Solid Mechanics and Its Applications

Volume 256

Aims and Scope of the Series

The fundamental questions arising in mechanics are: *Why*?, *How*?, and *How much*? The aim of this series is to provide lucid accounts written by authoritative researchers giving vision and insight in answering these questions on the subject of mechanics as it relates to solids.

The scope of the series covers the entire spectrum of solid mechanics. Thus it includes the foundation of mechanics; variational formulations; computational mechanics; statics, kinematics and dynamics of rigid and elastic bodies: vibrations of solids and structures; dynamical systems and chaos; the theories of elasticity, plasticity and viscoelasticity; composite materials; rods, beams, shells and membranes; structural control and stability; soils, rocks and geomechanics; fracture; tribology; experimental mechanics; biomechanics and machine design.

The median level of presentation is to the first year graduate student. Some texts are monographs defining the current state of the field; others are accessible to final year undergraduates; but essentially the emphasis is on readability and clarity.

More information about this series at http://www.springer.com/series/6557

Christopher G. Provatidis

Precursors of Isogeometric Analysis

Finite Elements, Boundary Elements, and Collocation Methods

Springer

Christopher G. Provatidis
School of Mechanical Engineering
National Technical University of Athens
Athens, Greece

ISSN 0925-0042 ISSN 2214-7764 (electronic)
Solid Mechanics and Its Applications
ISBN 978-3-030-03888-5 ISBN 978-3-030-03889-2 (eBook)
https://doi.org/10.1007/978-3-030-03889-2

Library of Congress Control Number: 2018960750

This Springer imprint is published by the registered company Springer Nature Switzerland AG
The registered company address is: Gewerbestrasse 11, 6330 Cham, Switzerland

Dedicated to
my Parents: Gabriel and Elisabeth
and
my Family: Eleni, Gabriel, Athanasios, Paul

Preface

This book was written with the purpose to be a self-contained text which thoroughly summarizes the attempts toward "CAD/CAE integration" so as to shorten the gap between Computer-Aided Engineering (CAE) and Computer-Aided Design (CAD) communities. In order to make this topic fully clear, besides the theory, a lot of benchmark tests, ranged between potential problems and elasticity problems, are solved in an instructive way and are thoroughly commented.

Having contributed with more than 60 scientific papers on this topic since 1980s, I felt the need to communicate my overall findings on an idea that first appeared in 1970s in industrial research centers (such as General Motors) and continues till nowadays in the framework of the contemporary nonuniform rational B-splines (NURBS)-based *isogeometric analysis* (IGA). Although our overall attempt was substantially developed well before the date on which IGA appeared, the reality of this numerous community has determined the title of the book, i.e., "*Precursors of Isogeometric Analysis*" with emphasis given in engineering analysis. Summarizing most of that useful knowledge associated with the CAGD interpolations which preceded the NURBS interpolation and formed the basis for the development of relevant *CAD-based macroelements*, it will become clear that the global approximation involved in all CAD-based methods (including the NURBS-based IGA) is the actual cause for the high accuracy of the numerical solution.

Close contact with young researchers has given me the impression that a lot of basic knowledge has not been fully absorbed, and this book highly contributes on the completion of the dialogue between finite element analysis (FEA) and the CAD communities. The book deals with three computer methods, i.e., the *Galerkin–Ritz* (finite element method: FEM), the *Global Collocation* as well as the *Boundary Element Method* (BEM) using CAD-based macroelements (with global approximation). This is the reason why this book refers to "CAD/CAE" instead of only "CAD/FEA" integration.

The extensive numerical examples of this book refer to both static and dynamic analysis. Except of Laplace and Poisson equation, particular attention was paid on the eigenvalue extraction of acoustic cavities for which closed-form analytical solutions exist. Therefore, the accuracy of the proposed single macroelements for rectangular and circular cavities refers to an analytically known exact value, a fact that is not possible in free vibration analysis of elastic structures (where the exact solution is not accurately known and therefore is approximated). And since due to the *Helmholtz decomposition* ($\mathbf{u} = grad\Phi + curl\mathbf{\Psi}$), the partial differential equation of elastic waves splits into *two* wave propagation equations, the first for the scalar potential Φ and the second for the vector potential Ψ, the study of acoustics may be a "crash test" for foreseeing the accuracy of any computational method in elastodynamics. This fact is the philosophy of this book: We start with acoustics and then continue with elastodynamics.

In the summer of 2018, my published work on this particular topic included:

- 38 papers in conjunction with Global Galerkin–Ritz
- 16 papers in conjunction with the Global Collocation method
- 5 papers in conjunction with the Global Boundary Element Method.

Synthesizing the numerical findings of the abovementioned papers as a whole, the book in hand consists of the following fourteen chapters:

Chapter 1 is a summary of earlier activities in CAD/CAE integration and offers the background or a refreshment of basic knowledge to the reader.

Chapter 2 refers to the basics of function interpolation/approximation and elements of CAGD. We start from univariate and then extend to multivariate functions. The main idea is that since in a 2D problem the solution $U(x, y)$ represents a surface, then several possible CAGD interpolations can be used to interpolate it. In other words, CAD surface interpolations of this surface are useful for further analysis. Following the historical evolution, Coons, Gordon, Barnhill, Bezier, B-splines, and NURBS interpolations are summarized.

In the sequence, Chaps. 3–8 refer to the application of global interpolation in conjunction with the Galerkin–Ritz formulation, whereas other methods are developed in Chaps. 11 (collocation) and 12 (boundary elements). In more details:

Chapter 3 refers to Coons interpolation, which is chronologically the first one in CAD theory. Large Coons elements are developed and a lot of engineering problems are solved.

Chapter 4 refers to Gordon transfinite interpolation, which is chronologically the second one in CAD theory. Large Gordon elements are developed and a lot of engineering problems are solved.

Chapter 5 refers to Barnhill interpolation applicable to triangular patches. Large Barnhill elements are developed and some engineering problems are solved.

Chapter 6 refers to nonrational Bézier curves and surfaces. The equivalency with elements of Lagrangian type is pointed out. The shortcoming of these elements in conjunction with conic sections is explained.

Chapter 7 refers to B-splines, covering the older truncated power formulation (before Schoenberg 1946) and the "contemporary" Curry–Schoenberg (1966) enriched by de Boor (1972) and Cox (1972) recursive algorithms. One- and two-dimensional problems are solved.

Chapter 8 refers to rational Bézier and rational B-splines (NURBS) surfaces.

Chapter 9 refers to the C^1-continuity problems and focuses on plate bending macroelements.

Chapter 10 refers to volume subregions (volume blocks) of hexahedral shape, which in this book are called "superbricks."

Chapter 11 refers to the global collocation method in conjunction with CAD-based macroelements.

Chapter 12 refers to the Boundary Element Method in conjunction with CAD-based patches.

Chapter 13 refers to domain decomposition problems.

Chapter 14 is a review of the overall performance of CAD-based macroelements in engineering analysis.

Hoping that this textbook will substantially contribute in the CAD/CAE dialogue, bridging the older with the contemporary ideas.

Athens, Greece
August 2018

Christopher G. Provatidis, Ph.D.
Professor of Computer Methods at NTUA

Contents

Abbreviations

General Note
Other books: In most CAD textbooks and papers, the intrinsic (dimensionless or normalized) coordinates, and sometimes the area coordinates of a point in a triangle, are denoted by (u, v, w). On the other point of view (concerning CAE), the symbols u, v, w represent the displacement components in the theory of elasticity or the velocity components in fluid mechanics.

This book. In order to avoid confusion as much as possible, this book:

(1) Uses the symbols (ξ, η, ζ) for the natural (normalized) coordinates in the unit square or the unit cube.
(2) Uses the symbols (u, v, w) for the normalized coordinates, useful in triangular Barnhill patches (see Chap. 5).
(3) Uses the symbols (u_x, u_y, u_z) for the displacement components in most elasticity problems. There are only some exceptions, where the traditional symbols (u, v, w) are used for the displacement components as well, but this rarely happens in specific places where there is no case the reader to be confused with the dimensionless coordinates.
(4) Uses the symbol (U) to describe the scalar potential (temperature, pressure, etc) in potential problems.

The most important symbols are as follows:

a_i	Generalized coefficients
$\{\mathbf{a}\}$ or $\mathbf{a}$	Column vector including the generalized coefficients a_i
$B_{i,n}$	Bernstein–Bézier polynomial
E_0, E_1	Blending functions
$H_{i,n}$	Hermite polynomial
$[\mathbf{K}]$ or $\mathbf{K}$	Stiffness matrix
$L_{i,n}$ or $L_i^{(n)}$	Lagrange polynomial of degree n
$[\mathbf{M}]$ or $\mathbf{M}$	Mass matrix
N_i	Shape function
$P_\xi, P_\eta, P_{\xi\eta}$	Projections

r, s, t	Normalized coordinates ($-1 \leq r, s, t \leq 1$)
t	Time
U_i	Nodal displacements or nodal potentials
$\{\mathbf{u}\}$ or $\mathbf{u}$	Column vector including the generalized coefficients U_i
x, y, z	Cartesian coordinates
u, v, w	Specific area coordinates in Barnhill triangular patch
u_x, u_y, u_z	Components of displacement vector

Greek Letters

ξ, η, ζ	Normalized (dimensionless) coordinates ($0 \leq \xi, \eta, \zeta \leq 1$)

Chapter 1
Initial Attempts on CAD/CAE Integration

Abstract This chapter discusses the meaning of the conventional "integrated CAD/CAE systems," which is contradicted from the "*CAD/CAE integration*" (under the umbrella of isogeometric analysis) adopted throughout this book. The history of several important CAD interpolations since 1964 is outlined. Five precursors of the NURBS-based isogeometric analysis are discussed. The general boundary value problem is posed. In order to solve it, three computational methods, i.e., the finite element method, the Boundary Element Method, and the collocation method are presented in brief. The implementation of Coons and Gordon interpolation formulas in mesh generation is discussed. Moreover, the utilization of the closely related transfinite elements in engineering analysis in conjunction with the aforementioned three major computational methods is discussed.

Keywords CAD/CAE integration · Finite element · Boundary element
Collocation method · Coons interpolation · Transfinite element

1.1 The Conventional Meaning of Integrated CAD/CAE Systems

Even in 1981, it had been written that "the field of finite elements and its relationship to Computer-Aided Design is a continuous development" [31]. Within this context, the term "CAD/CAE integration" has several meanings. For example, the older IDEAS® was a FEM package developed by SDRC company, which was later expanded by involving CAD features, thus creating Unigraphics® and currently the Siemens NX® PLM software. On the contrary, Pro/Engineer® was initially a commercial CAD system, which absorbed the FEA package Mechanica®, and today is known by the trademark Creo® PTC system. A third CAD system which preserves its name since 1993 is SolidWorks®, although at a certain time it absorbed the FEA package COSMOS/M®. Similarly, AutoCAD® absorbed the FEM package ALGOR® and created the CAD/CAE product Inventor®.

C. G. Provatidis, *Precursors of Isogeometric Analysis*, Solid Mechanics and Its Applications 256, https://doi.org/10.1007/978-3-030-03889-2_1

Among others, the abovementioned four software systems have been called CAD/CAE systems, and since also a CAM module can be always added (which may produce G-codes to feed CNC machines), the usual name has been "*integrated CAD/CAM/CAE*" systems (see, for instance, [80]). In every CAD/CAM/CAE system, the gap between geometry and analysis can somehow close. For example, during the design cycle of a bicycle, when the designer draws a hollow beam of its skeleton, the mechanical properties (like the cross-sectional area and the second moment of inertia) are automatically calculated, thus releasing him/her from performing time-consuming and error-prone calculations.

Later, these integrated CAD/CAM/CAE systems were extended to *Product Lifecycle Management* (PLM) ones, which means that these tools are not only useful for the process of initial design and manufacturing but also later, particularly, when the equipment may require continuous further changes and maintenance (e.g., in a nuclear plant). And the evolution still continues as a search on Internet or a visit in an international fair may reveal. Nowadays, most FEA packages, such as MSc/NASTRAN, MARC, ABAQUS, have been linked into corresponding PLM systems. For example, since some years ago, ABAQUS works in the context of SIMULIA PLM system. What is worthy to mention is that most of the aforementioned CAD/CAE systems are based on conventional (small sized) finite elements. Currently, intelligent management systems (such as 3DExperience® Platform, Dassault Systèmes) are developed, in which the FEA is a small part of the whole design or PLM process.

1.2 The Meaning of CAD/CAE Integration Adopted in This Book

Despite the commercial reality reviewed in Sect. 1.1, this book will not be concerned with the abovementioned classical CAD/CAE or PLM systems and therefore not will be involved in particular issues that refer to the reliability of data transfer from the CAD (geometry) model to the analysis (FEA) model. In contrast, the book will be concerned with potential and existing computational packages that implement the same interpolation for the CAD and the CAE models. A lot of researchers may have heard about the isogeometric analysis (IGA), which was introduced in 2005 by Hughes et al. [71]. Today, at least one FEA package, LS-DYNA, has been influenced by the IGA concept in their preprocessor and solver [66], but other attempts continue (e.g., [146]).

But even this really happens and probably is going to be extended to other systems as well, there is some knowledge still missing. In more details, the NURBS-based IGA is not the first and only way to integrate CAD with FEA, in the sense that all previous (than NURBS) CAGD interpolations have the same capability to perform this integration. For example, according to Cavendish and Hall [38], Gordon transfinite interpolation was included in the software environment GMSOLID (at General

Motors) for which a description has been reported earlier by Boyse and Rosen [30]. However, we have to admit that NURBS is the standard of *currently* used CAD packages. And obviously, in principle, the same philosophy of "CAD/CAE integration" stands for any possible *novel* CAD interpolation of the future.

This book will show that all CAD interpolations developed between 1964 and 2005 (i.e., before IGA appeared) are capable of building large CAD-based macroelements (subregions), which can be treated with (i) specific FEA (based on Galerkin–Ritz formulation) but also with (ii) CAD-based global collocation or (iii) CAD-based boundary elements. This comment justifies the term "*Precursors of IGA*", which is the title of this book.

1.3 CAD Interpolation

Classical textbooks in Computer-Aided Design (CAD) suggest that the first important achievement in CAD theory is *Coons interpolation formula* [44], which performs a reasonable approximation within a curvilinear patch ABCD (see Fig. 1.1) in terms of numerical data on its boundaries [54, 55]. The same general formula is applicable to the interpolation within a curvilinear volume block ABCDEFGH as well (see Fig. 10.1, Chap. 10).

Steven A. Coons (1912–1979) was an early pioneer in the field of computer graphical methods, who worked as a consultant of Boeing Co. and also a MIT Professor (see photograph in Fig. 1.2 and may refer to https://en.wikipedia.org/wiki/Steven_Anson_Coons). A critical reading of his published initial Technical Report (see [44]) reveals that it is so general and affirmative that it can cover the most general case of interpolation, including hypersurfaces and hypervolumes. In the same Report, in addition to quadrilateral patches and cuboid/hexahedral volume blocks, he also treated the case of triangular (or tetrahedral) patches by merely proposing the degeneration of one edge (or surface) into a single point. Although Steven Coons did not publish much in the open domain, his "shadow" continued to cover other activi-

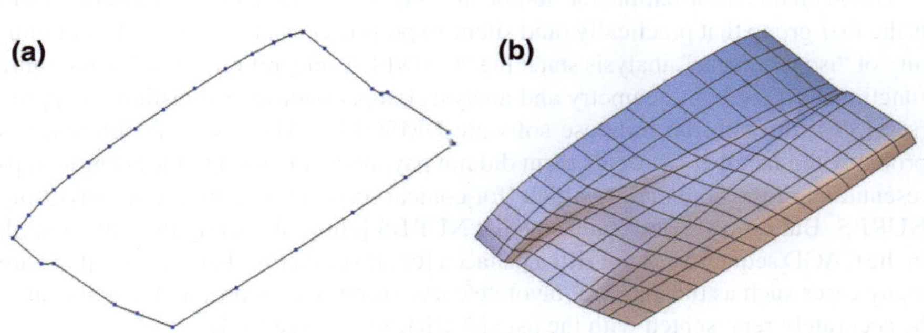

Fig. 1.1 Coons patch for a panel: **a** contour and **b** isoplots. *Source* Internet

Fig. 1.2 Professor Steven Anson Coons, MIT, [44]. *Source* http://photos1.blogger.com/img/155/960/1024/Steve%20Coons.jpg

ties up to the supervision of Versprille's PhD thesis (see [142]) in rational B-spline approximation form; it was published four years before his death. Having the author worked for many years with Coons formula (since 1982), he could say that in some sense this great achievement essentially is "nothing more" than the extension of the well-known linear interpolation from 1D intervals (lines) to 2D dimensions (surface patches) as well as to 3D dimensions (volume blocks).

The second important station in CAD theory is the work of William (Bill) Gordon at the Mathematical Department of General Motors, Detroit, USA (see his photograph in Fig. 1.3). Gordon himself worked much with the idea of considering interboundaries within the previously published Coons patches [58]. The motivation was the need to deal with more complex shapes based on information related to additional data associated to the interior of a domain. To give an example, if the boundary in the base of a hill (considered as a Coons patch ABCD) is given, the determination of its height requires additional data in the interior (the reader may imagine an infinite sequence of hills of *variable* height possessing the *same* given patch ABCD). Although Bill Gordon also worked alone in conjunction with spline-blended surface interpolation through curve networks (see [57]), in cooperation with his associates, Cavendish and Hall, he introduced the idea of decomposing a structure into large macroelements in which transfinite interpolation is performed [36–38, 60, 61].

Based on this information, the author strongly believes that General Motors' team is the *first* group that practically (and silently) proposed whatever today the community of "isogeometric" analysis state: the "CAD/FEA integration" by using the same functional set for both geometry and analysis [isoparametric or transfinite mapping (see [36])] through the in-house software GMSOLID. The essential difference is probably the fact that Gordon's team did not pay much attention to the accurate representation of the boundaries, which (for conical sections) is achieved mostly using NURBS. But even so, one could say that NURBS [chronologically the fifth or sixth in the CAGD sequence] is not still a panacea for all geometric shapes, since there are many cases such as the classical involute curve (used in gear profiles) which cannot be accurately represented with the usual NURBS as is (see [67]).

A third important station in CAD theory is the work of Pierre Bézier (1910–1999) (see his photograph in Fig. 1.4). From 1933 to 1975 Bézier worked for Renault, where

Fig. 1.3 William (Bill) J. Gordon was a professor of mathematics and computer science at Drexel University in Philadelphia since 1979. In addition, he was director of Drexel's center for interactive computer graphics. Prior to joining Drexel, he was a visiting scientist at IBM's T. J. Watson Research Center, spent 11 years at the General Motors Research Laboratories, and from 1976 to 1978, served as a liaison scientist for the London branch of the US Office of Naval Research. Gordon has taught at the University of Utah, Syracuse University, and the University of Detroit. His research interests included the mathematical foundations of geometry modeling. Gordon received his DSc degree from Brown University in 1965 [From Gordon [59] with permission]

he would ultimately develop his UNISURF CAD/CAM system (https://en.wikipedia.org/wiki/Pierre_B%C3%A9zier). Pierre has published a large number of papers (see literature list fom Bézier [11–29], Rabut [132]), and his main contribution is that he made clear the great importance of control points (previously proposed in the form of generators by de Casteljau [49, 50]) in both curves and surfaces (see https://en.wikipedia.org/wiki/Paul_de_Casteljau). It should be noticed that the nonrational Bézier interpolation is equivalent to the basis of Lagrange polynomials, but even so it has the advantage of the *control points*. Simply speaking, a nonrational Bézier tensor product merely "replaces" the classical tensor product of Lagrange polynomials. Between the two bases, i.e., nonrational Bézier and Lagrange polynomials, a linear dependency occurs (see [130]). Moreover, concerning the so-called rational Bézier interpolation, it belongs to the class of rational curves in CAGD and may be traced back to Coons [44, 45] and Forrest [56] (see [54], p. 231).

A fourth station in CAD theory is the collection of works by many researchers concerning spline interpolation and particularly the B-spline one. Although B-spline is an interpolation of general usefulness, tensor-product B-splines is an important way to interpolate a curvilinear patch or even a curvilinear volume block [54, 55]. Simply speaking, B-splines tensor product is an *analogue* of the classical tensor products of Lagrange polynomials. The idea of using the aforementioned tensor product as a series expansion for further analysis was documented in a CAD handbook by Höllig [68], whereas the first monograph on B-splines-based finite element methods was due to Höllig [69]. Interestingly, not only rectangular hexahedrals but also a whole sphere can be (approximately) handled using a unique B-spline tensor product [131].

A fifth station in CAD theory is the collection of works by many researchers concerning the so-called *NonUniform Rational B-splines* (NURBS). NURBS was developed by Versprille [142] and later by several other researchers (intensively

Fig. 1.4 Pierre Bézier in 1958. *Source* https://alchetron.com

Fig. 1.5 Robert E. Barnhill near 2018 (his Courtesy). He was a professor of mathematics at the University of Utah

within software companies in about 1985), but it became a commercial product in 1990. It is worth mentioning that ten years before the publication of Höllig's book, other researchers had already proposed the concept of using NURBS in analysis, in both the finite element method [137] and the Boundary Element Method [138]. In 2005, American [71] and Japanese [72] teams proposed the use of NURBS for analysis and shape design. There is no doubt that the excellent team of Professor Hughes attracted the major attention and managed to get funds to support PhD theses (Cottrell et al. [46]), including many collaborating teams in Germany and Italy, among other countries, and continue their successful research till today.

Turning from quadrilateral to triangular patches, a sixth important station in CAD theory is Barnhill's interpolation (see his photograph in Fig. 1.5). Historically, it was developed at the same period with Gordon interpolation, and Gordon himself contributed in this research [4, 5, 7]. More details in the form of a review paper were later presented by Barnhill himself ([6], among others).

The above six CAD-based interpolation formulas are not the only ones that exist in CAD theory (e.g. see [10]), but here these were reported first as the author happens to have worked with them since 1982 until today. Furthermore, as also mentioned in Cottrell et al. [46], "there are many other CAD-based formulations that may play a role in the future of isogeometric analysis such as N-sided Gregory patches [63], S-patches [82], and A-patches [3]." I agree and would also like to make an additional important remark. Not only the previous ones, but also any new CAD/CAGD formula that may be proposed in the future will produce novel macroelements because it probably will include a new set of basis functions. Need for new interpolations is not out of demand. As an example, we may refer to the case of *involute* curves that appear in all classical gear profiles and which are not still accurately represented using NURBS [67].

1.4 CAE Methods

As mentioned in Sect. 1.2, the term Computer-Aided Engineering (CAE) refers to engineering analysis which is performed using either of the traditional methods (finite elements, boundary elements, collocation methods) or newest methods such as mesh-free, mesh-less, point methods, etc. In this book, we will extensively refer to the traditional ones only.

In brief, the engineering analysis is posed as follows. The structural component (or mathematically the problem domain) is denoted by Ω and its surroundings by the boundary Γ (sometimes denoted by $\partial\Omega$). Within the domain Ω, the governing partial differential equation (PDE) is generally described by (the *strong form*):

$$D(U(\mathbf{x})) = 0 \quad \text{in } \Omega, \tag{1.1}$$

with D denoting a differential operator, whereas the boundary conditions (BCs) are (see Fig. 1.6):

$$U = \bar{U} \quad \text{in } \Gamma_1 (\text{Dirichlet BC}), \quad \text{and } \frac{\partial U}{\partial n} = \bar{q} \quad \text{in } \Gamma_2 \text{ (Neumann BC)}. \tag{1.2}$$

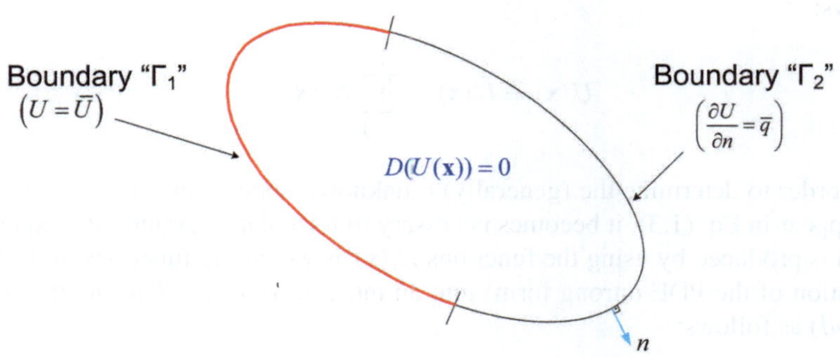

Fig. 1.6 Problem domain and boundary conditions

In elasticity problems, the flux $\bar{q}$ that appears in Eq. (1.2) is replaced by the vector of tractions $\mathbf{t}$.

1.4.1 Finite Element Method (FEM)

The *finite element method* (FEM) and the relevant finite element analysis (FEA) are of major importance for the numerical solution of boundary value problems [given partial differential equation under given boundary or/and initial conditions]. It is accepted that the origin of this method is mainly due to Argyris [1], whereas other pioneers were Clough [140] and Zienkiewicz [147]. The method is based on the discretization of the problem domain into small elements which are called "*finite elements*" and all together form a computational *mesh*. Each finite element consists of "*nodal points*" (or merely "*nodes*"), which form simple shapes such as triangles and quadrilaterals (in 2D problems), as well as tetrahedrals, hexahedrals, and prisms (in 3D problems). The size of the finite elements is generally small and generally should become relatively *smaller* (i) near the position of the external loads, (ii) near the position of the supports, as well as (iii) at places where the geometry along the boundary rapidly changes (e.g., when a notch appears). In other words, the size of the elements should be comparably small in places of stress concentration or when the gradient of the problem variable U is high. Three classical FEM books are Zienkiewicz and Taylor[148, 149], Bathe [8], Hughes [70], while many others cover particular issues (elasticity, plate bending, nonlinear, heat transfer, electromagnetics, etc.).

Within the context of this book, the FEM is implemented to macroelements of *large* size, which sometimes may include an entire *primitive* shape such as a rectangle, a circle, an ellipse. Although it is not always restrictive, each macroelement should be of preferably convex shape, as it happens with the classical finite elements.

For the purpose of completeness, below the very basics of FEM are exposed for static analysis (steady-state problems). At any position $\mathbf{x}$ of the domain, the unknown accurate function, $U(\mathbf{x})$ involved in Eq. (1.1), is approximated by the approximate solution, $\tilde{U}(\mathbf{x})$, which is written in the form of a series expansion of n terms, as follows:

$$U(\mathbf{x}) \cong \tilde{U}(\mathbf{x}) = \sum_{j=1}^{n} N_j(\mathbf{x}) U_j \tag{1.3}$$

In order to determine the (generally) n unknown nodal values U_j, $j = 1, \ldots, n$ that appear in Eq. (1.3), it becomes necessary to formulate a system of n equations, which is produced by using the functions $N_i(\mathbf{x})$ as weighting functions in the transformation of the PDE (strong form) into an integral from (*weak form*, or *Galerkin method*) as follows:

$$\int_{\Omega} [N_i \cdot D(U)] \mathrm{d}\Omega = 0, \quad i = 1, 2, \ldots, n. \tag{1.4}$$

Alternatively, the Galerkin method shown in Eq. (1.4) can be equivalently written in one equation only, as follows:

$$\int_{\Omega} [\delta U \cdot D(U)] \mathrm{d}\Omega = 0, \tag{1.5}$$

which is the generalization of the "*principle of virtual work*" or "*principle of virtual displacements.*"

The application of Eq. (1.4) for all the n weighting functions leads to the matrix form (equation of equilibrium)

$$[\mathbf{K}]\{\mathbf{u}\} = \{\mathbf{f}\}, \tag{1.6}$$

where $[\mathbf{K}]$ is the stiffness matrix (of dimensions $n \times n$), $\{\mathbf{u}\}$ is the vector of nodal displacements, and $\{\mathbf{f}\}$ is the vector of external nodal forces. In some sense, Eq. (1.6) generalizes the axial loading of a linear spring, from one degree of freedom to many ones.

Note-1 It is worth mentioning that the two abovementioned methods, i.e., (i) Galerkin method (Eq. 1.4), and (ii) principle of virtual work (Eq. 1.5), as well as the rest three (iii) minimum total potential energy Π, (iv) Hamilton's principle, and (v) Lagrange's equations, are all *equivalent* to one another.

Note-2 It should become clear that in either of Bézier, B-splines, and NURBS interpolation, the abovementioned displacement (or potential) vector $\{\mathbf{u}\}$ is not associated to nodal points but rather to control points and is substituted by the vector $\{\mathbf{a}\}$ of generalized constants (coefficients).

1.4.2 Boundary Element Method (BEM)

The *Boundary Element Method* (BEM), which also is met under the name Boundary Integral Element (or Equation) Method (BIEM), is an integral method competitive to the domain-based FEM. When the problem is linear, BEM avoids the integrals over the domain Ω and therefore deals only with its boundary Γ (see Fig. 1.7b). The most popular formulation of the Boundary Element Method is the direct one (DBEM), which is based either on the Betti-Maxwell principle or Green's identity. In contrast to FEM where the displacement is the primary variable and the stresses are produced by differentiation, in DBEM both the boundary displacements ($\mathbf{u}$) and the tractions ($\mathbf{t}$) are of equal importance.

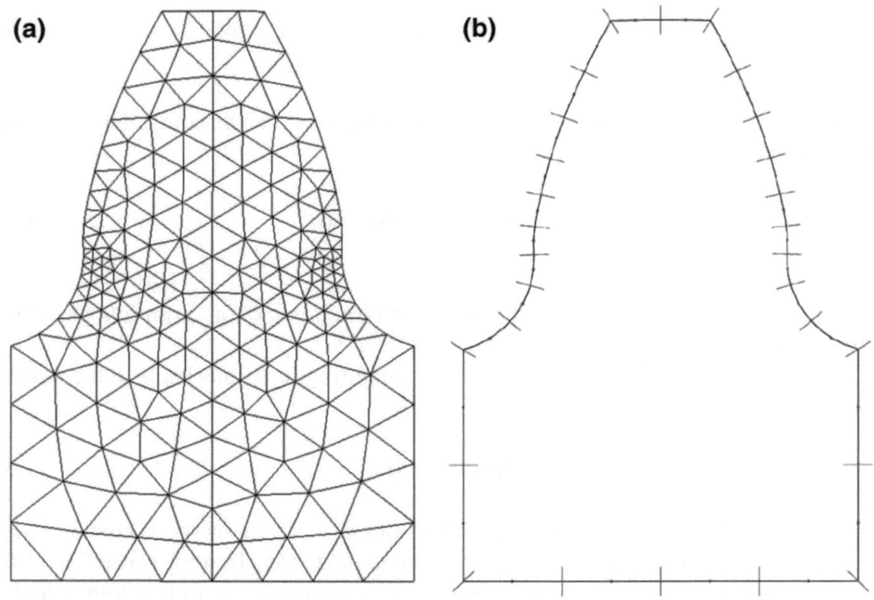

Fig. 1.7 **a** Finite elements versus **b** boundary elements

A concise form of DBEM is (see [32]):

$$c_i u_i + \int_{\Gamma} t^* u \, \mathrm{d}\Gamma = \int_{\Gamma} u^* t \, \mathrm{d}\Gamma, \tag{1.7}$$

where u^* and t^* are the fundamental displacement and the fundamental traction, respectively.

In matrix form, Eq. (1.7) is written as follows:

$$[\mathbf{H}]\{\mathbf{u}\} = [\mathbf{G}]\{\mathbf{t}\}, \tag{1.8}$$

where [**H**] and [**G**] are influence matrices of the boundary vector of displacements {**u**} and the boundary vector of tractions {**t**}.

After the imposition of the boundary conditions, the solution of the system, described by Eq. (1.8), determines the unknown values of the displacements and the tractions along the boundary. The displacement and the stresses at any arbitrary internal point of the structure are calculated afterward by setting $c_i = 1$ in Eq. (1.7). Since in nonlinear problems it becomes necessary to insert auxiliary internal cells, the practical usefulness of BEM is rather the analysis of external problems (e.g., acoustics), but it is also useful in particular internal problems such as the determination of crack intensity factors. This topic is covered in detail in the classical book of Brebbia and Dominguez [32], among others. In our book, this issue is studied in Chap. 12.

1.4.3 Collocation Method

The *collocation method* is an old technique [79], which was later revived in the framework of CAD-based analysis ([2, 115–119, 141]). In brief, the method is a subclass of the Galerkin methods, where the weighting function N_i in Eq. (1.4) is replaced by the Dirac delta function, Δ_i. This means that the PDE is fulfilled (collocated) at n discrete points in the domain Ω (generally not excluding the boundary). In this way, a system of n equations is formed, and if it is properly solved, the unknown displacements (potentials), $\mathbf{u}$, are calculated. Both linear and nonlinear boundary value problems as well as eigenvalue problems can be solved. A classical book including collocation methods is due to Collatz [42], where a thorough overview of collocation methods is given in Chap. 11 of the book in hands. Most of the books refer to the boundary conditions of Dirichlet type, whereas the proper treatment of the Neumann-type ones is still an open problem, particularly when the domain is subdivided into subregions.

1.5 Modules of CAD/CAE Integration

It is reasonable that the first achievements of CAD/CAE integration were in conjunction with the finite element method. Gordon's work in early 1970s was developed primarily in the context of General Motors' research until later, near 1980, he moved to Drexel University. His research was concerned with a methodology for generating structured FEM meshes based on Coons interpolation (Sect. 1.5.1) as well as with innovative transfinite macroelements that use the *same* global approximation for both the geometry and the variable (Sect. 1.5.2).

1.5.1 Mesh Generation

Initially, Gordon and Hall [60] applied the standard Coons formula in order to establish a mapping between the domain of the actual problem, Ω, and a unit square ABCD, Ω_e, as shown in Fig. 1.8. The procedure starts with the creation of a structured mesh within the unit square, which is further mapped into the actual domain. For example, the mapping of the nodal points of a pre-constructed uniform mesh of 3×2 subdivisions (shown in Fig. 1.8b) gives the desired computational mesh into the circular ring in Fig. 1.8a.

It is worth mentioning that the mesh generated using Coons interpolation formula is generally influenced by the *parameterization* of the domain, which essentially is the

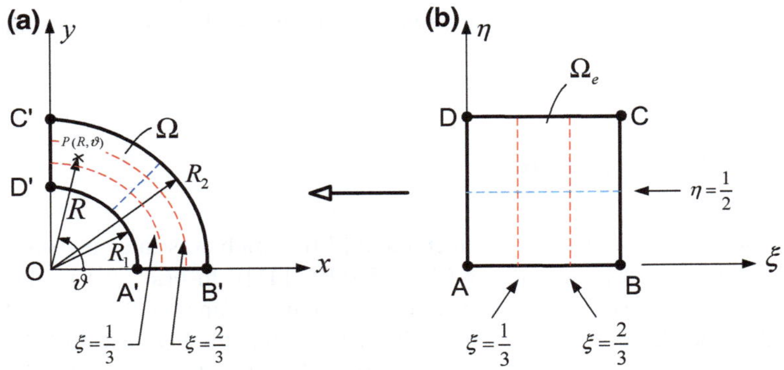

Fig. 1.8 Mapping **a** to a circular ring **b** from the unit square, using Coons interpolation formula

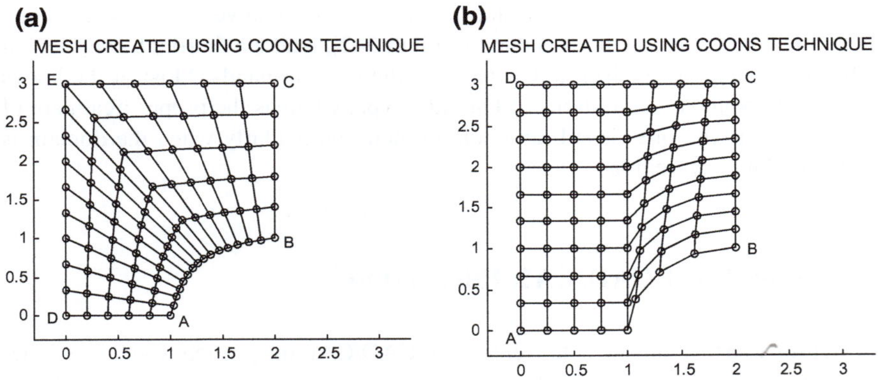

Fig. 1.9 Mesh generated in a domain with five discrete boundary segments, for two different parameterizations (choices of the corner points A, B, C, D of the reference square)

choice of the corresponding corner points A′, B′, C′, and D′ on the boundary $\Gamma \equiv \partial\Omega$ of the actual domain Ω. For example, in Fig. 1.9a, one may observe a generalized quadrilateral in which the edge AB is uniformly divided into 15 segments, whereas the edge CD is composite and consists of six segments along CE and another nine along ED (totally 15 along CD). In contrast, Fig. 1.9b shows another case in which the new edge AB is composite (totally divided into eight segments) and the opposite edge CD is a straight segment also divided into eight segments as well.

Other similar activities concerned with the application of *bivariate* Coons interpolation to the development of two-dimensional automatic mesh generators are [64, 65, 74, 134, 143, 144]. Moreover, *trivariate* Coons interpolation is capable of describing any boxlike solid component on the basis of its boundary ([44], p. 41), and this formula was used by Cook [43] to develop a generator for three-dimensional computational mesh, useful for further FEM analysis. One of the latest Gordon's works on

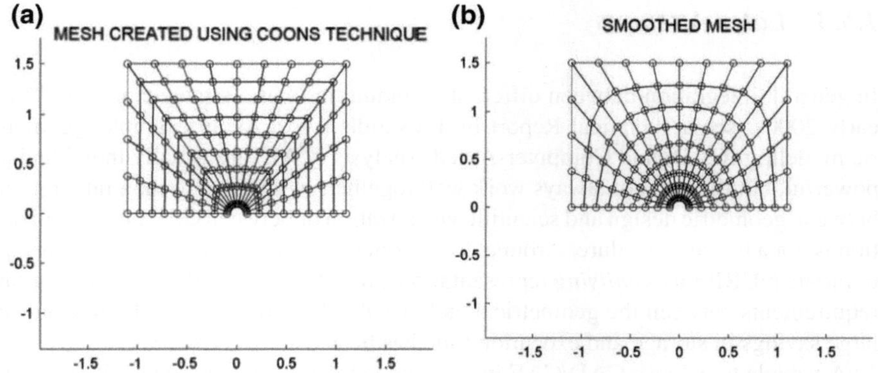

Fig. 1.10 Mesh generation as produced using Coons interpolation, **a** before smoothing and **b** after smoothing

mesh generation, after he moved to Drexel University, is Gordon and Thiel [62]. The interested reader may also consult the book of Beer and Watson [9], where details are provided for mesh generation in both finite elements (FEM) and boundary element (BEM) applications.

Moreover, it is worth mentioning that even if the mesh quality is somehow acceptable, as that shown in Fig. 1.9a, or its duplication shown in Fig. 1.10a, there is still space to make it better, for example, applying a simple relaxation scheme by averaging the eight surrounding nodes [107]; a typical smoothed mesh is shown in Fig. 1.10b. A similar smoothing of the internal nodes by averaging their twenty-six surrounding nodes in each iteration was applied to 3D structures as well [105].

1.5.2 Transfinite Elements

Below we shall show that William Gordon was aware of the topic of CAD/CAE integration as well. Actually, at the same time with the automatic mesh generation, Gordon and Hall [61] proposed the transfinite or blending-function elements which in fact are global interpolations within a whole patch. Later, Cavendish et al. [36] applied these elements in conjunction with the Galerkin–Ritz procedure (finite elements). The idea is to subdivide the structure in a certain number of macroelements (see [37]) and apply the global approximation in each of them. This procedure is characterized by high-order interpolation, and therefore, eventually it has some similarities with the technique proposed later by Babuska–Szabo's alternative approach [139]. After Gordon moved to the academia, the rest two collaborators (in the General Motors Lab) continued publishing toward this direction using the in-house CAD/CAE environment GMSOLID [38].

1.5.3 Later Attempts

In general, integration between different communities was a strategic aim until the early 2000s (see a Technical Report by Provatidis [95]). As an example, geometric modeling (CAD) and Computer-Aided Analysis (CAE) are usually individually powerful, but they do not always work well together. In addition to that, integration between geometric design and scientific visualization or between CAE and visualization is not a trivial procedure. Around 2000, some solutions were proposed by using trivariate NURBS as a *unifying* representation. Also, the topic of different resolution requirements between the geometrical and visualization models in order to achieve large savings in storage and execution time has been discussed [83, 85].

A vehicle to achieve CAD/CAE integration is to apply a common basis function for geometric modeling and representation of the multiphysics field (temperature, displacement, etc.). Casale [33] and his coworkers [34, 35] proposed trimmed surface patches as boundary elements. At the same time, Kanarachos and Deriziotis [77], influenced by the ideas of Gordon and Hall [61], developed a Coons-based method using cubic *natural* B-splines interpolation along the edges of the patch and applied it to solve 2D boundary value potential problems. In that primary work, it was found that their method leads to better results than FEM and BEM for both static and dynamic analyses. Previously, a Coons patch method including boundary derivatives had been proposed to differential equations with predominant lower-order derivatives [76]. Similar ideas were also applied to plate bending and shell problems (El-Zafrany and Cookson [52, 53], [145]), however, without presenting any numerical results.

Despite the fact that a lot of additional research had been conducted and reported at the National Technical University of Athens (NTUA) between 1982 and 1990, in the expectation of newly coming PhD students to fill the gap of an interrupted dissertation, after 1989 the NTUA group did not publish until only in 1999 [78]. Then, in a video recorded session which was chaired by Professor Robert Taylor (Univ. Berkeley), and in front of many distinguished professors and researchers (Zienkiewicz, Samuelson, etc.), the whole idea on CAD/CAE integration was disclosed during the ECCM'99 Conference in Munich. Although no fund was available, the author of this book contributed to the promotion and the further extension of the Coons macroelements as much as he could ([78], [88–131]). In its first stage, the aforementioned study includes 2D and axisymmetric potential and elasticity problems (static and dynamic analyses). Before going on with the rest activities at NTUA, where the author was involved (e.g., 3D analysis among others), below we continue with other attempts published between 1999 and 2004.

A monograph that covers FEM applications using B-splines till 2002 is due to Höllig [69]. Moreover, Kagan and Fisher [73] developed a B-spline-based finite element scheme. Renken and Subbarayan [133] used NURBS to represent the shape of droplets by integrating surface energy coefficients over appropriate surfaces.

A search in 2003 had revealed several other contributions toward the abovementioned CAD/CAE integration. Clark and Anderson [40, 41] proposed a penalty boundary method for performing finite element analysis using a regular overlapping mesh that does not have to coincide with the geometric boundaries. Previously, Charlesworth et al. [39] have proposed a "domain decomposition" technique aiming to relax or remove the restriction on the mesh to conform globally to the domain geometry. In addition, Natekar et al. [84] proposed an NURBS-based analysis methodology that is procedurally analogous to the Constructive Solid Geometry (CSG) integrating design and analysis and thereby enabling efficient optimal design. This method was applied to two-dimensional problems only.

According to Natekar et al. [84], bivariate NURBS representation is applied to derive shape functions $N_I(\xi, \eta)$ that are based on the set of Ith control points defining the system geometry. The same shape functions are also used to approximate the displacement field within the domain. It is remarkable that at any point within the domain, the sum of these shape functions equals to the unity but the value of $N_I(\xi, \eta)$ at the Ith control point is not unity to that node, i.e., $N_I(\xi_I, \eta_I) \neq 1$. Therefore, "*even if the control point were to be coincident with the location of the boundary condition, direct application of the boundary condition is not possible since the specified field value will be distributed to control points influencing the point under consideration*" [84]. At this point it is worthy to mention that some progress on this shortcoming appeared many after twelve years, using rational Lagrange polynomials [81, 135].

Having commented on the state-of-the-art until 2004, it is now time to come back and present some additional details concerning the philosophy of our NTUA CAD/CAE group. I have to claim that, in 1990, we were aware of the advantages of the new-coming NURBS and the closely related older Bézier approximation (note that we had a six-week certified training on the CAD system EUCLID_IS, at the premises of MATRADATAVISION, Palaiseau-Paris, France). We were also aware of the property of compact support inherent in the de Boor–Cox formulation of the B-spline interpolation [47, 48]. Despite these facts and in order to avoid difficulties with the imposition of boundary conditions, we preferred to follow the standard FEM procedure (using the well-known source code FEAP [147]), in which *cardinal* shape functions are used (i.e., functions of [1,0]-type, which means that they are equal to the unity at the associated nodal points and vanish at all the rest nodes), although they lack the advantage of compact support. As it will be shown in Chap. 2, the aforementioned cardinality can be ensured when using *natural* cubic B-splines in conjunction with the initial B-spline formulation (truncated power series, known before Schoenberg [136]. Most of the initial works were based on this type of approximation ([77, 78, 88, 92], among others). This fact makes an essential difference between Coons elements and NURBS-based techniques such as that in the above paragraph mentioned within quotation marks [84]. As a result, the initially proposed Coons patch methodology was directly applicable using all standard finite element procedures in both static and dynamics regimes, without being necessary to use penalty methods [40, 41] and related Lagrange multipliers techniques as occurs in NURBS-based meshless methods (Natekar et al. [84]). Note that only recently this shortcoming was resolved in the context of isogeometric analysis [81, 135]. In contrast, with our choice to use natural cubic B-splines, we did not even have to deal with nodeless coefficients, and this issue will be discussed and criticized in Chap. 7. By making our self-criticism, we must admit that until 2002, we had the mistaken opinion that Coons elements

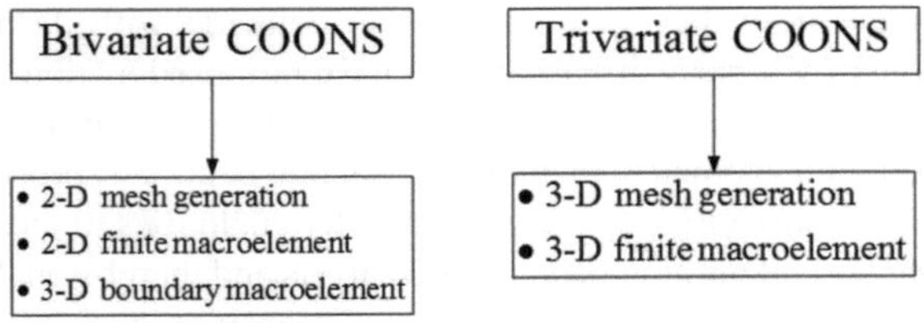

Fig. 1.11 Alternative utilization of Coons interpolation

could work without any internal nodes per Coons patch, an issue that was definitely resolved within the Ph.D. thesis of Dimitriou [51], in which the need of internal nodes was established.

To summarize, the overall applicability of Coons interpolation is schematically shown in Fig. 1.11, and some explanations are given below.

Bivariate Coons can be applied to produce 2D mesh [60], which can be easily smoothed [107]. Also, it can be used to construct 2D and axisymmetric arbitrary-noded Coons-based "C-elements" of which the nodal points belong to the boundary only [77, 88]. Finally, it can be used to construct boundary elements of which the boundary is divided into a certain number of Coons patches. As a special case of the latter, a hexahedral (superbrick) could be discretized in six Coons patches (in the limit, one per face), thus using its twelve edges only [91, 94, 99, 100]. Nevertheless, generally the boundary may be divided into any number of Coons patches, which in the extreme case can degenerate to the conventional boundary elements.

Trivariate Coons interpolation can be applied to produce 3D mesh [43], which can be easily smoothed if we wish [105]. Moreover, it can be used in order to derive large superbricks of which the nodal points are arranged along the twelve edges of the hexahedral [108, 109, 114]. As it was previously mentioned, this methodology appears two major advantages. First, it operates directly on the geometric modeling representation, and second, it preserves the FEM formulation in both static and dynamic analyses. Especially, in dynamic analysis, the mass matrix-based BEM has several shortcomings [75, 86, 87, 102], and therefore an alternative solution is needed.

Concerning boxlike structures, they consist of a curvilinear paralleloidal (cuboid) made of *six* surfaces, *eight* corners, and *twelve* edges. Therefore, the above trivariate Coons approach offers to designers a powerful tool in order to reduce data preparation and shape optimization costs, as they will have to handle with only a limited number of parameters, which are the control points along the twelve sides of the mechanical component. Obviously, for complicated components, it will be necessary to use "control" nodal points above the whole surface or to divide the model into a small number of large size boxlike regions.

Since the term "*Coons patch*" clearly describes curvilinear surfaces, in the sequence, the term "*superbrick*" will be introduced to describe boxlike volumes. Again, the term "superbrick" generally refers to a curvilinear paralleloidal (cuboid) made of six surfaces, eight corners, and twelve edges. Following to above, within this book the term "superbrick" will refer to either CAD (geometry representation) or

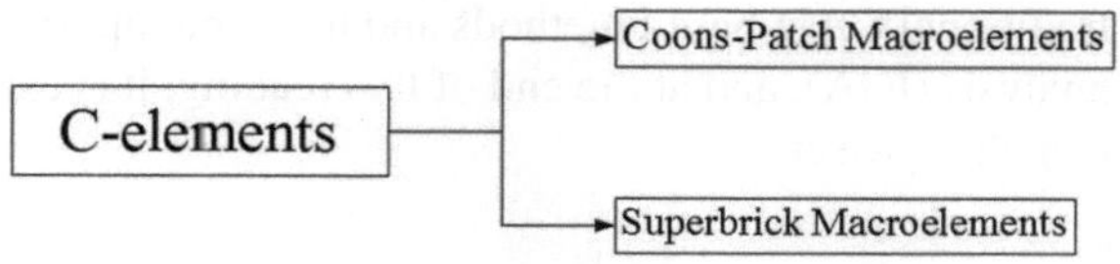

Fig. 1.12 Two categories of "C-elements" for 2D and 3D problems

CAE (trivariate attribute model) activities. In summary, the initial term "C-element" [77] is used for both the "Coons patch" and "superbrick" elements (Fig. 1.12).

1.6 Recapitulation and Some Historical Remarks

In the previous subsections it was clearly shown that, in mid 1970s, William (Bill) Gordon and his associates (Cavendish and Hall) at General Motors Lab had applied large transfinite macroelements for the solution of boundary value problems using the Galerkin–Ritz (finite element) method. These isoparametric elements used the same interpolation for both the geometry and the unknown variable. Nevertheless, they did not present a large amount of results (such as the usual benchmark test cases) in order to justify the performance of this method in comparison to other formulations. The only material they presented are the features of the software GMSOLID in which they linked their software. This gap was somehow completed by the research conducted by our CAD/CAE Group in the Department of Mechanical Engineering at NTUA, of which the first paper on the topic of CAD/FEA integration was published in 1989 [77], whereas—for reasons explained in Sect. 1.5.3—a second one was published with intentional delay in 1999 [78], and the publication procedure continues from 2000 [88, 107] and rather is completed today [131]. A small part of the activity of the NTUA group in Coons–Gordon interpolation is mentioned in the classical IGA book (see [46], p. 8).

Some of the above thoughts had been expressed "as is" in an early Report of 2002 [95]. Unfortunately, the lack of funding was the main reason that seven to eight PhD candidates at NTUA discontinued their dissertations, included the first one who started in 1984.

Since early 2000, the author of this book tried to promote the idea of CAD/FEA integration in the form of a compact paper, but either CAD journals found the papers outside their scope or they asked for a major revision or even a rejection. A positive event was a small grant of 10,000 Euros, as a result of a proposal that was submitted in May 2003 entitled "*DIDYMO: Investigation on the possibility of CAD/CAE integration*" (Project No. 65/1388, officially commenced in June 1, 2004, NTUA, Greece) [106]; the proposal took the high score of 93 out of 100 credit points.

The overall attempt of the author in this topic is about 60 papers in journals and conference proceedings with full papers, and this experience will be properly classified in an instructive way within the next chapters of this book. An overview will be given in Chap. 14. We believe that this book closes the gap between the older traditional CAD-based methods and the contemporary NURBS-based isogeometric analysis (IGA), and at the end of this reading, it is expected that the reader will feel somehow wiser.

References

1. Argyris JH (1955) Energy theorems and structural analysis: a generalized discourse with applications on energy principles of structural analysis including the effects of temperature and non-linear stress strain relations. Aircr Eng 26(1):347–356
2. Auricchio F, Beirão da Veiga L, Hughes TJR, Reali A, Sangalli G (2010) Isogeometric collocation methods. Math Models Methods Appl Sci 20(11):2075–2107
3. Bajaj C, Chen J, Xu G (1995) Modeling with cubic A-patches. ACM Trans Graph 14:103–133
4. Barnhill RE, Birkhoff G, Gordon WJ (1973) Smooth interpolation on triangles. J Approx Theory 8:114–128
5. Barnhill RE, Mansfield L (1974) Error bounds for smooth interpolation on triangles. J Approx Theory 11:306–318
6. Barnhill RE (1985) A survey of patch methods. NASA Report. Available at: https://ntrs.nasa.gov/search.jsp?R=19850020255
7. Barnhill RE, Gregory JA (1975) Compatible smooth interpolation on triangles. J Approx Theory 15:214–225
8. Bathe KJ (1996) Finite element procedures. Prentice-Hall, Upper Saddle River, New Jersey
9. Beer G, Watson JO (2002) Introduction to finite and boundary element methods for engineers. Wiley, Chichester, Chapter 11, pp 357–377
10. Bercovier M, Shilat E (1993) Enhancement of Gordon-Coons interpolations by 'bubble functions'. Comput Aided Des 10:253–265
11. Bezier P (1966) Définition numérique des courbes et surfaces I. Automatisme 11:625–632
12. Bezier P (1967) Définition numérique des courbes et surfaces II. Automatisme 12:17–21
13. Bezier P (1968) Procédé de définition numérique des courbes et surfaces non mathématiques. Automatisme 13(5):189–196
14. Bezier P (1968b) How Renault uses numerical control for car body design and tooling. SAE Paper 680010
15. Bezier PE (1971) Example of an existing system in motor industry: the UNISURF system. Proc R Soc Lond Ser A 321:207–218
16. Bezier P (1971) Numerical control and foundries. FOND-FR 26(299):77
17. Bezier P (1972) Numerical control: mathematics and applications. Wiley (translated by A. R. Forrest)
18. Bezier P (1973) UNISURF system: principles, program, language. In: Harvany J (ed)Proceedings of 1973 PROLAMAT conference, Budapest, North Holland Publ. Co., Amsterdam
19. Bezier P (1974) Mathematical and practical possibilities of UNISURF. In: Barnhill R, Riesenfeld R (eds) Computer aided geometric design. Academic Press, Cambridge
20. Bezier P (1977) Essay de définition numérique des courbes et des surfaces expérimentales. Ph.D. thesis, University of Paris VI
21. Bezier P (1978) General distortion of an ensemble of biparametric surfaces. Comput Aided Des 10(2):116–120
22. Bezier PE (1981) A view of CAD-CAM. Comput Aided Des 13(4):207–209
23. Bezier PE, Sioussiou S (1983) Semi-automatic system for defining free-form curves and surfaces. Comput Aided Des 15(2):65–72
24. Bezier PE (1983) UNISURF, from styling to tool-shop. Comput Ind 4(2):115–126
25. Bezier PE (1984) CADCAM—past, requirements, trends. Comput Aided Des 16(2):102

26. Bezier P (1986) The mathematical basis of the UNISURF CAD system. Butterworths, London
27. Bezier P (1989) 1st steps of CAD. Comput Aided Des 21(5):259–261
28. Bezier P (1990) Style, mathematics and NC. Comput Aided Des 22(9):524–526
29. Bezier P (1998) A view of the CAD/CAM development period. IEEE Ann Hist Comput 20(2):37–40
30. Boyse JW, Rosen JM (1981) GMSOLID—a system for interactive design and analysis of solids. SAE Trans, 90, Section 1: 810010-810234, pp 847–857
31. Brebbia CA (1982) Finite element systems: a handbook, 2nd edn. Springer, Berlin
32. Brebbia CA, Dominguez J (1992) Boundary elements: an introductory course. Computational Mechanics Publications, McGraw-Hill Book Company, Southampton
33. Casale MS (1989) Integration of geometric analysis and structural analysis using trimmed patches. Ph.D. thesis, University of California, Irvine
34. Casale MS, Bobrow JE (1989) The analysis of solids without mesh generation using trimmed patch boundary elements. Eng Comput 5 249–257
35. Casale MS, Bobrow JE, Underwood R (1992) Trimmed-patch boundary elements: bridging the gap between solid modeling and engineering analysis. Comput Aided Des 24:193–199
36. Cavendish JC, Gordon WJ, Hall CA (1976) Ritz-Galerkin approximations in blending function spaces. Numer Math 26:155–178
37. Cavendish JC, Gordon WJ, Hall CA (1977) Substructured macro elements based on locally blending interpolation. Int J Numer Meth Eng 11:1405–1421
38. Cavendish JC, Hall CA (1984) A new class of transitional blended finite elements for the analysis of solid structures. Int J Numer Meth Eng 28:241–253
39. Charlesworth WW, Cox JJ, Anderson DC (1994) The domain decomposition method applied to Poisson's equation in two dimensions. Int J Solids Struct 37(18):3093–3115
40. Clark BW, Anderson DC (2003) The penalty boundary method for combining meshes and solid models in finite element analysis. Eng Comput 20(4):344–365
41. Clark BW, Anderson DC (2003) The penalty boundary method. Finite Elem Anal Des 39:387–401
42. Collatz L (1960) The numerical treatment of differential equations, 3rd edn. Springer-Verlag, Berlin
43. Cook WA (1974) Body oriented (natural) co-ordinates for generating three-dimensional meshes. Int J Numer Meth Eng 8:27–43
44. Coons SA (1964) Surfaces for computer aided design of space form, Project MAC, MIT (1964), revised for MAC-TR-41 (1967), Springfield, VA 22161, USA. Available as AD 663 504 from the National Technical Information Service (CFSTI), Sills Building, 5285 Port Royal Road. Now, online available at: http://publications.csail.mit.edu/lcs/pubs/pdf/MIT-LCS-TR-041.pdf
45. Coons S (1968) Rational bibubic surface patches. Technical report, MIT, Project MAC
46. Cottrell JA, Hughes TJR, Bazilevs Y (2009) Isogeometric analysis: towards integration of CAD and FEA. Wiley, Chichester
47. Cox MG (1972) The numerical evaluation of B-splines. J Inst Math Its Appl 10:134–149
48. De Boor C (1972) On calculating with B-splines. J Approx Theory 6:50–62
49. De Casteljau P (1959) Courbes à poles. National Industrial Property Institute (INPI, France)
50. De Casteljau PF (1999) De Casteljau's autobiography: my life at Citroën. Comput Aided Geom Des 16:583–586
51. Dimitriou V (2004) Adaptive finite elements and related meshes, Ph.D. Dissertation (advisor: Prof. Andreas E. Kanarachos), National Technical University of Athens, School of Mechanical Engineering, Athens, August, 2004
52. El-Zafrany A, Cookson RA (1986) Derivation of Lagrangian and Hermitian shape functions for triangular elements. Int J Numer Meth Eng 23(2):275–285
53. El-Zafrany A, Cookson RA (1986) Derivation of Lagrangian and Hermitian shape functions for quadrilateral elements. Int J Numer Meth Eng 23(10):1939–1958
54. Farin G (1990) Curves and surfaces for computer aided geometric design: a practical guide. Academic Press, Boston

55. Farin G, Hoschek J, Kim MS (2002) Handbook of computer aided geometric design. Elsevier, North-Holland
56. Forrest AR (1968) Curves and surfaces for computer-aided design, Ph.D. dissertation, Cambridge University, Cambridge, UK
57. Gordon WJ (1969) Spline-blended surface interpolation through curve networks. Indiana Univ Math J 18, 10, pp 931–952. Online available at: http://www.iumj.indiana.edu/IUMJ/FULLTEXT/1969/18/18068
58. Gordon WJ (1971) Blending functions methods of bivariate and multivariate interpolation and approximation. SIAM J Numer Anal 8:158–177
59. Gordon WJ (1983) An operator calculus for surface and volume modeling. IEEE Comput Graph Appl 3:18–22
60. Gordon WJ, Hall CA (1973) Construction of curvilinear co-ordinate systems and application to mesh generation. Int J Numer Meth Eng 7:461–477
61. Gordon WJ, Hall CA (1973) Transfinite element methods: Blending-function interpolation over arbitrary curved element domains. Numer Math 21:109–112
62. Gordon WJ, Thiel LC (1982) Transfinite mappings and their application to grid generation. In: Thompson JP (ed) Numerical grid generation. Applied mathematics and computation, vol 11–12. Elsevier, pp. 171–233
63. Gregory JA (1983) N-sided surface patches. In: Gregory JA (ed) Math Surf. Clarendon Press, Oxford, pp 217–232
64. Haber RB, Shephard MS, Abel JF, Gallagher RH, Greensberg DP (1981) A general three-dimensional graphical finite element preprocessor utilizing discrete transfinite mappings. Int J Numer Meth Eng 17:1015–1044
65. Haber RB, Abel JF (1982) Discrete transfinite mappings for the description and meshing of three-dimensional surfaces using interactive computer graphics. Int J Numer Meth Eng 18:41–66
66. Hartmann S, Benson D, Nagy A (2016) Isogeometric analysis with LS-DYNA. J Phys Conf Ser 734:032125. https://doi.org/10.1088/1742-6596/734/3/032125
67. Higuchi F, Gofuku S, Maekawa T, Mukundan H, Patrikalakis NM (2007) Approximation of involute curves for CAD-system processing. Eng Comput 23(3):207–214
68. Höllig K (2002) Finite element approximation with splines. In: Farin G, Hoschek J, Kim MS (eds) Handbook of computer aided geometric design, North-Holland, Amsterdam, Chapter 11, pp 283–307
69. Höllig K (2003) Finite element methods with B-splines. SIAM, Philadelphia
70. Hughes TJR (2000), The finite element method: Linear static and dynamic finite element analysis. Dover Publications, Mineola, NY
71. Hughes TJR, Cottrell JA, Bazilevs Y (2005) Isogeometric analysis: CAD, finite elements, NURBS, exact geometry and mesh refinement. Comput Methods Appl Mech Eng 194:4135–4195
72. Inoue K, Kikuchi Y, Masuyama T (2005) A NURBS finite element method for product shape design. J Eng Des 16(2):157–174
73. Kagan P, Fischer A (2000) Integrated mechanical based CAE systems using B-spline based finite elements. Comput Aided Des 32(8–9):539–552
74. Kanarachos A, Röper O (1979) Rechnerunterstützte Netzgenerierung mit Hilfe der Coonsschen Abbildung. VDI-Z 121:297–303
75. Kanarachos A, Provatidis Ch (1987) Performance of mass matrices for the BEM dynamic analysis of wave propagation problems. Comput Methods Appl Mech Eng 63:155–165
76. Kanarachos AE, Spentzas CN (1988) Comparison of four finite element solutions of self-adjoint and non-self-adjoint problems governed by differential equations with predominant lower order derivatives. Eng Comput 5:71–74
77. Kanarachos A, Deriziotis D (1989) On the solution of Laplace and wave propagation problems using 'C-elements'. Finite Elem Anal Des 5:97–109
78. Kanarachos A, Provatidis C, Deriziotis D, Foteas N (1999),A new approach of the FEM analysis of two-dimensional elastic structures using global (Coons's) interpolation functions.

In: Wunderlich J (ed) CD proceedings first european conference on computational mechanics, München-Germany, August–September, 1999
79. Lanczos C (1938) Trigonometric interpolation of empirical and analytical functions. Stud Appl Math 17(1–4):123–199
80. Lee K (1999) Principles of CAD/CAM/CAE systems. Addison-Wesley, Reading, MA
81. Liu B, Xing Y, Wang Z, Lu X, Sun H (2017) Non-uniform rational Lagrange functions and its applications to isogeometric analysis of in-plane and flexural vibration of thin plates. Comput Methods Appl Mech Eng 321:173–208
82. Loop CT, DeRose TD (1989) A multisided generalization of Bezier surfaces. ACM Trans Graph 8:204–234
83. Martin W, Cohen E (2001) Representation and extraction of volumetric attributes using trivariate splines: a mathematical framework. In: Proceedings of the solid modelling symposium (http://www.cs.utah.edu)
84. Natekar D, Zhang X, Subbarayan G (2004) Constructive solid analysis: a hierarchical, geometry-based meshless analysis procedure for integrated design and analysis. Comput Aided Des 36(5):473–486
85. Park S, Lee K (1997) High-performance trivariate NURBS representation for analyzing and visualizing fluid flow data. Comput Graph 21:473–482
86. Provatidis C (1987) On the application of the boundary element method in the analysis of field and dynamic problems, Doctoral Dissertation, National Technical University of Athens, Mechanical Engineering Department, Greece, November 1987 (in Greek)
87. Provatidis CG, Kanarachos AE (1995) Further research on the performance of consistent mass matrices using BEM for symmetric/nonsymmetric formulations. Comput Mech 16:197–207
88. Provatidis C, Kanarachos A (2001) Performance of a macro-FEM approach using global interpolation (Coons') functions in axisymmetric potential problems. Comput Struct 79(19):1769–1779
89. Provatidis C (2001b) A global approximation technique in noise-control. In: Tsahalis DT (ed) Proceedings 4th European conference on noise control (EuroNoise-2001, 14–17 January 2001, Patras, Greece), vol I, pp 23–31
90. Provatidis C (2001c) Acoustic analysis of two-dimensional mufflers using large finite elements derived from Coons' interpolation. In: Drakatos PA (ed) CD proceedings of ASME—Greek Section, September 17–20, 2001, Patras, Greece (Paper ANG1/P099, pages 1–6)
91. Provatidis C (2001d) Stress analysis of 3D solid structures using large boundary elements derived from 2D-Coons' interpolation. In: Drakatos PA (ed) CD proceedings of ASME—-Greek Section, September 17–20, 2001, Patras, Greece (Paper ANG1/P129, pages 1–6)
92. Provatidis C (2002) Coons-patch macroelements in potential Robin problems. Forsch Ingenieurwes 67(1):19–26
93. Provatidis C (2002b) A comparative study between Coons-patch macroelements and boundary elements in two-dimensional potential problems. In: Tsahalis DT (ed) Proceedings 4th GRACM congress on computational mechanics, 27–29 June, 2002, Patras, Greece, pp 43–50
94. Provatidis C (2002c) Analysis of three-dimensional sound radiation problems using trimmed patch boundary elements. In: Tsahalis DT (ed) Proceedings 4th GRACM Congress on Computational Mechanics, 27–29 June, 2002, Patras, Greece, pp. 402-409
95. Provatidis CG (2002d) CAD-FEA integration using Coons interpolation. Technical Report MD&CS 01–2002, National Technical University of Athens, Mechanical Engineering Department; also submitted to "Engineering with Computers" (EWC03-019, Sept 12, 2003)
96. Provatidis C (2003) Analysis of axisymmetric structures using Coons' interpolation. Finite Elem Anal Des 39:535–558
97. Provatidis C (2003b) Frequency analysis and transient response of two-dimensional structures using Coons-patch macroelements. In: Brennan MJ, Ferman MA, Petersson BAT, Rizzi SA, Wentz K (eds) Proceedings of the VIII international conference on recent advances in structural dynamics, 14–16 July 2003, Southampton, UK (paper No. 24)
98. Provatidis C (2003c) Free vibrations of two-dimensional structures using Coons-patch macroelements, FEM and BEM. In: Atluri SN, Beskos DE, Polyzos D (eds) Proceedings from

ICCES 2003, advances in computational & experimental engineering & sciences, 24–29 July 2003, Corfu, Greece, Chapter 5, Paper No. 249
99. Provatidis C, Zafiropoulos N (2003d) Free-vibration analysis of three-dimensional solids using Coons-patch Boundary Superelements. In: Aifantis E (ed) Proceedings 5th European solids mechanics conference, Thessaloniki, 17–22 August 2003
100. Provatidis C, Zafiropoulos N (2003e) Determination of eigenfrequencies in three-dimensional acoustic cavities using Coons-patch Boundary Superelements. In: Aifantis E (ed) Proceedings 5th European solids mechanics conference, Thessaloniki, 17–22 August 2003
101. Provatidis CG (2004) Coons-patch macroelements in two-dimensional eigenvalue and scalar wave propagation problems. Comput Struct 82(4–5):383–395
102. Provatidis CG (2004) On DR/BEM for eigenvalue analysis of 2-D acoustics. Comput Mech 35:41–53
103. Provatidis CG (2004) Solution of two-dimensional Poisson problems in quadrilateral domains using transfinite Coons interpolation. Commun Numer Methods Eng 20(7):521–533
104. Provatidis CG (2005) Performance of large Gordon-Coons finite elements in 2-D potential problems. In: Proceedings GRACM 05, Limassol, Cyprus, 29 June–1 July, 2005
105. Provatidis CG, Vossou CG, Theodorou EG (2006) On the CAD/CAE integration using Coons interpolation. In: Proceedings 2nd international conference "from scientific computing to computational engineering", Athens, Greece, 5–8 July, 2006
106. Provatidis CG (2007) Final report for the project: PROTAGORAS (DIDYMO: Investigation on CAD/CAE integration), EDEIL Contract No. 65/1388-June 2004, February, 2007
107. Provatidis C, Kanarachos A (2000) On the use of Coons' interpolation in CAD/CAE systems. In: Mastorakis N (ed) Systems and control: theory and applications. World Scientific and Engineering Society Press, pp 343–348. Online available on: http://www.wseas.us/e-library/conferences/athens2000/Papers2000/547.pdf
108. Provatidis CG (2005) Three-dimensional Coons macroelements in Laplace and acoustic problems. Comput Struct 83:1572–1583
109. Provatidis CG (2005) Analysis of box-like structures using 3-D Coons' interpolation. Commun Numer Methods Eng 21:443–456
110. Provatidis CG (2005c) Performance of several RBFs in DR/BEM eigenvalue analysis of 2-D structures. GRACM 05, Limassol, Cyprus, 29 June–1 July 2005
111. Provatidis CG (2006) Coons-patch macroelements in two-dimensional parabolic problems. Appl Math Model 30:319–351
112. Provatidis CG (2006) Free vibration analysis of two-dimensional structures using Coons-patch macroelements. Finite Elem Anal Des 42(6):18–531
113. Provatidis CG (2006) Transient elastodynamic analysis of two-dimensional structures using Coons-patch macroelements. Int J Solids Struct 43(22–23):6688–6706
114. Provatidis CG (2006) Three-dimensional Coons macroelements: application to eigenvalue and scalar wave propagation problems. Int J Numer Meth Eng 65:111–134
115. Provatidis CG (2007) Performance of a Lagrange based global finite element collocation method for eigenvalue structural analysis. In: Proceedings 8th HSTAM international congress on mechanics, Patras, 12–14 July, 2007
116. Provatidis CG (2008) Free vibration analysis of elastic rods using global collocation. Arch Appl Mech 78(4):241–250
117. Provatidis CG (2008) Time- and frequency-domain analysis using lumped mass global collocation. Arch Appl Mech 78(11):909–920
118. Provatidis CG (2008) Global collocation method for 2-D rectangular domains. J Mech Mater Struct 3(1):185–194
119. Provatidis CG (2009) Integration-free Coons macroelements for the solution of 2-D Poisson problems. Int J Numer Meth Eng 77:536–557
120. Provatidis CG (2009) Eigenanalysis of two-dimensional acoustic cavities using transfinite interpolation. J Algorithms Comput Technol 3(4):477–502
121. Provatidis CG (2009c) Higher order Galerkin/Ritz approximations in 1-D eigenval problems. In: Tsahalis D (ed) Proceedings 3rd international conference on experiments/process/system modeling/simulation & optimization (3rd IC-EpsMsO), Athens, 8–11 July, 2009

122. Provatidis CG, Isidorou S (2009) Comparison of advanced collocation methods for the solution of ordinary differential equations. In: Tsahalis D (ed) Proceedings 3rd international conference on experiments/process/system modeling/simulation & optimization (3rd IC-EpsMsO), Athens, 8–11 July, 2009
123. Provatidis CG, Ioannou KS (2010) Static analysis of two-dimensional elastic structures using global collocation. Arch Appl Mech 80(4):389–400
124. Provatidis CG (2011) Equivalent finite element formulations for the calculation of eigenvalues using higher-order polynomials. Appl Math 1(1):13–23
125. Provatidis CG (2011b) Global versus local interpolation in the FEM free vibration analysis of prismatic bars. In: Proceedings 7th GRACM international congress in mechanics, Athens, 30 June–2 July, 2011
126. Provatidis CG (2011c) Some issues on CAD/CAE integration: global interpolation using isoparametric and isogeometric techniques. In: Proceedings 7th GRACM international congress in mechanics, Athens, 30 June-2 July, 2011
127. Provatidis CG, Isidorou SK (2011) B-splines collocation eigenvalue analysis of 1-D problems. In: Proceedings 7th GRACM international congress in mechanics, Athens, 30 June–2 July, 2011
128. Provatidis CG (2012) Two-dimensional elastostatic analysis using Coons-Gordon interpolation. Meccanica 47(4):951–967
129. Provatidis C (2013) A review on attempts towards CAD/CAE integration using macroelements. Comput Res 1(3):61–84
130. Provatidis CG (2014) Bezier versus Lagrange polynomials-based finite element analysis of 2-D potential problems. Adv Eng Softw 73:22–34
131. Provatidis CG (2018) Engineering analysis with CAD-based macroelements. Arch Appl Mech 88:121–140
132. Rabut C (2002) On Pierre Bezier's life and motivations. Comput Aided Des 34(7):493–510
133. Renken FP, Subbarayan G (2000) NURBS-based solution to inverse boundary problems in droplet shape prediction. Comput Methods Appl Mech Eng 190(11):1391–1406
134. Röper O (1978) Ein Geometrieprozessor für die rechnerunterstützte Auslegung von Maschinenbauteilen mit Hilfe der Methode der Finite Elemente, Dissertation, Ruhr-Universität Bochum, Mai 1978
135. Schillinger D, Ruthala PK, Nguyen LH (2016) Lagrange extraction and projection for NURBS basis functions: a direct link between isogeometric and standard nodal finite element formulations. Int J Numer Meth Eng 108(6):515–534
136. Schoenberg IJ (1946) Contributions to the problem of approximation of equidistant data by analytic functions. Q Appl Math 4:45–99
137. Schramm U, Pilkey WD (1993) The coupling of geometric descriptions and finite element using NURBS—a study in shape optimization. Finite Elem Anal Des 15:11–34
138. Schramm U, Pilkey WW (1994) Higher order boundary elements for shape optimization using rational B-splines. Eng Anal Boundary Elem 14(3):255–266
139. Szabó B, Babuška I (1991) Finite element analysis. Wiley, New York
140. Turner MJ, Clough RW, Martin HC, Topp LT (1956) Stiffness and deflection analysis of complex structures. Journal of the Aeronautical Sciences 23 (9):805–823
141. Van Blerk JJ, Botha JF (1993) Numerical solution of partial differential equations on curved domains by collocation. Numer Methods Partial Differ Equat 9:357–371
142. Versprille KJ (1975) Computer-aided design applications of the rational B-spline approximation form, Ph.D. Dissertation, Syracuse University, USA, 1975, http://surface.syr.edu/eecs_etd/262 (advisor: Prof. S. A. Coons)
143. Wu SC, Abel JF (1979) Representation and discretization of arbitrary surfaces for finite element shell analysis. Int J Numer Meth Eng 14:813–836
144. Yildir YB, Wexler A (1983) MANDAP—a FEM/BEM preparation package. IEEE Trans Magn 19:2562–2565
145. Zhaobei Z, Zhiqiang X (1987) Coons' surface method for formulation of finite element of plates and shells. Comput Struct 27(1):79–88

146. Zhang YJ (2017) Integrating CAD with Abaqus: a practical isogeometric analysis software platform for industrial applications. In: IACM 19th international conference on finite elements in flow problems—FEF 2017, 5–7 April 2017, Rome, Italy (paper: a79.pdf)
147. Zienkiewicz OC (1977) The finite element method, 3rd edn. McGraw-Hill, London
148. Zienkiewicz OC, Taylor RL (1988) The finite element method, 4rd edn, vol 1: Basic formulation and linear problems, and vol 2: Solid and fluid mechanics, dynamics and non-linearity. McGraw-Hill, London
149. Zienkiewicz OC, Taylor RL (2000) The finite element method, 5rd edn, 3 volume set. Butterworth-Heinemann, Elsevier, Kidlington, Oxford

Chapter 2
Elements of Approximation and Computational Geometry

Abstract In this chapter, we deal with several important formulas for approximation and interpolation. First we start with the one-dimensional problem and then we extend to the two-dimensional case. In addition to the classical Lagrange and Hermite interpolation, we also focus on some other interpolations which appear in CAGD theory. An easy way to understand the relationship between approximation and CAGD formulas is to consider the graph of the smooth solution $U(x, y)$ in a boundary value problem (or the graph of the eigenvector in an eigenvalue problem) as a surface patch described by the function $z = U(x, y)$. Then it is reasonable to approximate the variable U within this patch using any kind of known CAGD surface interpolation formulas. Fifteen exercises clarify the most important issues of the theory.

Keywords Approximation · Linear interpolation · Blending function
Lagrange polynomial · Hermite interpolation · Bernstein polynomial
Bézier interpolation · Control point · B-splines · Truncated power form
Curry–Schoenberg formulation · Natural cubic B-splines · Coons · Gordon
Rational Bézier · Lagrange–Bézier equivalency · NURBS · Barnhill

2.1 Description of Curve's Geometry as Well as of a Physical Quantity Along It

Interpolation means that one matches the given data exactly. *Approximation* is a more general term, which means that one matches the data. This dichotomy is discussed at some length in Davis book [13]. Böhm et al. [6] and Barnhill [2] have discussed this issue from the Computer-Aided Geometric Design (CAGD) point of view.

Within the context of this book, there are two cases in which we need to approximate a univariate function. In the first case, we need to represent a great variety of spatial curves $\mathbf{C}(\xi) \equiv [x(\xi), y(\xi), z(\xi)]^{\mathrm{T}}$, for which the Cartesian coordinates (x, y, z) are given in terms of an independent parameter ξ. In the second case, we need to find an approximate solution $U(\xi)$ of a two-point boundary value problem

C. G. Provatidis, *Precursors of Isogeometric Analysis*, Solid Mechanics and Its Applications 256, https://doi.org/10.1007/978-3-030-03889-2_2

governed by an ordinary differential equation (ODE), $D(U) = f(\xi)$ under given boundary conditions (BCs). In the former case, we deal with the *geometry* of the model (CAD), whereas in the latter we deal with the *analysis* of the model (CAE: Computer-Aided Analysis).

Below we remind of some basic knowledge which is useful from the point of view of both the geometry and the analysis.

2.2 Definitions

Definition 1 A function of a single variable (e.g., $U(\xi)$) is called a *univariate* function.

Definition 2 A function of more than one variables (e.g., $U(\xi,\eta)$, $U(\xi,\eta,\zeta)$, etc.) is called a *multivariate* function.

Definition 3 When a function $U(\xi)$ is *continuous* at a point c, this means that for every $\varepsilon > 0$ there is one $\delta > 0$ such as for all points ξ:

$$|\xi - c| < \delta \Rightarrow |U(\xi) - U(c)| < \varepsilon. \tag{2.1}$$

2.3 Approximation Using Univariate Functions

2.3.1 *Taylor Power Series*

A univariate function $U(x)$ can be expanded according to a Taylor series as follows

$$U(x) = U(0) + \frac{x}{1!}U'(0) + \frac{x^2}{2!}U''(0) + \frac{x^3}{3!}U'''(0) + \cdots \tag{2.2}$$

Introducing the following constants

$$a_0 = U(0), \quad a_1 = \frac{U'(0)}{1!}, \quad a_2 = \frac{U''(0)}{2!}, \quad a_3 = \frac{U'''(0)}{3!}, \ldots, \tag{2.3}$$

an alternative way to write Eq. (2.2) is:

$$U(x) = a_0 + a_1 x + a_2 x^2 + a_3 x^3 + \cdots = \sum_{j=0}^{\infty} a_j x^j \tag{2.4}$$

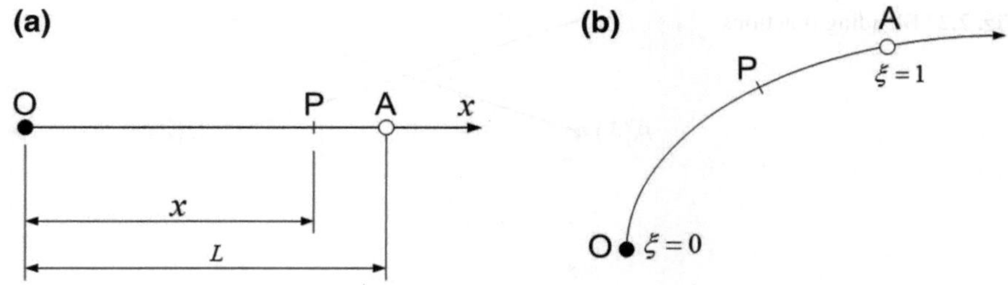

Fig. 2.1 **a** Straight and **b** curvilinear curves

Equation (2.4) constitutes the *power series form* of the univariate function $U(x)$ but a similar expression may be also used for the representation of a curve $\mathbf{C}(x)$. Here, the quantity x may represent either the Cartesian coordinate or the relevant normalized coordinate ξ (see Fig. 2.1). While for a straight segment OA the determination of a point P is given by a trivial formula such as $\xi = x/L$ (Fig. 2.1a), in the case a curvilinear segment OA the coordinate x could be replaced for example by the length s of the curvilinear segment OP (Fig. 2.1b) but the definition of ξ is not restricted to the ratio of the curved lengths (OP)/(AO) only.

Starting from the well-known *linear interpolation* and other interpolation schemes of a univariate function $U(x)$, we will later present some useful interpolations in CAD patches (surface portions). Moreover, volume blocks and relevant macroelements will be separately treated in Chap. 10.

2.3.2 One-Dimensional Linear Interpolation

Let us assume a univariate function $U(x)$ from which we isolate a part between the points x_1 and x_2 at which the function takes the values U_1 and U_2, respectively, i.e.,

$$U(x_1) = U_1 \quad \text{and} \quad U(x_2) = U_2 \tag{2.5}$$

A linear interpolation between the points x_1 and x_2 is equivalent with the assumption

$$U(x) = \alpha + \beta x \tag{2.6}$$

If Eq. (2.6) is applied to both points x_1 and x_2, a linear system of two equations is derived, from which we finally obtain

$$U(x) = \hat{E}_0(x) \cdot u_1 + \hat{E}_1(x) \cdot u_2 \tag{2.7}$$

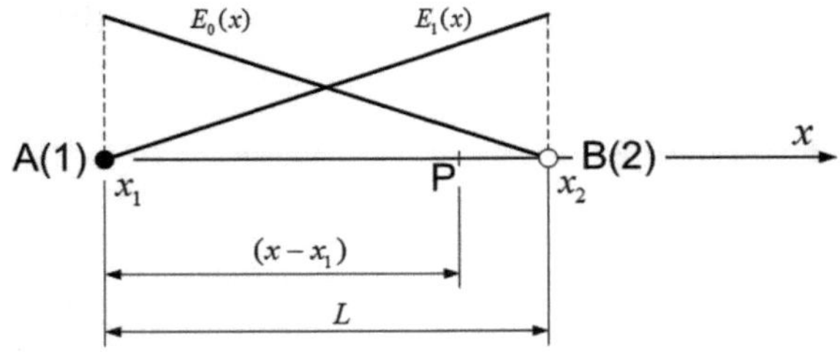

Fig. 2.2 Blending functions

with

$$\hat{E}_0(x) = 1 - \frac{(x - x_1)}{(x_2 - x_1)} \quad \text{and} \quad \hat{E}_1(x) = \frac{(x - x_1)}{(x_2 - x_1)} \tag{2.8}$$

The functions $\hat{E}_0(x)$ and $\hat{E}_1(x)$ in Eq. (2.7) are called *blending* functions and are associated with the endpoints 1 and 2 at x_1 and x_2, respectively, as shown in Fig. 2.2. The reason of using the subscripts "0" and "1", instead of the more reasonable "1" and "2" which better corresponds to the points x_1 and x_2, is explained below (after Eq. 2.11).

Alternatively, introducing the normalized coordinate

$$\xi = \frac{x - x_1}{x_2 - x_1}, \quad 0 \leq \xi \leq 1, \tag{2.9}$$

the blending functions take the normalized form:

$$E_0(\xi) = 1 - \xi \quad \text{and} \quad E_1(\xi) = \xi, \tag{2.10}$$

whereas Eq. (2.7) becomes:

$$U(\xi) = E_0(\xi) \cdot U_1 + E_1(\xi) \cdot U_2 \tag{2.11}$$

Since now $U_1 = U(\xi = 0) \equiv U(0)$ and $U_2 = U(\xi = 1) \equiv U(1)$, we can understand that the blending function $E_0(\xi)$ is associated to the node "1" that is found at the position ($\xi = 0$), whereas the blending function $E_1(\xi)$ is associated to the node "2" that is found at the position ($\xi = 1$). Thus, the subscript ("0" or "1") in Eq. (2.11) refers to the associated node of which the normalized coordinate ξ equals to the subscript.

We stretch the attention of the reader to the fact that Eq. (2.11) is valid for the interpolation of a physical quantity U associated to any point P with a coordinate x (or a normalized coordinate ξ) along the axis. But also the same happens with the coordinate itself. Indeed, given the coordinates x_1 and x_2 at two points along a straight line, any point in its interior (including the ends) is given by

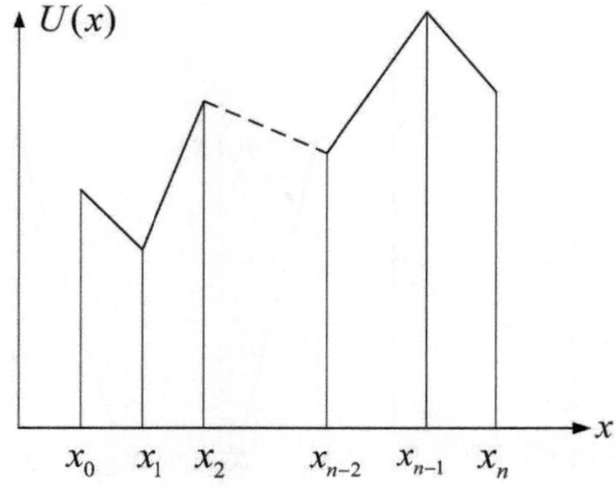

Fig. 2.3 Piecewise linear interpolation of the function $U(x)$

$$x(\xi) = E_0(\xi) \cdot x_1 + E_1(\xi) \cdot x_2, \quad 0 \leq \xi \leq 1 \tag{2.12}$$

In other words, both the variable u and the coordinate x itself are linearly interpolated in the same manner. While this remark is trivial for a straight segment, the transfer of the same idea to a curvilinear line segment constitutes the essence of the so-called *isoparametric* interpolation, which is a mapping between the reference domain $\xi \in [0, 1]$ and the actual one $x \in [x_1, x_2]$.

2.3.3 One-Dimensional (1D) Piecewise Linear Interpolation

Based on the abovementioned linear interpolation between any two successive points, we introduce the idea of *piecewise linear* interpolation. Concretely, between the successive points x_1 and x_2, the function $U(x)$ varies linearly according to Eq. (2.12). More generally, let $x_0, x_1, \ldots, x_{n-1}, x_n$ be a nondecreasing sequence along an axis (or a line), for which the corresponding values of the function $U(x)$ are $U_0, U_1, \ldots, U_{n-1}, U_n$.

Then the piecewise approximation of function $U(x)$ over the interval $x_0 \leq x \leq x_n$, as shown in Fig. 2.3, is

$$\tilde{U}(x) = \sum_{i=0}^{n} \varphi_i(x) \cdot U_i, \tag{2.13}$$

where (see, e.g., [35]):

$$\varphi_0(x) = \begin{cases} \frac{x_1 - x}{x_1 - x_0} & (x_0 \leq x \leq x_1) \\ 0 & (x_1 \leq x \leq x_n) \end{cases} \tag{2.14}$$

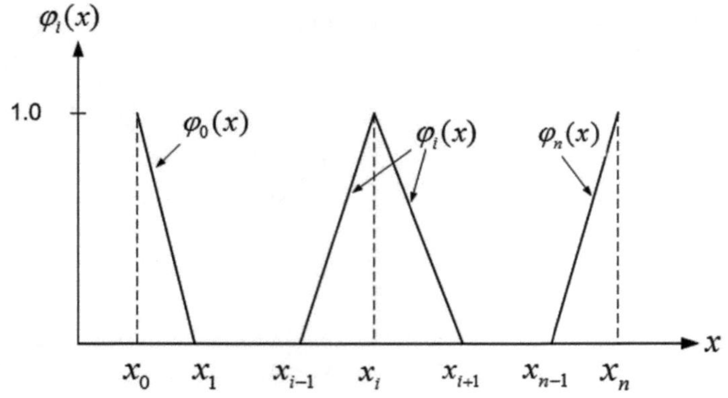

Fig. 2.4 Piecewise linear (hat) functions

also

$$\varphi_i(x) = \begin{cases} 0 & (x_0 \le x \le x_{i-1}) \\ \left(\frac{x-x_{i-1}}{x_i-x_{i-1}}\right) & (x_{i-1} \le x \le x_i) \\ \left(\frac{x_{i+1}-x}{x_{i+1}-x_i}\right) & (x_i \le x \le x_{i+1}) \\ 0 & (x_{i+1} \le x \le x_n) \end{cases} \tag{2.15}$$

and

$$\varphi_n(x) = \begin{cases} 0 & (x_0 \le x \le x_{n-1}) \\ \frac{x-x_{n-1}}{x_n-x_{n-1}} & (x_{n-1} \le x \le x_n) \end{cases} \tag{2.16}$$

are pyramid (hat) functions illustrated in Fig. 2.4.

2.3.4 One-Dimensional (1D) Lagrange and Power Series Interpolation

In order to interpolate $(n + 1)$ given points $(x_i, U_i), i = 0, 1, \ldots, n$, such that $x_0 < x_1 < \cdots < x_n$, we generally need a polynomial of degree n, which involves $(n + 1)$ coefficients and which may be expressed as

$$U_L(x) = \sum_{i=0}^{n} L_i(x) U_i, \tag{2.17}$$

where

$$L_i(x) = \frac{(x - x_0)(x - x_1)\dots(x - x_{i-1})(x - x_{i+1})\dots(x - x_n)}{(x_i - x_0)(x_i - x_1)\dots(x_i - x_{i-1})(x_i - x_{i+1})\dots(x_i - x_n)}$$
$$\equiv \prod_{\substack{j=0 \\ j \neq i}}^{n} \frac{(x - x_j)}{(x_i - x_j)} \tag{2.18}$$

This is *Lagrange interpolation formula*, and the expressions $L_i(x)$ are known as *Lagrange polynomials*. As can be seen, they are all polynomials of degree n in x, as in general will be $U_L(x)$ itself. Obviously, if it happens that the data points lie on a polynomial line of degree p lower than $n(p < n)$, then despite the fact that all Lagrange polynomials in Eq. (2.17) are of degree n, the involved monomials $x^{p+1}, \dots, x^n$ vanish.

Example 2.1 (Lagrange polynomials and series expansion) Let us take the function $U(x) = x$ which is defined in the interval [0, 1]. Supposing we do not still know the degree of the actual polynomial, we try to determine it using three data points, at the ends and the middle of the interval, i.e., $(x_0, U_0) = (0, 0)$, $(x_1, U_1) = (0.5, 0.5)$ and $(x_2, U_2) = (1, 1)$. Based on the aforementioned three values $(x_0, x_1, x_2) = (0, 0.5, 1)$, Eq. (2.17) gives the Lagrange polynomials: $L_0(x) = (2x - 1)(x - 1)$, $L_1(x) = 4x(1 - x)$, and $L_2(x) = x(2x - 1)$.

Therefore, Eq. (2.17) becomes:

$$\begin{aligned} U_L(x) &= L_0(x)\,U_0 + L_1(x)\,U_1 + L_2(x)\,U_2 \\ &= (2x-1)(x-1)\cdot 0 + 4x(1-x)\cdot\frac{1}{2} + x(2x-1)\cdot 1 \\ &= 2x - 2x^2 + 2x^2 - x \\ &= x \end{aligned}$$

The above relationship shows that despite the fact that we used quadratic Lagrange polynomials, the combination of the U_i's finally led to a linear function (i.e., $U_L(x) = x$). ■

In principle, Lagrange polynomials seem to be sufficient for interpolation and approximation. Nevertheless, there are some cases in which they suffer from numerical oscillations as it has been widely written. Below we shall demonstrate some of these cases.

Example 2.2 (Misleading behavior of Lagrange polynomials) In the beginning, we shall show that sometimes Lagrange interpolation may suffer from numerical oscillations when the function $U(x)$ is not of a polynomial form. One characteristic

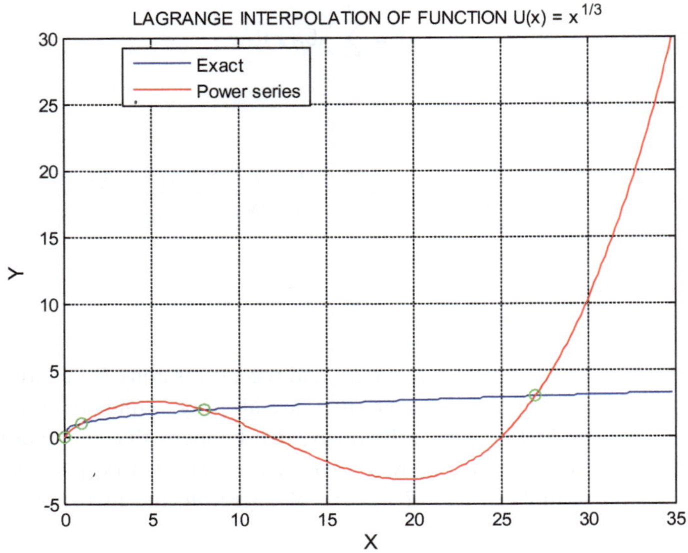

Fig. 2.5 Misleading behavior of Lagrange interpolation in a particular problem

(admittedly extreme) example can be found in a standard the textbook (see [22]) which refers to the function $U(x) = x^{1/3}$. Concretely, let us choose the four positions $x_0 = 0, x_1 = 1, x_2 = 8, x_3 = 27$, to which the corresponding values of the function $y_i = U(x_i) = x_i^{1/3}$ are $y_0 = 0, y_1 = 1, y_2 = 2, y_3 = 3$. Based on the aforementioned four pairs $(x_i, y_i), i = 0, 1, 2, 3$, the Lagrange interpolation is exceptionally inaccurate (gives quite misleading results) compared with the exact function $U(x) = x^{1/3}$, as shown in Fig. 2.5.

A critical explanation of this event is that this disparity arises because we are using a polynomial to approximate a function which does not behave like a polynomial. Actually, by taking the Taylor series around the point $x = 0$, the first term (i.e., $U(0)$) vanished, whereas all derivatives are of the form $U^{(k)}(x) \propto x^{(1/3-k)}$ and obviously tend to infinity when $x \to 0$; thus, a power series cannot be established.

On the other point of view, it is very easy to validate that the accurate solution fulfills the ODE $U'' + 2/(9U^5) = 0$, which is not known to represent any important phenomenon in engineering practice.

It is interesting to mention that, as Faux and Pratt [22] say "*more accurate results would be achieved by using linear (straight line) interpolation between successive pairs of data points in this particular case,*" which in this book is the aforementioned piecewise linear approximation (see §2.3.3). ■

Example 2.3 (The Runge example) Another case, labeled as "Runge example" (see de Boor [18], p. 17), is the function $U(x) = 1/(1+25x^2)$ at the following n uniformly spaced sites in $[-1 \ldots 1]$:

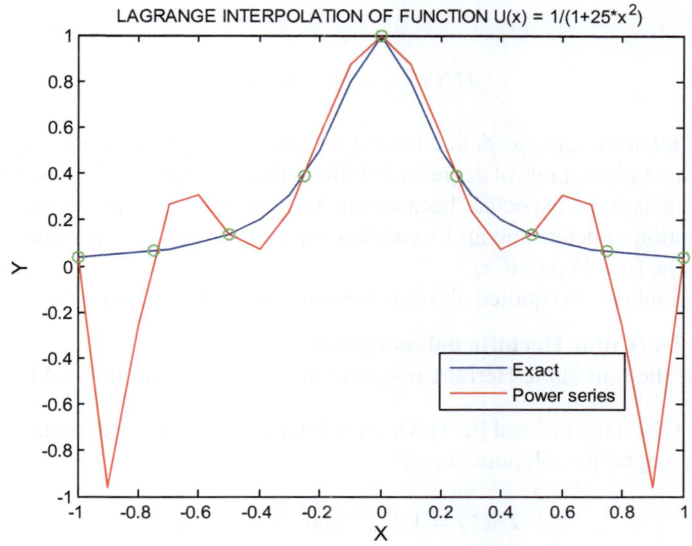

Fig. 2.6 Misleading Lagrange polynomials for the Runge example ($p = 8, n = 8$)

$$x_i = ih - 1, \quad i = 0, \dots, n, \quad \text{with } h = 2/n.$$

Using a power series, or equivalently, nine Lagrange polynomials each of eighth degree ($p = 8, n = 8$), the quality of interpolation through nine given points (green circles) is shown in Fig. 2.6. ■

2.3.5 *One-Dimensional (1D) Hermite Interpolation*

Suppose now that at $(n + 1)$ points x_i, $i = 0, 1, 2, \dots, n$, such that $x_0 < x_1 < \dots < x_n$, we are given not only the function values U_i but also the derivatives U_i'. We may determine the unique polynomial of lowest degree having these same properties at all the points x_i by using the formula

$$U_H(x) = \sum_{i=0}^{n} H_i(x)U_i + \sum_{i=0}^{n} H_i^*(x)U_i', \tag{2.19}$$

in which

$$H_i(x) = \left[1 - 2L_i'(x_i) \cdot (x - x_i)\right] L_i^2(x) \tag{2.20}$$

and

$$H_i^*(x) = (x - x_i)L_i^2(x), \tag{2.21}$$

with the functions $L_i(x)$ as defined in Eq. (2.18), i.e., Lagrange polynomials. Since the $L_i(x)$ are polynomials of degree n, it follows that $U_H(x)$ is a polynomial of degree $2n + 1$. This is to be expected, because such as polynomial requires $(2n + 2)$ items of information to determine all its coefficients, and we use known values of both U and U' at the $(n + 1)$ points x_i.

The formula (2.19) quoted above is *Hermite* interpolation formula.

Exercise 2.4 (Cubic Hermite polynomials)
Determine the four cubic Hermite polynomials within the unit interval [0, 1].

Solution Given the interval [0, 1] which is determined by the endpoints 1 and 2, the two linear Lagrange polynomials are:

$$L_0(\xi) = 1 - \xi \quad \text{and} \quad L_1(\xi) = \xi$$

Applying Eq. (2.19), one receives

$$H_0(\xi) = (1 + 2\xi)(1 - \xi)^2 \equiv 1 - 3\xi^2 + 2\xi^3 \text{ and}$$
$$H_1(\xi) = [1 - 2(\xi - 1)]\xi^2 \equiv 3\xi^2 - 2\xi^3$$

Moreover, applying Eq. (2.21) one receives

$$H_0^*(\xi) = (\xi - 0)(1 - \xi)^2 \equiv \xi - 2\xi^2 + \xi^3 \text{ and}$$
$$H_1^*(\xi) = [(\xi - 1)]\xi^2 \equiv -\xi^2 + \xi^3$$

One may observe that the above Hermite polynomials of third degree (*cubic Hermites*) are those which one can meet as the shape functions in *beam bending* finite element modeling (see Fig. 2.7). As will be mentioned later in Chap. 9, H_0 and H_1 are also noted by E_0^0 and E_1^0, respectively, and correspond to the translational degrees of freedom (partition of unity: $E_0^0 + E_1^0 \equiv 1$). Also, H_0^* and H_1^* are also noted by E_0^1 and E_1^1, respectively, and they correspond to the rotational degrees of freedom; their first derivatives fulfill the conditions $\left[\mathrm{d}E_0^1(x)/\mathrm{d}x\right]_{x=x_i} = \delta_{0i}$ (=Kronecker delta) and $\left[\mathrm{d}E_1^1(x)/\mathrm{d}x\right]_{x=x_i} = \delta_{1i}$, $i = 0, 1$. For example, the first derivative of $H_0(x) \equiv E_0^0(x)$ at $x = x_0 \equiv 0$ equals to the unity, whereas at $x = x_1 \equiv 1$ vanishes (see Fig. 2.7).

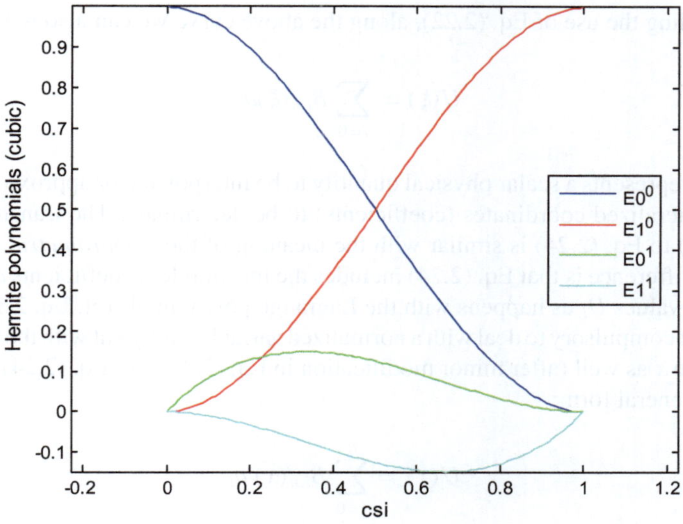

Fig. 2.7 Cubic Hermite polynomials

2.3.6 *One-Dimensional (1D) Nonrational Bézier Interpolation*

The standard form of a *nonrational* Bézier curve $\mathbf{C} = \mathbf{C}(\xi)$ in terms of the normalized coordinate ξ is

$$\mathbf{x}(\xi) = \sum_{i=0}^{n} B_{i,n}(\xi)\mathbf{x}_i\,, \tag{2.22}$$

where

$$B_{i,n}(\xi) \equiv \binom{n}{i} \xi^i (1-\xi)^{n-i} = \frac{n!}{i!\,(n-i)!}\xi^i (1-\xi)^{n-i} \tag{2.23}$$

represents a Bernstein polynomial of degree n and $\mathbf{x}_i$ are the coordinates of the $(n+1)$ control points. It is clarified that the quantity $\mathbf{x}(\xi)$ in Eq. (2.22) is the vector of the Cartesian coordinates of an arbitrary point along the curve $\mathbf{C} = \mathbf{C}(\xi)$. In general, the curve may be of any dimensionality, i.e., $\mathbf{x}(\xi) = x(\xi)$, $\mathbf{x}(\xi) = [x(\xi), y(\xi)]^{\mathrm{T}}$, or $\mathbf{x}(\xi) = [x(\xi), y(\xi), z(\xi)]^{\mathrm{T}}$. However, in this subsection the curve refers to the one-dimensional interval $[0, L]$ or $[a, b]$, along the axis of real numbers.

Extending the use of Eq. (2.22), along the above curve we can also write that

$$U(\xi) = \sum_{i=0}^{n} B_{i,n}(\xi) a_i, \tag{2.24}$$

where U represents a scalar physical quantity to be interpolated or approximated and a_i are generalized coordinates (coefficients) to be determined. The transition from Eq. (2.22) to Eq. (2.24) is similar with the meaning of the *isoparametric* mapping; the only difference is that Eq. (2.24) includes the meaningless coefficients a_i and *not* the nodal values U_i as happens with the Lagrange polynomials (cf. Eq. 2.17).

It is not compulsory to deal with a normalized variable ξ only but with the Cartesian coordinate x as well (after minor modification in Eq. (2.23)), so Eq. (2.24) may also take the general form:

$$U(x) = \sum_{i=0}^{n} B_{i,n}(x)\, a_i \tag{2.25}$$

In this case, Eq. (2.25) is used to interpolate $(n + 1)$ given points (x_i, U_i), $i = 0, 1, 2, \ldots, n$, such that $x_0 < x_1 < \cdots < x_n$, by determining $(n + 1)$ unknown coefficients a_i.

It is noted that the essential difference between Lagrange polynomials and nonrational Bézier (Bernstein) polynomials is that the first are associated with nodal values U_i, whereas the second are associated to control points through rather meaningless coefficients a_i.

Note: It can be easily proven that *nonrational* Bézier interpolation is an equivalent form of the Lagrange polynomial interpolation, just because both expansions span the same functional space. This general rule is also shown in the next exercise, while the discussion is extended in Section "Polynomial Expansion—Lagrange Polynomials and Bézier Curves Under the Prism of B-splines".

Exercise 2.5 (Runge example with Bernstein polynomials)
Repeat the same approximation with Exercise 2.3 (Runge example), but now using nonrational Bézier polynomials of eighth degree. Plot both functions in a common graph and compare them.

Hint Equation (2.24) is applied to all the nine points ($x_0, \ldots, x_n$, with $n = 8$); i.e., the green circles in Fig. 2.6 which uniformly divide the interval [–1,1] into eight equal segments. The obtained matrix equation is

$$\begin{Bmatrix} U_0 \\ \vdots \\ U_n \end{Bmatrix} = \begin{bmatrix} B_{0,n}(x_0) & \ldots & B_{n,n}(x_0) \\ \vdots & \ldots & \vdots \\ B_{0,n}(x_n) & & B_{n,n}(x_n) \end{bmatrix} \begin{Bmatrix} a_0 \\ \vdots \\ a_n \end{Bmatrix} \tag{2.26}$$

Inverting the arithmetically known matrix in the right-hand side of Eq. (2.26), the vector including the unknown coefficients $a_i, i = 0, \ldots, n$ is calculated. Then, for every value of the independent variable x of the domain $[-1, 1]$, the substitution of the aforementioned a_i's into Eq. (2.24) determines the interpolating value of the function $U(x)$. By programming this procedure, the reader will observe no difference in the approximation between the current nonrational Bézier and the previous Lagrange polynomial approximation.

Exercise 2.6 (Control points in univariate nonrational cubic Bézier interpolation along a straight segment)
Determine the control points for a cubic Bézier interpolation which is defined in the interval [0, 1].

Solution Given the interval [0, 1], the first control point is P_0 at the position $x = 0$ (left end), whereas the fourth one (noted by P_3) is located at $x = 1$ (right end of the domain). One way to determine the rest two control points, P_1 and P_2, is to apply Eq. (2.22) at two test points within the interval [0, 1]. In this case, for an arbitrary pair of points $(\xi_1, \xi_2) \in (0, 1) \times (0, 1)$, we have

$$\begin{Bmatrix} \xi_1 \\ \xi_2 \end{Bmatrix} = \left[\begin{array}{c|cc|c} B_{0,3}(\xi_1) & B_{1,3}(\xi_1) & B_{2,3}(\xi_1) & B_{3,3}(\xi_1) \\ B_{0,3}(\xi_2) & B_{0,3}(\xi_2) & B_{0,3}(\xi_2) & B_{0,3}(\xi_2) \end{array}\right] \begin{Bmatrix} x_{P_0} \\ \hline x_{P_1} \\ x_{P_2} \\ \hline x_{P_3} \end{Bmatrix} \tag{2.27}$$

Due to Eq. (2.23), the cubic Bernstein polynomials are $B_{0,3} = (1-\xi)^3$, $B_{1,3} = 3\xi(1-\xi)^2$, $B_{2,3} = 3\xi^2(1-\xi)$ and $B_{3,3} = \xi^3$. Separating the unknown quantities in the left hand and all the known ones in the right-hand side, Eq. (2.27) becomes:

$$\begin{bmatrix} B_{1,3}(\xi_1) & B_{2,3}(\xi_1) \\ B_{1,3}(\xi_2) & B_{2,3}(\xi_2) \end{bmatrix} \begin{Bmatrix} x_{P_1} \\ x_{P_2} \end{Bmatrix} = \begin{Bmatrix} \xi_1 \\ \xi_2 \end{Bmatrix} - \begin{Bmatrix} B_{0,3}(\xi_1) \cdot x_{P_0} + B_{3,3}(\xi_1) \cdot x_{P_3} \\ B_{0,3}(\xi_2) \cdot x_{P_0} + B_{3,3}(\xi_2) \cdot x_{P_3} \end{Bmatrix} \tag{2.28}$$

Considering that $x_{P_0} = 0$, $x_{P_3} = 1$, and taking arbitrary values such as $\xi_1 = \frac{1}{4}$ and $\xi_2 = \frac{3}{4}$, Eq. (2.28) becomes:

$$\begin{bmatrix} 3 \cdot \frac{1}{4} \cdot (1-\frac{1}{4})^2 & 3 \cdot (\frac{1}{4})^2 \cdot (1-\frac{1}{4}) \\ 3 \cdot \frac{3}{4} \cdot (1-\frac{3}{4})^2 & 3 \cdot (\frac{3}{4})^2 \cdot (1-\frac{3}{4}) \end{bmatrix} \begin{Bmatrix} x_{P_1} \\ x_{P_2} \end{Bmatrix} = \begin{Bmatrix} \frac{1}{4} \\ \frac{3}{4} \end{Bmatrix} - \begin{Bmatrix} (1-\frac{1}{4})^3 \cdot 0 + (\frac{1}{4})^3 \cdot 1 \\ (1-\frac{3}{4})^3 \cdot 0 + (\frac{3}{4})^3 \cdot 1 \end{Bmatrix}. \tag{2.29}$$

Solving the system, we finally get $x_{P_1} = \frac{1}{3}$ and $x_{P_2} = \frac{2}{3}$. Therefore, the four control points in the cubic Bézier interpolation *uniformly* divide the interval [0, 1]. The reader can easily check that the same numerical finding is obtained with any other pair of values (ξ_1, ξ_2).

Alternatively, one can apply Eq. (2.22) at any four test points within the interval [0, 1]. Then he/she will immediately find that the extreme control points P_0 and P_3 are indeed coincident with the endpoints of [0, 1], whereas the rest two uniformly divide the domain, thus $\{P_0, P_1, P_2, P_3\} = \{0, \frac{1}{3}, \frac{2}{3}, 1\}$.

Note: A quite different manner is to begin with a uniform location of the above four control points and then, applying the *de Casteljau* rule [19, 20] (see later in Chap. 6), to show that for each normalized parameter ξ the relevant point on the "Bézier curve" is given as $x_B = \xi L$ where L is the length of the general domain $x \in [0, L]$. ■

Example 2.7 (Uniform control points) Validate that in a straight segment $P_0 P_n$ and for any polynomial degree, n, the control points $P_1 \ldots P_{n-1}$ of the nonrational Bézier interpolation belong to the segment and uniformly subdivide it.

Hint Generalizing the procedure outlined in Exercise 2.6, the reader is advised to develop a computer program that performs the requested determination of the control points. The full mathematical proof is left to the reader, whereas a geometrical proof based on de Casteljau rule is given in Chap. 6.

Some remarks on piecewise interpolations

Remark 1 As was previously mentioned, when referring to the case of the function $U(x) = x^{1/3}$, dealt in Example 2.2, it would be more effective to use even a piecewise linear interpolation. Going on with similar thoughts, we could apply piecewise quadratic interpolation, piecewise cubic interpolation and so on. This class of interpolations ensures C^0-continuity only between any two adjacent segments (elements) of piecewise interpolation.

Remark 2 Therefore, if instead of the abovementioned piecewise (linear, quadratic, etc.) interpolation strictly related to a C^0-continuity, we demand a higher-order continuity at the internal nodes (breakpoints), then we have to consider the theory of the forthcoming Sect. 2.3.7 regarding interpolation using B-splines. In the latter, for a piecewise polynomial of degree p, a C^{p-1}-continuity may be achieved. Moreover, if we wish, it is possible to lower the continuity to C^{p-2}, C^{p-3}, and so on.

2.3.7 One-Dimensional (1D) Interpolation Using B-Splines

Interpolation using B-splines is capable of overcoming the possible numerical oscillations which, as we saw above (e.g., in Fig. 2.6), may appear when a power series, a set of Lagrange polynomials or of nonrational Bernstein–Bézier ones are used. The meaning of B-spline was documented by Schoenberg [32–34], Curry and Schoenberg [11, 12], whereas a fast recursive and robust computation was proposed later by Carl de Boor [17] and Maurice Cox [10], independently. Some papers refer also to the contribution of Lois Mansfield on the same issue at the same time, particularly an identity mentioned by Carl de Boor as well (see [17], p. 52).

Initially, I had formed the mistaken opinion that the truncated power form of a B-spline was due to Isaac Jacob Schoeneberg. However, after a private communication I had with Professor Carl de Boor, I finally understood that while B-splines can be found in the literature well before Schoenberg, it was Schoenberg who named them. According to de Boor [15], the form in which B-splines appear today (e.g., in MATLAB and elsewhere) is rightly associated with Curry and Schoenberg [12]. De Boor's contribution to that formulation is that he was the first to use B-splines for the numerical solution of ODEs, and, much more importantly, he found a stable algorithm for the evaluation of the B-form(ulation) that made use of that formulation practical. He also, through various papers, particularly the paper "*Splines as linear combinations of B-splines*" (de Boor [15]), and the book on splines [18] complete with FORTRAN programs, helped popularize the use of B-splines.

It is worthy to note that B-splines provide a convenient basis for various spline spaces. Therefore, computer analysts want to use "spline interpolation" and "spline approximation" accordingly. In Computer-Aided Geometric Design (CAGD) a "B-spline" is what a mathematician would call a spline and so researchers in this area use "B-spline basis function" for what Schoenberg was the first to call a "B-spline" [15]. Obviously, the expression "B-spline basis function" is particularly ridiculous since the B is an abbreviation for "Basis," so it says "Basis spline basis function." Therefore, if in some places I have used bad expressions such as "B-spline interpolation" or "B-spline approximation," I am asking for your understanding.

While most contemporary textbooks and papers rely on the *knot vector* and the relevant *control points* approach [12], we believe that for the benefit of the reader it is more instructive to start the presentation of the "*truncated power form*" of a B-spline. As soon as the original formulation is understood, then we pass on the most "modern" view.

2.3.7.1 Practical Understanding of B-Splines

Rephrasing Wikipedia (https://en.wikipedia.org/wiki/Flat_spline), a *spline* is a long flexible strip passing through a number of points (called breakpoints) at which it relaxes to form and hold a smooth curve for the purpose of transferring that curve to another material. The elasticity of the spline material combined with the constraint of the breakpoints would cause the strip to take the shape that minimized the energy required for bending it between the fixed points (breakpoints, shown by two arrows in Fig. 2.8), this being the smoothest possible shape. If we recall the beam theory in the finite element analysis (FEA), the shape of a beam between its two ends is a cubic polynomial (i.e., of third degree). Thus, the first spline functions that appeared for the first time were the *cubic* ones.

Moreover, if one assumes that the strip freely passes through the breakpoints, without being stuck on them, then the breakpoints do not exert any external moments (but only shear forces, if frictionless), and thus moment continuity is ensured. And

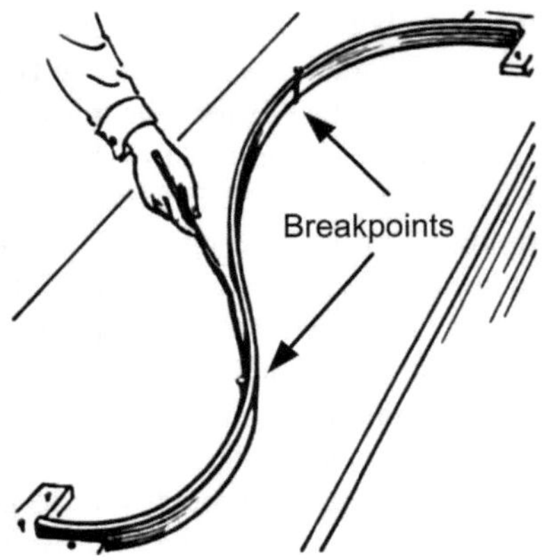

Fig. 2.8 Hand tool spline. *Source* https://en.wikipedia.org/wiki/Flat_spline, Archives of Pearson Scott Foresman, after modification

since the equation for the neutral axis is $EIw'' = -M(x)$, the aforementioned continuity in $M(x)$ implies the continuity in the curvature w'' (second derivative of deflection). Therefore, since between two successive breakpoints the strip (beam) is a piecewise cubic polynomial, it possesses a C^2-continuity. In the particular case that the bending moment at the ends vanishes ($w'' = 0$), the spline is called *natural* spline.

In 1946, the mathematician I. J. Schoenberg gave the spline function its name after its resemblance to the mechanical spline used by draftsmen (see [34]. The word "spline" characterizes smooth and *piecewise polynomial* approximations. A short history is given by Peddie [26].

One interesting source for the CAD history (including the creation of companies) is due to David E. Weisberg who wrote a book with over 650 pages, which may be found online at: http://www.cadhistory.net/. Spline tools were used by ship manufacturers and technicians in aircraft industry for drawing smooth shapes before electronic computers had been invented. In the prologue of the book by Bartels et al. [3], A. Robin Forrest (a former PhD student of W. J. Gordon) uses the term "lofting" to characterize a technique used by British aircraft industry during the Second Civil War in order to construct aircraft templates passing through thin wooden strips (called "splines") through points that were lying on the floor of a large design area, a technique lent from the design of ship structures.

The utilization of splines in the modeling of automotive chassis seems to have been done by many people. Particular recognition was given to Paul de Casteljau working for Citroën [20], Pierre Bézier in Renault (see [4, 29]), Garabedian, and de Boor in General Motors (see [5]). Finally, from the manuscripts of Paul de Casteljau only one was published, and even not completely, in 1959 [19], while in contrast Pierre Bézier published a lot of papers (most of them are cited by Rabut [29]). Perhaps this is the reason that the scientific contribution of de Casteljau was rather ignored compared to that of Pierre Bézier. According to Davis [14], the work of de Boor in General Motors led to a great number of papers published at the beginning of 1960s, including part of his fundamental research on B-spline [17]. More generally, the work in General Motors is described in a smart way in the texts of Birkhoff [4] and Young [36]. Birkhoff's contribution may be found online at (http://history.siam.org/birkhoff.htm).

Much research has been conducted in the industry Pratt & Whitney Aircraft, wherefrom Ahlberg, Nilson, and Walsh published their first research on splines in the form of a book (see [1]).

After listing the above historical facts, let us now focus on the contents of this chapter. Even if the reader happens to have accumulated considerable knowledge about splines from the subject of *Numerical Analysis*, we believe that it is necessary to complete this knowledge in relationship to the Computational Geometry and the so-called isogeometric methods that have obtained a high fame [8, 9, 24].

In any case, in order to introduce the meaning of a spline curve and demonstrate their important properties, there are many different ways, such as those of divided differences, of truncated monomials [11, 32], of connect-the-dots approach [30], and the recursive formula which is attributed to de Boor and Cox [10, 17] and Lois Mansfield as well. In the sequence, according to current needs, the most suitable methodology will be followed. This topic is thoroughly covered in older and contemporary textbooks about CAD (see [21, 22, 23, 27, 31], among others).

2.3.7.2 Truncated Power Form of a B-Spline

Introduction

From the previous subsection, it has become clear that the mathematical expression that describes the B-spline interpolation between two successive breakpoints changes from one subinterval to the next one. Also, we saw that in the cubic spline interpolation, we can construct the polynomials so as to have continuity until the *second* derivative.

Despite the great importance of cubic B-spline, for the sake of brevity in the sequence we refer to the general case of piecewise polynomial degree $p \geq 3$.

It is important to remember that, if the piecewise polynomials are of degree p, then we would like to ensure continuity until the $(p-1)$th derivative, which is labeled as C^{p-1}-continuity.

Let us assume that, on the plane OxU or equivalently $O\xi U$, we wish to interpolate the pairs of values $(\xi_0, U_0), (\xi_1, U_1), \ldots, (\xi_n, U_n)$ with $\xi_0 < \xi_1 < \ldots < \xi_n$, using a composite function $U(\xi)$. In other words, we wish to draw a curve (or to determine a function that cannot be described with a unified mathematical expression), which has the following properties:

(i) In every interval $\xi_{i-1} \leq \xi \leq \xi_i$ with $i = 1, 2, \ldots, n$, the function $U(\xi)$ is a polynomial of degree p.

(ii) The function $U(\xi)$ as well as its first and second derivative, and so on, until the $(p-1)$th derivative are continuous at the given points.

The points $\xi_0, \xi_1, \ldots, \xi_n$ are called *breakpoints*.

Continuity in truncated form of a spline

We wish to interpolate the data pairs $(\xi_0, U_0), (\xi_1, U_1), \ldots, (\xi_n, U_n)$ and determine the function $U(\xi)$, which is a B-spline of degree p, so that it belongs to the class of $C^{p-\lambda}$-continuity, which means that its derivatives $U', U'', \ldots, U^{(p-\lambda)}$ exist and are continuous.

Introducing the meaning of the *truncated* polynomial (where q is an integer):

$$\langle \xi - \xi_k \rangle^q = \begin{cases} 0, & \xi \leq \xi_k \\ (\xi - \xi_k)^q, & \xi > \xi_k, \end{cases} \tag{2.30}$$

in correspondence with the continuity class C^m, a number of terms (different than n) are involved in a series expansion of the function $U(\xi)$. Here it is sufficient to assume that the integer m (involved in C^m) is written as $m = p - \lambda$, and thus, the function $U(\xi)$ belongs to the continuity class $C^{p-\lambda}$, which is accomplished by properly selecting the *multiplicity* λ, of which the definition will be given in the next subsection. In brief, we distinguish the following cases:

A. C^{p-1}-*continuity* ($\lambda = 1$):

$$U(\xi) = a_0 + a_1\xi + a_2\xi^2 + \cdots + a_p\xi^p + \sum_{i=1}^{n-1} b_i \langle \xi - \xi_i \rangle^p \tag{2.31a}$$

B. C^{p-2}-*continuity* ($\lambda = 2$):

$$U(\xi) = a_0 + a_1\xi + a_2\xi^2 + \cdots + a_p\xi^p + \sum_{i=1}^{n-1} b_i \langle \xi - \xi_i \rangle^p + \sum_{i=1}^{n-1} c_i \langle \xi - \xi_i \rangle^{p-1} \tag{2.31b}$$

C. C^{p-3}-*continuity* ($\lambda = 3$)

$$U(\xi) = a_0 + a_1\xi + a_2\xi^2 + \cdots + a_p\xi^p + \sum_{i=1}^{n-1} b_i \langle \xi - \xi_i \rangle^p + \sum_{i=1}^{n-1} c_i \langle \xi - \xi_i \rangle^{p-1} + \sum_{i=1}^{n-1} d_i \langle \xi - \xi_i \rangle^{p-2}, \tag{2.31c}$$

and so on.

In the above expressions of Eq. (2.31a–2.31c), we can easily validate the claim that the function $U(\xi)$ and its derivatives are continuous at the breakpoints. Therefore, when the continuity is $C^{p-\lambda}$, then the integer λ denotes the *number of coefficients* $(a_i, b_i, c_i, \ldots)$ that correspond to every internal breakpoint $(i = 1, \ldots, n-1)$.

Obviously, the number of coefficients $(a_i, b_i, c_i, \ldots)$ that contribute to the B-spline expansion is:

$$n_{\text{coef}} = (n-1)\lambda + (p+1) \tag{2.32}$$

2.3.7.3 Curry–Schoenberg Formulation of B-Spline

Formulation

The Curry–Schoenberg formulation [12] is that which dominates today and can be found in academic and commercial codes such as MATLAB® among others. Carl de Boor [17] developed a robust numerical procedure and a FORTRAN 90 library which evaluates piecewise polynomial functions, including cubic splines may be found online at http://www.netlib.org/pppack/. The description and details on the subroutines are given in his latest textbook [18].

Based on the interpolation data at the breakpoints, and depending on the choice of continuity $(C^{p-\lambda})$, we construct the so-called *knot vector*, which consists of the two ends of the interval with a multiplicity $(p+1)$ for each of them, plus the $(n-1)$ inner breakpoints of the interval, each of multiplicity λ.

Therefore, for the breakpoints $\xi_0, \xi_1, \ldots, \xi_n$, the corresponding *knot vector* will be (see, also, [27]):

$$\hat{\mathbf{U}} = \left\{ \underbrace{\xi_0, \ldots, \xi_0}_{p+1}, \underbrace{\xi_1, \ldots, \xi_1}_{\lambda}, \ldots, \underbrace{\xi_{n-1}, \ldots, \xi_{n-1}}_{\lambda}, \underbrace{\xi_n, \ldots, \xi_n}_{p+1} \right\}, \tag{2.33}$$

and consists of

$$m = 2(p+1) + \lambda(n-1) \tag{2.34}$$

elements.

Comparing Eqs. (2.32) and (2.34), one gets

$$m = n_{\text{coef}} + (p+1), \tag{2.35a}$$

or, equivalently:

$$n_{\text{coef}} = m - (p+1) \tag{2.35b}$$

Important remark Instead of working with the n_{coef} polynomial coefficients involved in the truncated power form relationships (2.31a–2.31c), in the Curry–Schoenberg (de Boor) formulation $n_p + 1 = n_{\mathrm{coef}}$ *control points* are introduced, in a way analogous with the control points we previously saw in Bézier curves. The advantage of controlling the shape of a curve (under design) through these controls points, either in the Bézier interpolation or in the interpolation using B-splines, is obvious.

Therefore, if $0 \leq x \leq L$, the aforementioned breakpoints $x_0, x_1, \ldots, x_n$ obtain a secondary importance since the new consideration consists in the parameterization of the geometric curve [that is of the function $U(x)$] with the introduction of the control points, $P_i (i = 0, \ldots, n_p)$, of which the number equals to the number of polynomial coefficients mentioned in the earlier subsections. Given the $(n_p + 1)$ control points, for every normalized coordinate, $\xi = x/L$ (where $L = x_n - x_0$: length of interval), the position of the corresponding point of the interpolated function is given by the relationship:

$$x(\xi) = \sum_{i=0}^{n_p} \tilde{B}_{i,n_p}(\xi) x_{P_i}, \quad 0 \leq \xi \leq 1, \tag{2.36}$$

where $\tilde{B}_{i,n_p}(\xi)$ are the normalized B-splines. In other words, the interpolation of coordinates of the control points gives accurately the coordinate of the point under consideration that has the given normalized coordinate, ξ.

On the contrary, if we consider the values of an arbitrary function $U(\xi)$ at the $(n_p + 1)$ control points, we cannot apply the relationship (2.36) as is, but we have to determine proper coefficients, $a_i, i = 0, \ldots, n_p$, based on which we can interpolate, with adequate accuracy, the current functions, that is:

$$U(\xi) = \sum_{i=0}^{n_p} \tilde{B}_{i,n_p}(\xi) \cdot a_i \tag{2.37}$$

Obviously, these coefficients are not directly related to the values of the function $U(\xi)$ (more exactly they are associated to some values of U, but at some *unknown* positions in the neighborhood of the control points), except of the two extreme values for which we have:

$$\begin{aligned} a_0 &= U_1 = U(\xi = 0) \\ a_n &= U_n = U(\xi = 1) \end{aligned} \tag{2.38}$$

Basis Functions

There are many ways to define the B-splines and then demonstrate their important properties. At this point, we use the recursive formula as it is the most effective in computer implementation. The notation is according to the book by Piegl and Tiller [27], which includes valuable documented information.

Let us assume that $\widehat{\mathbf{U}} = \{\xi_0, \ldots, \xi_m\}$ is a nondecreasing sequence of real numbers, with $\xi_i \le \xi_{i+1}, i = 0, \ldots, m-1$. The ξ_i are called *knots*, and $\widehat{\mathbf{U}}$ is the *knot vector*. The ith B-spline of piecewise polynomial degree p (order $p+1$), which is denoted by $N_{i,p}(\xi)$, is defined as:

$$N_{i,0}(\xi) = \begin{cases} 1 \text{ if } \ \xi_i \le \xi < \xi_{i+1} \\ 0 \ \text{ elsewhere} \end{cases}$$
$$N_{i,p}(\xi) = \frac{\xi - \xi_i}{\xi_{i+p} - \xi_i} N_{i,p-1}(\xi) + \frac{\xi_{i+p+1} - \xi}{\xi_{i+p+1} - \xi_{i+1}} N_{i+1,p-1}(\xi) \tag{2.39}$$

Note that:

- The function $N_{i,0}(\xi)$ is a step function and is equal to zero always except of the semi-open interval $\xi \in [\xi_i, \xi_{i+1})$.
- If $p > 0$, then $N_{i,p}(\xi)$ is a linear combination of two basis functions of degree $(p-1)$.
- The computation of a set of basis functions requires the determination of the knot vector, $\widehat{\mathbf{U}}$, and the degree, p.
- Equation (2.39) includes the quotient $0/0$ which is defined equal to zero.
- The functions $N_{i,p}(\xi)$ are piecewise polynomials, which are defined along the entire axis of the real values. In general, they are of interest only within the interval $[\xi_0, \xi_m]$.
- The semi-open interval $[\xi_i, \xi_{i+1})$ is called the ith knot span. It can be of zero length, as it is not required that the knots are discrete.
- When the knot vector is of the form $\hat{\mathbf{U}} = \left\{ \underbrace{0, \ldots, 0}_{(p+1)\text{ terms}}, \underbrace{1, \ldots, 1}_{(p+1)\text{ terms}} \right\}$, then the produced B-spline $N_{i,p}$ coincides with the Bernstein polynomial of degree p, which are involved in Bézier curve.
- It is $N_{i,p}(\xi) = 0$ if the variable ξ is outside the interval $\left[\xi_i, \xi_{i+p+1}\right)$ (this is called the *local support property*).
- Within a given knot interval, $\left[\xi_j, \xi_{j+1}\right)$, at maximum $p+1$ functions $N_{i,p}$ are different than zero, and concretely the $N_{j-p,p}, \ldots, N_{j,p}$.
- $N_{i,p}(\xi) \ge 0$ for all i, p, and ξ (nonnegative). It is proven using mathematical induction with respect to p.
- For an arbitrary knot vector, $[\xi_i, \xi_{i+1})$, it is $\sum_{j=i-p}^{i} N_{j,p}(\xi) = 1$ for all $\xi \in [\xi_i, \xi_{i+1})$ (this is called the *partition of unity* property).

- All derivatives of $N_{i,p}(\xi)$ appear in the interior of a knot span (in which it is a perfect polynomial). At a node, $N_{i,p}(\xi)$ is $(p-k)$ times differentiable, where k is the *multiplicity* of the node.
- Except of the case $p=0$, $N_{i,p}(\xi)$ takes exactly a maximum value.

Derivatives of B-spline Functions

Using mathematical induction, it can be proven that the derivative of a B-spline is given by the recursive relationship:

$$N'_{i,p} = \frac{p}{\xi_{i+p}-\xi_i} N_{i,p-1}(\xi) - \frac{p}{\xi_{i+p+1}-\xi_{i+1}} N_{i+1,p-1}(\xi) \tag{2.40}$$

Generally, if $N_{i,p}^{(k)}$ denotes the k-order derivative of $N_{i,p}(\xi)$, successive differentiations of Eq. (2.40) give the following expression:

$$N_{i,p}^{(k)}(\xi) = p\left(\frac{N_{i,p-1}^{(k-1)}}{\xi_{i+p}-\xi_i} - \frac{N_{i+1,p-1}^{(k-1)}}{\xi_{i+p+1}-\xi_{i+1}}\right) \tag{2.41}$$

An alternative way to calculate the derivative of k-order, in terms of the functions $N_{i,p-k}, \ldots, N_{i+k,p-k}$, is the following one:

$$N_{i,p}^{(k)} = \frac{p!}{(p-k)!}\sum_{j=0}^{k} a_{k,j} N_{i+j,p-k}, \tag{2.42}$$

where

$$\begin{aligned}
a_{0,0} &= 1 \\
a_{k,0} &= \frac{a_{k-1,0}}{\xi_{i+p-k+1}-\xi_i} \\
a_{k,j} &= \frac{a_{k-1,j}-a_{k-1,j-1}}{\xi_{i+p+j-k+1}-\xi_{i+j}} \qquad j = 1, \ldots, k-1 \\
a_{k,k} &= \frac{-a_{k-1,k-1}}{\xi_{i+p+1}-\xi_{i+k}}
\end{aligned}$$

For more details, the interested reader may consult the book by Piegl and Tiller [27].

Polynomial Expansion—Lagrange Polynomials and Bézier Curves Under the Prism of B-splines

It can be easily proven that the expansion of a function $U(\xi)$ in a polynomial of degree p is equivalent with the expansion in $(p+1)$ Lagrange polynomials again of degree p, that is:

$$\begin{aligned} U(\xi) &= a_0 + a_1\xi + \cdots + a_p\xi^p \\ &\equiv L_0^{(p)}(\xi)\cdot U_0 + L_1^{(p)}(\xi)\cdot U_1 + \cdots + L_p^{(p)}(\xi)\cdot U_p \end{aligned} \tag{2.43}$$

In addition, the above expressions are equivalent with the expansion in nonrational Bézier functions (Bernstein polynomials):

$$\begin{aligned} U(\xi) &= a_0 + a_1\xi + \cdots + a_p\xi^p \\ &= B_{0,p}(\xi)\cdot\tilde{a}_0 + B_{1,p}(\xi)\cdot\tilde{a}_1 + \cdots + B_{p,p}(\xi)\cdot\tilde{a}_p \end{aligned} \tag{2.44}$$

Since the polynomial expansion covers both sets of basis functions (Lagrange and nonrational Bernstein–Bézier polynomials), given the fact that the polynomial of degree p constitutes the basic part of the expansion using B-splines (see Eq. (2.31a)) without the existence of truncated polynomials, it is immediately understood that:

The interpolation which is introduced by a nonrational Bézier curve is equivalent with that produced by the knot vector:

$$\widehat{\mathbf{U}}_{\text{Bezier}} = \left\{ \underbrace{0,\ldots,0}_{p+1}, \underbrace{1,\ldots,1}_{p+1} \right\} \tag{2.45}$$

Taking into consideration the above conclusion, we can handle Bézier curves or nonrational Bézier interpolation *with exactly the same way as we handle B-splines*. The only difference is that in Bézier interpolation *there are no inner breakpoints*. Therefore, since Bézier interpolation is a special case of B-splines, the subroutines used for the latter are also applicable to the former (e.g., function `spcol` in MATLAB®).

Cardinal Global Shape Functions in Natural Cubic B-Splines

Except of the control points at the ends of the domain interval, the classical basis functions $N_{i,p}(\xi)$ of Eq. (2.39) are *noncardinal* functions (i.e., *not* of [1-0]-type). In contrast, this subsection will focus on the derivation of *cardinal* B-spline-based shape functions which are directly associated to the corresponding nodal value U_i of the interpolated function (as is known in the classical finite element method, e.g., [35]), and not to the generalized arbitrary coefficients a_i (cf. Eq. 2.37). The main reason for referring to this issue is that in the past these were extensively used in the numerical solution of boundary value and eigenvalue problems based on the Galerkin–Ritz formulation or the collocation method ([25, 28], and papers therein).

Generally, the approximate solution of an ordinary differential equation is expressed as an expansion series of the form:

$$U(\xi) = \sum_{j=0}^{n_p} B_j(\xi)\hat{a}_j = \begin{bmatrix} B_0(\xi) \cdots B_{n_p}(\xi) \end{bmatrix} \cdot \begin{bmatrix} \hat{a}_0 \\ \vdots \\ \hat{a}_{n_p} \end{bmatrix}, \tag{2.46}$$

where the well-known basis functions, i.e., the monomials $B_j(\xi)$, are given by one of Eqs. (2.31a–2.31c) and $\hat{a}_j$ are the therein $(n_p + 1)$ unknown coefficients. In order to introduce the boundary conditions, it is useful to modify Eq. (2.46) in the form:

$$U(\xi) = \sum_{j=0}^{n_p} \phi_j(\xi) U_j \tag{2.47}$$

where U_j is the nodal value and $\phi_j(\xi)$ are the shape functions.

The general process is as follows. In the beginning, we apply Eq. (2.46) at all nodes (breakpoints) involved in the domain $[\xi_0, \xi_n]$ where the interpolation takes place. In addition, we take the first or the second derivative of Eq. (2.46) at the two endpoints. For example, in the case of first derivative at both ends, we have:

$$\underbrace{\begin{Bmatrix} U(\xi_0) \\ \vdots \\ U(\xi_n) \\ U'(\xi_0) \\ U'(\xi_n) \end{Bmatrix}}_{\{U\}} = \underbrace{\begin{bmatrix} B_0(\xi_0) \cdots B_{n_p}(\xi_0) \\ \vdots \quad \vdots \quad \vdots \\ B_0(\xi_n) \cdots B_{n_p}(\xi_n) \\ B_0'(\xi_0) \cdots B_{n_p}'(\xi_0) \\ B_0'(\xi_n) \cdots B_{n_p}'(\xi_n) \end{bmatrix}}_{[\mathbf{A}]} \cdot \underbrace{\begin{Bmatrix} \hat{a}_0 \\ \vdots \\ \hat{a}_{n_p} \end{Bmatrix}}_{\{a\}} \Rightarrow \{\hat{a}\} = [\mathbf{A}]^{-1}\{U\} \tag{2.48}$$

Introducing Eq. (2.48) into Eq. (2.46), the latter is written as follows:

$$U(\xi) = \underbrace{\begin{bmatrix} B_0(\xi) \cdots B_{n_p}(\xi) \end{bmatrix}}_{1\times(n_p+1)} \cdot \underbrace{[\mathbf{A}]^{-1}}_{(n_p+1)\times(n_p+1)} \cdot \underbrace{\{U\}}_{(n_p+1)\times 1} = \begin{bmatrix} \phi_0(\xi) \cdots \phi_{n_p}(\xi) \end{bmatrix} \cdot \begin{bmatrix} U_0 \\ \vdots \\ U_n \\ U_0' \\ U_n' \end{bmatrix} \tag{2.49}$$

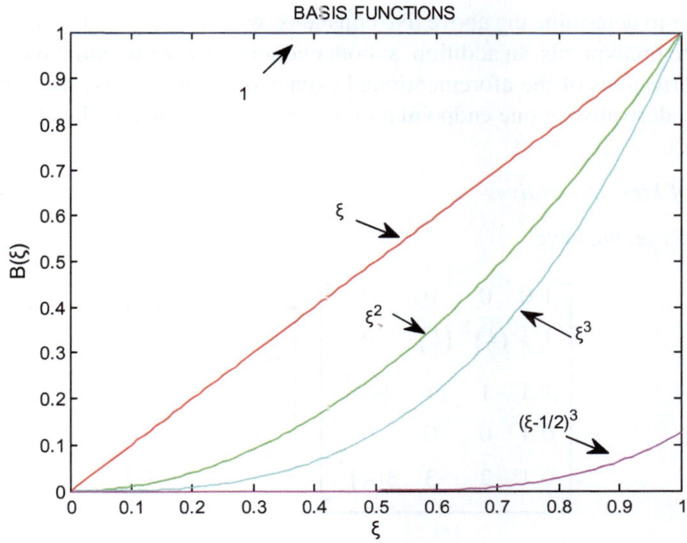

Fig. 2.9 Basis functions of truncated power terms (prior to Schoenberg) for three breakpoints ($p = 3$)

Equation (2.49) shows that the multiplication of the row vector by the matrix $[\mathbf{A}]^{-1}$, of dimensions $(n_p + 1) \times (n_p + 1)$, defines a new row vector of dimensions $1 \times (n_p + 1)$, which consists of the desired shape functions. Finally, each shape function is produced by the multiplication of the row vector $\left[\, B_0(\xi) \cdots B_{n_p}(\xi) \,\right]$ by the corresponding column of the matrix $[\mathbf{A}]^{-1}$. The application of this procedure is demonstrated in the following exercises.

Exercise 2.8 (Natural cubic B-splines for two breakpoint spans)
Derive the basis functions as well as the shape functions according to truncated power series formulation (prior to Schoenberg), which correspond to a *cubic B-splines* with full C^2-continuity, defined in the interval [0, 1]. The three given breakpoints are the two endpoints plus the midpoint.

Solution The given breakpoints are: $x_0 = 0, x_1 = \frac{1}{2}$, and $x_2 = 1$. According to prior-to-Schoenberg approach in conjunction with cubic splines, and taking one knot per inner breakpoint (multiplicity $\lambda = 1$, thus Eq. (2.31a) holds), the involved five basis functions are:
$\hat{B}_0(x) = 1$, $\hat{B}_1(x) = x$, $\hat{B}_2(x) = x^2$, $\hat{B}_3(x) = x^3$, $\hat{B}_4(x) = \left\langle x - \frac{1}{2} \right\rangle_+^3$ (see Fig. 2.9).

Therefore, the cubic B-spline interpolation is according to the formula:

$$U(\xi) = \begin{cases} a_0 + a_1\xi + a_2\xi^2 + a_3\xi^3 & 0 \le \xi \le \frac{1}{2} \\ a_0 + a_1\xi + a_2\xi^2 + a_3\xi^3 + b_1(\xi - \frac{1}{2})^3, & \frac{1}{2} \le \xi \le 1 \end{cases} \tag{2.50a}$$

In order to determine the above five constants, we fulfill Eq. (2.50a) at each of the given three breakpoints. In addition, at both endpoints we write either the first or the second derivatives of the aforementioned expansion (if necessary, the combination of the first derivative at one endpoint and the second derivative at the other endpoint is allowed).

A. *Use of First Derivatives*

In this case, we have:

$$\underbrace{\begin{bmatrix} 1 & 0 & 0 & 0 & 0 \\ 1 & \frac{1}{2} & \left(\frac{1}{2}\right)^2 & \left(\frac{1}{2}\right)^3 & 0 \\ 1 & 1 & 1 & 1 & \left(\frac{1}{2}\right)^3 \\ 0 & 1 & 0 & 0 & 0 \\ 0 & 1 & 2 & 3 & 3\left(\frac{1}{2}\right)^2 \end{bmatrix}}_{[\mathbf{A}]} \cdot \begin{bmatrix} a_0 \\ a_1 \\ a_2 \\ a_3 \\ b_1 \end{bmatrix} = \begin{bmatrix} U_0 \\ U_1 \\ U_2 \\ U_0' \\ U_2' \end{bmatrix}, \tag{2.50b}$$

whence the constants are given as

$$\begin{bmatrix} a_0 \\ a_1 \\ a_2 \\ a_3 \\ b_1 \end{bmatrix} = [\mathbf{A}]^{-1} \cdot \begin{bmatrix} U_0 \\ U_1 \\ U_2 \\ U_0' \\ U_2' \end{bmatrix}, \tag{2.50c}$$

with the inverse matrix given by:

$$[\mathbf{A}]^{-1} = \begin{bmatrix} 1 & 0 & 0 & 0 & 0 \\ 0 & 0 & 0 & 1 & 0 \\ -9 & 12 & -3 & -3.5 & 0.5 \\ 10 & -16 & 6 & 3 & -1 \\ -16 & 32 & -16 & -4 & 4 \end{bmatrix}. \tag{2.50d}$$

In matrix form, the relationship (2.50a) is written as:

$$U(\xi) = \begin{bmatrix} \underbrace{1 \ \ \xi \ \ \xi^2 \ \ \xi^3 \ \ \left\langle \xi - \frac{1}{2} \right\rangle_+^3}_{1\times 5} \end{bmatrix} \cdot \underbrace{\begin{bmatrix} a_0 \\ a_1 \\ a_2 \\ a_3 \\ b_1 \end{bmatrix}}_{5\times 1} \tag{2.50e}$$

Substituting Eq. (2.50c) into Eq. (2.50e), the latter is written as:

$$U(\xi) = \underbrace{\underbrace{\begin{bmatrix} 1 & \xi & \xi^2 & \xi^3 & \left\langle \xi - \frac{1}{2} \right\rangle_+^3 \end{bmatrix}}_{1\times 5} \cdot \underbrace{[\mathbf{A}]^{-1}}_{5\times 5}}_{\text{Shape Functions}} \cdot \begin{bmatrix} U_0 \\ U_1 \\ U_2 \\ U_0' \\ U_2' \end{bmatrix} \tag{2.50f}$$

The product of the first two terms in the right-hand side of Eq. (2.50f) gives the row vector of the shape functions. Therefore, substituting Eq. (2.50d) into Eq. (2.50f), the shape functions are given in a unified form as follows:

$$\begin{aligned} &\begin{bmatrix} N_1 & N_2 & N_3 & N_4 & N_5 \end{bmatrix} \\ &= \begin{bmatrix} 1 & \xi & \xi^2 & \xi^3 & \left\langle \xi - \frac{1}{2} \right\rangle_+^3 \end{bmatrix} \cdot \begin{bmatrix} 1 & 0 & 0 & 0 & 0 \\ 0 & 0 & 0 & 1 & 0 \\ -9 & 12 & -3 & -3.5 & 0.5 \\ 10 & -16 & 6 & 3 & -1 \\ -16 & 32 & -16 & -4 & 4 \end{bmatrix} \\ &= \begin{bmatrix} 1 - 9\xi^2 + 10\xi^3 - 16\left\langle \xi - \frac{1}{2} \right\rangle_+^3 \\ 12\xi^2 - 16\xi^3 + 32\left\langle \xi - \frac{1}{2} \right\rangle_+^3 \\ -3\xi^2 + 6\xi^3 - 16\left\langle \xi - \frac{1}{2} \right\rangle_+^3 \\ \xi - \frac{7}{2}\xi^2 + 3\xi^3 - 4\left\langle \xi - \frac{1}{2} \right\rangle_+^3 \\ \frac{1}{2}\xi^2 - \xi^3 + 4\left\langle \xi - \frac{1}{2} \right\rangle_+^3 \end{bmatrix}^T. \end{aligned} \tag{2.50g}$$

The graph of the five shape functions appearing in Eq. (2.50g) is given in Fig. 2.10. One may observe that the first three are cardinal, whereas the rest two (associated to the first derivatives at the ends) are not.

B. *Use of Second Derivatives*

In this case, we have:

$$\underbrace{\begin{bmatrix} 1 & 0 & 0 & 0 & 0 \\ 1 & \frac{1}{2} & \left(\frac{1}{2}\right)^2 & \left(\frac{1}{2}\right)^3 & 0 \\ 1 & 1 & 1 & 1 & \left(\frac{1}{2}\right)^3 \\ 0 & 0 & 2 & 0 & 0 \\ 0 & 0 & 2 & 6 & 3 \end{bmatrix}}_{[\tilde{\mathbf{A}}]} \cdot \begin{bmatrix} \tilde{a}_0 \\ \tilde{a}_1 \\ \tilde{a}_2 \\ \tilde{a}_3 \\ \tilde{b}_1 \end{bmatrix} = \begin{bmatrix} U_0 \\ U_1 \\ U_2 \\ U_0'' \\ U_2'' \end{bmatrix}, \tag{2.50h}$$

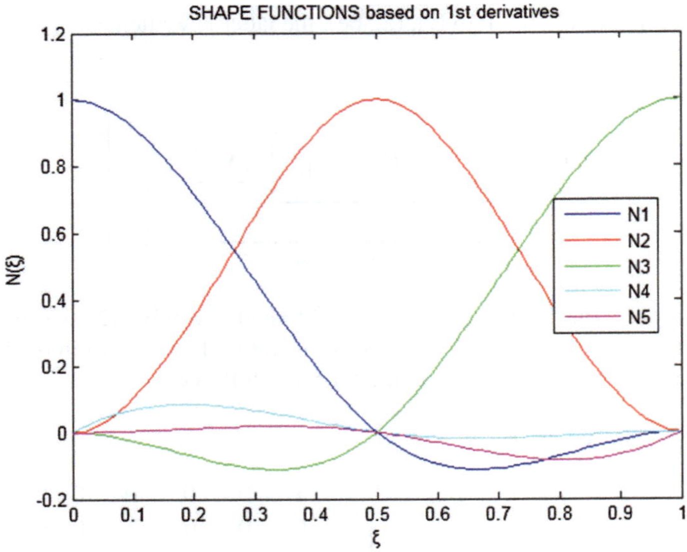

Fig. 2.10 Cubic ($p = 3$) shape functions (using truncated power form) for three breakpoints, when the additional boundary conditions are concerned with the first derivatives at the endpoints

whence the inverse matrix is:

$$[\mathbf{A}]^{-1} = \begin{bmatrix} 1 & 0 & 0 & 0 & 0 \\ -2.5 & 3 & -0.5 & -7/48 & 1/48 \\ 0 & 0 & 0 & 0.5 & 0 \\ 2 & -4 & 2 & -10/24 & -1/12 \\ -4 & 8 & -4 & 0.5 & 0.5 \end{bmatrix}$$

$$= \begin{bmatrix} 1 & 0 & 0 & 0 & 0 \\ -2.5 & 3 & -0.5 & -0.1458 & 0.0208 \\ 0 & 0 & 0 & 0.5 & 0 \\ 2 & -4 & 2 & -0.4167 & -0.0833 \\ -4 & 8 & -4 & 0.5 & 0.5 \end{bmatrix},$$

and therefore the five shape functions are given in a unified form by:

$$
\underbrace{\phi(\xi)}_{1\times5} = \left[\underbrace{1\ \xi\ \xi^2\ \langle\xi-0\rangle_+^3\ \langle\xi-1/2\rangle_+^3}_{1\times5}\right]\cdot\underbrace{[\mathbf{A}]^{-1}}_{5\times5}
$$

$$
= \left[1\ \xi\ \xi^2\ \xi^3\ \left\langle\xi-\tfrac{1}{2}\right\rangle_+^3\right]\cdot\begin{bmatrix} 1 & 0 & 0 & 0 & 0 \\ -2.5 & 3 & -0.5 & -7/48 & 1/48 \\ 0 & 0 & 0 & 0.5 & 0 \\ 2 & -4 & 2 & -1/24 & -1/12 \\ -4 & 8 & -4 & 0.5 & 0.5 \end{bmatrix}
$$

$$
= \begin{bmatrix} 1-\frac{5}{2}\xi+2\xi^3-4\left\langle\xi-\frac{1}{2}\right\rangle_+^3 \\ 3\xi-4\xi^3+8\left\langle\xi-\frac{1}{2}\right\rangle_+^3 \\ -\frac{1}{2}\xi+2\xi^3-4\left\langle\xi-\frac{1}{2}\right\rangle_+^3 \\ \frac{7}{48}\xi+\frac{1}{2}\xi^2-\frac{1}{24}\xi^3+\frac{1}{2}\left\langle\xi-\frac{1}{2}\right\rangle_+^3 \\ \frac{1}{48}\xi-\frac{1}{12}\xi^3+\frac{1}{2}\left\langle\xi-\frac{1}{2}\right\rangle_+^3 \end{bmatrix}^{T}. \qquad (2.50\text{i})
$$

The graph of the above five shape functions is illustrated in Fig. 2.11. Again, one may observe that the first three are cardinal, whereas the rest two (associated to the second derivatives at the ends) are not.

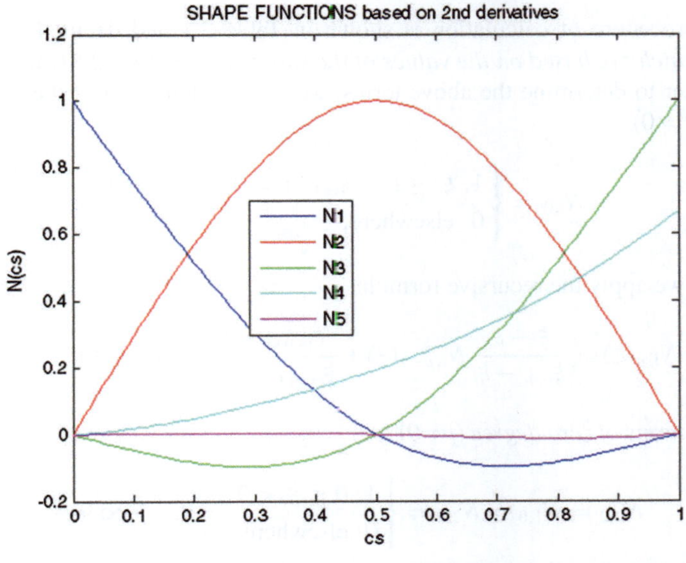

Fig. 2.11 Cubic ($p=3$) shape functions (using truncated power form) for three breakpoints, when the additional boundary conditions are concerned with the second derivatives at the endpoints

Remarks on Exercise 2.8:

(1) Comparing Eq. (2.50g) with Eq. (2.50i), we see that the cardinal shape functions that correspond to the first three (translational) degrees of freedom are not identical in both sets but they depend on the choice of the rotational DOFs at the ends (first or second derivative).
(2) However, in both cases the sum of the first three shape functions equal to the unity (*partition of unity*, or *rigid body* property).

Exercise 2.9 (Curry–Schoenberg (de Boor) cubic B-splines for two breakpoint spans)
Repeat the previous Exercise 2.8 but this time applying Curry–Schoenberg (de Boor) formulation. Calculate the position of the control points.

Solution If we take three breakpoints at the positions, $x_0 = 0, x_1 = 1/2, x_2 = 1$ and we select a cubic polynomial, $p = 3$, as well as a multiplicity of inner knots equal to "1", then the relevant knot vector consists of nine elements ($m = 8$):

$$\widehat{\mathbf{U}} = \left\{0, 0, 0, 0, \tfrac{1}{2}, 1, 1, 1, 1\right\} = \{\xi_0, \xi_1, \xi_2, \xi_3, \xi_4, \xi_5, \xi_6, \xi_7, \xi_8\} \tag{2.51a}$$

Therefore, there will be $n = m-(p+1)$ control points $P_0, P_1, \ldots, P_n$. The position of these five control points is determined in a way quite similar to that applied in Exercise 2.6. The difference is that here we do not know a priori the analytical expression (as was known in the case of Bézier curves) but we must follow a tedious process to determine the basis functions.

The procedure of calculation is shown in Table 2.1 and detailed calculations follow, *which are based on the values of the knot vector* in Eq. (2.51a).

In order to determine the above terms, we use the definition of the zero degree terms ($p = 0$)

$$N_{i,0} = \begin{cases} 1, \ \xi_i \le \xi < \xi_{i+1}, \ i = 0, \ldots, m-1 \\ 0 \ \text{elsewhere}, \end{cases} \tag{2.51b}$$

and then we apply the recursive formula:

$$N_{i,p}(\xi) = \frac{\xi - \xi_i}{\xi_{i+p} - \xi_i} N_{i,p-1}(\xi) + \frac{\xi_{i+p+1} - \xi}{\xi_{i+p+1} - \xi_{i+1}} N_{i+1,p-1}(\xi) \tag{2.51c}$$

(1) *The terms of zero degree (p=0) are:*

$$N_{0,0} = N_{1,0} = N_{2,0} = \begin{cases} 1, \ 0 \le \xi < 0 \\ 0 \ \text{elsewhere} \end{cases} = 0, \quad -\infty < \xi < \infty$$

$$N_{3,0} = \begin{cases} 1, \ 0 \le \xi < 1/2 \\ 0, \ \text{elsewhere} \end{cases},$$

Table 2.1 Demonstration of the recursive computation of basis functions (Exercise 2.9)

$N_{0,0}$						
	$\nwarrow$					
		$N_{0,1}$				
	$\swarrow$		$\nwarrow$			
$N_{1,0}$				$N_{0,2}$		
	$\nwarrow$		$\swarrow$		$\nwarrow$	
		$N_{1,1}$				$N_{0,3}$
	$\swarrow$		$\nwarrow$		$\swarrow$	
$N_{2,0}$				$N_{1,2}$		
	$\nwarrow$		$\swarrow$		$\nwarrow$	
		$N_{2,1}$				$N_{1,3}$
	$\swarrow$		$\nwarrow$		$\swarrow$	
$N_{3,0}$				$N_{2,2}$		
	$\nwarrow$		$\swarrow$		$\nwarrow$	
		$N_{3,1}$				$N_{2,3}$
	$\swarrow$		$\nwarrow$		$\swarrow$	
$N_{4,0}$				$N_{3,2}$		
	$\nwarrow$		$\swarrow$		$\nwarrow$	
		$N_{4,1}$				$N_{3,3}$
	$\swarrow$		$\nwarrow$		$\swarrow$	
$N_{5,0}$				$N_{4,2}$		
	$\nwarrow$		$\swarrow$		$\nwarrow$	
		$N_{5,1}$				$N_{4,3}$
	$\swarrow$		$\nwarrow$		$\swarrow$	
$N_{6,0}$				$N_{5,2}$		
	$\nwarrow$		$\swarrow$			
		$N_{6,1}$				
	$\swarrow$					
$N_{7,0}$						

$$N_{4,0} = \begin{cases} 1, & 1/2 \leq \xi < 1 \\ 0, & \text{elsewhere} \end{cases}$$

$$N_{5,0} = N_{6,0} = N_{7,0} = \begin{cases} 1, & 1 \leq \xi < 1 \\ 0 & \text{elsewhere} \end{cases} = 0, -\infty < \xi < \infty$$

(2) *The terms of first degree (p = 1) are:*

$$i=0 \Rightarrow N_{0,1}=\frac{\xi-0}{0-0}\cancel{N_{0,0}}+\frac{\xi_2-\xi}{\xi_2-\xi_1}\cancel{N_{1,0}}=0$$

$$i=1 \Rightarrow N_{1,1}=\frac{\xi-\xi_1}{\xi_2-\xi_1}\cancel{N_{1,0}}+\frac{\xi_3-\xi}{\xi_3-\xi_2}\cancel{N_{2,0}}=0$$

$$i=2 \Rightarrow N_{2,1}=\frac{\xi-\xi_2}{\xi_3-\xi_2}\cancel{N_{2,0}}+\frac{\xi_4-\xi}{\xi_4-\xi_3}N_{3,0}=\frac{\frac{1}{2}-\xi}{\frac{1}{2}-0}\cdot\begin{cases}1, & 0\le\xi<\frac{1}{2}\\ 0, & \text{elsewhere}\end{cases}=\begin{cases}(1-2\xi), & 0\le\xi<\frac{1}{2}\\ 0, & \text{elsewhere}\end{cases}$$

$$i=3 \Rightarrow N_{3,1}=\frac{\xi-\xi_3}{\xi_4-\xi_3}N_{3,0}+\frac{\xi_5-\xi}{\xi_5-\xi_4}N_{4,0}=\frac{\xi-0}{\frac{1}{2}-0}\cdot\begin{cases}1, & 0\le\xi<\frac{1}{2}\\ 0, & \text{elsewhere}\end{cases}+\frac{1-\xi}{1-\frac{1}{2}}\cdot\begin{cases}1, & \frac{1}{2}\le\xi<1\\ 0, & \text{elsewhere}\end{cases}$$
$$=\begin{cases}2\xi, & 0\le\xi<\frac{1}{2}\\ 2(1-\xi), & \frac{1}{2}\le\xi<1\end{cases}$$

$$i=4 \Rightarrow N_{4,1}=\frac{\xi-\xi_4}{\xi_5-\xi_4}N_{4,0}+\frac{\xi_6-\xi}{\xi_6-\xi_5}N_{5,0}=\frac{\xi-\frac{1}{2}}{1-\frac{1}{2}}\cdot\begin{cases}1, & 0\le\xi<\frac{1}{2}\\ 0, & \text{elsewhere}\end{cases}+\frac{1-\xi}{1-1}\cdot\cancel{N_{5,0}}$$
$$=\begin{cases}0, & 0\le u<\frac{1}{2}\\ (2\xi-1), & \frac{1}{2}\le\xi<1\end{cases}$$

$$i=5 \Rightarrow N_{5,1}=\frac{\xi-\xi_5}{\xi_6-\xi_5}\cancel{N_{5,0}}+\frac{\xi_7-\xi}{\xi_7-\xi_6}\cancel{N_{6,0}}=0$$

$$i=6 \Rightarrow N_{6,1}=\frac{\xi-\xi_6}{\xi_7-\xi_6}\cancel{N_{6,0}}+\frac{\xi_8-\xi}{\xi_8-\xi_7}\cancel{N_{7,0}}=0$$

(3) *The terms of second degree (p=2) are:*

$$i=0 \Rightarrow N_{0,2}=\frac{\xi-\xi_0}{\xi_2-\xi_0}\cancel{N_{0,1}}+\frac{\xi_3-\xi}{\xi_3-\xi_1}\cancel{N_{1,1}}=0,\ -\infty<\xi<\infty$$

$$i=1 \Rightarrow N_{1,2}=\frac{\xi-\xi_1}{\xi_3-\xi_1}\cancel{N_{1,1}}+\frac{\xi_4-\xi}{\xi_4-\xi_2}N_{2,1}=\frac{1/2-\xi}{1/2-0}\cdot\begin{cases}(1-2\xi), & 0\le\xi<1/2\\ 0, & \text{elsewhere}\end{cases}$$
$$=\begin{cases}(1-2\xi)^2, & 0\le\xi<1/2\\ 0, & \text{elsewhere}\end{cases}$$

$$i=2 \Rightarrow N_{2,2}=\frac{\xi-\xi_2}{\xi_4-\xi_2}N_{2,1}+\frac{\xi_5-\xi}{\xi_5-\xi_3}N_{3,0}=\frac{\xi-0}{\frac{1}{2}-0}\cdot\begin{cases}(1-2\xi), & 0\le\xi<\frac{1}{2}\\ 0, & \text{elsewhere}\end{cases}+\frac{1-\xi}{1-0}\cdot\begin{cases}2\xi, & 0\le\xi<\frac{1}{2}\\ 2(1-\xi), & \text{elsewhere}\end{cases}$$
$$=\begin{cases}2\xi(2-3\xi), & 0\le\xi<\frac{1}{2}\\ 2(1-\xi)^2, & \frac{1}{2}\le\xi<1\end{cases}$$

$$i=3 \Rightarrow N_{3,2}=\frac{\xi-\xi_3}{\xi_5-\xi_3}N_{3,1}+\frac{\xi_6-\xi}{\xi_6-\xi_4}N_{4,1}=\frac{\xi-0}{1-0}\cdot\begin{cases}2\xi, & 0\le\xi<\frac{1}{2}\\ 2(1-\xi), & \frac{1}{2}\le\xi<1\end{cases}+\frac{1-\xi}{1-\frac{1}{2}}\cdot\begin{cases}0, & 0\le\xi<\frac{1}{2}\\ 2\xi-1, & \frac{1}{2}\le\xi<1\end{cases}$$
$$=\begin{cases}2\xi^2, & 0\le\xi<\frac{1}{2}\\ 2(1-\xi)(3\xi-1), & \frac{1}{2}\le\xi<1\end{cases}$$

$$i=4 \Rightarrow N_{4,2}=\frac{\xi-\xi_4}{\xi_6-\xi_4}N_{4,1}+\frac{\xi_7-\xi}{\xi_7-\xi_5}\cancel{N_{5,1}}=\frac{\xi-\frac{1}{2}}{1-\frac{1}{2}}\cdot\begin{cases}0, & 0\le\xi<1/2\\ (2\xi-1), & \text{elsewhere}\end{cases}+0=\begin{cases}0, & 0\le\xi<1-\frac{1}{2}\\ (2\xi-1)^2, & \frac{1}{2}\le\xi<1\end{cases}$$

$$i=5 \Rightarrow N_{5,2}=\frac{\xi-\xi_5}{\xi_7-\xi_5}\cancel{N_{5,1}}+\frac{\xi_8-\xi}{\xi_8-\xi_6}\cancel{N_{6,1}}=0$$

Table 2.2 B-splines according to Curry–Schoenberg (de Boor), $N_{i,p}(\xi)$, for two intervals (5 control points: $P_0, \ldots, P_4$, i.e., $n_p = 4$) and one knot per inner breakpoint

For $0 < \xi < \frac{1}{2}$	For $\frac{1}{2} < \xi < 1$
$N_{0,3}(\xi) = -8\xi^3 + 12\xi^2 - 6\xi + 1 = (1 - 2\xi)^3$	$N_{0,3}(\xi) = 0$
$N_{1,3}(\xi) = 14\xi^3 - 18\xi^2 + 6\xi$	$N_{1,3}(\xi) = -2\xi^3 + 6\xi^2 - 6\xi + 2$
$N_{2,3}(\xi) = -8\xi^3 + 6\xi^2$	$N_{2,3}(\xi) = 8\xi^3 - 18\xi^2 + 12\xi - 2 = 2(1 - \xi)^2(4\xi - 1)$
$N_{3,3}(\xi) = 2\xi^3$	$N_{3,3}(\xi) = -14\xi^3 + 24\xi^2 - 12\xi + 2$
$N_{4,3}(\xi) = 0$	$N_{4,3}(\xi) = 8\xi^3 - 12\xi^2 + 6\xi - 1 = (2\xi - 1)^3$
$\sum_{i=0}^{4} N_{i,3}(\xi) = 1$	$\sum_{i=0}^{4} N_{i,3}(\xi) = 1$

(4) *The terms of third degree (p=3) are:*

$$i=0 \Rightarrow N_{0,3} = \frac{\xi-\xi_0}{\xi_3-\xi_0}\cancel{N_{0,2}} + \frac{\xi_4-\xi}{\xi_4-\xi_1}N_{1,2} = \begin{cases}(1-2\xi)^3, & 0 \le \xi < \frac{1}{2}\\ 0, & \frac{1}{2} \le \xi < 1\end{cases}$$

$$i=1 \Rightarrow N_{1,3} = \frac{\xi-\xi_1}{\xi_4-\xi_1}N_{1,2} + \frac{\xi_5-\xi}{\xi_5-\xi_2}N_{2,2} = \frac{\xi-0}{\frac{1}{2}-0}\cdot\begin{cases}(1-2\xi)^2, & 0 \le \xi < \frac{1}{2}\\ 0, & \text{elsewhere}\end{cases}$$

$$+\frac{1-\xi}{1-0}\cdot\begin{cases}2\xi(2-3\xi), & 0 \le \xi < \frac{1}{2}\\ 2(1-\xi)^2, & \text{elsewhere}\end{cases} = \begin{cases}14\xi^3-18\xi^2+6\xi, & 0 \le \xi < \frac{1}{2}\\ 2(1-\xi)^3, & \frac{1}{2} \le \xi < 1\end{cases}$$

$$i=2 \Rightarrow N_{2,3} = \frac{\xi-\xi_2}{\xi_5-\xi_2}N_{2,2} + \frac{\xi_6-\xi}{\xi_6-\xi_3}N_{3,2} = \frac{\xi-0}{1-0}\cdot\begin{cases}2\xi(2-3\xi), & 0 \le \xi < \frac{1}{2}\\ 2(1-\xi)^2, & \text{elsewhere}\end{cases}$$

$$+\frac{1-\xi}{1-0}\cdot\begin{cases}2\xi^2, & 0 \le \xi < \frac{1}{2}\\ 2(1-\xi)(3\xi-1), & \text{elsewhere}\end{cases} = \begin{cases}-8\xi^3+6\xi^2, & 0 \le \xi < \frac{1}{2}\\ 8\xi^3-18\xi^2+12\xi-2, & \frac{1}{2} \le \xi < 1\end{cases}$$

$$i=3 \Rightarrow N_{3,3} = \frac{\xi-\xi_3}{\xi_6-\xi_3}N_{3,2} + \frac{\xi_7-\xi}{\xi_7-\xi_4}N_{4,2} = \frac{\xi-0}{1-0}\cdot\begin{cases}2\xi^2, & 0 \le \xi < 1/2\\ 2(1-\xi)(3\xi-1), & \frac{1}{2} \le \xi < 1\end{cases}$$

$$+\frac{1-\xi}{1-1/2}\cdot\begin{cases}0, & 0 \le \xi < \frac{1}{2}\\ (2\xi-1)^2, & \frac{1}{2} \le \xi < 1\end{cases} = \begin{cases}2\xi^3 & 0 \le \xi < \frac{1}{2}\\ -14\xi^3+24\xi^2-12\xi+2, & \frac{1}{2} \le \xi < 1\end{cases}$$

$$i=4 \Rightarrow N_{4,3} = \frac{\xi-\xi_4}{\xi_7-\xi_4}N_{4,2} + \frac{\xi_8-\xi}{\xi_8-\xi_5}\cancel{N_{5,2}} = \frac{\xi-\frac{1}{2}}{1-\frac{1}{2}}\cdot\begin{cases}0, & 0 \le \xi < \frac{1}{2}\\ (2\xi-1)^2, & \text{elsewhere}\end{cases} + 0$$

$$= \begin{cases}0 & , 0 \le \xi < \frac{1}{2}\\ (2\xi-1)^3, & \frac{1}{2} \le \xi < 1\end{cases}$$

The above basis functions are shown in Fig. 2.12.

In more detail, the same shape functions are also shown in Table 2.2.

One may observe that in both subintervals the sum of the basis functions equals to the unity (*partition of unity* property). Also, in every subinterval only four or five functions are different than zero.

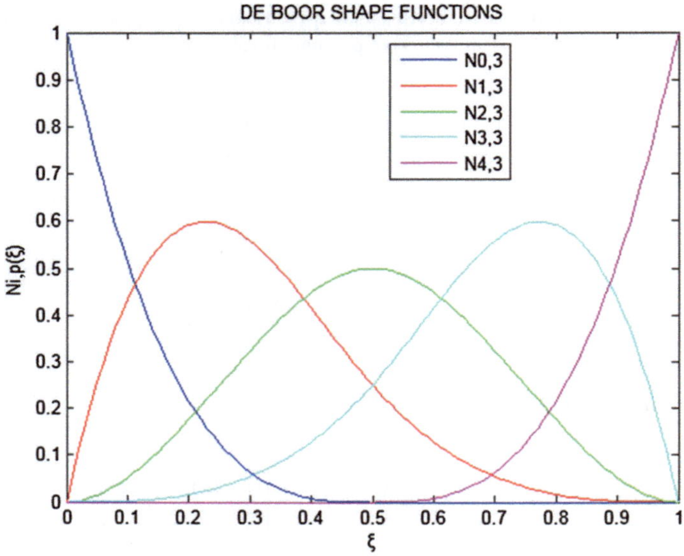

Fig. 2.12 Cubic ($p = 3$) basis functions according to Curry–Schoenberg (de Boor) approach, for three breakpoints, for multiplicity equal to the unity

The position of the control points is determined from the fulfillment of the relationship:

$$x(\xi) = \sum_{i=0}^{n} N_{i,p}(\xi) \cdot x_{\mathbf{P}_i} \tag{2.51d}$$

at $(n+1)$ test points in the domain. As the analytical expressions of the basis functions are now known, we repeat the procedure we had earlier applied (see Exercise 2.6), and then the position of the control points is determined as follows:

$$x_{P_0} = 0, \quad x_{P_1} = \tfrac{1}{6}, \quad x_{P_2} = \tfrac{1}{2}, \quad x_{P_3} = \tfrac{5}{6}, \quad x_{P_4} = 1. \tag{2.51e}$$

General conclusion Equation (2.51e) suggests that in the case of interpolating a straight segment $[0, L]$ by *cubic* B-splines, the control points include all used uniform breakpoints plus *two extra* points, which divide the two extreme subintervals ($[x_0, x_1]$ and $[x_{n-1}, x_n]$) in a ratio equal to 1:3, measured from the positions of P_0 and P_n, respectively.

Exercise 2.10 (Cubic B-splines with C^1-continuity)
Describe and fully discuss the interpolation using cubic B-splines with C^1-continuity in one dimension, as follows:

(A) Applying simple thoughts, calculate the number of involved coefficients regarding a given sequence of breakpoints $\{x_0, x_1, \ldots, x_n\}$.
(B) Fully explain the expression in Eq (2.31b), in terms of truncated power series (prior to Schoenberg) and Curry–Schoenberg (de Boor) formulation.

Solution **(A)** In the case of a function $U(x)$ with C^1-continuity, at each breakpoint x_i we know the value of the function, i.e., the pair (x_i, U_i), as well as its first derivative, i.e., the pair (x_i, U_i'), $i = 0, \ldots, n$. Therefore, in the first interval (i.e., $[x_0, x_1]$) we know the value $U_0 = U(x_0)$ and its first derivative $U_0' = U'(x_0)$ at the point x_0 as well as the similar quantities (U_1, U_1') at the point x_1; therefore, we are able to construct a unique polynomial of third degree $p_1(x)$ that fulfills all these four conditions. In the sequence, knowing the quantities (U_1, U_1') and (U_2, U_2'), we construct a second piecewise cubic polynomial in the interval $x_1 \leq x \leq x_2$. Due to the C^1-continuity, at the point (x_1, U_1), only two out of the four coefficients of the second polynomial are independent variables. In other words, running in the domain from left to the right, the full definition of the spline curve until the completion of the second interval of breakpoints requires six independent coefficients, $a_0, a_1, a_2, a_3, a_4, a_5$. Repeating the same thoughts until the nth interval of breakpoints is completed (where $(n-1)$ intervals exist after the first one), we conclude that the full definition of the spline curve requires $4 + 2(n-1)$ piecewise polynomial coefficients. *Therefore, in the case of cubic B-splines the crowd of these* $2(n+1)$ *polynomial coefficients is double of the number of the* $(n+1)$ *breakpoints.*

(B-1) Initially, we explain the expression in Eq. (2.31b) in the context of *truncated power series* B-spline formulation. If we consider that the above polynomial $p_1(x)$ extends in the second interval (i.e., $[x_1, x_2]$), then it should generally not pass through the point (x_2, U_2), or/and it will not have the desired derivative; therefore, the aforementioned deviation will be a polynomial error function $q(x) = p_2(x) - p_1(x) \neq 0$. In order to achieve the desired correction and obtain the correct polynomial in the form $p_2(x) = p_1(x) + q(x)$, we must construct the polynomial $q(x)$ so as to fulfill the conditions:

$$q(x_1) = q'(x_1) = 0 \text{ (first order continuity)} \tag{2.52a}$$

and

$$\left.\begin{aligned} q(x_2) &= f(x_2) - p_1(x_2) \\ q'(x_2) &= f'(x_2) - p_1'(x_2) \end{aligned}\right\}. \tag{2.52b}$$

It is apparent that the abovementioned polynomial $q(x)$ is unique, because it is of third degree (as a difference of two polynomials each of third degree) and fulfills four conditions. In Eq. (2.31b), we claimed that the unique polynomial is of the form

$q(x) = b_1(x - x_1)^3 + c_1(x - x_1)^2$, $x_1 \leq x \leq x_2$. Actually, this polynomial obviously fulfills both relationships involved in Eq. (2.52a) and on top of this, according to Eq. (2.52b) the constants are calculated as the unique solution of the linear equations system:

$$\left.\begin{aligned} b_1(x - x_1)^3 + c_1(x - x_1)^2 &= f(x_2) - p_1(x_2) \\ 3b_1(x - x_1)^2 + 2c_1(x - x_1) &= f'(x_2) - p_1'(x_2) \end{aligned}\right\}. \tag{2.52c}$$

Therefore, the spline curve within the second interval is given by the relationship:

$$p_2(x) = a_0 + a_1 x + a_2 x^2 + a_3 x^3 + b_1(x - x_1)^3 + c_1(x - x_1)^2, \quad x_1 \leq x \leq x_2 \tag{2.52d}$$

Introducing the *quadratic truncated monomial*:

$$\langle x - x_i \rangle_+^2 = \begin{cases} 0, \; x \leq x_i \\ (x - x_i)^2, \quad x > x_i, \end{cases} \tag{2.52e}$$

according to the definition (2.52e), for the first two intervals of breakpoints the spline curve is written in a unified way as:

$$p_2(x) = a_0 + a_1 x + a_2 x^2 + a_3 x^3 + b_1 \langle x - x_1 \rangle_+^3 + c_1 \langle x - x_1 \rangle_+^2, \quad x_0 \leq x \leq x_2 \tag{2.52f}$$

The above thoughts can be extended to a third interval, where now the position of $p_1(x)$ is occupied by the polynomial $p_2(x)$. Similarly, in order $p_2(x)$ pass through the point (x_3, U_3) and obtain the given derivative it requires a corrective polynomial of the form $b_2(x - x_2)^3 + c_2(x - x_2)^2$, $x_2 \leq x \leq x_3$, and so on. Finally, the general expression for the n involved intervals is given by the expression:

$$f(x) = a_0 + a_1 x + a_2 x^2 + a_3 x^3 + \sum_{j=1}^{n-1} b_j \langle x - x_j \rangle_+^3 + \sum_{j=1}^{n-1} c_j \langle x - x_j \rangle_+^2 \tag{2.52g}$$

Obviously, the relationship (2.52g) includes $(2n + 2)$ terms, as exactly we had concluded in question (A), and ensures the C^1-continuity.

Finally, the basis functions are the following three sets of monomials:

(B-2) In the sequence, we study the same issue following the *Curry–Schoenberg* (de Boor) formulation. Within this context, in the general case of $(n + 1)$ breakpoints $(x_0, x_1, \ldots, x_n)$, we construct the so-called knot vector sequence, which generally consists of $(m + 1)$ elements:

$$\hat{\mathbf{U}} = \left\{ \hat{U}_0, \hat{U}_1, \ldots, \hat{U}_m \right\} \tag{2.52h}$$

Set No. 1	Set No. 2	Set No. 3
$\begin{cases} \hat{B}_0(x) = 1 \\ \hat{B}_1(x) = x \\ \hat{B}_2(x) = x^2 \\ \hat{B}_3(x) = x^3 \end{cases}$	$\begin{cases} \hat{B}_4(x) = \langle x - x_1 \rangle_+^3 \\ \vdots \\ \hat{B}_{n+2}(x) = \langle x - x_{n-1} \rangle_+^3 \end{cases}$	$\begin{cases} \hat{B}_{n+3}(x) = \langle x - x_1 \rangle_+^2 \\ \vdots \\ \hat{B}_{2n+1}(x) = \langle x - x_{n-1} \rangle_+^2 \end{cases}$

Given the polynomial degree p of the spline curve, the general relationship that correlates the number of elements in the knot vector $\hat{\mathbf{U}}$ with the number of control points is:

$$m = n_p + p + 1 \tag{2.52i}$$

When we wish C^1-continuity, at each breakpoint we associate two knots, and then the knot vector becomes:

$$\widehat{\mathbf{U}} = \left\{ \underbrace{x_0, \ldots, x_0}_{(p+1) \text{ terms}}, \underbrace{x_1, x_1}_{2 \text{ terms}}, \ldots, \underbrace{x_{n-1}, x_{n-1}}_{2 \text{ terms}}, \underbrace{x_n, \ldots, x_n}_{(p+1) \text{ terms}} \right\}, \tag{2.52j}$$

whence the number of its terms (see relationship 2.52j) is $m'+1 = 2(p+1)+2(n-1) = 2n + 2p$. Then, according to the relationship (2.52i), the number of control points should be $n'_p = m' - p - 1 = (2n + 2p - 1) - p - 1 = 2n + p - 2$.

Actually, if the interpolating polynomial is *cubic* ($p = 3$), then the above relationship gives $n''_p = 2n + 3 - 2 = 2n + 1$, whence the number of control points $(P_0, P_1, \ldots, P_{2n+1})$ is $(2n + 2)$, which is *exactly double* than that of the breakpoints.

Exercise 2.11 [Cubic B-splines, in the case of C^1-continuity (cont'd)]
Find the normalized shape functions that are obtained from the interpolation using cubic B-splines, in the case of C^1-continuity, for two subintervals $\left(0 \leq x \leq \frac{1}{2}, \frac{1}{2} \leq x \leq 1\right)$. Apply both known approaches (prior-to-Schoenberg truncated power series, and Curry–Schoenberg [12] and de Boor [17]).

Solution This exercise differs from Exercise 2.8 in the fact that it here refers to C^1 instead of the previous C^2-continuity. In the case of two subintervals, which means three breakpoints (two endpoints and one middle point), we need six coefficients and six control points ($n_p = 5$) or six constants.

(A) ***Truncated power basis (prior-to-Schoenberg approach)***

The procedure is as follows. For each of the three breakpoints $\{0, \frac{1}{2}, 1\}$, we write the expansion of $U(x)$ in a series of six terms:

$$U(x) = a_0 + a_1 x + a_2 x^2 + a_3 x^3 + b_1 \left\langle x - \tfrac{1}{2}\right\rangle_+^3 + c_1 \left\langle x - \tfrac{1}{2}\right\rangle_+^2 \tag{2.53a}$$

In the sequence, also for each of the three breakpoints we write the first derivatives, $U'(x)$, of the aforementioned expansion,

$$U'(x) = 0 + a_1 \cdot 1 + 2a_2 x + 3a_3 x^2 + 3b_1 \left\langle x - \tfrac{1}{2}\right\rangle_+^2 + 2c_1 \left\langle x - \tfrac{1}{2}\right\rangle_+ . \tag{2.53b}$$

Therefore, the following linear system is derived:

$$\underbrace{\begin{bmatrix} 1 & 0 & 0 & 0 & 0 & 0 \\ 1 & 1/2 & 1/4 & 1/8 & 0 & 0 \\ 1 & 1 & 1 & 1 & 1/8 & 1/4 \\ 0 & 1 & 0 & 0 & 0 & 0 \\ 0 & 1 & 1 & 3/4 & 0 & 0 \\ 0 & 1 & 2 & 3 & 3/4 & 1 \end{bmatrix}}_{[\mathbf{A}]} \cdot \begin{bmatrix} a_0 \\ a_1 \\ a_2 \\ a_3 \\ b_1 \\ c_1 \end{bmatrix} = \begin{bmatrix} U_0 \\ U_1 \\ U_2 \\ U'_0 \\ U'_1 \\ U'_2 \end{bmatrix} . \tag{2.53c}$$

The inverse of the matrix $[\mathbf{A}]$ that appears in Eq. (2.53c) is:

$$[\mathbf{A}]^{-1} = \begin{bmatrix} 1 & 0 & 0 & 0 & 0 & 0 \\ 0 & 0 & 0 & 1 & 0 & 0 \\ -12 & 12 & 0 & -4 & -2 & 0 \\ 16 & -16 & 0 & 4 & 4 & 0 \\ -16 & 32 & -16 & -4 & 0 & 4 \\ -12 & 0 & 12 & -2 & -8 & -2 \end{bmatrix},$$

and therefore the shape functions, $\mathbf{\Phi} = [\phi_1(x), \phi_2(x), \phi_3(x), \phi_4(x), \phi_5(x), \phi_6(x)]^{\mathrm{T}}$, are given in a unified expression as follows:

$$\underbrace{\mathbf{\Phi}(x)}_{1\times 6} = \left[\underbrace{1 \; x \; x^2 \; x^3 \; \langle x-1/2\rangle_+^3 \; \langle x-1/2\rangle_+^2}_{1\times 6} \right] \cdot \underbrace{[\mathbf{A}]^{-1}}_{6\times 6}$$

$$= \left[1 \; x \; x^2 \; x^3 \; \langle x-1/2\rangle_+^3 \; \langle x-1/2\rangle_+^2 \right] \cdot \begin{bmatrix} 1 & 0 & 0 & 0 & 0 & 0 \\ 0 & 0 & 0 & 1 & 0 & 0 \\ -12 & 12 & 0 & -4 & -2 & 0 \\ 16 & -16 & 0 & 4 & 4 & 0 \\ -16 & 32 & -16 & -4 & 0 & 4 \\ -12 & 0 & 12 & -2 & -8 & -2 \end{bmatrix}$$

$$= \begin{bmatrix} 1 - 12x^2 + 16x^3 - 16\langle x-1/2\rangle_+^3 - 12\langle x-1/2\rangle_+^2 \\ 12x^2 - 16x^3 + 32\langle x-1/2\rangle_+^3 \\ -16\langle x-1/2\rangle_+^3 + 12\langle x-1/2\rangle_+^2 \\ x - 4x^2 + 4x^3 - 4\langle x-1/2\rangle_+^3 - 2\langle x-1/2\rangle_+^2 \\ -2x^2 + 4x^3 - 8\langle x-1/2\rangle_+^2 \\ 4\langle x-1/2\rangle_+^3 - 2\langle x-1/2\rangle_+^2 \end{bmatrix} \tag{2.53d}$$

Remark At this point, we consider the auxiliary shape functions of a beam with length L, in bending (e.g., [37]):

$$\left.\begin{aligned} N_1(w) &= 1 - 3w^2 + 2w^3 \\ N_2(w) &= (w - 2w^2 + w^3) \cdot L \\ N_3(w) &= 3w^2 - 2w^3 \\ N_4(w) &= (-w^2 + w^3) \cdot L \end{aligned}\right\} \tag{2.53e}$$

Now we will attempt to compare the basis functions in Eq. (2.53d) with those of the bending beam in Eq. (2.53e), in the interval $0 \le x \le 1/2$. That is, we consider a beam starting from the left end and ending to the middle of the interval [0, 1]. Since the length of the interval is $L = 1/2$, then the apparent relationship between the normalized coordinate $w(\triangleq x/L)$ and the global coordinate x will be

$$w = 2x. \tag{2.53f}$$

Substituting Eq. (2.53f) into Eq. (2.53d), for the interval $0 \le x \le 1/2$ the following conclusions may be derived:

(1) The function $\phi_1(x) = 1 - 12x^2 + 16x^3$ becomes exactly equal to the function $N_1(w) = 1 - 3w^2 + 2w^3$ of Eq. (2.53e), which is the distribution of the flexure in the beam due to a unit displacement at the *left* end of the interval [0, 1].

Table 2.3 B-spline basis functions according to the notation in the book of Piegl and Tiller [27], $N_{i,p}(\xi)$, for two uniform subdivisions (six control points: $n_p = 5$) and two knots per breakpoint

For $0 < \xi < \frac{1}{2}$	For $\frac{1}{2} < \xi < 1$
$N_{0,3}(\xi) = -8\xi^3 + 12\xi^2 - 6\xi + 1 = (1 - 2\xi)^3$	$N_{0,3}(\xi) = 0$
$N_{1,3}(\xi) = 24\xi^3 - 24\xi^2 + 6\xi$	$N_{1,3}(\xi) = 0$
$N_{2,3}(\xi) = -20\xi^3 + 12\xi^2$	$N_{2,3}(\xi) = -4\xi^3 + 12\xi^2 - 12\xi + 4$
$N_{3,3}(\xi) = 4\xi^3$	$N_{3,3}(u) = 20\xi^3 - 48\xi^2 + 36\xi - 8$
$N_{4,3}(\xi) = 0$	$N_{4,3}(u) = -24\xi^3 + 48\xi^2 - 30\xi + 6$
$N_{5,3}(\xi) = 0$	$N_{5,3}(u) = 8\xi^3 - 12\xi^2 + 6\xi - 1$
$\sum_{i=0}^{5} N_{i,3}(\xi) = 1$	$\sum_{i=0}^{5} N_{i,3}(\xi) = 1$

(2) The function $\phi_2(x) = 12x^2 - 16x^3$ becomes exactly equal to the function $N_3(w) = 3w^2 - 2w^3$ of Eq. (2.53e), which is the flexure of the *middle* breakpoint in the interval [0, 1].
(3) The function $\phi_4(x) = x - 4x^2 + 4x^3$ becomes equal to the function $\frac{1}{2}N_2(w) = w - 2w^2 + w^3$ of Eq. (2.53e), that is *half* of the rotational angle (slope) at the *left* end of the interval [0, 1].
(4) The function $\phi_5(x) = -2x^2 + 4x^3$ becomes equal to the function $2N_4(w) = -w^2 + w^3$ of Eq. (2.53e), which is *double* of the rotational angle at the *middle* breakpoint in the interval [0, 1].

Conclusion Therefore, despite the small differentiation in the coefficients (1/2 or 2), in the first subinterval of the B-spline interpolation $[0, \frac{1}{2}]$, essentially a beam of length $l_e = \frac{1}{2}$ appears in bending, which means that the supposed global interpolation is practically a set of four piecewise local Hermite polynomials for this half interval $[0, \frac{1}{2}]$ and another set for $[\frac{1}{2}, 1]$ (by mirroring with respect to the midpoint)!

(B) ***Curry–Schoenberg (de Boor) approach***

Following either the original recursive procedure developed by de Boor [17], or even performing reverse engineering (using the relevant MATLAB® functions: `augknt`, `spcol`, etc.), the obtained six basis functions are tabulated in the form of composite expressions in Table 2.3.

As also in the previous case (Exercise 2.8), we observe that again in both subintervals the sum of basis functions equals to the unity (partition of unity property). Also, in each subinterval only four out of the six functions are different than zero.

Exercise 2.12 (Natural cubic B-splines in three breakpoint spans)
In the case of subdividing the unit length [0, 1] into three equal segments, compare the cubic Bézier curve with that of the *natural* B-spline (for multiplicity equal to the unity).

Solution Now we have 4 breakpoints (2 endpoints and 2 intermediate ones: $[0, \frac{1}{3}, \frac{2}{3}, 1]$). Therefore, considering the unit interval, the given breakpoints are: $\xi_0 = 0, \xi_1 = \frac{1}{3}, \xi_2 = \frac{2}{3}, \xi_3 = 1$.

Since the multiplicity equals to one, in the general formulation of cubic splines according to Schoenberg, we will take one knot at every breakpoint, and therefore we will have six shape functions. From these functions, four are concerned with the values of the function ("translational" degrees of freedom), whereas the rest two may refer either to the first or to the second derivatives at the ends of the interval [0, 1].

In the particular case we study, the interpolation becomes:

$$U(\xi) = b_0 + b_1\xi + b_2\xi^2 + \sum_{i=0}^{2} a_i \langle \xi - \xi_i \rangle_+^3 \tag{2.54a}$$

The application of Eq. (2.54a) to the four nodes gives:

$$\begin{aligned}
\hat{d}_1 &= b_0 + b_1(0) + b_2(0)^2 + a_1(0) + a_2(0) + a_3(0) \\
\hat{d}_2 &= b_0 + b_1(1/3) + b_2(1/3)^2 + a_1(1/3 - 0)^3 + a_2(0) + a_3(0) \\
\hat{d}_3 &= b_0 + b_1(2/3) + b_2(2/3)^2 + a_1(2/3 - 0)^3 + a_2(2/3 - 1/3)^3 + a_3(0) \\
\hat{d}_4 &= b_0 + b_1(1) + b_2(1)^2 + a_1(1 - 0)^3 + a_2(1 - 1/3)^3 + a_3(1 - 2/3)^3,
\end{aligned} \tag{2.54b}$$

where $\hat{d}_i, i = 1, \ldots, 4$ correspond to the degrees of freedom.

Obviously, the above four equations are incapable of determining the six unknowns $(b_0, b_1, b_2, a_1, a_2, a_3)$ involved in Eq. (2.54a). Therefore, two more equations are required, which are produced, e.g., by vanishing the second derivatives at the two ends of the domain (vanishing bending moment).

The first derivative of Eq. (2.54a) becomes:

$$U'(\xi) = b_1 + 2b_2\xi + 3\left[a_1\langle \xi - 0 \rangle_+^2 + a_2\langle \xi - 1/3 \rangle_+^2 + a_3\langle \xi - 2/3 \rangle_+^2\right], \tag{2.54c}$$

whence the second derivative, in which we are interested, becomes:

$$U''(\xi) = 2b_2 + 6\left[a_1\langle \xi - 0 \rangle_+^1 + a_2\langle \xi - 1/3 \rangle_+^1 + a_3\langle \xi - 2/3 \rangle_+^1\right] \tag{2.54d}$$

Therefore, for $\xi = 0$ and $\xi = 1$, from Eq. (2.54d) we take:

$$\left.\begin{aligned}\hat{d}_5 &= 0 = U''(0) = 2b_2 = 0\\ \hat{d}_6 &= 0 = U''(0) = 2b_2 + 6[a_1(1-0) + a_2(1-1/3) + a_3(1-2/3)] = 0\end{aligned}\right\} \tag{2.54e}$$

Therefore, the matrix involved in Eq. (2.54b) (in the form $\mathbf{Aa} = \mathbf{u}$, $\mathbf{a}$ = unknown constants, $\mathbf{u}$ = nodal values) is:

$$\mathbf{A} = \begin{bmatrix} 1 & 0 & 0 & 0 & 0 & 0\\ 1 & 1/3 & (1/3)^2 & (1/3)^3 & 0 & 0\\ 1 & 2/3 & (2/3)^2 & (2/3)^3 & (1/3)^3 & 0\\ 1 & 1 & 1 & 1 & (2/3)^3 & (2/3)^3\\ 0 & 0 & 2 & 0 & 0 & 0\\ 0 & 0 & 2 & 6 & 4 & 2 \end{bmatrix},$$

and its inverse:

$$\mathbf{A}^{-1} = \begin{bmatrix} 1 & 0 & 0 & 0 & 0 & 0\\ -3.8 & 4.8 & -1.2 & 0.2 & -0.0963 & -0.0037\\ 0 & 0 & 0 & 0 & 0.5 & 0\\ 7.2 & -16.2 & 10.8 & -1.8 & -0.6333 & 1/30\\ -16.2 & 43.2 & -37.8 & 10.8 & 0.8 & -0.2\\ 10.8 & -37.8 & 43.2 & -16.2 & -0.2 & 0.8 \end{bmatrix}$$

Ignoring the last two columns, which correspond to vanishing rotational DOFs (curvature) at the ends due to the assumed *natural* B-splines, the following matrix is derived:

$$\mathbf{B}_{11} = \begin{bmatrix} 1 & 0 & 0 & 0\\ -3.8 & 4.8 & -1.2 & 0.2\\ 0 & 0 & 0 & 0\\ 7.2 & -16.2 & 10.8 & -1.8\\ -16.2 & 43.2 & -37.8 & 10.8\\ 10.8 & -37.8 & 43.2 & -16.2 \end{bmatrix}$$

Therefore, the shape functions become:

$$\underbrace{\mathbf{\Phi}(\xi)}_{1\times4} = \underbrace{\left[1\ \xi\ \xi^2\ \langle\xi-0\rangle_+^3\ \langle\xi-1/3\rangle_+^3\ \langle\xi-2/3\rangle_+^3\right]}_{1\times6} \cdot \underbrace{(\mathbf{B}_{11})}_{6\times4}$$

or

$$\underbrace{\boldsymbol{\Phi}(\xi)}_{1\times 4} = \left[1\ \xi\ \xi^2\ \langle\xi-0\rangle_+^3\ \langle\xi-1/3\rangle_+^3\ \langle\xi-2/3\rangle_+^3\right]\cdot\begin{bmatrix} 1 & 0 & 0 & 0 \\ -3.8 & 4.8 & -1.2 & 0.2 \\ 0 & 0 & 0 & 0 \\ 7.2 & -16.2 & 10.8 & -1.8 \\ -16.2 & 43.2 & -37.8 & 10.8 \\ 10.8 & -37.8 & 43.2 & -16.2 \end{bmatrix}$$

More analytically, the shape functions within the three subintervals are defined in terms of the breakpoints and are given as follows:

For the interval $0 \le \xi \le 1/3$:

$$\begin{aligned}
\phi_1(\xi) &= 1 - 3.8\xi + 7.2\xi^3 \\
\phi_2(\xi) &= 4.8\xi - 16.2\xi^3 \\
\phi_3(\xi) &= -1.2\xi + 10.8\xi^3 \\
\phi_4(\xi) &= 0.2\xi - 1.8\xi^3 \\
\phi_1(\xi) + \phi_2(\xi) &+ \phi_3(\xi) + \phi_4(\xi) \equiv 1
\end{aligned}$$

For the interval $1/3 \le \xi \le 2/3$:

$$\begin{aligned}
\phi_1(\xi) &= 1 - 3.8\xi + 7.2\xi^3 - 16.2(\xi - 1/3)^3 \\
\phi_2(\xi) &= 4.8\xi - 16.2\xi^3 + 43.2(\xi - 1/3)^3 \\
\phi_3(\xi) &= -1.2\xi + 10.8\xi^3 - 37.8(\xi - 1/3)^3 \\
\phi_4(\xi) &= 0.2\xi - 1.8\xi^3 + 10.8(\xi - 1/3)^3 \\
\phi_1(\xi) + \phi_2(\xi) &+ \phi_3(\xi) + \phi_4(\xi) \equiv 1
\end{aligned}$$

Finally, *for the interval* $2/3 \le \xi \le 1$:

$$\begin{aligned}
\phi_1(\xi) &= 1 - 3.8\xi + 7.2\xi^3 - 16.2(\xi - 1/3)^3 + 10.8(\xi - 2/3)^3 \\
\phi_2(\xi) &= 4.8\xi - 16.2\xi^3 + 43.2(\xi - 1/3)^3 - 37.8(\xi - 2/3)^3 \\
\phi_3(\xi) &= -1.2\xi + 10.8\xi^3 - 37.8(\xi - 1/3)^3 + 43.2(\xi - 2/3)^3 \\
\phi_4(\xi) &= 0.2\xi - 1.8\xi^3 + 10.8(\xi - 1/3)^3 - 16.2(\xi - 2/3)^3 \\
\phi_1(\xi) + \phi_2(\xi) &+ \phi_3(\xi) + \phi_4(\xi) \equiv 1
\end{aligned}$$

Remark Obviously, if instead of the submatrix $\mathbf{B}_{11}$ we had used the whole matrix $\mathbf{A}^{-1}$, then in addition we would have taken the shape functions that correspond to the "rotational" degrees of freedom, i.e., those related to the second derivative (curvature) at the two endpoints of the interval.

Table 2.4 Comparison between basis and shape functions

Function	Cubic Bezier ($n = p = 3$) (basis function)	Cubic B-spline ($0 \le \xi \le 1/3$) (shape function)
$N_1(\xi)$	$B_{0,3}(\xi) = (1-\xi)^3$	$1 - 3.8\xi + 7.2\xi^3$
$N_2(\xi)$	$B_{1,3}(\xi) = 3\xi(1-\xi)^2$	$4.8\xi - 16.2\xi^3$
$N_3(\xi)$	$B_{2,3}(\xi) = 3\xi^2(1-\xi)$	$-1.2\xi + 10.8\xi^3$
$N_4(\xi)$	$B_{3,3}(\xi) = \xi^3$	$0.2\xi - 1.8\xi^3$
$\sum_{i=1}^4 N_i(\xi)$	1	1

After the above extensive calculations, let us now restrict in the interval between the first two breakpoints, i.e., $0 \le \xi \le 1/3$.

In Table 2.4, we see that cubic **Bézier curve** is a continuous function that is described by a *unified formula* in the whole interval [0, 1]. On the contrary, the natural cubic **B-splines** requires a *piecewise definition,* from which we present only the interval $0 \le \xi = x/L \le 1/3$.

From Table 2.4, we see that the functions $N_{i,p}(\xi) = B_{i,3}(\xi)$ related to *Bernstein–Bézier* polynomials are *basis* functions, as exactly their name denotes, while in contrast the *natural cubic B-splines* are *shape* functions [they possess the property: $N_i(\xi_j) = \delta_{ij}$]. The basis functions take a unit value only at the extreme endpoints ($\xi = 0, 1$), whereas those that are associated to intermediate control points have a maximum (but not unit value) at them.

A comparative contradiction is given in Figs. 2.13 and 2.14, from which it becomes clear that again there is *no relationship* between these two approximations.

Concretely, both approximations are of polynomial form until the third degree, but in Table 2.4, it is shown that in natural cubic B-spline formulation the term ξ^2 does *not* appear in the shape functions. From a mathematical point of view, this fact is explained considering the first term of Eq. (2.54d) which—in conjunction with $U''(0) = 0$—implies $b_2 = 0$. Thus considering Eq. (2.54a) the quadratic term vanishes, and this fact shows that the natural cubic B-splines is a *reduced* set of basis functions. Of course, this fact would not occur if instead of the *natural* spline (vanishing moment at the ends, here we are concerned with the condition $U''(0) = 0$ only) another arbitrary value had been considered.

In contrast, in the Bézier formulation all powers of ξ appear accordingly, and concretely the term ξ^2 appears in three places, i.e., in the basis functions $B_{0,3}(\xi)$, $B_{1,3}(\xi)$ and $B_{2,3}(\xi)$.

Interim Remark: At this point of Chap. 2, since the most important issues of the *univariate* interpolations and approximations have been thoroughly presented, it is the proper time to turn from 1D to 2D interpolations.

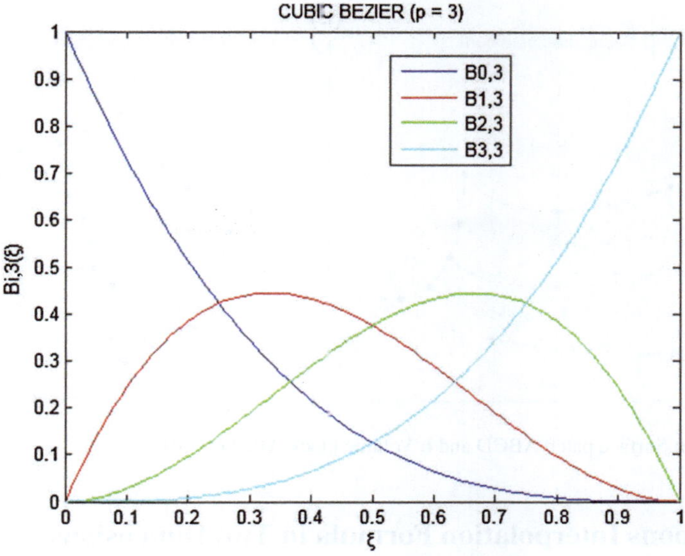

Fig. 2.13 Basis functions involved in the cubic Bézier curve (Bernstein polynomials), which are also provided in Table 2.4

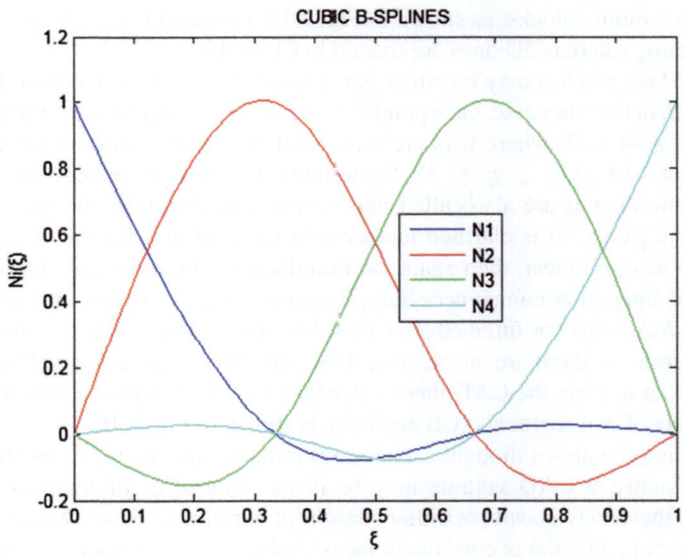

Fig. 2.14 Shape functions of the natural cubic B-splines (the region $0 < \xi < \frac{1}{3}$ is also shown in Table 2.4)

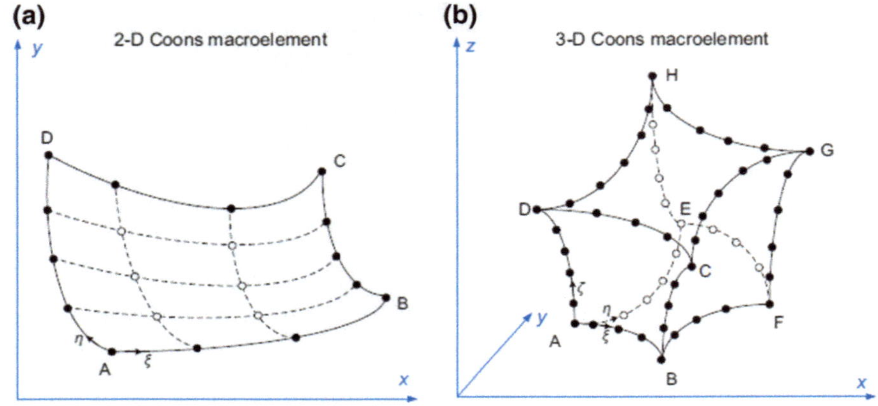

Fig. 2.15 **a** Surface patch ABCD and **b** Volume block ABCDEFGH

2.4 Coons Interpolation Formula in Two Dimensions

The theory of Computer-Aided Geometric Design (CAGD) refers either to two-dimensional (2D) parametric patches (surface portions) or to three-dimensional (3D) parametric volume blocks, as shown in Fig. 2.15. In the sequence, we proceed with 2D problems, whereas 3D ones are treated in Chap. 10.

The surface patches may be either flat (planar), thus lying on a plane Oxy or be curvilinear. In the latter case, each point P(x, y) [or P(x, y, z)] is defined in a curvilinear system O$\xi\eta$ as well, where it is preferred to deal with normalized coordinates in the unit interval ($0 \leq \xi, \eta \leq 1$). Sometimes the aforementioned normalized or intrinsic coordinates are also called *parameters,* and therefore, the patch is called "parametric patch." It is clarified that even in the case of a flat patch, if its edges happen to be curvilinear, then again the introduction of normalized coordinates $\xi\eta$ in the unit interval becomes necessary. The usual theory covers patches of either four (*quadrilaterals*) or three edges (*triangles*). In the general case of more actual edges, if some of them are successive, these may be unified in a single composite edge in order to meet the CAD theory. A characteristic example, in which the edge CD consists of two segments (CE and ED), is shown in Fig. 2.16.

Extending the above thoughts, analogous remarks may be made for the volume blocks. Usually, a CAD volume may be of *hexahedral* or of *tetrahedral* shape. Similarly, these may consist of either straight of curvilinear edges. Moreover, these may consist of either flat of curvilinear faces (called external surfaces or boundaries).

Essentially, *Coons interpolation formula* [7] extends the univariate linear interpolation given by Eq. (2.11), from 1D segments to 2D quadrilateral patches. The actual curvilinear patches are mapped to 2D squares of unit edge on the $\xi\eta$-plane. A typical Coons patch is shown in Fig. 2.17. Given the distribution of a function $U(\xi, \eta)$ along the boundary of the patch, Coons interpolation formula offers estimation for

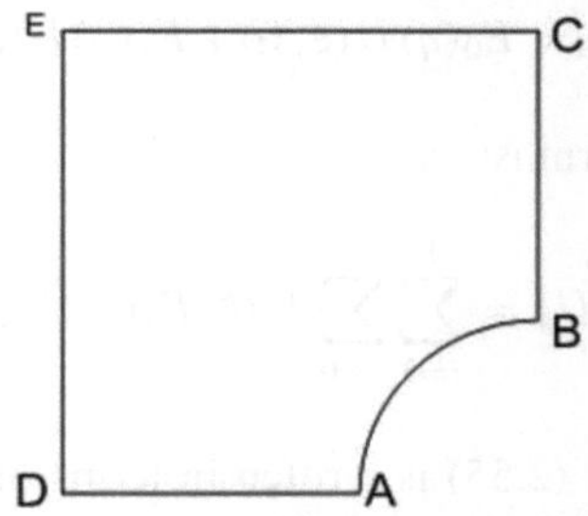

Fig. 2.16 A patch consisting of more than four discrete boundary segments (here five)

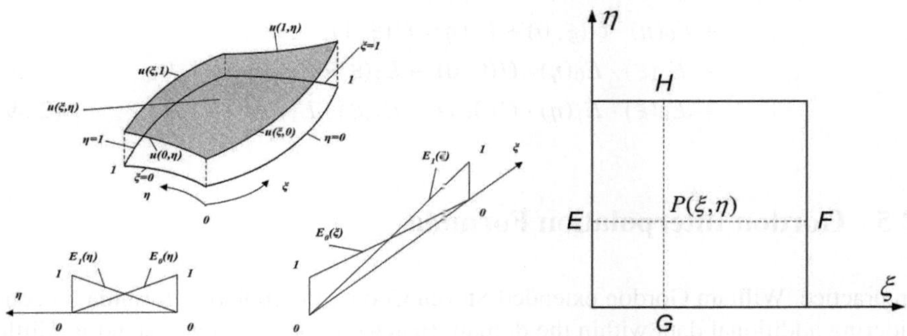

Fig. 2.17 Coons interpolation in a quadrilateral patch

a reasonable distribution of this function within the patch provided the boundary conditions are fulfilled.

Every point $P(\xi, \eta)$ in the interior of the patch ABCD (Fig. 2.17, on the right) can be considered to be linearly interpolated either toward the ξ-direction (between the points E and F) or in the η-direction (between the points G and H). But if these two interpolations are merely added, then their sum becomes unreasonably double, for example, at the four corners, thus a proper correction in the aforementioned sum is required.

After a series of simple thoughts, Coons interpolation formula takes the general form [7]:

$$U(\xi, \eta) = P_\xi[U] + P_\eta[U] - P_\xi P_\eta[U], \tag{2.55}$$

where the projection in the ξ-direction is:

$$P_\xi[U] = E_0(\xi)\, U(0, \eta) + E_1(\xi)\, U(1, \eta), \tag{2.56}$$

the projection in the η-direction is:

$$P_\eta[U] = E_0(\eta)\,U(\xi,0) + E_1(\eta)\,U(\xi,1), \tag{2.57}$$

and finally the corrective term is:

$$P_\xi P_\eta[U] = \sum_{i=0}^{1}\sum_{j=0}^{1} E_i(\xi)E_j(\eta)U(\xi_i,\eta_j) \tag{2.58}$$

In a complete form, Eq. (2.55) is written in terms of the four boundaries (i.e., $U(0,\eta)$, $U(1,\eta)$, $U(\xi,0)$, $U(\xi,1)$), as follows:

$$\begin{aligned} U(\xi,\eta) = {} & E_0(\xi)\cdot U(0,\eta) + E_1(\xi)\cdot U(1,\eta) \\ & + E_0(\eta)\cdot U(\xi,0) + E_1(\eta)\cdot U(\xi,1) \\ & - E_0(\xi)\cdot E_0(\eta)\cdot U(0,0) - E_1(\xi)\cdot E_0(\eta)\cdot U(1,0) \\ & - E_0(\xi)\cdot E_1(\eta)\cdot U(0,1) - E_1(\xi)\cdot E_1(\eta)\cdot U(1,1) \end{aligned} \tag{2.59}$$

2.5 Gordon Interpolation Formula

In practice, William Gordon extended Steven Coons' interpolation formula by considering additional data within the domain (patch). In order to understand it a little better, let us first consider a series of planes normal to the ξ-axis. Then, except of the extreme positions AD and BC (normal to the ξ-axis) on which the function $U(\xi,\eta)$ is given (with $\xi = 0$ and $\xi = 1$), we have to consider a greater number of functions $U(\xi_i,\eta)$ with i denoting the ascending number of the cross section normal to the ξ-axis (Fig. 2.18).

The common matter between Gordon's and Coons' interpolation is that in both of them the variable U is expressed as an algebraic sum of three projections in the ξ- and η-directions, according to Eq. (2.55). The difference lies on the particular form each of the three aforementioned projections should take in each interpolation. In brief, Gordon's interpolation considers more than two blending functions; each of them equals to the unity on the associated section and vanishes at all the other sections.

In summary, Eqs. (2.55)–(2.58) are again valid, but now the involved projections have a broader meaning, as internal nodes are now included. In more details, information is given along a series of positions, which are stored in the vector $\boldsymbol{\xi} = [0 \equiv \xi_0, \xi_1, \ldots, \xi_{n-1}, \xi_n \equiv 1]$ in the ξ-direction, and another vector $\boldsymbol{\eta} = [0 \equiv \eta_0, \eta_1, \ldots, \eta_{m-1}, \eta_m \equiv 1]$ in the η-direction. Now, instead of the previous four corner nodes met in Coons interpolation, a crowd of $(n+1)\times(m+1)$ intersection points usually appear in the corrective term, and this is the number of terms involved in Eq. (2.58) $[P_{\xi\eta}]$ that has to be properly modified.

Additional information is provided in Chaps. 4 and 13.

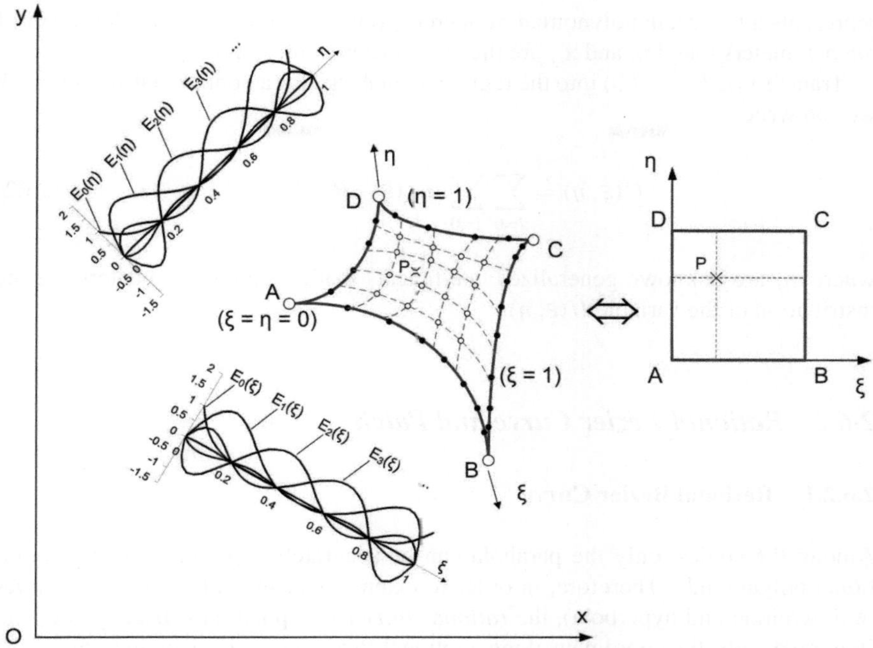

Fig. 2.18 Gordon interpolation in a quadrilateral patch

2.6 Bézier Interpolation Formulas

2.6.1 Nonrational Bézier Patch (Surface)

For 1D problems, this topic has been discussed in Eqs. (2.22) to (2.25) and is related to the Bernstein polynomials $B_{i,n}(\xi)$.

For 2D problems, we can use different polynomial degrees per direction, for example, n for the ξ- and m for the η-direction, respectively. Therefore, the parametric expression of the coordinate vector over a nonrational Bézier patch is given in terms of the normalized coordinates ξ and η by

$$\mathbf{x}(\xi, \eta) = \sum_{i=0}^{n} \sum_{j=0}^{m} B_{i,n}(\xi) \cdot B_{j,m}(\eta)\, \mathbf{x}_{ij}, \tag{2.60}$$

where

$$B_{i,q}(w) \equiv \binom{q}{i} w^i (1-w)^{q-i} = \frac{q!}{i!\,(q-i)!} w^i (1-w)^{q-i} \tag{2.61}$$

represents a Bernstein polynomial of degree q (either n or m), w stands for either of the parameters ξ and η, and $\mathbf{x}_{ij}$ are the coordinates of the control points.

Transferring Eq. (2.60) into the regime of analysis, for a nonrational Bézier patch we can write:

$$U(\xi, \eta) = \sum_{i=0}^{n} \sum_{j=0}^{m} B_{i,n}(\xi) \cdot B_{j,m}(\eta)\, a_{ij}, \tag{2.62}$$

where a_{ij} are unknown generalized coefficients that are required to determine the distribution of the variable $U(\xi, \eta)$.

2.6.2 *Rational Bézier Curve and Patch*

2.6.2.1 Rational Bézier Curve

Among the conics, only the parabola can be accurately represented using nonrational polynomials. Therefore, in order to extend this capability to the rest curves (ellipse/circle and hyperbola), the *rational* Bézier interpolation is usually adopted. In more details, the coordinate along a rational Bézier curve is given in terms of the normalized coordinate ξ by

$$\mathbf{x}(\xi) = \sum_{i=0}^{n} R_{i,n}(\xi)\, \mathbf{x}_i \tag{2.63}$$

with

$$R_{i,n}(\xi) = \frac{B_{i,n}(\xi)\, w_i}{\sum_{j=0}^{n} B_{i,n}(\xi)\, w_j}, \tag{2.64}$$

where w_i and w_j are proper weighting factors.

2.6.2.2 Rational Bézier Patch and Approximation

Extending Eq. (2.63) to a 2D pach ABCD, we can write

$$\mathbf{x}(\xi, \eta) = \sum_{i=0}^{n} \sum_{j=0}^{m} \tilde{R}_{ij}(\xi, \eta) \cdot \mathbf{x}_{ij}, \tag{2.65}$$

where

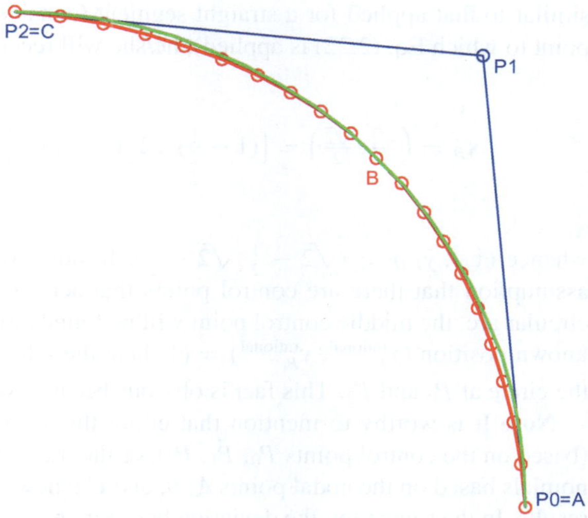

Fig. 2.19 Erroneous representation of the circular segment (in green) using nonrational Bézier (solid redline) and Lagrange polynomials (red circlets), both of second degree

$$\tilde{R}_{ij}(\xi,\eta)=\frac{B_{i,n}(\xi)\cdot w_{ij}\cdot B_{j,m}(\eta)}{\sum_{k=0}^{n}\sum_{l=0}^{m}B_{k,n}(\xi)\cdot w_{kl}\cdot B_{l,m}(\eta)}, \tag{2.66}$$

is the bivariate rational Bézier polynomial, which interpolates the Cartesian coordinates $\mathbf{x}_{ij}$ at the position (ξ_i,η_j), whereas the weighting factor is usually taken as $w_{ij}=w_i w_j$.

Extending Eq. (2.65) from the geometric to the analysis model, we can approximate the problem variable $U(\xi,\eta)$ by

$$U(\xi,\eta)=\sum_{i=0}^{n}\sum_{j=0}^{m}\tilde{R}_{ij}(\xi,\eta)\cdot a_{ij}, \tag{2.67}$$

where now a_{ij} are the generalized coefficients to be determined.

Example 2.13 (Circular arc using rational Bernstein polynomials) To make this point more clear, let us consider a circular arc, for example, the first quadrant of a unit circle (Fig. 2.19). Defining the variable ξ in the counterclockwise direction, this portion of the circle is geometrically determined by the two extreme points, i.e., $A(1,0)$ at $\xi=0$ and $C(0,1)$ at $\xi=1$, plus the middle point $B\left(\frac{\sqrt{2}}{2},\frac{\sqrt{2}}{2}\right)$ of the circular arc at $\xi=\frac{1}{2}$.

A. *Nonrational approach*

As usual, two control points are associated with the ends of the segment. Therefore, considering a quadratic interpolation ($n=2$), the extreme control points (P_0 and P_2) will be $\mathbf{x}_{P_0}=(1,0)$ and $\mathbf{x}_{P_2}=(0,1)$. Moreover, if one attempts to determine the third (intermediate) control point P_1 of the nonrational Bézier interpolation in a way similar to that applied for a straight segment (considering the midpoint B as a test point to which Eq. (2.22) is applied), he/she will receive the system

$$\mathbf{x}_B = \left(\tfrac{\sqrt{2}}{2}, \tfrac{\sqrt{2}}{2}\right) = \left[(1-\tfrac{1}{2})^2, 2\tfrac{1}{2}(1-\tfrac{1}{2}), (\tfrac{1}{2})^2\right] \cdot \left\{ \begin{matrix} (1,0) \\ (\tilde{x}_{P_1}, \tilde{y}_{P_1}) \\ (0,1) \end{matrix} \right\},$$

whence $(\tilde{x}_{P_1}, \tilde{y}_{P_1}) = (\sqrt{2}-\frac{1}{2}, \sqrt{2}-\frac{1}{2})$. In other words, if one made the wrong assumption that there are control points that achieve to accurately interpolate the circular arc, the middle control point will be found closer to the circle than the well-known position $(x_{P_1}^{\text{rational}}, y_{P_1}^{\text{rational}}) = (1, 1)$ at the intersection of the tangent lines of the circle at P_0 and P_2. This fact is obvious because $\sqrt{2}-\frac{1}{2} \cong 0.9142 < 1.0$.

Note It is worthy to mention that either this (erroneous) nonrational approach (based on the control points $P_0, \tilde{P}_1, P_2$), or the traditional quadratic Lagrange polynomials based on the nodal points A, B, and C (shown in Fig. 2.19) lead to *identical* results. In the same plot, the deviation between the circular arc (green color) and the parabola (red color) is distinguishable. Since the above middle point B was considered as a test point, "by construction" the nonrational Bézier curve eventually passes through it.

The above subsection clearly shows the incapability of the nonrational Bézier formulation to represent a circular segment, which therefore demands the use of rational polynomials.

B. *Rational approach*

In the case of rational Bézier interpolation, the control points are:

$$(x_{0p}, y_{0p}) = (1,0), (x_{1p}, y_{1p}) = (1,1) \text{ and } (x_{2p}, y_{2p}) = (0,1). \tag{2.68}$$

This is so because the extreme control points (P_0 and P_2) always coincide with the extreme points of a curve ($P_0 \equiv A$, $P_2 \equiv$ C), whereas the middle control point is at the intersection of the two tangents at P_0 and P_2. In other words, the generator of the arc is the polygonal line $P_0P_1P_2$, as shown in Fig. 2.20.

Below, we shall show that the rational Bézier formulation ensures the accurate representation of a circular arc.

Actually, the quadratic Bézier (Bernstein) polynomials are:

$$B_0 = (1-\xi)^2, \quad B_1 = 2(1-\xi)\xi, \quad B_2 = \xi^2 \tag{2.69}$$

Between several choices, here we take symmetric weighting functions, as follows:

$$w_0 = 1, \quad w_1 = \frac{1}{\sqrt{2}}, \quad w_2 = 1 \tag{2.70}$$

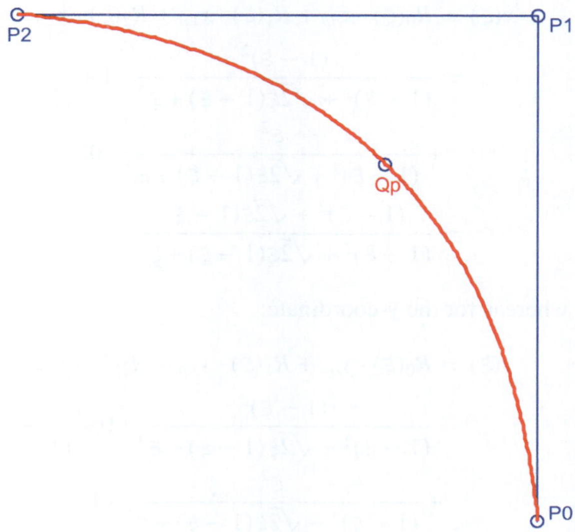

Fig. 2.20 Contour of the first quadrant of a unit circle

Therefore, the three rational functions will be:

$$R_0(\xi) = \frac{B_0(\xi)\cdot w_0}{B_0(\xi)\cdot w_0 + B_1(\xi)\cdot w_1 + B_2(\xi)\cdot w_2}$$
$$R_1(\xi) = \frac{B_1(\xi)\cdot w_1}{B_0(\xi)\cdot w_0 + B_1(\xi)\cdot w_1 + B_2(\xi)\cdot w_2}$$
$$R_2(\xi) = \frac{B_2(\xi)\cdot w_2}{B_0(\xi)\cdot w_0 + B_1(\xi)\cdot w_1 + B_2(\xi)\cdot w_2} \tag{2.71}$$

Substituting Eqs. (2.68) to Eq. (2.70) into Eq. (2.71), the analytical expression of the rational basis functions is found as follows:

$$R_0(\xi) = \frac{(1-\xi)^2}{(1-\xi)^2 + \sqrt{2}\xi(1-\xi) + \xi^2}$$
$$R_1(\xi) = \frac{\sqrt{2}\xi(1-\xi)}{(1-\xi)^2 + \sqrt{2}\xi(1-\xi) + \xi^2}$$
$$R_2(\xi) = \frac{\xi^2}{(1-\xi)^2 + \sqrt{2}\xi(1-\xi) + \xi^2} \tag{2.72}$$

The parametric expression of the x-coordinate is given as

$$
\begin{aligned}
x(\xi) &= R_0(\xi)\cdot x_{op} + R_1(\xi)\cdot x_{1p} + R_2(\xi)\cdot x_{2p}\\
&= \frac{(1-\xi)^2}{(1-\xi)^2+\sqrt{2}\xi(1-\xi)+\xi^2}\cdot 1 + \frac{\sqrt{2}\xi(1-\xi)}{(1-\xi)^2+\sqrt{2}\xi(1-\xi)+\xi^2}\cdot 1\\
&\quad + \frac{\xi^2}{(1-\xi)^2+\sqrt{2}\xi(1-\xi)+\xi^2}\cdot 0\\
&= \frac{(1-\xi)^2+\sqrt{2}\xi(1-\xi)}{(1-\xi)^2+\sqrt{2}\xi(1-\xi)+\xi^2}
\end{aligned}
\tag{2.73}
$$

whereas for the y-coordinate:

$$
\begin{aligned}
y(\xi) &= R_0(\xi)\cdot y_{op} + R_1(\xi)\cdot y_{1p} + R_2(\xi)\cdot y_{2p}\\
&= \frac{(1-\xi)^2}{(1-\xi)^2+\sqrt{2}\xi(1-\xi)+\xi^2}\cdot 0 + \frac{\sqrt{2}\xi(1-\xi)}{(1-\xi)^2+\sqrt{2}\xi(1-\xi)+\xi^2}\cdot 1\\
&\quad + \frac{\xi^2}{(1-\xi)^2+\sqrt{2}\xi(1-\xi)+\xi^2}\cdot 1\\
&= \frac{\sqrt{2}\xi(1-\xi)+\xi^2}{(1-\xi)^2+\sqrt{2}\xi(1-\xi)+\xi^2}
\end{aligned}
\tag{2.74}
$$

Therefore, it can be easily shown that:

$$
[x(\xi)]^2 + [y(\xi)]^2 = \frac{\left[(1-\xi)^2+\sqrt{2}\xi(1-\xi)\right]^2 + \left[\sqrt{2}\xi(1-\xi)+\xi^2\right]^2}{\left[(1-\xi)^2+\sqrt{2}\xi(1-\xi)+\xi^2\right]^2} \equiv 1 \tag{2.75}
$$

Equation (2.75) shows that for every point ξ along the segment the standard equation of the circle is fulfilled ($x^2 + y^2 = 1$), which is the mathematical proof that rational Bézier interpolation using quadratic polynomials is capable of accurately representing a circular arc of 90°.

Remarks

(1) The reader may easily check that: $B_0(\xi) + B_1(\xi) + B_2(\xi) \equiv 1$
(2) The reader may easily check that: $R_0(\xi) + R_1(\xi) + R_2(\xi) \equiv 1$

Exercise 2.14 (Equivalency between Bernstein–Bézier and Lagrange quadratic polynomials)
Show the equivalency of a quadratic Bézier curve with the *quadratic* Lagrange polynomials and determine the linear transformation that relates the two sets of basis functions.

Solution Let us assume the domain $0 \le x \le 1$, and the breakpoints $x_0 = 0, x_1 = 1/2, x_2 = 1$, whence $n = 2$ is the number of parameters minus one.

The corresponding interpolation between the three above breakpoints using Lagrange polynomials is:

$$\mathbf{L} = \left[\, L_{0,2}(x)\; L_{1,2}(x)\; L_{2,2}(x) \,\right]^t$$

where

$$L_{0,2}(x) = 2x^2 - 3x + 1, \quad L_{1,2}(x) = -4x^2 + 4x, \quad L_{2,2}(x) = 2x^2 - x \tag{2.76a}$$

Also, the corresponding Bernstein pclynomials are:

$$\mathbf{B} = \left[\, B_{0,2}(x)\; B_{1,2}(x)\; B_{2,2}(x) \,\right]^t$$

where

$$B_{0,2}(x) = x^2 - 2x + 1, \quad B_{1,2}(x) = -2x^2 + 2x, \quad B_{2,2}(x) = x^2 \tag{2.76b}$$

Based on Lagrange polynomials interpolation, the variable is written as:

$$\begin{aligned} U_L(x) &= L_{0,2}(x)U_0 + L_{1,2}(x)U_1 + L_{2,2}(x)U_2 \\ &= \left(2x^2 - 3x + 1\right)U_0 + \left(-4x^2 + 4x\right)U_1 + \left(2x^2 - x\right)U_2 \\ &= x^2(2U_0 - 4U_1 + 2U_2) + x(-3U_0 + 4U_1 - U_2) + U_0 \end{aligned} \tag{2.76c}$$

Also, based on Bézier interpolation, the same variable is written as:

$$\begin{aligned} U_B(x) &= B_{0,2}(x)a_0 + B_{1,2}(x)a_1 + B_{2,2}(x)a_2 \\ &= \left(x^2 - 2x + 1\right)a_0 + \left(-2x^2 + 2x\right)a_1 + \left(x^2\right)a_2 \\ &= x^2(a_0 - 2a_1 + a_2) + x(-2a_0 + 2a_1) + a_0 \end{aligned} \tag{2.76d}$$

Demanding that $U_L(x) \equiv U_B(x)$, the coefficients of the three monomials (i.e., 1, x, and x^2) that appear in Eqs. (2.76c) and (2.76d) must be equal one another, and therefore:

$$\left.\begin{aligned} &2U_0 - 4U_1 + 2U_2 = a_0 - 2a_1 + a_2 \\ &-3U_0 + 4U_1 - U_2 = -2a_0 + 2a_1 \\ &U_0 = a_0 \end{aligned}\right\} \tag{2.76e}$$

Solving the linear system of Eqs. (2.76e) in $\mathbf{a} = [a_0, a_1, a_2]^{\mathrm{T}}$, we take:

$$\mathbf{a} = \begin{bmatrix} a_0 \\ a_1 \\ a_2 \end{bmatrix} = \underbrace{\begin{bmatrix} 1 & 0 & 0 \\ -1/2 & 2 & -1/2 \\ 0 & 0 & 1 \end{bmatrix}}_{\mathbf{T}} \cdot \begin{bmatrix} U_0 \\ U_1 \\ U_2 \end{bmatrix} = \mathbf{Tu}, \tag{2.76f}$$

Then, substituting Eq. (2.76f) into Eq. (2.76d), the latter is written as:

$$
\begin{aligned}
U_B(x) &= B_{0,2}(x)a_0 + B_{1,2}(x)a_1 + B_{2,2}(x)a_2 \\
&= B_{0,2}(x)U_0 + B_{1,2}(x)\left(-\frac{1}{2}U_0 + 2U_1 - \frac{1}{2}U_2\right) + B_{2,2}(x)U_2 \\
&= \left[B_{0,2}(x) - \frac{1}{2}B_{1,2}(x)\right]U_0 + \left[2B_{1,2}(x)\right]U_1 + \left[-\frac{1}{2}B_{1,2}(x) + B_{2,2}(x)\right]U_2
\end{aligned} \tag{2.76g}
$$

Comparing Eq. (2.76g) with its equivalent Eq. (2.76c), the unique representation in terms of the constant **U**, leads to:

$$
\begin{aligned}
L_{0,2}(x) &= B_{0,2}(x) - \frac{1}{2}B_{1,2}(x) \\
L_{1,2}(x) &= 2B_{1,2}(x) \\
L_{2,2}(x) &= -\frac{1}{2}B_{1,2}(x) + B_{2,2}(x)
\end{aligned} \tag{2.76h}
$$

Therefore, the two sets of functions are interrelated through a matrix transformation, as follows:

$$
\underbrace{\begin{bmatrix} L_{0,2}(x) \\ L_{1,2}(x) \\ L_{2,2}(x) \end{bmatrix}}_{\mathbf{L}} = \underbrace{\begin{bmatrix} 1 & -1/2 & 0 \\ 0 & 2 & 0 \\ 0 & -1/2 & 1 \end{bmatrix}}_{\mathbf{T}^t} \cdot \underbrace{\begin{bmatrix} B_{0,2}(x) \\ B_{1,2}(x) \\ B_{2,2}(x) \end{bmatrix}}_{\mathbf{B}}, \tag{2.76i}
$$

and

$$
\begin{bmatrix} B_{0,2}(x) \\ B_{1,2}(x) \\ B_{2,2}(x) \end{bmatrix} = \underbrace{\begin{bmatrix} 1 & 1/4 & 0 \\ 0 & 1/2 & 0 \\ 0 & 1/4 & 1 \end{bmatrix}}_{(\mathbf{T}^t)^{-1}} \cdot \begin{bmatrix} L_{0,2}(x) \\ L_{1,2}(x) \\ L_{2,2}(x) \end{bmatrix}. \tag{2.76j}
$$

Obviousy, from Eq. (2.76i) one may immediately understand the relationship $\mathbf{L} = \mathbf{T}^t\mathbf{B}$ that connects the two functional sets, whereas Eq. (2.76j) expresses the inverse relationship $\mathbf{B} = (\mathbf{T}^t)^{-1}\mathbf{L}$.

Exercise 2.15 (Circular arc using rational Lagrange polynomials)
Show that it is possible to modify the traditional rational Bézier approximation that was presented in Exercise 2.13, by shifting the control point P_1 to the middle point B and at the same time by changing the weighting factors.

Solution It will be shown that the answer is accomplished by creating proper rational Lagrange polynomials, which are capable of accurately representing the entire

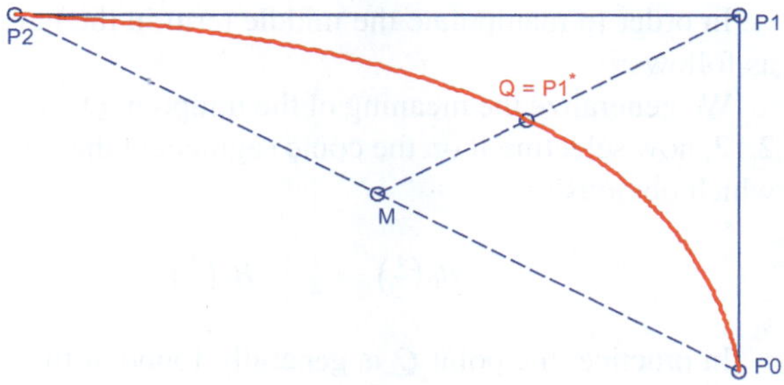

Fig. 2.21 Determination of the intermediate control point $Q \equiv P_1^*$ on the conic segment, at which the intermediate control point P_1 is shifted

circular arc. Below we generalize the procedure to any conical section (ellipse, circle, hyperbola).

A. *Linear dependency between quadratic Bézier and Lagrange polynomials*

In Exercise 2.14, it was shown that in case of three control points (P_0, P_1, and P_2 with Cartesian coordinates $\mathbf{x}_{P_0}, \mathbf{x}_{P_1}, \mathbf{x}_{P_2}$, respectively), the vector including the Bernstein polynomials $[B_0, B_1, B_2]^{\mathrm{T}} \equiv \left[(1-\xi)^2, 2\xi(1-\xi), \xi^2\right]^{\mathrm{T}}$ is related to the vector including the Lagrange polynomials $[L_0, L_1, L_2]^{\mathrm{T}} \equiv [(2\xi-1)(\xi-1), 4\xi(1-\xi), \xi(2\xi-1)]^{\mathrm{T}}$, through the linear transformation:

$$\begin{bmatrix} B_0 \\ B_1 \\ B_2 \end{bmatrix} = \begin{bmatrix} 1 & 1/4 & 0 \\ 0 & 1/2 & 0 \\ 0 & 1/4 & 1 \end{bmatrix} \begin{bmatrix} L_0 \\ L_1 \\ L_2 \end{bmatrix} \tag{2.77}$$

B. *Nonuniform rational Lagrange polynomials*

Let us now consider a general conical segment, which is determined again by the classical three control points (P_0, P_1, and P_2) as previously shown in Fig. 2.20. Below it will be shown that instead of using the aforementioned standard control points P_0, P_1, and P_2 in conjunction with the standard weighting factors $w_0 = w_2 = 1$ and w_1 of a proper given value (depending on the angle $\measuredangle(OP_0, OP_2) = 2\alpha : w_1 = \tan\alpha$), it is possible to make all the three control points P_0, P_1^* (updated shown in Fig. 2.21) and P_2 belong to the conic segment.

Actually, substituting Eq. (2.77) into the classical rational expression,

$$\mathbf{x}(\xi) = \frac{\sum_{i=0}^{2} w_i B_i(\xi)\mathbf{x}_{Pi}}{\sum_{i=0}^{2} w_i B_i(\xi)}, \quad 0 \le \xi \le 1, \tag{2.78}$$

after some manipulation one obtains:

$$\mathbf{x}(\xi) = \frac{[w_0 \cdot L_0(\xi)]\mathbf{x}_{P0} + \left[\frac{1}{4}w_0\mathbf{x}_{P0} + \frac{1}{2}w_1\mathbf{x}_{P1} + \frac{1}{4}w_2\mathbf{x}_{P2}\right]L_1(\xi) + [w_2 \cdot L_2(\xi)]\mathbf{x}_{P2}}{w_0 \cdot L_0(\xi) + \left(\frac{1}{4}w_0 + \frac{1}{2}w_1 + \frac{1}{4}w_2\right) \cdot L_1(\xi) + w_2 \cdot L_2(\xi)}, \tag{2.79}$$

In order to manipulate the middle term in the numerator of Eq. (2.79), we work as follows:

We generalize the meaning of the midpoint Q_p (or B) which was used in Exercise 2.12, now selecting it on the conic segment at the position $\xi = \frac{1}{2}$ (see Fig. 2.21), for which obviously:

$$B_0\left(\tfrac{1}{2}\right) = \tfrac{1}{4}, \quad B_1\left(\tfrac{1}{2}\right) = \tfrac{1}{2}, \quad B_2\left(\tfrac{1}{2}\right) = \tfrac{1}{4}. \tag{2.80}$$

In practice, the point Q is generally found at the intersection of the straight line (P_1M) and the conical segment, where M is the middle of the straight segment P_0P_2 (see Fig. 2.21).

Therefore, at the position of the point Q, Eq. (2.78) progressively yields

$$\begin{aligned}\mathbf{x}_Q \equiv \mathbf{x}\left(\tfrac{1}{2}\right) &= \frac{\sum_{i=0}^{2} w_i B_i(\frac{1}{2})\mathbf{x}_{Pi}}{\sum_{i=0}^{2} w_i B_i(\frac{1}{2})} = \frac{w_0 B_0(\frac{1}{2})\mathbf{x}_{P0} + w_1 B_1(\frac{1}{2})\mathbf{x}_{P1} + w_2 B_2(\frac{1}{2})\mathbf{x}_{P2}}{w_0 B_0(\frac{1}{2}) + w_1 B_1(\frac{1}{2}) + w_2 B_2(\frac{1}{2})} \\ &\equiv \frac{\frac{1}{4}w_0\mathbf{x}_{P0} + \frac{1}{2}w_1\mathbf{x}_{P1} + \frac{1}{4}w_2\mathbf{x}_{P2}}{\frac{1}{4}w_0 + \frac{1}{2}w_1 + \frac{1}{4}w_2}\end{aligned} \tag{2.81}$$

Equation (2.81) depicts that the point Q can be selected as the new control point, P_1^*, and then Eq. (2.79) becomes:

$$\mathbf{x}(\xi) = \frac{[w_0 \cdot L_0(\xi)]\mathbf{x}_{P0} + \left[(\frac{1}{4}w_0 + \frac{1}{2}w_1 + \frac{1}{4}w_2) \cdot L_1(\xi)\right]\mathbf{x}_{P1*} + [w_2 \cdot L_2(\xi)]\mathbf{x}_{P2}}{w_0 \cdot L_0(\xi) + (\frac{1}{4}w_0 + \frac{1}{2}w_1 + \frac{1}{4}w_2) \cdot L_1(\xi) + w_2 \cdot L_2(\xi)}. \tag{2.82}$$

Moreover, setting a modified weighting factor

$$w_1^* \equiv \tfrac{1}{4}w_0 + \tfrac{1}{2}w_1 + \tfrac{1}{4}w_2, \tag{2.83}$$

then Eq. (2.82) takes the smart form:

$$\mathbf{x}(\xi) = \frac{[w_0 \cdot L_0(\xi)]\mathbf{x}_{P0} + \left[w_1^* \cdot L_1(\xi)\right]\mathbf{x}_{P1*} + [w_2 \cdot L_2(\xi)]\mathbf{x}_{P2}}{w_0 \cdot L_0(\xi) + w_1^* \cdot L_1(\xi) + w_2 \cdot L_2(\xi)}. \tag{2.84}$$

Therefore, Eq. (2.84) clearly shows that the Cartesian coordinates along the conic are given in terms of rational Lagrange polynomials associated to the nodal points P_0, P_1^* and P_2, which all are located along the conic segment.

Obviously, the rational Lagrange basis functions which are involved in Eq. (2.84) are given by

$$S_0(\xi) = \frac{w_0 \cdot L_0(\xi)}{w_0 \cdot L_0(\xi) + w_1^* \cdot L_1(\xi) + w_2 \cdot L_2(\xi)}$$
$$S_1(\xi) = \frac{w_1^* \cdot L_1(\xi)}{w_0 \cdot L_0(\xi) + w_1^* \cdot L_1(\xi) + w_2 \cdot L_2(\xi)},$$
$$S_2(\xi) = \frac{w_2 \cdot L_2(\xi)}{w_0 \cdot L_0(\xi) + w_1^* \cdot L_1(\xi) + w_2 \cdot L_2(\xi)} \quad (2.85)$$

Due to the fact that the standard Lagrange polynomials possess the property of cardinality, i.e., $L_i(\xi_j) = \delta_{ij}$ (Kronecker delta), it is trivial to validate that the rational basis functions of Eq. (2.85) are also cardinal, i.e., $S_i(\xi_j) = \delta_{ij}$ (Kronecker delta).

Application Example

Let us consider the first quadrant of an ellipse with semi-axes a and b along the x- and y-directions, respectively, with standard control points $P_0(a,0)$, $P_1(a, b)$, and $P_2(0, b)$. In this particular case, the standard rational Bézier parametric form requires the weights $w_0 = 1$, $w_1 = 1/\sqrt{2}$, and $w_2 = 1$.

In the proposed form, the control points $P_0(a, 0)$ and $P_2(0, b)$ remain as they are, that is at the intersections of the positive semi-axes with the boundary of the ellipse. Moreover, fulfilling Eq. (2.78) for $\xi = ½$, the new intermediate point takes the position $\left(x_{P1^*} = a\sqrt{2}/2,\ y_{P1^*} = b\sqrt{2}/2\right)$. Also, Eq. (2.83) implies that the associated weight becomes $w_1^* = \frac{1}{2} + \frac{\sqrt{2}}{4}$.

Based on the above calculations, the plot of the three rational Lagrange basis functions is given in Fig. 2.22 (bold lines), where one may observe their cardinality. One may also observe that the usual Lagrange polynomials (dashed lines) are very close to the rational ones.

2.7 Interpolation Formula Using NURBS

Similarly to the rational Bézier curves, rational B-splines can be produced from usual B-splines. Below we choose the de Boor form, where the basis functions involved in the approximation using B-splines are denoted by $N_{i,p}$. Extension to 2D-NURBS patches is analogous to the abovementioned rational Bézier formulas.

2.7.1 *NURBS Curve*

A NURBS curve is represented by:

$$\mathbf{x}(\xi) = \sum_{i=0}^{n} \widehat{R}_{i,p}(\xi)\mathbf{x}_i \quad (2.86)$$

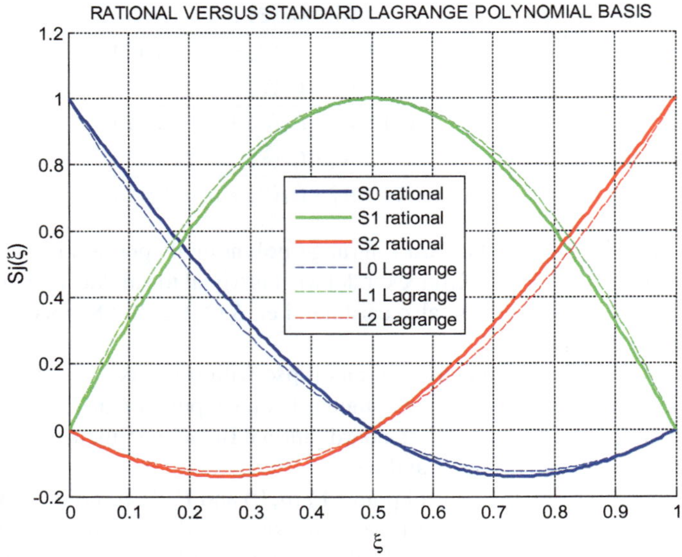

Fig. 2.22 Three nonuniform rational Lagrange polynomials (bold lines) versus the usual standard ones (dashed lines) in the reference domain $\xi \in [0, 1]$

with

$$\widehat{R}_{i,p}(\xi) = \frac{N_{i,p}(\xi)\, w_i}{\sum_{j=0}^{n} N_{j,p}(\xi)\, w_j}, \tag{2.87}$$

In Eq. (2.86), $\mathbf{x}$ represents the coordinates vector $\mathbf{x}(\xi) = [x(\xi), y(\xi)]^t$, whereas $\mathbf{x}_i = [x_i, y_i]^t,\ i = 0, 1, \ldots, n$ are the control points.

For a univariate function $U(\xi)$, the NURBS-based interpolation becomes:

$$U(\xi) = \sum_{i=0}^{n} \widehat{R}_{i,p}(\xi)\, a_i, \tag{2.88}$$

where a_i, $i = 0, 1, \ldots, n$ are generalized coordinates or coefficients.

2.7.2 *NURBS Patch*

Extending Eq. (2.86) to a 2D patch ABCD, we can write

$$\mathbf{x}(\xi, \eta) = \sum_{i=0}^{n} \sum_{j=0}^{m} \widehat{R}_{ij}(\xi, \eta) \cdot \mathbf{x}_{ij}, \tag{2.89}$$

where

$$\widehat{R}_{ij}(\xi,\eta) = \frac{N_{i,n}(\xi)\cdot w_{ij}\cdot N_{j,m}(\eta)}{\sum_{k=0}^{n}\sum_{l=0}^{m} N_{k,n}(\xi)\cdot w_{kl}\cdot N_{l,m}(\eta)}, \tag{2.90}$$

is the bivariate NURBS, which interpolates the Cartesian coordinates $\mathbf{x}_{ij}$ at the position (ξ_i, η_j), whereas the weighting factor is usually taken as $w_{ij} = w_i w_j$.

Extending Eq. (2.89) from the geometric to the analysis model, we can approximate the problem variable $U(\xi, \eta)$ by

$$U(\xi,\eta) = \sum_{i=0}^{n}\sum_{j=0}^{m} \widehat{R}_{ij}(\xi,\eta)\cdot a_{ij}, \tag{2.91}$$

where now a_{ij} are the generalized coefficients to be determined.

2.8 Barnhill's Interpolation Formula in a Parametric Triangular Domain

This is a quite new interpolation scheme that is not based on the material of the book presented so far. As previously was noted, Coons proposed a degeneration of the quadrilateral patch ABCD into a triangle ABC. Despite this fact, the degenerated Coons formula is not preferred because it gives a preference to the degenerated point C (with which D coincides), whereas Barnhill's formula treats a triangular patch with *affine invariance*. Barnhill and Gordon cooperated and proposed a new formula that uses boundary data of the variable U and determines the functions $U(x, y)$ at any internal point of the triangular patch ABC.

2.8.1 Area Coordinates

In the most trivial case of a triangle *ABC* with straight edges, each point $P(x, y)$ inside *ABC* is fully determined by its two Cartesian coordinates, x and y. Alternatively, if the vertices *A*, *B*, and *C* are replaced by the numbers 1, 2, and 3, respectively, and then A_1, A_2 and A_3 are designated as the areas of the three subtriangles PBC (opposite to vertex 1 or *A*), *PCA* (opposite to vertex 2 or *B*), and *PAB* (opposite to vertex 2 or B), respectively, then the *area coordinates* are defined by (see Fig. 2.23a):

$$\xi_1 = A_1/A, \quad \xi_2 = A_2/A, \quad \xi_3 = A_3/A \tag{2.92}$$

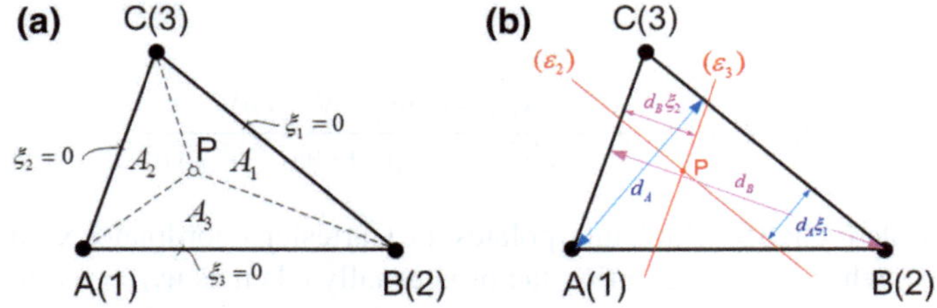

Fig. 2.23 **a** Definition of area coordinates and **b** determination of an internal point for a given pair (ξ_1, ξ_2)

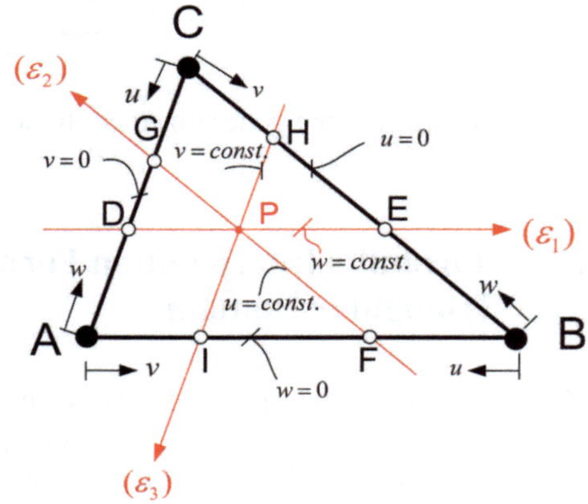

Fig. 2.24 Barnhill's parameters in a triangular patch ABC

Obviously,

$$\xi_1 + \xi_2 + \xi_3 = \frac{A_1 + A_2 + A_3}{A} = \frac{A}{A} \equiv 1 \tag{2.93}$$

Details are given below.

Therefore, instead of two Cartesian coordinates it is sufficient to know two area coordinates, e.g., ξ_1 and ξ_2, since the third will be expressed in terms of the other two ($\xi_3 = 1 - \xi_1 - \xi_2$).

A given constant value of one area coordinate, e.g., of ξ_1, means that it lies along a straight line (ε_2) parallel to BC, at a distance $d = \xi_1 d_A$, where d_A is the distance of the point A from the straight-line segment BC (height of the triangle from A) as shown in Fig. 2.23b. Given the pair (ξ_1, ξ_2), the intersection of the lines (ε_2) and (ε_3) parallel to BC and AC, respectively, at given distances ($d_{P,BC} = \xi_1 d_A, d_{P,AC} = \xi_2 d_B$), determines the position of the point P (see Fig. 2.23b).

Furthermore, in order to be consistent with classical CAD textbooks (e.g., [21]), we introduce three variables (u, v, w) which are coincident with the area coordinates as shown in Fig. 2.24 ($0 \leq u, v, w \leq 1$). Therefore, for any point P of the triangle we can write:

$$u = \frac{A_1}{A} = \frac{(P\overset{\Delta}{B}C)}{(A\overset{\Delta}{B}C)} = \xi_1 \tag{2.94a}$$

$$v = \frac{A_2}{A} = \frac{(P\overset{\Delta}{C}A)}{(A\overset{\Delta}{B}C)} = \xi_2 \tag{2.94b}$$

$$w = \frac{A_3}{A} = \frac{(P\overset{\Delta}{A}B)}{(A\overset{\Delta}{B}C)} = \xi_3 \tag{2.94c}$$

According to the above definitions, for all points along the edge AB $w = 0$, along edge BC $u = 0$, and finally along the edge CA $v = 0$. Closely related, all points along the line (ε_1) correspond to a constant $w = c_1$, similarly the line (ε_2) to another constant $u = c_2$, and finally the line (ε_3) to a third constant $v = c_3$. Obviously, these constants are related by $c_1 + c_2 + c_3 = 1$.

2.8.2 Derivation of Barnhill's Formula

Now it is time to proceed with the derivation of Barnhill's formula. For the sake of brevity, here only the particular case of C^0-continuity will be discussed in detail. At any arbitrary point P in the interior of the triangle ABC, we consider the lines (ε_1), (ε_2) and (ε_3), parallel to the edges AB, BC, and CA, respectively. The choice of the indexes in ε_k, $k = 1, 2, 3$ is so that the vector sequence $(\overrightarrow{AB}, \overrightarrow{BC}, \overrightarrow{CA})$ is in the anticlockwise direction. Furthermore, the positive orientation of the lines is toward the directions $\overrightarrow{AB}, \overrightarrow{BC}, \overrightarrow{CA}$, respectively.

In the direction of $\overrightarrow{AB}$, the line ε_1 intersects the triangle at the points D and E. Similarly, in the direction of $\overrightarrow{BC}$ the line ε_2 intersects the triangle at the points F and G. Finally, in the direction of $\overrightarrow{CA}$ the line ε_3 intersects the triangle at the points H and I.

In each of the abovementioned three directions $(\overrightarrow{AB}, \overrightarrow{BC}, \overrightarrow{CA})$, the value of the function at P can be calculated as a linear interpolation between the values at the corresponding intersections. For example, if the point P is considered to be along the line $\left(\varepsilon_1 \middle\| \overrightarrow{AB}\right)$ with intersections D and E, then the value at P can be easily calculated in terms of the values of the function at the points D and E. This interpolation is called *projection* P_{AB} toward the $\overrightarrow{AB}$ direction and is given in terms of the linear blending functions (E_0, E_1) by:

$$P_{AB} = E_0 U_D + E_1 U_E \tag{2.95}$$

"Thales theorem" (of Euclidian Geometry) yields:

$$\frac{DE}{\left|\overrightarrow{AB}\right|} = \frac{CE}{BC} = 1 - \frac{BE}{BC} = 1 - w,$$

whence

$$DE = (1 - w) \cdot \left|\overrightarrow{AB}\right| \tag{2.96}$$

Also,

$$PE = FB \equiv u \cdot \left|\overrightarrow{AB}\right| \text{ and } DP = AI \equiv v \cdot \left|\overrightarrow{AB}\right| \tag{2.97}$$

By virtue of Eq. (2.96) and (2.97), the blending functions become:

$$E_0 = \frac{PE}{DE} = \frac{u \cdot \left|\overrightarrow{\mathrm{AB}}\right|}{(1-w)\cdot\left|\overrightarrow{\mathrm{AB}}\right|} = \frac{u}{(1-w)}$$
$$E_1 = \frac{DP}{DE} = \frac{u \cdot \left|\overrightarrow{\mathrm{AB}}\right|}{(1-w)\cdot\left|\overrightarrow{\mathrm{AB}}\right|} = \frac{v}{(1-w)} \tag{2.98}$$

Substituting Eq. (2.98) into Eq. (2.95), one obtains:

$$P_{AB} = \frac{u}{(1-w)}U_D + \frac{v}{(1-w)}U_E \tag{2.99}$$

By cyclic symmetry, the two rest projections are given by:

$$P_{BC} = \frac{v}{(1-u)}U_F + \frac{w}{(1-u)}U_G \tag{2.100}$$

and

$$P_{CA} = \frac{w}{(1-v)}U_H + \frac{u}{(1-v)}U_I \tag{2.101}$$

The superposition of the three above projections, as the sum of Eqs. (2.99)–(2.101), leads to an erroneous value of the function $U(u, v, w)$ at the point P. For example, if $U(u, v, w) \equiv U_0$ for the entire boundary, then the superposition gives the erroneous result $S = P_{AB} + P_{BC} + P_{CA} = 3U_0$. Therefore, in order to settle this error, we resort to the particular case in which the point P lies, for example, along the edge AB. Then the first project becomes $P_{AB} = uf(0) + vg(0)$ is the *linear interpolation* between the values $f(0)$ and $g(0)$ which correspond to the corners*A* and*B*, respectively, whereas the rest two projections give $P_{BC} = P_{CA} = U_P$

with U_P denoting the *actual* value of the function at the point P. It should become clear that the latter can be assigned an arbitrary value, so in other words it is not compulsorily a linear interpolation of the aforementioned values at A and B. Therefore, if one subtracts the linear interpolation P_{AB} between the values at A and B from the erroneous sum $S = P_{AB} + P_{BC} + P_{CA}$, the remaining quantity becomes $2U_P$, and thus is should be divided by 2 in order to obtain the correct value U_P.

In the above case in which P was taken along the edge AB, since $w = 0$, the corrective term (linear interpolation between A and B) in the abovementioned subtraction $S - C_t$ can be written in an expanded form as follows:

$$C_t = uf(0) + vg(0) + wh(0) \tag{2.102}$$

It is obvious that Eq. (2.102) is valid along all three edges, i.e., AB, BC, and CA. Based on the above skeptic, a first expression for Barnhill's interpolation is:

$$U_P(u, v, w) = \frac{1}{2}\left[\frac{v}{1-w}U_E + \frac{u}{1-w}U_D + \frac{w}{1-u}U_G + \frac{v}{1-u}U_F + \frac{u}{1-v}U_I + \frac{w}{1-v}U_H - uf(0) - vg(0) - wh(0)\right] \tag{2.103}$$

The given functions f, g, h (along the edges AB, BC, and CA, respectively), as well as their interpolation $U_P(u, v, w)$ in the interior of the triangular patch, may represent (i) geometric coordinates of the position $\mathbf{x} = [x, y, z]^{\mathrm{T}}$, but equally well may represent (ii) any mechanical quantity U such as temperature, displacement, which is known along the boundary.

Based on Fig. 2.23, the six values of the function $U(u, v, w)$ involved in Eq. (2.103) are given by

$$\left.\begin{aligned} U_D &= h(1-w),\ U_E = g(w) \\ U_F &= f(1-u),\ U_G = h(u) \\ U_H &= g(1-v),\ U_I = f(v). \end{aligned}\right\} \tag{2.104}$$

Therefore, the triangular patch ABC can be mapped in the parametric space through the formula:

$$\begin{aligned} U(u, v, w) = \frac{1}{2}\Bigg[&\frac{vg(w)}{1-w} + \frac{uh(1-w)}{1-w} + \frac{wh(u)}{1-u} \\ &+ \frac{vf(1-u)}{1-u} + \frac{uf(v)}{1-v} + \frac{wg(1-v)}{1-v} \\ &- uf(0) - vg(0) - wh(0)\Bigg] \end{aligned} \tag{2.105}$$

2.9 Recapitulation

In this chapter, we saw several ways to parametrically describe either a curve $\mathbf{C}(\xi)$ or a univariate function $U(\xi)$. Except of the power series which is closely related to the Taylor expansion, we also dealt with Lagrange and Hermite polynomials, nonrational and rational Bernstein–Bézier polynomials, B-splines (in older truncated power form and "modern" Curry–Schoenberg/de Boor formulations) and NURBS. Extending the well-known linear interpolation toward a certain direction, we saw that proper projections are useful to interpolate a quantity in two dimensions (within a patch) as well. While Coons and Gordon (transfinite) interpolation within a quadrilateral patch require two projections minus a corrective term, Barnhill's interpolation in triangles requires three projections minus a corrective term and division by the factor two. Moreover, it was mentioned that B-spline, nonrational Bézier, and NURBS tensor products are easily applicable in quadrilaterals. For the sake of brevity, some other particular formulas for triangles were omitted. Again, it should become clear that this chapter covered the most important CAGD expressions, whereas some others, although they may be of interest, were not covered. Nevertheless, we believe that if the reader absorbs the material of this chapter, then he/she can easily continue the study with the following chapters, as well as with any older or any future CAGD formula in the wider literature.

References

1. Ahlberg JH, Nilson EN, Walsh JL (1967) The theory of splines and their applications. Academic Press, NY
2. Barnhill RE (1985) Surfaces in computer aided geometric design: A survey with new results. Computer Aided Geometric Design 2:1–17
3. Bartels RH, Beatty JC, Barsky BA (1987) An introduction to splines for use in computer graphics and geometric modeling. Morgan Kaufmann, Los Altos
4. Birkhoff G (1990) Fluid dynamics, reactor computations, and surface representation. In: Nash S (ed) A history of scientific computation
5. Birkhoff G, De Boor C (1965) Piecewise polynomial interpolation and approximation. In: Garabedian HL (ed) Proceedings General Motors symposium of 1964. Elsevier, New York, pp 164–190
6. Böhm W, Farin G, Rahmann J (1984) A survey of curve and surface methods in CAGD. Comput Aided Geom Design 1:1–60
7. Coons SA (1964) Surfaces for computer aided design of space form, Project MAC, MIT (1964), revised for MAC-TR-41 (1967), Springfield, VA 22161, USA: Available as AD 663 504 from the National Technical Information Service (CFSTI), Sills Building, 5285 Port Royal Road. Available online: http://publications.csail.mit.edu/lcs/pubs/pdf/MIT-LCS-TR-041.pdf
8. Cottrell JA, Hughes TJR, Bazilevs Y (2009) Isogeometric analysis: Towards integration of CAD and FEA. Wiley, Chichester
9. Cottrell JA, Reali A, Bazilevs Y, Hughes TJR (2006) Isogeometric analysis of structural vibrations. Comput Methods Appl Mech Eng 195:5257–5296
10. Cox MG (1972) The numerical evaluation of B-splines. J Inst Math Its Appl 10:134–149
11. Curry HB, Schoenberg IJ (1947) On spline distributions and their limits: the Pólya distribution functions. Abstract 380t, Bull Am Math Soc 53:109

12. Curry HB, Schoenberg IJ (1966) On Pólya frequency functions IV: the fundamental spline functions and their limits. Journal d'Analyse Mathématique 17(1):71–107
13. Davis PJ (1975) Interpolation and approximation. Dover, New York
14. Davis P (1997) B-splines and geometric design. SIAM News 29(5)
15. De Boor C (2018) Private communication
16. De Boor C (1976) Splines as linear combinations of B-splines. In: Lorentz GG, Chui CK, Schumaker LL (eds) Approximation theory, II. Academic Press, New York, pp 1–47. Also available on (http://www.dtic.mil/docs/citations/ADA031944, while an emended version (in 1986) can be found at (ftp://cs.wisc.edu/Approx/survey76.pdf)
17. De Boor C (1972) On calculating with B-splines. J Approx Theory 6:50–62
18. De Boor C (2001) A practical guide to splines, Revised edition. Springer, New York (first edition in 1978)
19. De Casteljau P (1959) Courbes à poles. National Industrial Property Institute (INPI), France
20. De Casteljau P (1999) De Casteljau's autobiography: my life at Citroën. Comput Aided Geom Design 16:583–586
21. Farin G (1990) Curves and surfaces for computer aided geometric design: a practical guide. Academic Press, Boston
22. Faux ID, Pratt MJ (1979) Computational geometry for design and manufacture. Ellis Horwood, Chichester
23. Gordon WJ, Riesenfeld RF (1974) B-spline curves and surfaces. In: Barnhill RE, Riesenfeld RF (eds) Computer aided geometric design. Academic Press, New York
24. Hughes TJR, Cottrell JA, Bazilevs Y (2005) Isogeometric analysis: CAD, finite elements, NURBS, exact geometry and mesh refinement. Comput Methods Appl Mech Eng 194:4135–4195
25. Kanarachos AE, Deriziotis DG (1989) On the solution of Laplace and wave propagation problems using "C-elements". Finite Elem Anal Des 5:97–109
26. Peddie J (2013) The history of visual magic in computers. Springer, London, pp 43–44
27. Piegl L, Tiller W (1997) The NURBS book. 2nd edn. Springer-Verlag, Berlin
28. Provatidis C, Kanarachos A (2001) Performance of a macro-FEM approach using global interpolation (Coons') functions in axisymmetric potential problems. Comput Struct 79(19):1769–1779
29. Rabut C (2002) On Pierre Bézier's life and motivations. Comput Aided Des 34:493–510
30. Ramshaw L (1987) Blossoming: a connect-the-dots—approach to splines. Report 19, Digital, Systems Research Center, Palo Alto, CA
31. Riesenfeld RF (1973) Applications of B-spline approximation to geometric problems of computer-aided design, Ph.D. dissertation, Syracuse University, USA
32. Schoenberg IJ (1946) Contributions to the problem of approximation of equidistant data by analytic functions. Q Appl Math 4:45–99, 112–141
33. Schoenberg IJ, Whitney A (1953) On Polya frequency functions III: the positivity of translation determinants with an application to the interpolation problem by splines curves. Trans Am Math Soc 74:246–259
34. Schoenberg IJ (1964) Spline functions and the problem of graduation. Proc Nat Acad Sci USA Nat Acad Sci 52(4):947–950
35. Wait R, Mitchell AR (1985) Finite element analysis and applications. Wiley, Chichester
36. Young DM (1997) Garrett Birkhoff and applied mathematics. Not AMS 44(11):1446–1449
37. Zienkiewicz OC (1977) The finite element method. McGraw-Hill, London

Chapter 3
COONS' Interpolation as a Vehicle to Derive Large Isoparametric Elements

Abstract This chapter explains that Coons interpolation, which chronologically is the first formula for the mathematical representation of surface patches in Computational Geometry, can be used to derive the closed-form analytical expressions of shape functions that appear in classical isoparametric finite elements of the Serendipity family. Moreover, it is shown that not only Lagrange polynomials but also reduced natural cardinal cubic B-splines as well as most of other interpolations discussed in Chap. 2 can be also used as trial functions along the four edges of a Coons patch macroelement ("C-element"). The latter is generally a large isoparametric finite element with nodal points along the boundary of the patch. In the case of using Curry-Schoenberg (de Boor) B-splines or NURBS along each edge in the Coons patch, the nodal points are merely replaced by the control points. The performance of all these elements is thoroughly investigated through ten examples in two-dimensional and axisymmetric potential and elasticity problems.

Keywords COONS' interpolation · Blending functions · Trial functions
C-element · Macroelement · Global shape functions · Boundary value problem
Potential · Elasticity · Solved examples

3.1 General

In Chap. 2 it was explained that Coons interpolation formula [7] within a patch $ABCD$ is given by

$$U(\xi,\eta) = P_\xi[U] + P_\eta[U] - P_\xi P_\eta[U], \tag{3.1}$$

where the three involved projections are:

$$\text{Towards } \xi\text{-direction}: \; P_\xi[U] = E_0(\xi)U(0,\eta) + E_1(\xi)U(1,\eta) \tag{3.2}$$

$$\text{Towards } \eta\text{-direction}: \; P_\eta[U] = E_0(\eta)U(\xi,0) + E_1(\eta)U(\xi,1) \tag{3.3}$$

C. G. Provatidis, *Precursors of Isogeometric Analysis*, Solid Mechanics and Its Applications 256, https://doi.org/10.1007/978-3-030-03889-2_3

and

$$\text{The corrective term:} P_\xi P_\eta[U] = \sum_{i=0}^{1}\sum_{j=0}^{1} E_i(\xi)E_j(\eta)\, U(\xi_i, \eta_j) \tag{3.4}$$

By virtue of Eqs. (3.2)–(3.4), the Coons interpolation formula (3.1) takes the form:

$$\begin{aligned} U(\xi, \eta) = {} & E_0(\xi) \cdot U(0, \eta) + E_1(\xi) \cdot U(1, \eta) \\ & + E_0(\eta) \cdot U(\xi, 0) + E_1(\eta) \cdot U(\xi, 1) \\ & - E_0(\xi) \cdot E_0(\eta) \cdot U(0, 0) - E_1(\xi) \cdot E_0(\eta) \cdot U(1, 0) \\ & - E_0(\xi) \cdot E_1(\eta) \cdot U(0, 1) - E_1(\xi) \cdot E_1(\eta) \cdot U(1, 1), \end{aligned} \tag{3.5}$$

where the involved (linear) blending functions are given by:

$$\begin{aligned} E_0(\xi) = 1 - \xi, \quad E_1(\xi) = \xi \\ E_0(\eta) = 1 - \eta, \quad E_1(\eta) = \eta \end{aligned} \tag{3.6}$$

Based on the above general interpolation formula (3.5), it will be shown that

- classical finite elements may be derived
- a generalization of Serendipity family is possible (Fig. 3.1).

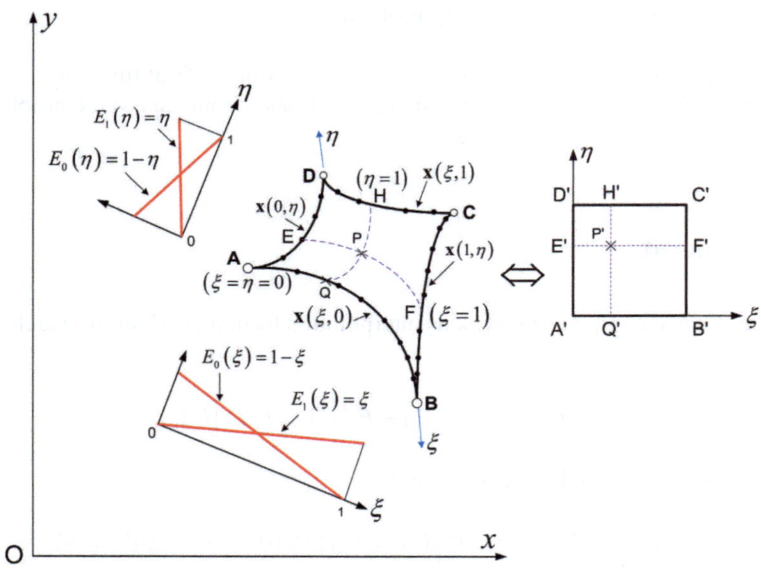

Fig. 3.1 Coons macroelement (C-element)

3.2 Small-Size Classical Finite Elements

For reasons of better understanding the formulation of the general "C-element", we start with the implementation of Coons formula in conjunction with a standard procedure that eventually leads to the derivation of the well-known *Serendipity* elements. For the definition of the Serendipity family the reader is referred to the classical book ([39], p. 155) or the newest editions. The most popular elements of this family are the 4-noded and the 8-noded isoparametric elements shown in Fig. 3.2.

3.2.1 Four-Noded Element

For this particular patch *ABCD*, which is identical with the element 1234 (Fig. 3.2a), we assume that the variable U along an edge varies linearly between any two associated successive nodes, i.e. we perform linear interpolations along the four edges of the patch *ABCD* as follows:

$$\begin{aligned} U_{AD} &\equiv U(0,\eta) = (1-\eta)\cdot U_A + \eta\cdot U_D \\ U_{BC} &\equiv U(1,\eta) = (1-\eta)\cdot U_B + \eta\cdot U_C \\ U_{AB} &\equiv U(\xi,0) = (1-\xi)\cdot U_A + \xi\cdot U_B \\ U_{CD} &\equiv U(\xi,1) = (1-\xi)\cdot U_D + \xi\cdot U_C \end{aligned} \tag{3.7}$$

Substituting Eqs. (3.6) and (3.7) into Eq. (3.5) one obtains:

$$\begin{aligned} U(\xi,\eta) = {} & (1-\xi)\cdot[(1-\eta)\cdot U_A + \eta\cdot U_D] + \xi\cdot[(1-\eta)\cdot U_B + \eta\cdot U_C] \\ & + (1-\eta)\cdot[(1-\xi)\cdot U_A + \xi\cdot U_B] + \eta\cdot[(1-\xi)\cdot U_D + \xi\cdot U_C] \\ & - (1-\xi)\cdot(1-\eta)\cdot U_A - \xi\cdot(1-\eta)\cdot U_B \\ & - (1-\xi)\cdot\eta\cdot U_D - \xi\cdot\eta\cdot U_C \end{aligned} \tag{3.8}$$

After the obvious reduction of similar terms, one obtains:

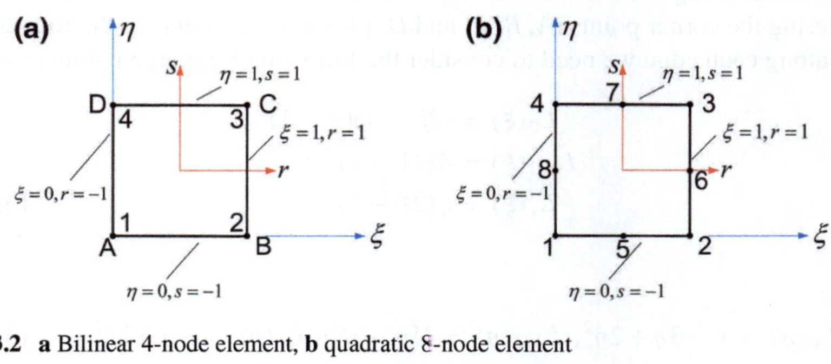

Fig. 3.2 **a** Bilinear 4-node element, **b** quadratic 8-node element

$$\boxed{U(\xi,\eta) = U_A(1-\xi)(1-\eta) + U_B\xi(1-\eta) + U_C\xi\eta + U_D(1-\xi)\eta} \tag{3.9}$$

One may observe that the factor $N_A = (1-\xi)(1-\eta)$ associated to the nodal value U_A in the right-hand-side of Eq. (3.9), also $N_B = \xi(1-\eta)$ to U_B, $N_C = \xi\eta$ to U_C, and $N_D = (1-\xi)\eta$ to U_D, are identical with the well-known shape functions of the four-noded bilinear finite element. Indeed, the classical shape functions of the bilinear element, found in all standard FEM textbooks such as [1, 39], are $N(r,s) = (1+r_i r)(1+s_i s)/4$.

For example, for the corner A (node 1):

$$\widehat{N}_1(r,s) = (1-r)(1-s)/4, \quad -1 \le r, s \le 1. \tag{3.10}$$

Furthermore, considering that

$$\xi = (1+r)/2 \text{ and } \eta = (1+s)/2, \tag{3.11a}$$

or equivalently

$$r = 2\xi - 1 \text{ and } s = 2\eta - 1, \tag{3.11b}$$

the function $\widehat{N}_1$ in Eq. (3.10) takes exactly the form of the function $N_A = (1-\xi)(1-\eta)$, that is $N_A \equiv \widehat{N}_1$. Therefore, the shape function that was produced by Coons interpolation is *identical* with that coming from isoparametric considerations (in classical FEM theory).

3.2.2 Eight-Noded Element

The 8-noded element occupies the entire patch *ABCD* and its numbering is shown in Fig. 3.2b. Equations (3.5) and (3.6) will be again the starting point of shape function derivation. The basic difference with the above procedure is that now we use quadratic trial functions along each of the four edges of the patch *ABCD*. This is implemented considering the corner points (*A*, *B*, *C*, and *D*) plus the mid-points of the four edges. Then, along each edge we need to consider the following Lagrange polynomials:

$$\begin{aligned} L_0(\xi) &= (2\xi-1)(\xi-1) \\ L_{1/2}(\xi) &= 4\xi(1-\xi) \\ L_1(\xi) &= \xi(2\xi-1) \end{aligned} \tag{3.12a}$$

and

$$L_0(\eta) = 1 - 3\eta + 2\eta^2,\ L_{1/2}(\eta) = 4\left(\eta - \eta^2\right),\ L_1(\eta) = -\eta + 2\eta^2 \tag{3.12b}$$

Using Eqs. (3.12), the four quadratic interpolations along the edges *AD*, *BC*, *AB* and *CD* will be:

$$\begin{aligned}
U_{AD} &\equiv U(0,\eta) = L_0(\eta)\cdot U_1 + L_{1/2}(\eta)\cdot U_8 + L_1(\eta)\cdot U_4\\
U_{BC} &\equiv U(1,\eta) = L_0(\eta)\cdot U_2 + L_{1/2}(\eta)\cdot U_6 + L_1(\eta)\cdot U_3\\
U_{AB} &\equiv U(\xi,0) = L_0(\xi)\cdot U_1 + L_{1/2}(\xi)\cdot U_5 + L_1(\xi)\cdot U_2\\
U_{CD} &\equiv U(\xi,1) = L_0(\xi)\cdot U_4 + L_{1/2}(\xi)\cdot U_7 + L_1(\xi)\cdot U_3
\end{aligned} \tag{3.13}$$

Then, substituting (3.6) and (3.13) into Eq. (3.5), one receives

$$\begin{aligned}
U(\xi,\eta) = {} & E_0(\xi)\cdot\left[L_0(\eta)U_1 + L_{1/2}(\eta)U_8 + L_1(\eta)U_4\right]\\
& + E_1(\xi)\cdot\left[L_0(\eta)U_2 + L_{1/2}(\eta)U_6 + L_1(\eta)U_3\right]\\
& + E_0(\eta)\cdot\left[L_0(\xi)U_1 + L_{1/2}(\xi)U_5 + L_1(\xi)U_2\right]\\
& + E_1(\eta)\cdot\left[L_0(\xi)U_4 + L_{1/2}(\xi)U_7 + L_1(\xi)U_3\right]\\
& - E_0(\xi)E_0(\eta)U_1 - E_1(\xi)E_0(\eta)U_2\\
& - E_0(\xi)E_1(\eta)U_4 - E_1(\xi)E_1(\eta)U_3
\end{aligned} \tag{3.14}$$

After manipulation and reduction of equal terms, the latter becomes

$$\begin{aligned}
U(\xi,\eta) = {} & U_1\cdot[E_0(\xi)L_0(\eta) + L_0(\xi)E_0(\eta) - E_0(\xi)E_0(\eta)]\\
& + U_2\cdot[E_1(\xi)L_0(\eta) + L_1(\xi)E_0(\eta) - E_1(\xi)E_0(\eta)]\\
& + U_3\cdot[E_1(\xi)L_1(\eta) + L_1(\xi)E_1(\eta) - E_1(\xi)E_1(\eta)]\\
& + U_4\cdot[E_0(\xi)L_1(\eta) + L_0(\xi)E_1(\eta) - E_0(\xi)E_1(\eta)]\\
& + U_5\cdot\left[L_{1/2}(\xi)E_0(\eta)\right]\\
& + U_6\cdot\left[E_1(\xi)L_{1/2}(\eta)\right]\\
& + U_7\cdot\left[L_{1/2}(\xi)E_0(\eta)\right]\\
& + U_8\cdot\left[E_0(\xi)L_{1/2}(\eta)\right]
\end{aligned} \tag{3.15}$$

Again, the terms in the brackets [...] of Eq. (3.15) are the shape functions associated to the nodal points 1–8 of the eight-noded element.

Now we shall prove that the shape functions involved in Eq. (3.15) are identical with those of the classical isoparametric quadratic elements, which are as follows [1, 39]:

$$\text{Corner: } N_i = \frac{1}{4}(1+r_0)(1+s_0)(r_0+s_0-1) \tag{3.16}$$

$$\begin{aligned}
&\text{Middle:}\\
&r_i = 0, \quad N_i = \tfrac{1}{2}\left(1-r^2\right)(1+s_0)\\
&s_i = 0, \quad N_i = \tfrac{1}{2}(1+r_0)\left(1-s^2\right)\\
&r_0 = r\,r_i,\ s_0 = s\,s_i
\end{aligned} \tag{3.17}$$

Actually, substituting Eqs. (3.6) and (3.12) into Eq. (3.15) and finally considering Eq. (3.11), (3.16) and (3.17) can be derived. This is mostly obvious for a middle nodal point, for example, the node "5" along the edge AB, for which $\widehat{N}_5 = L_{1/2}(\xi)E_0(\eta) = 4\xi(1-\xi)(1-\eta)$. By virtue of Eq. (3.11a), the latter becomes $\widehat{N}_5 = \frac{1}{2}(1-r^2)(1-s)$, which is identical with the first relationship in the group of Eq. (3.17) with $s_0 = s(-1) = -s$. A similar equivalency is found for all the rest shape functions, in both Eqs. (3.16) and (3.17).

Interim Remark In other words, Coons interpolation is capable of directly producing the shape functions of both the 4-node and the 8-node isoparametric elements, without resorting to inspection. This partial conclusion was the initiation for the construction of the (arbitrary-noded) *generalized Coons macroelement*.

3.3 Global Shape Functions of Arbitrary Noded Coons Macroelement ("C-Element")

3.3.1 General Formulation

Below we shall extend Eq. (3.15) to arbitrary noded elements. To this purpose, let us denote by q_1 the number of the nodal points along the edge AB. Similarly, q_2, q_3 and q_4 denote the number of nodes along the edges BC, CD and DA, respectively. In general, q_i's are independent one another whereas no need for uniform distribution along each edge of the patch $ABCD$ (see, Fig. 3.3).

Under these definitions, Eq. (3.15) is generalized as follows:

Corner nodes:

$$\begin{aligned} N_A(\xi,\eta) &= E_0(\xi)B_{q_4}^{DA}(\eta) + E_0(\eta)B_1^{AB}(\xi) - E_0(\xi)E_0(\eta) \\ N_B(\xi,\eta) &= E_1(\xi)B_1^{BC}(\eta) + E_0(\eta)B_{q_1}^{AB}(\xi) - E_1(\xi)E_0(\eta) \\ N_C(\xi,\eta) &= E_1(\xi)B_{q_2}^{BC}(\eta) + E_1(\eta)B_1^{CD}(\xi) - E_1(\xi)E_1(\eta) \\ N_D(\xi,\eta) &= E_0(\xi)B_1^{DA}(\eta) + E_1(\eta)B_{q_3}^{CD}(\xi) - E_0(\xi)E_1(\eta) \end{aligned} \tag{3.18}$$

Intermediate nodes along the edge AB:

$$N_j(\xi,\eta) = E_0(\eta)B_j^{AB}(\xi), \quad 2 \le j \le q_1 - 1 \tag{3.19}$$

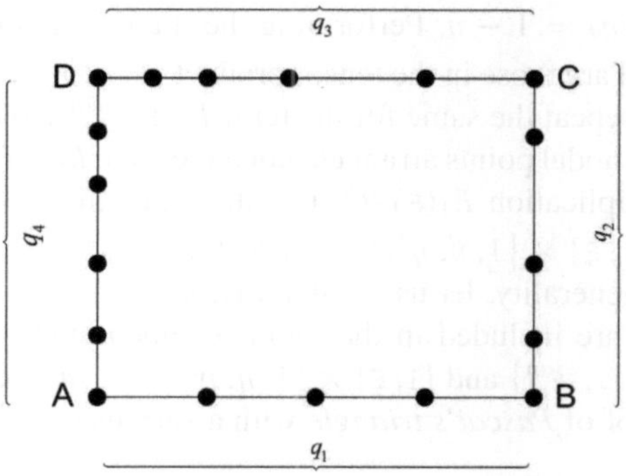

Fig. 3.3 General numbering of the boundary-only Coons macroelement (here $q_1 = 5, q_2 = 5, q_3 = 6, q_4 = 6$)

Intermediate nodes along the edge BC:

$$N_j(\xi,\eta) = E_1(\xi)B_j^{BC}(\eta), \quad 2 \le j \le q_2 - 1 \tag{3.20}$$

Intermediate nodes along the edge CD:

$$N_j(\xi,\eta) = E_1(\eta)B_j^{CD}(\xi), \quad 2 \le j \le q_3 - 1 \tag{3.21}$$

Intermediate nodes along the edge DA:

$$N_j(\xi,\eta) = E_0(\xi)B_j^{DA}(\eta), \quad 2 \le j \le q_4 - 1 \tag{3.22}$$

Regarding the indexes associated to the abovementioned trial functions, let us take the example of the first function that appears in Eq. (3.18). There, $B_{q_4}^{DA}(\eta)$ refers to the edge $\overrightarrow{DA}$ of the Coons patch and is associated to the last nodal point along it (which is the q_4 th nodal point in local numbering, i.e. the corner point A). Similar remarks hold for the rest trial functions, which all are set in the form $B_{\text{node}}^{\text{EDGE}}$.

3.4 A Deeper Examination of Coons Interpolation and Relevant Coons Macroelements

Let us consider the term $E_0(\eta)\, B_j^{AB}(\xi)$ of Eq. (3.19), which corresponds to those nodal points arranged along the edge AB of the patch. Among several alternatives, we take the case in which the trial functions B_j are Lagrange polynomials of degree $(q_1 - 1)$ as dictated by the corresponding q_1 nodes, whereas the blending function is the linear one $E_0(\eta) = 1 - \eta$. Performing the multiplication $E_0(\eta)\, B_j^{AB}(\xi)$, the monomials produced are those in the tensor product $\{1, \eta\} \times \left\{1, \xi, \xi^2, \ldots, \xi^{q_1-1}\right\}$.

Moreover, let us repeat the same for the term $E_1(\xi)\, B_j^{BC}(\eta)$ of Eq. (3.20), which corresponds to those nodal points arranged along the edge BC of the patch. Similarly, performing the multiplication $E_1(\xi)\, B_j^{BC}(\eta)$, the monomials produced are those in the tensor product $\{1, \xi\} \times \left\{1, \eta, \eta^2, \ldots, \eta^{q_2-1}\right\}$.

Without loss of generality, let us assume that $q_1 - 1 = q_2 - 1 = m$. Then, all the monomials that are included in the two abovementioned tensor products, i.e. $\{1, \eta\} \times \left\{1, \xi, \xi^2, \ldots, \xi^{\mathrm{m}}\right\}$ and $\{1, \xi\} \times \left\{1, \eta, \eta^2, \ldots, \eta^{\mathrm{m}}\right\}$ can be schematically arranged into the roof of *Pascal's triangle* with a surplus of *two* terms, as shown in Fig. 3.4.

But this set of monomials is exactly what is included into the classical *Serendipity* family (see for example, (Zienkiewicz [39], p. 155)).

Remark In other words, it was shown that a Coons macroelement based on trial functions of Lagrange polynomials type, is *identical* with an element of Serendipity family.

Nevertheless, the advantage of the Coons interpolation (with respect to the older Serendipity elements) lies on the following facts:

1. It allows for dealing with different number of nodal points along the four edges. Not only any two successive edges may have a different number of nodes ($q_1 \neq q_2, q_2 \neq q_3, q_3 \neq q_4$ and $q_4 \neq q_1$) but also the opposite edges may have different number of nodes ($q_1 \neq q_3$ and $q_2 \neq q_4$).
2. It allows for using any kind of trial functions B_j along each of the four edges. Typical examples are the above-mentioned Lagrange polynomials, but also other

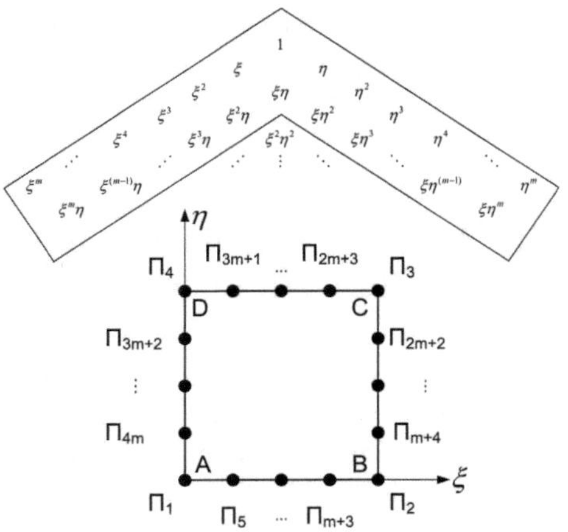

Fig. 3.4 The roof of Pascal's triangle with a surplus of two including all monomials relevant with the edges AB and BC (toward the axes ξ and η, respectively)

type of *univariate* interpolation $B_j(\xi)$ or $B_j(\eta)$ accordingly (see Chap. 2) can be used. For example, piecewise-linear, piecewise-quadratic, B-splines, Bézier, NURBS (see an example in Sect. 3.8), etc.

3.5 Trial Functions

3.5.1 Piecewise Linear Trial Functions

The piecewise linear univariate trial functions, along each of the four edges of the patch *ABCD*, have three advantages:

(i) Obviously, they have compact support.
(ii) They preserve the classical FEM formulation, which is based on nodal value U_j.
(iii) They can approximate polygonal chains with high accuracy.

First, the *compact support* is clearly shown in Fig. 3.5 (for a 16-node Coons element), for one corner (Fig. 3.5a) and another intermediate (Fig. 3.5b) nodal point. Both global shape functions are *cardinal*, that is of [1, 0]-type:

$$N_i(x_j, y_j) = \delta_{ij} = \begin{cases} 1, & i = j \\ 0, & i \neq j \end{cases} \tag{3.23}$$

Second, the *preservation of the FEM formulation* is concerned with the fact that the produced stiffness and mass matrices are multiplied by the vector of the nodal displacements, $\{\mathbf{u}\}$, and the nodal accelerations, $\{\ddot{\mathbf{u}}\}$, respectively. This fact makes the implementation of the boundary conditions a trivial task, as exactly in the conventional finite elements.

Third, let us now examine the case of representing *polygonal chains* by other trial functions such as Lagrange polynomials and natural cubic B-splines.

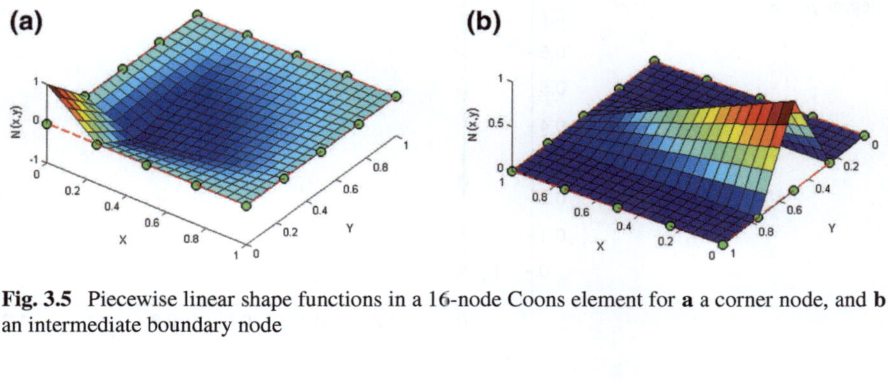

Fig. 3.5 Piecewise linear shape functions in a 16-node Coons element for **a** a corner node, and **b** an intermediate boundary node

3.5.2 Lagrange Polynomials

The Lagrange polynomials have been introduced in Sect. 2.3.4. Now we shall see whether these are applicable to polygonal lines. The simplest case is that of a line EFG that consists of two straight line segments of equal length $L = 1$, at a rectangle angle $\alpha = 90°$ (see, Fig. 3.6). Each line segment (*EF* and *FG*) is divided into four equal subintervals, thus creating nine nodal points along the polygonal chain curve *EFG*. In order to show the inconvenience of the Lagrange polynomials in representing the entire curve *EFG* as a whole, we assume the axis origin at point *E* ($\xi = 0$) whereas the ending point ($\xi = 1$) is at point *G*. Considering the nine Lagrange polynomials $L_1(\xi), L_2(\xi), \ldots, L_9(\xi)$, we apply the isoparametric interpolation

$$x(\xi) = \sum_{i=1}^{9} L_i(\xi) \cdot x_i, \quad y(\xi) = \sum_{i=1}^{9} L_i(\xi) \cdot y_i \tag{3.24}$$

The result is shown in Fig. 3.6. The thick blue line is the polygon chain with the nodal points at the small circles, whereas the red line is according to Eq. (3.24). The result of the isoparametric interpolation is entirely symmetric with respect to the corner point *F*. Clearly the reason of the bad quality of the red line is not the inability of Lagrange polynomials but the global character of the approximation due to the appearance of the corner at *F* (similar results occur even when using natural cubic B-splines). The involved Lagrange polynomials of all of 8th degree.

The quality of the representation becomes still worse when the polygonal chain becomes more complicated like the Π-shape shown in Fig. 3.7. There, the red line [according to Eq. (3.24) where now the index in the sum extends from $i = 1$ to $i = 13$ thus including Lagrange polynomials to 12th degree : $p = 12$], is far away from the real line at both vertical portions of the Π-shape.

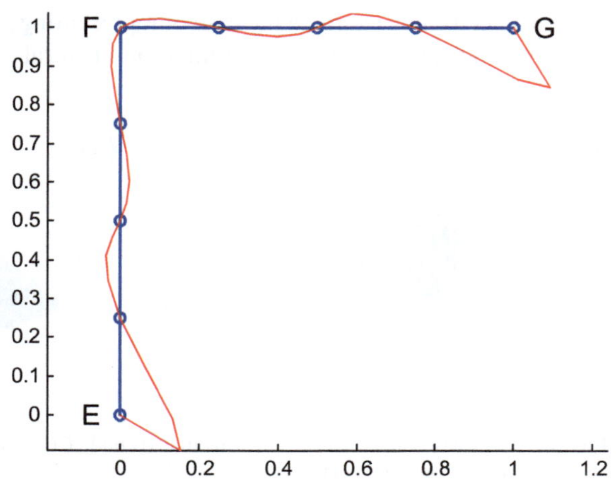

Fig. 3.6 Quality in the representation of the a polygon chain EFG using Lagrange polynomials of degree $p = 8$

Therefore, in both cases above, the mapping from the unit square to the actual domain may lead to points outside the problem domain.

The conclusion from the plots shown in Figs. 3.6 and 3.7 is that it is not possible to construct a Coons macroelement of concave shape (e.g. of *L*-shape) using *Lagrange polynomials* as trial functions (see, Fig. 3.8a). Still worse is the case of a Π-shaped macroelement (see, Fig. 3.8b) in conjunction with Lagrange polynomials. This fact can be easily checked into the computer code by calculating the total area *A* of the domain using the formula $A = \int_{\Omega} \det \mathbf{J}\, dxdy$, where **J** is the Jacobian matrix; in both cases of Fig. 3.8 the deviation between the calculated and the exact area is high.

In contrast, *piecewise* linear polynomials do *not* have any problem and perform well provided a proper parameterization of the Coons patch has been implemented. This fact has been shown in a number of applications including Π-shaped domains (see for example, (Provatidis [34], p. 347; 35, p. 6701)).

Below we continue our investigation on the performance of natural cubic B-splines as candidate trial functions.

3.5.3 Natural Cubic B-Splines as Trial Functions

The basic theory has been reported in Chap. 2. Here we shall point out some important topics.

In the beginning, below we refer to a computer code that was written in 1984 in FORTRAN77 (published by Kanarachos and Deriziotis [12]). Here it was rewritten in MATLAB (named Program P3-1) and refers to the interval $[x_{\min}, x_{\max}]$ of length $L = x_{\max} - x_{\min}$. The most important variables are explained within the code, while some critical ones are tabulated below.

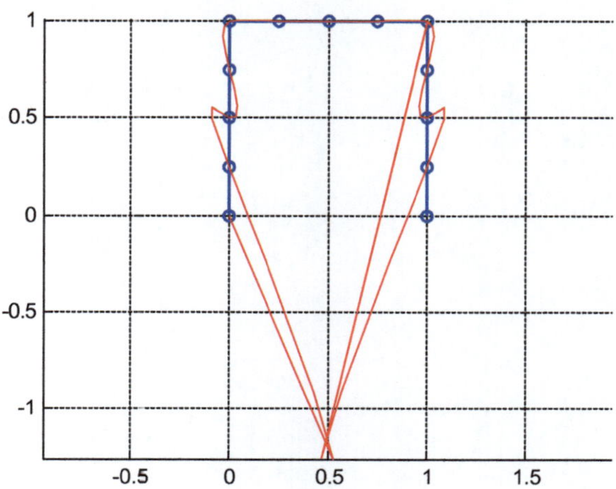

Fig. 3.7 Quality in the representation of the a polygon chain EFGH using Lagrange polynomials of degree $p = 12$

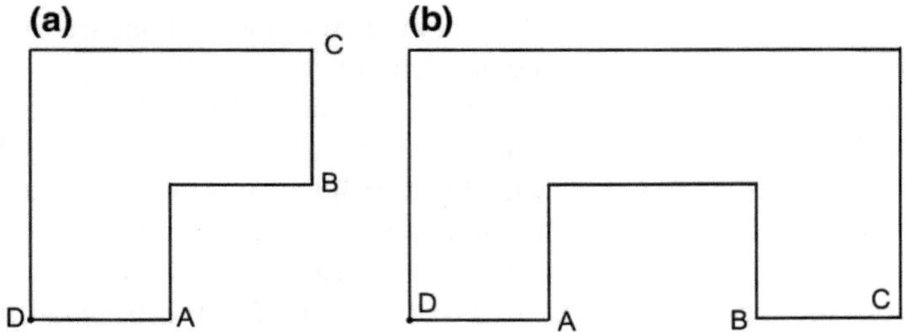

Fig. 3.8 Coons macroelements of **a** L-shape, and **b** Π–shape

Variable	Meaning
`nele`	Number of elements (subdivisions) in the interval
`nodes`	Number of nodes (nodal points)
`rcur`	Vector including the normalized coordinates $\xi = x/L$ of nodal points
`curv1, curv2`	Curvatures $U''(x_{\min})$, $U''(x_{\max})$ at the ends of the interval
`ncolumns`	Number of columns of matrix [**A**]
`a`	The matrix [**A**] (see Exercise 3.1, below)
`ainv`	Inverse of matrix [**A**]
`nsteps`	Number of segments in which the domain is divided. At the endpoints of these segments the values of shape functions are calculated (for plotting purposes)
`r`	Normalized coordinate at the endpoints of the above `nsteps` segments

```
    %----------------------------------------------------------------
    %
    %*** PROGRAM P3-1
    %
    %...Main Program to apply truncated power series cubic B-splines interpolation
    %   It calculates and plots the cardinal shape functions (natural B-splines).
    % ***
    %...Professor C. G. Provatidis, NTUA
    % ***
    %-------------------------------------------------------------------------
    %...DATA: number of elements / segments:
        nele=4;          % number of elements / segments
    %
        nodes=nele+1;    %number of nodes including the two ends
    %...Normalized co-ordinates (isodistantly):
        rcur=linspace(0,1,nodes);      %uniform subdivision of interval [0,1]
    %
%*********************************************************************
        curv1=0;     %given curvature at left end
        curv2=0;     %given curvature at right end
    %
        ncolumns = nodes+2;  %number of columns = number of unknown coefficients (a's,b's)
    %
    %...Build-up the matrix [A] of coefficients:
        a = spline1d(ncolumns,rcur);
    %...Calculate the inverse matrix:
        ainv=inv(a);
    %...Calculate the values of shape functions along the domain at 'nsteps' equidistant points:
    %
        nsteps=40;
        dr=1/nsteps; %normalized step
        for icou=1:nsteps+1
            r=(icou-1)*dr; %reference coordinate
            csi(icou)=r;   %store for plot purposes
            for j=1:nodes+2
               [shp,dershp]=bsplines_int(ainv,x,nodes,j,r);  %shape functions and first deriva-
tive.
               Nj(icou,j)=shp;
               derNj(icou,j)=dershp;
            end
        end
    %...Plot the shape functions (Nj,j=1,...,nodes):
        for j=1:nodes
            plot(csi,Nj(:,j),'b')
            hold on
        end
        for j=nodes+1:nodes+2
            plot(csi,100*Nj(:,j),'r')
            hold on
        end
    %--------- END-OF-MAIN-PROGRAM ------------------------------------
```

The involved function `spline1d` determines the monomials in the B-spline expansion. Its input `rcur` is the vector that includes all the normalized coordinates $\xi = x/L$ of nodal points, whereas its output is the matrix `a`, which corresponds to the matrix [**A**] (see Exercise 3.1, below).

```
function am = spline1d(ncolumns,xcur)
%...Computes the matrix that multiplies spline-coefficients (a0,a1,a2,bj,...)
      am=0; nm=ncolumns;
      nm1=nm-1; nm2=nm-2; nm3=nm-3;
%...assign normalized coordinates at nodes:
      for m=1:nm2
        w(m)=xcur(m);
      end
%...The first column corresponds to the constant term:
      for i=1:nm2
        am(i,1)=1;
      end
%...The last two rows(correspond to the curvature):
      for i=nm1:nm
        for j=1:2
         am(i,j)=0;
        end
        am(i,3)=2;
      end
      for i=4:nm
        am(nm1,i)=0;
      end
%...
      for i=1:nm3
         ih1=i+3;
         for j=ih1:nm
           am(i,j)=0;
         end
      end
%...second column (linear term), and third column (quadratic term):
      for i=1:nm2
        for j=2:3
          ih3=j-1;
          am(i,j)=w(i)^ih3;
        end
      end
%...cubic terms:
      [id1,ihdif1,ideik1,id2,ihdif2,ideik2,iter] = findconst(nm);
%
      for j=1:iter
        for i=id1(j):nm2
          am(i,ideik1(j))=(w(i)-w(ihdif1(j)))^3;
        end
        for i=id2(j):nm2
          am(i,ideik2(j))=(w(i)-w(ihdif2(j)))^3;
        end
      end
%...last row and columns related to cubic terms:
      for i=4:nm
        iha=i-3;
        am(nm,i)=6*(w(nm2)-w(iha));
      end
%---end-of-function ------------------------------------------------------%
```

The subroutine that finds the constants (`findconst`) is given below.

```
%
function [id1,ih1,idk1,id2,ih2,idk2,iter] = findconst(nm)
%
      iter=(nm-2)/2;
      nmd(1)=5;
       if(iter~=1)
            for i=2:iter
              i1=i-1;
              nmd(i)=nmd(i1)+2;
            end
       end
      for i=1:iter
       id1(i)  = nmd(i)-3;
       ih1(i)  = nmd(i)-4;
       idk1(i)= nmd(i)-1;
       id2(i)  = nmd(i)-2;
       ih2(i)  = nmd(i)-3;
       idk2(i)= nmd(i);
      end
```

Finally, the involved function `bsplines_int` calculates the cardinal shape functions and their derivative.

```
%
    function [yvalue,dervalue] = bsplines_int(c0,x,n,i,xvalue)
    yvalue  =   splCP(xvalue,c0,n+2,x,i);
    dervalue=dersplCP(xvalue,c0,n+2,x,i);
%
```

The above task is performed using the series expansion of splines (`spl_cp`).

```
%
    function spl_cp = splCP(ww,ams,nd,wcur,ncol)
    idyn=3;
    spl_cp=ams(1,ncol)+ams(2,ncol)*ww+ams(3,ncol)*ww^2;
    for i=4:nd
        j=i-3;
      spl_cp=spl_cp+ams(i,ncol)*qfun(j,ww,wcur,idyn);
    end
%
```

3.5.3.1 Exercise 3.1: Based on the Above Computer Program, Derive the Cardinal Shape Functions Based on the Cubic Natural B-spline Interpolation

The unit length interval [0, 1] is divided into 4 equal parts (`nele = 4`) and for each of them we assume a piecewise cubic polynomial. As explained earlier in Chap. 2, the number of coefficients is $5+2=7$, which corresponds to five shape functions associated to the five nodal points (translational DOFs) plus two extreme nodes which refer to the second derivative (curvature) thus related to the rotational DOFs at the ends. Using the abovementioned code (Program P3-1), the obtained result is shown in Fig. 3.9, where the blue line corresponds to the nodes (translational)

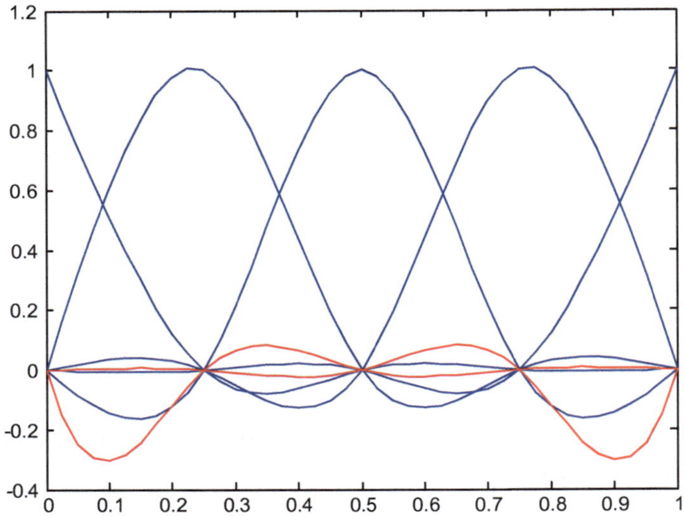

Fig. 3.9 Univariate shape functions (nele = 4 subdivisions, i.e. 5 nodes)

while the red to the rotational DOFs. We may observe that the blue-coloured shape functions are *cardinal* (due to the constant term), that is of [1, 0]-type, whereas the red (rotational) are *not* cardinal.

It is noted that the abovementioned natural cubic B-spline interpolation allows for an enormous number of nodal points. For example, Fig. 3.10 shows the result obtained for a number of `nele` = 20 subdivisions, i.e. 21 nodal points.

The property of cardinality inherent in the abovementioned functions (i.e., [0, 1]-type: $N_i(\mathbf{x}_j) = \delta_{ij}$), in conjunction with the existing partition of unity (rigid-body) property, was the reason that the CAD/CAE team at NTUA started in 1980s using "natural cubic B-splines" as global shape functions [12, 14]. Concretely, in 1984, it seemed easier for us to deal directly with *nodal* values U_i and stick to the standard computer program FEAP ([39], Chap. 24 by Robert L. Taylor), than developing a new computer code based on control points and associated *nodeless* coefficients a_i (followed, twenty years later, by the isogeometric community).

If now we study again the case of a polygonal chain made of two equal segments at a right angle, then the quality of geometry isoparametric approximation is shown in Fig. 3.11. One may observe that it is better than the Lagrange interpolation but it has a substantial error in the neighourhood of the corner point.

A deviation of the same quality appears in case of a Π-shaped polygonal line, as shown in Fig. 3.12.

The results illustrated in Figs. 3.11 and 3.12 suggest that if such a non-convex Coons patch (such as those shown in Fig. 3.8) is modeled using natural cubic B-spline trial functions along the polygonal edges, a certain deviation occurs. For example,

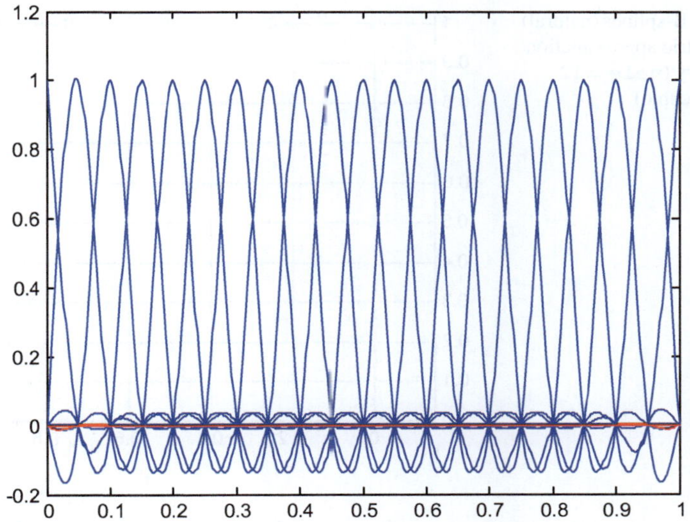

Fig. 3.10 Shape functions for nele = 20 subdivisions (natural cubic B-splines)

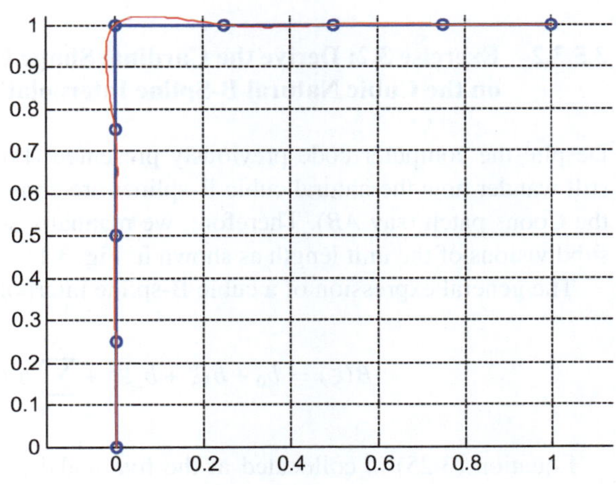

Fig. 3.11 B-splines (natural) isoparametric approximation of geometry (`nele` = 8, cubic)

even in case of a polygon with straight lines the calculated area of the patch differs from its accurate value.

Let us now see the two-dimensional global shape functions ought to natural cubic B-spline interpolation along the edges (a separate *natural* B-spline along each separate edge of the quadrilateral patch *ABCD*). In Fig. 3.13 we present the natural cubic B-spline based global shape functions which are associated to the nodal points 1, 2, and 3 on the edge *AB* in a 20-noded Coons macroelement, along either the edge "ξ" or the diagonal "*d*" direction shown at the top.

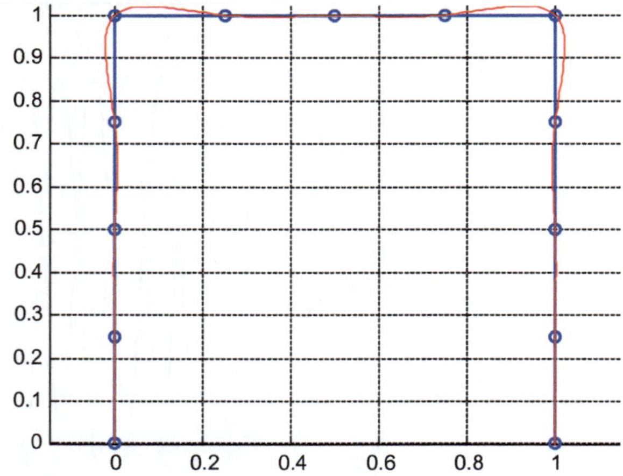

Fig. 3.12 B-splines (natural) isoparametric approximation of geometry (`nele` = 12 segments, cubic)

A 3D visualization for a C-macroelement of dimensions 3.0 × 1.0, with still more nodal points (24-noded) is illustrated in Fig. 3.14.

3.5.3.2 Exercise 3.2: Derive the Cardinal Shape Functions Based on the Cubic Natural B-Spline Interpolation

Despite the computer code previously presented (Program P3-1), the reader may still wander how the natural cubic B-splines are constructed along an entire side of the Coons patch (say *AB*). Therefore, we manually study the case of four uniform subdivisions of the unit length as shown in Fig. 3.15.

The general expression of a cubic B-spline interpolation is:

$$B(\xi) = b_0 + b_1\xi + b_2\xi^2 + \sum_{i=0}^{3} a_i \langle \xi - \xi_i \rangle_+^3 \tag{3.25}$$

Equation (3.25) is collocated at the five nodal points shown in Fig. 3.15 thus receiving the first five degrees of freedom as follows:

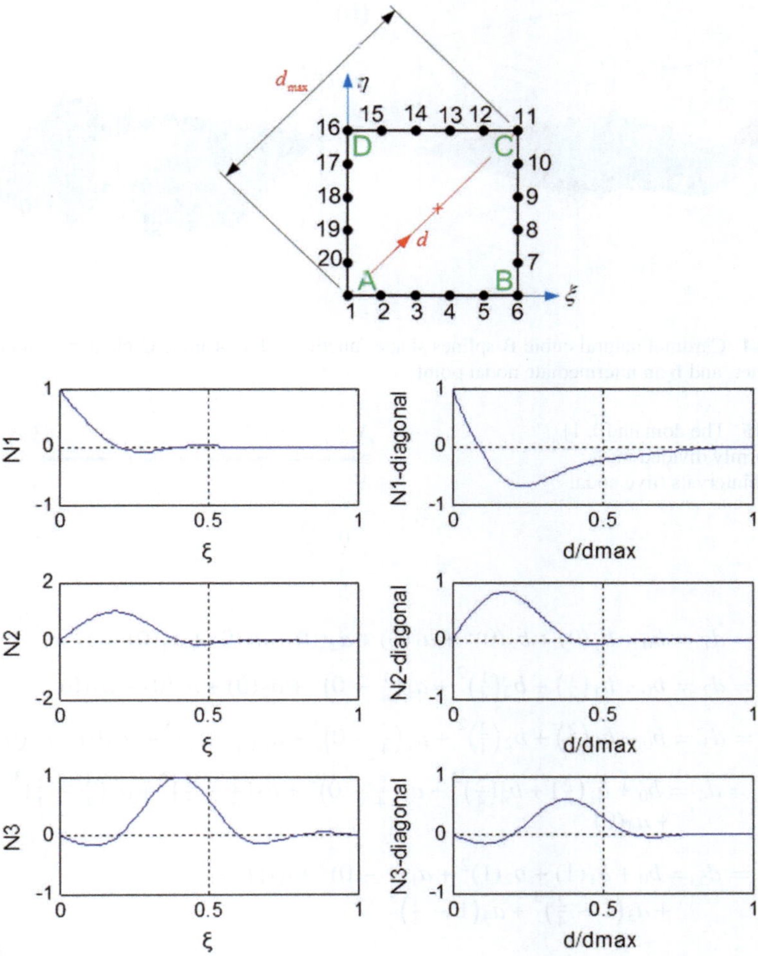

Fig. 3.13 Global shape functions N_i associated to the nodal points $i = 1$, 2, and 3 (the left column refers to ξ whereas the right to the diagonal d direction, shown on the top)

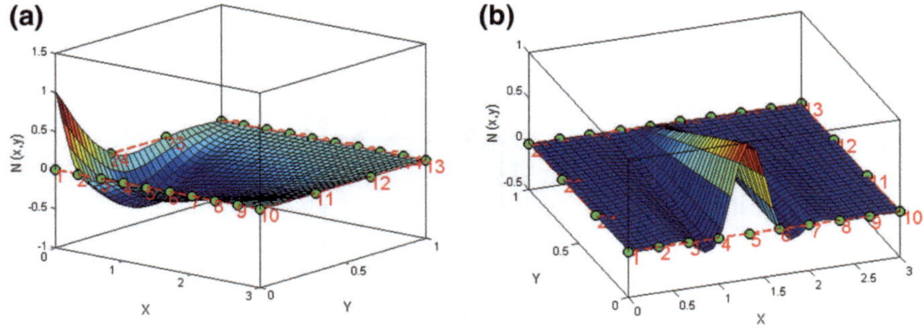

Fig. 3.14 Cardinal natural cubic B-splines shape functions of a 24-node C-element associated to **a** a corner, and **b** an intermediate nodal point

Fig. 3.15 The domain [0, 1] is uniformly divided into four subintervals (five nodal points)

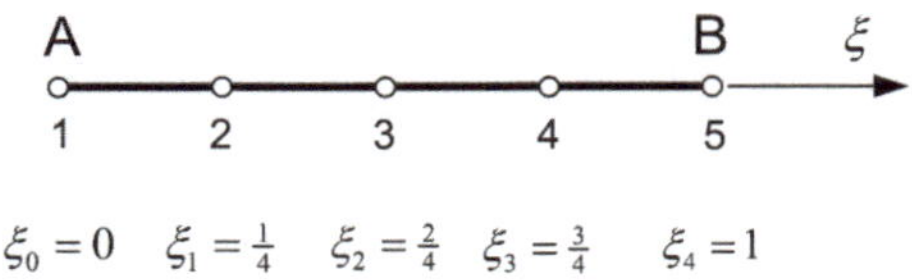

$$\left.\begin{aligned}
\hat{d}_1 &= d_1 = b_0 + b_1(0) + b_2(0)^2 + a_1(0) + a_2(0) + a_3(0) + a_4(0) \\
\hat{d}_2 &= d_2 = b_0 + b_1\left(\tfrac{1}{4}\right) + b_2\left(\tfrac{1}{4}\right)^2 + a_1\left(\tfrac{1}{4} - 0\right)^3 + a_2(0) + a_3(0) + a_4(0) \\
\hat{d}_3 &= d_3 = b_0 + b_1\left(\tfrac{2}{4}\right) + b_2\left(\tfrac{2}{4}\right)^2 + a_1\left(\tfrac{2}{4} - 0\right)^3 + a_2\left(\tfrac{2}{4} - \tfrac{1}{4}\right)^3 + a_3(0) + a_4(0) \\
\hat{d}_4 &= d_4 = b_0 + b_1\left(\tfrac{3}{4}\right) + b_2\left(\tfrac{3}{4}\right)^2 + a_1\left(\tfrac{3}{4} - 0\right)^3 + a_2\left(\tfrac{3}{4} - \tfrac{1}{4}\right)^3 + a_3\left(\tfrac{3}{4} - \tfrac{2}{4}\right)^3 \\
&\quad + a_4(0) \\
\hat{d}_5 &= d_5 = b_0 + b_1(1) + b_2(1)^2 + a_1(1 - 0)^3 + a_2\left(1 - \tfrac{1}{4}\right)^3 \\
&\quad + a_3\left(1 - \tfrac{2}{4}\right)^3 + a_4\left(1 - \tfrac{3}{4}\right)^3
\end{aligned}\right\} \tag{3.26}$$

The second derivative of the function $B(\xi)$ given by Eq. (3.25) is:

$$B''(\xi) = 2b_2 + 6\left[a_1\langle\xi - 0\rangle_+^1 + a_2\left\langle\xi - \tfrac{1}{4}\right\rangle_+^1 + a_3\left\langle\xi - \tfrac{2}{4}\right\rangle_+^1 + a_4\left\langle\xi - \tfrac{3}{4}\right\rangle_+^1\right] \tag{3.27}$$

Collocating Eq. (3.27) at the two ends of the interval [0, 1] we obtain:

$$\left.\begin{aligned}
\hat{d}_6 &= 0 = B''(0) = 2b_2 = 0 \\
\hat{d}_7 &= 0 = B''(1) = 2b_2 + 6\left[a_1(1 - 0) + a_2\left(1 - \tfrac{1}{4}\right)\right. \\
&\qquad \left. + a_3\left(1 - \tfrac{2}{4}\right) + a_4\left(1 - \tfrac{3}{4}\right)\right] = 0
\end{aligned}\right\} \tag{3.28}$$

Summarizing Eqs. (3.26) and (3.28) in one linear equations system, we have:

$$\mathbf{d} = \mathbf{A} \cdot \mathbf{a} \tag{3.29}$$

with

$$\mathbf{A} = \begin{bmatrix} 1 & 0 & 0 & 0 & 0 & 0 & 0 \\ 1 & \frac{1}{4} & \left(\frac{1}{4}\right)^2 & \left(\frac{1}{4}\right)^3 & 0 & 0 & 0 \\ 1 & \frac{2}{4} & \left(\frac{2}{4}\right)^2 & \left(\frac{2}{4}\right)^3 & \left(\frac{1}{4}\right)^3 & 0 & 0 \\ 1 & \frac{3}{4} & \left(\frac{3}{4}\right)^2 & \left(\frac{3}{4}\right)^3 & \left(\frac{2}{4}\right)^3 & \left(\frac{1}{4}\right)^3 & 0 \\ 1 & 1 & 1 & 1 & \left(\frac{3}{4}\right)^3 & \left(\frac{2}{4}\right)^3 & \left(\frac{1}{4}\right)^3 \\ 0 & 0 & 2 & 0 & 0 & 0 & 0 \\ 0 & 0 & 2 & 6 & \frac{18}{4} & \frac{12}{4} & \frac{6}{4} \end{bmatrix} \tag{3.30}$$

where the vector of the DOFs is:

$$\mathbf{d} = [d_1, d_2, d_3, d_4, d_5, d_6, d_7]^{\mathrm{T}}, \tag{3.31}$$

and the vector of the coefficients:

$$\mathbf{d} = [b_0, b_1, b_2, a_1, a_2, a_3, a_4]^{\mathrm{T}} \tag{3.32}$$

Based on Eq. (3.29), the vector of the coefficients is given by

$$\mathbf{a} = \mathbf{A}^{-1} \cdot \mathbf{d} \tag{3.33}$$

with

$$\mathbf{A}^{-1} = \begin{bmatrix} 1 & 0 & 0 & 0 & 0 & 0 & 0 \\ -5.0714 & 6.4286 & -1.7143 & 0.4286 & -0.0714 & -0.0722 & 0.0007 \\ 0 & 0 & 0 & 0 & 0 & 0.5000 & 0 \\ 17.1429 & -38.8571 & 27.4286 & -6.8571 & 1.1429 & -0.8452 & -0.0119 \\ -38.8571 & 105.1429 & -100.5714 & 41.1429 & -6.8571 & 1.0714 & 0.0714 \\ 27.4286 & -100.5714 & 146.2857 & -100.5714 & 27.4286 & -0.2857 & -0.2857 \\ -6.8571 & 41.1429 & -100.5714 & 105.1429 & -38.8571 & 0.0714 & 1.0714 \end{bmatrix} \tag{3.34}$$

Following the standard FEM procedure ([39], p. 151), the problem variable is given in terms of the basis functions by

$$U(\xi) = [1, \xi, \xi^2, \langle \xi - \xi_0 \rangle_+^3, \langle \xi - \xi_1 \rangle_+^3, \langle \xi - \xi_2 \rangle_+^3, \langle \xi - \xi_3 \rangle_+^3] \cdot \mathbf{a}, \tag{3.35}$$

Substituting Eq. (3.34) into Eq. (3.33), and then into Eq. (3.35), we obtain:

$$U(\xi) = [1, \xi, \xi^2, \langle \xi - \xi_0 \rangle_+^3, \langle \xi - \xi_1 \rangle_+^3, \langle \xi - \xi_2 \rangle_+^3, \langle \xi - \xi_3 \rangle_+^3] \cdot \mathbf{A}^{-1} \cdot \mathbf{d}, \tag{3.36}$$

Performing the multiplications between the row vector and the matrix in the right hand side of Eq. (3.36), the variable is expressed in terms of seven shape functions:

$$U(\xi) = [N_1(\xi), N_2(\xi), N_3(\xi), N_4(\xi), N_5(\xi), N_6(\xi), N_7(\xi)] \cdot \mathbf{d}, \tag{3.37}$$

where the shape functions are given by:

$$\left.\begin{array}{l} N_1(\xi) = 1 \cdot \mathbf{d},\, N_2(\xi) = \xi \cdot \mathbf{d},\, N_3(\xi) = \xi^2 \cdot \mathbf{d}, \\ N_4(\xi) = \langle \xi - \xi_0 \rangle_+^3 \cdot \mathbf{d},\, N_5(\xi) = \langle \xi - \xi_1 \rangle_+^3 \cdot \mathbf{d}, \\ N_6(\xi) = \langle \xi - \xi_2 \rangle_+^3 \cdot \mathbf{d},\, N_7(\xi) = \langle \xi - \xi_3 \rangle_+^3 \cdot \mathbf{d} \end{array}\right\} \tag{3.38}$$

We notice that the number of nodes is *five* (see Fig. 3.15) whereas the number of shape functions is *seven*. Actually, this is generally so when the second derivative at the two ends is different than zero, which is the usual *complete* cubic B-spline. But since here we deal with a *natural* B-spline interpolation (reduced basis), which is characterized by the conditions $\hat{d}_5 = \hat{d}_6 = 0$, the shape functions $N_6(\xi)$ and $N_7(\xi)$ are useless because they are multiplied by null. In other words, since we need only the first *five* DOF of the vector $\mathbf{d}$, in Eq. (3.33) we do not need the whole inverse matrix $\mathbf{A}^{-1}$ but it is sufficient to consider its first *five* columns which constitute the submatrix:

$$\mathbf{B}_{11} = \begin{bmatrix} 1 & 0 & 0 & 0 & 0 \\ -5.0714 & 6.4286 & -1.7143 & 0.4286 & -0.0714 \\ 0 & 0 & 0 & 0 & 0 \\ 17.1429 & -38.8571 & 27.4286 & -6.8571 & 1.1429 \\ -38.8571 & 105.1429 & -100.5714 & 41.1429 & -6.8571 \\ 27.4286 & -100.5714 & 146.2857 & -100.5714 & 27.4286 \\ -6.8571 & 41.1429 & -100.5714 & 105.1429 & -38.8571 \end{bmatrix} \tag{3.39}$$

Therefore, the *five* useful shape functions involved in the natural cubic B-spline (incomplete or reduced set, see Chap. 7) compose the vector of the shape functions given by:

$$\underbrace{\mathbf{\Phi}(\xi)}_{1\times 5} = \begin{Bmatrix} \varphi_1(\xi) \\ \varphi_2(\xi) \\ \varphi_3(\xi) \\ \varphi_4(\xi) \\ \varphi_5(\xi) \end{Bmatrix}^{\mathrm{T}} = \underbrace{\left[1\ \xi\ \xi^2\ \langle \xi - 0 \rangle_+^3\ \langle \xi - \tfrac{1}{4} \rangle_+^3\ \langle \xi - \tfrac{2}{4} \rangle_+^3, \langle \xi - \tfrac{3}{4} \rangle_+^3 \right]}_{1\times 7} \cdot \underbrace{(\mathbf{B}_{11})}_{7\times 5} \tag{3.40}$$

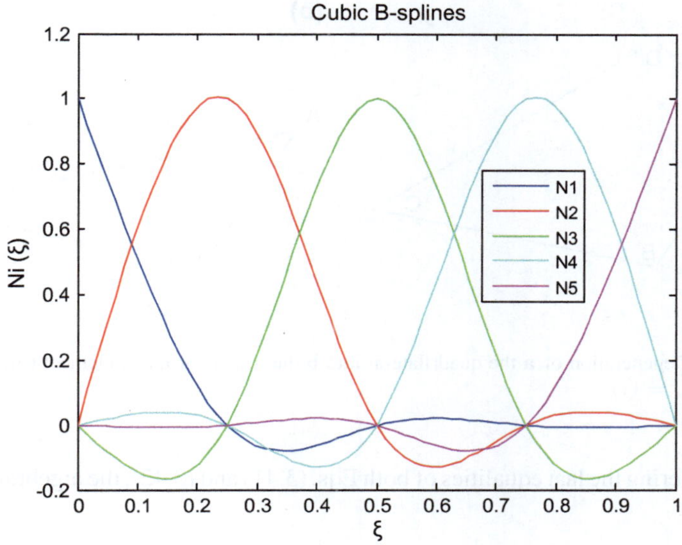

Fig. 3.16 The five shape functions involved in the natural cubic B-spline (of cardinal [0, 1]-type) that are associated to the five nodal points shown in Fig. 3.15

The graphical representation of the abovementioned four shape functions is shown in Fig. 3.16. The same results were also derived using the program P3-1 in the Sect. 3.5.3.

3.6 Degeneration of Quadrilateral into a Triangular Patch

According to the famous Coons' Report ([7], p. 3), a triangular patch ABC can be produced by degenerating the whole side AD into a single vertex A (see, Fig. 3.17). Then, since obviously $U_{AD} = U_A$ and $U_{CD} = U_{CA}$, the sum of the two projections progressively becomes:

$$\begin{aligned} P_\xi + P_\eta &= (1-\xi)U_{AD} + \xi U_{BC} + (1-\eta)U_{AB} + \eta U_{CD} \\ &= (1-\xi)U_A + \xi U_{BC} + (1-\eta)U_{AB} + \eta U_{CA} \end{aligned} \tag{3.41}$$

Also, the corrective term progressively becomes:

$$\begin{aligned} P_{\xi\eta} &= (1-\xi)(1-\eta)U_A + \xi(1-\eta)U_B + \xi\eta U_C + (1-\xi)\eta U_D \\ &= (1-\xi)(1-\eta)U_A + \xi(1-\eta)U_B + \xi\eta U_C + (1-\xi)\eta U_A \\ &= (1-\xi)U_A + \xi(1-\eta)U_B + \xi\eta U_C \end{aligned} \tag{3.42}$$

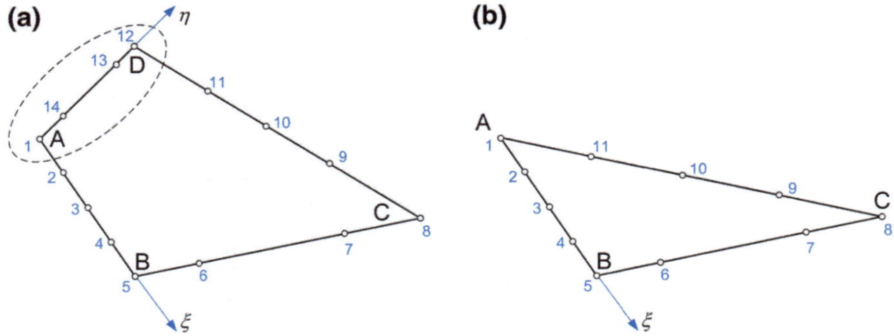

Fig. 3.17 Degeneration of **a** the quadrilateral into **b** the triangular macroelement ($q_1 = 4, q_2 = 5, q_3 = 5; q_e = 11$)

Considering the last equalities of both Eqs. (3.41) and (3.42), the algebraic expression

$$U(\xi, \eta) = P_\xi + P_\eta - P_{\xi\eta}, \tag{3.43a}$$

takes its final form:

$$\begin{aligned} U(\xi, \eta) = U_{AB}(\xi) \cdot (1 - \eta) + U_{BC}(\eta) \cdot \xi + U_{CA}(\xi) \cdot \eta \\ - \xi \cdot [(1 - \eta)U_B + \eta U_C] \end{aligned} \tag{3.43b}$$

Remark One may observe that Eq. (3.43b) has *not* a symmetry, neither in the variables ξ and η, nor in the vertexes *A*, *B*, and *C*. In contrast, affine invariance is achieved using *Barnhill's interpolation*, which was mentioned in Chap. 2 whereas relevant macroelements are studied in Chap. 5.

If we assume that the three sides (*AB*, *BC*, and *CA*) consist of q_1, q_2 and q_3 nodes, respectively, the total number of nodes becomes:

$$q_e = (q_1 + q_2 + q_3) - 3 \tag{3.44}$$

Approximating the variable U along the three sides of the triangular macroelement by:

$$U_{AB} = \sum_{j=1}^{q_1} B_j^{AB} U_j, \quad U_{BC} = \sum_{j=1}^{q_2} B_j^{BC} U_j, \quad U_{CA} = \sum_{j=1}^{q_3} B_j^{CA} U_j, \tag{3.45}$$

and further substituting into Eq. (3.43b), after rearrangement of the terms one obtains:

$$U(\xi, \eta) = \sum_{j=1}^{q_e} N_j(\xi, \eta) U_j \tag{3.46}$$

where the shape functions are given as follows:

Corner nodes

$$\left.\begin{aligned} N_A(\xi, \eta) &= B_1^{AB}(\xi) \cdot (1 - \eta) + B_{q_3}^{CA}(\xi) \cdot \eta, \text{ degenerated} \\ N_B(\xi, \eta) &= B_{q_1}^{AB}(\xi) \cdot (1 - \eta) + B_1^{BC}(\eta) \cdot \xi - \xi(1 - \eta) \\ N_C(\xi, \eta) &= B_{q_2}^{BC}(\eta) \cdot \xi + B_1^{CA}(\xi) \cdot \eta - \xi\eta \end{aligned}\right\} \tag{3.47}$$

Side nodes

$$\text{Side } AB\text{: } N_i(\xi, \eta) = B_i^{AB}(\xi) \cdot (1 - \eta), \quad 2 \le i \le q_1 - 1 \text{ (local numbering)} \tag{3.48}$$

$$\text{Side } BC\text{: } N_i(\xi, \eta) = B_i^{BC}(\eta) \cdot \xi, \quad 2 \le i \le q_2 - 1 \text{ (local numbering)} \tag{3.49}$$

$$\text{Side } BC\text{: } N_i(\xi, \eta) = B_i^{CA}(\xi) \cdot \eta, \quad 2 \le i \le q_3 - 1 \text{(local numbering)} \tag{3.50}$$

One may observe in the first line of Eq. (3.47) that the shape function which is associated to the degenerated node (here is A) consists of two terms, whereas the rest ones (here are B and C) consist of three terms.

3.7 Test Cases

In order to demonstrate the performance of "C-elements", a lot of engineering examples will be presented. First we apply the C-elements to potential problems and then to elasticity ones.

Between many possible cases, we restrict our study to the following *three* selected classes of trial functions (along each of the edges of a quadrilateral patch $ABCD$):

(i) Piecewise linear interpolation (model C1 or M1)
(ii) Cardinal natural cubic B-splines (model C2 or M2)
(iii) Lagrange polynomials (model C3 or M3).

Note The abovementioned symbols are justified as follows. The letter "M" stands for the word "model" whereas M1 means model No. 1, and so on. Alternatively, the letter "C" stands for the word "Coons". We mention both of them, because it happened to have been used in our journal papers.

In many cases the obtained results will be also compared with conventional finite elements of the same mesh density along the boundary.

3.7.1 Potential Problems

In this Chapter numerical results will restrict to 2D problems only. The general potential problem is governed by a Partial Differential Equation (PDE) $D(U) - f = 0$, where $U(x, y)$ is the problem variable, $f(x, y)$ represents the "sources" at an arbitrary point with Cartesian coordinates (x, y), and $D(\ldots)$ is the differential operator that corresponds to the (physical) problem under consideration. The problem domain is noted by the symbol Ω, which generally corresponds either to a surface (2*D* problem) or to a volume (3*D* problem). The boundary surrounding the domain Ω is noted by the symbol Γ. A part of the boundary, Γ_1, is under boundary conditions of *Dirichlet* type, whereas the rest, Γ_2, is under *Neumann* ones, thus we can write that $\Gamma = \Gamma_1 \cup \Gamma_2$. Nevertheless, more generally, we usually distinguish three types of linear boundary conditions:

- *Dirichlet* type of boundary conditions (BCs) are sometimes called BCs of *first kind* or *essential* BCs. They describe given (prescribed) values of the primary variable $U(x, y)$ at the boundary, and then we put a "bar" over it, $\Gamma_1 : U = \bar{U}$.
- *Dirichlet* type of boundary conditions (BCs) are sometimes called BCs of *first kind* or *natural* BCs. They refer to a given (prescribed) derivative of the primary variable $U(x, y)$ such as the gradient (∇U) or the normal derivative (towards the outward normal unit vector *n*), for example, $\Gamma_2 : \partial U/\partial n \equiv \mathbf{n}\nabla U = \bar{q}$.
- *Robin* type of boundary conditions (BCs) are sometimes called BCs of *third kind* or *mixed* BCs. They are a mixture, usually a linear combination, between the primary variable U and its derivative $q = \partial U/\partial n$. A typical case is the relationship $\Gamma_3 : \alpha U + \beta q = \gamma$, where the scalar quantities (α, β, γ) may be constants.

Then, we can generally write that $\Gamma = \Gamma_1 \cup \Gamma_2 \cup \Gamma_3$.

Usually, we distinguish two types of problems, the boundary value problem (steady-state or transient), and the eigenvalue problem.

- In the *boundary value problem*, the arithmetic value of the *BC*s plays a significant role for the numerical solution of the problem. The aforementioned *BC*s may be independent (elliptic problem) or time dependent (parabolic, hyperbolic problems). Typical cases of time independent BCs in potential problems are the LAPLACE and POISSON equations, which describe steady-state thermal problems, inviscid fluid flow and electrostatics, among others. Typical cases of time dependent problems are the transient heat transfer (parabolic problem) and the wave propagation (hyperbolic problem). In time dependent problems, in addition to the boundary conditions it is necessary to know the *initial* conditions, $U(x, y; t = 0) = \bar{U}_0(x, y)$, as well.
- In the *eigenvalue problem*, not the value but only the type of boundary conditions (Dirichlet or Neumann) are of importance (Fig. 3.18).

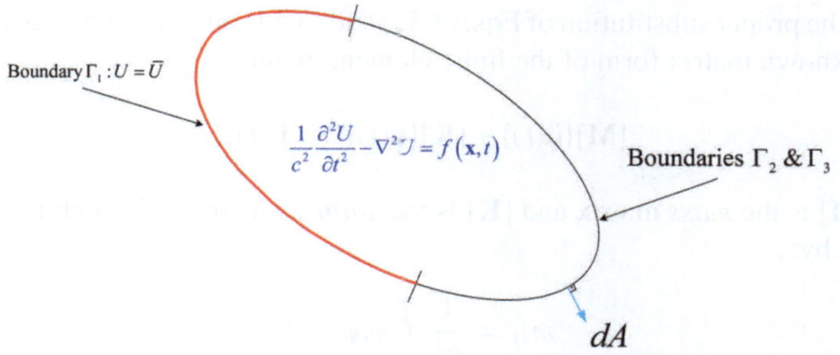

Fig. 3.18 Domain and boundary conditions

A typical PDE of high value is the wave propagation equation:

$$\left(\frac{1}{c^2}\right)\frac{\partial^2 U}{\partial t^2} - \nabla^2 U = f, \tag{3.51}$$

where c is the wave velocity (speed) and t is the time.

The above-mentioned high importance of the above equation is due to the fact it includes both the Laplace and the Poisson equation, merely leaving out the first (time dependent) term. In addition, it covers the eigenvalue acoustic problem as well as the transient wave propagation.

Finally, the mass and stiffness finite element matrices are almost those necessary to perform transient thermal analysis, with the major difference that the second temporal derivative $\partial^2 U/\partial t^2$ in Eq. (3.51) is merely replaced by the first one $\partial U/\partial t$.

Applying the weighted Galerkin procedure using as a weighting function the variation δU gives the equivalent weak form of Eq. (3.51):

$$\int_{\Omega} \delta U\left[\left(\frac{1}{c^2}\right)\frac{\partial^2 U}{\partial t^2} - \nabla^2 U - f\right] \mathrm{d}\Omega = 0. \tag{3.52}$$

Then approximating the problem variable $U(x, y; t)$ in terms of the global shape functions:

$$U(x, y; t) = \sum_{j=1}^{q_e} \varphi_j(x, y) \cdot U_j(t), \tag{3.53}$$

implies that

$$\delta U = \sum_{j=1}^{q_e} \varphi_j(x, y) \cdot \delta U_j(t). \tag{3.54}$$

After the proper substitution of Eqs. (3.53) and (3.54) into Eq. (3.52), one receives the well known matrix form of the finite element method:

$$[\mathbf{M}]\{\ddot{\mathbf{u}}(t)\} + [\mathbf{K}]\{\mathbf{u}(t)\} = \{\mathbf{F}(t)\}, \tag{3.55}$$

where [**M**] is the *mass* matrix and [**K**] is the *stiffness* matrix, of which the elements are given by:

$$m_{ij} = \frac{1}{c^2} \int_{\Omega} \varphi_i \varphi_j \mathrm{d}\Omega \tag{3.56}$$

and

$$k_{ij} = \int_{\Omega} \nabla\varphi_i \nabla\varphi_j \mathrm{d}\Omega, \quad i, j = 1, \ldots, q_e. \tag{3.57}$$

Despite the fact the above matrices are of dimensions $q_e \times q_e$, the standard finite element procedures are directly applicable.

In the sequence, typical potential problems (static and time dependent) will be solved, in order to demonstrate the performance of Coons macroelements.

Application Example 3.1 (Steady-state conduction in a cylindrical wall) This is the first test case on which the natural cubic B-splines based C-element (model M2) was successfully applied since many years ago [12]. Here, minor corrections are made on the previous results and additional original results are presented for two more sets of trial functions, i.e. piecewise-linear (model M1) and Lagrange polynomial (model M3). All models were programmed in a unified MATLAB® environment, without any manual calculation (as was done in the past).

The boundary value problem refers to a uniform long hollow cylinder, subject to a uniform inner surface temperature $T_i = 1000$ °C and a uniform outer surface temperature $T_o = 0$ °C. The cylinder under consideration (with internal radius $R_i = 1$ and outer one $R_o = 32$) is assumed insulated so as to prevent heat flow toward the axis of symmetry (z). According to Carslaw and Jaeger [4], at any radial position ($R_o \le R \le R_i$) the steady state temperature distribution is given by:

$$T(R) = T_i + \left[\frac{(T_o - T_i)}{\ln(R_o/R_i)}\right] \ln(R/R_i). \tag{3.58}$$

The model is shown in Fig. 3.19 and consists of the half cross section, on which a portion is under Neumann type boundary conditions.

Two mesh densities have been tested. The coarse mesh uses 10 radial and 6 circumferential subdivisions (Fig. 3.20a, b), whereas the fine one uses 14 radial and 16 circumferential ones (Fig. 3.20c, d). The same mesh density (per direction) has been applied to both the assemblage of conventional bilinear 4-noded finite elements and the C-element.

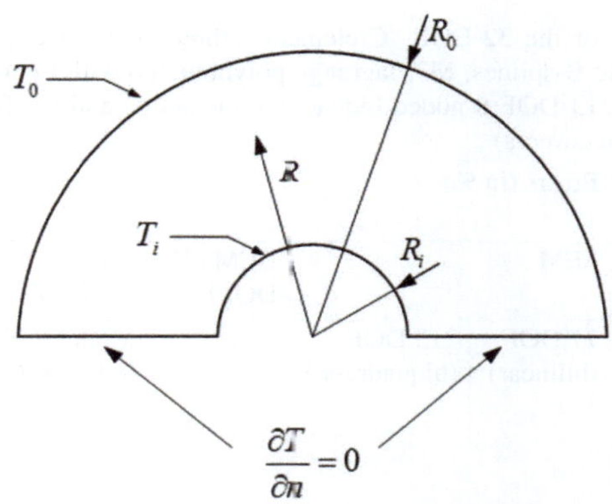

Fig. 3.19 Temperature distribution in a long hollow cylinder with $R_o = 32$, $R_i = 1$ and radial heat flow

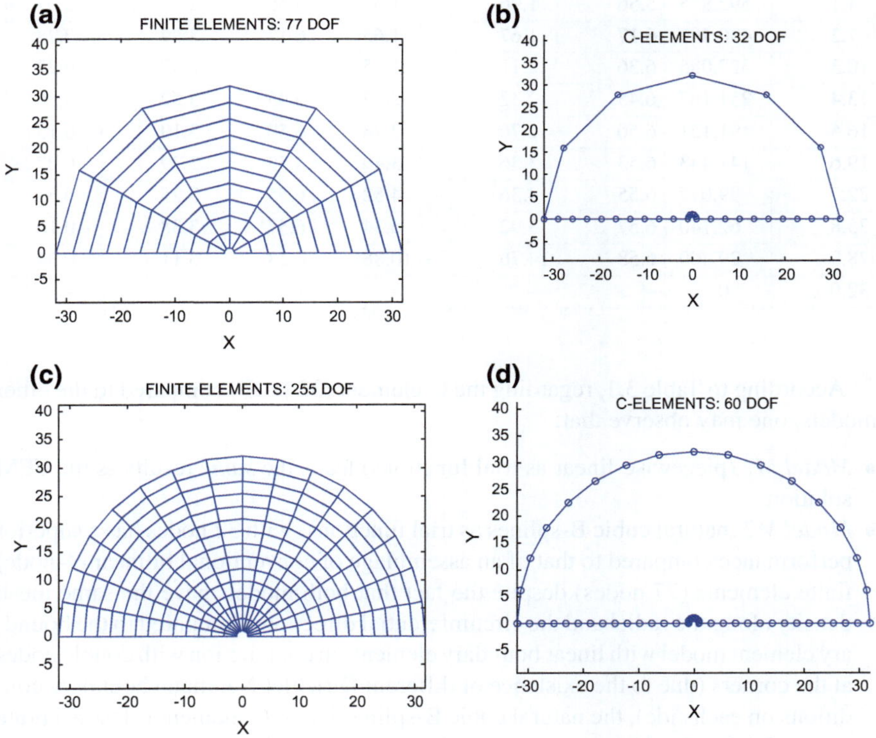

Fig. 3.20 Comparison of the C-element with the conventional FEM of same density: **a** 77-DOF bilinear finite elements (also 213-DOF biquadratic finite elements, both for 60 elements), **b** 32-DOF "C-element", **c** 255-DOF bilinear finite elements, **d** 60-DOF "C-element"

Table 3.1 Comparison of the 32-DOF "C-element" (three alternative models, M1: piecewise-linear, M2: natural cubic B-splines, M3: lagrange polynomials) with the FEM (77-DOF 4-noded bilinear elements, and 213-DOF 8-noded biquadratic elements) and the BEM (32-DOF element with double nodes in the corners)

Radius (R)	Exact solution	Errors (in %)					
		FEM		BEM (32 DOF)	C-element (32 DOF)		
		77 DOF (bilinear)	213 DOF (biquadratic)		Model M1 (2×2 gauss)	Model M2 (2×2 gauss) Kanarachos and Deriziotis [12]	Model M3
1.0	1000.0	–	–	–	–	–	–
4.1	592.875	5.56	1.96	1.37	5.56	2.73	−0.07
7.2	430.400	6.17	1.67	1.65	6.17	3.89	0.28
10.3	327.085	6.36	2.11	1.95	6.36	2.97	−0.22
13.4	251.167	6.45	2.32	2.33	6.45	3.52	0.36
16.5	191.121	6.50	2.76	2.78	6.50	3.19	0.13
19.6	141.443	6.53	3.36	3.49	6.53	3.39	−0.27
22.7	99.075	6.55	4.36	4.52	6.55	3.25	0.53
25.8	62.140	6.57	5.92	6.34	6.57	3.38	−0.16
28.9	29.400	6.58	14.76	10.88	6.58	3.11	0.34
32.0	0	–	–	–	–	–	–

According to Table 3.1, regarding the C-element (32 DOF) compared to the other models, one may observe that:

- *Model M1* (piecewise-linear as trial functions) leads the *same* results as the FEM solution.
- *Model M2* (natural cubic B-splines as trial functions) with 32 nodes has a superior performance compared to that of an assemblage of conventional bilinear (4-node) finite elements (77 nodes) despite the fact that both models have the same mesh density along the radial and the circumferential direction. Compared to the boundary element model with linear boundary elements in conjunction with double nodes at the corners (due to the existence of different Dirichlet/Neumann boundary conditions on each side), the natural cubic B-splines based C-element is less accurate due to the fact that its own shape functions do not satisfy the differential equation as the BEM solution does.
- *Model M3* (Lagrange polynomials as trial functions) is the overall best model. It performs better even than the FEM model using 8-node biquadratic elements of 213 DOFs, as well as better than the BEM model.

Table 3.2 Comparison of the 60-DOF "C-element" (in three formulations) with the FEM (255-DOF, 4-noded bilinear elements)

Radius (R)	Exact solution	Errors (in %)			
		FEM (255 DOF)	C-element (60 DOF)		
			Model M1 (2 × 2 gauss)	Model M2 (2 × 2 gauss) Kanarachos and Deriziotis [12]	Model M3
1.0	1000.0	–	–	–	–
3.2143	663.1003	3.23	3.23	1.32	0.05
5.4285	511.8854	3.67	3.67	2.13	−0.04
7.6428	413.1776	3.82	3.82	1.54	0.03
9.8571	339.7661	3.89	3.89	1.87	0.02
12.0714	281.2951	3.92	3.92	1.69	−0.05
14.2857	232.6997	3.95	3.95	1.78	0.06
16.5000	191.1211	3.97	3.97	1.73	0.01
18.7143	154.7864	3.98	3.98	1.76	−0.06
20.9285	122.5196	3.99	3.99	1.74	0.07
23.1428	93.5009	4.00	3.99	1.75	−0.00
25.3571	67.1359	4.00	4.00	1.74	−0.06
27.5714	42.9795	4.00	4.00	1.78	0.13
29.7857	20.6902	4.01	4.01	1.62	−0.17
32.0	0	–	–	–	–

Interestingly, the superior performance of the C-element compared to that of an assemblage of conventional finite elements is also observed in computational models with more degrees of freedom. Referring to the models shown in Fig. 3.20c, d, Table 3.2 shows the results obtained using (i) a 60-DOF C-element (Fig. 3.20d), and (ii) a 255-DOF FEM model (Fig. 3.20c), which do seem to sustain the above statements. It is remarkable that, although the shape functions are polynomials of high degree (sixteenth in ξ and fourteenth in η), the numerical solution is stable and excellently outperforms, as the last column of Table 3.2 depicts.

Example 3.2 (Rectangular adiabatic plate [thermal analysis]) This is the second test case where the C-element was applied [12]. Here, minor corrections are made and many new results are presented. A thin rectangular plate, free of heat sources, insulated at the top and bottom surfaces is considered (Fig. 3.21a). The dimensions of the entire plate are: $L = 2a = 6$ and $b = 12$ in the horizontal and the vertical direction, respectively. The analytical solution of the problem is given by (see, Kreith [15]):

$$T(x, y) = T_m[\sinh(\pi y/L)/\sinh(\pi b/L)]\cos(\pi x/L). \quad (3.59)$$

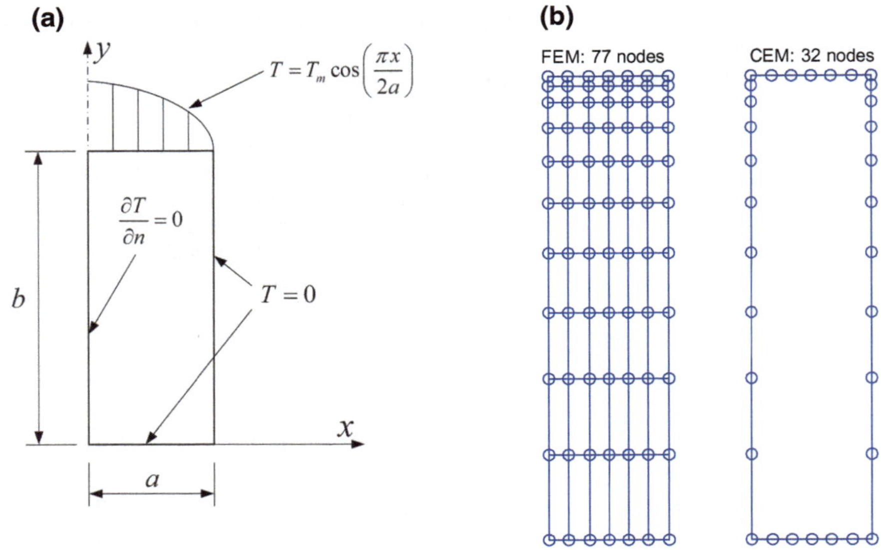

Fig. 3.21 **a** Temperature distribution in a rectangular adiabatic plate with $a = 3.0$ and $b = 12.0$, solved by the following computational models: **b** FEM (77-DOF, 4-noded bilinear elements) as well as C-element (32-DOF) and the BEM (32-DOF element with double nodes in the corners)

The maximum temperature was taken equal to $T_m = 100$ °C. Due to the symmetry of the problem with respect to the y-axis, only half the domain (of dimensions $a \times b = 3 \times 12$) was analyzed.

The mesh was constructed using six uniform subdivisions in the x-direction, whereas ten unequal segments of which the width follows an arithmetic progression were used in the y-direction. Given the same mesh (shown in Fig. 3.21b), we distinguish two cases, the first of the classical usual linear blending functions, whereas the second of trigonometrical ones.

A. *Considering the usual Linear Blending functions*

Using the standard linear blending functions according to Eq. (3.6), the obtained numerical results at the nodal points along the axis of symmetry are shown in Table 3.3. One may observe that all three models (M1, M2, and M3) are inferior to the FEM solution for the same mesh density (on the boundary). The smaller error appears in the natural cubic B-splines element (introduced by Kanarachos and Deriziotis [12]) but its superiority is not that critical. It is noted that the different results between the M2 model of this book as well as Kanarachos and Deriziotis [12] may be attributed to the way the numerical integration was performed. It is clarified that all three models (M1, M2 and M3) shown in Table 3.3 were based on integration cells (of unequal size in the y-direction) that are determined by the isolines passing through the boundary nodes. Thus, Model M1 (piecewise-linear) requires 2×2 quadrature, whereas model M2 (natural cubic B-splines) needs a 3×3 Gauss integration scheme. In contrast, since the integrand of the stiffness matrix in model M3 (Lagrange polynomials of 6th and tenth degree in x and y, respectively) includes monomials up to the degree $(x^{2\times 5}y^{2\times 10})$, a global Gaussian scheme 6×11 is adequate to handle the current static problem.

The interesting reader may wish to extend the above calculations by testing full B-splines ([9]) which include the missing rotational terms at the two ends of each side (see details in Chap. 7), and not the reduced natural cubic ones (model M2) shown in Table 3.3. The answer is given in the Sect. 3.8 of this chapter.

B. *Considering cosine-like blending functions*

The poor results tabulated in the abovementioned Table 3.3 was the motivation that a second option of Coons formulation was proposed by Provatidis [25]. This choice is influenced by the particular boundary conditions, which suggest linear blending functions in y-direction (as usual) but *cosinusoidal* ones in x-direction:

$$\left.\begin{array}{l} E_0(\xi) = \cos\left(\frac{\pi\xi}{2}\right), \quad E_1(\xi) = 1 - \cos\left(\frac{\pi\xi}{2}\right) \\ E_0(\eta) = 1 - \eta, \quad E_1(\eta) = \eta \end{array}\right\} \tag{3.60}$$

The partially cosinusoidal blending functions of Eq. (3.60) *dramatically* reduce the numerical error as shown in Table 3.4 (bottom area, where the mean average error

Table 3.3 Comparison of the 32-DOF C-element (three alternative models) with the FEM (77-DOF, 4-nodes bilinear elements)

y coordinate (at $x = 0$)	Exact solution	Errors (in %)			
		FEM (77 DOF)	C-element (32 DOF) with trial functions		
			M1 (piecewise-linear)	M2 (natural cubic B-splines)	M3 (lagrange polynomials)
12.0	100.0	–	–	–	–
11.781	89.2034	−0.09	−1.24	−1.15	−1.17
11.345	70.9852	−0.32	−3.73	−3.45	−3.46
10.691	50.5861	−0.76	−7.45	−6.80	−6.79
9.818	31.9046	−1.51	−12.41	−11.04	−11.06
8.727	18.0200	−2.73	−18.56	−16.10	−16.13
7.418	9.0768	−4.60	−25.81	−21.76	−21.81
5.891	4.0729	−7.30	−33.99	−27.85	−27.92
4.145	1.6152	−10.91	−42.68	−34.01	−34.13
2.182	0.5257	−14.96	−50.66	−39.46	−39.63
0	0	–	–	–	–

Table 3.4 Temperature distribution in a rectangular adiabatic plate with $L = 6$ ($L = 2a$) and $b = 12$. Comparison between the 32-node CP-element (three formulations) with the FEM(77-node four-node bilinear elements) and the BEM (32-nodes)

y	Exact solution	Errors in %					
		FEM 77 DOF	BEM 32 DOF	Macro-element 32-DOF (Kanarachos and Deriziotis [12], p. 105)	CP-element 32-DOF *Cosine-like*		
				Natural B-splines	M1 (piecewise-linear)	M2 (Provatidis [25])	M3 (lagrange polynomials)
12.000	100.0000	–	–	–	–	–	–
11.781	89.2034	0.09	0.32	0.34	−0.00	0.02	0.00
11.345	70.9852	0.32	0.40	0.84	−0.03	0.02	−0.00
10.691	50.3861	0.75	0.49	3.08	−0.12	0.04	0.00
9.818	31.9046	1.51	0.68	6.33	−0.28	0.10	0.00
8.727	18.0200	2.73	1.04	10.00	−0.49	0.19	−0.00
7.418	9.0768	4.59	1.77	13.80	−0.58	0.37	0.00
5.891	4.0729	7.29	3.25	16.58	−0.06	0.69	−0.00
4.145	1.6152	10.91	6.30	16.65	1.88	1.26	0.00
2.182	0.5257	14.97	12.95	10.00	5.57	2.06	−0.01
0	0	–	–	–	–	–	–
Mean average absolute error (%)		4.80	3.02	8.62	**1.00**	**0.52**	**0.00**

Bold numbers correspond to the findings using the novel CP-element, and merely emphasize its superiority

varies between 0.00 and 1.00%). In particular, model M3 (Lagrange polynomials) appears an extremely low error.

Example 3.3 (Rectangular Acoustic Cavity) This third example is concerned with a linear (acoustic) wave propagation problem in a rectangular cavity of dimensions $a = 2.5$ in the x-direction and $b = 1.1$ in the y-direction, under Neumann boundary conditions (Fig. 3.22, top). The exact eigenvalues of the cavity ω^2 (see, Courant and Hilbert [8], p. 301) are:

$$\omega^2 = \pi^2 c^2 \left[(m/a)^2 + (n/b)^2 \right], \quad m,\ n = 0,\ 1,\ 2,\ \ldots \tag{3.61}$$

and the corresponding eigenforms are:

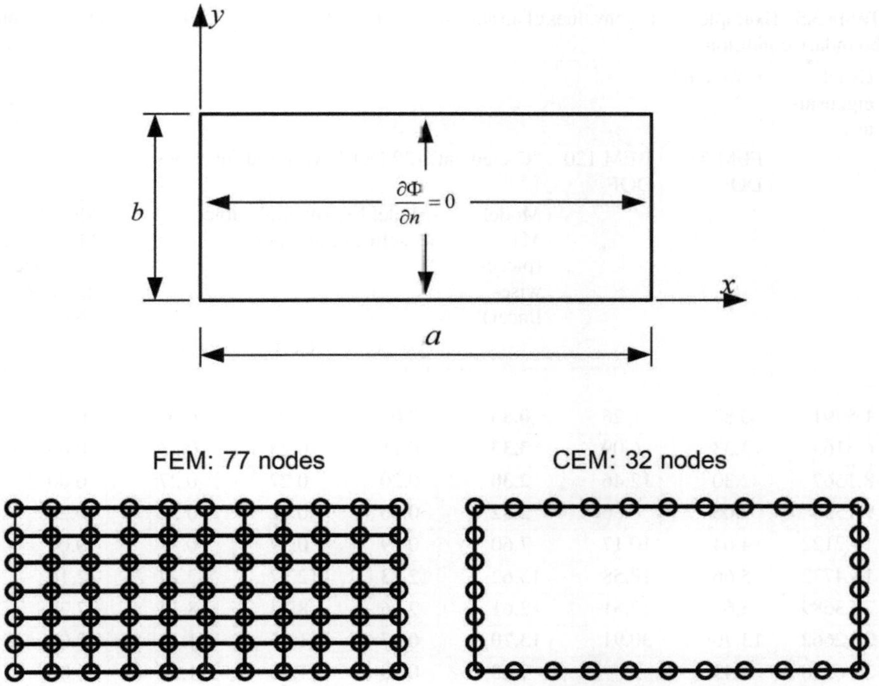

Fig. 3.22 Acoustic cavity with $c = 1$, $a = 2.5$, $b = 1.1$ and Neumann boundary conditions. Comparison of the 32-DOF C-element with the FEM (77-DOF, 4-node bilinear elements) and the BEM method (120-DOF, not shown)

$$u_{mn}(x, y) = \cos(m\pi x/a)\cos(n\pi y/b), \quad m, \; n = 0, \; 1, \; 2, \; \ldots \tag{3.62}$$

with c denoting the sound velocity.

In this eigenvalue problem, the "C-element" with 32 DOFs is compared to (i) the Nardini-Brebbia formulation [18, 19] using an adequately fine mesh 120 constant boundary elements DOFs (an analysis taken from [26], and (ii) the FEM using 60 four-node bilinear elements with a total of 77 DOFs (having the same mesh density, i.e. subdivisions per direction, with the C-element). Some of these meshes are shown in the bottom of Fig. 3.22. In all cases, the eigenvalue problem is formulated as follows: $\det(\mathbf{K} - \omega^2\mathbf{M}) = 0$, where $\mathbf{M}$ is the mass matrix and $\mathbf{K}$ is the stiffness matrix. In the BEM eigensolution the stiffness matrix $\mathbf{K}$ is replaced by the first influence matrix $\mathbf{H}$ (see definitions and further explanations in Chap. 12).

From Table 3.5 it is observed that the performance of the C-element in this case is superior to that of the mass matrix based BEM (Nardini-Brebbia formulation) as well, although almost one-fourth (32:120) of the DOFs are used. The reason for this poor BEM solution in the dynamic case, is certainly associated with the approximation accuracy of both the inertia and stiffness terms of the formulation of Nardini-Brebbia

Table 3.5 Example 3.3: Eigenvalues of an acoustic cavity with $c = 1, a = 2.5, b = 1.1$ and Neumann boundary conditions

Exact eigenvalues	Errors in %						
	FEM 77 DOF	BEM 120 DOF	"C-element" (32 DOF) with trial functions:				
			Model M1 (piecewise-linear)	Model M2 (natural cubic B-spline) gauss points			Model M3 (lagrange polynomials)
				2×2	3×3	4×4	
–	–	–	–	–	–	–	–
1.5791	0.83	1.26	**0.83**	0.04	0.05	0.05	0.00
6.3165	3.33	6.09	**3.33**	0.15	0.23	0.23	0.00
8.1567	2.30	12.46	**2.30**	0.20	0.27	0.27	0.00
9.7358	2.06	6.86	2.32	0.46	0.52	0.52	0.28
14.2122	4.64	10.17	7.60	0.29	0.57	0.57	0.00
14.4732	5.66	18.58	13.62	12.23	12.27	12.27	12.16
22.3689	5.67	12.31	12.61	7.96	8.14	8.14	7.78
25.2662	13.70	30.91	**13.70**	0.37	1.16	1.15	0.00
32.6268	9.43	–	**9.43**	0.70	1.20	1.20	0.00

Comparison of the 32-DOF C-element with the FEM (77-DOF, 4-nodes bilinear elements) and the BEM method (120-DOF)
Bold values refers to novel results using the M1-model compared with traditional FEM. Bold values at those figures where the M1-model coincides with the FEM results on a left column

[18, 19] on which a dissertation has been conducted [26, 13], while further research is [23, 27].

Furthermore, one may observe that:

- Model M3 (Lagrange polynomials as trial functions) outperforms and has an excellent behavior.
- Model M1 (piecewise-linear trial functions) gives *some* eigenvalues that coincide with the FEM solution (the shadowed ones in the fourth column of Table 3.5).

3.7.2 Application to Plane Elastostatics and Eigenanalysis

The first public reports using the same approximation for *both* the geometry and the displacement field were released from the General Motors Research Laboratories (Warren, Michigan, USA) but they did not refer to the accuracy achieved by the transfinite elements ([5, 11, 6] in GMSOLID). To the best of our knowledge, in chronological order the first numerical results concerning boundary-only Coons macroelements in stress analysis were published by Kanarachos et al. [14] as well as Provatidis and Kanarachos [28]. The research continued with axisymmetric problems [29], adaptive schemes [10], three-dimensional elasticity [21], whereas (after many revisions) a detailed report in 2D stress analysis was published with delay in 2012 [22].

The "C-element" based eigenvalue analysis was presented for the first time by Kanarachos and Deriziotis [12] in acoustics, and further findings were reported by Provatidis [24, 30], among others. Concerning eigenvalue extraction of elastic bodies, the first published work is by Kanarachos et al. [14], a paper that was orally presented by the author of this book in ECCM'99. Extended work and several formulations in structural eigenanalysis were published by Provatidis [31, 32], among others.

3.7.2.1 Governing Equations and Matrix Formulation

As usual (e.g. Zienkiewicz and Taylor [40]), the governing equations of elasticity are written in compact form as:

$$\text{Stress-equilibrium: } \mathbf{D}^{\mathrm{T}}\boldsymbol{\sigma} + \mathbf{b} = 0 \tag{3.63}$$

$$\text{Hooke's law: } \boldsymbol{\sigma} = \mathbf{E}\boldsymbol{\varepsilon} \tag{3.64}$$

$$\text{Strain vector: } \boldsymbol{\varepsilon} = \mathbf{D}\mathbf{u} \tag{3.65}$$

In Eq. (3.63) the operator $\mathbf{D}$ for the 2-D problem is given by

$$\mathbf{D} = \begin{bmatrix} \frac{\partial}{\partial x} & 0 \\ 0 & \frac{\partial}{\partial y} \\ \frac{\partial}{\partial y} & \frac{\partial}{\partial x} \end{bmatrix}, \tag{3.66}$$

and it is applied to the stress vector $\{\boldsymbol{\sigma}\} = \left[\sigma_x, \sigma_y, \tau_{xy}\right]^{\mathrm{T}}$, whereas $\{\mathbf{b}\} = \left[b_x, b_y\right]^{\mathrm{T}}$ is the vector of body forces.

In Eq. (3.64), $\mathbf{E}$ denotes the elasticity matrix, which for 2-D problems (either plane stress or plane strain isotropic case) has the general form:

$$\mathbf{E} = \begin{bmatrix} E_{11} & E_{12} & 0 \\ E_{12} & E_{22} & 0 \\ 0 & 0 & E_{33} \end{bmatrix}, \tag{3.67}$$

whereas $\{\boldsymbol{\varepsilon}\} = \left[\varepsilon_x, \varepsilon_y, \gamma_{xy}\right]^{\mathrm{T}}$ is the vector of strains.

Now, the global shape functions of the previous section relate the displacement at every point inside the macroelement with the nodal values as follows:

$$\mathbf{u}(x, y) = \mathbf{N}(x, y)\mathbf{a} \tag{3.68}$$

where $\mathbf{a}$ is a vector denoting the nodal displacements at the boundary of the Coons macroelement. Substituting Eqs. (3.68) into (3.65) one obtains (with $\mathbf{B} \equiv \mathbf{DN}$):

$$\boldsymbol{\varepsilon} = \mathbf{DNa} = \mathbf{Ba} \tag{3.69}$$

The application of the principle of virtual work (of inertial, elastic and external forces):

$$\delta W_m + \delta W_k = \delta W_{\text{ext}}, \tag{3.70}$$

with

$$\delta W_m = \int_\Omega \delta \mathbf{u}^{\mathrm{T}}(\rho \ddot{\mathbf{u}}) \mathrm{d}\Omega = \delta \mathbf{a}^{\mathrm{T}} \left(\int_\Omega \mathbf{N}^{\mathrm{T}} \rho \, \mathbf{N} \mathrm{d}\Omega \right) \ddot{\mathbf{a}}, \tag{3.71}$$

$$\delta W_k = \int_\Omega \delta \boldsymbol{\varepsilon}^{\mathrm{T}} \boldsymbol{\sigma} \, \mathrm{d}\Omega = \delta \mathbf{a}^{\mathrm{T}} \left(\int_\Omega \mathbf{B}^{\mathrm{T}} \mathbf{E} \, \mathbf{B} \, \mathrm{d}\Omega \right) \mathbf{a}, \tag{3.72}$$

and

$$\delta W_{\text{ext}} = \int_\Omega \delta \mathbf{u}^{\mathrm{T}} \mathbf{f} \mathrm{d}\Omega = \delta \mathbf{a}^{\mathrm{T}} \int_\Gamma \mathbf{N}^{\mathrm{T}} \boldsymbol{\sigma}_\Gamma \mathrm{d}\Gamma + \delta \mathbf{a}^{\mathrm{T}} \int_\Omega \mathbf{N}^{\mathrm{T}} \mathbf{b} \, \mathrm{d}\Omega, \tag{3.73}$$

leads to the matrix formulation:

$$[\mathbf{M}]\{\ddot{\mathbf{a}}(t)\} + [\mathbf{K}]\{\mathbf{a}(t)\} = \{\mathbf{f}(t)\}, \tag{3.74}$$

where the mass matrix $[\mathbf{M}]$ and the stiffness matrix, $[\mathbf{K}]$, are given by

$$[\mathbf{M}] = \int_\Omega \mathbf{N}^{\mathrm{T}} \rho \, \mathbf{N} \, \mathrm{d}\Omega, \tag{3.75}$$

and

$$[\mathbf{K}] = \int_\Omega \mathbf{B}^{\mathrm{T}} \mathbf{E} \mathbf{B} \, \mathrm{d}\Omega, \tag{3.76}$$

whereas the vector of the external force, $\{\mathbf{f}\}$, by

$$\{\mathbf{f}\} = \int_\Gamma \mathbf{N}^{\mathrm{T}} \boldsymbol{\sigma}_\Gamma \mathrm{d}\Gamma + \int_\Omega \mathbf{N}^{\mathrm{T}} \mathbf{b} \, \mathrm{d}\Omega \tag{3.77}$$

Given the numbers (q_1, q_2, q_3, q_4) of nodal points along each of the four edges (see, Fig. 3.3), the corresponding numbers of nodal segments along them are

$$n_1 = q_1 - 1, n_2 = q_2 - 1, n_3 = q_3 - 1, \text{and} \quad n_4 = q_4 - 1. \tag{3.78}$$

Therefore, the total number of nodes of the C-macroelement is $q_e = (n_1 + n_2 + n_3 + n_4)$ or:

$$q_e = (q_1 + q_2 + q_3 + q_4) - 4 \tag{3.79}$$

Obviously, for the abovementioned number of nodal points along the boundary of the macroelement, the matrix [**K**] (of size $2q_e \times 2q_e$) is both symmetric and, in general, fully occupied (except in case of Model C1, which is defined below).

3.7.2.2 Models Applied

Concerning boundary-only Coons-patch macroelements (notated by "C-element" or "CPM"), again three different formulations are studied below as follows:

Model C1 (or "M1"): C-element using piecewise-linear interpolation along each side.

Model C2 (or "M2"): C-element using natural cubic B-splines interpolation along each side.

Model C3 (or "M3"): C-element using Lagrange interpolation along each side.

Besides, one more formulation is occasionally applied for comparison purposes as follows:

Model FEM: Conventional Finite Element formulation using four-node bilinear elements.

3.7.2.3 Numerical Integration

All three above-mentioned CPM models for a single "C-element" (C1, C2 and C3) can be basically treated in the same manner. Considering the aforementioned n_1, n_2, n_3 and n_4 segments along the sides *AB*, *BC*, *CD* and *DA*, respectively, the maximum number of subdivisions per each one of the opposite couples, i.e., $n = \max(n_1, n_3)$ and $m = \max(n_2, n_4)$, is determined. In the general case where $n_1 \neq n_3$ or/and $n_2 \neq n_4$, a uniform mesh of $n \times m$ integration cells could be automatically used. Otherwise, the domain is divided into $n \times m$ integration cells by $\xi =$ constant and $\eta =$ constant families of the introduced DOFs. In each integration cell, a standard Gaussian quadrature such as 2×2 or 3×3 can be applied.

Although the above integration scheme always works, improvements are possible, as follows:

(1) The piecewise-linear global shape functions (model M1 or C1) are characterized by *compact support* and are reduced to small strips being vertical to the side where the associated node belongs. In more detail, if the ith node lies between the corners A and B, the corresponding shape function $N_i(\xi, \eta)$ is nonzero only within the strip given by $\xi_{i-1} \leq \xi \leq \xi_{i+1}$. Due to the linear approximation, a representative term such as $\partial N_i/\partial \xi$ [involved in matrix **B**, cf. Equation (3.69)] is a constant. Therefore, in order to calculate the corresponding elements, $k_{IJ}(I = 2i - 1, 2i$ and $J = 2j - 1, 2j)$, of the stiffness matrix, in case of a *rectangular* domain it is only necessary to use *one* Gauss point within the former strip (otherwise, a 2×2 Gaussian quadrature is adequate).

(2) The natural cubic B-splines (model M2 or C2) are continuous functions only within two successive nodes. By recalling that n_g Gauss points are necessary to accurately calculate polynomial integrands of maximum $2n_g - 1$ degree, the accurate calculation of the stiffness matrix (it consists of fourth degree polynomials, e.g., $k_{2i,2j}$ includes the product $\partial N_i/\partial \xi \cdot \partial N_j/\partial \xi$ and each term is a polynomial of second degree), would require a 3×3 Gaussian quadrature per integration cell. Nevertheless, many numerical tests have shown that no practical improvement is achieved between a reduced 2×2 and an accurate 3×3 Gaussian quadrature. Therefore, model C2 practically requires 4 nm Gauss points for the calculation of each stiffness matrix element.

(3) The Lagrange-polynomial interpolation (model M3 or C3) leads to global shape functions that are of nth degree in ξ and mth degree in η direction. Consequently, the integrands involved in the *mass* matrix are of $(3n - 1)$-th degree in ξ and $(3m - 1)$-th degree in η; in more details, the terms N_i, N_j and $\det \mathbf{J}$ are of degree n, n and $(n - 1)$, respectively. Also, the integrands involved in the *stiffness* matrix are polynomials of final degree $3(n - 1)$ in ξ and degree $3(m - 1)$ in η direction. Therefore, the accurate numerical integration to service both matrices (**M, K**) would require $n_{g,\xi} = \text{ceil}(\frac{3}{2}n)$ and $n_{g,\eta} = \text{ceil}(\frac{3}{2}m)$ Gauss points in ξ and η directions, respectively.

3.7.2.4 Numerical Results

In order to assess the accuracy of the proposed procedure in 2D elasticity, below some two-dimensional test cases will be worked out using a single C-element (or a very small number of them) and in some cases the conventional FEM technique as well as closed form analytical expressions, for comparison purposes.

Error norms refer to the L_2 norm of the displacement error:

$$L_2 = \left\| \left(\mathbf{u}^h - \mathbf{u}_{\text{exact}}\right) \right\| = \left\{ \int_{\Omega} \left(\mathbf{u}^h - \mathbf{u}_{\text{exact}}\right)^{\mathrm{T}} \left(\mathbf{u}^h - \mathbf{u}_{\text{exact}}\right) \mathrm{d}\Omega \right\}^{1/2}, \tag{3.80}$$

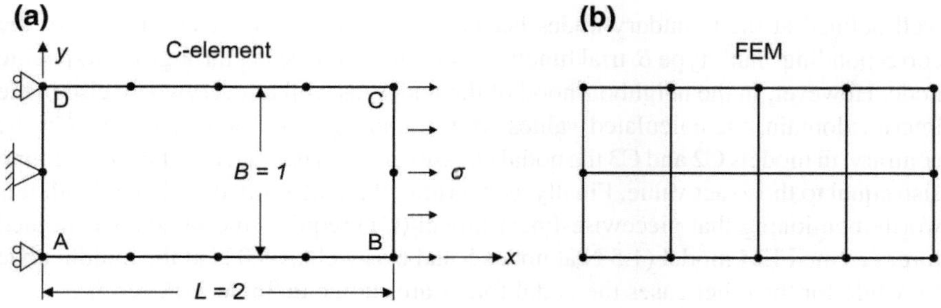

Fig. 3.23 Two-dimensional bar of dimensions 2 m × 1 m subject to pure tension (σ =6 N/m^2) analyzed using **a** One Coons macroelement (12 nodes) and **b** Eight 4-node finite elements (15 nodes)

where $\mathbf{u}^h$ is the numerical solution, $\mathbf{u}_{\text{exact}}$ is the analytical solution, and Ω is the domain.

Besides, two additional normalized error norms (in %) are defined, as follows:

$$\left.\begin{aligned} L_u &= \frac{L_2}{\left\{\int_\Omega \mathbf{u}_{\text{exact}}^{\mathrm{T}}\cdot\mathbf{u}_{\text{exact}}\,\mathrm{d}\Omega\right\}^{1/2}} \times 100\% \\ L_\sigma &= \frac{\left\{\int_\Omega (\boldsymbol{\sigma}^h-\boldsymbol{\sigma}_{\text{exact}})^{\mathrm{T}}\cdot(\boldsymbol{\sigma}^h-\boldsymbol{\sigma}_{\text{exact}})\,\mathrm{d}\Omega\right\}^{1/2}}{\left\{\int_\Omega \boldsymbol{\sigma}_{\text{exact}}^{\mathrm{T}}\cdot\boldsymbol{\sigma}_{\text{exact}}\,\mathrm{d}\Omega\right\}^{1/2}} \times 100\% \end{aligned}\right\} \tag{3.81}$$

where $\boldsymbol{\sigma}^h$ and $\boldsymbol{\sigma}_{\text{exact}}$ are the approximate and the exact stress vectors, respectively.

Example 3.4 (Beam in tension) This problem consists of a short two-dimensional beam of length $L = 2$ and height $B = 1$, shown in Fig. 3.23, with Young's modulus $E = 1000$ and Poisson's ratio $\nu = 0.3$. Boundary conditions consist of a fully restrained node in the middle of the edge DA and frictionless sliding on the other nodes along the same side. In this way, the situation is one-dimensional but this model problem offers important information for the overall efficiency of the proposed method.

The analytical solution is obviously given by:

$$\left.\begin{aligned} u_x(x,y) &= \tfrac{\sigma}{E}x, & 0 \le x \le L \\ u_y(x,y) &= -\tfrac{\nu\sigma}{E}\left(y - \tfrac{B}{2},\right) & 0 \le y \le B \end{aligned}\right\} \tag{3.82}$$

The Coons macroelement occupies the entire patch and includes twelve nodes, using uniform subdivision along the four sides of the rectangle $ABCD$ as shown in Fig. 3.23a. For comparison, a FEM model with the same number of nodes along the boundary is considered as shown in Fig. 3.23b. In all cases, the nodal displacements were found equal to the exact ones. With respect to the calculated stresses, it is worth mentioning that in case of model C1 (piecewise-linear) the boundary stresses are not well defined at the boundary nodes because of the discontinuous derivative of the corresponding "hat"-type B-trial functions (cf. Fig. 2.4) passing through each specific node. However, in the neighbourhood of the nodes, as well as everywhere else in the interior domain, the calculated values were found equal to the exact ones. On the contrary, in models C2 and C3 the nodal stresses are well defined and they were found also equal to the exact value. Finally, concerning the uniformly distributed load, it is worth mentioning that piecewise-linear model (C1) requires the same concentrated forces as the FEM model (1.5 N at nodes 5 and 6 as well as 3.0 N at the middle node 6), while for the other cases the nodal forces are shown in Table 3.6.

Example 3.5 (Bar under linear body forces) This is practically a one-dimensional example taken from Beissel and Belytschko [2]. Here, it is modified and applied on a rectangular sheet of dimensions $L \times B = 10 \times 1$ under plane stress conditions. The bar with fixed boundaries,

$$\bar{u}_x(0, y) = 0, \quad \bar{u}_x(10, y) = 0, \quad 0 \le y \le B \tag{3.83}$$

is loaded by a linear body force,

$$\mathbf{b} = \begin{Bmatrix} 10x \\ 0 \end{Bmatrix} \tag{3.84}$$

It is easy to validate that the analytical solution is given by

$$\left.\begin{aligned} u_x(x, y) &= \tfrac{10}{6E}\left(-x^3 + 100x\right) \\ u_y(x, y) &= -\tfrac{10\nu y}{6E}\left(-3x^2 + 100\right) \end{aligned}\right\} \tag{3.85}$$

The single Coons macroelement ("C-element") consists of a uniform mesh of eleven nodes along the length and three nodes along the width, which are totally twenty-four nodes as shown in Fig. 3.24a. For comparison, the same problem is also solved using conventional FEM with 33 nodes and 20 four-node finite elements (Fig. 3.24b). In both cases, the number of nodes along the boundary is the same

Table 3.6 Transformation of the uniform stress of magnitude $\sigma = 6$ N/m^2 applied on the straight segment of unit length BC (shown in Fig. 3.23a) into three concentrated forces at the ends (B, C) and the middle. The total force equals to 6 N

Model	Concentrated nodal forces (N)	
	Edge nodes	Middle node
Piecewise-linear (C1)	1.5	3.0
Natural Cubic B-splines (C2)	1.125	3.750
Lagrange (C3)	1.0	4.0
FEM	1.5	3.0

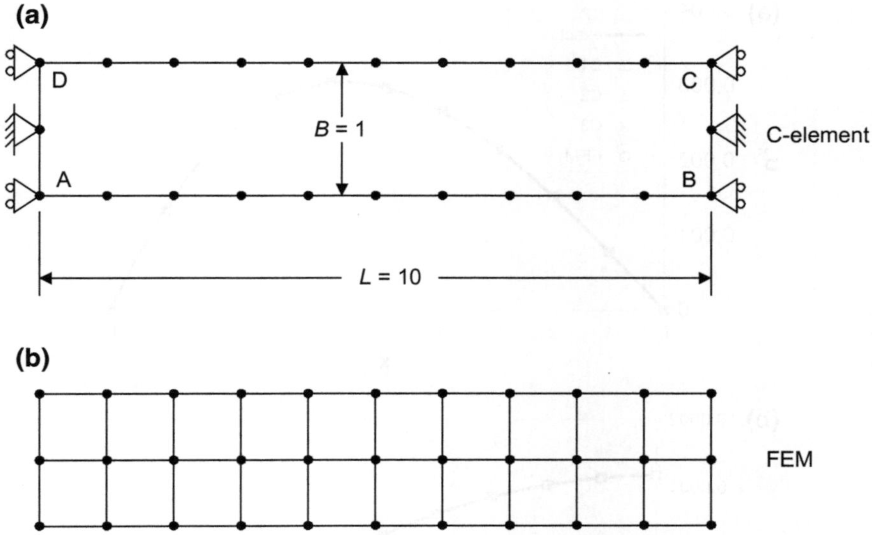

Fig. 3.24 Bar of dimensions 10 × 1 under uniaxial body forces ($b = 10x$) towards x-direction

(twenty-four). Numerical solution is successfully compared with analytical one in Fig. 3.25. One can notice that:

- The piecewise-linear (model C1) leads to results very similar to the FEM solution.
- All four models approximate the u_x-displacement with excellent accuracy.
- Piecewise-linear (model C1), as well as FEM solution, appear a significant error of the vertical displacement, u_y, at the end $x = 10$, that is −10.4 and −9.7%, respectively.
- Lagrange interpolation (model C3), when applied using a 2 × 2 quadrature scheme within each of the 20 integration cells (80 integration points in total), appears an error of +15.6% in the vertical displacement, u_y, at the end $x = 0$, as shown in Fig. 3.25b. In contrast, by using a *global* Gaussian 10 × 2 quadrature applied to the entire domain (20 integration points in total), the error reduces to only 0.24%.
- Cubic B-splines interpolation (model C2) appears an error in the vertical displacement, u_y, of only 0.13% at the end $x = 0$ and −5.94% at $x = 10$.

Example 3.6 (Patch tests) Two patch tests will be performed below. They refer to a square domain of dimensions 2 × 2, which is uniformly divided into sixteen nodes, in total, as shown in Fig. 3.26. The test is performed at twenty internal points, the same with those in the paper of Beissel and Belytschko [2].

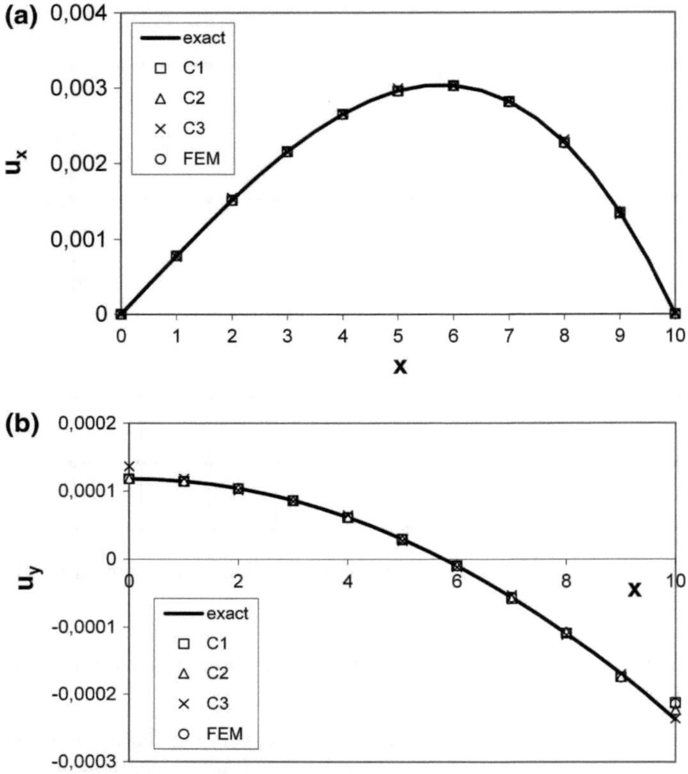

Fig. 3.25 Calculated displacements at nodal points towards **a** x- and **b** y-direction

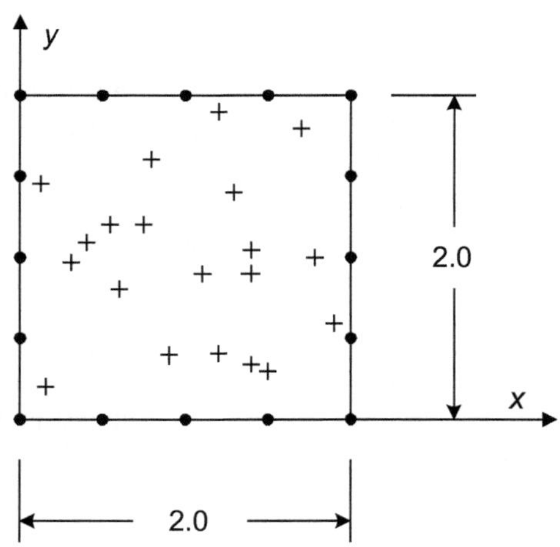

Fig. 3.26 Square of dimensions 2×2 discretized into sixteen boundary nodes. Linear and quadratic patch test is performed at the twenty internal points (+)

(i) ***Linear patch***

The displacement field was described at the boundaries as

$$\bar{\mathbf{u}} = \begin{Bmatrix} \bar{u}_x \\ \bar{u}_y \end{Bmatrix} = \begin{Bmatrix} 0.1 + 0.1x + 0.2y \\ 0.05 + 0.15x + 0.1y \end{Bmatrix} \tag{3.86}$$

The exact solution is then $\mathbf{u} = \bar{\mathbf{u}}$ when no body forces are applied. Since this is a problem of first kind (with Dirichlet boundary conditions), the "C-element" is applicable by performing a simple interpolation according to Coons' interpolation formula. For all three "C-elements" (models: C1, C2 and C3), the calculated displacements at the abovementioned twenty nodes were found to be equal to the exact values.

(ii) ***Quadratic patch***

The displacement field was described at the boundaries as in Beissel and Belytschko [2]:

$$\bar{\mathbf{u}} = \begin{Bmatrix} \bar{u}_x \\ \bar{u}_y \end{Bmatrix} = \begin{Bmatrix} 0.1x^2 + 0.1xy + 0.2y^2 \\ 0.15x^2 + 0.29xy + 0.06y^2 \end{Bmatrix} \tag{3.87}$$

The corresponding exact solution is $\mathbf{u} = \bar{\mathbf{u}}$ when the following body forces are applied:

$$\mathbf{b} = \begin{Bmatrix} -0.20E_{11} - 0.29E_{12} - 0.40E_{33} \\ -0.12E_{11} - 0.10E_{12} - 0.69E_{33} \end{Bmatrix} \tag{3.88}$$

where E_{ij} represents the i, jth component of the matrix of elastic moduli in Eq. (3.67).

Surprisingly, although neither the constitutive equations nor the body forces have been taken into consideration in the "C-element", it is clearly shown in Table 3.7 that the boundary-only Coons interpolation (linear blending) using piecewise-linear (model C1) and cubic B-splines (model C2) leads to acceptable errors excepting the case of the second node at $(x, y) = (0.15, 0.20)$ where the high error is probably due to the corresponding small value. It is noted that while model C1 (piecewise-linear) and model C2 (cubic B-spline) lead errors ($L_2 = 3.05\text{E}{-02}$, $L_u = 1.63\%$) and ($L_2 = 6.91\text{E}{-03}$, $L_u = 0.37\%$), respectively, when utilizing Lagrange polynomials (model C3) the errors completely vanish not only at all twenty internal points shown in Table 3.7 but also everywhere in the domain.

Remark It is also noted that for the same mesh density (25 nodes, 16 bilinear elements) the conventional FEM formulation leads to errors ($L_2 = 7.81\text{E}{-02}$, $L_u = 4.17\%$), which is more than twice of that obtained using the model C1. In other words, for this particular problem, the macroelement based on boundary-only Coons

Table 3.7 Patch test: Calculated displacement components at twenty internal points in the interior of the square shown in Fig. 3.26, using a single "C-element"

Internal point	x	y	Errors (in %) of calculated displacements					
			u_x displacement component			u_y displacement component		
			Model			Model		
			C1	C2	C3	C1	C2	C3
1	0.12	1.45	2.06	0.35	0.00	4.58	1.70	0.00
2	0.15	0.2	130.19	53.77	0.00	79.27	34.04	0.00
3	0.3	0.98	3.44	0.81	0.00	6.12	1.81	0.00
4	0.4	1.1	3.97	0.23	0.00	3.75	0.58	0.00
5	0.55	1.2	3.71	−0.25	0.00	2.16	−0.20	0.00
6	0.6	0.8	7.55	−0.54	0.00	4.15	−0.38	0.00
7	0.75	1.2	4.20	−0.26	0.00	3.01	−0.20	0.00
8	0.8	1.6	1.99	0.21	0.00	1.84	0.01	0.00
9	0.9	0.4	8.05	1.15	0.00	3.57	0.16	0.00
10	1.1	0.9	3.14	−0.09	0.00	1.62	−0.05	0.00
11	1.2	0.4	6.25	0.66	0.00	3.12	0.01	0.00
12	1.2	1.9	1.28	0.32	0.00	1.04	0.06	0.00
13	1.3	1.4	1.88	−0.19	0.00	1.27	−0.12	0.00
14	1.4	0.34	5.58	0.97	0.00	2.11	0.05	0.00
15	1.4	0.9	2.48	−0.14	0.00	1.19	−0.11	0.00
16	1.4	1.05	1.51	−0.09	0.00	0.93	−0.09	0.00
17	1.5	0.3	4.17	1.31	0.00	0.76	0.24	0.00
18	1.7	1.8	1.45	0.54	0.00	0.83	0.28	0.00
19	1.78	1.0	0.89	0.34	0.00	0.88	0.34	0.00
20	1.9	0.6	2.19	0.19	0.00	0.94	0.30	0.00

Results are presented in errors %

interpolation is better than the ensemble of standard elements defined at the interior of the domain.

Example 3.7 (A rectangular cantilever subject to shear force) This is a particular example in elastostatics where the boundary-only formulation is applicable extremely well. Consider a cantilever beam subject to end load as shown in Fig. 3.27a. The following parameters were used: Young's modulus $E = 3.0 \times 107$ kPa; Poisson's ratio $\nu = 0.30$, diameter $D = 12$ m, length $L = 48$ m, shear force $P = 1000$ kN, and plane stress conditions.

The exact solution is given by Timoshenko and Goodier [36] as

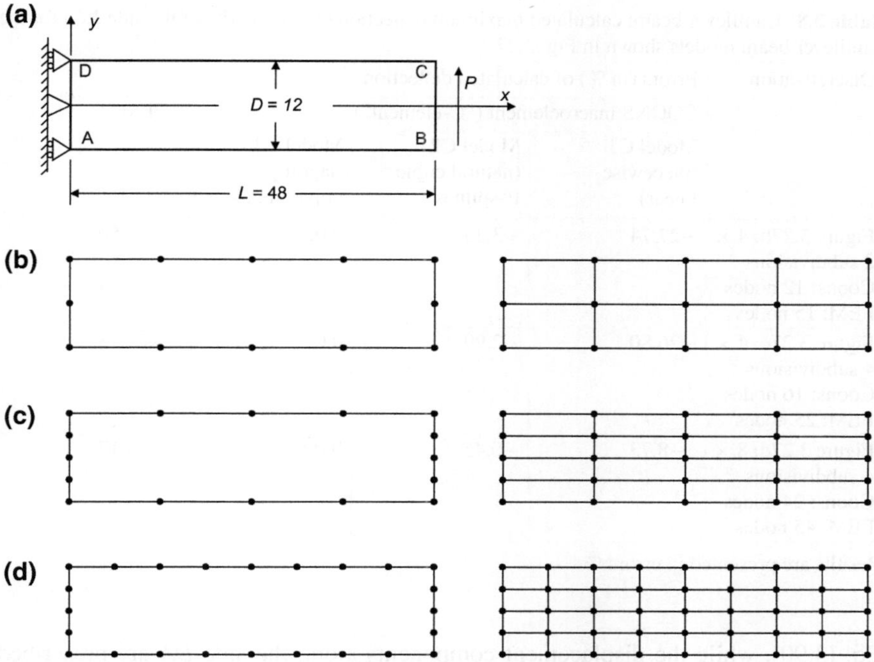

Fig. 3.27 Cantilever beam subject to a parabolically distributed shear force P, and its discretizations

$$\left.\begin{aligned} u_x &= \tfrac{-Py}{6EI}\left[(6L-3x)x+(2+\nu)\left(y^2-\tfrac{D^2}{4}\right)\right] \\ u_y &= \tfrac{P}{6EI}\left[3\nu y^2(L-x)+(4+5\nu)\tfrac{D^2x}{4}+(3L-x)x^2\right] \end{aligned}\right\} \tag{3.89}$$

The stresses are given by

$$\left.\begin{aligned} \sigma_{xx} &= \tfrac{P(L-x)y}{I} \\ \sigma_{yy} &= 0 \\ \sigma_{xy} &= \tfrac{P}{2I}\left(\tfrac{D^2}{4}-y^2\right) \end{aligned}\right\} \tag{3.90}$$

where I is the moment of inertia and for a beam with rectangular cross-section and unit thickness it is given by

$$I = \frac{D^3}{12} \tag{3.91}$$

Clearly, the top (CD) and bottom (AB) edges of the beam are considered to be traction free; the traction along BC is prescribed by the parabolic shear stress of

Table 3.8 Cantilever beam: calculated maximum deflection (at the middle of the side BC) for the cantilever beam models shown in Fig. 3.27

Discretization	Errors (in %) of calculated deflection			
	COONS macroelement ("C-element")			FEM
	Model C1 (piecewise-linear)	Model C2 (natural cubic B-splines)	Model C3 (lagrange polynomials)	
Figure 3.27b: 4 × 2 subdivisions Coons: 12 nodes FEM: 15 nodes	−27.74	−3.33	0.02	−28.50
Figure 3.27c: 4 × 4 subdivisions Coons: 16 nodes FEM: 25 nodes	−26.80	−2.99	0.00	−27.73
Figure 3.27d: 8 × 4 subdivisions Coons: 24 nodes FEM: 45 nodes	−8.73	−0.45	0.00	−9.17

Results are presented in errors %

Eq. (3.90), while the displacement components along the side DA are prescribed according to Eq. (3.89).

For the models shown in Fig. 3.27, the error of the calculated maximum deflection is shown in Table 3.8.

A careful inspection of both displacement components in Eq. (3.89) reveals that the analytical solution includes the monomials/binomials $\{x, x^2, x^3, y, y^2, y^3, xy, xy^2, x^2y\}$, which all belong to the functional space of the Serendipity elements. This fact explains why the boundary-only "C-element" (based on Lagrange polynomials, i.e. model C3) converges toward the exact solution.

Let us now see some more results with different meshes. The long sides (AB, CD) were discretized using four equal subdivisions while the short ones (BC, DA) using only one subdivision. Then, the subdivisions of the short sides progressively increased to 2, 3, 4 and 5 while those along the long sides increased to 8, 12, 16 and 20, respectively, as shown in Fig. 3.28. In this way, the aspect ratio of the boundary segments in the "C"-element was achieved to be unity.

For the same number and location of boundary nodes, FEM solution was also obtained for comparison purposes; obviously the corresponding finite element mesh consists of perfect squares, a discretization also adopted by previous investigators ([2], p. 71). Results are shown in Fig. 3.29 where one may observe that the piecewise-linear formulation (model C1) is of the same quality with the FEM solution (inappreciably better), the natural cubic B-splines (model C2) is significantly better, whereas the Lagrange-polynomial (model C3) rapidly converges and *coincides* with the exact solution when the short side is discretized using *three* subdivisions only.

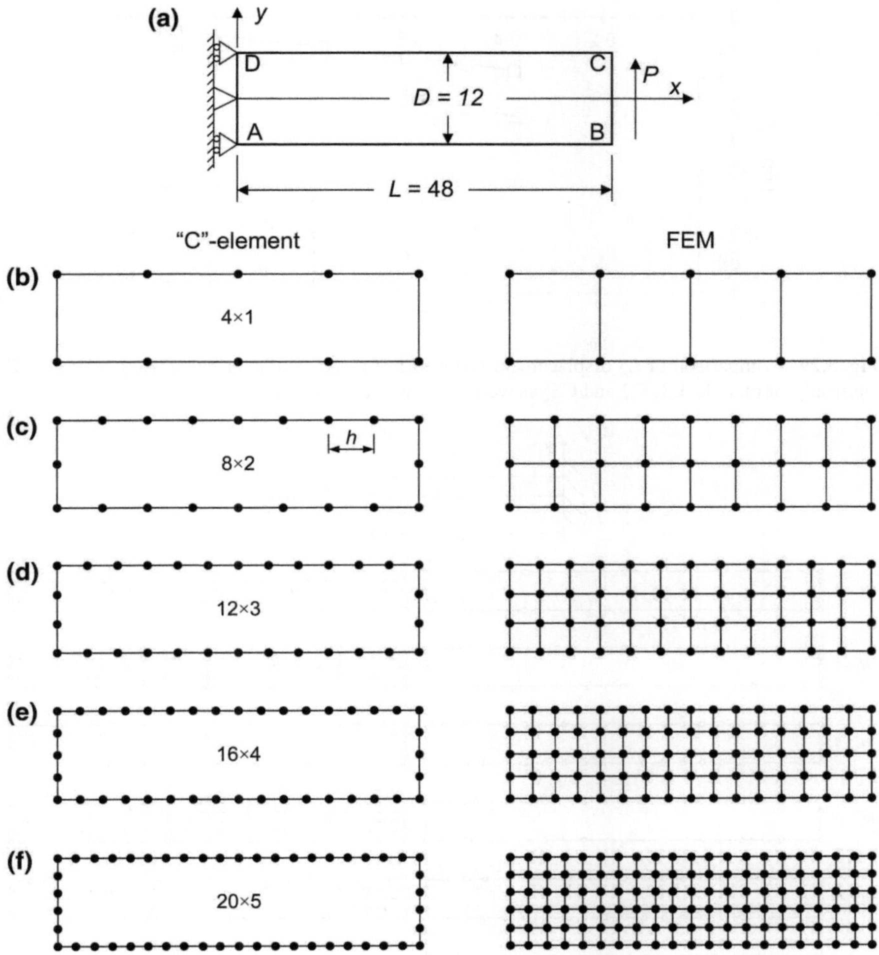

Fig. 3.28 Alternative discretization of the cantilever beam

Finally, all sides were discretized using the *same* number of subdivisions, thus leading to aspect ratios equal to four. It was again found that model C3 achieves the exact solution within the entire domain in case of the critical discretization 3 × 3 (which coincides with the 12-node conventional cubic Serendipity element) and further. This finding is fully justified in the light of the abovementioned monomials $\{x, x^2, x^3, y, y^2, y^3, xy, xy^2, x^2y\}$, as the highest degree in both x and y is three, which equals to the number of subdivisions along each side of the patch.

Therefore, in this example the single COONS macroelement based on boundary-only Coons interpolation, was better than the ensemble of standard finite elements

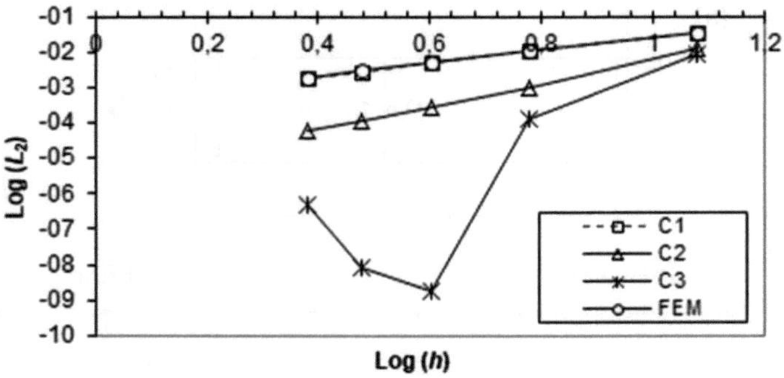

Fig. 3.29 Comparison of L_2 displacement-error norm for the cantilever beam using a single "C-element" (of models: C1, C2 and C3) as well conventional FEM solution

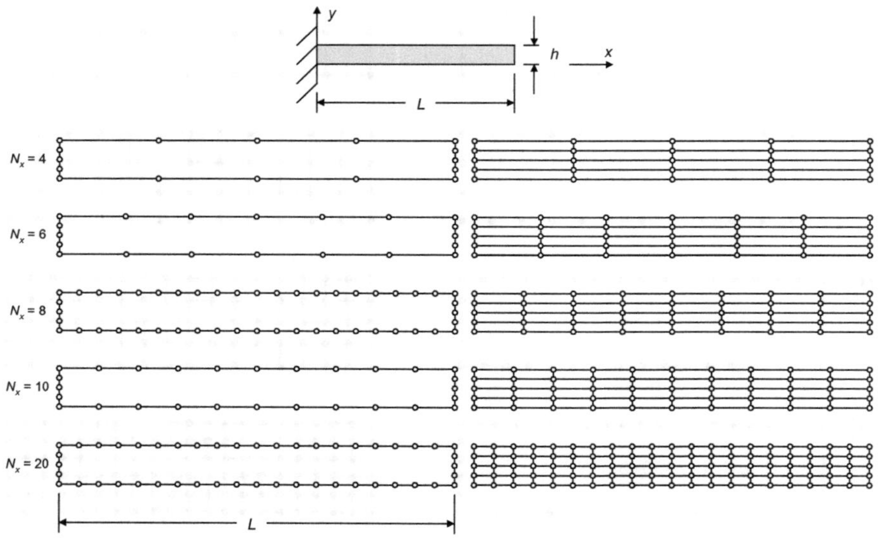

Fig. 3.30 Cantilever beam and several models using C-element (left) and usual FEM (right)

defined at the interior of its domain, because the true solution is best approximated by the Coons' function.

Example 3.8 (Long cantilever beam) This problem is concerned with a long two-dimensional cantilever beam of length $L = 10$ and height $h = 1$. The beam with Young's modulus $E = 10{,}000$, Poisson's ratio $\nu = 0$, mass density $\rho = 1$ and thickness $t = 1$, is analyzed using a single C-element and FEM models of the same mesh density, as shown in Fig. 3.30. The reason for the selection of Poisson's ratio equal to zero is that it abandons contraction in the y-direction, and therefore the plane stress deformation becomes very close to the one-dimensional beam model.

Shortly, the beam undertakes either transverse or longitudinal deformations for which analytical expressions exist (Weaver et al. [38]).

Within this context, given the material and geometric data (modulus of elasticity E, mass density ρ, cross sectional area A, and second moment of inertia I), the theoretical *transverse* eigenvalues of the one-dimensional model are given by:

$$\omega_i^2 = \left(\frac{\lambda_i^{\text{TRANS}}}{L}\right)^4 \cdot \left(\frac{EI}{\rho A}\right), \quad i = 1, 2, \ldots \tag{3.92}$$

The smallest (transverse) eigenvalue is determined for $\lambda_1^{\text{TRANS}} = 1.87510407$, whereas greater constants are also known.

Also, the *longitudinal* eigenvalues of the one-dimensional model are given by:

$$\omega_i^2 = \left(\frac{\lambda_i^{\text{LONG}}}{L}\right)^2 \cdot \left(\frac{E}{\rho}\right), \quad i = 1, 2, \ldots \tag{3.93}$$

where the analytical values are $\lambda_i^{\text{LONG}} = (2i-1)\pi/2, \quad i = 1, 2, 3, \ldots$.

In general, the free vibration of a cantilever beam is the combination of the transverse and the longitudinal vibrations in ascending order.

Following Provatidis [32], first a FEM convergence analysis with conventional 4-node (bilinear) and rectangular finite elements was performed. In more details, beam's length, L, was progressively divided into $n_x = 4, 6, 8, 10, 20, 40, 60, 80, 100, 120, 200, 240$ and 300 uniform segments while the height, h, into $n_y = 4$ segments. In the sequence, by preserving the 300 longitudinal subdivisions, the discretization of the height was increased to $n_y = 8$ segments. The eigenvalues which were obtained using the aforementioned finite element mesh of $n_x \times n_y = 300 \times 8$ subdivisions (2709 nodes, 2400 elements) were considered to be the '*exact*' solution. This choice is supported by the fact that, the relative error of the first eigenvalue in the previously mentioned slightly coarser mesh of 240×4 subdivisions (1205 nodes, 1200 elements) with respect to the 'exact mesh' differs less than 0.06%, while the differences in the next four eigenvalues are very small as shown in Table 3.9. In some later paper regarding collocation (Chap. 11), a still higher number was taken but it has no serious influence to the overall results. Since the FEM-convergence is usually achieved from greater to smaller values, it was judged *fairer* to compare with the 'exact' FE solution than the 1-*D* theoretical one. Major differences between 1-*D* theory and 2-*D* FE solution were found for the fourth and fifth mode. Concerning the fifth eigenvalue, it is clear that the "C-element" based solution is far from the "exact" solution, and therefore the introduction of internal nodes becomes necessary.

In the sequence, the performance of the single "C-element" is compared with the FE solution, in both cases using the *same* number of *boundary* nodes. Therefore, the beam length was progressively divided into n_x=4, 6, 8, 10 and 20 segments while the height constantly into n_y=4 ones, as shown in Fig. 3.30. The convergence behavior of the first four calculated eigenvalues is shown in Fig. 3.31, while similar findings were also obtained for the fifth one. It can be there noticed that the two

Table 3.9 Long cantilever beam (Fig. 3.30)

Analytical eigenvalue ω^2 [s^{-2}] 1D-theory	Mode	Calculated eigenvalues ω^2 [s^{-2}]							
		Fine FEM models with uniform subdivisions $n_x \times n_y$				Coarse "C"-elements (boundary-only nodes) $n_x \times n_y$			
		200 x 4 (1005 nodes)	**240 x 4** (1205 nodes)	**300 x 4** (1505 nodes)	**300 x 8** (2709 nodes) **'Exact'**	**4 x 4** (16 nodes)	**6 x 4** (20 nodes)	**8 x 4** (24 nodes)	**10 x 4** (24 nodes)
1.030	Transverse	1.019	1.018	1.018	**1.018**	1.018	1.018	1.018	1.018
40.460	Transverse	37.239	37.224	37.212	**37.132**	37.980	37.436	37.434	37.434
246.740	Longitudinal	246.741	246.741	246.741	**246.741**	246.740	246.740	246.740	246.740
–	Transverse	263.486	263.369	263.273	**262.088**	619.332	273.747	266.858	266.802
–	Transverse	888.966	888.514	888.144	**881.772**	2232.176	995.885	909.671	907.345

The "C-element", based on Lagrange interpolation (model C3), is compared with the analytical one-dimensional (1-D) eigenvalues, ω^2 (s^{-2}), and converging plane stress FEM models using $n_x \times n_y$ uniform subdivisions in the longitudinal and transverse directions, respectively

Bold values on the column of Exact-Solution are compared by all other numbers in the table

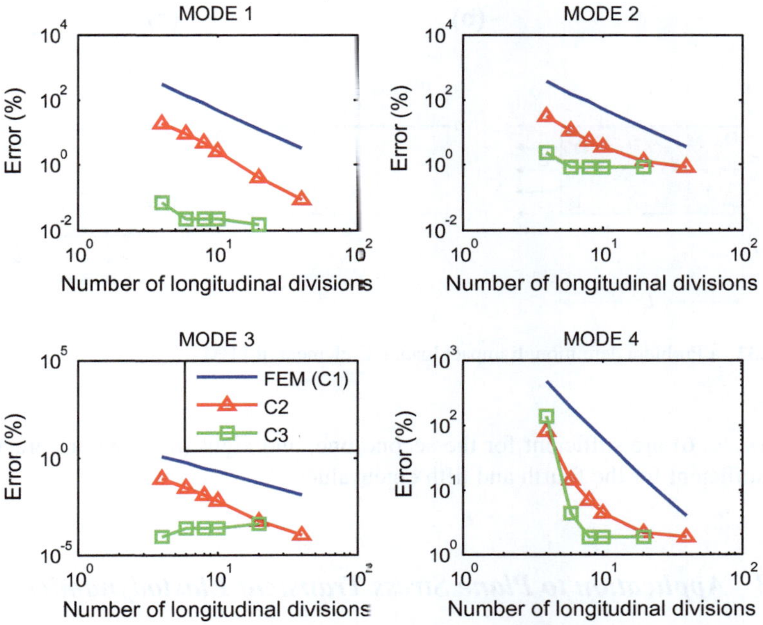

Fig. 3.31 Long cantilever beam: Convergence quality of the first four calculated eigenvalues, ω^2 (s^{-2}), for the "C-element" and FEM models shown in Fig. 3.30 (model C2 = natural cubic B-splines, model C3 = Lagrange polynomials)

"C-element" models (C2: natural cubic B-splines and C3: Lagrange polynomials) are characterized by an excellent convergence towards the 'exact' solution. In contrast, model C1 (piecewise-linear interpolation) practically coincides with the FE solution (not shown as it could not be visually distinguished), despite the fact that it refers to a quite different formulation. Based on these findings, the following remarks can be made:

(1) The one-dimensional character of this example is probably the reason for the numerical coincidence between the single "C-element"-model C1 (piecewise-linear trail functions) and the FE solution. In general, these two formulations are somehow close but they do not coincide.
(2) Concerning the third mode, all four models, i.e. FEM and "C-element" (models C1, C2 and C3) have an excellent behavior due to the fact that is refers to longitudinal vibration and thus only a small number of longitudinal nodes is sufficient to approximate its sinusoidal mode shape.
(3) For the rest modes which refer to transverse vibrations, the model C3 (Lagrange polynomials) outperforms. For this model, one may observe in Fig. 3.31 and Table 3.9 that the discretization of the length into only four subdivisions ($n_x = 4$) is sufficient to practically achieve convergence for the first eigenvalue, six

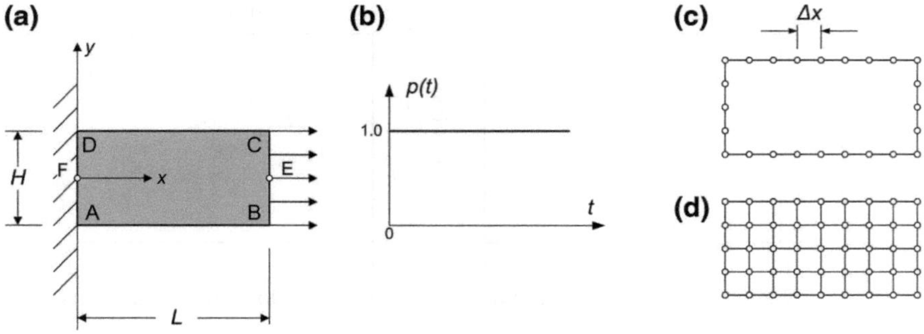

Fig. 3.32 **a** Problem definition, **b** impact load, **c** C-element, **d** FEM

($n_x = 6$) are sufficient for the second one, and eight ($n_x = 8$) are practically sufficient for the fourth and fifth eigenvalues.

3.7.3 Application to Plane Stress Transient Elastodynamics

The performance of C-elements in transient analysis will be studied below.

The first report regarding the performance of a single Coons macroelement in transient elastodynamics is due to Provatidis [31]. At the same time C-elements performed well in scalar wave propagation problems [30], while final 2D reports on elastodynamics were published by Provatidis [35, 33]. The necessity of using internal nodes was definitely shown by Provatidis [32].

In general, similarly with the standard FEM procedure, the solution of the equations of motion, $\mathbf{M}\ddot{\mathbf{u}} + \mathbf{K}\mathbf{u} = f(t)$, is performed using direct or indirect time integration methods. For the definitions, see Bathe [1].

Example 3.9 (Clamped bean under suddenly applied tensile impact) A two-dimensional rod clamped at one end and free at the other, is subjected to a suddenly applied axial tension of a constant value in time, i.e. pressure $p(t) = 1$, as shown in Fig. 3.32a, b. The material properties are $E = 10^4$, $\rho = 1$, $\nu = 0$, length of the rod $L = 4$ and height $H = 2$. For the transient analysis, the model has been discretized with only twenty four boundary nodes along the boundary, as shown in Fig. 3.32c, d.

The analytical solution of this problem may be found in textbooks of mechanics (e.g. Timoshenko and Goodier [36], as well as mathematical physics (e.g., (Pipes and Harvill [20], pp. 494–496)).

The longitudinal wave propagation velocity becomes $c_L = \sqrt{E/\rho} = 100$ m/s. Based on the latter quantity, the analytical expressions of the axial displacements are:

$$\left.\begin{aligned} U(L,t) &= \tfrac{pc_L}{E}t, \quad 0 \le t \le \tfrac{2L}{c_L} \equiv T \\ U(L,t) &= \tfrac{pc_L}{E}t - \frac{2p\,c_L\left(t-\frac{2L}{c_L}\right)}{E}, \quad \tfrac{2L}{c_L} \le t \le \tfrac{4L}{c_L}, \text{etc.} \end{aligned}\right\} \tag{3.94}$$

The time between the shock and the return of the longitudinal wave is given as

$$T = \frac{2L}{c_L} = 0.08\,\mathrm{s} \tag{3.95}$$

According to Eq. (3.94), at the abovementioned time ($t = T$), the axial displacement of the loaded end takes the value $U_{\mathrm{dyn}} = 0.08$ m, which is twice the value of the static extension, $U_{\mathrm{static}} = 0.04$ m.

The eight subdivisions along the x-axis (length $L = 4$) define a spatial step $\Delta x = 0.5$ m. Moreover, according to the CFL-criterion, the time step must be less than the critical value $\Delta t_{\mathrm{cr}}^{\mathrm{CFL}} = \Delta x / c_L = 5 \times 10^{-3}$s. Nevertheless, this is a very crude estimation that requires further adjustment to be of practical use, according to the adopted numerical integration scheme, as follows.

1. ***Central-difference method***

According to standard knowledge concerning the explicit central difference scheme (e.g. Bathe [1] , p. 772), this integration method requires that the time step Δt is smaller than a critical value, Δt_{cr}, that is

$$\Delta t \le \Delta t_{\mathrm{cr}} = \frac{T_n}{\pi} = \frac{2}{\omega_n} \tag{3.96}$$

where $T_n(\omega_n)$ is the smallest period (cyclic frequency) of the finite element assemblage with n degrees of freedom. Since C-element (CPM) is actually a finite element, it is hypothesized that the above receipt may be applied equal well.

The minimum and maximum eigenvalues calculated using FEM and CPM are given in Table 3.10. Also, the fourth column shows the critical time-step calculated through Eq. (3.96) while the last column shows the scaling factor of the aforementioned step to the critical value of time step based on the CFL-criterion.

By virtue of the findings in Table 3.10, we illustrate the time response exactly at the critical time step according to Eq. (3.96).

First we compare the C-element (model M1-piecewise linear, 24 nodes, Fig. 3.32c) with the competitive FEM solution (45 nodes, Fig. 3.32d). Each graph refers to its separate critical time step shown in Table 3.10. The time response (axial displacement) at the loaded side and the reaction stress in the axial direction are shown in Figs. 3.33 and 3.34, respectively. Interestingly, although FEM and CEM-model M1 (piecewise linear) are two different formulations, they almost lead to the same numerical results. A similar observation was made in a past study concerning Poisson's ratio different than zero [35]. One may also observe (in Fig. 3.34) that the reaction stress obtains *double* the value of the imposed stress, but this occurs during half the

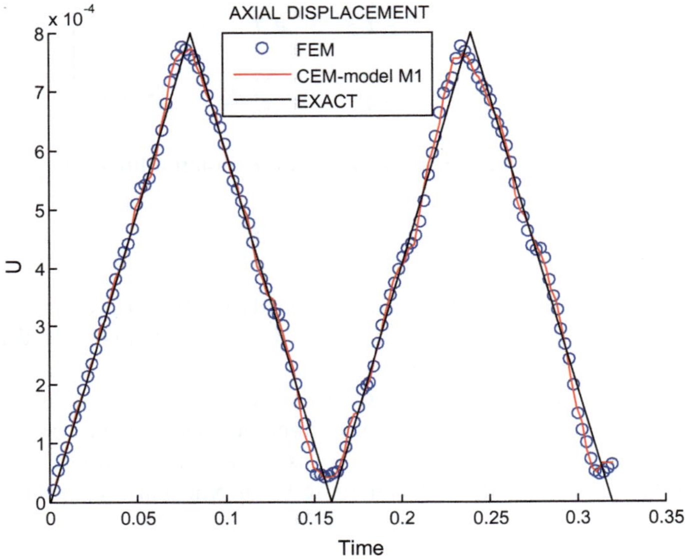

Fig. 3.33 Time response at the middle of the loaded side (point E in Fig. 3.32a)

period of time (of course the impulse is preserved in the system). This fact justifies the usual safety factor involved in impulsive loadings, which equals to $S = 2$.

Now, we compare all three CEM models (M1, M2, and M3). Figure 3.35 depicts that model M1 (piecewise-linear) is inferior to the rest two since it slightly underestimates the peak value of the axial displacement. Also, Fig. 3.36 shows that model M1 appears a relatively high overshoot and exceeds the reaction stress, more than the two other models. Between models M2 (natural cubic B-splines) and M3 (Lagrange polynomials) it is not clear which of them performs better.

If the time step decreases (e.g. by a factor of five) then the approximation becomes better, as shown in Fig. 3.37.

Table 3.10 Beam under tensile impact

Model	Eigenvalues		Critical time step according to Eq. (3.96) Δt (s)	Ratio $\left(\frac{\Delta t}{\Delta t_{cr}^{CFL}}\right)$
	Minimum (s^{-2})	Maximum (s^{-2})		
FEM	1.22681e+02	7.14064e+05	2.36680e−03	4.73360e−01
CPM–model M1	1.27733e+02	4.81834e+05	2.88125e−03	5.76250e−01
CPM–model M2	1.25106e+02	4.78074e+05	2.89256e−03	5.78512e−01
CPM–model M3	1.24894e+02	1.71488e+06	1.52726e−03	3.05452e−01

Minimum and maximum calculated eigenvalues and critical time step using all four alternative formulations

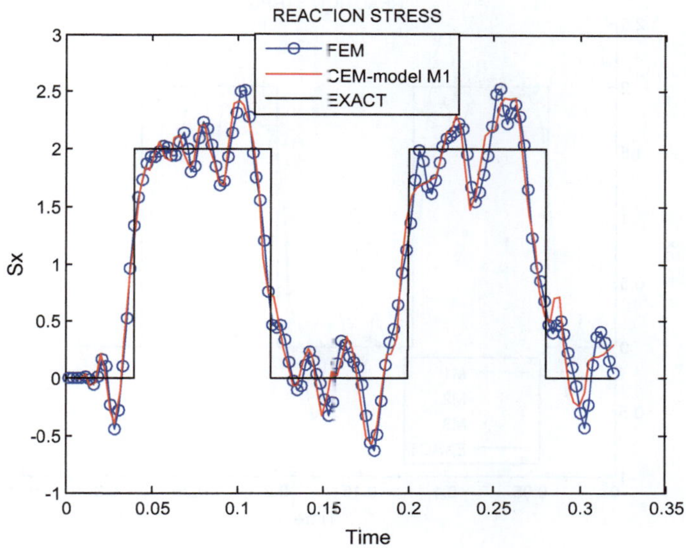

Fig. 3.34 Reaction stress (point *A* in Fig. 3.32a)

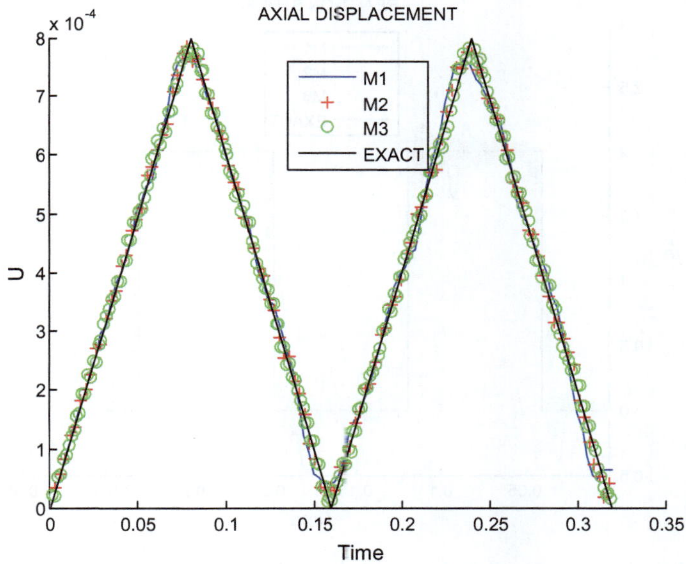

Fig. 3.35 Time response at the middle of the loaded side for all models of "C-element"

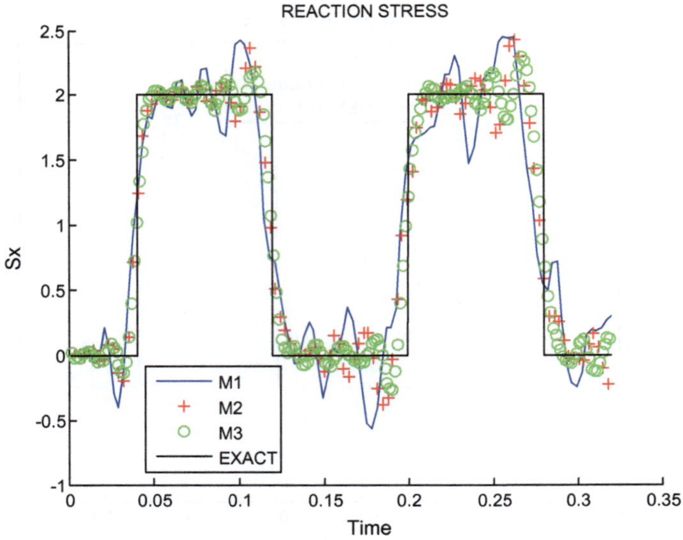

Fig. 3.36 Reaction stress for all models of "C-element"

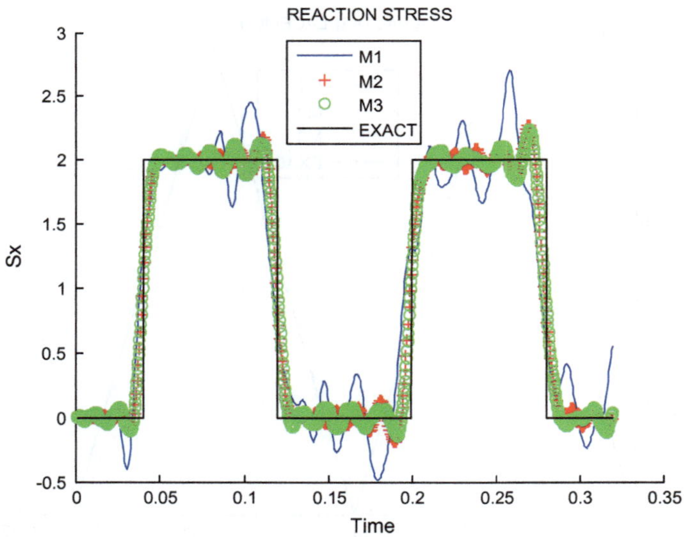

Fig. 3.37 Reaction stress for all models of "C-element" using one-fifth of the time step shown in Table 3.10

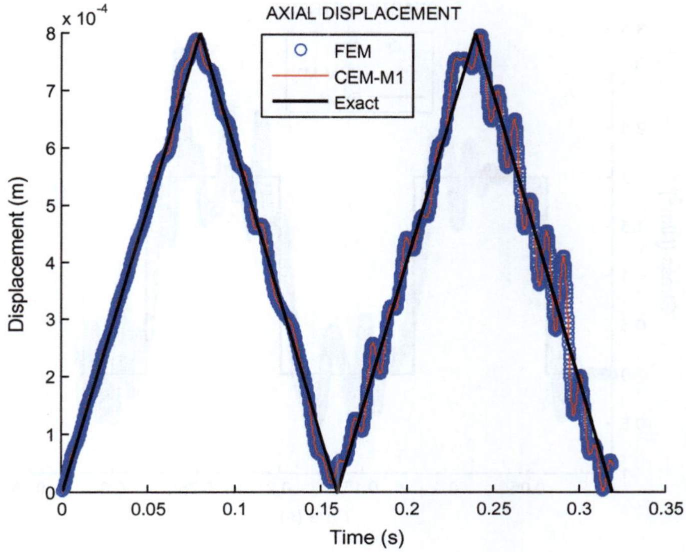

Fig. 3.38 Displacement history for point "*E*" in a clamped bar under impact, using FEM solution and CEM (model M1) formulation. Time step: $\Delta t = 0.025\ \Delta t_{\mathrm{cr}}^{\mathrm{CFL}} = 1.2500\mathrm{E}{-}04\mathrm{s}$ in conjunction with the θ-Wilson scheme

2. ***θ*-Wilson method**

In order to investigate the performance of the C-elements to implicit time-integration methods, the θ-Wilson has been chosen as a representative. As an example, the corresponding results using a time step: $\Delta t = 0.025\ \Delta t_{\mathrm{cr}}^{\mathrm{CFL}} = 1.2500\mathrm{E}{-}04$ s are shown in Figs. 3.38 and 3.39.

Once more, both the displacements (Fig. 3.38) and also the reaction stresses (Fig. 3.39) are equal between the FEM and C-element (model M1) solution.

Moreover, for a very similar problem, the interested reader can find additional results regarding models M2 and M3 in Provatidis [35].

3.7.4 Application to Problems of Axisymmetric Elasticity

A report on the use of CAD-based Coons macroelements in axisymmetric potential problems, following the spirit of the previous paper by Kanarachos and Deriziotis [13], was published with delay in 2001 by Provatidis and Kanarachos [24]. For an initial publication with many examples and detailed explanations in linear axisymmetric elasticity, where the C-element is successfully compared with an assemblage of conventional finite elements of the same mesh density, the reader is referred to Provatidis [29] full of theoretical details and many test cases. In this subsection of the book, only cumulative original results will be presented.

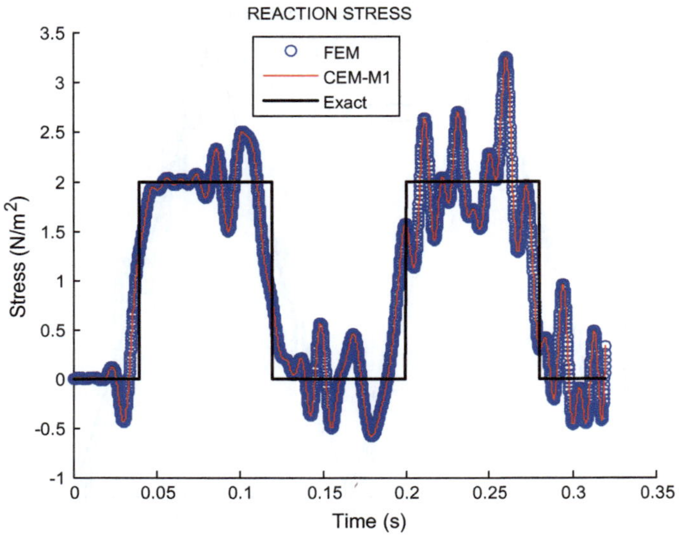

Fig. 3.39 Reaction stress in a bar under impact, using FEM solution and CEM (model M1) formulation. Time step: $\Delta t = 0.025\ \Delta t_{\mathrm{cr}}^{\mathrm{CFL}} = 1.2500\mathrm{E}{-}04$ s in conjunction with the θ-Wilson scheme

Axisymmetric elasticity was the reason to consider piecewise-linear interpolation in the Coons macroelement for the first time. Until the year 2002, the author used only the "reduced basis" of *natural cubic B-splines* as univariate trial functions along each of the four sides in the Coons patch. But for this particular approximation the accuracy of the calculated stresses near the boundaries was poor (in contrast to the accurately calculated boundary displacements). Since the calculated displacements were known along the boundary only and not around the point of interest (as happens in the conventional FEM), with the purpose to achieve a better estimation of the stresses, different shape functions were utilized for the stresses than those used for the interpolation of the displacements. Details are provided below.

3.7.4.1 General Equations

It is well known that axisymmetric structures may be solved on an axial *section*. The entire or a portion of this section is considered to be idealized as a four-sided Coons patch *ABCD*, which refers to the cylindrical curvilinear co-ordinates (r, z), as shown in Fig. 3.40. Following the standard FEM procedures, symmetric and arbitrary boundary conditions may be considered as well. In the latter case a Fourier series (or a FFT) of the loading is required. For the sake of simplicity, here it is assumed that the boundary conditions are also *axisymmetric* and the structure does not undertake any torsion. Therefore, the angular direction (ϑ) does not participate in the following formulation.

The non-vanishing strains are expressed in terms of the radial (u_r) and axial (u_z) component of the displacement vector, $\mathbf{u} = \left[u_r, u_z\right]^{\mathrm{T}}$, and are given by

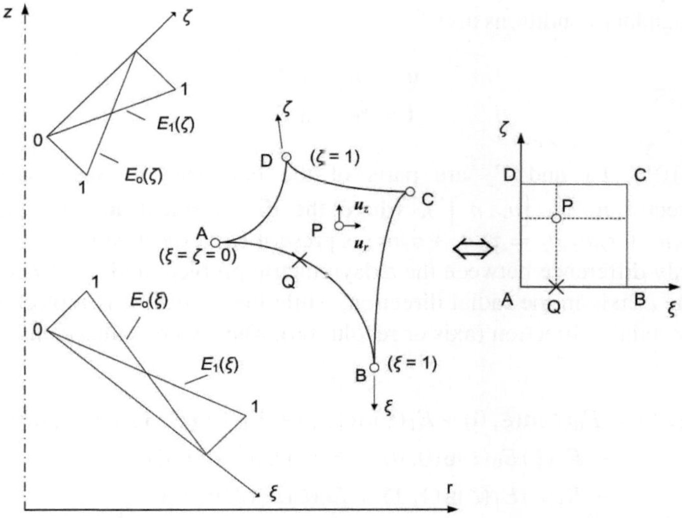

Fig. 3.40 Axial section in the form of a four-sided patch *ABCD* in polar (*r*, *z*) and curvilinear co-ordinates (ξ, ζ) in the parametric square

$$\varepsilon_r = \frac{\partial u_r}{\partial r}, \quad \varepsilon_\vartheta = \frac{u_r}{r}, \varepsilon_z = \frac{\partial u_z}{\partial z}, \quad \gamma_{zr} = \frac{\partial u_r}{\partial z} + \frac{\partial u_z}{\partial r} \tag{3.97}$$

Equation (3.97) can be written in matrix form, as follows:

$$\boldsymbol{\varepsilon} = \begin{Bmatrix} \varepsilon_r \\ \varepsilon_\vartheta \\ \varepsilon_z \\ \gamma_{zr} \end{Bmatrix} = [\mathbf{L}]\mathbf{u}, \quad \text{where} \quad [\mathbf{L}] = \begin{bmatrix} \partial/\partial r & 0 \\ 1/r & 0 \\ 0 & \partial/\partial z \\ \partial/\partial z & \partial/\partial r \end{bmatrix} \tag{3.98}$$

Also, the non-vanishing stresses are related to the strains through *Hooke*'s law:

$$\boldsymbol{\sigma} = \begin{Bmatrix} \sigma_r \\ \sigma_\theta \\ \sigma_z \\ \tau_{zr} \end{Bmatrix} = \frac{(1-\nu)E}{(1+\nu)(1-2\nu)} \begin{bmatrix} 1 & f & f & 0 \\ f & 1 & f & 0 \\ f & f & 1 & 0 \\ 0 & 0 & 0 & g \end{bmatrix} \begin{Bmatrix} \varepsilon_r \\ \varepsilon_\theta \\ \varepsilon_z \\ \gamma_{zr} \end{Bmatrix} = [\mathbf{E}]\boldsymbol{\varepsilon}, \tag{3.99}$$

in which the appearing constants are:

$$f = \frac{\nu}{1-\nu} \quad \text{and} \quad g = \frac{1-2\nu}{2(1-\nu)}, \tag{3.100}$$

with ν and E the Poisson's ratio and the Young's modulus, respectively.

The boundary conditions are:

$$\begin{aligned} \mathbf{u} &= \mathbf{u}_\Gamma \quad \text{on } \Gamma_1 \\ \mathbf{t} &= \mathbf{t}_\Gamma \quad \text{on } \Gamma_2 \end{aligned} \tag{3.101}$$

In (3.101), Γ_1 and Γ_2 are parts of the boundary Γ (with outward unit normal vector $n = [n_r, n_z]^{\mathrm{T}}$), where the displacement $\mathbf{u}_\Gamma$ and the traction $\mathbf{t}_\Gamma(t_r = \sigma_r n_r + \tau_{zr} n_z, t_z = \tau_{zr} n_r + \sigma_z n_z)$ is prescribed, respectively.

The only difference between the axisymmetric problem and the 2-*D* one, is that we use the ξ axis in the radial direction, while the vertical η is replaced by the ζ axis in the axial z-direction (axis or revolution). Then, Coons interpolation formula becomes:

$$\begin{aligned} \mathbf{u}(\xi, \zeta) = {} & E_0(\zeta)\mathbf{u}(\xi, 0) + E_1(\xi)\mathbf{u}(1, \zeta) + E_1(\zeta)\mathbf{u}(\xi, 1) + E_0(\xi)\mathbf{u}(0, \zeta) \\ & - E_0(\xi)E_0(\zeta)\mathbf{u}(0, 0) - E_1(\xi)E_0(\zeta)\mathbf{u}(1, 0) \\ & - E_1(\xi)E_1(\zeta)\mathbf{u}(1, 1) - E_0(\xi)E_1(\zeta)\mathbf{u}(0, 1) \end{aligned} \tag{3.102}$$

Furthermore, the univariate boundary displacements $\mathbf{u}(\xi, 0)$, $\mathbf{u}(1, \zeta)$, $\mathbf{u}(\xi, 1)$ and $\mathbf{u}(0, \zeta)$ in Eq. (3.102) are interpolated by a suitable set of trial functions. Following 2D problems, three particular cases of alternative trial functions are of major interest: (i) natural cardinal cubic B-splines (ii) linear interpolants (piecewise-linear) and (iii) Lagrange polynomials. The first and the third one have been extensively used earlier in this Chapter (in potential and plane elasticity problems), whereas the linear interpolants play a certain additional role in the estimation of the boundary (primarily) and internal strains and stresses.

3.7.4.2 Linear Interpolants

Piecewise linear interpolation was introduced in Chap. 2 and was applied earlier (as global shape functions, i.e. in Model "M1" or "C1") to 2D potential and 2D elasticity problems. As happened in the 2D problems, despite the same *essential* boundary conditions with those of conventional bilinear finite elements, the piecewise-linear global shape functions do not only influence the neighbourhood of the nodal point '*i*' but a *wider* area (i.e., into a strip vertical to the boundary) of the patch *ABCD*. This fact is clearly illustrated in a typical graph for a quadrilateral patch in Fig. 3.5 of a previous subsection.

It is reminded that sometimes (but not always) the obtained results may be very similar to those obtained by the FEM solution, and it is under question if this observation is valid to axisymmetric problems as well.

Another reason that the piecewise-linear interpolation is useful is the fact that the boundary stresses computed using the natural cubic B-splines are not that accurate, despite the accurate values of the calculated boundary displacements. This shortcoming has been resolved applying piecewise-linear interpolation (instead of natural

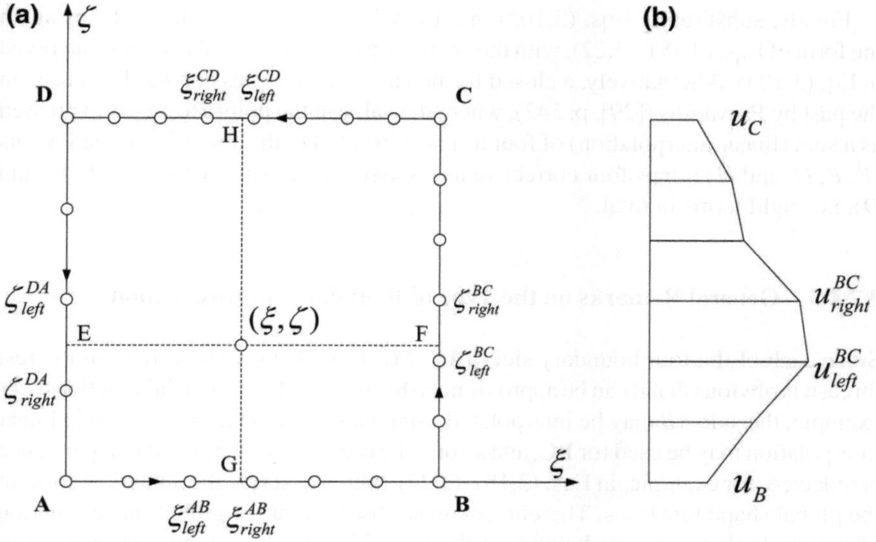

Fig. 3.41 **a** Sketch for the derivation of Coons' formula when using linear interpolation along the sides of the patch *ABCD*, and **b** piecewise-linear interpolation along the side *BC* of the patch

cubic B-splines) on the displacement calculated using the aforementioned natural cubic B-splines (for more details the reader is referred to Provatidis [29]).

In brief, for an arbitrary point $P(r, z)$ in the interior of the Coons patch, the lines of constant ξ and ζ values intersect the boundaries at four points (E, F, G and H) between certain corresponding pairs of nodal boundary points as shown in Fig. 3.41. The values of the displacement vector $\mathbf{u}$ at the aforementioned four intermediate points (E, F, G and H) as well as of the corners (A, B, C, and D), i.e. eight terms, are involved in Coons interpolation. But each of the intermediate points (E, F, G and H) generally lies between two boundary nodes as shown in Fig. 3.41, and the corresponding value is found using linear interpolation between the aforementioned two points. Regarding the notation in Fig. 3.41, moving on the boundary towards the anti-clock wise direction, the first (*left*, at the entrance) and the second (*right*, at the exit of the arrow) nodal points met on the boundary *AB* (where the intermediate point G is found) are denoted as ξ^{AB}_{left} and ξ^{AB}_{right}, respectively, and so on. The linear interpolation of the boundary displacements $\mathbf{u}(\xi, 0)$, $\mathbf{u}(1, \zeta)$, $\mathbf{u}(\xi, 1)$ and $\mathbf{u}(0, \zeta)$, involved in Eq. (3.102), between the left and the right nodal points, can be expressed in terms of the two "hat" functions, as follows:

$$\mathbf{u}(\xi, 0) = N_{\text{left}}(\xi)\mathbf{u}^{AB}_{\text{left}} + N_{\text{right}}(\xi)\mathbf{u}^{AB}_{\text{right}}, \quad \xi^{AB}_{\text{left}} \le \xi \le \xi^{AB}_{\text{right}} \tag{3.103}$$

Similar expressions with Eq. (3.103) can be written for the approximation of the univariate functions $\mathbf{u}(1, \zeta)$, $\mathbf{u}(\xi, 1)$ and $\mathbf{u}(0, \zeta)$.

Finally, substituting Eqs. (3.103) into (3.102), *Coons*' interpolation takes again the form of Eqs. (3.18)–(3.22), with the proper meaning of the trial functions involved in Eq. (3.103). Alternatively, a closed formed analytical expression has been used in the past by Provatidis ([29], p. 542), where the value at the point $P(\xi, \zeta)$ is expressed as a sum (linear interpolation) of four terms associated to the abovementioned points (E, F, G, and H) minus four corrective terms associated to the corners (A, B, C, and D), i.e. eight terms in total.

3.7.4.3 General Remarks on the Type of Boundary Approximation

Since each of the four boundary sides (AB, BC, CD, DA) is independent of the rest three, it is obvious that it can be approximated by *any* suitable set of trial functions. For example, the side AB may be interpolated using natural cubic B-splines, while linear interpolation may be used for BC, and so on. The previously mentioned independence is reflected, for example, in Eqs. (3.18)–(3.22), which describe the analytical form of the global shape functions. Therein, one may observe that the global shape function of any node that does not belong on the mth side vanishes along that, exactly as happens with the conventional finite elements. This property allows us to assemble macro-elements along their common sides, where displacement continuity is per se fulfilled.

From the computational point of view, the set of Eqs. (3.18)–(3.22) can be used once using a certain set (e.g., natural B-splines or Lagrange polynomials) for the estimation of the stiffness matrix, and a second time using piecewise-linear interpolation for the estimation of the stresses.

3.7.4.4 Stiffness Matrix and Force Vector

It is obvious that both stiffness matrix and force vector depend on the choice of the global shape functions $N_j (1 \le j \le q_e)$, of which typical cases were previously presented. Similarly to the 2D problems, the displacement vector with the C-element is written in matrix form as

$$\{\mathbf{u}\} = [\mathbf{N}_u]\{\mathbf{d}\} \tag{3.104}$$

where $\{\mathbf{u}\}$ is the displacement vector at the boundaries of the macro-element which includes q_e nodal points, $[\mathbf{N}_u]$ is the matrix of the global shape functions and $\{\mathbf{d}\}$ is the vector of the nodal displacements.

The equilibrium equations are:

$$[\mathbf{K}]\{\mathbf{d}\} = \{\mathbf{f}\}, \tag{3.105}$$

with the stiffness matrix given by

$$[\mathbf{K}] = \int_{\Omega} [\mathbf{B}]^T[\mathbf{E}][\mathbf{B}]\mathrm{d}\Omega = \int_{A}\int_{-\pi}^{+\pi} [\mathbf{B}]^T[\mathbf{E}][\mathbf{B}]r\,\mathrm{d}\vartheta\,\mathrm{d}A = 2\pi\int_{A}[\mathbf{B}]^T[\mathbf{E}][\mathbf{B}]r\quad\mathrm{d}A$$

$$(2q_e \times 2q_e) \tag{3.106}$$

In Eq. (3.106), A is the cross-sectional area of the macroelement (with $\mathrm{d}A = \mathrm{d}r\,\mathrm{d}z$), whereas the strain operator is

$$[\mathbf{B}] = [\mathbf{L}][\mathbf{N}_u]. \tag{3.107}$$

Obviously, the obtained stiffness matrix [**K**] is symmetric and fully populated (unless piecewise-linear approximation, or de Boor B-splines is used), whereas the vector of the external force (**f**) is given by

$$[\mathbf{f}]_i = \int_{\Gamma} N_i\mathbf{t}_{\Gamma}(2\pi r)\mathrm{d}\Gamma, \tag{3.108}$$

where $\mathbf{t}_{\Gamma}$ is the traction vector along the boundary Γ of the macro-element.

3.7.4.5 Internal Strains and Stresses

After the solution of the equilibrium Eq. (3.105), the displacement field **u** is known along the entire boundary of the macro-element. Then, one can evaluate the internal strains within the macro-element in terms of the basis of the *global* functional set as follows:

$$\boldsymbol{\varepsilon} = [\mathbf{L}]\mathbf{u} = [\mathbf{L}][\mathbf{N}_u]\mathbf{d} \tag{3.109}$$

Working with many test problems, local oscillations of the stresses were noticed in the particular case that [**K**] was calculated using natural cubic B-splines and then the same functions were also used to calculate the strains by differentiating Eqs. (3.18)–(3.22). Since the displacement field is known, a different functional set than Eqs. (3.18)–(3.22) can be used. Using global linear interpolants instead, the $\boldsymbol{\varepsilon}'$-field produced by the differentiation in Eq. (3.107) will be *discontinuous* at the boundary nodal points $(\xi_i, \zeta_i), i = 1, \ldots, q_e$, but in many cases it will be more accurate than that obtained using the natural cubic B-splines. In other words, the alternative functional set that may be used for the determination of the discontinuous strains leads to a new matrix $[\mathbf{N}_\sigma]$ of *global* shape functions, so as

$$\boldsymbol{\varepsilon}' = [\mathbf{L}][\mathbf{N}_\sigma]\mathbf{d} \tag{3.110}$$

In general, the associated stresses $\boldsymbol{\sigma}$ are calculated using the *global* shape functions by substituting Eq. (3.109) or Eq. (3.110) in Hooke's law, as follows:

$$\boldsymbol{\sigma} = \begin{cases} [\mathbf{E}]\boldsymbol{\varepsilon} = [\mathbf{E}][\mathbf{L}][\mathbf{N}_u]\mathbf{d}, & \text{continuous stresses} \\ [\mathbf{E}]\boldsymbol{\varepsilon}' = [\mathbf{E}][\mathbf{L}][\mathbf{N}_\sigma]\mathbf{d}, & \text{discontinuous stresses} \end{cases} \tag{3.111}$$

The interested reader is referred to Provatidis [29], where he/she can find many results at the *nodal points* of the C-element. Due to the major difference between the continuous stress field in the C-element and the discontinuous one in the assemblage of the finite elements, the FEM solution included averaged values based on "*Adjacent Corner Points*" (ACP) and also "*Superconvergent Mid Points*" (SMP).

In contrast, below we present overall error norms (in percent), of the displacement field (L_u) and the stress filed (L_σ), according to Eq. (3.81), which will be reported for the first time. For the purpose of completeness, in this book we present results using all three models, i.e.:

- Model M1: Piecewise-linear
- Model M2: Natural cubic B-splines
- Model M3: Lagrange polynomials.

Particularly for Model M2, not only the consistent stress field but also the piecewise-linear based (mentioned in the Sect. 3.7.4.2) will be reported.

Caution should be paid on the fact that the abovementioned C-element models (M1, M2, and M3) used below are harmonized with the previous part of Chap. 3, and therefore are different (and more) than those used in the paper by Provatidis [29].

Example 3.10 (Hollow cylinder [axisymmetric structure]) Consider a thick-walled hollow cylinder of inner radius R_1, outer radius R_2 and height H, with pressure P prescribed on the inner surface (Fig. 3.42). The ends of the cylinder are constrained to move in the radial direction only (i.e. $\varepsilon_z = 0$), so that the results can be compared with the analytical plane strain Lamé solutions [16] . The numerical values employed are: $R_1 = 3.0$, $R_2 = 6.0$, $H = 2.0$ and $P = 1.0$, with material properties $E = 1.0$ and $\nu = 0.3$. The Lamé solutions for hollow cylinders can be written as (see, Timoshenko and Goodier [36], pp. 69–71):

$$\begin{aligned} u_r &= C_1 r + \frac{C_2}{r} \\ \sigma_r &= C_3 - \frac{C_4}{r^2} \\ \sigma_\vartheta &= C_3 + \frac{C_4}{r^2} \end{aligned} \tag{3.112}$$

where C_1–C_4 are constants depending on the numerical values of R_1, R_2, P, E and ν. Note that in Eq. (3.112), r is the distance of a point to the axis of revolution, i.e. the abscissa in the Orz Cartesian system shown in Fig. 3.40.

Imposing the boundary conditions the analytical expression for the abovementioned constants are:

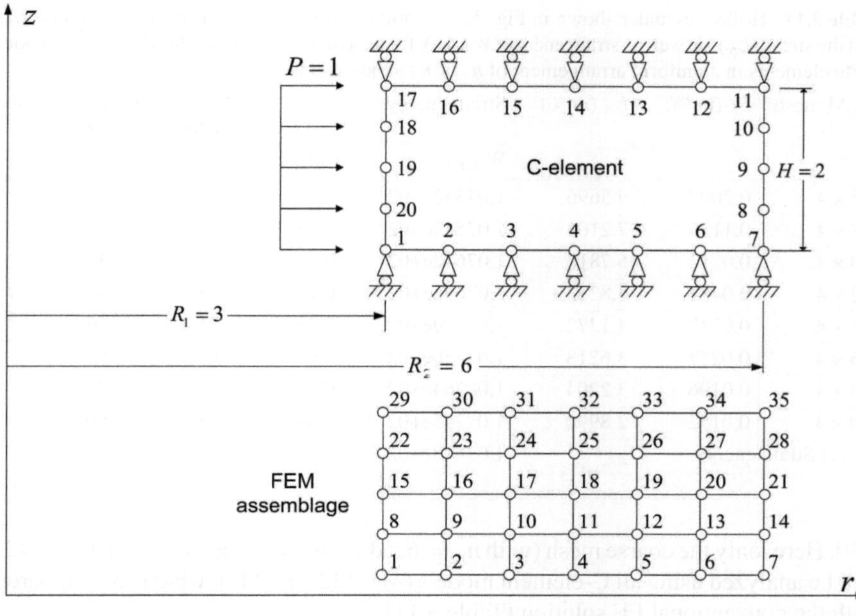

Fig. 3.42 Hollow cylinder under internal pressure $P = 1$ discretized as a 20-noded macro-element (top) or a set of 24 conventional linear elements based on 35 nodes (bottom)

$$\left.\begin{aligned} C_1 &= \frac{P(1+\nu)(1-2\nu)}{E\left[\left(\frac{R_2}{R_1}\right)^2-1\right]}, \quad C_2 = \frac{P(1+\nu)}{E\left[\left(\frac{R_2}{R_1}\right)^2-1\right]}, \\ C_3 &= \frac{P}{\left(\frac{R_2}{R_1}\right)^2-1}, \quad C_4 = \frac{P}{\frac{1}{R_1^2}-\frac{1}{R_2^2}} \end{aligned}\right\} \tag{3.113}$$

In the present example (with $P = 1 > 0$), they take the numerical values $C_1 = 0.1733$, $C_2 = 15.6$, $C_3 = 0.3333$, $C_4 = 12.0$.

Based on Eq. (3.112), the strain energy in the entire volume of the hollow cylinder is found to be

$$W_{\text{strain}} = \tfrac{1}{2} \cdot 2\pi PH\left(R_2^2 - R_1^2\right)\left[C_1 C_3 + \frac{C_2 C_4}{R_1^2 R_2^2}\right] \tag{3.114}$$

Similarly to the conventional FEM, the uniform internal pressure should be transformed into equivalent nodal forces along the nodes at $R_1 = 3$. Using piecewise-linear approximation (model M1), the value of the concentrated nodal forces are the same with used in the FEM solution. For the rest models (M2: natural cubic B-splines, M3: Lagrange polynomials) they differ.

The previous numerical study includes five different meshes using $N_r = 6$, 12, 18 and 24 boundary segments in r-direction and constantly four segments in z-direction

Table 3.11 Hollow cylinder shown in Fig. 3.42: Normalized L_2-norms for the displacement (L_u) and the stress (L_σ), as well as strain energy (W_{strain}), for an assemblage of standard bilinear (4-node) finite elements in a uniform arrangement of $n_x \times n_y$ subdivisions

FEM mesh	L_u (in %)	L_σ (in %)	Strain energy		Number of nodes	Number of elements
			W_{strain}	Error (%)		
6 × 4	0.2007	9.5696	1.07332e+02	−0.45	35	24
8 × 4	0.1126	7.2106	1.07543e+02	−0.25	45	32
10 × 4	0.0712	5.7810	1.07642e+02	−0.16	55	40
12 × 4	0.0484	4.8233	1.07696e+02	−0.11	65	48
14 × 4	0.0347	4.1372	1.07729e+02	−0.08	75	56
16 × 4	0.0257	3.6218	1.07750e+02	−0.06	85	64
18 × 4	0.0196	3.2204	1.07764e+02	−0.05	95	72
20 × 4	0.0152	2.8990	1.07775e+02	−0.04	105	80
Exact Strain energy			1.07818e+02			

[29]. Here, only the coarse mesh (with $n_r = 6$ radial subdivisions, shown in Fig. 3.42) will be analyzed using all C-element models (M1, M2, and M3), which are compared with the conventional FE solution (Table 3.11).

In the sequence, we present similar results using a single "C"-element (in literature, the alternative term "Coons Patch Macroelement"-CPM has been extensively used). For the three models (M1, M2, and M3), the corresponding results are shown in Tables 3.12, 3.13, and 3.14, respectively.

One may observe that all three models of the single "C-element" converge to the exact value, similarly to the FEM solution. In this problem no need appeared to replace the natural cubic B-splines interpolation with the piecewise-linear one (as happened regarding the boundary nodes, in [29]).

Example 3.11 (Hollow sphere [axisymmetric model]) Consider a thick-walled hollow sphere of inner radius R_1 and outer radius R_2, subjected to pressures P_1 and P_2 at the inner and outer surfaces, respectively (Fig. 3.43). The numerical values employed are: $R_1 = 1.0$, $R_2 = 2.0$, $P_1 = 5.0$ and $P_2 = 3.0$ with material properties $E = 1.0$ and $\nu = 0.3$. The exact analytical solutions for this problem is due to Lamé and are given by (see, Timoshenko and Goodier [36], pp. 394–395):

$$\begin{aligned} u_R &= \frac{c_1}{R^2} + c_2 R \\ \sigma_R &= \frac{c_3}{R^3} + c_4 \\ \sigma_\theta &= \frac{c_5}{R^3} + c_6 \end{aligned} \tag{3.115}$$

It should be clarified that the radius R in Eq. (3.120) refers to the position vector of the arbitrary point $P(r, z)$ with respect to the center O of the sphere. In practice, we work on the axial section $Orz \equiv Oxz$, where the center O is also the axis origin

Table 3.12 Hollow cylinder shown in Fig. 3.42: Normalized L_2-norms for the displacement (L_u) and the stress (L_σ), as well as strain energy (W_{strain}), for a single Coons macroelement ("C-element"), based on piecewise-linear trial functions (model M1)

"C-element" (Piecewise-linear, model M1)	L_u (in %)	L_σ (in %)	Strain energy		Number of nodes	Number of elements
			W_{strain}	Error (%)		
6 × 4	0.2232	9.5792	1.07333e+02	−0.45	20	1
8 × 4	0.1254	7.2147	1.07543e+02	−0.25	24	1
10 × 4	0.0795	5.7832	1.07642e+02	−0.16	28	1
12 × 4	0.0543	4.8245	1.07696e+02	−0.11	32	1
14 × 4	0.0391	4.1380	1.07729e+02	−0.08	36	1
16 × 4	0.0292	3.6223	1.07750e+02	−0.06	40	1
18 × 4	0.0224	3.2207	1.07764e+02	−0.05	44	1
20 × 4	0.0176	2.8993	1.07775e+02	−0.04	48	1
Exact strain energy			1.07818e+02			

The boundary is uniformly discretized using n_x and n_y subdivisions in the directions x- and y, respectively, which is denoted as $n_x \times n_y$

Table 3.13 Hollow cylinder shown in Fig. 3.42: Normalized L_2-norms for the displacement (L_u) and the stress (L_σ), as well as strain energy (W_{strain}), for a single Coons macroelement ("C-element"), based on natural cubic B-splines trial functions (model M2)

"C-element" (natural cubic B-splines, model M2)	L_u (in %)	L_σ (in %)	Strain energy		Number of nodes	Number of elements
			W_{strain}	Error (%)		
6 × 4	0.0709	3.3843	1.07780e+02	−0.0347	20	1
8 × 4	0.0343	2.2113	1.07803e+02	−0.0136	24	1
10 × 4	0.0194	1.5839	1.07811e+02	−0.0060	28	1
12 × 4	0.0123	1.2046	1.07815e+02	−0.0027	32	1
14 × 4	0.0085	0.9553	1.07817e+02	−0.0010	36	1
16 × 4	0.0064	0.7814	1.07817e+02	−0.0001	40	1
18 × 4	0.0052	0.6544	1.07818e+02	0.0005	44	1
20 × 4	0.0045	0.5584	1.07818e+02	0.0008	48	1
Exact strain energy			1.07818e+02			

The boundary is uniformly discretized using n_x and n_y subdivisions in the directions x- and y, respectively, which is denoted as $n_x \times n_y$

Table 3.14 Hollow cylinder shown in Fig. 3.42: Normalized L_2-norms for the displacement (L_u) and the stress (L_σ), as well as strain energy (W_{strain}), for a single Coons macroelement ("C-element"), based on Lagrange polynomials as trial functions (model M3)

"C-element" (Lagrange polynomials, model M3)	L_u (in %)	L_σ (in %)	Strain energy		Number of nodes	Number of elements
			W_{strain}	Error (%)		
4 × 4	0.0241	0.1336	1.07817e+02	-0.0003	16	1
5 × 4	0.0054	0.0268	1.07819e+02	0.0017	18	1
6 × 4	0.0037	0.0010	1.07819e+02	0.0017	20	1
8 × 4	0.0037	3.6714e−05	1.07819e+02	0.0017	24	1
10 × 4	0.0037	0.1336	1.07819e+02	0.0017	28	1
Exact strain energy			1.07818e+02			

The boundary is uniformly discretized using n_x and n_y subdivisions in the directions x- and y, respectively, which is denoted as $n_x \times n_y$

of the Cartesian system, thus $R = \sqrt{r^2 + z^2}$. Therefore, if the radial vector through the point $P(r, z)$ is $\vec{n} = n_x\vec{i} + n_z\vec{k}$, the abscissa and the ordinate will be $r = n_x R$ and $z = n_z R$, respectively.

The c_i's in Eq. (3.115) are constants depending on the geometry and boundary conditions. In this example, $c_1 = 1.4857, c_2 = -1.0857, c_3 = -2.2857, c_4 = -2.7143$, $c_5 = 1.1429$ and $c_6 = -2.7143$.

Analytical integration using Eq. (3.115) gives the strain energy for the half of the hollow sphere (i.e. that produced by the body shown in Fig. 3.42, when it is rotated by 360° about the z-axis), as

$$W_{\text{strain}} = \frac{1}{2} \cdot 2\pi \cdot \left[\frac{A}{3}\left(R_2^3 - R_1^3\right) + B \ln\left(\frac{R_2}{R_1}\right) - \frac{C}{3}\left(\frac{1}{R_2^3} - \frac{1}{R_1^3}\right)\right] \tag{3.116}$$

where

$$\left.\begin{aligned} A &= c_2(c_4 + 2c_6), \\ B &= c_2(c_3 + 2c_5) + 2c_1(-c_4 + c_6), \\ C &= 2c_1(c_5 - c_3). \end{aligned}\right\} \tag{3.117}$$

This problem has been previously solved by Provatidis [29], where the results and the discussion focused on the nodal values along the boundary. As the interested reader can find valuable information therein, here we will present only new results related to the normalized error norms of displacement (L_u) and stress (L_σ) field.

Concerning the comparison between the calculated FEM stress field and the analytical solution, it is worthy to mention that while the exact σ_θ (circumferential or

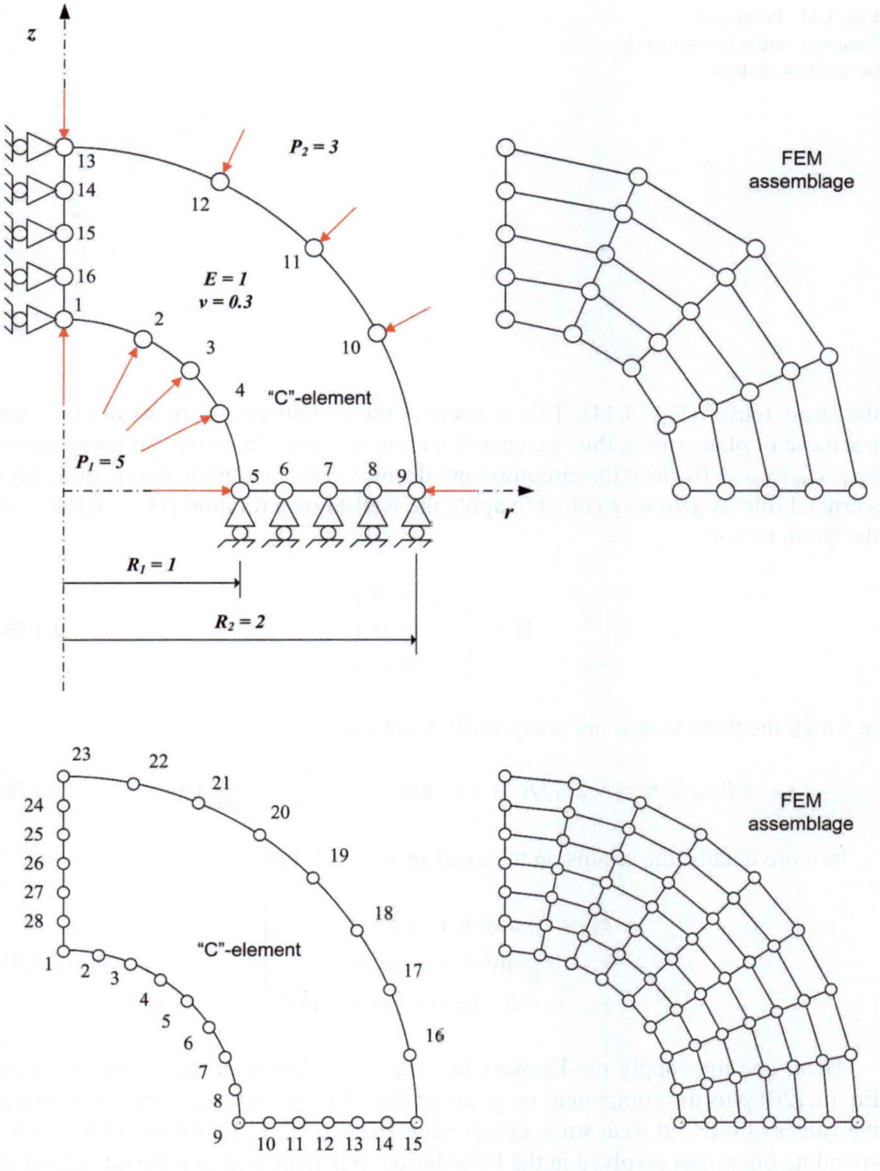

Fig. 3.43 Hollow sphere using a coarse (top) and a fine (bottom) discretization

tangential) component given by Eq. (3.115) is conceptually the same quantity with that involved in FEM analysis (only its numerical value may differ, depending on the mesh density), the reader may probably find a difficulty to determine the accurate stress field in order to compare with the corresponding (σ_x, σ_z, τ_{xz}) FEM values of

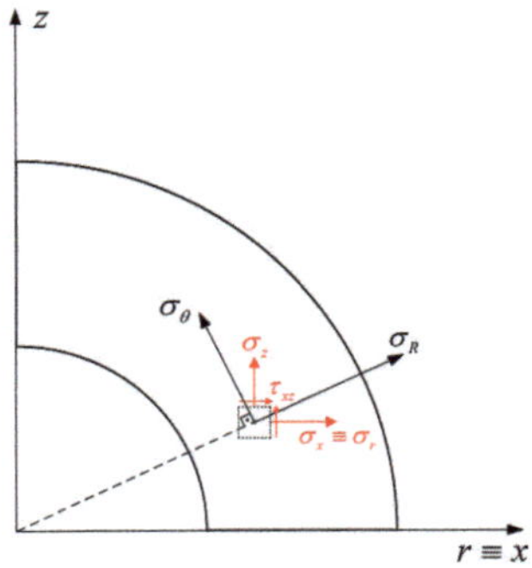

Fig. 3.44 Polar and Cartesian stress components on an axial section

the stress tensor (Fig. 3.44). This is because the conditions on the axial plane are not those of plane stress, thus we cannot merely rotate the known polar components $(\sigma_{rr}, \sigma_{\theta\theta}, \tau_{r\theta} = 0)$ since the circumferential stress in the ϕ-spherical direction plays a crucial role. A safe way is first to apply the well-known rotation $\left(\mathbf{H}' = \mathbf{RHR}^{\mathrm{t}}\right)$ on the strain tensor:

$$\mathbf{H} = \begin{bmatrix} \varepsilon_R & 0 & 0 \\ 0 & \varepsilon_\theta & 0 \\ 0 & 0 & 0 \end{bmatrix}, \tag{3.118}$$

in which the three strains are analytically known as:

$$\varepsilon_R = \partial u_R / \partial R = -2c_1/R^3 + c_2 \quad \text{and} \quad \varepsilon_\theta = \frac{u_R}{R} = \frac{c_1}{R^3} + c_2. \tag{3.119}$$

In more details, the strains on the axial section will be:

$$\left.\begin{aligned} \varepsilon_x &= \varepsilon_R \cos^2\theta + \varepsilon_\theta \sin^2\theta \\ \varepsilon_z &= \varepsilon_R \sin^2\theta + \varepsilon_\theta \cos^2\theta \\ \gamma_{xz} &= -2(-\varepsilon_R + \varepsilon_\theta)\cos\theta\sin\theta \end{aligned}\right\} \tag{3.120}$$

Then, one may apply the Hooke's law [i.e., Eq. (3.99)] on the strains given by Eq. (3.120) plus the component ε_θ given by Eq. (3.119), and therefore to determine the values of the Cartesian stress components $[\sigma_r, \sigma_\theta, \sigma_z, \tau_{zr}]^{\mathrm{T}}$, which are the corresponding quantities involved in the FE solution. It is reminded that the numerator in the error L_σ will be the L^2-norm:

$$L^2 - \text{norm} = \int_A (\tilde{\sigma}_r - \sigma_r)^2 + (\tilde{\sigma}_\theta - \sigma_\theta)^2 + (\tilde{\sigma}_z - \sigma_z)^2 + (\tilde{\tau}_{zr} - \tau_{zr})^2 r \, \mathrm{d}A \tag{3.121}$$

Table 3.15 (Hollow sphere: Fig. 3.43). Normalized L-two error norms for the displacement (L_u) and stress (L_σ), as well as strain energy (W_{strain}) value and error (in %), using an assemblage of bilinear finite elements arranged in a uniform mesh of $n_x \times n_y$ subdivisions

FEM mesh	L_u (in %)	L_σ (in %)	Strain energy		Number of nodes	Number of elements
			W_{strain}	Error (%)		
4 × 4	4.4056	1.5811	7.03041e+01	−5.1749	25	16
8 × 6	1.1531	0.4763	7.29658e+01	−1.5849	63	48
10 × 8	0.7230	0.2938	7.34167e+01	−0.9767	99	80
20 × 16	0.1755	0.0691	7.39716e+01	−0.2282	357	320
40 × 32	0.0447	0.0174	7.40990e+01	−0.0564	1353	1280
60 × 48	0.0212	0.0083	7.41212e+01	−0.0265	2989	2880
100 × 80	0.0084	0.0030	7.41342e+01	−0.0089	8181	8000
Exact strain energy			7.41408e+01			

with $(\tilde{\sigma}_r, \tilde{\sigma}_\theta, \tilde{\sigma}_z, \tilde{\tau}_{zr})$ representing the FEM solution, whereas the exact stress components are calculated in terms of the known pair $(\sigma_R, \sigma_\theta)$ [cf. Eq. (3.115)] by the final formulas:

$$\left.\begin{aligned} \sigma_r \equiv \sigma_x &= \sigma_R \cos^2\theta + \sigma_\theta \sin^2\theta \\ \sigma_z &= \sigma_R \sin^2\theta + \sigma_\theta \cos^2\theta \\ \tau_{xz} = G\gamma_{xz} &= -\tfrac{6c_1}{R^3}\cos\theta\sin\theta \end{aligned}\right\} \tag{3.122}$$

Note that the shear stress τ_{xz} in Eq. (3.122) differs than the value produced by an erroneous rotation on the axial plane (i.e., $\tau_{xz} \neq -(\sigma_R - \sigma_\theta)\cos\theta\sin\theta$).

Based on the abovementioned analytical formulas as the accurate exact solution of the problem of the hollow sphere, Table 3.15 shows the convergence quality of the FEM solution, which one may observe that is monotonic in all three quantities (displacement, stress, and strain energy).

In the sequence, the results obtained with the three models (M1, M2, an M3) of a single Coons macroelement ("C-element") are shown in Tables 3.16, 3.17 and 3.18.

One may observe that model M1 converges to an erroneous value, which however (from the engineering point of view) could be acceptable as a first order approximate solution.

Now, in Table 3.17 we present the same calculations using model M2 (natural cubic B-splines as trial functions). One may observe that, again, the convergence is towards displacement and stress values with a permanent error around 2.0 or 2.35%.

Finally, Lagrange polynomials (model M3) were implemented as univariate trial functions along each side of the Coons patch. The results are shown in Table 3.18, and again show a similar behavior as the two previous models.

Table 3.16 (Hollow sphere: Fig. 3.43). Normalized L-two error norms for the displacement (L_u), the stress (L_σ) as well as the strain energy (W_{strain}) using a single Coons macroelement (model M1: piecewise-linear trial functions) in several discretizations of $n_x \times n_y$ subdivisions

Model M1	L_u (in %)	L_σ (in %)	Strain energy		Number of nodes	Number of elements
			W_{strain}	Error (%)		
4×4	3.8147	6.2771	7.09098e+01	−4.3580	16	1
8×6	1.4721	4.2533	7.32379e+01	−1.2179	28	1
10×8	1.2074	3.3374	7.35629e+01	−0.7795	36	1
20×16	1.0340	2.0265	7.39768e+01	−0.2213	72	1
40×32	1.0243	1.5226	7.40808e+01	−0.0810	144	1
60×48	1.0242	1.4091	7.41001e+01	−0.0550	216	1
100×80	1.0243	1.3473	7.41100e+01	−0.0416	360	1
Exact strain energy			7.41408e+01			

Table 3.17 (Hollow sphere: Fig. 3.43). Normalized L-two error norms for the displacement (L_u), stress (L_σ) and strain energy (W_{strain}), using a single Coons macroelement (model M2: natural cubic B-splines) in several discretizations of $n_x \times n_y$ subdivisions

Model M2	L_u (in %)	L_σ (in %)	Strain energy		Number of nodes	Number of elements
			W_{strain}	Error (%)		
4×4	3.6541	3.7292	7.41210e+01	−0.0267	16	1
8×6	2.1549	2.7984	7.40412e+01	−0.1344	28	1
10×8	2.0649	2.5449	7.40461e+01	−0.1278	36	1
20×16	1.9915	2.3722	7.40473e+01	−0.1262	72	1
40×32	1.9829	2.3505	7.40475e+01	−0.1259	144	1
Exact strain energy			7.41408e+01			

Table 3.18 (Hollow sphere: Fig. 3.43). Normalized L-two error norms for the displacement (L_u), stress (L_σ) and strain energy (W_{strain}), using a single Coons macroelement (model M3: Lagrange polynomials as trial functions) in several discretizations of $n_x \times n_y$ subdivisions

Model M3	L_u (in %)	L_σ (in %)	Strain energy		Number of nodes	Number of elements
			W_{strain}	Error (%)		
4×4	1.9560	2.4248	7.40424e+01	−0.1328	16	1
8×6	1.9817	2.3477	7.40475e+01	−0.1259	28	1
10×8	1.9817	2.3475	7.40475e+01	−0.1259	36	1
20×16	1.9818	2.3475	7.40475e+01	−0.1259	72	1
Exact strain energy			7.41408e+01			

Table 3.19 Deviation between vertical (at corner *A*) and horizontal (at corner *B*) radial displacements for a dense mesh using 20×16 subdivisions

Method	Radial displacement (u_r)	
	Corner *A*	Corner *B*
FEM	0.4045501	0.3957292
C-element (model M1)	0.5025899	0.4396046
C-element (model M2)	0.4808382	0.4543561
C-element (model M3)	0.4792125	0.4549158
Exact	0.4000	0.4000

Remark 1 The conclusion from this example (hollow sphere) is that the boundary-only Coons macroelements ("C-elements"), in all its three modes (M1, M2, and M3) fails to converge to the exact solution. Nevertheless, the level of deviation is not that high (less than 2.5 and 2.0% in stress and displacements, respectively).

Remark 2 Compared to the previous example of the hollow cylinder, we can comment as follows. In the hollow cylinder the solution was quite uniform because the *z*-axis did not influence it at all. Here (hollow sphere), the solution is supposed to be uniform with respect to the radial direction (*R*), however even in the FEM solution it is not, as shown in Table 3.19 for an adequately dense mesh using 20×16 subdivisions.

3.8 COONS Macroelement as a NURBS

In general, there is no restriction in the choice of the univariate trial functions B_i^{SIDE} along a certain side (say *AB*) of the quadrilateral Coons patch *ABCD*. Whatever the choice of B_i^{SIDE} is, the global shape functions of the Coons macroelement are always given by Eqs. (3.18)–(3.22).

In order to be productive, as an example below we study the case in which, instead of the abovementioned *reduced* natural cubic B-splines (model M2 or C2), approximation using *full* Curry-Schoenberg (de Boor) cubic splines is considered (see for details in Chap. 7). Based on the findings of Exercise 3.2, de Boor's subroutine could be skipped simply by considering *all* the columns of the matrix **A**, thus implicitly creating two *additional* rotational DOF (curvatures) at the ends of each side of *ABCD*. However, the de Boor subroutines (`spcol` in MATLAB) have the advantage of robustness and compact support, while the latter is not ensured in the use of the inverse matrix $\mathbf{A}^{-1}$ left-multiplied by the truncated power series. As a result, both ways eventually lead to the same numerical values (see further discussion in Chap. 7).

Nevertheless, whatever formulation the analyst choices, in the case of *full* cubic B-splines with breakpoints that divide the sides of the patch into n_x and n_y segments, the number of degrees of freedom (DOF) in the Coons-patch macroelement increases from $n_{\mathrm{DOF}} = 2(n_x + n_y)$ to $n'_{\mathrm{DOF}} = 2(n_x + n_y) + 8$. This is due to the fact that two extra control points appear near the two endpoints in each side of the quadrilateral Coons patch *ABCD*, thus creating eight additional DOF in total.

The computational procedure in the full B-spline based Coons patch macroelement consists of the following steps:

- *Step*-1: Select $(n_x + 3)$ *test* points, preferably uniformly distributed, along each of the "opposite" sides *AB* and *CD* ("parallel" to the x-axis).
- *Step*-2: Find the control points along each of the sides AB and CD, by implementing the relationship $\mathbf{x} = \sum_{j=1}^{n_x+3} N_{i,p}\mathbf{x}_{P_i}$ at the aforementioned test points. This is easily accomplished by the inversion of the produced linear system which eventually gives the unknown Cartesian coordinates of the control points $\mathbf{x}_{P_i}$, $i = 1, \ldots, n_x + 3$ along them.
- *Step*-3: Repeat the above Step-2, now along the sides *BC* and *DA* ("parallel" to the y-axis). Determine the control points $\mathbf{x}_{P_j}$, $j = 1, \ldots, n_y + 3$ along these two sides of the patch.
- *Step*-4: Form the unified vector of *all*, $n_{\mathrm{cntrl}} = 2(n_x + n_y) + 8$, control points in the counter-wise direction, from the corner *A* until the entire boundary is completed. This constant vector feeds the subroutine that calculates the shape functions and their first derivatives, at each integration point.
- *Step*-5: Estimate the stiffness matrix implementing the usual Gauss integration in each of the $(n_x n_y)$ domain cells, which are determined by the breakpoints.
- *Step*-6: Impose the boundary conditions and create the right-hand-side of the equations system. Solve this system and determine the unknown coefficients.
- *Step*-7: Calculate the basis functions $N_{i,p}$ (B-splines) at the same positions where a previous uniform mesh of $(n_x \times n_y)$ subdivisions had been used, and calculate the true values of the physical quantity (e.g., temperature).

As an example, we refer again in the application Example 3.1 that refers to the steady-state conduction in a cylindrical wall. In this problem, for the discretization $(n_x,\ n_y) = (6,\ 10)$ the previous natural cubic B-splines based C-element includes 32 DOF, whereas the de Boor B-splines based C-element includes 40 DOF (eight more than previously) associated to control points (see Fig. 3.45). The numerical results at the eleven equidistant positions along the radius are given in the third column of Table 3.20, where they are compared with those previously obtained using nodal based shape functions (32 DOF). One may observe that the full cubic B-spline formulation (40 DOF) is much superior to the reduced one (Model 2, with 32 DOF), but it is not clear whether it outperforms with respect to the Lagrange polynomials based numerical solution (sixth column in Table 3.20) despite the fact that less parameters are used in the latter (i.e. 32 DOF).

Remark 1 An *equivalent* numerical solution with "C-element" Model 3 is obtained when *nonrational* using Bernstein-Bézier polynomials are used instead of the Lagrange polynomials. Since the set of Lagrange polynomials is mathematically equivalent with a set of nonrational Bernstein-Bézier polynomials of the same degree (details may be found in Chap. 6), the latter can easily replace the former ones.

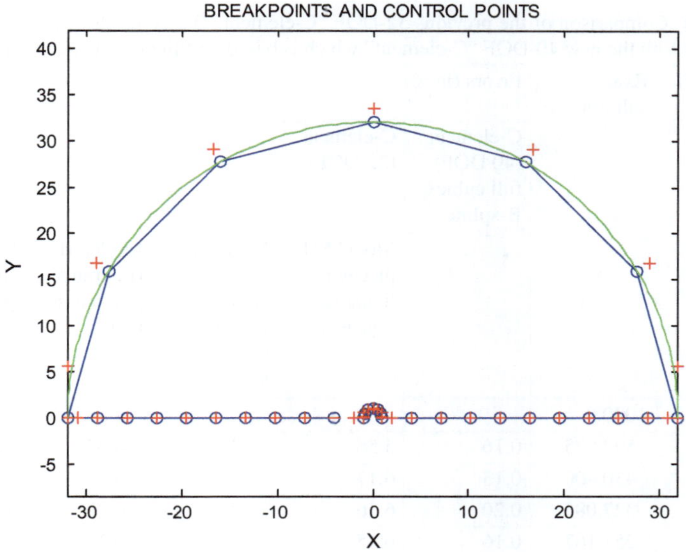

Fig. 3.45 Nine control points (+) and seven breakpoints (o) along the four sides of the *nonrational* B-splines based Coons patch macroelement ($n_x = 6,\ n_y = 10$)

Then, Eqs. (3.18)–(3.22) are again valid, but now the involved B_i^{SIDE} functions are Bernstein-Bézier polynomials of degree $(q_1 - 1)$, $(q_2 - 1)$, $(q_3 - 1)$ and $(q_4 - 1)$, along the sides *AB*, *BC*, *CD* and *DA*, respectively.

Then, the interpolation within a Coons macroelement becomes:

$$U(\xi,\eta) = \sum_{j=1}^{q_e} N_j(\xi,\eta)\cdot a_j \tag{3.123}$$

It is noted that the coefficients a_j in the above expression are nodeless quantities (coefficients) associated to the q_e control points, except of those associated to the four corners (*A*, *B*, *C*, and *D*) for which they are the nodal values:

$$a_1 = U_A,\quad a_{q_1} = U_B,\quad a_{q_1+q_2-1} = U_C,\quad a_{q_1+q_2+q_3-2} = U_D \tag{3.124}$$

Obviously, the numerical results obtained using nonrational Bernstein-Bézier polynomials along the sides of a Coons macroelement will be identical to those obtained using Lagrange polynomials, as shown in the last column of Table 3.20. The position of the control points are shown in Fig. 3.46.

Remark 2 Both of the abovementioned formulations, i.e. the Lagrange and the nonrational Bernstein-Bézier based Coons elements are *different faces* of the Serendipity element. Particularly, the Lagrange polynomials based Coons macroelement is iden-

Table 3.20 Comparison of the previous 32-DOF "C-element" (based on three plus one new formulations) with the new 40-DOF "C-element" which is based on full cubic B-splines (de Boor)

Radius (R)	Exact solution	Errors (in %)				
		C-element (40 DOF) full cubic B-spline	C-element (32 DOF)			
			Model M1 piecewise-linear (2 × 2 gauss)	Model M2 reduced cubic B-splines (2 × 2 gauss)	Model M3 (Lagrange polynomials)	Bernstein-Bézier polynomials
1.0	1000.0	–	–	–	–	–
4.1	592.875	0.16	5.56	2.73	−0.07	−0.07
7.2	430.400	0.13	6.17	3.89	0.28	0.28
10.3	327.085	0.20	6.36	2.97	−0.22	−0.22
13.4	251.167	0.16	6.45	3.52	0.36	0.36
16.5	191.121	0.18	6.50	3.19	0.13	0.13
19.6	141.443	0.17	6.53	3.39	−0.27	−0.27
22.7	99.075	0.17	6.55	3.25	0.53	0.53
25.8	62.140	0.17	6.57	3.38	−0.16	−0.16
28.9	29.400	0.17	6.58	3.11	0.34	0.34
32.0	0.0	–	–	–	–	

tical to the Serendipity element without needed to perform any linear transformation at all.

Remark 3 Moreover, it is obvious that there is no problem to replace the abovementioned trial functions B_i^{SIDE} in the boundary function $[U(\xi, 0)$, $U(1, \eta)$, $U(\xi, 1)$ and $U(0, \eta)]$ with *rational* Bézier polynomials or even with NURBS, either for all sides of the Coons patch *ABCD* or for some of it. Such "mixed" Coons elements can be coupled partially (along some edges) with conventional FEM or other nodal-value based (U-based) elements, and partially with isogeometric elements (a-based).

In the past, the matter of considering Coons interpolation as a general NURBS problem has attracted the interest of CAGD researchers [17, 37]. Nevertheless, in the above discussion we did not need to recall new knowledge out of these papers.

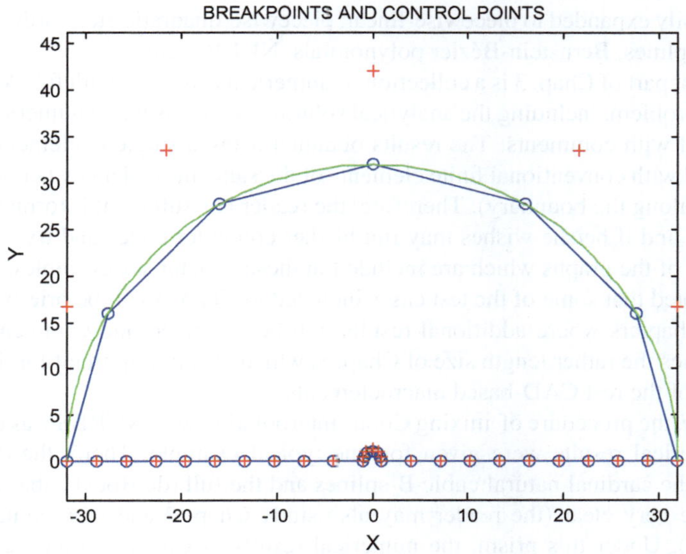

Fig. 3.46 Seven control points (+) and seven breakpoints (o) along the four sides of the *nonrational* Bézier based Coons patch macroelement ($n_x = 5,\ n_y = 10$)

3.9 Recapitulation

In Chap. 3 we dealt with the first type of CAD-based macroelements, called "C-elements" (where C stands for the name of Coons). These elements are based on Coons interpolation formula, which uses data only along the boundary of a Coons patch. Therefore, all analytical expressions concerning the global shape functions refer to a macroelement with nodal points arranged along the four edges (sides) of a quadrilateral patch *ABCD*, or a degenerated triangular patch *ABC*.

We saw that in the particular case the trial functions are Lagrange polynomials and the blending functions are linear functions, the Coons macroelements are identical with the classical elements of "Serendipity family". Of particular interest is the finding that classical isoparametric elements such as the 4-node (bilinear) and the 8-node (biquadratic) are directly produced using the receipt of Coons interpolation formula.

Although closed form expressions are found in all finite element textbooks regarding small-size elements (4-node, 8-node), here generalized expressions have been presented for larger elements as well. For example, in C-elements we can deal with different number of nodes along the opposite edges of a quadrilateral patch. Even the latter general case can be easily programmed in a single subroutine of general use.

But the most novel point is that, the above mentioned general expressions concerning the global shape functions do not restrict to Lagrange polynomials only, but can be easily expanded to piecewise-linear, piecewise-quadratic etc., cardinal natural cubic B-splines, Bernstein-Bézier polynomials, NURBS, etc.

The last part of Chap. 3 is a collection of numerical examples with full description for each problem, including the analytical solution as well as many numerical results associated with comments. The results obtained using a single C-element are also compared with conventional finite elements of the same mesh density (same number of nodes along the boundary). Therefore, the reader has sufficient information to be based on, and if he/she wishes may run his/her computer codes and try to validate the tables of the graphs which are included in these benchmark examples.

It is noted that some of the test cases included in Chap. 3 will be briefly repeated in next Chapters where additional results will be presented and commented. This fact justifies the rather length size of Chap. 3, which is very important for the understanding of the rest CAD-based macroelements.

Finally, the procedure of mixing Coons interpolation with NURBS was discussed and numerical results were given for one typical example. There, the difference between the cardinal natural cubic B-splines and the full (de Boor) cubic B-splines was made very clear (the reader may also study Chap. 7 and then come back to this topic). Under this prism, the numerical results of Chap. 3 could be repeated using coefficients a for all the intermediate control points (in the Bézier, B-splines or NURBS univariate approximations along each edge), whereas the four nodal values of the displacement (or potential) remain invariant at the four corners of the Coons patch.

References

1. Bathe KJ (1990) finite element procedures. Prentice-Hall, Upper Saddle River, New Jersey
2. Beissel S, Belytschko T (1996) Nodal integration of the element-free Galerkin method. Comput Methods Appl Mech Eng 139(1–4):49–74
3. Blevins RD (1979) Formulas for natural frequency and mode shape. Van Nostrand Reinhold, New York
4. Carslaw HS, Jaeger JC (1986) Conduction of heat in solids, 2nd edn. Oxford Science Publications, Oxford
5. Cavendish JC, Gordon WJ, Hall CA (1977) Substructured macro elements based on locally blended interpolation. Int J Numer Meth Eng 11:1405–1421
6. Cavendish JC, Hall CA (1984) A new class of transitional blended finite elements for the analysis of solid structures. Int J Numer Meth Eng 20:241–253
7. Coons SA (1964) Surfaces for computer aided design of space form. Project MAC, MIT (1964), revised for MAC-TR-41 (1967). Available as AD 663 504 from the National Technical Information Service (CFSTI), Sills Building, 5285 Port Royal Road, Springfield, VA 22161, USA. Available online: http://publications.csail.mit.edu/lcs/pubs/pdf/MIT-LCS-TR-041.pdf
8. Courant R, Hilbert D (1966) Methods of mathematical physics, vol I, 1st English edn (translated and revised from the German original). InterScience Publishers, New York
9. De Boor C (1972) On calculating with B-splines. J Appl Theory 6(1):50–62
10. Dimitriou V (2004) Adaptive finite elements and related meshes, Ph.D. Dissertation (advisor: Prof. Andreas E. Kanarachos), National Technical University of Athens, School of Mechanical Engineering, Athens, August, 2004 (in Greek). Online available from: http://thesis.ekt.gr/thesisBookReader/id/16430#page/132/mode/2up

11. Gordon WJ, Hall CA (1973) Transfinite element methods: blending-function interpolation over arbitrary curved element domains. Numer Math 21:109–112
12. Kanarachos A, Deriziotis D (1989) On the solution of Laplace and wave propagation problems using 'C-elements'. Finite Elem Anal Des 5:97–109
13. Kanarachos A, Provatidis C (1987) Performance of mass matrices for the BEM dynamic analysis of wave propagation problems. Comput Methods Appl Mech Eng 63:155–165
14. Kanarachos A, Provatidis C, Deriziotis D, Foteas N (1999) A new approach of the FEM analysis of two-dimensional elastic structures using global (Coons's) interpolation functions. In: Wunderlich J (ed) CD Proceedings first european conference on computational mechanics (ECCM'99): (Paper 763.pdf: pp 1–19), Munich-Germany, 28 Aug–3 Sept 1999. Available online at: http://users.ntua.gr/cprovat/yliko/763.pdf
15. Kreith F (1968) Principles of heat transfer, 2nd edn. International Textbook Company, Pennsylvania
16. Lamé G (1852) Leçons sur la Théorie Mathématique de l' Élasticité des Corps Solides. Gautheir-Villars, Paris
17. Lin F, Hewitt WT (1994) Expressing coons-gordon surfaces as NURBS. Comput Aided Des 26(2):145–155
18. Nardini D, Brebbia CA (1983) A new approach to free vibration analysis using boundary elements. Appl Math Model 7:157–162
19. Nardini D, Brebbia CA (1985) Boundary integral formulation of mass matrices for dynamic analysis. In: Brebbia CA (ed) Topics in boundary element research, vol 2. Springer, Berlin, pp 191–208
20. Pipes LA, Harvill LR (1981) Applied mathematics for engineers and physicists, 3rd edn. McGraw-Hill International Book Company, International Student Edition
21. Provatidis CG (2005) Analysis of box-like structures using 3-D Coons' interpolation. Commun Numer Methods Eng 21:443–456
22. Provatidis CG (2012) Two-dimensional elastostatic analysis using Coons-Gordon interpolation. Meccanica 47:951–967
23. Provatidis CG, Kanarachos AE (1995) Further research on the performance of consistent mass matrices using BEM for symmetric/nonsymmetric formulations. Comput Mech 16:197–207
24. Provatidis C, Kanarachos A (2001) Performance of a macro-FEM approach using global interpolation (Coons') functions in axisymmetric potential problems. Comput Struct 79:1769–1779
25. Provatidis CG (2002) A comparative study between Coons-patch macroelements and boundary elements in two-dimensional potential problems. In: Tsahalis D (ed) Proceedings 4th GRACM congress on computational mechanics, Patras, 27–29 June, 2002, pp 43–50
26. Provatidis C (1987) On the application of the boundary element method in the analysis of field and dynamic problems. Doctoral Dissertation, National Technical University of Athens, Mechanical Engineering Department, Greece, November 1987 (in Greek). Available online at: https://www.didaktorika.gr/eadd/handle/10442/0616
27. Provatidis CG (2004) On DR/BEM for eigenvalue analysis of 2-D acoustics. Comput Mech 35:41–53
28. Provatidis C, Kanarachos A (2000) On the use of Coons' interpolation in CAD/CAE systems. In: Mastorakis N (ed.) Systems and control: theory and applications. World Scientific and Engineering Society Press, pp 343–348; Second international conference on mathematics and computers in mechanical engineering (MCME 2000), pp 10–15 July 2000, Vouliagmeni, Greece. (ISBN: 960-8052-11-4). Online available on: http://www.wseas.us/e-library/conferences/athens2000/Papers2000/547.pdf
29. Provatidis C G (2003) Analysis of axisymmetric structures using Coons' interpolation. Finite Elem Anal Des 39:535–558
30. Provatidis CG (2004) Coons-patch macroelements in two-dimensional eigenvalue and scalar wave propagation problems. Comput Struct 82(4–5):383–395
31. Provatidis C (2003) Frequency analysis and transient response of two-dimensional structures using Coons-patch macroelements. In: Brennan MJ, Ferman MA, Rizzi SA, Wenz K (eds)

Proceedings of the VIII international conference on recent advances in structural dynamics. Organized by the Institute of Sound & Vibration Research University of Southampton, Southampton, UK, 14–16 July 2003
32. Provatidis CG (2006) Free vibration analysis of two-dimensional structures using Coons-patch macroelements. Finite Elem Anal Des 42(6):18–531
33. Provatidis CG (2006) Transient elastodynamic analysis of plane structures using coons-patch macroelements and modal superposition. In: The eighth international conference on computational structures technology, 12–15 Sept 2006, Las Palmas de Gran Canaria—Spain (https://doi.org/10.4203/ccp.83.197. You can find your paper webpage in the repository by following the link http://dx.doi.org/10.4203/ccp.83.197)
34. Provatidis CG (2006) Coons-patch macroelements in two-dimensional parabolic problems. Appl Math Model 30:319–351
35. Provatidis CG (2006) Transient elastodynamic analysis of two-dimensional structures using Coons-patch macroelements. Int J Solids Struct 43(22–23):6688-6706
36. Timoshenko S, Goodier JN (1970) Theory of elasticity, 3rd edn. McGraw-Hill, New York
37. Wang L, Zhu X, Tang Z (1997) COONS type blended B-spline (CNSBS) surface and its conversion to NURBS surface. Comput Graph 21(3):297–303
38. Weaver W Jr, Timoshenko SP, Young DH (1990) Vibration problems in engineering, 5th edn. Wiley, New York
39. Zienkiewicz OC (1977) The finite element method. McGraw-Hill, London
40. Zienkiewicz OC, Taylor RL (1988) The finite element method. In: Basic Formulation and Linear Problems, 4th edn, vol 1. McGraw-Hill, London

Chapter 4
GORDON's Transfinite Macroelements

Abstract This chapter discusses transfinite macroelements, which are based on Gordon interpolation formula. The latter extends Coons interpolation formula (see Chap. 3) considering internal nodes as well. It will be shown that the standard tensor-product elements of Lagrange family constitute a subclass of transfinite elements, while one may generally use more or less internal nodes in several configurations. Moreover, true transfinite elements with different pattern in the arrangement of the internal nodes, as well as degenerated triangular macroelements, are discussed. A class of **Cij** macroelements is introduced, by influencing the trial functions as well as the blending functions. This class is so wide that can include even an assemblage of conventional bilinear elements in a structured $n_\xi \times n_\eta$ arrangement. A careful programming of the shape functions and their global partial derivatives resulted in a single subroutine that includes all twelve combinations. The theory is supported by several test cases that refer to potential and elasticity problems in simple domains of primitive shapes where a single macroelement is used. In a couple of cases, somehow more complex domains are successfully treated using two or three Gordon macroelements.

Keywords GORDON interpolation · Transfinite macroelement
Global shape functions · Test cases

4.1 General Formulation

In Chap. 2, we presented Gordon interpolation, which is an enhancement of Coons interpolation where inter-boundaries have been posed. Therefore, the essential difference of Gordon's compared to Coons interpolation is that, in the former each projection depends on boundary data as well as on internal data along lines of $\xi =$ const and $\eta =$ const.

If we consider a patch *ABCD* (see Fig. 4.1) in which the normalized parameters ξ and η are directed toward the edges *AB* and *AD*, respectively, whereas $(n_\xi - 1)$

C. G. Provatidis, *Precursors of Isogeometric Analysis*, Solid Mechanics and Its Applications 256, https://doi.org/10.1007/978-3-030-03889-2_4

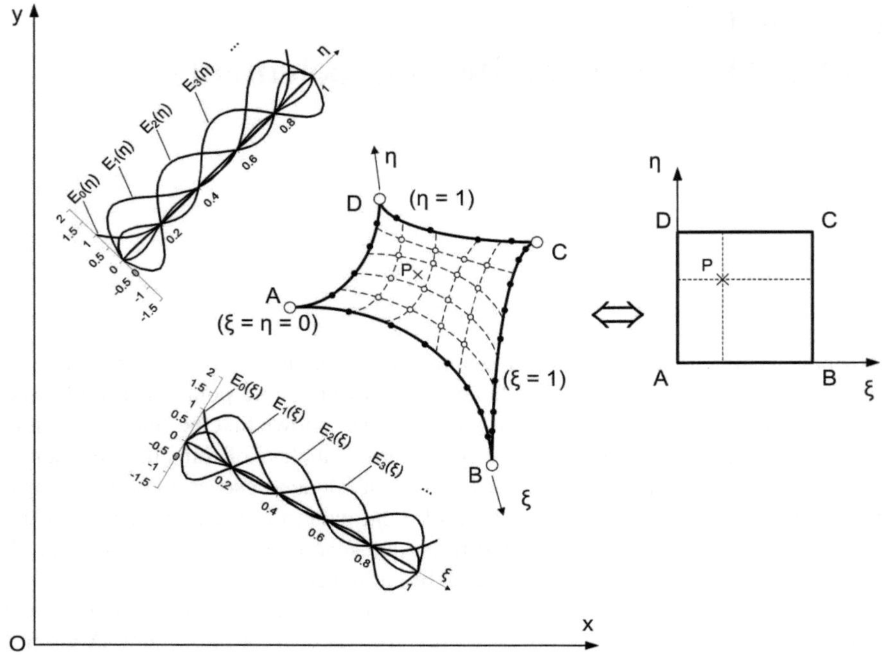

Fig. 4.1 Gordon transfinite macroelement

internal lines of constant value ξ and $(n_\eta - 1)$ internal lines of constant value η are added, Gordon *transfinite* interpolation formula is:

$$U(\xi, \eta) = P_\xi(U) + P_\eta(U) - P_{\xi\eta}(U) \tag{4.1}$$

with

$$\left.\begin{aligned} P_\xi(\mathbf{w}) &= \sum_{i=0}^{n_\xi} \mathbf{w}(\xi_i, \eta) \cdot E_i(\xi), \\ P_\eta(\mathbf{w}) &= \sum_{j=0}^{n_\eta} \mathbf{w}\big(\xi, \eta_j\big) \cdot E_j(\eta) \\ P_{\xi\eta}(\mathbf{w}) &= P_\xi P_\eta(\mathbf{w}) = \sum_{i=0}^{n_\xi} \sum_{j=0}^{n_\eta} \mathbf{w}\big(\xi_i, \eta_j\big) \cdot E_i(\xi) \cdot E_j(\eta). \end{aligned}\right\} \tag{4.2}$$

The position of the lines "perpendicular" to the ξ-axis (including the edges AD and BC, see Fig. 4.1) is:

$$\big[\boldsymbol{\xi}\big] = \big[0 = \xi_0, \xi_1, \ldots, \xi_{n_\xi} = 1\big], \tag{4.3}$$

whereas the position of the rest lines "perpendicular" to the η-axis (including the edges *AB* and *CD*, see Fig. 4.1) is:

$$[\boldsymbol{\eta}] = \left[0 = \eta_0, \eta_1, \ldots, \eta_{n_\eta} = 1\right] \tag{4.4}$$

In Eq. (4.2), E_i denote the blending functions, which are no more linear, and fulfill the condition:

$$E_i(\xi_m) = \delta_{im} \tag{4.5}$$

Based on the formulation of Eq. (4.1), it becomes possible to construct large finite elements, called *transfinite* elements, with both boundary and internal nodes [11]. We recall that Coons interpolation produces boundary-only macroelements, which means that the nodal points are arranged along the boundaries of the patch. Later, El-Zafrany and Cookson [9, 10] proposed similar elements using Lagrange (C^0-continuity) and Hermitian (C^1-continuity) concepts, from a quite different departure. Nevertheless, no numerical results were presented by them and this was a motivation to include many of them in this book.

Example 4.1 (The 5-node transfinite element) The way of deriving the global shape functions will become clear after the Gordon transfinite formula has been implemented to the 5-node isoparametric element shown in Fig. 4.2 (see, Provatidis [17]).

It is clarified that toward one direction (say the ξ-direction), the data are known along three positions: $\xi = 0, \frac{1}{2}, 1$ which correspond to the lines *AD*, *GH*, and *BC*, respectively (Fig. 4.2). Therefore, the aforementioned three positions of ξ determine blending functions $E_i(\xi)$ which are Lagrange polynomials of second degree. This is consistent with the fact that the number of formed subdivisions by the aforementioned three iso-ξ lines (due to the internal node "5") is $n_\xi = 2$:

$$E_0(\xi) = (2\xi - 1)(\xi - 1),\ E_{1/2}(\xi) = 4\xi(1 - \xi),\ E_1(\xi) = \xi(2\xi - 1). \tag{4.6a}$$

Similarly, in the η-direction data are given along $\eta = 0, \frac{1}{2}, 1$. This fact justifies the use of quadratic blending functions $E_j(\eta)$:

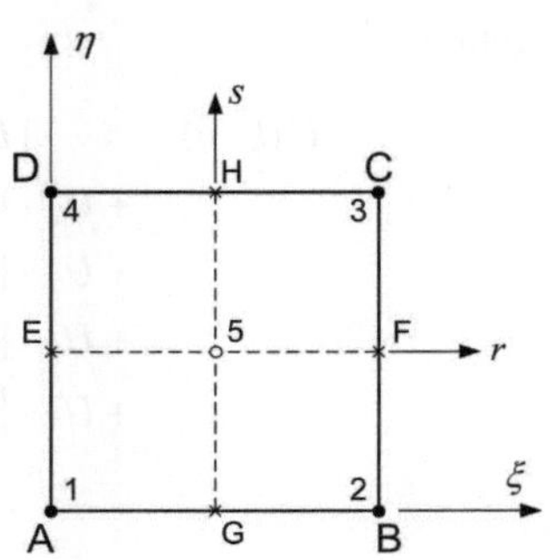

Fig. 4.2 Five-node isoparametric element

$$E_0(\eta) = (2\eta - 1)(\eta - 1),\ E_{1/2}(\eta) = 4\eta(1 - \eta),\ E_1(\eta) = \eta(2\eta - 1). \tag{4.6b}$$

Based on the abovementioned blending functions, the starting point is Eq. (4.1), in which the three lofting operators are given by:

$$P_\xi(U) = E_0(\xi) \cdot U(0, \eta) + E_{1/2}(\xi) \cdot U(1/2, \eta) + E_1(\xi) \cdot U(1, \eta) \tag{4.7a}$$

$$P_\eta(u) = E_0(\eta) \cdot u(\xi, 0) + E_{1/2}(\eta) \cdot u(\xi, 1/2) + E_1(\eta) \cdot u(\xi, 1) \tag{4.7b}$$

$$\begin{aligned} P_{\xi\eta}(U) = {} & E_0(\xi)E_0(\eta) \cdot U(0, 0) + E_1(\xi)E_0(\eta) \cdot U(1, 0) + E_1(\xi)E_1(\eta) \cdot U(1, 1) \\ & + E_0(\xi)E_1(\eta) \cdot U(0, 1) + E_{1/2}(\xi)E_0(\eta) \cdot U_G + E_1(\xi)E_I(\eta) \cdot U_F \\ & + E_{1/2}(\xi)E_1(\eta) \cdot U_H + E_0(\xi)E_{1/2}(\eta) \cdot U_E + E_{1/2}(\xi)E_{1/2}(\eta) \cdot U_I, \end{aligned} \tag{4.7c}$$

One may observe that $P_{\xi\eta}$ involves nine terms, which are the intersections of the two sets of isolines: $\boldsymbol{\xi} = [0, \frac{1}{2}, 1]$ and $\boldsymbol{\eta} = [0, \frac{1}{2}, 1]$.

We further consider that along each edge (*AB*, *BD*, *CD*, and *DA*) the variable U is interpolated using Lagrange polynomials (*trial* functions) of first degree:

$$L_0(\xi) = 1 - \xi,\ L_1(\xi) = \xi \quad \text{(similarly for } \eta), \tag{4.8}$$

whereas along the lines *EF* and *GH* the functions $U(\xi, \frac{1}{2})$ and $U(\frac{1}{2}, \eta)$ are interpolated in terms of quadratic Lagrange polynomials which obviously coincide with the blending functions of Eqs. (4.6a, b).

Therefore:

$$\begin{aligned} U(0, \eta) &= L_0(\eta)U_1 + L_1(\eta)U_4 \\ U\left(\tfrac{1}{2}, \eta\right) &= E_0(\eta)U_G + E_I(\eta)U_5 + E_1(\eta)U_H \\ U(1, \eta) &= L_0(\eta)U_2 + L_1(\eta)U_3 \\ U(\xi, 0) &= L_0(\xi)U_1 + L_1(\xi)U_2 \\ U\left(\xi, \tfrac{1}{2}\right) &= E_0(\xi)U_E + E_I(\xi)U_5 + E_1(\xi)U_F \\ U(\xi, 1) &= L_0(\xi)U_4 + L_1(\xi)U_3. \end{aligned} \tag{4.9}$$

Substituting Eq. (4.9) into Eqs. (4.7a–c) and then into Eq. (4.1), one eventually obtains:

$$\begin{aligned} U(\xi, \eta) = {} & U_1 \cdot [L_0(\xi)E_0(\eta) + E_0(\xi)L_0(\eta) - E_0(\xi)E_0(\eta)] \\ & + U_2 \cdot [L_1(\xi)E_0(\eta) + E_1(\xi)L_0(\eta) - E_1(\xi)E_0(\eta)] \\ & + U_3 \cdot [L_1(\xi)E_1(\eta) + E_1(\xi)L_1(\eta) - E_1(\xi)E_1(\eta)] \\ & + U_4 \cdot [L_0(\xi)E_1(\eta) + E_0(\xi)L_1(\eta) - E_0(\xi)E_1(\eta)] \\ & + U_5 \cdot \left[E_{1/2}(\xi)E_{1/2}(\eta)\right]. \end{aligned} \tag{4.10}$$

Obviously, the five quantities inside the brackets of Eq. (4.10) represent the five *shape* functions of the 5-node element which were obtained through the transfinite Gordon interpolation.

Let us now see whether Eq. (4.10) is relevant to the classical FEM theory. Actually, according to Bathe [1, p. 344], the conventional shape functions for a 5-node isoparametric element are given by the intuitively derived formulas:

$$
\begin{aligned}
N_1(r,s) &= h_1 - \frac{1}{4}h_5 = \frac{(1-r)(1-s)}{4} - \frac{1}{4}(1-r^2)(1-s^2) \\
N_2(r,s) &= h_2 - \frac{1}{4}h_5 = \frac{(1+r)(1-s)}{4} - \frac{1}{4}(1-r^2)(1-s^2) \\
N_3(r,s) &= h_3 - \frac{1}{4}h_5 = \frac{(1+r)(1+s)}{4} - \frac{1}{4}(1-r^2)(1-s^2) \\
N_4(r,s) &= h_4 - \frac{1}{4}h_5 = \frac{(1-r)(1+s)}{4} - \frac{1}{4}(1-r^2)(1-s^2) \\
N_5(r,s) &= h_5 = \frac{1}{4}(1-r^2)(1-s^2)
\end{aligned}
\tag{4.11}
$$

where h_1, h_2, h_3 and h_4 denote the shape functions of the 4-node (bilinear) element and h_5 is the "bubble" function $\frac{1}{4}(1-r^2)(1-s^2)$. Using the transformation $r = 2\xi - 1, s = 2\eta - 1 (-1 \le r, s \le 1$, and $0 \le \xi, \eta \le 1)$, it is trivial to prove that the shape functions which were derived in (4.10) through transfinite Gordon–Coons interpolation *coincide* with those of the well-known finite element theory (4.11).

4.2 General Formulation

Now, analytical expressions of the global shape functions will be derived. Therefore, let us assume that the four edges (*AB*, *BC*, *CD*, and *DA*) of the macroelement *ABCD* are discretized into $\hat{n}_1$, $\hat{n}_2$, $\hat{n}_3$ and $\hat{n}_4$ segments, so that the corresponding nodal points per side will be q_1, q_2, q_3 and q_4, respectively. Obviously, $q_i = \hat{n}_i + 1, i = 1, 2, 3, 4$. Since the corner points (*A*, *B*, *C*, and *D*) appear twice in two successive boundary sides, the total number of the *boundary* nodal points becomes

$$q_{\mathrm{B}} = (q_1 + q_2 + q_3 + q_4) - 4. \tag{4.12a}$$

In addition to the abovementioned boundary nodes which are compulsory, a mesh of internal nodes generally exists. For the assumed (n_ξ, n_η) subdivisions, the number of *internal* nodes is:

$$q_{\mathrm{I}} = (n_\xi - 1) \times (n_\eta - 1) \tag{4.12b}$$

Therefore, the *total* number of nodal points in this typical Gordon macroelement is:

$$q_{\mathrm{e}} = q_{\mathrm{B}} + q_{\mathrm{I}}. \tag{4.12c}$$

In this chapter, the numbering of the boundary nodes begins from the corner A and continues in the counterclockwise direction until the boundary closes. Then, the internal nodes follow, from the $(q_{\mathrm{B}} + 1)$th until the q_{e}th node.

If the boundary values $U(\xi, 0)$, $U(\xi, 1)$, $U(0, \eta)$ and $U(1, \eta)$ in Eq. (4.1) are interpolated by any set of trial functions $B_j(w)$ (where w is either of ξ or η; also the upper index in B_j below corresponds to the pertinent edge):

$$\text{Edge } AB: \quad U(\xi, 0) = \sum_{j=1}^{q_1} B_j^{AB}(\xi) U(\xi_j, 0) \tag{4.13a}$$

$$\text{Edge } BC: \quad U(1, \eta) = \sum_{j=1}^{q_2} B_j^{BC}(\eta) U(1, \eta_j) \tag{4.13b}$$

$$\text{Edge } CD: \quad U(\xi, 1) = \sum_{j=1}^{q_3} B_j^{CD}(\xi) U(\xi_j, 1) \tag{4.13c}$$

$$\text{Edge } DA: \quad U(0, \eta) = \sum_{j=1}^{q_4} B_j^{DA}(\eta) U(0, \eta_j) \tag{4.13d}$$

and then they are introduced into Eqs. (4.1)–(4.2), exactly as was performed in Example 4.1, the global shape functions associated to all nodes of the reference macroelement through equation:

$$U(\xi, \eta) = \sum_{j=1}^{q_{\mathrm{e}}} N_j(\xi, \eta) U_j \tag{4.14}$$

are derived as follows:

(i) *Corner nodes*:

$$\begin{aligned}
N_A(\xi, \eta) &= E_0(\xi) B_{q_4}^{DA}(\eta) + E_0(\eta) B_1^{AB}(\xi) - E_0(\xi) E_0(\eta) \\
N_B(\xi, \eta) &= E_1(\xi) B_1^{BC}(\eta) + E_0(\eta) B_{q_1}^{AB}(\xi) - E_1(\xi) E_0(\eta) \\
N_C(\xi, \eta) &= E_1(\xi) B_{q_2}^{BC}(\eta) + E_1(\eta) B_1^{CD}(\xi) - E_1(\xi) E_1(\eta) \\
N_D(\xi, \eta) &= E_0(\xi) B_1^{DA}(\eta) + E_1(\eta) B_{q_3}^{CD}(\xi) - E_0(\xi) E_1(\eta)
\end{aligned} \tag{4.15}$$

(ii) *Intermediate nodes along the edge AB (local numbering)*:

$$N_j(\xi, \eta) = E_0(\eta) B_j^{AB}(\xi), \quad 2 \le j \le q_1 - 1 \tag{4.16}$$

(iii) *Intermediate nodes along the edge BC (local numbering)*:

$$N_j(\xi,\eta) = E_1(\xi)B_j^{BC}(\eta), \quad 2 \le j \le q_2 - 1 \tag{4.17}$$

(iv) *Intermediate nodes along the edge CD (local numbering)*:

$$N_j(\xi,\eta) = E_1(\eta)B_j^{CD}(\xi), \quad 2 \le j \le q_3 - 1 \tag{4.18}$$

(v) *Intermediate nodes along the edge DA (local numbering)*:

$$N_j(\xi,\eta) = E_0(\xi)B_j^{DA}(\eta), \quad 2 \le j \le q_4 - 1 \tag{4.19}$$

(vi) *Internal nodes (local numbering)*:

$$N_j(\xi,\eta) = E_k(\xi)E_l(\eta); \quad k = 1,\dots,n_\xi - 1,\ l = 1,\dots,n_\eta - 1. \tag{4.20}$$

4.3 A Structured Mesh of Bilinear Elements as a Special Case of Gordon Macroelement

Lemma 4.1 *Let us assume that ($n_1 = n_3 = n_\xi$, $n_2 = n_4 = n_\eta$, i.e., equal number of subdivisions along the opposite edges of the Gordon patch) and all nodes (boundary and internal ones) are distributed along lines of constant ξ- and η-values (see Fig. 4.3). Then, when using piecewise-linear interpolation for both the trial ($B_j(\hat{\xi})$) and the blending functions ($E_i(\hat{\xi})$), Eqs. (4.15) up to (4.30) produce a mesh of $n_\xi \times n_\eta$ conventional four-node bilinear finite elements.*

Proof Considering the compact support of the piecewise-linear functions (according to Fig. 2.4) and substituting into either of Eqs. (4.15)–(4.30), the proof is quite trivial. ■

As an example, the assemblage of the thirty bilinear (4-node) finite elements shown in Fig. 4.3 is mathematically equivalent to a single 42-node transfinite macroelement, which is defined by:

- Piecewise-linear *trial* functions, $B_i(\xi)$, $i = 1,\dots,7$ along the ξ-direction (shown in Fig. 4.3), and $B_j(\eta)$, $j = 1,\dots,6$ along the η-direction (not shown).
- Piecewise-linear *blending* functions, $E_i(\xi)$, $i = 1,\dots,7$ along the ξ-direction (shown in Fig. 4.3), and $E_j(\eta)$, $j = 1,\dots,6$ along the η-direction (not shown).

For the better understanding, the trial function $B_3(\xi)$ and the blending function $E_1(\xi)$ have been drawn by a thicker line (colored in the Internet version).

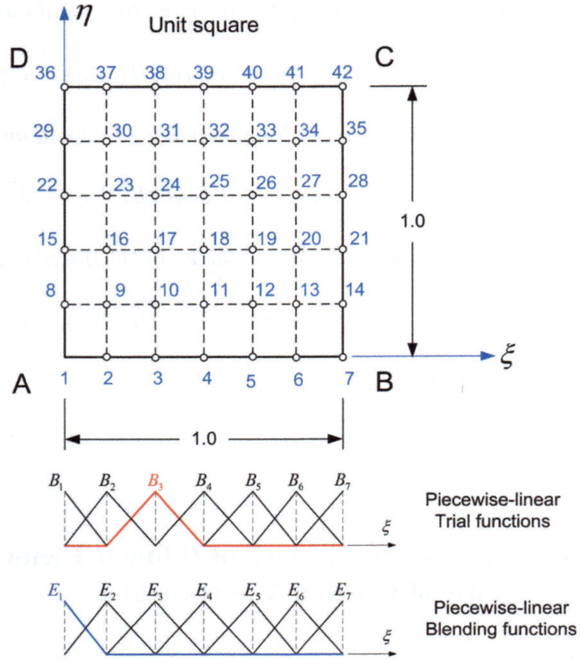

Fig. 4.3 Uniform mesh of conventional bilinear finite elements, considered as a transfinite macroelement in the unit square (parametric domain)

4.4 Lagrangian-Type Elements as a Special Case of Gordon Macroelement

Theorem 4.1 *In the particular case of using Lagrange polynomials for both the trial* (B_i) *and the blending* (E_j) *functions, in conjunction with uniformly arranged internal nodes so that their projection on the four edges of the reference square coincide with all the intermediate boundary nodes (see Fig. 4.4a), the Gordon macroelement becomes identical with the classical finite element of Lagrange type.*

Proof From the posed conditions, it is obvious that the blending functions $E_i(\xi)$ and $E_j(\eta)$ are *identical* with the trial functions $B_i(\xi)$ and $B_j(\eta)$, respectively, as they share the same number and relative position of nodes toward ξ- and η-direction. Introducing the symbol $L_i(w)$ of a Lagrangian polynomial in the w-direction, we have: $B_i(\xi) = E_i(\xi) = L_i(\xi)$ and $B_i(\eta) = E_i(\eta) = L_i(\eta)$. We shall prove that the coincidence between the Gordon macroelement and the classical Lagrangian element is valid for all three characteristic cases, i.e., (i) for an internal node, (ii) for an intermediate along an edge, and (iii) for one node at a corner.

First, concerning the internal nodes Eq. (4.20) yields that for this particular Gordon macroelement:

$$N_j(\xi,\eta) = L_k(\xi)L_l(\eta);\quad k = 1,\ldots,n_\xi - 1,\ l = 1,\ldots,n_\eta - 1, \tag{4.21}$$

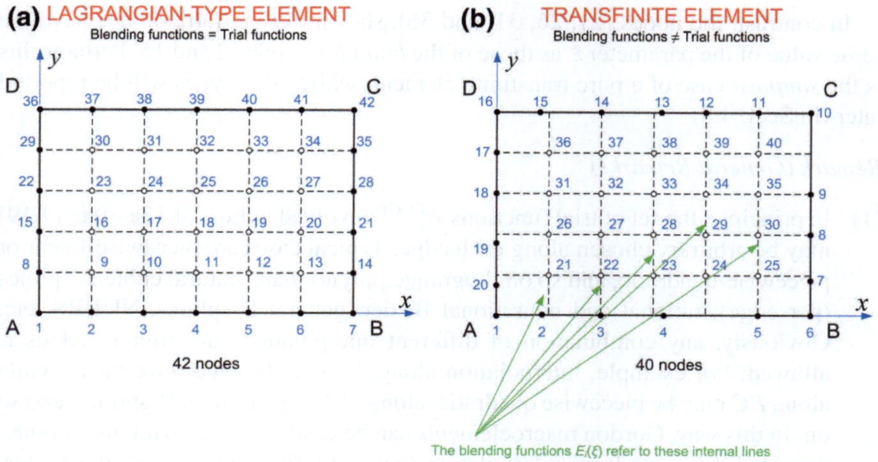

Fig. 4.4 Generalized Gordon macroelements: **a** classical Lagrange and **b** fully transfinite element

which is exactly the expression for the classical element of Lagrange family.

Second, for an intermediate point along an edge, say AB, Eq. (4.16) becomes:

$$N_j(\xi,\eta) = L_j(\xi)L_1(\eta), \quad 2 \le j \le q_1 - 1, \tag{4.22}$$

which again is exactly the same expression with the classical Lagrangian element.

Third, for a corner node such as A, the first relationship in Eq. (4.15) includes successively the terms $E_0(\xi)B_{q_4}^{DA}(\eta) = L_1(\xi)L_1(\eta)$, $E_0(\eta)B_1^{AB}(\xi) = L_1(\eta)L_1(\xi)$, and $-E_0(\xi)E_0(\eta) = -L_1(\xi)L_1(\eta)$, of which the sum is obviously $N_A(\xi,\eta) = L_1(\xi)L_1(\eta)$, as was anticipated.

Since the coincidence of this particular Gordon macroelement with the classical finite element of Lagrange type was shown for all three possible positions of a nodal point, the proof of the Theorem 4.1 was completed. ■

Within this context, the above theorem concerns a structured transfinite macroelement (i.e., of Lagrange family) like that shown in Fig. 4.4a. One may observe that any sequence of nodal points perpendicular to the ξ-axis (such as the nodes 2, 9, 16, 23, 30, and 37) have the *same* value of the parameter ξ (normalized coordinate). A similar observation can be made for any sequence of nodes parallel to the η-direction (e.g., nodes 8–14).

In contrast, the nodes (21, 26, 31, and 36), shown in Fig. 4.4b, do *not* have the same value of the parameter ξ as those of the *boundary* nodes 2 and 15. Perhaps this is the *simplest* case of a pure transfinite element, while other types will be reported later (in Sect. 4.8).

Remark (General Remarks)

(1) In principle, the set of trial functions B_j^{EDGE} involved in Eqs. (4.13a–d) to (4.19) may be arbitrary chosen along each edge. Typical cases are piecewise-linear or piecewise-quadratic, and so on, Lagrange polynomials, natural cubic B-splines (per edge), rational and nonrational Bézier, general B-splines, NURBS, etc. Obviously, any combination of different interpolations at different edges is allowed. For example, interpolation along AB may be piecewise linear while along BC may be piecewise quadratic, along CD may be cubic B-splines, and so on. In this way, Gordon macroelements can be easily coupled with one another, provided the *same* local interpolation (compatibility) exists along their interface. The case of mismatched Gordon transfinite macroelements is covered in Chap. 13.
(2) Also, the Gordon macroelement can be easily coupled with conventional finite elements. For example, piecewise-linear macroelements can be coupled with bilinear (three- and four-node) finite elements while piecewise-quadratic macroelements can be coupled with quadratic (six-node triangular as well as eight-node or/and nine-node isoparametric) finite elements. Although the Gordon macroelement can stand alone as novel software, the aforementioned Gordon/FEM coupling may become necessary at a specific instance during the product development cycle when adding a new Coons macroelement on a pre-existing mesh of conventional finite elements representing a certain structural component that we do not want to modify. In other words, these macroelements can enrich the library of any commercial FEM code. Quite similarly, in case of combined interior/exterior domains such as those requiring vibro-acoustic analysis, a structural Gordon macroelement can be combined with conventional acoustic boundary elements of the same interpolation (i.e., linear, quadratic, splines).
(3) Since different edges of the Gordon patch may be interpolated using either Lagrange polynomials or general NURBS, it becomes evident that a Gordon macroelement may be connected at the same time with a region of FEM elements (along one side) and NURBS elements along another side.
(4) As previously mentioned, Eqs. (4.15)–(4.19) are valid in conjunction with higher-order blending functions (E_i), while the shape functions associated to internal nodes are tensor products of the aforementioned blending functions (Eq. 4.20). In general, the trial functions may be piecewise-linear polynomials, cubic B-splines, Lagrange polynomials, etc. Also, the blending functions should be *cardinal* functions (i.e., of [0, 1] type: $E_i(\xi_j) = \delta_{ij}$) such as Lagrange polynomials (the most applicable), natural cubic B-splines, etc. Thus a large number of possible Gordon macroelement classes may arise.

(5) Following to the above comment, and in accordance with previous definitions where several classes of Coons elements had been characterized by the symbol "**C***ij*" (see, Provatidis [22]), these macroelements will be denoted by the symbol "$\mathbf{G_{ij}}$", where "*G*" stands for "Gordon". The first index $i = 1, 2, 3$, etc., stands for the type of trial functions (B_k), whereas the second index $j = 0, 1, 2, 3$ stands for the type of the blending functions (E_l). In more detail, for both indexes (*i*, *j*), "1" refers to *piecewise linear*, "2" to *natural cubic B-splines*, and "3" to *Lagrange polynomials* interpolation, whereas $j = 0$ (model $\mathbf{G}_{i0}$) refers to the *boundary-only* formulation (absence of internal nodes).

(6) The first index *i* may be extended so as to include nonrational Bernstein–Bézier ($i = 4$) and rational Bernstein–Bézier polynomials ($i = 5$) as well as NURBS ($i = 6$). Moreover, piecewise quadratic, etc., polynomials can be added, although all these could be also considered as special cases of NURBS.

(7) In addition, conventional finite element formulation using four-node bilinear elements, denoted by FEM, is applied for comparison purposes. Recalling the above remark as well as Lemma 4.1, for a structured mesh $(n_1 = n_3 = n_\xi, n_2 = n_4 = n_\eta)$ it becomes obvious that "the FEM model (of $n \times m$ bilinear elements) theoretically coincides with the abovementioned model **G**11."

(8) In all cases, in principle there is no restriction for the relative location of the boundary nodes, so that both uniform and nonuniform meshes can be used.

4.5 State-of-the-Art

Historically, Ritz–Galerkin-based transfinite elements were proposed for the first time by Gordon and Hall [11] at a period William (Bill) Gordon was at General Motors Research Laboratories (Warren, Michigan, USA). In that pioneering paper (of very mathematical character), the same interpolation was used for both the geometry and the approximation of the problem variable U. The implementation of these elements in analysis is due to James Cavendish (also at General Motors Research Laboratories), who published on this topic jointly with Gordon [4, 5]. James Cavendish published also on the external problem jointly with Hall and Zienkiewicz [6] and finally extended the theory jointly with Hall [7] using the in-house CAD/CAE system GMSOLID at General Motors. Regarding Bill Gordon, since 1979 he had moved to Drexel University where he also published similarly in transfinite methods jointly with Linda Thiel [12] (see, photograph in Fig. 4.5). In my view, the aforementioned papers are the pioneering precursors of CAD/FEA integration and therefore of the so-called isogeometric analysis (IGA) of today.

Nevertheless, not many numerical results were reported on benchmark engineering problems with full detail, and therefore, the performance of these elements still remains unclear. The next Sect. (4.6) is going to cover this gap at a certain degree.

Furthermore, in the textbook of Zienkiewicz [25, pp. 157, 176] it is reported that in the context of an undergraduate project, a quartic element with a *central* node

Fig. 4.5 Linda Thiel in 2018 (with her Courtesy)

was developed by Scott [23] (see, Zienkiewicz [25], pp. 157, 176). Probably, this could be considered as an early transfinite element (in the lack of information, it is hypothesized that it was derived by inspection). Also, in Chap. 3 we presented several test cases in which a single Coons macroelemenet ("C-element") with all its nodal points along its boundaries was used. Moreover, we discussed in detail about the relationship between a "C-element" and the Serendipity family. In addition, the capability of Coons interpolation to represent the family of Serendipity elements was noticed also by T. Sederberg (see in the "Acknowledgment" of the paper by Bercovier and Shilat [2, p. 264]).

Applying Gordon interpolation formula, a class of Poisson problems was solved using one internal node only, which is allowed to be located at an arbitrary position [16]. Among others, the latter paper included two examples in a circular domain and one example in an elliptical domain, all solved using a *single* Gordon transfinite macroelement. In all these cases, the *parameterization* (of the circle or the ellipse) was established as follows. The boundary was divided into four equal parts that were mapped onto the four sides of a unit square *ABCD*. Each of the aforementioned parts (say *AB*) was uniformly divided into n_s segments, thus creating $4n_s$ boundary nodes (although a different subdivision is allowed for each particular side). Then Coons interpolation was blindly applied, thus creating a large isoparametric element. Although that paper used piecewise-linear and Lagrange polynomials as univariate trial functions, if the reader of this book has already read and understood the theory in Chap. 2, he/she will immediately understand that any other reasonable approximation (e.g., piecewise quadratic, natural cubic B-splines, de Boor B-splines) will work as

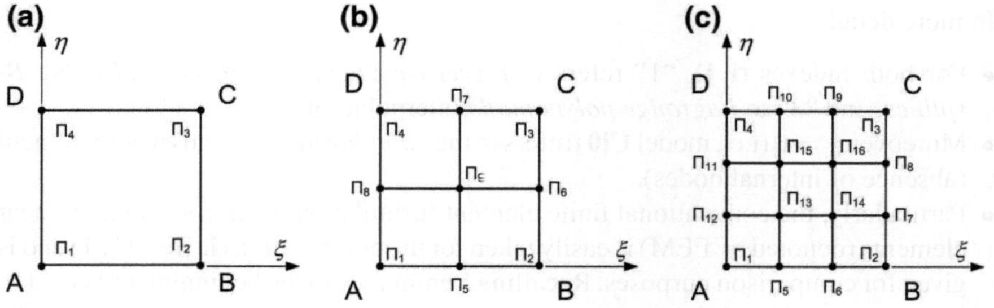

Fig. 4.6 Gordon and equivalent Lagrange elements **a** linear, **b** quadratic, **c** cubic

well. This work was followed by Provatidis [17], where not only elliptic but also parabolic problems were successfully solved.

The coincidence of a structured Gordon element with the (tensor product) Lagrangian type one was discussed in detail for the first time by Provatidis ([18], see therein the "Remark" on pp. 327–328). A special case in which the number of subdivisions is equal in both directions ($n = m$) is shown in Fig. 4.6.

The significance of Gordon transfinite elements, which are based on internal nodes as well, was also reported by Dimitriou [8], while Provatidis definitely demonstrated this fact (I should admit with considerable delay) in a couple of works [21, 22].

A presentation of many other possible transfinite elements will be shown later in Sect. 4.8, where we will come back after the presentation of some typical test cases in engineering analysis.

4.6 Test Problems

Several benchmark problems will be solved and the performance of Gordon transfinite elements will be commented. We start with potential problems and then continue with problems in elasticity.

Standardized transfinite models (Cij)

A general purpose subroutine was written, which covers all *twelve* models of Gordon–Coons transfinite macroelements. Following Provatidis [22], these combinations are denoted by "**Cij**", where "**C**" stands for the word "Coons", the first index (**i** = 1, 2, 3) stands for the type of trial functions B_i, while the second index (**j** = 0, 1, 2, 3) stands for the type of the blending functions E_j.

In more detail:

- For both indexes $(\mathbf{i}, \mathbf{j})$, "1" refers to *piecewise linear*, "2" to *natural cubic B-splines*, and "3" to *Lagrange polynomials* interpolation.
- Moreover, $\mathbf{j} = 0$ (i.e., model **Ci0**) refers to the *boundary-only* Coons macroelement (absence of internal nodes).
- Particularly, the conventional finite element formulation using four-node bilinear elements (denoted by FEM) is easily taken for the combination $(\mathbf{i}, \mathbf{j}) = (1, 1)$ and is given for comparison purposes. Recalling Lemma 4.1 in the beginning of Sect. 4.4, for a structured mesh $\left(n_1 = n_3 = n_\xi, n_2 = n_4 = n_\eta\right)$, it becomes obvious that the FEM model theoretically coincides with this particular transfinite model.

Example 4.2 (Elliptical section of a prismatic beam in torsion) This example concerns a potential problem within a curved domain, in which the Poisson's equation ($\nabla^2 U = -2$, with $U = 0$ along the circumference) governs the torsion of a uniform beam with an elliptic cross section of dimensions $a \times b = 2 \times 1$ (a and b denote the semi-axes). The numerical solution is performed in many ways, using conventional 3-node triangular elements, 4-node quadrilateral (bilinear) elements, as well as a *single* macroelement of Lagrange type (higher-order) having the same number of nodes as the previously mentioned assemblages of small size conventional elements.

The accurate solution is given by:

$$U(x, y) = 0.8\left(1 - \frac{x^2}{4} - y^2\right) \tag{4.23}$$

The overall accuracy of the numerical solution is perfectly described by the natural magnitude of the *torsional rigidity*

$$J = 2G \int_A U \,\mathrm{d}A \tag{4.24}$$

Taking for simplicity $G = 1$, the analytical expression of the torsional rigidity is

$$J_{\text{exact}} = G\frac{\pi a^3 b^3}{(a^2 + b^2)} \cong 5.027 \tag{4.25}$$

The boundary of the ellipse is divided into four equal parts (defined between the two main Cartesian axes) thus defining the four points A, B, C, and D, which correspond to the corners of the unit square (see the general case in Fig. 4.1). Then, each of the four edges, let us take AB for instance, is uniformly subdivided into n_ξ segments of equal length, thus leading to $4n_\xi$ boundary points. In the sequence, applying the boundary-only Coons interpolation we derive the internal points, and this procedure leads to a mesh of totally $(n_\xi + 1)^2$ nodes. Figure 4.7 shows the particular case of $n_\xi = 8$, including the numbering of the macroelement (the corner nodes A, B, C, and D are replaced by the nodal points 1, 9, 81, and 73, respectively). Based on the aforementioned mesh, we can produce the following discretization:

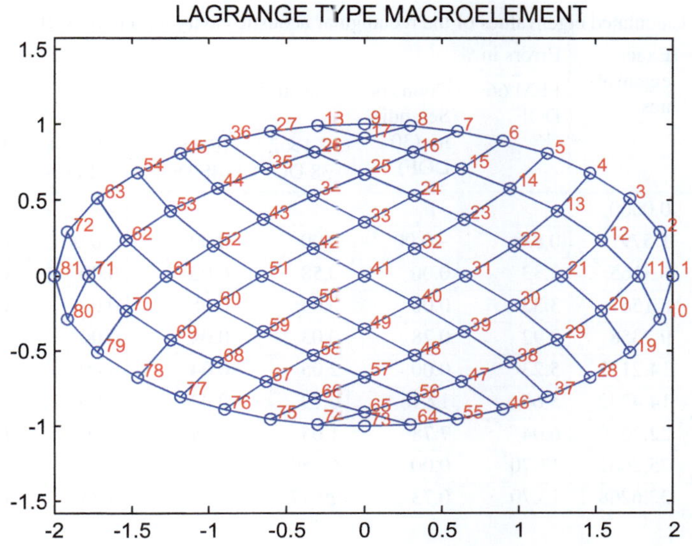

Fig. 4.7 Discretization of an ellipse ($a = 2, b = 1$) using $n_\xi = 8$ subdivisions per edge, implementing Lagrangian-type formulation

Table 4.1 Calculated torsional rigidity and area of elliptic section using several types of elements

Quantity	Error in %		
	Lagrange element	Four node	Three node
J (rigidity)	0.00092	−3.33	−2.04
A (area)	0.00786	−0.76	−0.76
Number of equations	49	49	49

- n_ξ^2 quadrilateral elements.
- $2n_\xi^2$ triangular elements.
- a single Lagrangian macroelement that consists of $(n_\xi + 1)^2$ nodes.

For the specific Dirichlet-type conditions of this problem, the number of final equations will be $n_{eq} = (n_\xi - 1)^2$ for all the above three schemes.

Based on the mesh shown in Fig. 4.7, the obtained results are shown in Table 4.1. One may observe that the Gordon–Lagrange macroelement (model **C33** of degree $p = 8$) is far superior compared to either the quadrilateral (4-node) bilinear elements or the triangular (3-node) ones, for the same total number of nodes involved. The aforementioned superiority does not restrict to the accurate calculation of the rigidity J but also extends to the estimation of the cross-sectional area A.

Again, all the above three models share the same nodal points, in position and number. In other studies in which 3-node and 6-node triangular elements have been

Table 4.2 Calculated eigenvalues of the rectangular acoustic cavity with hard walls

Mode	Exact eigenvalues	Errors in %					
		FEM 66 DOF	Coons or Serendipity (30 DOF)	Transfinite			Lagrange type
				IP 4×2 (38 DOF)	IP 6×3 (48 DOF)	IP 8×4 (62 DOF)	IP 9×4 (66 DOF)
1	0.0000	–	–	–	–	–	–
2	1.5791	0.83	−0.00	0.00	0.00	−0.00	0.00
3	6.3165	3.33	0.00	0.58	0.00	0.00	0.00
4	8.1567	3.33	0.00	0.05	0.05	0.00	0.00
5	9.7358	2.92	0.28	0.03	0.04	0.00	0.00
6	14.2122	5.23	0.00	2.05	0.04	0.00	0.00
7	14.4732	5.66	12.16	0.29	0.03	0.00	0.00
8	22.3689	6.04	7.78	1.05	0.04	0.00	0.00
9	25.2662	13.70	0.00	61.59	5.62	0.20	0.00
10	32.6268	13.70	0.73	29.77	0.73	0.73	0.73

used, average errors of the order 30 and 5%, respectively, have been previously reported (Brebbia and Ferrante [3, pp. 287, 328–335]).

Example 4.3 (Rectangular acoustic cavity with hard walls) We deal with the same cavity (of dimensions 2.5×1.1) as that analyzed previously in Example 3.3. The study includes several models which are shown in Fig. 4.8, where the symbol IP refers to the arrangement of the Internal Points (shown by filled black circles). Therefore, the transfinite element IP 9×4 is conceptually equivalent with the classical 66-node element of Lagrange family. This problem has been previously studied in detail by Provatidis [21] but here more results are provided.

The overall performance of a single Gordon macroelement is shown in Table 4.2. For the sake of brevity, all the univariate trial functions $\left(B_i^{\text{SIDE}}\right)$ along the four sides of the Gordon–Coons patch were chosen to be Lagrange polynomials. One may observe that the classical Lagrangian element (66 DOF) is the most accurate model, whereas the transfinite macroelement also performs well following the rule "the smallest the deviation from the Lagrangian element the highest the accuracy."

For different boundary conditions of the same acoustic cavity, the interested reader may consult Provatidis [21].

Example 4.4 (Circular acoustic cavity with hard walls) We deal with a circular cavity of unit radius ($a = 1$) under Neumann boundary conditions ($\partial u/\partial n = 0$). In this problem, the exact wavenumbers are calculated as the roots of Bessel functions:

$$J_m'(ka) = 0, \quad m = 0, 1, 2, \ldots \tag{4.26}$$

whereas the eigenvector is given by

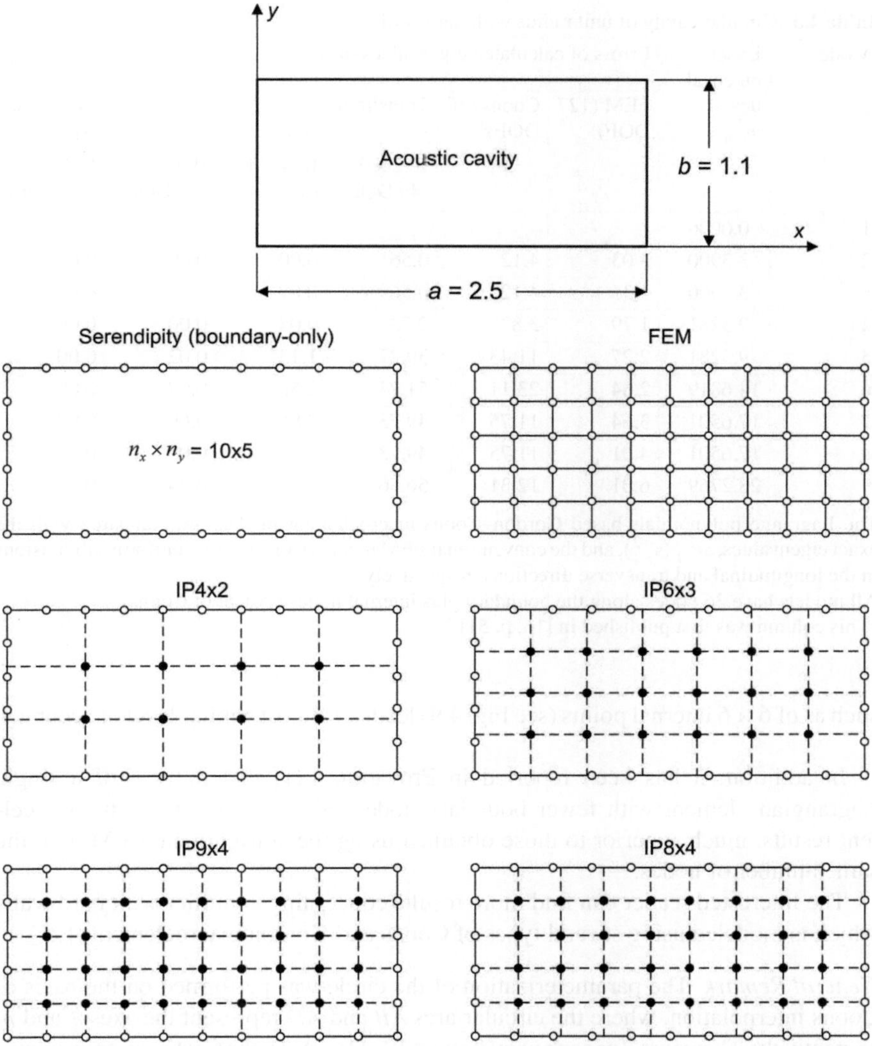

Fig. 4.8 Rectangular acoustic cavity of dimensions $a \times b = 2.5\,\text{m} \times 1.1\,\text{m}$, discretized using several Gordon–Coons models (IP: internal points), as well as a FEM model with the same mesh density

$$P_m(r, \vartheta) = J_m(kr/a)\sin m\vartheta \tag{4.27}$$

Using a single Gordon macroelement with 36 nodal points along the boundary and a variable number of internal nodes, the results are shown in Table 4.3. One may observe that the Lagrangian element (totally 81 nodes of which the internal ones are arranged in a 8×8 scheme) leads to the best results, but also a smaller arrangement

Table 4.3 Circular cavity of unit radius with hard walls

Mode	Exact eigenvalues ω_{mn}^2	Errors of calculated eigenvalues in %					
		FEM (127 DOF)	Coons (36 DOF)[a]	Transfinite			Lagrange type
				IP 2×2 (40 DOF)	IP 4×4 (52 DOF)	IP 6×6 (72 DOF)	IP 8×8 (66 DOF)
1	0.0000	–	–	–	–	–	–
2	3.3900	3.03	4.12	0.56	0.00	0.00	0.00
3	3.3900	1.35	4.12	0.56	0.00	0.00	0.00
4	9.3281	1.79	3.83	2.73	0.05	0.00	0.00
5	9.3281	2.27	11.43	39.27	1.13	0.02	0.00
6	14.6819	2.64	23.11	51.74	2.58	0.07	0.00
7	17.6501	3.64	11.75	49.73	2.86	0.09	0.00
8	17.6501	4.01	11.75	49.73	2.86	0.09	0.00
9	28.2769	6.01	12.31	56.56	5.32	0.24	0.00

The Lagrange polynomials-based Gordon–Coons macroelement models are compared with the exact eigenvalues, ω_{mn}^2 (s^{-2}), and the conventional FEM models using $n_x \times n_y$ uniform subdivisions in the longitudinal and transverse directions, respectively

All models have 36 nodes along the boundary plus internal nodes (except of Coons)

[a]This column was first published in [16, p. 51]

such as of 6×6 internal points (see Fig. 4.9) leads to almost similar level of accuracy.

In addition, it has been reported in Provatidis [21] that the use of a single Lagrangian element with fewer boundary nodes (32, 28, and 24) leads to excellent results, much superior to those obtained using the conventional FEM with the same number of nodes.

The interested reader can find more results concerning a muffler with hard walls which is modeled using several types of Coons and Gordon macroelements [21].

General Remark The parameterization of the circle was performed on the basis of Coons interpolation, where the circular arcs *AB* and *AD* represent the axes ξ and η, respectively. Therefore, for each pair $(\xi, \eta) \in [0, 1] \times [0, 1]$, the Cartesian coordinates (x, y) can be analytically found; a closed-form expression is given in Chap. 8 (see, Eq. 8.7). Regarding Fig. 4.9, it is important to understand that the nodes 37–42 lie on the isoline $\xi = \frac{1}{7}$, the nodes 43–48 belong to the isoline $\xi = \frac{2}{7}$, and so on. Similarly, the nodes 37, 43, 49, 55, 61, and 67 lie on the isoline $\eta = \frac{1}{7}$, and so on.

Example 4.5 (Bending of a beam by uniform load) This example depicts the necessity of using internal nodes in the Gordon model and was taken from the literature (Timoshenko and Goodier [24, pp. 46–50]). A beam *ABCD* of narrow rectangular cross section of unit width, supported at the ends, is bent by a uniformly distributed load of density q, as shown in Fig. 4.10a. The following parameters were used: $E = 3.0 \times 10^7$ kPa; $v = 0.30$, $c = 6$ m, $L = 24$ m, $q = 1000$ kN/m, and plane stress

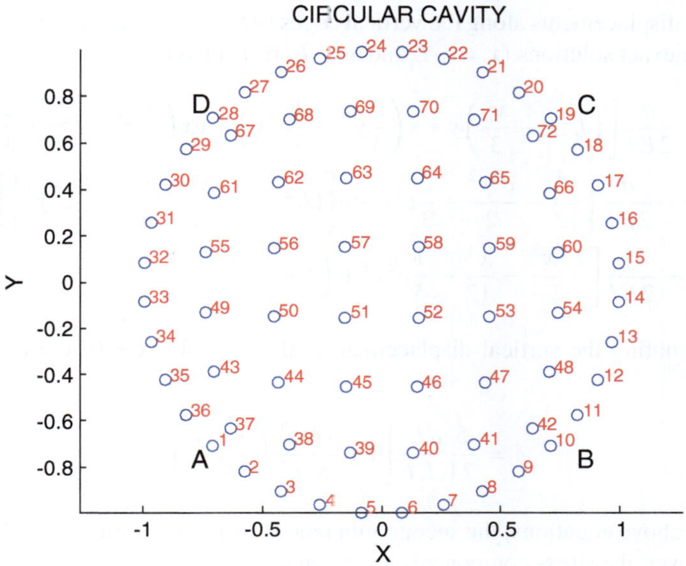

Fig. 4.9 Discretization of a circular acoustic cavity using the transfinite element IP 6×6

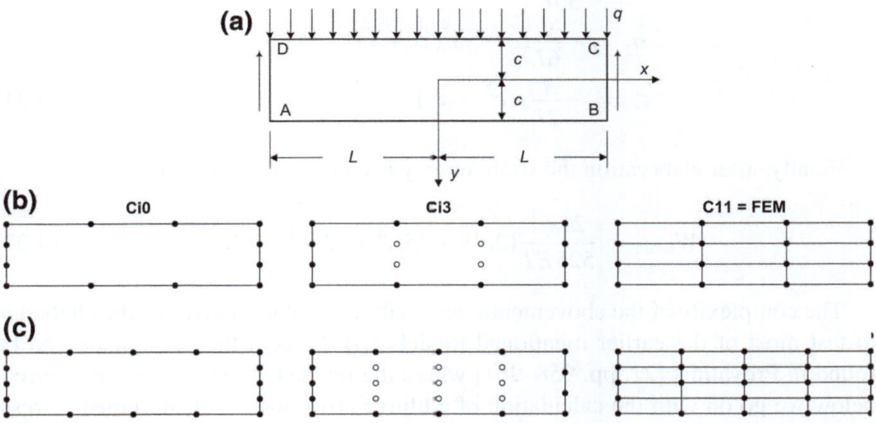

Fig. 4.10 **a** Cantilever beam subjected to a uniformly distributed load q at topside, using several models with **b** $n_x = n_y = 3$ and **c** $n_x = n_y = 4$ subdivisions per side

conditions. The given conditions at the upper (CD) and lower (AB) edges of the beam are:

$$\left(\tau_{xy}\right)_{y=\pm c} = 0 \quad \left(\sigma_{yy}\right)_{y=+c} = 0 \quad \left(\sigma_{yy}\right)_{y=-c} = -q, \tag{4.28}$$

while the displacements along the vertical edges (AD and BC) are prescribed according to the exact solutions ($x = -L$ and $x = L$, respectively):

$$
\begin{aligned}
u_x &= \frac{q}{2EI}\left[\left(L^2x - \frac{x^3}{3}\right)y + x\left(\frac{2}{3}y^3 - \frac{2}{5}c^2y\right) + \nu x\left(\frac{1}{3}y^3 - c^2y + \frac{2}{3}c^3\right)\right] \\
u_y &= -\frac{q}{2EI}\left\{\frac{y^4}{12} - \frac{c^2y^2}{2} + \frac{2}{3}c^3y + \nu\left[(L^2 - x^2)\frac{y^2}{2} + \frac{y^4}{6} - \frac{1}{5}c^2y^2\right]\right\} \\
&\quad - \frac{q}{2EI}\left[\frac{L^2x^2}{2} - \frac{x^4}{12} - \frac{1}{5}c^2x^2 + \left(1 + \frac{1}{2}\nu\right)c^2x^2\right] + \delta
\end{aligned}
\tag{4.29}
$$

with δ denoting the vertical displacement at the centroid ($x = 0$, $y = 0$), which is given by:

$$
\delta = \frac{5}{24}\frac{qL^4}{EI}\left[1 + \frac{12}{5}\frac{c^2}{L^2}\left(\frac{4}{5} + \frac{\nu}{2}\right)\right]
\tag{4.30}
$$

In the above equations, the second moment of inertia is given as $I = 2c^3/3$. Moreover, the stress components are given as:

$$
\begin{aligned}
\sigma_x &= \frac{q}{30I}y(15L^2 - 6c^2 - 15x^2 + 10y^2) \\
\sigma_y &= -\frac{q}{6I}(c - y)^2(2c + y) \\
\tau_{xy} &= -\frac{q}{2I}x(c^2 - y^2)
\end{aligned}
\tag{4.31}
$$

Finally, after elaboration the strain energy is analytically found as

$$
W_{\text{strain}} = \frac{2Lq^2}{525EI}(2c^4\nu + 35L^4 + 132c^4 + 70L^2c^2)
\tag{4.32}
$$

The complexity of the abovementioned analytical solution gives us the challenge to test most of the earlier mentioned models **Cij**. Since a thorough study can be found in Provatidis [22, pp. 958–961] where the interested reader may be referred, below we go on with the calculation of all three error norms (displacement, stress, and strain energy).

Each side of the Gordon–Coons patch ABCD is uniformly divided into n_s segments. In this way, $n_b = 4n_s$ nodal points appear along the boundary, which is the minimum number of required nodes. In general, an additional set of (uniformly arranged) $n_{I\xi} \times n_{I\eta}$ *internal* nodes will be used to examine the level at which they improve the numerical solution.

I. *Boundary-only solution*

Here we consider that no internal nodes are used, i.e., $n_{I\xi} = n_{I\eta} = 0$. In Table 4.4, it is definitely demonstrated that a single boundary-only Coons macroelement (models C10≡M1, C20≡M2, and C30≡M3, with no internal nodes) is *incapable* of accurately solving this particular problem. One may observe that all three models, C10 (M1), C20 (M2), and C30 (M3) converge to about $L_u \cong 1.5\%$ and $L_\sigma \cong 32.8\%$, whereas the strain energy deviates about 9% from its exact value. This finding definitely suggests the necessity for the enhancement of the functional space by introducing additional shape functions associated to some internal nodes.

Particular interest has the model C30, in which convergence to an erroneous value appears when a *coarse* mesh with $n_s = 4$ subdivisions per side is applied (notation 4×4, i.e., quartic element of the Serendipity family). This is a strong indication that internal points are needed.

For the completeness of the study (i.e., for comparison purposes), preserving the aspect ratio again equal to four (see in the right column of Fig. 4.10), the obtained results of the FEM solution (transfinite element **C11**) are shown in Table 4.5. There, one may observe the slow but monotonic convergence of all error norms.

II. *Introduction of internal nodes*

It is reminded that, according to Provatidis [22], a subclass of transfinite elements may be noted by the symbol **Cij**. Again, the letter "C" stands for Coons, the index "**i**" stands for the trial functions $[B_i^{AB}(\xi), B_i^{BC}(\eta), B_i^{CD}(\xi)$, and $B_i^{DA}(\eta)]$, whereas the index "**j**" stands for the blending functions $[E_j(\xi)$ and $E_j(\eta)]$ utilized by this macroelement. The index "**j**" takes the value zero, when the blending functions do *not* depend on internal nodes (i.e., indicates zero influence of internal nodes).

The practical question to be answered through this example is how to select the internal nodes within each Gordon–Coons (transfinite) macroelement. As has been previously reported, the boundary-only Coons macroelement (with $4n_s$ nodes appearing only along the boundary) which is on "one bank of the river" and the Lagrangian macroelement (with $(n_s + 1)^2$ nodes, in total) which is on the opposite "bank" are two *extreme* situations. This means that adding $n_I = (n_s - 1)^2$ internal nodes to the aforementioned Coons macroelement (with $4n_s$ boundary nodes, initially), the Lagrangian macroelement is obtained.

It should be clarified that when the univariate trial functions B_i^{SIDE} are *Lagrange polynomials*:

(1) The abovementioned boundary-only Coons macroelement (of model M3) is the classical finite element of *Serendipity* family (found in all FEM textbooks, e.g., Zienkiewicz [25, p. 160]). The latter includes all monomials along the

two legs of Pascal's triangle (i.e., $\{1, x, \ldots, x^{n_s}\}$ and $\{1, y, \ldots, y^{n_s}\}$) plus a surplus of one additional layer. Therefore, the corresponding basis functions are: $f_1 = 1$, plus the discrete monomials included in the products $\{1, x, \ldots, x^{n_s}\}y$ and $\{1, y, \ldots, y^{n_s}\}x$. The notation for a relevant Coons macroelement is **C30**, where "3" refers to the Lagrange polynomials (model M3) and "0" refers to the linear blending function which depends on the opposite boundaries only (i.e., "0" indicates the absence of internal node-based blending functions).

(2) If the blending functions are also *Lagrange polynomials*, and even identical to those used as trial functions (used for the approximation of the four boundary sides), the produced Gordon macroelement becomes identical to the finite element of Lagrange family (see, Zienkiewicz [25, p. 156]). In this case, the

Table 4.4 Normalized error norms, and error of maximum deflection, using a single boundary-only Coons macroelement of variable size and three alternative models of different trial functions

Arrangement of boundary nodes in Coons macroelement	Error (in %)				Number of nodes
	L_u	L_σ	W_{strain}	Maximum deflection	
Model **C10** *(≡M1: piecewise linear)*					
2 × 2	33.17	94.01	108.30	−18.16	8
3 × 3	19.11	68.83	49.80	−13.79	12
4 × 4	12.67	56.35	30.68	−9.31	16
5 × 5	9.14	49.23	22.31	−7.23	20
6 × 6	7.02	44.82	17.95	−5.91	24
10 × 10	3.61	37.41	11.90	−3.68	40
20 × 20	2.07	33.80	9.48	−2.61	80
40 × 40	1.68	32.84	8.88	−2.33	160
Model **C20** *(≡M2: natural cubic B-splines)*					
2 × 2	16.21	57.05	8.59	−14.75	8
3 × 3	2.66	36.91	9.21	−4.64	12
4 × 4	2.00	35.03	9.09	−1.92	16
5 × 5	1.71	33.65	8.82	−2.22	20
6 × 6	1.63	33.21	8.79	−2.31	24
10 × 10	1.56	32.66	8.71	−2.24	40
20 × 20	1.55	32.53	8.69	−2.23	80
Model **C30** *(≡M3: Lagrange polynomials)*					
2 × 2	15.84	51.97	−1.52	−18.16	8
3 × 3	7.28	35.75	4.51	−9.54	12
4 × 4	1.55	32.51	8.69	−2.23	16
5 × 5	1.55	32.51	8.69	−2.23	20
6 × 6	1.55	32.51	8.69	−2.23	24
10 × 10	1.55	32.51	8.69	−2.23	40

transfinite macroelement is noted by **C33** (where “3” indicates Lagrange polynomials for both the trial and the blending functions).

Remark 1 While the classical FEM theory covers mostly the two above situations, i.e., the Serendipity and Lagrange family (all based on Lagrange polynomials), any *smaller* or *greater* number of internal nodes than the abovementioned $n_I = (n_s - 1)^2$ ones is possible to be developed in terms of the transfinite (Gordon’s) interpolation formula (Eq. 4.1).

Remark 2 All issues mentioned above can be repeated replacing the Lagrange polynomials with (i) piecewise linear (**i** = 1) or (ii) natural cubic B-splines (**j** = 2), and similarly for **j** = 1, 2. This issue has not been covered by the classical FEM theory so far, because a *CAD-based* point of view is necessary to accomplish this task.

Remark 3 Continuing Remark 2, the abovementioned Lagrange polynomials may be also replaced by a lot of other trial functions (approximations) such as (iii) piecewise quadratic, cubic, quintic, etc., (iv) nonrational or rational Bernstein–Bézier polynomials, (v) full (de Boor) B-splines, (vi) NURBS with several multiplicities of inner knots, and other “nodeless” basis functions (e.g., Legendre polynomials like those used in the p-method, Chebyshev polynomials, among others). Thus the **Cij** family may include more than the previously mentioned (e.g., see Abstract) twelve combinations of different of macroelements.

Table 4.5 Normalized error norms in the FEM solution

Arrangement of nodes in FEM solution	L_u (in %)	L_σ (in %)	Strain energy W_{strain}	
			Calculated value	Error (in %)
2 × 2	42.1186	7.8353	1.86168e+02	−33.54
3 × 3	23.0459	12.6710	2.22023e+02	−20.74
4 × 4	14.6340	9.8082	2.39563e+02	−14.48
6 × 6	7.3202	5.6367	2.57733e+02	−7.99
8 × 8	4.3341	3.4981	2.66266e+02	−4.94
10 × 10	2.8477	2.3481	2.70806e+02	−3.32
20 × 20	0.7401	0.6273	2.77612e+02	−0.89
30 × 30	0.3316	0.2824	2.78987e+02	−0.40
40 × 40	0.1871	0.1595	2.79477e+02	−0.23
50 × 50	0.1199	0.1023	2.79707e+02	−0.15
100 × 100	0.0300	0.0256	2.80014e+02	−0.04
200 × 200	0.0075	0.0064	2.80091e+02	−0.01
			Exact value = 2.80117e+02	

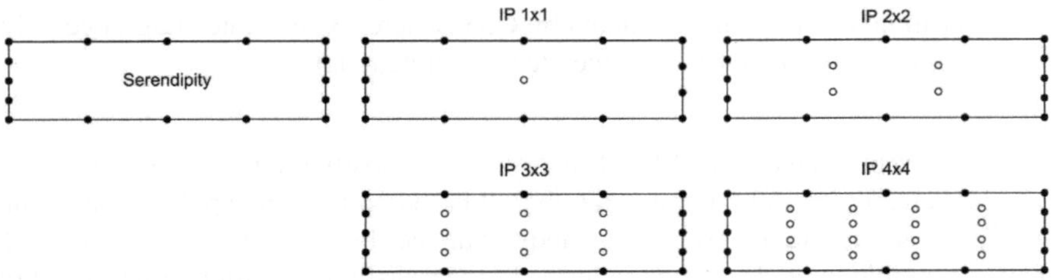

Fig. 4.11 Internal nodes in the model of the cantilever beam subjected to a uniformly distributed load at the top ($n_s = 4$ subdivisions per side)

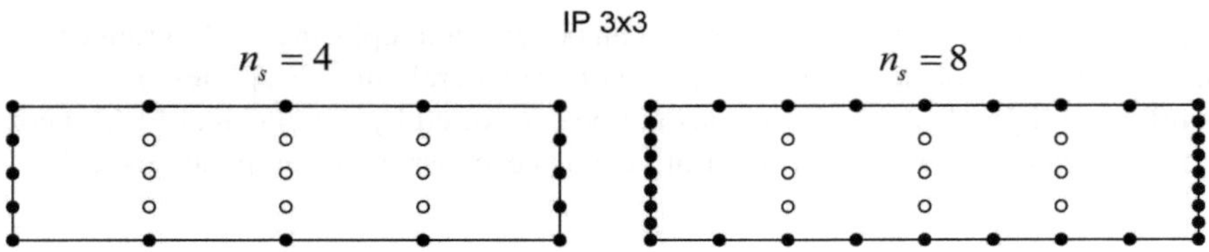

Fig. 4.12 Progressive increase of boundary nodes using piecewise-linear approximation, for a fixed set of nine internal nodes (IP 3 × 3 configuration)

From the computational point of view, all the abovementioned **Cij** models can be treated by a single subroutine (or a function in MATLAB®). This may have as an input the coordinates of the macroelement plus the parameters (ξ, η) of the integration (Gauss) point, and as an output the shape functions and their global derivatives as well as the determinant of the Jacobian matrix **J**.

Interestingly, the introduction of either a very small or a very large number of internal nodes does not always ensure that the numerical solution will improve. For the sake of brevity, here we shall restrict to the case of $n_s = 4$. Two cases will be studied, both using Lagrange polynomials as blending functions, whereas the trial functions will be either Lagrange polynomials (model **C33**) or piecewise linear (model **C13**). The configuration of the models studied is shown in Fig. 4.11, whereas the results are shown in Table 4.6. One may observe in the upper part of Table 4.6 that the boundary-only Coons solution (C30) is better than the transfinite one (C33 for IP 1 × 1 and IP 2 × 2), while the Lagrange element (C33 for IP 3 × 3) outperforms.

Table 4.6 Influence of the internal nodes in a coarse macroelement with $n_s = 4$ segments per side (see Fig. 4.11)

Lagrange polynomials as trial functions					
Type of macroelement	Errors (in %)				Number of nodes
	L_u	L_σ	W_{strain}	Maximum deflection (δ)	
Model **C33**					
Boundary-only (**C30**)	1.55	32.51	8.69	−2.23	16
IP 1 × 1 (C33)	6.47	17.51	−5.06	**−8.15**	17
IP 2 × 2 (C33)	6.81	17.44	−5.71	**−8.52**	20
IP 3 × 3 (C33)	0	0	0	0	25
IP 4 × 4 (C33)	0	0	0	0	32
Model **C13**					
Serendipity (**C10**)	12.67	56.35	30.68	−9.31	16
IP 1 × 1 (C13)	12.22	37.18	−2.59	**−14.11**	17
IP 2 × 2 (C13)	12.42	37.91	−2.84	**−14.73**	20
IP 3 × 3 (C13)	4.27	22.41	−0.37	−3.94	25
IP 4 × 4 (C13)	3.71	22.12	0.16	−3.15	32

Bold were set to attract the attention of the reader, as they correspond to results that are worse than one could anticipate

It is remarkable that when the quartic element of the Serendipity family is enhanced by nine uniformly arranged internal points (IP 3 × 3), it becomes a quartic element of Lagrange family and then it leads to no numerical error at all. Not only this, but the zero error appears for a higher number of sixteen internal points (IP 4 × 4) as well.

The reason that the (IP 3 × 3)-based **C33** model achieves the absolute accuracy, is due to the fact that all monomials involved in the analytical solution given by Eq. (4.29) belong to the tensor product of terms appearing in Pascal's triangle, i.e., elements of the following matrix **M** (where M stands for "monomials"):

$$\mathbf{M} = \begin{bmatrix} 1 \\ x \\ x^2 \\ x^3 \\ x^4 \end{bmatrix} \cdot \begin{bmatrix} 1 & y & y^2 & y^3 & y^4 \end{bmatrix} \tag{4.33}$$

Let us now see whether it is possible to approximate the accurate solution using different trial functions than Lagrange polynomials, again in conjunction with internal nodes, e.g., using the models **C1j** (piecewise linear) and **C2j** (natural cubic B-splines).

The weakest case is the class of **C1j** elements, because piecewise-linear approximation requires a large number of nodal points toward a certain direction. For simplicity, let us focus in the most favorite case which corresponds to a net of nine uniformly arranged internal points (IP 3 × 3).

Preserving the nine internal points in the setup of (IP 3 × 3), in Table 4.7 we study the influence of the number of the ($4n_s$) boundary points (Fig. 4.12).

A similar study using **C23** elements (natural cubic B-splines) is shown in Table 4.8. One may observe that the numerical solution slowly tends to the accurate one, but it requires a large number of DOF along each side of the patch. This is probably due to the piecewise nature of the natural cubic B-splines, accompanied by the fact that the monomials x^2 and y^2 are missing in the expansion series.

Table 4.7 Influence of the boundary discretization for a Gordon coarse macroelement of **C13** type, with nine internal nodes in the (IP 3 × 3) configuration, according to Fig. 4.12

Gordon macroelement **C13** (IP 3 × 3), with variable discretization per side: n_s	Errors (in %)				Number of nodes ($4n_s + 9$)
	L_u	L_σ	W_{strain}	Maximum deflection (δ)	
4	4.2698	22.4093	−0.3676	−3.9390	25
6	1.5583	12.8421	−0.3866	−1.4330	33
8	0.8180	9.5171	−0.2031	−0.7658	41
16	0.1860	4.8477	−0.0257	−0.1773	73
32	0.0448	2.4473	−0.0033	−0.0428	137

Table 4.8 Influence of the boundary discretization for a Gordon coarse macroelement of **C23** type, with nine internal nodes in the (IP 3 × 3) configuration, according to Fig. 4.12

Gordon macroelement **C23** (IP 3 × 3), with variable discretization per side: n_s	Errors (in %)				Number of nodes ($4n_s + 9$)
	L_u	L_σ	W_{strain}	Maximum deflection (δ)	
4	1.4545	10.2006	−1.6258	−1.0749	25
6	0.3115	4.6677	−0.3801	−0.0971	33
8	0.1302	2.8720	−0.1732	−0.0298	41
16	0.0178	0.9532	−0.0254	−0.0037	73
32	0.0025	0.3310	−0.0033	−0.0004	137

Table 4.9 Influence of the boundary discretization for a Gordon coarse macroelement of **C22** type, with nine internal nodes in the (IP 3 × 3) configuration, according to Fig. 4.12

Gordon macroelement **C22** (IP 3 × 3), with variable discretization per side: n_s	Errors (in %)				Number of nodes ($4n_s + 9$)
	L_u	L_σ	W_{strain}	Maximum deflection (δ)	
4	1.0483	13.5242	0.5481	0.4930	25
6	0.6174	11.0906	0.9003	0.3446	33
8	0.5797	10.2942	1.0097	0.3990	41
10	0.5765	10.0480	1.0584	0.4262	49
16	0.5779	9.7958	1.1119	0.4570	73
32	0.5783	9.6878	1.1229	0.4644	137

Furthermore, not only the trial functions but also the blending functions may be changed. For example, using again natural cubic B-splines for both the trial functions (as previously in Table 4.8) but now also for the blending functions as well, the new findings are shown in Table 4.9. One may observe the slow convergence toward a value near the exact solution.

For the sake of clarity, in Fig. 4.13 we visualize the convergence rate for only a few out of the twelve **Cij** transfinite models. The interested reader can find more detailed charts of the L_2-norm (for the displacement only) in Provatidis [22, pp. 960–961]. Also, the same study in elastostatics [22, p. 961] shows that in case that piecewise-linear trial functions in conjunction with blending functions in the form of Lagrange polynomials are used (model C13), the number of internal nodes should not be greater than $2n_s$–1 in each direction.

It is remarkable that the decrease of accuracy in the numerical solution when very few internal nodes (with respect to the boundary ones) are used has been also observed in the eigenvalue problem. As an example, for a boundary mesh of twenty nodes in a form 6 × 4 subdivisions, the introduction of one node (noted by IP 1 × 1) has led to worse results than the Serendipity element [19, p. 526].

Remark Of course, if the natural cubic B-splines are replaced by complete (de Boor) splines, it is anticipated that the numerical solution will improve. Particular improvement is anticipated when complete (de Boor) B-splines of fourth degree is implemented. In general, the tensor product of (de Boor) B-splines is a very good choice.

Example 4.6 (Infinite plate with a hole) While most of the previous examples refer to a rectangle or a quadrilateral-like domain, here the boundary consists of five discrete segments. Moreover, here a stress concentration arises.

We consider the problem of an infinite plate with a central circular hole of radius a. The plate is subjected to a uniform tension, σ_0, in the x-direction at infinity, where

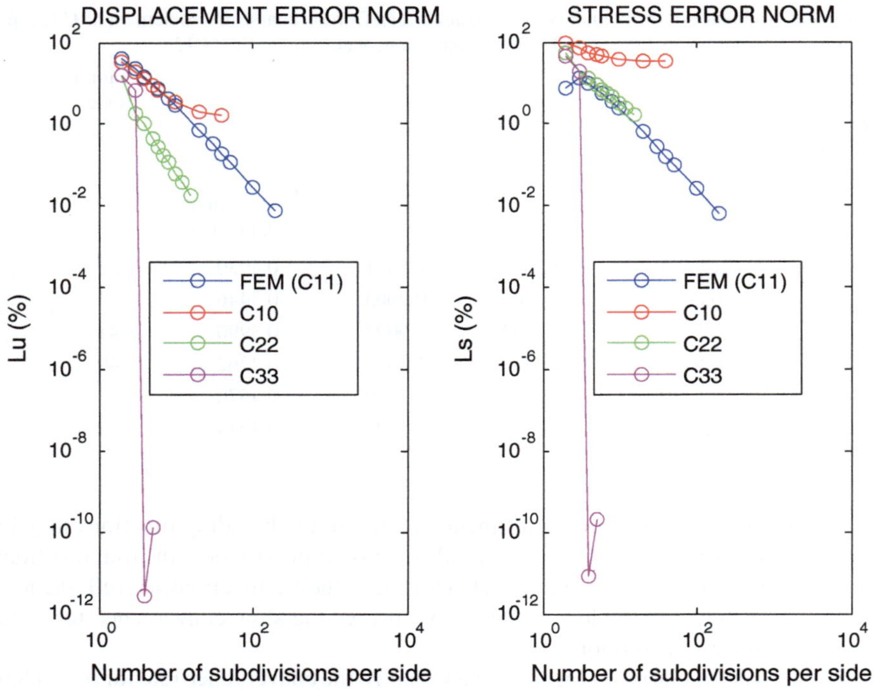

Fig. 4.13 Displacement and stress error norms (in %) for the beam under uniform load

the axis origin coincides with the center of the hole. The exact solution for the stress components is

$$\begin{aligned}
\sigma_{xx} &= \sigma_0\left\{1 - \frac{a^2}{r^2}\left[\frac{3}{2}\cos(2\theta) + \cos(4\theta)\right] + \frac{3a^4}{2r^4}\cos(4\theta)\right\}\\
\sigma_{yy} &= -\sigma_0\left\{\frac{a^2}{r^2}\left[\frac{1}{2}\cos(2\theta) - \cos(4\theta)\right] + \frac{3a^4}{2r^4}\cos(4\theta)\right\}\\
\sigma_{xy} &= -\sigma_0\left\{\frac{a^2}{r^2}\left[\frac{1}{2}\sin(2\theta) + \sin(4\theta)\right] - \frac{3a^4}{2r^4}\sin(4\theta)\right\}
\end{aligned} \tag{4.34}$$

Also, the exact polar components of the displacement vector are

$$\begin{aligned}
u_r &= \frac{\sigma_0}{4\mu}\left\{r\left[\frac{(\kappa-1)}{2} + \cos(2\theta)\right] + \frac{a^2}{r}[1 + (1+\kappa)\cos(2\theta)] - \frac{a^4}{r^3}\cos(2\theta)\right\}\\
u_\theta &= \frac{\sigma_0}{4\mu}\left[(1-\kappa)\frac{a^2}{r} - r - \frac{a^4}{r^3}\right]\sin(2\theta)
\end{aligned} \tag{4.35}$$

where

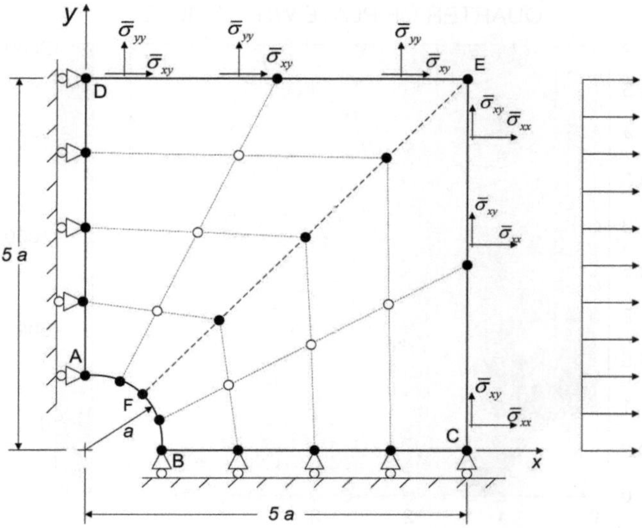

Fig. 4.14 Infinite plate with a hole subject to a uniform stress at infinity, and schematic decomposition in two transfinite elements

$$\mu = \frac{E}{2(1+\nu)} \qquad \kappa = \begin{cases} 3-4\nu & \text{plane strain} \\ \frac{3-\nu}{1+\nu} & \text{plane stress} \end{cases} \tag{4.36}$$

For the configuration of Fig. 4.14, the maximum stress concentration ratio appears at the point A and according to Eq. (4.34) it is $K = \sigma_{xx}^{\max}/\sigma_0 = 3$.

As has been shown in Chap. 2, it is not possible to define a smooth Lagrange polynomial for the entire part CED of the boundary. For this side of the patch, we have either to use piecewise-linear approximation or in the best case a first approximation for the segment CE and a second one for ED.

In the first case, using $n_b = 48$ boundary nodes (and no internal nodes) which are produced by a uniform discretization of each side in the patch $ABCD$ using $n_s = 12$ divisions, the mesh and the corresponding calculated stresses are shown in Fig. 4.15. The normalized displacement and stress error norms were found equal to $L_u = 2.84\%$ and $L_\sigma = 8.84\%$, respectively. At the corner A with $(\xi, \eta) = (0, 0)$, the stress concentration factor was determined equal to $K = 2.26$ which is far from the exact value 3.0.

Improved results for the stress concentration factor can be obtained by introducing internal nodes; the interested reader may consult Provatidis [22]. Furthermore, several scenarios can be made as follows: either considering again a single macroelement in which a structured net of internal nodes is added or splitting the domain into two or four subdomains. Each subregion is considered as a separate large transfinite element.

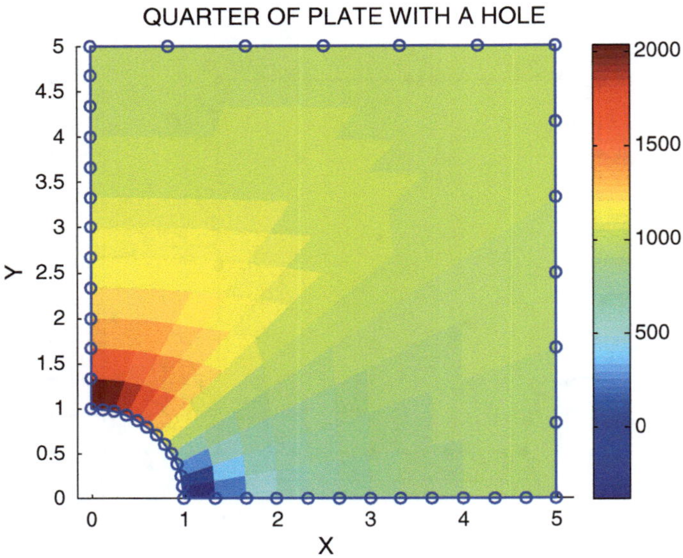

Fig. 4.15 Distribution of the stress component σ_{xx} around the hole

In the second case, uniform and nonuniform (close to the circular arc) arrangements of the internal nodes may be considered.

Example 4.7 (Free vibration analysis) Several examples have been successfully solved so far; many of them were retrieved from literature related to boundary-type methods [13, 14]. Free vibrations of a deep cantilever with a nonzero Poisson ratio ($\nu = 0.2$) and dimensions ($L = 24$, $h = 6$) have been studied by Provatidis [19]. In general, the first eigenvalue and eigenmode are calculated with adequate accuracy with respect to both the conventional FEM and the mass matrix-based BEM solution. Higher modes usually require the contribution of internal nodes, either in the single macroelement or through the creation of artificial subregions.

Within this context, starting from a single boundary-only Coons macroelement, the influence of internal nodes was thoroughly investigated (Fig. 4.16). In brief, a very small number of internal nodes disproportionate to the density of the boundary ones may somehow decrease the accuracy, and always there is a critical point that improves the numerical solution. For this rectangular shape and given boundary discretization, the classical tensor-product element using Lagrange polynomials was found to be a very good (perhaps the best) choice.

However, there are cases where a structured mesh is not offered. For example, each side of the Gordon macroelement may be already coupled with a given substructure. The arch-like structure shown in Fig. 4.17 is a typical case and has been also retrieved from literature related to boundary-type methods [15]. We can notice that the outer circular arc of the structure is divided into ten segments whereas the inner into six ones. While in the usual FEM analysis the so-called transition elements become

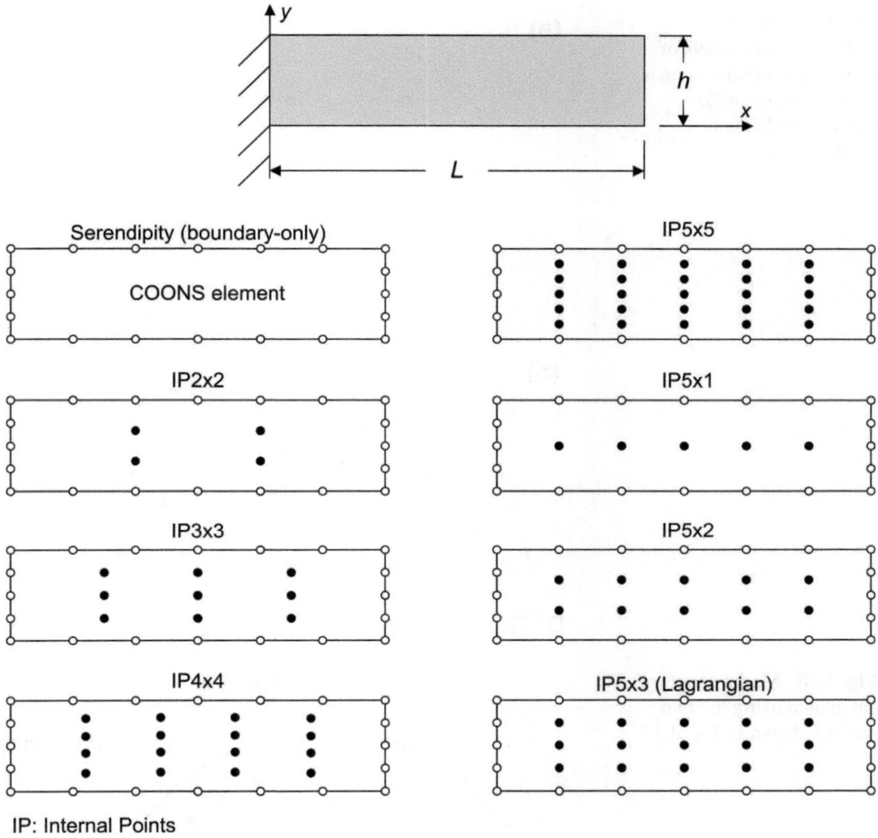

Fig. 4.16 Some of the transfinite elements studied in Provatidis (2003b)

necessary, if the difference of nodal spans between the opposite sides is small ($6 < 10$) it is possible to average these two discretizations and consider a uniform mesh of IP 7×4 internal nodes as shown in Fig. 4.17. In terms of the calculated eigenvalues, the result of the transfinite element is superior to the FEM solution, while Fig. 4.18 shows the small differences in the fourth eigenmode. Details of the relevant study may be found in the paper by Provatidis [19].

Example 4.8 (Pi-shaped domain) This test case deals with a "Π"-shaped nonconvex domain, as shown in Fig. 4.19a. The domain and the solution correspond to the well-known problem of the thick cylinder (see Example 3.1), where now the circular arcs have become straight segments. In order to obtain a steep numerical solution on purpose, the dimensions have to be chosen so as $a = 1$ and $b = 32$.

The computational mesh is shown in Fig. 4.20, where the boundary is discretized into sixty nodes (similar to the exaggerated Fig. 4.19a) plus an additional net of internal nodes (in configuration IP 5×5) is numbered therein (nodes 61–85). Due to

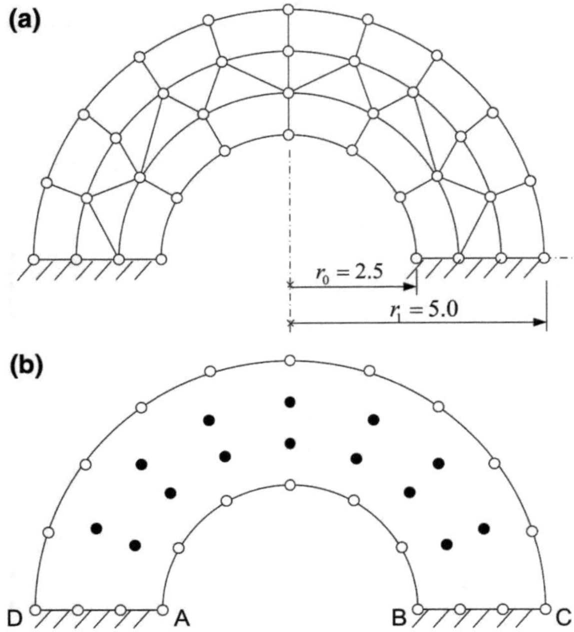

Fig. 4.17 Problem definition and models of arch-like structure using **a** finite element and **b** transfinite element formulations

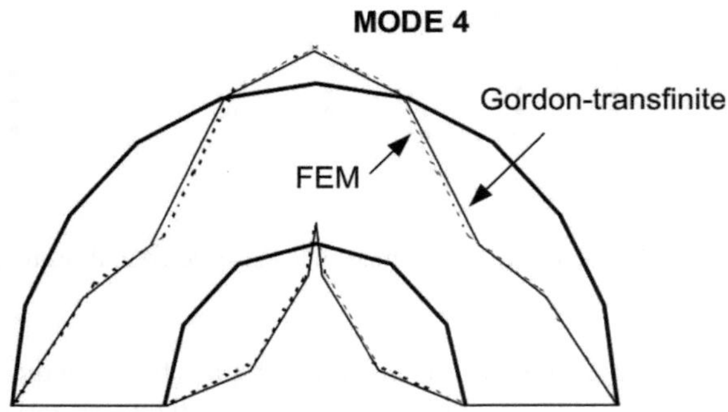

Fig. 4.18 Modal shape calculated using the two models shown in Fig. 4.17

the nonconvex shape, neither Lagrange polynomials nor the natural cubic B-splines cannot accurately describe the boundary as a unique function (see details in Chap. 2), and therefore, piecewise-linear trial functions have been chosen for the entire boundary. Moreover, the blending functions have been chosen to be Lagrange polynomials of sixth degree. This combination defines a 85-node Coons–Gordon macroelement of type **C13** (1—for piecewise linear, and 3—for Lagrange polynomials).

While the boundary-only Coons 60-node macroelement (model C10) gives an average error $L_u = 4.81\%$, the introduction of the abovementioned 25 internal nodes reduces it to $L_u = 3.29\%$ (see the results in Fig. 4.21).

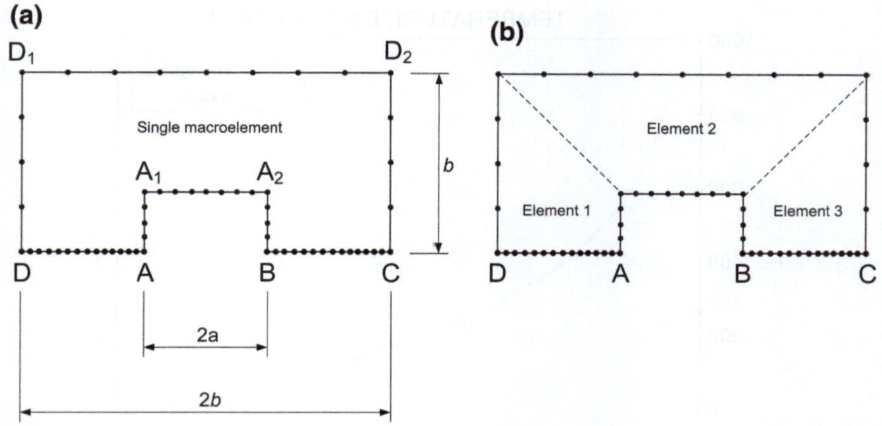

Fig. 4.19 **a** Π-shaped domain as a single macroelement (exaggerated for visual purposes), or **b** split into three convex domains

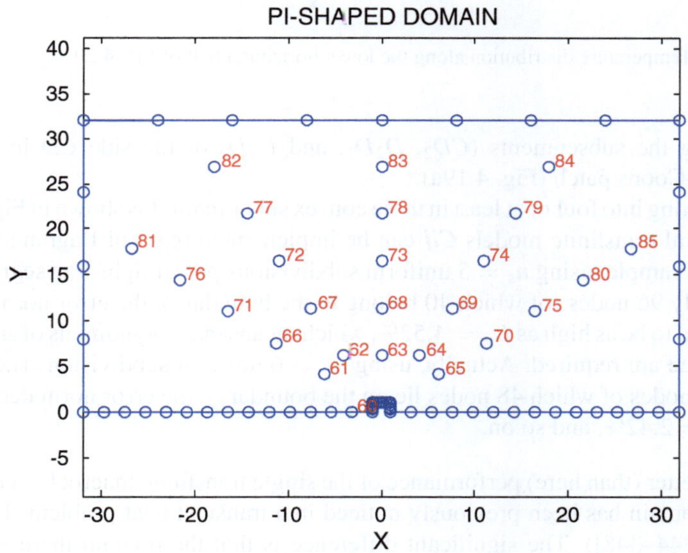

Fig. 4.20 Single Gordon macroelement (of **C13** type) for the entire domain

Remarks The accuracy of the numerical solution could be improved in many ways as follows:

(1) Using a more fine mesh of piecewise-linear approximation along the boundary.
(2) Using piecewise-quadratic, piecewise-cubic, and so on, interpolation along the boundary. For example, three different Lagrange polynomials could be used

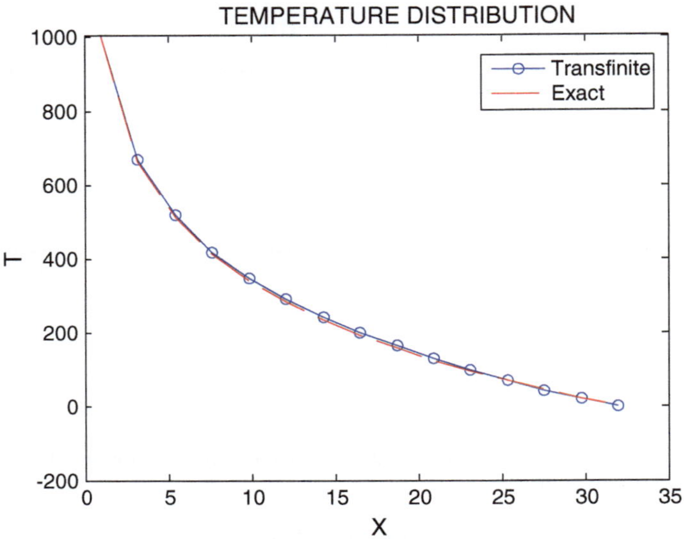

Fig. 4.21 Temperature distribution along the lower horizontal half of Fig. 4.20

along the subsegments (CD_2, D_2D_1, and D_1D) of the side CD in the Gordon–Coons patch (Fig. 4.19a).

(3) Splitting into four or at least in three convex subdomains (as shown in Fig. 4.19b), several transfinite models ***Cij*** can be implemented (e.g., of Lagrange family). For example, using $n_s = 5$ uniform subdivisions per straight-line segment, i.e., totally 96 nodes (of which 40 belong to the boundary), the error norm is again found to be as high as $L_u = 4.52\%$, which means that polynomials of still higher degree are required. Actually, using $n_s = 6$ uniform subdivisions (i.e., totally 133 nodes of which 48 nodes lie on the boundary), the error norm decreases to $L_u = 2.42\%$, and so on.

Note A better (than here) performance of the single transfinite macroelement in a Π-shaped domain has been previously noticed in a transient heat problem (Provatidis [18, pp. 344–348]). The significant difference is that the domain there was well-shaped, without the "singularity" that appears here in the neighborhood of the axes origin. The interested reader may repeat this example using a new ratio of dimensions such as $a = b/2$ or $a = b/4$. Moreover, the interested reader may also test *nonuniform* locations of the nodal points.

Also, another successful application of a single macroelement in the transient analysis of a Π-shaped elastic structure can be found in Provatidis [20, pp. 6700–6701]. The reader may also try to implement the abovementioned techniques in an L-shaped problem.

4.7 Degenerated Triangular Transfinite Elements

4.7.1 General

If a whole side of the quadrilateral macroelement *ABCD* degenerates in a single point, then a triangular macroelement is produced. Obviously, whatever has been written above for the quadrilateral is valid for the triangular as well.

Following the choice we made in Chap. 3, here again we deal with the degeneration of the entire side *AD* into a single point *A*. The procedure will become fully understood when a four-node element degenerated element is studied in Sect. 4.7.2.

4.7.2 Degenerated Quadrilateral with Four Nodes

We consider a 5-node quadrilateral with one internal node at its centroid (Fig. 4.22a). Given the blending functions, which are equal to the quadratic Lagrange polynomials:

$$\left.\begin{aligned} E_0(\xi) = (2\xi - 1)(\xi - 1),\ E_{\frac{1}{2}}(\xi) = 4\xi(1-\xi),\ E_1(\xi) = \xi(2\xi - 1), \\ E_0(\eta) = (2\eta - 1)(\eta - 1),\ E_{\frac{1}{2}}(\eta) = 4\eta(1-\eta),\ E_1(\eta) = \eta(2\eta - 1), \end{aligned}\right\} \tag{4.37}$$

the relevant transfinite shape is given by

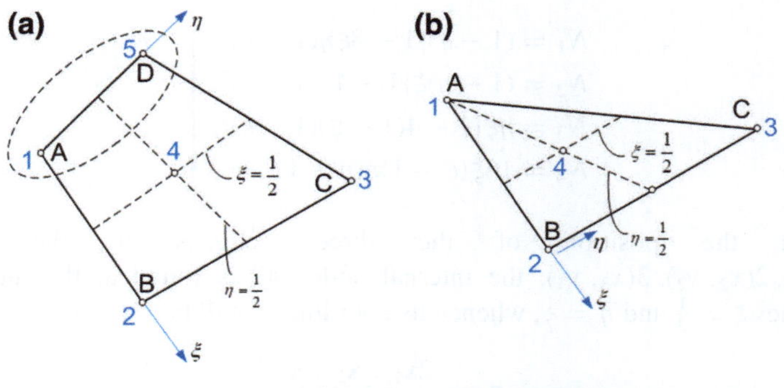

Fig. 4.22 Degeneration of a 5-node quadrilateral into a 4-node triangular element

$$\left.\begin{aligned}\hat{N}_1 &= L_0(\xi)\cdot E_0(\eta)+L_0(\eta)\cdot E_0(\xi)-E_0(\xi)\cdot E_0(\eta)\\ \hat{N}_2 &= L_1(\xi)\cdot E_0(\eta)+L_0(\eta)\cdot E_1(\xi)-E_1(\xi)\cdot E_0(\eta)\\ \hat{N}_3 &= L_1(\xi)\cdot E_1(\eta)+L_1(\eta)\cdot E_1(\xi)-E_1(\xi)\cdot E_1(\eta)\\ \hat{N}_5 &= L_0(\xi)\cdot E_1(\eta)+L_1(\eta)\cdot E_0(\xi)-E_0(\xi)\cdot E_1(\eta)\\ \hat{N}_4 &= E_{\frac{1}{2}}(\xi)\cdot E_{\frac{1}{2}}(\eta)\end{aligned}\right\} \tag{4.38}$$

where due to the linear interpolation along the boundary:

$$\left.\begin{aligned}L_0(\xi) &= 1-\xi, \quad L_1(\xi)=\xi,\\ L_0(\eta) &= 1-\eta, \quad L_1(\eta)=\eta.\end{aligned}\right\} \tag{4.39}$$

Assuming the entire edge *AD* is degenerated to a single point (*A* or 3), for the degenerated configuration shown in Fig. 4.22b the above expression becomes

$$\left.\begin{aligned}N_1 &= \hat{N}_1+\hat{N}_5\\ N_2 &= \hat{N}_2\\ N_3 &= \hat{N}_3\\ N_4 &= \hat{N}_4\end{aligned}\right\} \tag{4.40}$$

After manipulation, the shape functions of the degenerated 4-node element shown in Fig. 4.22b become:

$$\left.\begin{aligned}N_1 &= (1-\xi)[1-8\xi\eta(1-\eta)],\\ N_2 &= (1-\eta)\xi[1-4\eta(1-\xi)],\\ N_3 &= \eta\xi[1-4(1-\xi)(1-\eta)],\\ N_4 &= 16\xi(\xi-1)\eta(\eta-1).\end{aligned}\right\} \tag{4.41}$$

Given the position of the three vertexes in the triangle, $1(x_1, y_1), 2(x_2, y_2), 3(x_3, y_3)$, the internal node "4" is found at the intersection of the lines $\xi = \frac{1}{2}$ and $\eta = \frac{1}{2}$, whence its coordinates will be:

$$\mathbf{x}_4 = \frac{2\mathbf{x}_1+\mathbf{x}_2+\mathbf{x}_3}{4} \tag{4.42}$$

For an isosceles and rectangular triangle with vertexes 1(0, 0), 2(1, 0), and 3(1, 1), the shape functions are shown in Fig. 4.23.

4.7.3 Degenerated Quadrilateral with Seven Nodes

We consider a 9-node quadrilateral with one internal node at its centroid (Fig. 4.24a). In this case, both the trial and blending functions are the quadratic Lagrange polynomials given by Eq. (4.34), and therefore, the shape functions $\hat{N}_1$–$\hat{N}_9$ are given as tensor products.

Following the same procedure with that of Sect. 4.7.2, the shape functions of the degenerated 7-node triangular element of which the degenerated node is 1 (Fig. 4.24b) will be:

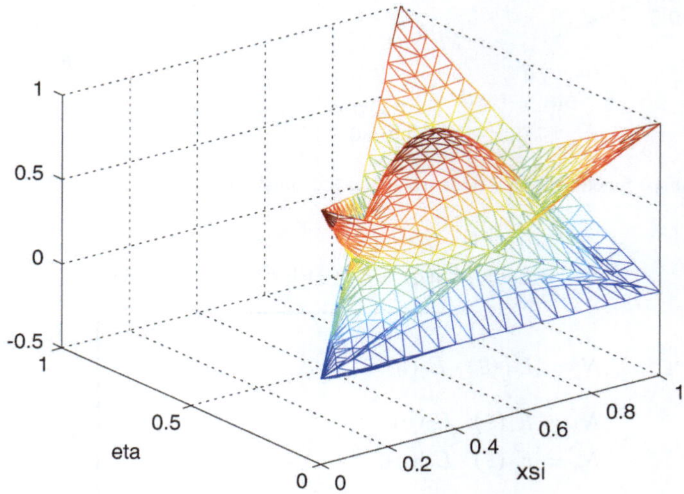

Fig. 4.23 Shape functions for a degenerated 4-node triangular element

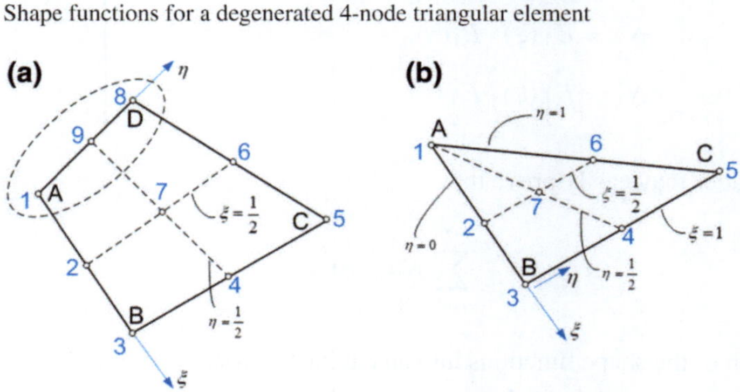

Fig. 4.24 Degeneration of a 9-node quadrilateral into a 7-node triangular element

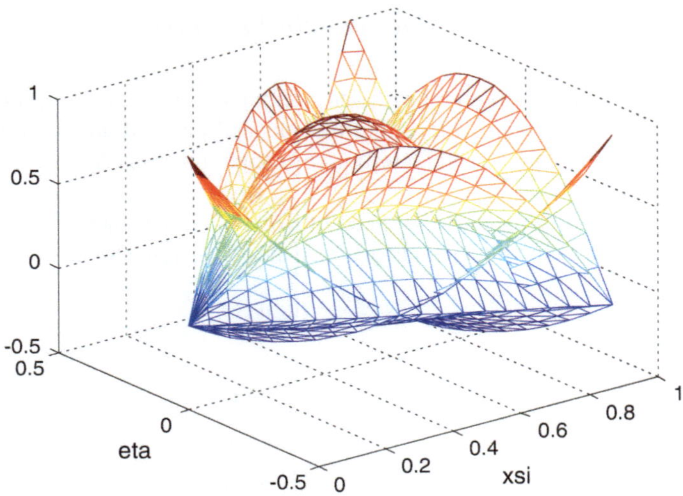

Fig. 4.25 Shape functions of a 7-node triangular transfinite element

$$\left.\begin{aligned}
N_1 &= E_0(\xi)\underbrace{[E_0(\eta)+E_{\frac{1}{2}}(\eta)+E_1(\eta)]}_{1} \equiv E_0(\xi)\\
N_2 &= E_{\frac{1}{2}}(\xi)\cdot E_0(\eta),\\
N_3 &= E_1(\xi)\cdot E_0(\eta),\\
N_4 &= E_1(\xi)\cdot E_{\frac{1}{2}}(\eta),\\
N_5 &= E_1(\xi)\cdot E_1(\eta),\\
N_6 &= E_{\frac{1}{2}}(\xi)\cdot E_1(\eta),\\
N_7 &= E_{\frac{1}{2}}(\xi)\cdot E_{\frac{1}{2}}(\eta).
\end{aligned}\right\} \tag{4.43}$$

The reader may easily prove that

$$\sum_{i=1}^{7} N_i(\xi,\eta) \equiv 1 \tag{4.44}$$

A graph of the shape functions into an equilateral triangle of unit length, with the vertices at the points $\mathbf{1}(0,0)$, $\mathbf{2}(\frac{\sqrt{3}}{2},-\frac{1}{2})$ and $\mathbf{3}(\frac{\sqrt{3}}{2},\frac{1}{2})$, is shown in Fig. 4.25.

4.7.4 Arbitrary-Node Degenerated Quadrilateral into Triangular Element

Let us consider that the sides *AB* and *CA* of the degenerated triangle are divided into n_ξ segments, whereas the side *BC* into n_η ones. Considering that all nodal points of the side *AD* degenerate to the single node "1" (see Fig. 4.26), the total number of nodes will be:

$$n_e = n_\xi(n_\eta + 1) + 1 \tag{4.45}$$

The shape function that is associated to the degenerated node "1" will be the sum of all relevant nodes along the degenerated side *DA*, that is:

$$N_1 = E_0(\xi) \cdot \sum_{j=1}^{n_\eta+1} E_j(\eta) = E_0(\xi) \cdot 1 \equiv E_0(\xi), \tag{4.46a}$$

where $E_i(\xi), i = 1, \ldots, n_\xi + 1$ are the Lagrange polynomials that correspond to the sequence $\{\xi_1, \ldots, \xi_{n_\xi+1}\}$, whereas $E_j(\xi), j = 1, \ldots, n_\eta + 1$ are the ones that correspond to the sequence $\{\eta_1, \ldots, \eta_{n_\eta+1}\}$.

Based on the same Lagrange polynomials, the shape functions associated to the rest are given by a simple expression as follows. If the node is found at the intersection of the perpendicular lines ξ_k (with $k = 1, \ldots, n_\xi + 1$) and η_l (with $l = 1, \ldots, n_\eta + 1$), the corresponding shape function will be:

$$N_i(\xi, \eta) = E_k(\xi) \cdot E_k(\eta), i = 2, \ldots, n_e \tag{4.46b}$$

Obviously, the above degenerated triangular elements are characterized by a preference in the η-direction, in the sense that all lines with constant η-value pass through the point 1 (degenerate vertex of the triangle).

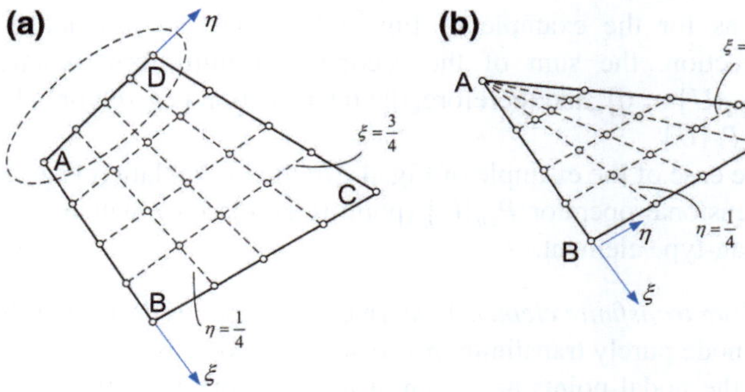

Fig. 4.26 General degenerated triangular element

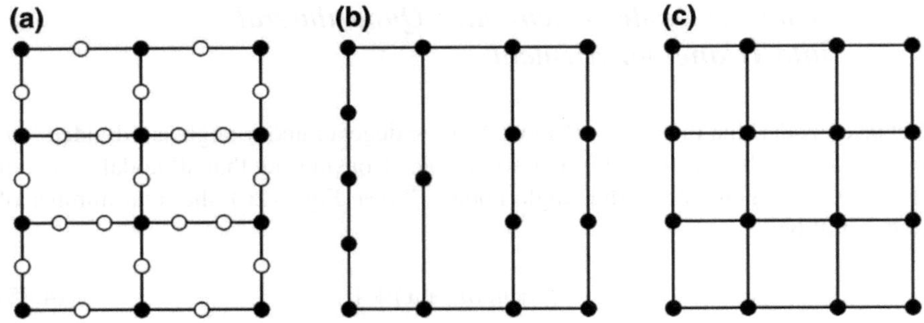

Fig. 4.27 Characteristic cases of Coons interpolation: **a** full formula, **b** blending in ξ-direction, **c** tensor product

Remark More generally, the procedure does not restrict to a uniform grid of internal nodes only (structured as in Fig. 4.27c), but it can include several patterns of internal nodes arranged as also mentioned in Sect. 4.7 (see Fig. 4.27a, b).

4.8 Special Cases of Transfinite Interpolation

The general expression of the transfinite interpolation is given by:

$$U(\xi, \eta) = P_\xi\{U\} + P_\eta\{U\} - P_{\xi\eta}\{U\} \tag{4.47}$$

Applying (4.47) to the examples of Fig. 4.27, one may make the following comments:

- For the example of Fig. 4.27a, the full interpolation formula (4.47) is necessary because internal nodes are arranged in *both* directions. The black-colored nodes are arranged at the intersection of horizontal and vertical lines whereas the white-colored ones do not.
- In contrast, as for the example of Fig. 4.27b there are no nodes arranged in the ξ-direction, the sum of the second and third term equals to zero $\left(P_\eta\{U\} - P_{\xi\eta}\{U\} = 0\right)$, and therefore, the interpolation is given only by the lofting operator $P_\xi\{U\}$.
- Finally, in the case of the example of Fig. 4.27c the interpolation is given only by the two-dimensional operator $P_{\xi\eta}\{U\}$ (point interpolation), which is the case of the Lagrangian-type element.

Example 4.9 (True transfinite element) As an example, let us derive the shape functions of the 21-node purely transfinite macroelement shown in Fig. 4.28.

The data at the nodal points are given at the positions $\boldsymbol{\xi} = \left[0, \frac{1}{4}, \frac{1}{2}, \frac{3}{4}, 1\right]$ and $\boldsymbol{\eta} = \left[0, \frac{1}{4}, \frac{1}{2}, \frac{3}{4}, 1\right]$. As previously was mentioned, the black-colored nodes are found at the intersection of vertical and horizontal lines, which are perpendicular to the ξ- and η-axes at the positions of the vectors $\boldsymbol{\xi}$ and $\boldsymbol{\eta}$, respectively.

The symmetric nature of the macroelement, with *four* subdivisions in each direction, implies the adoption of blending functions which are Lagrange polynomials of *fourth* degree, which are denoted by E_1, E_2, E_3, E_4 and E_5.

It is clear that at the boundary as well as along the middle lines, the approximation of the variable U is performed using trial functions, which again are Lagrange polynomials of fourth degree. For the sake of clarity, these are denoted by L_1^5, L_2^5, L_3^5, L_4^5 and L_5^5, although they coincide with the abovementioned E_1, E_2, E_3, E_4 and E_5. Therefore:

$$\left.\begin{aligned}
U(0,\eta) &= L_1^5(\eta)\cdot U_1 + L_2^5(\eta)\cdot U_{16} + L_3^5(\eta)\cdot U_{15} + L_4^5(\eta)\cdot U_{14} + L_5^5(\eta)\cdot U_{13}\\
U(\tfrac{1}{2},\eta) &= L_1^5(\eta)\cdot U_3 + L_2^5(\eta)\cdot U_{20} + L_3^5(\eta)\cdot U_{18} + L_4^5(\eta)\cdot U_{21} + L_5^5(\eta)\cdot U_{11}\\
U(1,\eta) &= L_1^5(\eta)\cdot U_5 + L_2^5(\eta)\cdot U_6 + L_3^5(\eta)\cdot U_7 + L_4^5(\eta)\cdot U_8 + L_5^5(\eta)\cdot U_9
\end{aligned}\right\} \tag{4.48}$$

and similarly

$$\left.\begin{aligned}
U(\xi,0) &= L_1^5(\eta)\cdot U_1 + L_2^5(\eta)\cdot U_2 + L_3^5(\eta)\cdot U_3 + L_4^5(\eta)\cdot U_4 + L_5^5(\eta)\cdot U_5\\
U(\xi,\tfrac{1}{2}) &= L_1^5(\eta)\cdot U_{15} + L_2^5(\eta)\cdot U_{17} + L_3^5(\eta)\cdot U_{18} + L_4^5(\eta)\cdot U_{19} + L_5^5(\eta)\cdot U_7\\
U(\xi,1) &= L_1^5(\eta)\cdot U_{13} + L_2^5(\eta)\cdot U_{12} + L_3^5(\eta)\cdot U_{11} + L_4^5(\eta)\cdot U_{10} + L_5^5(\eta)\cdot U_9
\end{aligned}\right\} \tag{4.49}$$

In contrast, at the places $\xi = \frac{1}{4}, \frac{3}{4}$ and $\eta = \frac{1}{4}, \frac{3}{4}$ it is necessary to use Lagrange polynomials of second degree, which are denoted by L_1^3, L_2^3 and L_3^3. Thus, the corresponding approximation is:

$$\left.\begin{aligned}
U(\tfrac{1}{4},\eta) &= L_1^3(\eta)\cdot U_2 + L_2^3(\eta)\cdot U_{17} + L_3^3(\eta)\cdot U_{12}\\
U(\tfrac{3}{4},\eta) &= L_1^3(\eta)\cdot U_4 + L_2^3(\eta)\cdot U_{19} + L_3^3(\eta)\cdot U_{10}
\end{aligned}\right\} \tag{4.50a}$$

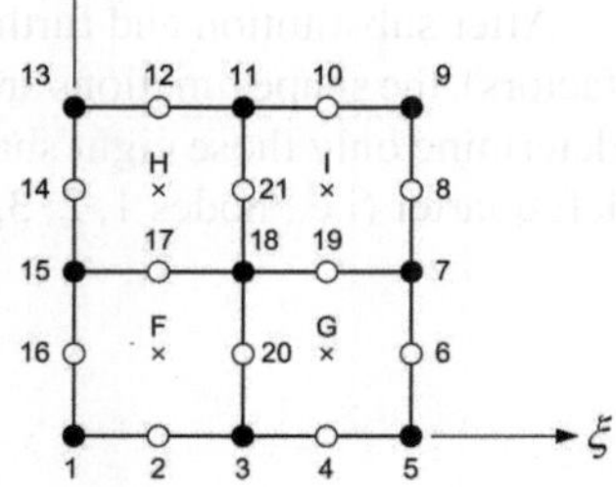

Fig. 4.28 Transfinite element with 21 nodes

and also

$$\left.\begin{aligned} U(\xi, \tfrac{1}{4}) &= L_1^3(\xi) \cdot U_{16} + L_2^3(\xi) \cdot U_{20} + L_3^3(\xi) \cdot U_6 \\ U(\xi, \tfrac{3}{4}) &= L_1^3(\xi) \cdot U_{14} + L_2^3(\xi) \cdot U_{21} + L_3^3(\xi) \cdot U_8 \end{aligned}\right\} \quad (4.50b)$$

The three involved projections are given in terms of the abovementioned approximations by:

$$\left.\begin{aligned} P_\xi &= E_1(\xi) \cdot U(0, \eta) + E_2(\xi) \cdot U(\tfrac{1}{4}, \eta) + E_3(\xi) \cdot U(\tfrac{1}{2}, \eta) \\ &\quad + E_4(\xi) \cdot U(\tfrac{3}{4}, \eta) + E_5(\xi) \cdot U(1, \eta) \\ P_\eta &= E_1(\eta) \cdot U(\xi, 0) + E_2(\eta) \cdot U(\xi, \tfrac{1}{4}) + E_3(\eta) \cdot U(\xi, \tfrac{1}{2}) \\ &\quad + E_4(\eta) \cdot U(\xi, \tfrac{3}{4}) + E_5(\eta) \cdot U(\xi, 1) \\ P_{\xi\eta} &= \sum_{i=1}^{5} \sum_{j=1}^{5} E_i(\xi) \cdot E_j(\eta) \end{aligned}\right\} \quad (4.51)$$

After substitution and extraction of common factors, it is found that the shape functions associated to the black-colored nodes are expressed as tensor product of the corresponding blending functions. A characteristic finding for this element (usual in fully transfinite ones) is that the remaining terms in $P_{\xi\eta}$ are

$$\begin{aligned} \tilde{P}_{\xi\eta} &= E_2(\xi) \cdot E_2(\eta) \cdot U_F + E_4(\xi) \cdot E_2(\eta) \cdot U_G \\ &\quad + E_2(\xi) \cdot E_4(\eta) \cdot U_H + E_4(\xi) \cdot E_4(\eta) \cdot U_I \end{aligned} \quad (4.52)$$

One may observe in Fig. 4.28 that these four auxiliary points (F, G, H, and I) are denoted by a cross ($\times$) and do not correspond to nodal points. In order to eliminate them, it is reasonable to average the sum of their particular corresponding projections, which is explicitly written, for example, for the point F, by

$$\begin{aligned} U_F &= \tfrac{1}{2}(P_\xi + P_\eta) \\ &= \tfrac{1}{2}\left[L_1^3(\tfrac{1}{4}) \cdot U_2 + L_2^3(\tfrac{1}{4}) \cdot U_{17} + L_3^3(\tfrac{1}{4}) \cdot U_{12}\right] \\ &\quad + \tfrac{1}{2}\left[L_1^3(\tfrac{1}{4}) \cdot U_{16} + L_2^3(\tfrac{1}{4}) \cdot U_{20} + L_3^3(\tfrac{1}{4}) \cdot U_6\right] \end{aligned} \quad (4.53)$$

Similar expression can be easily derived for the points G, H, and I as well.

After substitution and further manipulation (reduction and extraction of common factors), the shape functions are derived. Due to the double symmetry, it is sufficient to determine only those eight shape functions associated to the nodal points in bottom-left quarter (i.e., nodes 1, 2, 3, 20, 18, 17, 15, and 16), which are as follows:

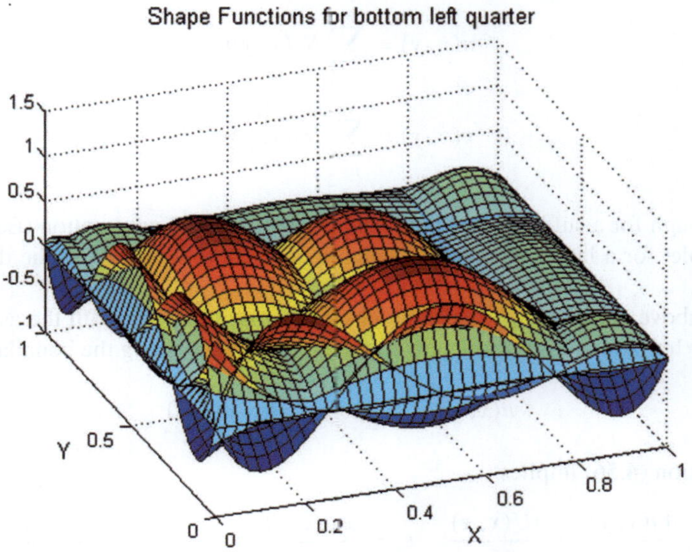

Fig. 4.29 Shape functions of the nodes that lie in the bottom-left quarter of the 21-node macroelement shown in Fig. 4.28

$$\left.\begin{aligned}
N_1(\xi,\eta) &= E_1(\xi)\cdot E_1(\eta)\\
N_2(\xi,\eta) &= E_2(\xi)\cdot L_1^3(\eta) - \tfrac{3}{16}E_2(\xi)\cdot E_2(\eta) + \tfrac{1}{16}E_2(\xi)\cdot E_4(\eta)\\
N_3(\xi,\eta) &= E_3(\xi)\cdot E_1(\eta)\\
N_{15}(\xi,\eta) &= E_1(\xi)\cdot E_3(\eta)\\
N_{16}(\xi,\eta) &= L_1^3(\xi)\cdot E_2(\eta) - \tfrac{3}{16}E_2(\xi)\cdot E_2(\eta) + \tfrac{1}{16}E_4(\xi)\cdot E_2(\eta)\\
N_{17}(\xi,\eta) &= E_2(\xi)\cdot L_2^3(\eta) - \tfrac{3}{8}E_2(\xi)\cdot E_2(\eta) - \tfrac{3}{8}E_2(\xi)\cdot E_4(\eta)\\
N_{18}(\xi,\eta) &= E_3(\xi)\cdot E_3(\eta)\\
N_{20}(\xi,\eta) &= L_2^3(\xi)\cdot E_2(\eta) - \tfrac{3}{8}E_2(\xi)\cdot E_2(\eta) - \tfrac{3}{8}E_4(\xi)\cdot E_2(\eta)
\end{aligned}\right\} \quad (4.54)$$

The graph of the abovementioned eight shape functions is shown in Fig. 4.29.

4.9 Potential Versus Elasticity Problems

In potential problems, Coons (see Chap. 3) and Gordon (this chapter) single macroelements seem to work somehow better than what they did in problems of elasticity. This is probably due to the assumption that nodal displacement components in one direction induce components in the same direction only as follows:

$$u(x, y) = \sum_{j=1}^{q_e} N_j(x, y)u_j$$
$$v(x, y) = \sum_{j=1}^{q_e} N_j(x, y)v_j \tag{4.55}$$

Although for a sufficiently small element this classical assumption (Eq. 4.55) is acceptable, for a larger element a deviation appears with respect to the theoretical solution.

The above statement will become obvious working in line with the assumption that only horizontal displacement components, u, appear along the boundary, that is

$$u(x, y) = U(x, y); \quad v(x, y) \equiv 0 \tag{4.56}$$

Equation (4.56) implies

$$\varepsilon_x = \frac{\partial u(x, y)}{\partial x} = \frac{\partial U(x, y)}{\partial x}; \quad \varepsilon_y = \frac{\partial v(x, y)}{\partial x} \equiv 0; \quad \gamma_{xy} = \frac{\partial u}{\partial y} + \frac{\partial v}{\partial x} \equiv \frac{\partial u}{\partial y} \tag{4.57}$$

Substituting Eq. (4.57) in Hooke law for plane stress conditions, the stress components are found as follows:

$$\sigma_x = \frac{E}{1-\nu^2}\frac{\partial U}{\partial x}; \quad \sigma_y = \frac{\nu E}{1-\nu^2}\frac{\partial U}{\partial x}; \quad \gamma_{xy} = \frac{E}{2(1+\nu)}\frac{\partial U}{\partial y}. \tag{4.58}$$

After trivial manipulation, the two stress equilibrium equations finally become:

$$\frac{\partial^2 U}{\partial x^2} + \frac{1-\nu}{2}\frac{\partial^2 U}{\partial y^2} = 0 \tag{4.59a}$$

and

$$\frac{\partial^2 U}{\partial x \partial y} \equiv 0 \tag{4.59b}$$

Obviously, Eq. (4.59b) demands that the function $U(x, y)$ should be a constant. Therefore, since the assumption that at any point of the elastic body the vertical displacement component vanishes ($v = 0$) leads to the conclusion that the horizontal displacement component (u) should be a constant, it is only the *rigid body* motion that is fully consistent with the decoupling assumed in Eq. (4.55).

4.10 Recapitulation

Gordon–Coons (transfinite) macroelements cover an entire spectrum that varies *between* the classical *Serendipity* and *Lagrangian* elements (including both of them). In Chap. 3, we saw that Coons elements in conjunction with Lagrange polynomials as trial functions coincide with the classical elements of Serendipity family. Furthermore, in Chap. 4 we continued the findings by showing that Gordon elements in conjunction with Lagrange polynomials as trial functions, and at the same time as blending functions, coincide with the classical elements of Lagrangian family.

Therefore, Gordon–Coons (transfinite) macroelements are capable of handling any intermediate situation between Serendipity and Lagrangian elements.

Moreover, transfinite elements are capable of handling *even more* nodes than those involved in the classical Lagrangian-type elements (however without achieving a higher accuracy). It was also shown that the arrangement of the internal nodes does not restrict to a structured net (like a tensor product) only but may be of more complex shape, as for example in Fig. 4.28.

In this book, the term "Coons macroelements" (or "C-elements") refers to the use of nodal points only along the boundary of the Coons patch *ABCD*. If internal nodes are inserted in the interior of the patch, then the term "Gordon macroelements" is preferred to make the difference more recognizable.

Using a single Gordon element, a lot of primitive two-dimensional domains such as rectangle, circular, and elliptical shape were successfully analyzed with high accuracy. In contrast to the boundary-only Coons macroelement, no difficulty appeared in the handling of Dirichlet-type boundary conditions.

In potential problems, Coons and Gordon single elements worked somehow better than what they did in problems of elasticity. This is probably due to the assumption that nodal displacement components in one direction (say x) induce components in the same direction (here x) only. Although for a sufficiently small element this classical assumption is acceptable, for a larger element a deviation appears with respect to the theoretical solution. Therefore in elasticity problems, the domain decomposition is more imperative than what in potential problems is.

References

1. Bathe KJ (1996) Finite element procedures. Prentice-Hall, Upper Saddle River, New Jersey
2. Bercovier M, Shilat E (1993) Enhancement of Gordon-Coons interpolations by 'bubble functions'. Comput Aided Geom Des 10(3–4):253–265
3. Brebbia CA, Ferrante AJ (1986) Computational methods for the solution of engineering problems, 3rd revised edn. Pentech Press, London
4. Cavendish JC, Gordon WJ, Hall CA (1976) Ritz-Galerkin approximations in blending function spaces. Numer Math 26:155–178
5. Cavendish JC, Gordon WJ, Hall CA (1977) Substructured macro elements based on locally blending interpolation. Int J Numer Meth Eng 11:1405–1421

6. Cavendish JC, Hall CA, Zienkiewicz OC (1978) Blended infinite elements for paraboblic boundary value problems. Int J Numer Meth Eng 12(12):1841–1851
7. Cavendish JC, Hall CA (1984) A new class of transitional blended finite elements for the analysis of solid structures. Int J Numer Meth Eng 28:241–253
8. Dimitriou V (2004) Adaptive finite elements and related meshes. Ph.D. Dissertation (advisor: Kanarachos AE), National Technical University of Athens, School of Mechanical Engineering, Athens, Aug 2004 (in Greek). Online available from: http://thesis.ekt.gr/thesisBookReader/id/16430#page/132/mode/2up
9. El-Zafrany A, Cookson RA (1986) Derivation of Lagrangian and Hermitian shape functions for triangular elements. Int J Numer Meth Eng 23(2):275–285
10. El-Zafrany A, Cookson RA (1986) Derivation of Lagrangian and Hermitian shape functions for quadrilateral elements. Int J Numer Meth Eng 23(10):1939–1958
11. Gordon WJ, Hall CA (1973) Transfinite element methods: blending-function interpolation over arbitrary curved element domains. Numer Math 21:109–112
12. Gordon WJ, Thiel LC (1982) Transfinite mappings and their application to grid generation. In: Thompson JP (ed) Numerical grid generation. Elsevier. Appl Math Comput 11–12:171–233
13. Nardini D, Brebbia CA (1983) A new approach to free vibration analysis using boundary elements. Appl Math Model 7:157–162
14. Nardini D, Brebbia CA (1985) Boundary integral formulation of mass matrices for dynamic analysis. In: Brebbia CA (ed) Topics in boundary element research, vol 2. Springer, Berlin, pp 191–208
15. Partridge PW, Brebbia CA, Wrobel LC (1992) The dual reciprocity boundary element method. Computational Mechanics Publications, Southampton; Elsevier, New York, pp. 236–238
16. Provatidis CG (2004) Solution of two-dimensional Poisson problems in quadrilateral domains using transfinite Coons interpolation. Commun Numer Methods Eng 20(7):521–533
17. Provatidis CG (2005) Performance of large Gordon-Coons finite elements in 2-D potential problems. In: Georgiou G, Papanastasiou P, Papadrakakis M (eds) Proceedings of the 5th GRACM international congress on computational mechanics, Limassol, Cuprus, 29 June–1 July 2005, vol 2, pp 805–812. Dedicated to the Memory of Professor John H. Argyris. Online available from: http://euclid.mas.ucy.ac.cy/~georgios/bookfiles/GRACM05v2.pdf
18. Provatidis CG (2006) Coons-patch macroelements in two-dimensional parabolic problems. Appl Math Model 30:319–351
19. Provatidis CG (2006) Free vibration analysis of two-dimensional structures using Coons-patch macroelements. Finite Elem Anal Des 42(6):18–531
20. Provatidis CG (2006) Transient elastodynamic analysis of two-dimensional structures using Coons-patch macroelements. Int J Solids Struct 43:6688–6706
21. Provatidis CG (2009) Eigenanalysis of two-dimensional acoustic cavities using transfinite interpolation. J Algorithms Comput Technol 3(4):477–502
22. Provatidis CG (2012) Two-dimensional elastostatic analysis using Coons-Gordon interpolation. Meccanica 47(4):951–967
23. Scott FC (1968) A quartic, two dimensional isoparametric element. Undergraduate project, Univ. of Wales, Swansea
24. Timoshenko S, Goodier JN (1970) Theory of elasticity, 3rd edn. McGraw-Hill, New York
25. Zienkiewicz OC (1977) The finite element method. McGraw-Hill, London

Chapter 5
BARNHILL's Interpolation and Relevant Isoparametric Elements in Triangular Patches

Abstract This chapter discusses the derivation of macroelements based on Barnhill's interpolation formula within triangular CAD patches. Particular attention is paid to the classical triangular elements up to the sixth degree. In both cases, not only boundary nodes but also internal ones will be considered. The performance of a single macroelement is tested in potential boundary value and eigenvalue problems (Laplace equation and acoustics). For the sake of brevity, the discussion restricts to C^0-continuity only.

Keywords BARNHILL interpolation · Triangular patch
Triangular macroelements · Test cases

5.1 Introduction

Given a triangular patch ABC, in which (i) a function $f(v)$ is given along the edge AB, (ii) a function $g(w)$ is given along BC, and (iii) a third one $h(u)$ is given along the edge CA (see Fig. 5.1), it was found in Chap. 2 that at every arbitrary triplet (u, v, w) of area coordinates, the function $U(u, v, w)$ is interpolated by the rational expression:

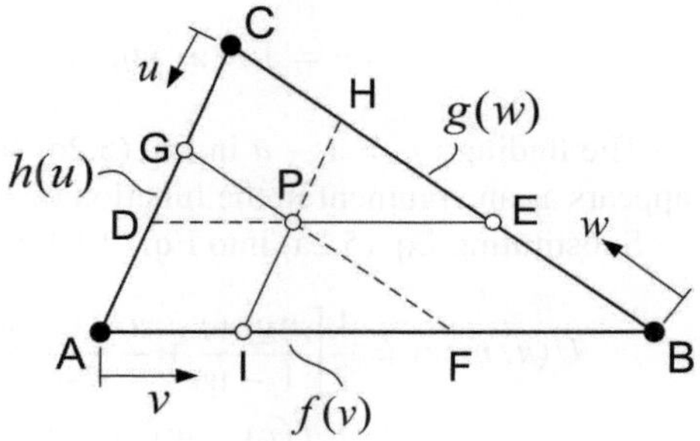

Fig. 5.1 Projections and area coordinates in the Barnhill patch

C. G. Provatidis, *Precursors of Isogeometric Analysis*, Solid Mechanics and Its Applications 256, https://doi.org/10.1007/978-3-030-03889-2_5

$$U(u,v,w)=\frac{1}{2}\left[\frac{v}{1-w}U_E+\frac{u}{1-w}U_D+\frac{w}{1-u}U_G+\frac{v}{1-u}U_F+\frac{u}{1-v}U_I\right.$$
$$\left.+\frac{w}{1-v}U_H-uf(0)-vg(0)-wh(0)\right] \tag{2.103}$$

In terms of the function U, the three involved univariate functions are written as

$$f(v)=U(u,v,0),\quad g(w)=U(0,v,w),\quad h(u)=U(u,0,w) \tag{5.1}$$

Given the fact that the *primary* projections of a point $P(u,v,w)$ on the boundary are the points **I** (at the position v measured from A), **E** (at the position w measured from B), and **G** (at the position u measured from C), whereas the *secondary* ones are the points (F, H, and D); the values of the function $U(u,v,w)$ at the six aforementioned boundary points (all involved in Eq. 2.103) are given by:

$$\left.\begin{array}{ll} U_E=g(w), & U_D=h(1-w),\\ U_G=h(u), & U_F=f(1-u),\\ U_I=f(v), & U_H=g(1-v).\end{array}\right\} \tag{5.2a}$$

A detailed explanation of Eq. (5.2a) is as follows. First, from Fig. 5.1, one may observe that (by definition) the area coordinates of the projections are:

$$v_I=v_H=v,\ w_E=w_D=w \text{ and } u_G=u_F=u. \tag{5.2b}$$

Second, let us now consider a concrete side of the triangle (say *AB*). The univariate reference coordinate of the point I is $v_I=v$, measured from A, thus, $U_I=f(v)$. In order to determine U_F, we need to find v_F. Since the point F belongs to the side *AB* where $w=0$, the general condition $u+v+w=1$ leads to $u_F+v_F=1$, whence $v_F=1-u_F$ and therefore $U_F=f(v_F)=f(1-u_F)$. Due to the last equality of Eq. (5.2b), $U_F=f(1-u)$, which is involved in the second row of Eq. (5.2a). Similar remarks may be made for the rest two sides, *BC* and *CA*. Totally, for the secondary projections, we can write:

$$v_F=1-u,\ w_H=1-v \text{ and } u_D=1-w. \tag{5.2c}$$

The finding $v_F=1-u$ in Eq. (5.2c) supports the reason why the term $(1-u)$ appears as an argument in the function of $U_F=f(1-u)$ in Eq. (5.2a), and so on.

Substituting Eq. (5.2a) into Eq. (2.103), one obtains:

$$U(u,v,w)=\frac{1}{2}\left[\frac{vg(w)}{1-w}+\frac{uh(1-w)}{1-w}+\frac{wh(u)}{1-u}+\frac{vf(1-u)}{1-u}\right.$$
$$\left.+\frac{uf(v)}{1-v}+\frac{wg(1-v)}{1-v}-uf(0)-vg(0)-wh(0)\right] \tag{5.3a}$$

Introducing the parameters:

$$\left.\begin{aligned} E_{vw} &= \tfrac{v}{1-w}, \; E_{uw} = \tfrac{u}{1-w}, \\ E_{wu} &= \tfrac{w}{1-u}, \; E_{vu} = \tfrac{v}{1-u}, \\ E_{uv} &= \tfrac{u}{1-v}, \; E_{wv} = \tfrac{w}{1-v}, \end{aligned}\right\} \tag{5.3b}$$

Equation (5.3a) takes now the more manageable form:

$$\begin{aligned} U(u,v,w) = \frac{1}{2}[&E_{vw}g(w) + E_{uw}h(1-w) + E_{wu}h(u) + E_{vu}f(1-u) \\ &+E_{uv}f(v) + E_{wv}g(1-v) - uf(0) - vg(0) - wh(0)] \end{aligned} \tag{5.3c}$$

5.2 Triangular Elements Under the Prism of Barnhill's Interpolation

The possibility of creating large triangular finite elements was understood very early, and error bounds were determined for Poisson's equation [6, 7]. Here we introduce this topic in an instructive way.

We remind that in Chaps. 3 and 4, we saw that a subclass of quadrilateral Coons and Gordon macroelements is the classical finite elements of the Serendipity and Lagrange families, respectively. In a similar way, below (Sects. 5.2.1 and 5.2.2) we shall see that a subclass of *Barnhill macroelements* includes the conventional triangular elements (3-node and 6-node), as also reported by Provatidis [34].

Afterward, the general formula will be given for the arbitrary-noded Barnhill macroelement in a triangular patch *ABC*, where the nodes are arranged along the boundary only (see Sect. 5.2.3). Later, the general methodology of introducing internal nodes will be discussed, first in the classical high-order element and later in the Barnhill macroelement.

5.2.1 *The Classical Linear Triangular Element*

Let us consider the triangular element $ABC \equiv 123$ as shown in Fig. 5.2a. Assuming the variation of the variable, U, along every side to be *linear*, the boundary terms shown in Fig. 5.2a can be written as follows:

$$f(v) \triangleq U_I = (1-v)U_A + vU_B \tag{5.4}$$

$$g(w) \triangleq U_E = (1-w)U_B + wU_C \tag{5.5}$$

$$h(u) \triangleq U_G = (1-u)U_C + uU_A \tag{5.6}$$

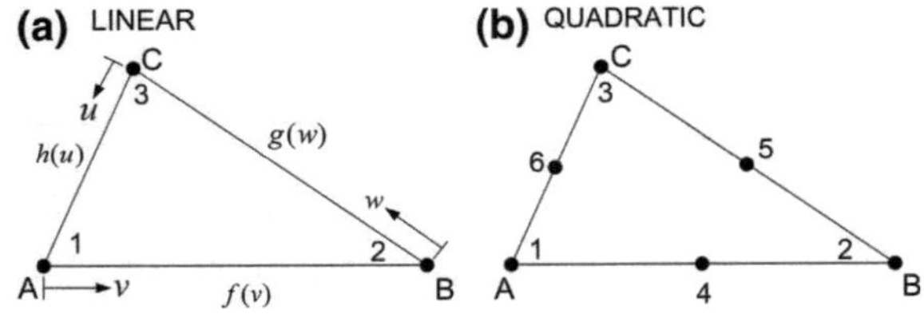

Fig. 5.2 Barnhill's formula applied to derive conventional triangular elements for **a** linear and **b** quadratic interpolations

Moreover, the values at the rest three boundary points involved in Eq. (5.3c) are:

$$f(1-u) \triangleq U_F = [1-(1-u)]U_A + (1-u)U_B = uU_A + (1-u)U_B \tag{5.7}$$

$$g(1-v) \triangleq U_H = [1-(1-v)]U_B + (1-v)U_C = vU_B + (1-v)U_C \tag{5.8}$$

$$h(1-w) \triangleq U_D = [1-(1-w)]U_C + (1-w)U_A = wU_C + (1-w)U_A \tag{5.9}$$

Substituting Eqs. (5.4)–(5.9) into Eq. (5.3a), after extensive manipulation and deduction, one obtains:

$$U(u, v, w) = u \cdot U_A + v \cdot U_B + w \cdot U_C \tag{5.10}$$

Taking into consideration that the *barycentric* (or areal) coordinates ($\xi_1,\ \xi_2,\ \xi_3$) are identical to the abovementioned coordinates (u, v, w), that is:

$$\xi_1 = u, \xi_2 = v, \xi_3 = w, \tag{5.11a}$$

with

$$u + v + w = 1, \tag{5.11b}$$

Equation (5.10) leads to the well-known formula:

$$U(u, v, w) = \xi_1 U_A + \xi_2 U_B + \xi_3 U_C, \tag{5.12}$$

which implies the classical shape functions:

$$N_i(u, v, w) = \xi_i, \quad i = 1, 2, 3. \tag{5.13}$$

5.2.2 *The Classical Quadratic Triangular Element*

Let us consider the triangular element $ABC \equiv 123456$, as shown in Fig. 5.2b. Since the variation of the variable, U, along the sth side, ($s = AB, BC, CA$), is *quadratic*, we introduce the Lagrange polynomials:

$$L_S^1(\xi) = (2\xi - 1)(\xi - 1),\ L_S^2(\xi) = 4\xi(1 - \xi),\ L_S^3(\xi) = \xi(2\xi - 1) \tag{5.14}$$

Due to the fact that each of the abcvementioned three Lagrange polynomials appears as a well-known symmetry with respect to the midpoint $\xi = 1/2$ (the half is a mirror of the rest half), we can write:

$$L_S^1(1 - \xi) = L_S^3(\xi), L_S^2(1 - \xi) = L_S^2(\xi), L_S^3(1 - \xi) = L_S^1(\xi) \tag{5.15}$$

Based on Eq. (5.14), the boundary terms shown in Fig. 5.2b can be written in terms of Lagrange polynomials as:

$$f(v) \triangleq U_I = L_{AB}^1(v)U_1 + L_{AB}^2(v)U_4 + L_{AB}^3(v)U_2 \tag{5.16}$$

$$g(w) \triangleq U_E = L_{BC}^1(w)U_2 + L_{BC}^2(w)U_5 + L_{BC}^3(w)U_3 \tag{5.17}$$

$$h(u) \triangleq U_G = L_{CA}^1(u)U_3 + L_{CA}^2(u)U_6 + L_{CA}^3(u)U_1 \tag{5.18}$$

Moreover, the values at the rest three boundary points are:

$$f(1 - u) \triangleq U_F = L_{AB}^1(1 - u) \cdot U_1 + L_{AB}^2(1 - u) \cdot U_4 + L_{AB}^3(1 - u) \cdot U_2 \tag{5.19a}$$

$$g(1 - v) \triangleq U_H = L_{BC}^1(1 - v) \cdot U_2 + L_{3C}^2(1 - v) \cdot U_5 + L_{BC}^3(1 - v) \cdot U_3 \tag{5.20a}$$

$$h(1 - w) \triangleq U_D = L_{CA}^1(1 - w) \cdot U_3 - L_{CA}^2(1 - w) \cdot U_6 + L_{CA}^3(1 - w) \cdot U_1 \tag{5.21a}$$

By virtue of Eq. (5.15), where the general variable ξ takes the values u, v, and w in turn, the above three equations become:

$$f(1 - u) = L_{AB}^3(u) \cdot U_1 + L_{AB}^2(u) \cdot U_4 + L_{AB}^1(u) \cdot U_2 \tag{5.19b}$$

$$g(1 - v) = L_{BC}^3(v) \cdot U_2 + L_{BC}^2(v) \cdot U_5 + L_{BC}^1(v) \cdot U_3 \tag{5.20b}$$

$$h(1 - w) = L_{CA}^3(w) \cdot U_3 + L_{CA}^2(w) \cdot U_6 + L_{CA}^1(w) \cdot U_1 \tag{5.21b}$$

Substituting Eqs. (5.16)–(5.18) as well as Eqs. (5.19b)–(5.21b) into Eq. (5.1) and then extracting common factors, one obtains:

$$
\begin{aligned}
U(u, v, w) = \frac{1}{2}\Bigg\{ & U_1 \cdot \left[\frac{vL^3_{AB}(u)}{1-u} + \frac{uL^1_{AB}(v)}{1-v} + \frac{uL^1_{CA}(w)}{1-w} + \frac{wL^3_{CA}(u)}{1-u} - u\right] \\
& + U_2 \cdot \left[\frac{vL^1_{BC}(w)}{1-w} + \frac{wL^3_{BC}(v)}{1-v} + \frac{vL^1_{AB}(u)}{1-u} + \frac{uL^3_{AB}(v)}{1-v} - v\right] \\
& + U_3 \cdot \left[\frac{vL^3_{BC}(w)}{1-w} + \frac{wL^1_{BC}(v)}{1-v} + \frac{vL^1_{CA}(w)}{1-w} + \frac{uL^3_{CA}(w)}{1-w} - w\right] \\
& + U_4 \cdot \left[\frac{vL^2_{AB}(u)}{1-u} + \frac{uL^2_{AB}(v)}{1-v}\right] \\
& + U_5 \cdot \left[\frac{vL^2_{BC}(w)}{1-w} + \frac{wL^2_{BC}(v)}{1-v}\right] \\
& + U_6 \cdot \left[\frac{uL^2_{CA}(w)}{1-w} + \frac{wL^2_{CA}(u)}{1-u}\right]\Bigg\}
\end{aligned}
\tag{5.22}
$$

It is evident that one-second of the functions into the brackets, which multiply the nodal values U_1, U_2, U_3, U_4, U_5, and U_6 in Eq. (5.22), are the shape functions of the six-node (quadratic) finite element.

Substituting Eq. (5.14) into Eq. (5.22), after manipulation the shape functions become:

$$N_1(u, v, w) = u(2u - 1) \tag{5.23a}$$

$$N_2(u, v, w) = v(2v - 1) \tag{5.23b}$$

$$N_3(u, v, w) = w(2w - 1) \tag{5.23c}$$

$$N_4(u, v, w) = 4uv \tag{5.23d}$$

$$N_5(u, v, w) = 4vw \tag{5.23e}$$

$$N_6(u, v, w) = 4uw \tag{5.23f}$$

Considering the mutual coincidence between the symbols (u, v, w) and (ξ_1, ξ_2, ξ_3), (see Eq. 5.11a), it is immediately concluded that Eq. (5.23) coincides with those mentioned in finite element textbooks, for example (Zienkiewicz [46, p. 167]; Bathe [9, p. 374]).

5.2.3 The General Arbitrarily Noded Boundary-Only Barnhill Triangular Element

Several possibilities of using Barnhill's interpolation in the finite element analysis have been early reported by Barnhill [3]. Here we shall somehow reduce and show that Barnhill's formula allows for the construction of *arbitrary*-noded triangular macroelements, in which the nodal points are arranged along the *boundary* (Fig. 5.3). In such as case, Eqs. (5.16)–(5.21) must be properly extended and then the whole procedure is easily programmed. The general procedure is presented below.

Let us assume that the edges AB, BC, and CA are divided into $(\hat{n}_1, \hat{n}_2, \hat{n}_3)$ segments, respectively, which leads to $(q_1 = \hat{n}_1 + 1, q_2 = \hat{n}_2 + 1, q_3 = \hat{n}_3 + 1)$ nodal points. Therefore, the number of boundary nodes can be written in two equivalent ways as follows:

$$q_B = \hat{n}_1 + \hat{n}_2 + \hat{n}_3 \equiv (q_1 + q_2 + q_3) - 3 \tag{5.24}$$

If the boundary values $U(u, v, 0) = f(v)$, $U(0, v, w) = g(w)$, and $U(u, 0, w) = h(u)$ in Eq. (5.3c) are interpolated by any set of trial functions B_j^{edge} as follows:

Edge AB:

$$U_I = \sum_{j=1}^{q_1} B_j^{AB}(v) \cdot U_j^{AB}, \; U_F = \sum_{j=1}^{q_1} B_j^{AB}(1-u) \cdot U_j^{AB} \tag{5.25a}$$

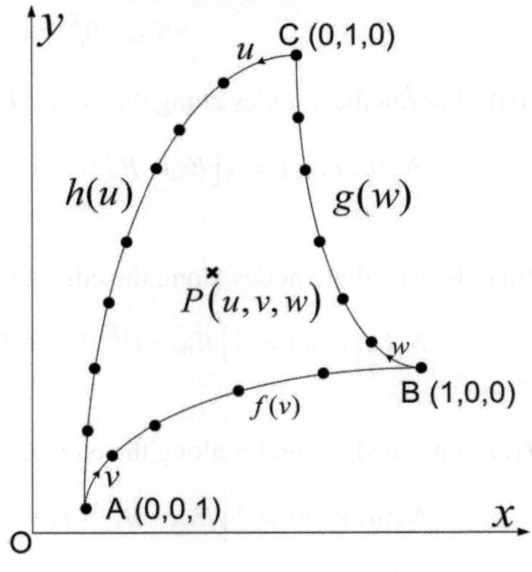

Fig. 5.3 Typical Barnhill macroelement

Edge BC:

$$U_E = \sum_{j=1}^{q_2} B_j^{BC}(w) \cdot U_j^{BC}, \; U_H = \sum_{j=1}^{q_2} B_j^{BC}(1-v) \cdot U_j^{BC} \tag{5.25b}$$

Edge CA:

$$U_G = \sum_{j=1}^{q_3} B_j^{CA}(u) \cdot U_j^{CA}, \; U_D = \sum_{j=1}^{q_3} B_j^{CA}(1-w) \cdot U_j^{CA}, \tag{5.25c}$$

and then they are introduced into Eq. (5.3c), exactly as was performed in Sects. 5.2.1 and 5.2.2, the global shape functions associated to all nodes of the reference macroelement through equation:

$$U(u,v,w) = \sum_{j=1}^{q_e} N_j(u,v,w) \cdot U_j \tag{5.26}$$

are derived as follows:

(i) Corner nodes (at vertices):

$$\begin{aligned} N_A(u,v,w) &= \tfrac{1}{2}\big[E_{uv} \cdot B_1^{AB}(v) + E_{vu} \cdot B_1^{AB}(1-u) \\ &\quad + E_{wu} \cdot B_{q_3}^{CA}(u) + E_{uw} \cdot B_{q_3}^{CA}(1-w) - u\big] \\ N_B(u,v,w) &= \tfrac{1}{2}\big[E_{vw} \cdot B_1^{BC}(w) + E_{wv} \cdot B_1^{BC}(1-v) \\ &\quad + E_{uv} \cdot B_{q_3}^{AB}(v) + E_{vu} \cdot B_{q_3}^{AB}(1-u) - v\big] \\ N_C(u,v,w) &= \tfrac{1}{2}\big[E_{wu} \cdot B_1^{CA}(u) + E_{uw} \cdot B_1^{CA}(1-w) \\ &\quad + E_{vw} \cdot B_{q_3}^{BC}(w) + E_{wv} \cdot B_{q_3}^{BC}(1-v) - w\big] \end{aligned} \tag{5.27}$$

(ii) Intermediate nodes along the edge AB (local numbering):

$$N_j(u,v,w) = \tfrac{1}{2}\big[E_{uv} \cdot B_j^{AB}(v) + E_{vu} \cdot B_j^{AB}(1-u)\big], \quad 2 \le j \le q_1 - 1 \tag{5.28}$$

(iii) Intermediate nodes along the edge BC (local numbering):

$$N_j(u,v,w) = \tfrac{1}{2}\big[E_{vw} \cdot B_j^{BC}(w) + E_{wv} \cdot B_j^{BC}(1-v)\big], \quad 2 \le j \le q_2 - 1 \tag{5.29}$$

(iv) Intermediate nodes along the edge CA (local numbering):

$$N_j(u,v,w) = \tfrac{1}{2}\big[E_{wu} \cdot B_j^{CA}(u) + E_{uw} \cdot B_j^{CA}(1-w)\big], \quad 2 \le j \le q_3 - 1 \tag{5.30}$$

One may observe the affine invariance that characterizes all the abovementioned shape functions, from Eqs. (5.27) to (5.30).

Remarks

(1) It is noted that the trial functions B_j^{edge} that appear in Eqs. (5.27)–(5.30) may be the well-known Lagrange polynomials, but they can also be cardinal natural cubic B-splines, piecewise-linear (hat functions), and so on. Even nonrational or rational Bézier, B-splines or NURBS may be applied as well. For the sake of brevity, below we shall deal with Lagrange polynomials only.
(2) The number of nodes along each side may generally be different one another $(q_1 \neq q_2 \neq q_3 \neq q_1)$.

Although the computer programming of these equations is a straightforward procedure, one may be interested in deriving closed-form analytical expressions. For example, the case of cubic approximation along each side of the triangle is shown in Exercise 5.1.

5.3 Introduction of Internal Nodes

As usual, the quality of the numerical solution improves when internal nodes are introduced, because in this way *complete* polynomials are created. The incremental step from the boundary-only formulation to that of using internal nodes in triangular patches would be anticipated to be analogous to the changes involved in the transit from Coons to Gordon formulation in quadrilaterals. Nevertheless, the transfer of the transfinite blending methodology from quadrilateral to triangular patches is not that simple, because the concept of tensor product is not applicable as is (for a discussion, see Barnhill [3]).

Before we apply Barnhill's interpolation in conjunction with internal nodes, for both reference and comparison reasons, below we deal first with conventional higher-order triangular elements.

5.3.1 *The General Classical Triangular Element*

We refer to the classical case in which each side of the triangle *ABC* is uniformly divided into *m* segments. Moreover, using Lagrange interpolation, the complete *m*th-order polynomial

$$\tilde{U}_m(x, y) = \sum_{k+l=0}^{m} a_{kl} x^k y^l \tag{5.31}$$

can be used to interpolate a function, say $U(x, y)$, at the $n_e = \frac{1}{2}(m+1)(m+2)$ symmetrically placed nodes in a triangle. In general, $n_M = 3m$ nodes are placed on the boundary of the triangle, whereas the rest $n_I = (m-1)(m-2)/2$ nodes appear in its interior; obviously, $n_e = n_B + n_I$.

For $m \geq 2$, the node numbering is according to Fig. 5.4. We start with the vertices (A, B, and C) at where the nodes 1, 2, and 3 are set. Then we continue with the nodes $4, 5, \ldots, (3+m-1)$ along the edge $AB \equiv 12$. In the sequence, we proceed with the nodes $(3+m), \ldots, (2m+1)$ along the edge BC. The boundary closes with the nodes $(2m+2), \ldots, 3m$ along the side CA. Finally, we arrange the nodes $3m+1, \ldots, n_e$ in the interior of the triangular element.

Instead of dealing with many separate cases found in previous textbooks (see Wait and Mitchell [43]; Strang and Fix [41], among others), starting from the linear case ($m = 1$), and then continuing with the quadratic ($m = 2$), cubic ($m = 3$), quartic ($m = 4$), quintic ($m = 5$), etc., we present here generalized expressions that were obtained by a careful *inspection* on the established simple cases.

(i) Corner nodes (vertices):

$$\begin{aligned} N_1 &= \frac{1}{(m-1)!}\xi_1(m\xi_1 - 1)\cdots(m\xi_1 - m + 1) \\ N_2 &= \frac{1}{(m-1)!}\xi_2(m\xi_2 - 1)\cdots(m\xi_2 - m + 1) \\ N_3 &= \frac{1}{(m-1)!}\xi_3(m\xi_3 - 1)\cdots(m\xi_3 - m + 1) \end{aligned} \tag{5.32}$$

(ii) Side nodes along the edge AB:

$$N_{3+k} = \frac{m^2}{(m-k)!\,k!}\xi_1\xi_2 \prod_{i=1}^{m-k-1}(m\xi_1 - i)\prod_{j=1}^{k-1}(m\xi_2 - j), \quad k = 1, \ldots, m-1 \tag{5.33}$$

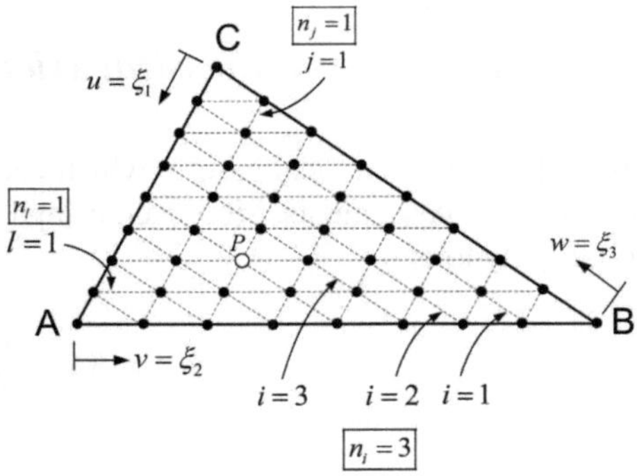

Fig. 5.4 Triangular element for ($m = 8$). Here, for the internal node at $P(\frac{4}{8}, \frac{2}{8}, \frac{2}{8})$, the number of interfering isolines is: $n_i = 3$, $n_j = 1$, and $n_l = 1$

(iii) Side nodes along the edge BC:

$$N_{2+m+k} = \frac{m^2}{(m-k)!\,k!}\xi_2\xi_3 \prod_{i=1}^{m-k-1}(m\xi_2 - i)\prod_{j=1}^{k-1}(m\xi_3 - j), \quad k = 1, \ldots, m-1 \tag{5.34}$$

(iv) Side nodes along the edge CA:

$$N_{2m+1+k} = \frac{m^2}{(m-k)!\,k!}\xi_3\xi_1 \prod_{i=1}^{m-k-1}(m\xi_3 - i)\prod_{j=1}^{k-1}(m\xi_1 - j), \; k = 1, \ldots, m-1 \tag{5.35}$$

(v) Internal nodes:

$$N_{3m+k} = \alpha_P\xi_1\xi_2\xi_3 \prod_{i=1}^{n_i}(m\xi_1 - i)\prod_{j=1}^{n_j}(m\xi_2 - j)\prod_{l=1}^{n_l}(m\xi_3 - l), \quad k = 1, \ldots, n_{\mathrm{I}} \tag{5.36}$$

Equation (5.36) requires further explanation as follows. The internal nodes are usually located at the intersections of isolines ($\xi_1 =$ constant, $\xi_2 =$ constant, $\xi_3 =$ constant). The latter isolines create a net of lines (inter-boundaries) where the data are known. Between a certain internal nodal point $P(\xi_1^P, \xi_2^P, \xi_3^P)$ and the boundary, some isolines interfere. In more detail, there are n_i isolines interfering between P and the side BC, n_j isolines between P and the side CA, and finally n_l isolines between P and the side AB (see Fig. 5.4). One may observe that the aforementioned (n_i, n_j, and n_l) isolines include all the internal nodes except the nodal point $P(\xi_1^P, \xi_2^P, \xi_3^P)$.

In order to ensure that the shape function of the nodal point P vanishes on the entire boundary, it becomes necessary to include the term "($\xi_1\xi_2\xi_3$)." Moreover, in order to ensure that they also vanish along the lines that interfere between P and BC, it becomes necessary to include the term $\prod_{i=1}^{n_i}(m\xi_1 - i)$.

Similarly, in order to ensure that they also vanish along the lines that interfere between P and CA and AB, it becomes necessary to include the terms $\prod_{j=1}^{n_j}(m\xi_2 - j)$ and $\prod_{l=1}^{n_l}(m\xi_3 - l)$, respectively.

Based on the above explanations, the expression of Eq. (5.36) becomes now obvious. The factor α_P has a different value for each particular internal node P and can be easily determined by the cardinality of the shape function of the node P to itself, i.e., by the condition $N_P(\xi_1^P, \xi_2^P, \xi_3^P) = 1$.

Note For all the internal nodes that lie on the outer surface of the internal net of isolines (within the boundary layer of the patch ABC), only two out of the three product sets will exist. For example, all the $(m-2)$ internal nodes on the isoline $\xi_1 = \frac{1}{m}$ (parallel to the side BC), where obviously $n_i = 0$, will not include the term $\prod_{i=1}^{n_i}(m\xi_1 - i)$.

Typical cases are as follows:

Linear element ($m = 1$)

$$N_1 = \xi_1, \quad N_2 = \xi_2, \quad N_3 = \xi_3. \tag{5.37}$$

Quadratic element ($m = 2$)

$$\begin{aligned} &N_1 = \xi_1(2\xi_1 - 1), \; N_2 = \xi_2(2\xi_2 - 1), \; N_3 = \xi_3(2\xi_3 - 1), \\ &N_4 = 4\xi_1\xi_2, \qquad N_5 = 4\xi_2\xi_3, \qquad N_6 = 4\xi_3\xi_1. \end{aligned} \tag{5.38}$$

Cubic element ($m = 3$)

$$\begin{aligned} &N_1 = \tfrac{1}{2}\xi_1(3\xi_1 - 1)(3\xi_1 - 2), \; N_2 = \tfrac{1}{2}\xi_2(3\xi_2 - 1)(3\xi_2 - 2), \\ &N_3 = \tfrac{1}{2}\xi_3(3\xi_3 - 1)(3\xi_3 - 2), \; N_4 = \tfrac{9}{2}\xi_1\xi_2(3\xi_1 - 1), \\ &N_5 = \tfrac{9}{2}\xi_1\xi_2(3\xi_2 - 1), \qquad N_6 = \tfrac{9}{2}\xi_2\xi_3(3\xi_2 - 1), \\ &N_7 = \tfrac{9}{2}\xi_2\xi_3(3\xi_3 - 1), \qquad N_8 = \tfrac{9}{2}\xi_3\xi_1(3\xi_3 - 1), \\ &N_9 = \tfrac{9}{2}\xi_3\xi_1(3\xi_1 - 1), \qquad N_{10} = 27\xi_1\xi_2\xi_3. \end{aligned} \tag{5.39}$$

Quartic element ($m = 4$)

$$\begin{aligned} &N_1 = \tfrac{1}{6}\xi_1(4\xi_1 - 1)(4\xi_1 - 2)(4\xi_1 - 3), \; N_2 = \tfrac{1}{6}\xi_2(4\xi_2 - 1)(4\xi_2 - 2)(4\xi_2 - 3), \\ &N_3 = \tfrac{1}{6}\xi_3(4\xi_3 - 1)(4\xi_3 - 2)(4\xi_3 - 3), \; N_4 = \tfrac{8}{3}\xi_1\xi_2(4\xi_1 - 1)(4\xi_1 - 2), \\ &N_5 = 4\xi_1\xi_2(4\xi_1 - 1)(4\xi_2 - 1), \qquad N_6 = \tfrac{8}{3}\xi_1\xi_2(4\xi_2 - 1)(4\xi_2 - 2), \\ &N_7 = \tfrac{8}{3}\xi_2\xi_3(4\xi_2 - 1)(4\xi_2 - 2), \qquad N_8 = 4\xi_2\xi_3(4\xi_2 - 1)(4\xi_3 - 1), \\ &N_9 = \tfrac{8}{3}\xi_2\xi_3(4\xi_3 - 1)(4\xi_3 - 2), \qquad N_{10} = \tfrac{8}{3}\xi_3\xi_1(4\xi_3 - 1)(4\xi_3 - 2), \\ &N_{11} = 4\xi_3\xi_1(4\xi_3 - 1)(4\xi_1 - 1), \qquad N_{12} = \tfrac{8}{3}\xi_3\xi_1(4\xi_1 - 1)(4\xi_1 - 2), \\ &N_{13} = 32\xi_1\xi_2\xi_3(4\xi_1 - 1), \qquad N_{14} = 32\xi_1\xi_2\xi_3(4\xi_2 - 1), \\ &N_{15} = 32\xi_1\xi_2\xi_3(4\xi_3 - 1). \end{aligned} \tag{5.40}$$

Quintic element ($m = 5$)

Corner nodes:

$$\begin{aligned} &N_1 = \tfrac{1}{24}\xi_1(5\xi_1 - 1)(5\xi_1 - 2)(5\xi_1 - 3)(5\xi_1 - 4), \\ &N_2 = \tfrac{1}{24}\xi_2(5\xi_2 - 1)(5\xi_2 - 2)(5\xi_2 - 3)(5\xi_2 - 4), \\ &N_3 = \tfrac{1}{24}\xi_3(5\xi_3 - 1)(5\xi_3 - 2)(5\xi_3 - 3)(5\xi_3 - 4). \end{aligned} \tag{5.41a}$$

Side AB ($\equiv 12$):

$$
\begin{aligned}
N_4 &= \tfrac{25}{24}\xi_1\xi_2(5\xi_1-1)(5\xi_1-2)(5\xi_1-3),\\
N_5 &= \tfrac{25}{12}\xi_1\xi_2(5\xi_1-1)(5\xi_1-2)(5\xi_2-1),\\
N_6 &= \tfrac{25}{12}\xi_1\xi_2(5\xi_2-1)(5\xi_2-2)(5\xi_1-1),\\
N_7 &= \tfrac{25}{24}\xi_1\xi_2(5\xi_2-1)(5\xi_2-2)(5\xi_2-2).
\end{aligned}
\tag{5.41b}
$$

Side BC ($\equiv 23$):

$$
\begin{aligned}
N_8 &= \tfrac{25}{24}\xi_2\xi_3(5\xi_2-1)(5\xi_2-2)(5\xi_2-3),\\
N_9 &= \tfrac{25}{12}\xi_2\xi_3(5\xi_2-1)(5\xi_2-2)(5\xi_3-1),\\
N_{10} &= \tfrac{25}{12}\xi_2\xi_3(5\xi_3-1)(5\xi_3-2)(5\xi_2-1),\\
N_{11} &= \tfrac{25}{24}\xi_2\xi_3(5\xi_3-1)(5\xi_3-2)(5\xi_3-3).
\end{aligned}
\tag{5.41c}
$$

Side CA ($\equiv 31$):

$$
\begin{aligned}
N_{12} &= \tfrac{25}{24}\xi_3\xi_1(5\xi_3-1)(5\xi_3-2)(5\xi_3-3),\\
N_{13} &= \tfrac{25}{12}\xi_3\xi_1(5\xi_3-1)(5\xi_3-2)(5\xi_1-1),\\
N_{14} &= \tfrac{25}{12}\xi_3\xi_1(5\xi_1-1)(5\xi_1-2)(5\xi_3-1),\\
N_{15} &= \tfrac{25}{24}\xi_3\xi_1(5\xi_1-1)(5\xi_1-2)(5\xi_1-3).
\end{aligned}
\tag{5.41d}
$$

Internal nodes:

$$
\begin{aligned}
N_{16} &= \tfrac{125}{6}\xi_1\xi_2\xi_3(5\xi_1-1)(5\xi_1-2), & N_{17} &= \tfrac{125}{4}\xi_1\xi_2\xi_3(5\xi_1-1)(5\xi_2-1),\\
N_{18} &= \tfrac{125}{6}\xi_1\xi_2\xi_3(5\xi_2-1)(5\xi_2-2), & N_{19} &= \tfrac{125}{4}\xi_1\xi_2\xi_3(5\xi_1-1)(5\xi_3-1),\\
N_{20} &= \tfrac{125}{4}\xi_1\xi_2\xi_3(5\xi_2-1)(5\xi_3-1), & N_{21} &= \tfrac{125}{6}\xi_1\xi_2\xi_3(5\xi_3-1)(5\xi_3-2).
\end{aligned}
\tag{5.41e}
$$

5.3.2 Internal Nodes in Nonclassical Triangular Elements

We begin with the easiest case of a linear triangular element in which we add an internal node. This element is not one of the classical ones (compare with those in Fig. 5.5) and can be treated in two ways. The first is to resort to the standard transfinite interpolation using degeneration of one edge, whereas the second is to introduce a new affine invariant formulation. Both of these procedures will be studied below.

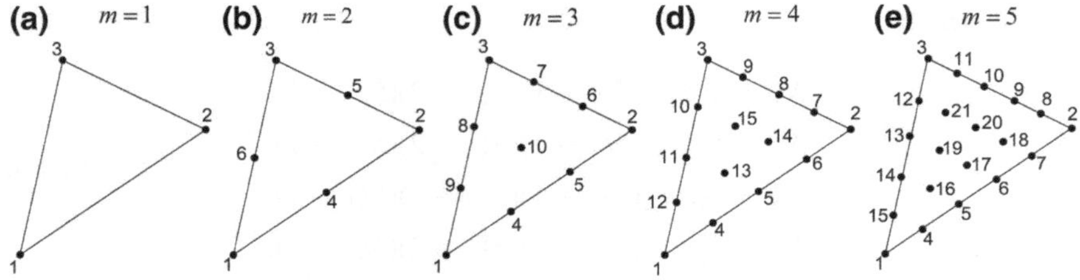

Fig. 5.5 Triangular elements: **a** linear: $m = 1$; **b** quadratic: $m = 2$; **c** cubic: $m = 3$; **d** quartic: $m = 4$; **e** quintic: $m = 5$

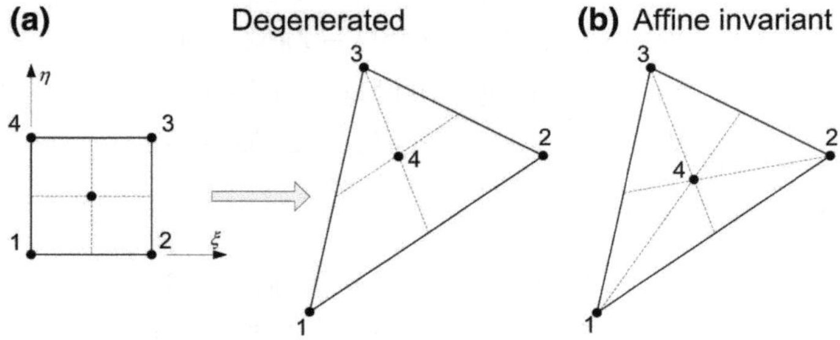

Fig. 5.6 **a** Degenerated versus **b** affine invariant (cyclically symmetric) four-node triangular element

5.3.2.1 Degenerated Quadrilateral into a Triangular Element

The degenerated four-node triangular element has been previously studied in Chap. 4 (see Sect. 4.7). According to Eq. (4.38), the shape functions of the degenerated four-node element are given by (see Fig. 5.6a):

$$\begin{aligned} N_1 &= (1-\xi)\cdot[1-8\xi\eta(1-\eta)],\\ N_2 &= (1-\eta)\xi\cdot[1-4\eta(1-\xi)],\\ N_3 &= \eta\xi\cdot[1-4(1-\xi)(1-\eta)],\\ N_4 &= 16\xi(\xi-1)\eta(\eta-1). \end{aligned} \tag{5.42a}$$

For the above straight triangular element, one can easily verify that the *area coordinates* are expressed in terms of (ξ, η) as follows:

$$\xi_1 = 1-\xi, \quad \xi_2 = (1-\eta)\xi, \quad \xi_3 = \eta\xi. \tag{5.42b}$$

Considering Eq. (5.42b), the three area coordinates, ξ_1, ξ_2, and ξ_3, can be easily recognized as leading factors in the three first lines of Eq. (5.42a), respectively. In contrast, the product $\xi_1\xi_2\xi_3$ does not appear in N_4, since instead of the term ξ^2 only the first power (i.e., ξ) exists therein. Thus, Eq. (5.42a) is rewritten as:

$$N_1 = \xi_1 \cdot [1 - 8\xi\eta(1-\eta)],$$
$$N_2 = \xi_2 \cdot [1 - 4\eta(1-\xi)],$$
$$N_3 = \xi_3 \cdot [1 - 4(1-\xi)(1-\eta)],$$
$$N_4 = 16\xi(\xi - 1)\eta(\eta - 1). \tag{5.42c}$$

For an equilateral triangle of unit side ($a = 1$) and vertices at ($x_1 = y_1 = 0, x_2 = \frac{\sqrt{3}}{2}a, y_2 = -\frac{1}{2}a, x_3 = \frac{\sqrt{3}}{2}a, y_3 = \frac{1}{2}a$), with *one* internal node at the position ($x = \frac{\sqrt{3}}{2}, y = 0$) according to Fig. 5.6a, the shape functions are shown in Fig. 5.7.

Note While in a classical linear triangle with three nodes the determinant of the Jacobian matrix $\mathbf{J}_3 = \begin{bmatrix} \partial x/\partial\xi_2 & \partial y/\partial\xi_2 \\ \partial x/\partial\xi_3 & \partial y/\partial\xi_3 \end{bmatrix}$ equals to twice its area (2*A*), the reader may validate that for the determinant of the Jacobian matrix $\mathbf{J}_4 = \begin{bmatrix} \partial x/\partial\xi & \partial y/\partial\xi \\ \partial x/\partial\eta & \partial y/\partial\eta \end{bmatrix}$ in a degenerated quadrilateral into a triangle with four nodes equals to $2A\xi$, with $0 \le \xi \le 1$. Therefore, this determinant vanishes at the degenerated vertex "1".

Hint In the classical linear (three-node) triangle, the coordinates of an internal point are given by $\mathbf{x} = \xi_1\mathbf{x}_1 + \xi_2\mathbf{x}_2 + \xi_3\mathbf{x}_3$, whereas in the four-node degenerated element, they are given by $\mathbf{x} = N_1\mathbf{x}_1 + N_2\mathbf{x}_2 + N_3\mathbf{x}_3 + N_4\mathbf{x}_4$, with the N_i's according to Eqs. (5.42a, c) and $\mathbf{x}_4 = (2\mathbf{x}_1 + \mathbf{x}_2 + \mathbf{x}_3)/4$ (see Fig. 5.6a).

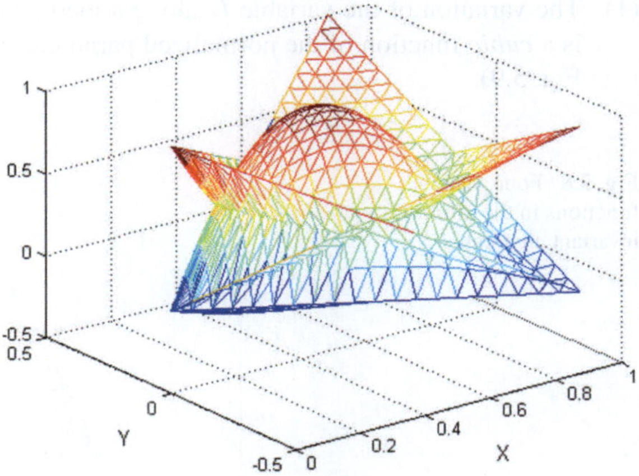

Fig. 5.7 Four shape functions in the degenerated element

5.3.2.2 Affine Invariant Triangular with One Internal Node (by Inspection)

Let us consider a straight triangle 123 with one additional node 4 at its *centroid* (see Fig. 5.6b), i.e.,

$$\mathbf{x}_4 = \frac{\mathbf{x}_1 + \mathbf{x}_2 + \mathbf{x}_3}{3} \tag{5.43}$$

In this particular case, intuition gives the following analytical expressions for the shape functions (shown in Fig. 5.8):

$$\begin{aligned} N_1 &= \xi_1(1 - 9\xi_2\xi_3), \\ N_2 &= \xi_2(1 - 9\xi_3\xi_1), \\ N_3 &= \xi_3(1 - 9\xi_1\xi_2), \\ N_4 &= 27\xi_1\xi_2\xi_3. \end{aligned} \tag{5.44}$$

Given the boundary values, U_1, U_2, and U_3, the shape of the interpolation or approximation is controlled nodal by the value U_4 at the centroid, through the well-known formula:

$$\begin{aligned} U(\xi_1, \xi_2, \xi_3) = N_1(\xi_1, \xi_2, \xi_3) \cdot U_1 + N_2(\xi_1, \xi_2, \xi_3) \cdot U_2 \\ + N_3(\xi_1, \xi_2, \xi_3) \cdot U_3 + N_4(\xi_1, \xi_2, \xi_3) \cdot U_4. \end{aligned} \tag{5.45}$$

Remarks In the affine invariant triangle, the interested reader may verify that:

(1) The variation of the variable U along a median (say CM) of the triangle ABC is a *cubic* function of the normalized parameter $t = CP/CM$, $0 \leq t \leq 1$ (see Fig. 5.9).

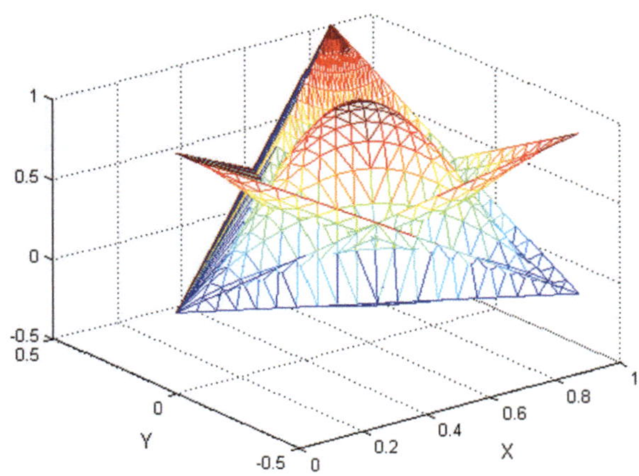

Fig. 5.8 Four shape functions in the affine invariant element

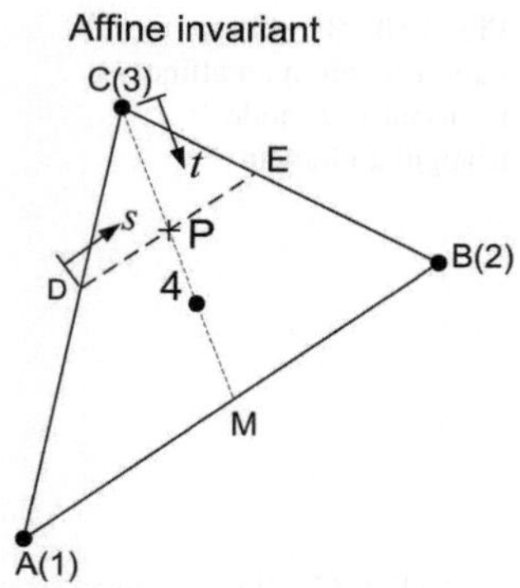

Fig. 5.9 Particular directions in the affine invariant four-node triangular element

(2) The variation of the variable U along the line DE, parallel to the edge AB, is a *quadratic* function of the normalized parameter $s = DP/DE, 0 \le s \le 1$ (see Fig. 5.9). Therefore, the (linear) projection P_1 along DE includes an intentional error, which is recovered later through the corrective term.

5.3.2.3 Extension of Barnhill's Interpolation Formula to Include One Internal Node

It should be clarified that Barnhill's formula is a smooth interpolant to boundary data only. Therefore, it is under question whether this formula can be extended to handle internal data as well.

Before we go on, we remind that in case of an internal node within a *quadrilateral* $ABCD$, the transfinite procedure was straightforward, whereas the shape function associated to it is merely the product of the blending functions in the two directions, i.e., $N = E_{1/2}(\xi)E_{1/2}(\eta)$. It is clarified that each of the two involved functions, i.e., $E_{1/2}(\xi)$ and $E_{1/2}(\eta)$, vanishes along the boundary.

Similarly, for a triangular element 1234 where 4 is at the center of mass (centroid), the shape function associated to 4 should be reasonably a multiple of the three area coordinates, so as the affine invariance is ensured. This means that we need an additional corrective term that (i) ensures the independent value U_4 at "4" and (ii) has no influence along the boundary.

The obvious value that ensures both conditions is the corrective term

$$\mathrm{CT} = 9uvw(U_1 + U_2 + U_3 - 3U_4), \tag{5.46}$$

which when subtracted gives the final expression:

$$\begin{aligned} U(u, v, w) = &\frac{1}{2}[P_1 + P_2 + P_3 - uf(0) - vg(0) - wh(0)] \\ &- 9uvw(U_1 + U_2 + U_3 - 3U_4) \end{aligned} \tag{5.47}$$

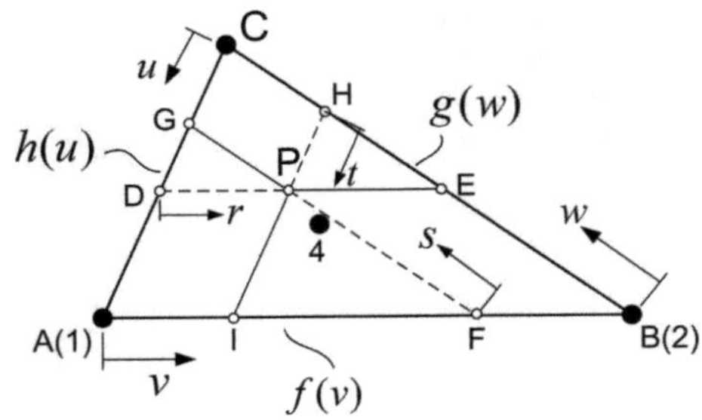

Fig. 5.10 Barnhill's interpolation in an affine invariant four-node triangular element

In Eq. (5.47), the three first projects are those well-known involved in the handling of boundary data (see Fig. 5.10):

$$P_1 = (1 - r)U_D + rU_E \tag{5.48a}$$

$$P_2 = (1 - s)U_F + sU_G \tag{5.48b}$$

$$P_3 = (1 - t)U_H + tU_I \tag{5.48c}$$

Substituting Eqs. (5.48a–c) into Eq. (5.47), and further manipulation, leads eventually to the previously obtained Eq. (5.44) by inspection.

Note The difference of Eq. (5.47) with the previous transfinite interpolations into quadrilaterals (which we met in Chap. 4) lies on the fact that in the triangular patch we first start with the rational interpolation based on *boundary* data and then we add the additional corrective term to match with internal data. In other words, if one applies the boundary-only formula (i.e., the first six terms of Eq. 5.47) will obtain the following erroneous value:

$$\tilde{U}_4 = \frac{U_1 + U_2 + U_3}{3}, \tag{5.49}$$

which generally differs from the *independent* U_4. Therefore, the additional corrective term described by Eq. (5.46) was identified and its opposite value should be inserted in Eq. (5.47) so as to remove this inconsistency.

5.3.2.4 Extension of Barnhill's Interpolation Formula to Include Three Internal Nodes

Let us now deal with the case where the approximation along the boundary remains linear, but three internal nodes are involved (shown in Fig. 5.11) like those three also

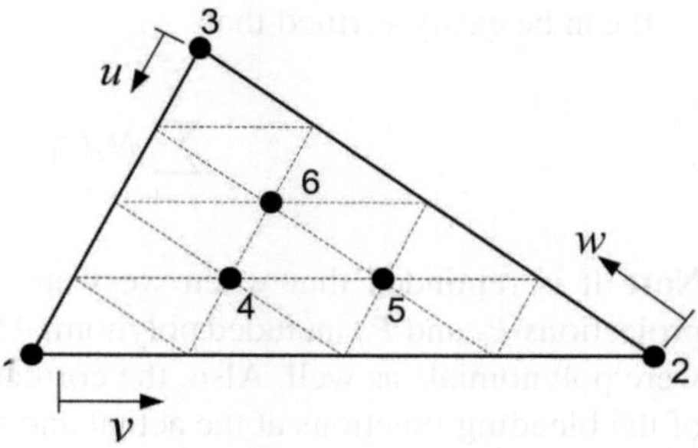

Fig. 5.11 Six-node triangular element with three internal nodes

in Fig. 5.5d (but not the intermediate ones on the boundary). Since the internal nodes have the area coordinates: $4(\frac{2}{4}, \frac{1}{4}, \frac{1}{4})$, $5(\frac{1}{4}, \frac{2}{4}, \frac{1}{4})$, and $6(\frac{1}{4}, \frac{1}{4}, \frac{2}{4})$, the boundary-only rational Barnhill interpolation (here a simple linear interpolation between the nodes 1, 2, and 3) gives the following deviations:

Node 4: $\Delta U_4 = U_4 - \left(\frac{2}{4}U_1 + \frac{1}{4}U_2 + \frac{1}{4}U_3\right)$
Node 5: $\Delta U_5 = U_5 - \left(\frac{1}{4}U_1 + \frac{2}{4}U_2 + \frac{1}{4}U_3\right)$
Node 6: $\Delta U_6 = U_6 - \left(\frac{1}{4}U_1 + \frac{1}{4}U_2 + \frac{2}{4}U_3\right)$

The obvious value that restores the abovementioned three deviations, and at the same time does not contribute to the boundary, is the corrective term:

$$\begin{aligned} \mathrm{CT} &= 32uvw(4u-1)\left[U_4 - ((\tfrac{2}{4}U_1 + \tfrac{1}{4}U_2 + \tfrac{1}{4}U_3)\right] \\ &\quad + 32uvw(4v-1)\left[U_5 - \left(\tfrac{1}{4}U_1 + \tfrac{2}{4}U_2 + \tfrac{1}{4}U_3\right)\right] \\ &\quad + 32uvw(4w-1)\left[U_6 - \left(\tfrac{1}{4}U_1 + \tfrac{1}{4}U_2 + \tfrac{2}{4}U_3\right)\right], \end{aligned} \tag{5.50}$$

When the above term is added to Barnhill's (boundary-only) formula, the final expression for the interpolation through the six points (1–6) becomes:

$$U(u, v, w) = \frac{1}{2}[P_1 + P_2 + P_3 - uf(0) - vg(0) - wh(0)] + \mathrm{CT} \tag{5.51}$$

Rearranging the terms by forming the factors of the six nodal values (U_1–U_6), one can easily derive the following shape functions:

$$\begin{aligned} N_1 &= \xi_1[1 - 32\xi_2\xi_3(4\xi_1 - 1)], \\ N_2 &= \xi_2[1 - 32\xi_3\xi_1(4\xi_2 - 1)], \\ N_3 &= \xi_3[1 - 32\xi_1\xi_2(4\xi_3 - 1)], \\ N_4 &= 32\xi_1\xi_2\xi_3(4\xi_1 - 1), \\ N_5 &= 32\xi_1\xi_2\xi_3(4\xi_2 - 1), \\ N_6 &= 32\xi_1\xi_2\xi_3(4\xi_3 - 1). \end{aligned} \tag{5.52}$$

It can be easily verified that

$$\sum_{j=1}^{6} N_j(\xi_1, \xi_2, \xi_3) \equiv 1 \tag{5.53}$$

Note It is reminded that when we were dealing with *quadrilateral* patches, the projections P_ξ and P_η included polynomial blending functions and the trial functions were polynomials as well. Also, the corrective term consisted of the tensor product of the blending functions at the actual and the artificial points.

In contrast, in *triangular* patches (according to the procedure we have followed in this book so far), the situation is somehow different. Clearly, we initially assumed the boundary-only smooth Barnhill's rational approximation, and afterward we imposed a corrective term to depress the deviations at the internal nodes. In a way analogous to the tensor product (applicable to quadrilaterals), the terms of the corrective projection $P_{\xi\eta}$ are associated to the uniformly arranged internal nodes into the triangles (generally shown as in Fig. 5.5) and are given by the general expression of Eq. (5.36).

5.4 An Overview on Approximation into Triangular Patches

I. *Triangular Coons patches*

In his original text, Steven Coons proposed the degeneration of a side in the quadrilateral patch *ABCD* (say *DA*) into a triangle *ABC* (Coons [14], p. 3). Moreover, triangular Coons-like patches owing affine invariance were initiated by Barnhill, Birkhoff, and Gordon [8] to handle boundary data only. Not only the C^0-continuity but also C^1-continuity was considered with the corresponding cubic Hermite projectors along parallel to each side, that is, P_1, is the cubic Hermite lofting interpolant along parallel to side 1, etc. The so-called BBG triangle (from author's initials) is a family of interpolants formed by taking Boolean sums of the three lofting interpolants P_1, P_2, and P_3, plus a corrective term depending on the values at the vertices. Twist incompatibilities inherent in all Boolean sums must again be resolved in order to produce suitable schemes. In 1978, Little made the very important step of generalizing the "BBG" schemes to an arbitrary triangle [5], a key concept being a calculus for the functions of barycentric coordinates.

Other transfinite triangular interpolants have subsequently been discovered, including Nielson's radial schemes [30], Gregory's symmetric schemes which are generalized to n-dimensional simplices [27], and Brown, Dube, and Little's convex combinations.

Later, Alfeld and Barnhill [1] constructed a C^2 BBG scheme, whereas Little has devised a trivariate C^1 BBG scheme [5].

The above works have been documented by Barnhill until 1985, and no standard way of dealing with arbitrary located internal nodes such as that of Eqs. (5.47) and (5.51) is known till today.

Instead, there are many variations of the so-called Bernstein–Bézier patches which are discussed below.

II. *Bernstein–Bézier patches*

As has been reported in many papers (e.g., De Casteljau [19]; Boehm and Müller [13]; Bieri and Prautzsch [11]), when Paul de Casteljau invented Bézier curve in 1959 (working at Citroën), he realized the need for the extension of the curve ideas to surfaces. According to Farin [26, p. 303], it is of interest that the *first* extension of de Casteljau algorithm from curves to surfaces [18, 16] was what we now call *Bézier triangles* (see later in Chap. 6). In other words, the triangular patches were a more "natural" generalization of Bézier curves than the tensor-product patches. Due to the fact that de Casteljau's work was never published (it was documented in two internal Citroën reports, De Casteljau [18, 17]), the relevant CAD curves and surfaces bring the name of Pierre Bézier because it was him who was publishing at least a paper per year (see Rabut [35]).

Thanks to Prof. Wolfgang Boehm of the Technical University of Braunschweig, the fundamental work of Paul de Casteljau finally became known to the world of specialists in 1976 (after seventeen years). There are also other researchers who relied on the need for triangular patches. Sabin [36], in his Ph.D. thesis, was aware of de Casteljau's work and worked in conjunction with Bernstein polynomials. Considering that piecewise surfaces were defined over regular triangulations, many researchers have contributed (Sablonnière [37–39]; Stancu [40], among others). Arbitrary triangulations were considered by Farin [23, 24], whose Ph.D. thesis was also related to triangular domains [22]. For the placement of triangular Coons patches (see above) as well as Bernstein–Bézier CAD patches, the reader is referred to the survey of Barnhill [4]. Moreover, he/she may consult two additional surveys on the particular field of triangular Bézier patches [25], De Boor [15]. A couple of recent works are due to Zieniuk and Szerszen [45] for triangles, as well as Xia and Qian [44] for tetrahedral (see also the list of papers therein).

A concise presentation of Bernstein–Bézier interpolation in triangular patches will be given in Chap. 6.

III. *Different approaches and recent works*

Starting from a different departure, an interesting theoretical work on triangular patches using Lagrange and Hermite polynomials was presented by El-Zafrany and Cookson [21], without numerical examples.

Moreover, from the point of view of spectral methods, triangular patches have been also treated using transfinite interpolation (see Pasquetti [31], and Perronnet [32, 33], among a few others).

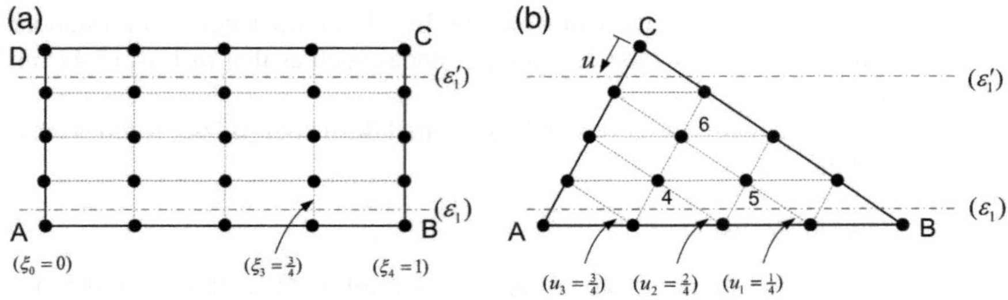

Fig. 5.12 Intersection of isolines with parallel lines to the side AB

IV. *Similarities and differences between triangular and quadrilateral patches*

Major *similarity*: For a certain number of m divisions per side, the unit square $ABCD$ is automatically subdivided into m^2 smaller subsquares. Similarly, for the same number of subdivisions, m^2 smaller subtriangles can be automatically produced within the triangular patch ABC. In Fig. 5.12, one may observe that for $m = 4$, the number of either subsquares (Fig. 5.12a) or subtriangles (Fig. 5.12b) is $m^2 = 16$.

Major *difference*: If a line (ε) passes through an arbitrary point P of the patch, and it is parallel to a side (say AB) of the patch, the aforementioned line (ε) will cut the isolines (perpendicular lines to the side AB at the points $\xi = 0, \frac{1}{m}, \frac{2}{m}, \ldots, \frac{m-1}{m}, 1$) at a certain number of points that depend on the type of the patch. In more details, if the line is close to the side AB of the triangle, it cuts $(m + 1)$ lines, whereas when it is at the highest level, it cuts only two lines (AC and BC). In the particular case it reaches the vertex C, it cuts only it (one point). This is an essential reason, for which we cannot work with a standard number of data along a certain projection. Therefore, we are forced to deal with three projections (parallel to the three sides of the triangle), where in each of them we use linear interpolation.

5.5 Applications

The performance of Barnhill macroelementis shown through two numerical examples.

Example 5.1 (Laplace equation) This is an artificial example, where (on purpose) the boundary-only rational Barnhill's interpolation is not capable of accurately approximating the function $U(x, y)$.

We consider a rectangular and isosceles triangle ABC (the right angle at B as shown in Fig. 5.13), in which the length of each of the equal sides equals to the unity ($AB = BC = 1$). The physical problem is governed by the Laplace equation, $\nabla^2 U = 0$, and the boundary conditions are:

Side AB: $U = 2 - x^4$

Fig. 5.13 Triangular domain governed by Laplace equation (the discretization is arbitrary), solved by a single macroelement (for example, here is of m = third degree)

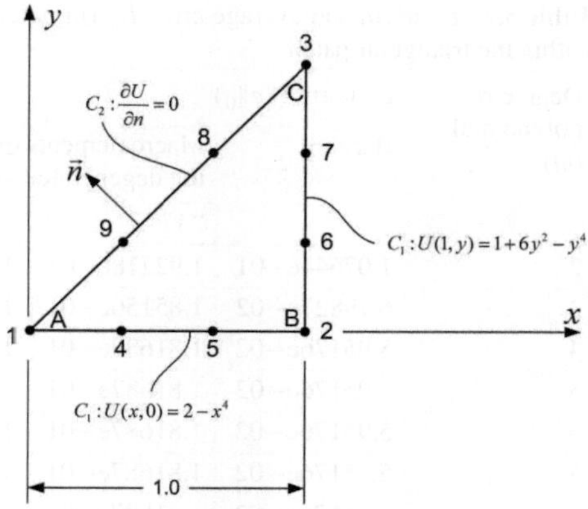

Side BC: $U = 1 + 6y^2 - y^4$
Side CA: $\frac{\partial U}{\partial n} = 0$

The exact solution is given by:

$$U_{\text{exact}}(x, y) = 6x^2y^2 + (1 - x^4) + (1 - y^4) \tag{5.54}$$

The reader can easily show that the function in Eq. (5.54) satisfies Laplace equation and the boundary conditions as well.

The quality of the numerical solution is measured in terms of the mean-square (L^2 or "L-two") norm of the error over the triangular domain and is defined by (for example, see Becker et al. [10, p. 35]):

$$\|e\|_0 = \left(\int_\Omega (U_{\text{calculated}} - U_{\text{exact}})^2 \mathrm{d}\Omega \right)^{1/2}, \tag{5.55}$$

where $U_{\text{calculated}}$ is the approximate (calculated) solution and U_{exact} is the abovementioned exact solution given by Eq. (5.54).

Moreover, we also present a more engineering measure (the mean average error in percent), which is a normalization of $\|e\|_0$, defined by:

$$L_u(\text{in}\%) = \frac{\left(\int_\Omega (U_{\text{calculated}} - U_{\text{exact}})^2 \mathrm{d}\Omega\right)^{1/2}}{\left(\int_\Omega (U_{\text{exact}})^2 \mathrm{d}\Omega\right)^{1/2}} \times 100, \tag{5.56}$$

The overall results using three alternatives of the degenerated Serendipity elements and the Barnhill one are given in Table 5.1, where they are also compared with an assemblage of conventional linear triangular elements of the same mesh density. From these results, it becomes evident that:

Table 5.1 L^2 norm and average error L_u (in percent) where the approximate solutions converge within the triangular patch

Degree of polynomial (m)	L^2 norm ($\|e\|_0$)				
	Barnhill	Macroelements of Serendipity family with the degenerated vertex at the corner point			FEM
		"A"	"B"	"C"	
2	1.07644e−01	1.92218e−01	1.69966e−01	9.73000e−02	2.12259e−01
3	6.19827e−02	1.85150e−01	1.48605e−01	3.20586e−02	9.24824e−02
4	5.95176e−02	1.81687e−01	1.48605e−01	3.34522e−02	5.16364e−02
5	5.95176e−02	1.81687e−01	1.54644e−01	3.34522e−02	3.29303e−02
6	5.95176e−02	1.81687e−01	1.54494e−01	3.34522e−02	2.28234e−02
7	5.95176e−02	1.81687e−01	1.54809e−01	3.34522e−02	1.67481e−02
8	5.95176e−02	1.81687e−01	1.54818e−01	3.34522e−02	1.28127e−02
9	5.95176e−02	1.81687e−01	1.54819e−01	3.34522e−02	1.01181e−02
Convergence at L_u (in %)	3.54	10.79	9.20	1.99	0.43 ÷ 8.92

(i) A single Barnhill macroelement(with $3m$ boundary nodes) converges very rapidly (at $m = 4$), not to the accurate solution but to another where a mean error equal to 3.54% appears.

(ii) A single degenerated Serendipity macroelement (with $3m$ boundary nodes) rapidly converges to different values with substantially higher or slightly smaller mean errors (between 1.99 and 10.79%). It is remarkable that the latter errors greatly depend on the choice of the particular vertex where the degeneration occurs (*A*, *B*, or *C*), and there is not a safe way to guess the best one.

(iii) In this particular problem, the FEM solution (which uses more nodes than the abovementioned boundary-only macroelements, i.e., totally $\frac{(m+1)(m+2)}{2}$ nodes, of which $3m$ belong to the boundary and the rest $\frac{(m-1)(m-2)}{2}$ are internal nodes) has an average error that starts from 8.92% and reduces to 0.43%.

The incapability of the Serendipity family, and at a certain degree of the Barnhill boundary-only approach as well (especially in this particular problem), induces the need to incorporate internal nodes. In this spirit, in Table 5.2, we present results for single macroelements with internal nodes as follows:

(i) For the complete triangular elements such as those in Fig. 5.5c–e (i.e., cubic: $m = 3$, quartic: $m = 4$, quintic: $m = 5$) as well as for the septic element ($m = 6$) similar to that shown in Fig. 5.15a.

(ii) For a degenerated triangular element with internal nodes as shown in Fig. 5.15b.

From the results of Table 5.2, it becomes evident that the introduction of internal nodes leads to excellent results in both the Barnhill triangular macroelement, as well as to the degenerated quadrilateral of Lagrangian type. The domain integration

Table 5.2 Quality of the calculated function $U(x, y)$ using a single macroelement with internal nodes

Degree of polynomial (m)	L^2 norm ($\|e\|_0$) and mean error L_u (in %)					
	Barnhill			Degenerated		
	L^2-norm	L_u (in %)	Total number of nodes	L^2-norm	L_u (in %)	Total number of nodes
2	1.07644e−01	6.40	6	8.44480e−02	5.02	7
3	1.42931e−02	0.85	10	1.24029e−02	0.74	13
4	1.94310e−11	0.00	15	2.76326e−12	0.00	21
5	9.06916e−10	0.00	21	2.85150e−14	0.00	31
6	3.53957e−09	0.00	28	1.41187e−12	0.00	43

was performed using 44-point (see Hussain et al. [28]) and tensor-product Gaussian quadrature (see Stroud and Secrest [42]), respectively.

Example 5.2 (Acoustic cavity) We consider a two-dimensional acoustic cavity which has the shape of an isosceles right triangle with the right sides equal to a (Fig. 5.14).

If Neumann boundary conditions ($\partial U/\partial n = 0$) are imposed along the entire boundary of the acoustic cavity (due to hard walls), the eigenvector is given by (see Blevins [12, p. 367]):

$$U_{nm}(x, y) = \cos\frac{n\pi x}{a}\cos\frac{m\pi y}{a} \pm \cos\frac{m\pi x}{a}\cos\frac{n\pi y}{a}, \quad n, m = 0, 1, 2, \ldots \tag{5.57}$$

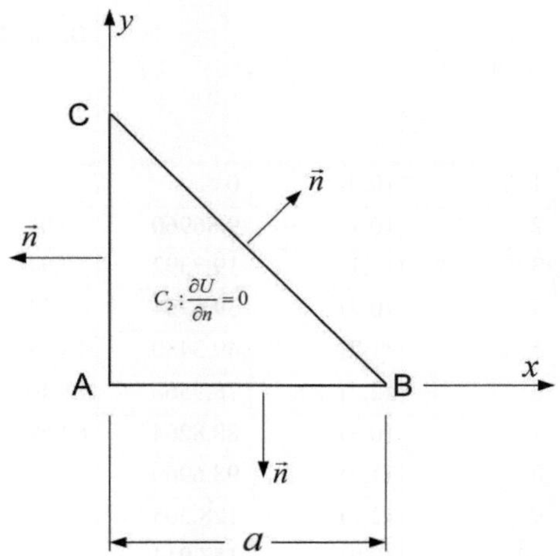

Fig. 5.14 Triangular acoustic cavity

The sign in Eq. (5.57) is plus (+) when $(n + m)$ is an even number and is minus (−) when $(n + m)$ is an odd one.

Substituting Eq. (5.57) into the Helmholtz equation, $\nabla^2 U_{nm} + k^2 U_{nm} = 0$, it can be found that the eigenvalues are given by

$$\omega_{nm}^2 = \pi^2 c^2 \frac{(n^2 + m^2)}{a^2} \tag{5.58}$$

Several types of Barnhill macroelements are tested and are compared with conventional triangular (three-node) elements of the same mesh density. The results are shown in Table 5.3.

One may observe that the degenerated macroelements (based on the Serendipity family) rather work, but they do *not* lead to better results than those obtained using linear triangular elements of the same mesh density and therefore can be recommended as a better solution. As also happened in Example 5.1, the numerical solution depends on the choice of the corner where the degeneration occurs.

In contrast, for the first five nonzero modes (particularly for mode 2), the boundary-only rational Barnhill macroelement is superior to the degenerated Serendipity macroelements as well as to the FEM solution. Particularly, the accuracy of the first nonzero calculated eigenvalue is excellent (with error equal to only 0.04%). Nevertheless, the smaller accuracy in the calculated higher modes suggests the use of internal points.

Table 5.3 Calculated eigenvalues for a triangular acoustic cavity using a single macroelement of polynomial degree $m = 6$ (18 boundary nodes only)

Mode	Harmonics (n, m)	Exact ω_{nm}^2	Errors (in %) of calculated eigenvalues			
			A single macroelement (18 nodes)			FEM (28 nodes, 36 triangular linear elements)
			Barnhill	Serendipity family with degenerated vertex at		
				Point *A* (90°)	Point *B* (45°)	
1	(0,0)	0	–	–	–	–
2	(0,1)	9.86960	0.04	2.54	1.99	2.23
3	(1,1)	19.7392	2.94	3.28	4.38	6.67
4	(0,2)	39.4784	5.74	39.43	10.91	9.03
5	(1,2)	49.3480	18.28	11.55	25.75	18.51
6	(2,2)	78.9568	21.46	24.83	24.09	24.31
7	(0,3)	88.8264	63.59	57.28	22.25	20.64
8	(1,3)	98.6960	–	100.32	57.19	30.07
9	(2,3)	128.305	–	60.40	48.74	52.21
10	(0,4)	157.914	–	55.70	45.84	34.46

First we study the influence of internal nodes in the case of the degenerated elements of Lagrangian family. For the sake of briefness, we restrict to the case of degeneration of the side *DA* at the corner *A*, and the obtained results are shown in Table 5.4.

Finally, we present the influence of the internal nodes for single Barnhill macroelements, which in this case are eventually identical with the complete ones shown in Fig. 5.5c–e, and so on. The results are shown in Table 5.5, where one may observe that, when the number m of subdivisions increases, for every mode, the error monotonically decreases tending to zero values. It is remarkable that for six divisions along every side of the triangle ($m = 6$), the complete Barnihill macroelement (28 nodes) is of higher quality for the degenerated Lagrange macroelement (31 nodes) for ($m = 5$), whereas it is of similar quality with the degenerated Lagrangian for ($m = 6$) (however, with 43 nodes).

The model ($m = 6$), not included in Fig. 5.4, at which the complete Barnhill macroelement is competitive with the degenerated Lagrangian one, is shown in Fig. 5.15.

5.6 Historical Note

According to a recent communication I had with Professor R. E. Barnhill, through the kind intervention of Prof. N. Sapidis, the former believes that it was the summer of 1969 that he spent at General Motors Research (GMR) Laboratories. During it, his interest in triangular patches melded with Bill Gordon "blending function" Coons

Table 5.4 Performance of degenerated elements (at the side *DA*) of Lagrange family

Mode	Exact ω_{nm}^2	Errors (in %) of calculated eigenvalues				
		Number of divisions per side				
		$m = 2$ (7 nodes)	$m = 3$ (13 nodes)	$m = 4$ (21 nodes)	$m = 5$ (31 nodes)	$m = 6$ (43 nodes)
1	0	–	–	–	–	–
2	9.86960	11.56	0.03	0.02	0.00	0.00
3	19.7392	14.07	7.22	0.12	0.10	0.00
4	39.4784	19.69	18.53	0.25	0.24	0.00
5	49.3480	120.86	30.38	15.91	0.74	0.63
6	78.9568	90.60	42.31	9.11	6.97	0.82
7	88.8264	304.99	35.04	30.30	0.98	0.80
8	98.6960	–	144.51	33.19	18.45	1.56
9	128.305	–	159.78	64.70	20.15	16.22
10	157.914	–	146.43	46.02	22.42	1.74

Table 5.5 Performance of classical higher-order triangular elements (Fig. 5.3)

Mode	Harmonics (n, m)	Exact ω_{nm}^2	Errors (in %) of calculated eigenvalues			
			Complete triangular elements (according to Fig. 5.4)			
			$m = 3$ (c): 10 nodes	$m = 4$ (d): 15 nodes	$m = 5$ (e): 21 nodes	$m = 6$ 28 nodes
1	(0,0)	0	–	–	–	–
2	(0,1)	9.86960	0.06	0.03	0.00	0.00
3	(1,1)	19.7392	7.69	0.17	0.11	0.00
4	(0,2)	39.4784	23.59	0.73	0.40	0.00
5	(1,2)	49.3480	34.84	16.92	0.87	0.65
6	(2,2)	78.9568	62.41	15.96	9.03	1.33
7	(0,3)	88.8264	91.53	37.43	2.70	1.43
8	(1,3)	98.6960	152.60	53.17	19.73	2.04
9	(2,3)	128.305	175.48	129.67	34.17	20.20
10	(0,4)	157.914	153.57	92.25	38.75	6.39

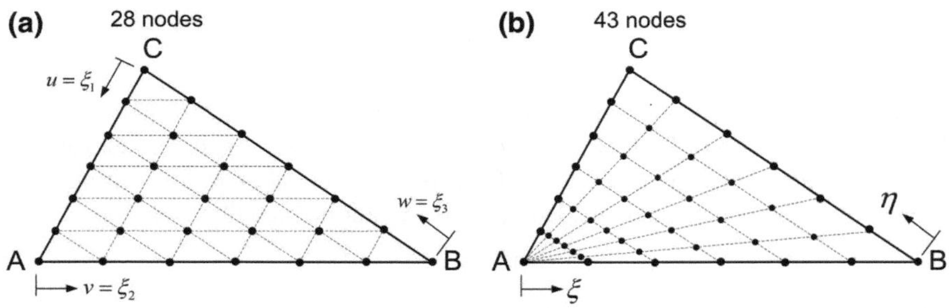

Fig. 5.15 General concept of **a** complete Barnhill and **b** degenerated Lagrangian septic macroelements ($m = 6$)

patches. Garrett Birkhoff was the very senior consultant who appeared monthly at GMR. It was an interesting trifecta combination of people!

When Barnhill went to Brunel University during 1971–1972, he met John Gregory and they sorted out the lack of commutativity of Gordon's projectors in the standard rectangular C^1 case, from which the solution, which he named Gregory's square, emerged. John Gregory came to Utah for a year, about this same time.

Steve Coons visited Utah for a semester or year, courtesy of his student Rich Riesenfeld. Pierre Bezier came to at least one of the Oberwolfach CAGD conferences—Coons and he were the first recipients later of the John Gregory Award there, later yet Prof. Barnhill was the third recipient.

Regarding research activities by the CAD/CAE Group at the National Technical University of Athens (NTUA) to which the author belongs, it should be mentioned that the performance of the *degenerated* triangular macroelements was studied in a certain extent by Dimitriou [20] and previously by Kanarachos and Dimitriou [29] in potential and elasticity problems, focusing on their applicability in adaptive procedures. In contrast, the performance of the *affine invariant* Barnhill elements was studied later, for 2D elasticity problems by Antoniou [2], theoretically by Provatidis [34], and eventually the study was extended to potential problems for which numerical results are presented for the first time in this book.

5.7 Exercises and Solutions

5.1. For the cubic Barnhill element shown in Fig. 5.16, find the shape functions.

Solution With respect to the point sequence $(t_1, t_2, t_3, t_4) = (0, \frac{1}{4}, \frac{2}{4}, \frac{3}{4}, 1)$, the four Lagrange polynomials along any side of the triangle are:

$$\left.\begin{aligned} B_1 &= L_{1,4} = \tfrac{1}{2}(3t-1)(3t-2)(1-t), & B_2 &= L_{2,4} = \tfrac{9}{2}t(3t-2)(t-1), \\ B_3 &= L_{3,4} = \tfrac{9}{2}t(3t-1)(1-t), & B_4 &= L_{4,4} = \tfrac{1}{2}t(3t-1)(3t-2). \end{aligned}\right\} \tag{5.59}$$

Substituting Eq. (5.59) into Eqs. (5.27)–(5.30), after extensive manipulation, one obtains the following closed-form analytical expressions:
Corner nodes:

$$\begin{aligned} N_1 &= \tfrac{1}{4}\xi_1[(3\xi_1-1)(3\xi_1-2)+(3\xi_2-1)(3\xi_2-2)+(3\xi_3-1)(3\xi_3-2)-2] \\ N_2 &= \tfrac{1}{4}\xi_2[(3\xi_1-1)(3\xi_1-2)+(3\xi_2-1)(3\xi_2-2)+(3\xi_3-1)(3\xi_3-2)-2] \\ N_3 &= \tfrac{1}{4}\xi_3[(3\xi_1-1)(3\xi_1-2)+(3\xi_2-1)(3\xi_2-2)+(3\xi_3-1)(3\xi_3-2)-2] \end{aligned} \tag{5.60}$$

For the intermediate nodes of the side AB:

$$N_4 = \tfrac{9}{4}\xi_1\xi_2(3\xi_1-3\xi_2+1), \quad N_5 = \tfrac{9}{4}\xi_1\xi_2(-3\xi_1+3\xi_2+1) \tag{5.61}$$

Similarly, by cyclic rotation, the nodes along BC are:

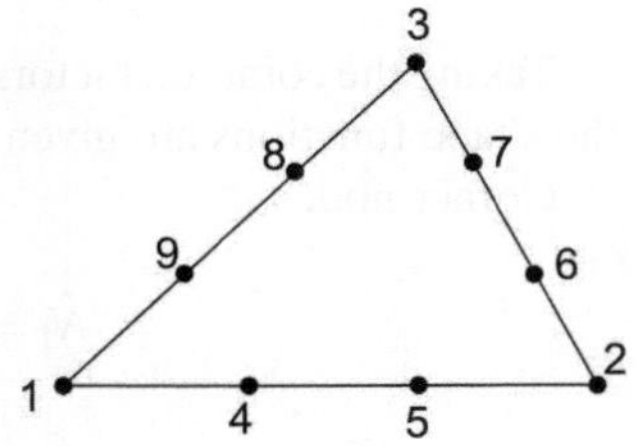

Fig. 5.16 Nine-node Barnhill element (without internal node)

$$N_6 = \tfrac{9}{4}\xi_2\xi_3(3\xi_2 - 3\xi_3 + 1), \quad N_7 = \tfrac{9}{4}\xi_2\xi_3(-3\xi_2 + 3\xi_3 + 1), \tag{5.62}$$

and finally along *CA*:

$$N_8 = \tfrac{9}{4}\xi_3\xi_1(3\xi_3 - 3\xi_1 + 1), \quad N_9 = \tfrac{9}{4}\xi_3\xi_1(-3\xi_3 + 3\xi_1 + 1). \tag{5.63}$$

We notice that despite the fact Barnhill's interpolation formula is of rational character, the final expression of the shape functions is not.

5.2. Extend the above Exercise 5.1 by introducing one internal node (No. 10) at the center of mass. Determine the proper corrective term to Barnhill boundary interpolation, and show that the result is identical with that given by Eq. (5.39).

Solution At the centroid $(\xi_1, \xi_2, \xi_3) = (\frac{1}{3}, \frac{1}{3}, \frac{1}{3})$, the abovementioned Eqs. (5.60)–(5.63), which are related to the boundary-only Barnhill interpolation, give:

$$N_1 = N_2 = N_3 = -\tfrac{1}{6} \text{ and } N_4 = N_5 = N_6 = N_7 = N_8 = N_9 = \tfrac{1}{4}, \tag{5.64}$$

and, therefore, the sum $\sum_1^9 N_i U_i$ is written as:

$$U_{\mathrm{B}} = -\tfrac{1}{6}(U_1 + U_2 + U_3) + \tfrac{1}{4}(U_4 + U_5 + U_6 + U_7 + U_8 + U_9) \tag{5.65}$$

where the subscript "B" stands for the boundary.

In order to ensure the value U_{10} at the abovementioned centroid, the approximation should include the corrective term $\mathrm{CT} = 27\xi_1\xi_2\xi_3(U_{10} - U_{\mathrm{B}})$, and therefore (Eq. 5.65) is updated to:

$$\begin{aligned} U(\xi_1, \xi_2, \xi_3) &= \underbrace{\sum_1^9 N_i U_i}_{\text{Boundary term}} + \underbrace{27\xi_1\xi_2\xi_3(U_{10} - U_{\mathrm{B}})}_{\text{Corrective term}} \\ &= \sum_1^9 N_i U_i + 27\xi_1\xi_2\xi_3\big[U_{10} + \tfrac{1}{6}(U_1 + U_2 + U_3) \\ &\quad - \tfrac{1}{4}(U_4 + U_5 + U_6 + U_7 + U_8 + U_9)\big] \end{aligned} \tag{5.66}$$

Taking the common factors of the nodal values in (5.66), $\sum_1^9 N_i U_i,\ i = 1, \ldots, 10$, the shape functions are given by the following formulas:

Corner nodes:

$$\hat{N}_i = N_i + \tfrac{9}{2}\xi_1\xi_2\xi_3, \quad i = 1, 2, 3. \tag{5.67}$$

Intermediate nodes:

$$\hat{N}_i = N_i - \tfrac{27}{4}\xi_1\xi_2\xi_3, \quad i = 4, 5, 6, 7, 8, 9. \tag{5.68}$$

Internal node:

$$\hat{N}_{10} = 27\xi_1\xi_2\xi_3 \tag{5.69}$$

Substituting Eqs. (5.60)–(5.63) into Eqs. (5.67) and (5.68), after trivial manipulations, the latter expressions are easily found to coincide with those in Eq. (5.39).

5.3. Apply Barnhill macroelement, as well as the degenerated quadrilateral of Lagrangian type, to the solution of an elastic circular disk compressed by two opposite and colinear forces F along its diameter. The single macroelement should refer to one-fourth of this disk (two vertical radii plus one-fourth of the circumference). Rolling boundary conditions have to be imposed along the two vertical radii. Consider that half of the force F is exerted on the macroelement.

5.8 Recapitulation

It was shown that Barnhill's interpolation in triangular patches is so general that it includes the shape functions which are involved in the conventional linear (three-node) and quadratic (six-node) finite elements of triangular shape. Furthermore, in analogy with the quadrilateral elements of the Serendipity family (Coons macroelements), arbitrary-noded *boundary-only* large elements (i.e., with all nodal points placed along the boundary, here called *Barnhill macroelements*) may be automatically constructed in triangular patches of arbitrary shape. The advantage of these elements is that they possess the property of affine invariance. Although this chapter reduced to C^0-continuity (for the sake of brevity), since Barnhill interpolation is applicable for C^1-continuity as well, it is evident that C^1-macroelements may be also constructed.

Regarding the consideration of *internal nodes*, which play a crucial role in the convergence of the numerical solution (completeness), in this book we have implemented the advanced Barnhill macroelements through the following automatic procedure (using a subroutine that calculates the shape functions and the derivatives). For each integration point, first the shape functions due to the boundary nodes are determined, and their values at the internal nodes are calculated. Second, the aforementioned shape functions are slightly updated so as to vanish at the position of the internal nodes. This procedure has been described with full detail in many places of this chapter. It is worthy to mention that if the $n_{\mathrm{I}} = (m-1)(m-2)/2$ internal nodes are symmetrically placed in a triangle and the trial functions are chosen to be Lagrange polynomials of degree m, the produced Barnhill macroelement becomes identical with the classical higher order triangular elements. But the theoretical advantage of

Barnhill' elements is that they are capable of handling more or less internal nodes than the aforementioned critical number n_{I}. Moreover, they can handle any number of nodes along the sides of the triangle (equal one another or not) and any type of trial functions.

Alternatively, large triangular elements may be constructed based on the degeneration of the reference square in which Gordon–Coons transfinite interpolation may be easily applied (see Chap. 4). Although these macroelements can be automatically constructed, probably slightly more easily than the Barnhill ones, they *lack* the property of *affine invariance*, and the numerical solution may highly depend on the choice (guess) of the degenerated corner.

The performance of Barnhill macroelement was shown through two test cases where the macroelements were successfully compared with degenerated triangles as well as with conventional triangular elements of small size for similar mesh densities.

To come to an end, it should become clear that Barnhill macroelements, either using boundary nodes only or in conjunction with internal nodes as well, are associated to nodal values U_i of the true variable $U(x, y)$. Obviously, if nonrational or rational Bernstein–Bézier polynomials (also B-splines or NURBS) are used as trial functions, generalized coefficients, α_i, will appear in the formulation. Moreover, the reader should be prepared to meet an alternative at the end of Chap. 6, where we shall study a substitute of the classical triangular elements with symmetrically posed $n_e = \frac{1}{2}(m+1)(m+2)$ nodes, by the so-called Bernstein–Bézier interpolation, which again refers to arbitrary coefficients α_i (instead of the aforementioned nodal values U_i).

References

1. Alfeld P, Barnhill RE (1984) A transfinite C^2 interpolant over triangles. Rocky Mt J Math 14:17–40
2. Antoniou D (2010) Barnhill's macroelements in plane elasticity problems. M.Sc. thesis, National Technical University of Athens, Interdepartmental Course in Computational Mechanics (supervisor: Prof. Provatidis CG)
3. Barnhill RE (1976) Blending function interpolation: a survey and some new results. In: Collatz L, Werner H, Meinardus G (eds) Numerische Methoden der Approximationstheorie, Band 3. Springer Basel AG, pp 43–89
4. Barnhill R (1985) Surfaces in computer aided geometric design: a survey with new results. Comput Aided Geom Des 2:1–17
5. Barnhill RE, Little FF (1984) Three- and four-dimensional surfaces. Rocky Mt J Math 14:77–102
6. Barnhill RE, Whiteman JR (1973) Error analysis of finite element methods with triangles for elliptic boundary value problems. In: Whiteman JR (ed) The mathematics of finite elements and applications. Academic Press, London, pp 83–122
7. Barnhill RE, Gregory JA, Whiteman JR (1972) The extension and application of Sard kernel theorems to compute finite element error bounds. In: Aziz AK (ed) The mathematical foundations of the finite element method with applications to partial differential equations. Academic Press, New York, pp 749–755
8. Barnhill R, Birkhoff G, Gordon W (1973) Smooth interpolation in triangles. J Approx Theory 8(2):114–128

9. Bathe K-J (1996) Finite element procedures. Prentice-Hall, Upper Saddle River, New Jersey
10. Becker EB, Carey GF, Oden JT (1981) Finite elements: an introduction, vol I. Prentice-Hall, Englewood Cliffs, New Jersey
11. Bieri HP, Prautzsch H (1999) Preface (of a special issue). Comput Aided Geom Des 16:579–581
12. Blevins RD (1979) Formulas for natural frequency and mode shape. Van Nostrand Reinhold Company, New York
13. Boehm W, Müller A (1999) On de Casteljau's algorithm. Comput Aided Geom Des 16(7):587–605
14. Coons SA (1964) Surfaces for computer aided design of space form. Project MAC, MIT (1964), revised for MAC-TR-41 (1967). Available as AD 663 504 from the National Technical Information Service (CFSTI), Sills Building, 5285 Port Royal Road, Springfield, VA 22161, USA. Online Available from http://publications.csail.mit.edu/lcs/pubs/pdf/MIT-LCS-TR-041.pdf
15. De Boor C (1987) B-form basics. In: Farin G (ed) Geometric modeling: algorithms and new trends. SIAM, Philadelphia, pp 131–148
16. De Casteljau P (1959) Outillages methods calcul. Technical report, A. Citroën, Paris
17. De Casteljau P (1959) Courbes à poles. National Industrial Property Institute, INPI, France
18. De Casteljau P (1963) Courbes et surface à poles. Technical report, A. Citroën, Paris
19. de De Casteljau PF (1999) De Casteljau's autobiography: my life at Citroën. Comput Aided Geom Des 16:583–586
20. Dimitriou V (2004) Adaptive finite elements and related meshes. Ph.D. dissertation (advisor: Prof. Kanarachos AE), National Technical University of Athens, School of Mechanical Engineering, Athens, Aug 2004 (in Greek). Online available from http://thesis.ekt.gr/thesisBookReader/id/16430#page/132/mode/2up
21. El-Zafrany A, Cookson RA (1986) Derivation of Lagrangian and Hermitian shape functions for triangular elements. Int J Numer Meth Eng 23(2):275–285
22. Farin G (1979) Subsplines über Dreiecken (transl: subsplines over triangles). Ph.D. thesis, Technical University of Braunschweig, Germany
23. Farin G (1980) Bézier polynomials over triangles and the construction of piecewise C^r polynomials. Technical report TR/91, Brunel University, Uxbridge, England
24. Farin G (1985) A modified Clough-Tocher interpolant. Comput Aided Geom Des 2(1–3):19–27
25. Farin G (1986) Triangular Bernstein-Bézier patches. Comput Aided Geom Des 3(2):83–128
26. Farin G (1990) Curves and surfaces for computer aided geometric design: a practical guide. Academic Press, Boston
27. Gregory JA (1985) Interpolation to boundary data on the simplex. Comput Aided Geom Des 2(1–3):43–52
28. Hussain F, Karim MS, Ahamad R (2012) Appropriate Gaussian quadrature formulae for triangles. Int J Appl Math Comput 4(1):24–38 (Journal homepage: www.darbose.in/ijamc)
29. Kanarachos A, Dimitriou V (2002) Adaptive FEM for plane elasticity problems using 'preconstructed' large finite elements. In: Tsahalis D (ed) Proceedings of 4th GRACM congress on computational mechanics, Patras Greece, vol I, pp 324–332
30. Nielson GM (1979) The side-vertex method for interpolation in triangles. J Approx Theory 25:318–336
31. Pasquetti R (2016) Comparison of some isoparametric mappings for curved triangular spectral elements. J Comput Phys 316:573–577
32. Perronnet A (1998) Triangle, tetrahedron, pentahedron transfinite interpolations. application to the generation of C^0 or G^1-continuous algebraic meshes. In: Proceedings of international conference on numerical grid generation in computational field simulations, Greenwich, England, pp 467–476
33. Perronnet A (1998) Interpolation transfinie sur le triangle, le tetraèdre et le pentaèdre. Application à la création de maillages et à la condition de Dirichlet. C R Acad Sci Paris 326(I):117–122
34. Provatidis CG (2013) A review on attempts towards CAD/CAE integration using macroelements. Comput Res 1(3):61–84
35. Rabut C (2002) On Pierre Bézier's life and motivations. Comput Aided Des 34(7):493–510

36. Sabin M (1976) The use of piecewise forms for the numerical representation of shape. Ph.D. thesis, Hungarian Academy of Sciences, Budapest, Hungary
37. Sablonnière P (1982) Bases de Bernstein et approximants splines. Ph.D. thesis, University of Lille
38. Sablonnière P (1982) Interpolation by quadratic splines on triangles and squares. Comput Ind 3:45–52
39. Sablonnière P (1985) Composite finite elements of class C^k. J Comput Appl Math 12 & 13:542–550
40. Stancu D (1960) Some Bernstein polynomials in two variables and their applications. Sov Math 1:1025–1028
41. Strang G, Fix G (2008) An analysis of the finite element method, 2nd edn. Wellesley-Cambridge Press and SIAM, Wellesley MA
42. Stroud AH, Secrest D (1966) Gaussian quadrature formulas. Prentice-Hall, Englewood Cliffs, New Jersey
43. Wait R, Mitchell AR (1985) Finite element analysis and applications. Wiley, Chichester
44. Xia S, Qian X (2017) Isogeometric analysis with Bézier tetrahedra. Comput Methods Appl Mech Eng 316:782–816
45. Zieniuk E, Szerszen K (2013) Triangular Bézier surface patches in modeling shape of boundary geometry for potential problems in 3D. Eng Comput 29:517–527
46. Zienkiewicz OC (1977) The finite element method. McGraw-Hill, London

Chapter 6
BEZIER Interpolation and Relevant Isoparametric Elements

Abstract Previously, in Chap. 2 the basics of Bézier interpolation have been exposed. This chapter continues with the construction of univariate (1D) and tensor-product (2D) Bézier-based macroelements. The reader is introduced to the fact that, due to a basis change, the nonrational Bézierian elements are equivalent with Lagrangian ones, in all dimensions. The aforementioned higher-order elements are also compared with the well-known p-method. It is shown that, particularly in the 1D problem, all three approaches (Lagrange polynomials, Bernstein–Bézier polynomials, and p-method) are equivalent, in the sense that each of them includes the same monomials. The theory is supported by eleven exercises. Also, the de Casteljau algorithm in curves and the Bernstein–Bézier triangles are explained in detail through three original Appendices.

Keywords BEZIER interpolation · Bézier curve · Casteljau algorithm
Lagrange versus Bézier polynomials · Bézierian macroelements · Equivalency
P-method · Rational Bézier element · Bézier triangle · Test cases

6.1 One-Dimensional Problems

First we begin with 1D problems, where the relationships become more apparent. As known (e.g., [10]), Bézier interpolation of a curve or a univariate function has the following form

$$\mathbf{C}(\xi) = \sum_{i=0}^{n} B_{i,n}(\xi)\,\mathbf{P}_i \quad 0 \le \xi \le 1, \tag{6.1}$$

with the Bernstein polynomials given by

$$B_{i,n}(\xi) = \frac{n!}{i!\,(n-i)!}\xi^i(1-\xi)^{n-i}, \tag{6.2}$$

whereas $\mathbf{P}_i$ are the $(n+1)$ control points.

C. G. Provatidis, *Precursors of Isogeometric Analysis*, Solid Mechanics and Its Applications 256, https://doi.org/10.1007/978-3-030-03889-2_6

The abovementioned dimensionless variable ξ is the normalized coordinate (parameter). In the particular case, the curve $\mathbf{C}(\xi)$ is a straight segment AB of length L, with the Ox-axis having its origin at A and directed toward B, $\xi \triangleq x/L$.

The simplest case is the *quadratic* Bézier curve which is shown in Fig. 6.1, where the variable "u" in the graph stands for the Greek parameter ξ. As also shown in this graph, the corresponding Bernstein polynomials are given by:

$$\left.\begin{aligned} B_{0,2}(\xi) &= \tfrac{2!}{0!(2-0)!}\xi^0(1-\xi)^{2-0} \equiv (1-\xi)^2 \\ B_{1,2}(\xi) &= \tfrac{2!}{1!(2-1)!}\xi^1(1-\xi)^{2-1} \equiv 2\xi(1-\xi) \\ B_{2,2}(\xi) &= \tfrac{2!}{2!(2-2)!}\xi^2(1-\xi)^{2-2} \equiv \xi^2 \end{aligned}\right\} \tag{6.3}$$

Another interesting case is the basis of the *cubic* Bézier (Bernstein) polynomials given by:

$$\left.\begin{aligned} B_{0,3}(\xi) &= \tfrac{3!}{0!(3-0)!}\xi^0(1-\xi)^{3-0} \equiv (1-\xi)^3 \\ B_{1,3}(\xi) &= \tfrac{3!}{1!(3-1)!}\xi^1(1-\xi)^{3-1} \equiv 3\xi(1-\xi)^2 \\ B_{2,3}(\xi) &= \tfrac{3!}{2!(3-2)!}\xi^2(1-\xi)^{3-2} \equiv 3\xi^2(1-\xi) \\ B_{3,3}(\xi) &= \tfrac{3!}{3!(3-3)!}\xi^3(1-\xi)^{3-3} \equiv \xi^3 \end{aligned}\right\} \tag{6.4}$$

The sketch in Fig. 6.2 shows how the position of the control points of Eq. (6.4) may influence the shape of the Bézier curve (thick line).

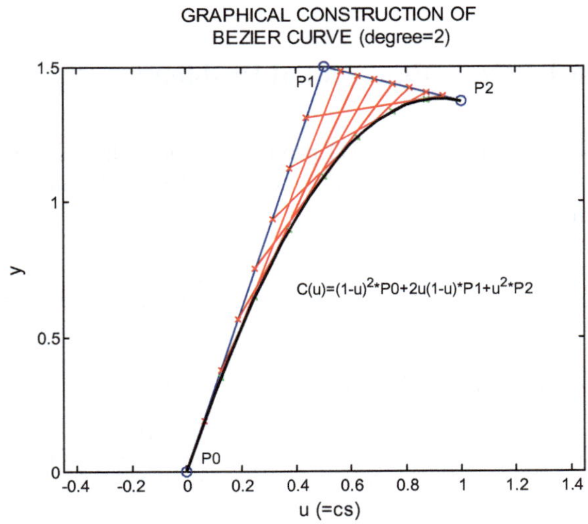

Fig. 6.1 Quadratic Bézier curve

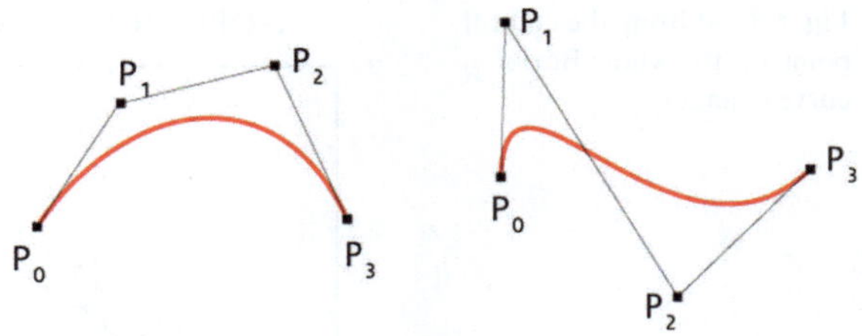

Fig. 6.2 Cubic Bézier curve

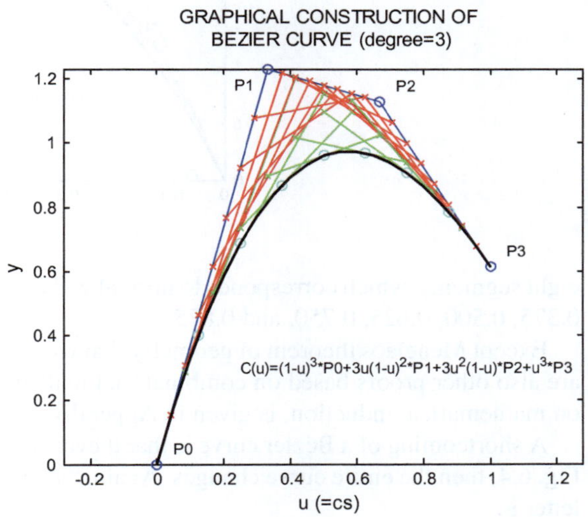

Fig. 6.3 Graphical construction of Bézier curve

The construction of the Bézier curves is based on the generator $P_0 \ldots P_n$, which is a polygonal line of which the vertices $(P_0, P_1, \ldots, P_n)$ are the so-called *control points*. Based on the generator, de Casteljau [4, 5] proposed a geometric procedure to produce points along the relevant Bézier curve. This procedure, which is shown in Figs. 6.1 and 6.3, consists of joining *corresponding* points along the edges of the polygon line. In mode details, all the middle points M_i of the edges in the initial polygon line L_0 are connected and thus a new polygon line L_1 is produced. Taking the midpoints of the polygon line L_1, a third polygon line L_2 is produced, and so on, until in the polygon L_{n-1}, the final polygon line becomes a single segment and then its middle C_M is a useful point, because it can be proven that belongs to the Bézier curve. According to Farin [7, pp. 22–24], de Casteljau algorithm is based and can be proven on the use of Menelaos theorem of Euclidian Geometry.

The procedure is repeated for a sequence of points, say A_i, for example at one-fourth of each polygon edge (measured from its start corner when moving from P_0 to P_n) until, finally, the useful corresponding point C_A is obtained along the new polygon L_{n-1}, thus producing a new point of the Bézier curve. In both cases shown in Figs. 6.1 and 6.3, each edge of the initial polygon line was uniformly divided into

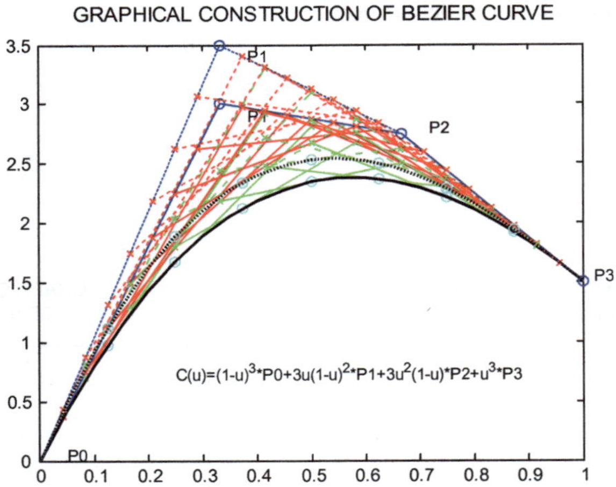

Fig. 6.4 Shifting the control point P_1, the whole Bézier curve changes

eight segments, which corresponds to normalized coordinates of values 0.125, 0.250, 0.375, 0.500, 0.625, 0.750, and 0.875.

Except Menelaos theorem of geometry that was utilized by de Casteljau [4], there are also other proofs based on combinatorial mathematics. A different proof, based on mathematical induction, is given in Appendix 1.

A shortcoming of a Bézier curve is that if even a control point is moved, say P_1 in Fig. 6.4, then the entire curve changes. Again, "u" in the graphs stands for the Greek letter ξ.

When comparing Bernstein–Bézier with Lagrange polynomials of the same degree, we see that the first ones have an upper limit equal to the unity and also they are nonnegative, whereas Lagrange polynomials do not. Figure 6.5 represents a case of a relatively low polynomial degree, seventh ($p = 7$), for which the peak values exceed unity but not that much. The horizontal axis refers to ξ.

Raising the degree of the polynomial to (say) $p = 14$, Bernstein polynomials continue to be smooth, whereas Lagrange polynomials appear substantially large values, particularly near the ends, as shown in Fig. 6.6.

6.2 Equivalency Between Lagrange and Nonrational Bézier Elements: 1D Case

From a first glance (see Figs. 6.5 and 6.6), Bernstein–Bézier polynomials seem to be quite different from Lagrange ones. Nevertheless, we shall show that they share the same set of basis functions (span the same functional space). Actually, let us consider the interval [0, 1] in which the independent variable x belongs. Moreover, let us take three breakpoints at the positions $x_0 = 0$, $x_1 = 1/2$, and $x_2 = 1$, thus

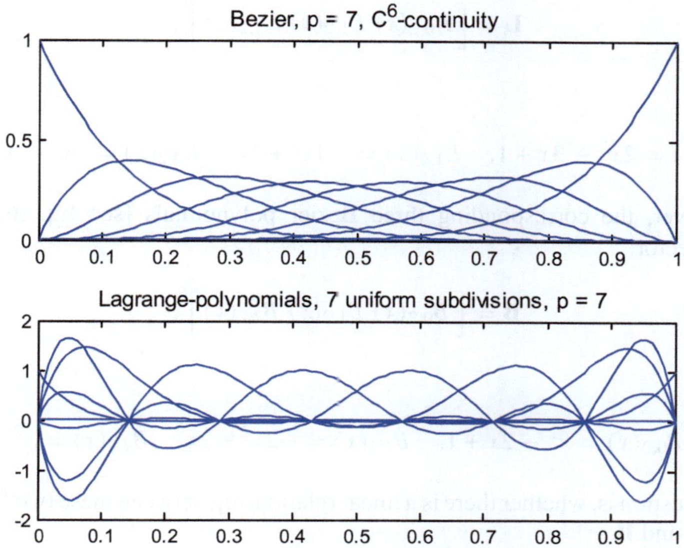

Fig. 6.5 Comparison of Bézier with Lagrange polynomials of seventh degree ($p = 7$)

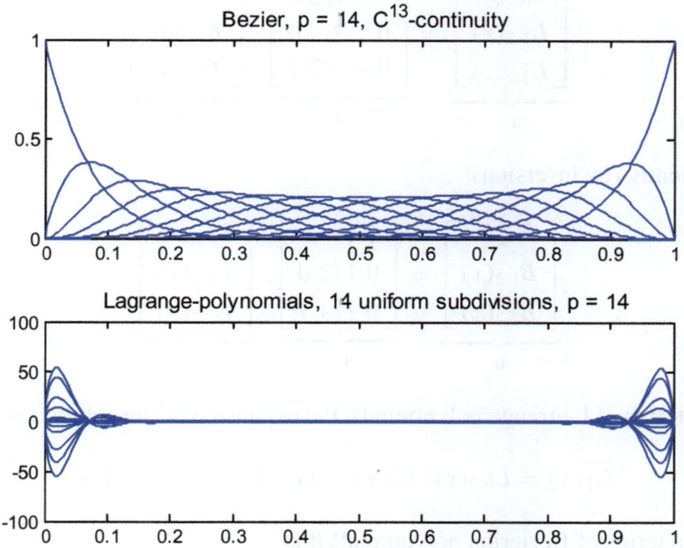

Fig. 6.6 Comparison of Bézier with Lagrange polynomials of fourteenth degree ($p = 14$)

establishing a quadratic interpolation. Based on these breakpoints, the corresponding three Lagrange polynomials form the vector:

$$\mathbf{L} = \left[L_{0,2}(x)\ L_{1,2}(x)\ L_{2,2}(x) \right]^t, \tag{6.5}$$

with

$$L_{0,2}(x) = 2x^2 - 3x + 1, \quad L_{1,2}(x) = -4x^2 + 4x, \quad L_{2,2}(x) = 2x^2 - x. \tag{6.6}$$

Moreover, the corresponding three Bezier polynomials [see Eq. (6.3)] form another vector:

$$\mathbf{B} = \left[B_{0,2}(x)\ B_{1,2}(x)\ B_{2,2}(x) \right]^t, \tag{6.7}$$

with

$$B_{0,2}(x) = x^2 - 2x + 1, \quad B_{1,2}(x) = -2x^2 + 2x, \quad B_{2,2}(x) = x^2. \tag{6.8}$$

The question is, whether there is a linear relationship between these two functional spaces, **L** and **B**.

Below, it will be proven that the relationship between **L** and **B** is:

$$\underbrace{\begin{bmatrix} L_{0,2}(x) \\ L_{1,2}(x) \\ L_{2,2}(x) \end{bmatrix}}_{\mathbf{L}} = \underbrace{\begin{bmatrix} 1 & -1/2 & 0 \\ 0 & 2 & 0 \\ 0 & -1/2 & 1 \end{bmatrix}}_{\mathbf{T}} \cdot \underbrace{\begin{bmatrix} B_{0,2}(x) \\ B_{1,2}(x) \\ B_{2,2}(x) \end{bmatrix}}_{\mathbf{B}}, \tag{6.9a}$$

or equivalently (by inversion):

$$\underbrace{\begin{bmatrix} B_{0,2}(x) \\ B_{1,2}(x) \\ B_{2,2}(x) \end{bmatrix}}_{\mathbf{B}} = \underbrace{\begin{bmatrix} 1 & 1/4 & 0 \\ 0 & 1/2 & 0 \\ 0 & 1/4 & 1 \end{bmatrix}}_{\mathbf{T}^{-1}} \cdot \underbrace{\begin{bmatrix} L_{0,2}(x) \\ L_{1,2}(x) \\ L_{2,2}(x) \end{bmatrix}}_{\mathbf{L}}, \tag{6.9b}$$

Proof In terms of Lagrange polynomials, the expansion of the solution is

$$U_L(x) = L_{0,2}(x) \cdot U_0 + L_{1,2}(x) \cdot U_1 + L_{2,2}(x) \cdot U_2, \tag{6.10a}$$

whereas in terms of Bézierian polynomials it is

$$U_B(x) = B_{0,2}(x) \cdot a_0 + B_{1,2}(x) \cdot a_1 + B_{2,2}(x) \cdot a_2, \tag{6.10b}$$

with $a_0 = U_0$ and $a_2 = U_2$.

Due to Eqs. (6.6) and (6.8), both expressions (6.10a) and (6.10b) are polynomials of second degree, and thus, each of them includes three coefficients. In order these polynomials to be equal to one another, it is sufficient to follow one of the following approaches, i.e., either

(i) to equalize the coefficients of the corresponding terms (of the same degree) in the polynomials
(ii) or to force them to coincide at three different positions.

Therefore, in the Lagrange interpolation Eq. (6.10a) becomes

$$U_L(x) = x^2(2U_0 - 4U_1 + 2U_2) + x(-3U_0 + 4U_1 - U_2) + U_0, \tag{6.10c}$$

whereas in the Bézier interpolation

$$U_B(x) = x^2(a_0 - 2a_1 + a_2) + x(-2a_0 + 2a_1) + a_0 \tag{6.10d}$$

Comparing the polynomials given by Eq. (6.10c) with Eq. (6.10d) and equalizing the corresponding coefficients of monomials, one obtains the following system of equations:

$$\left.\begin{aligned} &2U_0 - 4U_1 + 2U_2 = a_0 - 2a_1 + a_2 \\ &-3U_0 + 4U_1 - U_2 = -2a_0 + 2a_1 \\ &U_0 = a_0 \end{aligned}\right\} \tag{6.10e}$$

The third equality in Eq. (6.10e) is obvious, which when substituted in the first two equalities gives a linear system of two equations in a_1 and a_2. The solution of this system shows that the three coefficients in Eq. (6.10b) or Eq. (6.10d) can be expressed in terms of the nodal values (at the ends and the middle) as follows:

$$a_0 = U_0, \quad a_1 = -\frac{U_0}{2} + 2U_1 - \frac{U_2}{2}, \quad a_2 = U_2 \tag{6.10f}$$

Substituting Eq. (6.10f) into Eq. (6.10d) and equating with Eq. (6.10c), one obtains

$$\begin{aligned} U_B(x) &= B_{0,2}(x)\cdot a_0 + B_{1,2}(x)\cdot a_1 + B_{2,2}(x)\cdot a_2 \\ &= B_{0,2}(x)\cdot U_0 + B_{1,2}(x)\cdot\left(-\frac{U_0}{2} + 2U_1 - \frac{U_2}{2}\right) + B_{2,2}(x)\cdot U_2 \\ &= \left[B_{0,2}(x) - \frac{1}{2}B_{1,2}(x)\right]\cdot U_0 + \left[2B_{1,2}(x)\right]\cdot U_1 + \left[-\frac{1}{2}B_{1,2}(x) + B_{2,2}(x)\right]\cdot U_2 \end{aligned} \tag{6.10g}$$

Equating Eq. (6.10g) with Eq. (6.10c), since these two polynomials are identical for every combination of values (U_0, U_1, U_2), including the case in which two of them vanish, it yields

$$\left.\begin{aligned} L_{0,2}(x) &= B_{0,2}(x) - \tfrac{1}{2}B_{1,2}(x) \\ L_{1,2}(x) &= 2B_{1,2}(x) \\ L_{2,2}(x) &= -\tfrac{1}{2}B_{1,2}(x) + B_{2,2}(x) \end{aligned}\right\} \tag{6.10h}$$

Obviously, when Eq. (6.10h) is written in matrix form then it becomes identical with (6.9a). This statement completes the proof. ■

Conclusion Obviously, similar expressions with Eq. (6.9a, b) can be written straightforwardly for any polynomial degree n. Therefore, in general, *nonrational* Bernstein–Bézier and Lagrange polynomials are linearly dependent on one another.

Remark As it was pointed out, the fulfillment of the series expansion Eq. (6.10a) or Eq. (6.10b) at so many points as the number of the coefficients of a polynomial (i.e., $(n+1)$ locations) ensures a unique expression. If we are interested in calculating the control points, then it becomes necessary to fulfill

$$x(\xi) = B_0(\xi) \cdot x_0^C + B_1(\xi) \cdot x_1^C + B_2(\xi) \cdot x_2^C, \tag{6.10i}$$

at three arbitrary test points along the interval. Choosing these test points at the two ends plus the middle of the interval [0, 1], we have:

$$\begin{Bmatrix} x_0 \\ x_1 \\ x_2 \end{Bmatrix} = \begin{bmatrix} B_0(0) & B_1(0) & B_2(0) \\ B_0(1/2) & B_1(1/2) & B_2(1/2) \\ B_0(1) & B_1(1) & B_2(1) \end{bmatrix} \begin{Bmatrix} x_0^C \\ x_1^C \\ x_2^C \end{Bmatrix}. \tag{6.10j}$$

Substituting B_0, B_1, B_2 using Eq. (6.3), Eq. (6.10j) becomes:

$$\begin{Bmatrix} x_0 \\ x_1 \\ x_2 \end{Bmatrix} = \begin{bmatrix} 1 & 0 & 0 \\ 1/4 & 1/2 & 1/4 \\ 0 & 0 & 1 \end{bmatrix} \begin{Bmatrix} x_0^C \\ x_1^C \\ x_2^C \end{Bmatrix}, \tag{6.10k}$$

whence

$$\begin{Bmatrix} x_0^C \\ x_1^C \\ x_2^C \end{Bmatrix} = \begin{bmatrix} 1 & 0 & 0 \\ 1/4 & 1/2 & 1/4 \\ 0 & 0 & 1 \end{bmatrix}^{-1} \cdot \begin{Bmatrix} 0 \\ 1/2 \\ 1 \end{Bmatrix} = \begin{bmatrix} 1 & 0 & 0 \\ -1/2 & 2 & -1/2 \\ 0 & 0 & 1 \end{bmatrix} \cdot \begin{Bmatrix} 0 \\ 1/2 \\ 1 \end{Bmatrix} = \begin{Bmatrix} 0 \\ 1/2 \\ 1 \end{Bmatrix}. \tag{6.10l}$$

One may observe in Eq. (6.10l) that:

(i) The control points coincide with the *uniformly* arranged nodal points.
(ii) The transpose of the matrix $\mathbf{T}^{-1}$ involved in Eq. (6.9b) appears again in Eq. (6.11).

The reader is suggested to choose different triplets of test points and repeat the sequence of Eq. (6.10i) to Eq. (6.10l), so as to validate that the obtained control points remain *invariable*. Furthermore, he/she may validate that for any polynomial degree n, the control points of the relevant nonrational Bézier polynomials $B_{i,n}$, $i = 0, \ldots, n$ *uniformly* divide the domain [0, 1] into n equal segments (each of length $1/n$). A rigorous proof is given in Appendix 1.

Based on the abovementioned equivalent functional sets **L** and **B**, now we proceed with a computational model concerning one-dimensional acoustics.

Exercise 6.1 (Wave propagation in one dimension) *Concerning the one-dimensional wave equation in the interval* $[0, L]$,

$$\frac{1}{c^2} \cdot \frac{\partial^2 p}{\partial t^2} - \frac{\partial^2 p}{\partial x^2} = 0, \tag{6.11}$$

calculate the stiffness and mass matrices for a uniform quadratic element (see Fig. 6.7), using Lagrange and Bernstein–Bézier polynomials.

Solution If $\varphi_i(x), i = 0, 1, 2$ are the three basis functions in the corresponding formulation (using either Lagrange or Bernstein polynomials), the elements of the stiffness (**K**) and mass (**M**) matrices are given by:

$$k_{ij} = \int_0^L \varphi_i' \varphi_j' \mathrm{d}x \quad \text{and} \quad m_{ij} = \frac{1}{c^2} \int_0^L \varphi_i \varphi_j \mathrm{d}x \tag{6.12}$$

Therefore, according to the choice of the basis functions, after trivial manipulations on the integrals, we have:

1. Lagrange polynomials ($\varphi_i(x) = L_{0,i}(x), i = 0, 1, 2$) according to Eq. (6.6):

$$\mathbf{K}_u = \frac{1}{3\,l} \begin{bmatrix} 7 & -8 & 1 \\ -8 & 16 & -8 \\ 1 & -8 & 7 \end{bmatrix}, \quad \mathbf{M}_u = \frac{l}{30\,c^2} \begin{bmatrix} 4 & 2 & -1 \\ 2 & 16 & 2 \\ -1 & 2 & 4 \end{bmatrix} \tag{6.13}$$

2. Bernstein polynomials ($\varphi_i(x) = B_{0,i}(x), i = 0, 1, 2$) according to Eq. (6.8):

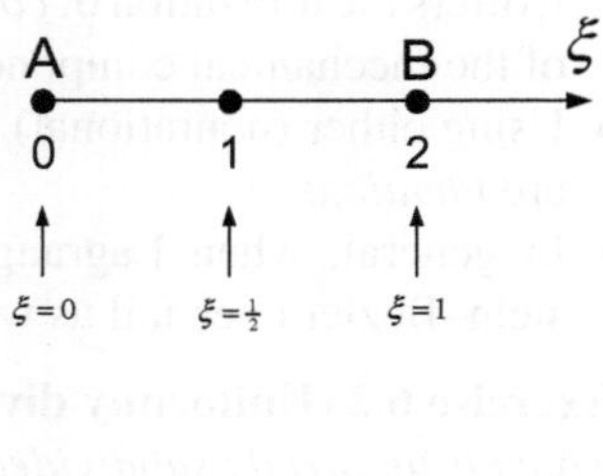

Fig. 6.7 Quadratic element in the reference domain [0, 1]

$$\mathbf{K}_a = \frac{1}{3l}\begin{bmatrix} 4 & -2 & -2 \\ -2 & 4 & -2 \\ -2 & -2 & 4 \end{bmatrix}, \quad \mathbf{M}_a = \frac{l}{30\,c^2}\begin{bmatrix} 6 & 3 & 1 \\ 3 & 4 & 3 \\ 1 & 3 & 6 \end{bmatrix}. \tag{6.14}$$

Furthermore, considering the transformation matrix **T** that appears in Eq. (6.9a), and starting—for example—from the Bernstein formulation (6.14), the Lagrange-based matrices are properly transformed and are found to be identical with the original ones [Eq. (6.13)] as shown below:

$$\mathbf{T}^t\mathbf{K}_a\mathbf{T} = \begin{bmatrix} 1 & 0 & 0 \\ -1/2 & 2 & -1/2 \\ 0 & 0 & 1 \end{bmatrix}^t \cdot \frac{1}{3l}\begin{bmatrix} 4 & -2 & -2 \\ -2 & 4 & -2 \\ -2 & -2 & 4 \end{bmatrix} \cdot \begin{bmatrix} 1 & 0 & 0 \\ -1/2 & 2 & -1/2 \\ 0 & 0 & 1 \end{bmatrix}$$

$$= \frac{1}{3l}\begin{bmatrix} 7 & -8 & 1 \\ -8 & 16 & -8 \\ 1 & -8 & 7 \end{bmatrix} = \mathbf{K}_u. \tag{6.15a}$$

Similarly,

$$\mathbf{T}^t\mathbf{M}_a\mathbf{T} = \begin{bmatrix} 1 & 0 & 0 \\ -1/2 & 2 & -1/2 \\ 0 & 0 & 1 \end{bmatrix}^t \cdot \frac{l}{30\,c^2}\begin{bmatrix} 6 & 3 & 1 \\ 3 & 4 & 3 \\ 1 & 3 & 6 \end{bmatrix} \cdot \begin{bmatrix} 1 & 0 & 0 \\ -1/2 & 2 & -1/2 \\ 0 & 0 & 1 \end{bmatrix}$$

$$= \frac{l}{30\,c^2}\begin{bmatrix} 4 & 2 & -1 \\ 2 & 16 & 2 \\ -1 & 2 & 4 \end{bmatrix} = \mathbf{M}_u. \tag{6.15b}$$

Therefore, the stiffness (**K**) and the mass (**M**) matrices that correspond to the Lagrange polynomials are produced (in quadratic form) by transformation of those produced by nonrational Bernstein polynomials. Either of these two classes of polynomials (Lagrange, Bernstein) can be considered as "local" whereas the other will be the "global," and vice versa.

Remarks

- Bézier interpolation is more convenient for the designer because, in a CAD system, it offers the utilization of *control points* to control and accomplish the desired shape of the mechanical component.
- Using either (nonrational) Bézier or Lagrange formulation, the numerical results are *identical*.
- In general, when Lagrange polynomials fail (due to truncation errors) Bernstein–Bézier ones fail as well.

Exercise 6.2 (Uniformly divided straight segment) *For a straight segment AB, which is uniformly subdivided into n equal segments (i.e., n + 1 breakpoints; e.g., in Fig. 6.7, n = 2):*

- *What is the polynomial degree p of the relevant equivalent nonrational Bézier interpolation?*
- *What about the position of the control points?*

Answer It is easy to prove that the polynomial degree p is equal to the number of segments n ($p \equiv n$). The position of the control points is exactly that of the $n + 1$ nodal points involved in a *uniform* discretization of n subintervals (see Appendix 1). In the particular case of Fig. 6.7, Eq. (6.10) shows that the number of subintervals is $n = 2$, whence $p = n = 2$ as well. Furthermore, in the case of Fig. 6.8, the number of subintervals is $n = 3$, whence $p = n = 3$ as well, and so on. The points 1, 2, 3, and 4 represent either the *nodal* points in the Lagrange formulation or the *control* points (P_0, P_1, P_2, and P_3) in the Bézier formulation.

Note The abovementioned coincidence between nodal points and control points is valid only for straight segments.

Exercise 6.3 (Transformation matrix) *Determine the transformation matrices between Lagrange and Bézier polynomials of third degree (see Fig. 6.8).*

Answer Let us define the Lagrange polynomial-based vectors

$$\mathbf{L} = \left[L_{0,3}(x)\ L_{1,3}(x)\ L_{2,3}(x)\ L_{3,3}(x) \right]^t, \tag{6.16}$$

with

$$\left.\begin{aligned} L_{0,3}(x) &= \tfrac{1}{2}(3x-1)(3x-2)(1-x) \equiv -\tfrac{9}{2}x^3 + 9x^2 - \tfrac{11}{2}x + 1,\\ L_{1,3}(x) &= \tfrac{9}{2}x(3x-2)(x-1) \equiv \tfrac{27}{2}x^3 - \tfrac{45}{2}x^2 + 9x,\\ L_{2,3}(x) &= \tfrac{9}{2}x(3x-1)(1-x) \equiv -\tfrac{27}{2}x^3 + 18x^2 - \tfrac{9}{2}x,\\ L_{3,3}(x) &= \tfrac{1}{2}x(3x-1)(3x-2) \equiv \tfrac{9}{2}x^3 - \tfrac{9}{2}x^2 + x. \end{aligned}\right\} \tag{6.17}$$

and also the Bezier ones:

$$\mathbf{B} = \left[B_{0,3}(x)\ B_{1,3}(x)\ B_{2,3}(x)\ B_{3,3}(x) \right]^t, \tag{6.18}$$

with

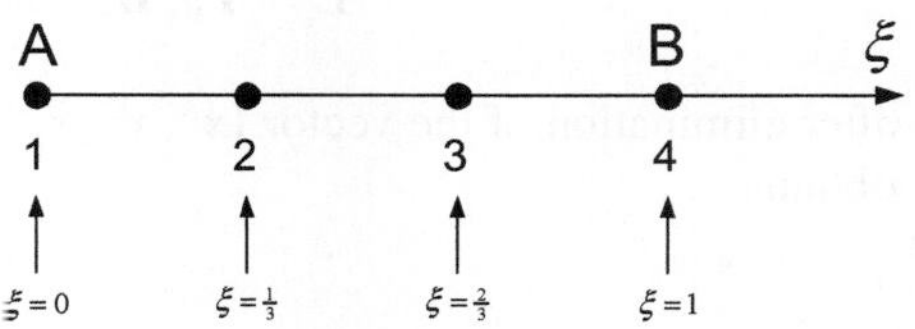

Fig. 6.8 Uniform mesh of three segments (four nodes or control points)

$$\left.\begin{aligned} B_{0,3}(x) &= (1-x)^3 = -x^3 + 3x^2 - 3x + 1, \\ B_{1,3}(x) &= 3x(1-x)^2 = 3x^3 - 6x^2 + 3x, \\ B_{2,3}(x) &= 3x^2(1-x) = -3x^3 + 3x^2, \\ B_{3,3}(x) &= x^3. \end{aligned}\right\} \tag{6.19}$$

Both functional sets, given by Eqs. (6.17) and (6.19), respectively, refer to the discretization shown in Fig. 6.8.

In order to easily determine the transformation between the above two sets, we proceed using matrix manipulations as follows.

Equation (6.17) is written as

$$\begin{Bmatrix} L_{0,3} \\ L_{1,3} \\ L_{2,3} \\ L_{3,3} \end{Bmatrix} = \begin{bmatrix} -\frac{9}{2} & 9 & -\frac{11}{2} & 1 \\ \frac{27}{2} & -\frac{45}{2} & 9 & 0 \\ -\frac{27}{2} & 18 & -\frac{9}{2} & 0 \\ \frac{9}{2} & -\frac{9}{2} & 1 & 0 \end{bmatrix} \begin{Bmatrix} x^3 \\ x^2 \\ x \\ 1 \end{Bmatrix}, \tag{6.20a}$$

or, in short notation, like

$$\mathbf{L} = \mathbf{A}_L \left[x^3, x^2, x, 1\right]^T, \tag{6.20b}$$

whereas Eq. (6.19) is written as

$$\begin{Bmatrix} B_{0,3} \\ B_{1,3} \\ B_{2,3} \\ B_{3,3} \end{Bmatrix} = \begin{bmatrix} -1 & 3 & -3 & 1 \\ 3 & -6 & 3 & 0 \\ -3 & 3 & 0 & 0 \\ 1 & 0 & 0 & 0 \end{bmatrix} \begin{Bmatrix} x^3 \\ x^2 \\ x \\ 1 \end{Bmatrix}, \tag{6.21a}$$

or, in short notation, like

$$\mathbf{B} = \mathbf{A}_B \left[x^3, x^2, x, 1\right]^T, \tag{6.21b}$$

Asking the transformation matrices to fulfill the following conditions:

$$\mathbf{L} = \mathbf{T}_{BL}\mathbf{B}, \quad \text{and} \quad \mathbf{B} = \mathbf{T}_{LB}\mathbf{L}, \tag{6.22}$$

after elimination of the vector $\{x^3, x^2, x, 1\}^t$ between Eqs. (6.20a) and (6.21a), one obtains

$$\mathbf{T}_{BL} = \begin{bmatrix} 1 & -\frac{5}{6} & \frac{1}{3} & 0 \\ 0 & 3 & -\frac{3}{2} & 0 \\ 0 & -\frac{3}{2} & 3 & 0 \\ 0 & \frac{1}{3} & -\frac{5}{6} & 1 \end{bmatrix}, \quad \text{and} \quad \mathbf{T}_{LB} = \begin{bmatrix} 1 & \frac{8}{27} & \frac{1}{27} & 0 \\ 0 & \frac{4}{9} & \frac{2}{9} & 0 \\ 0 & \frac{2}{9} & \frac{4}{9} & 0 \\ 0 & \frac{1}{27} & \frac{8}{27} & 1 \end{bmatrix} \tag{6.23}$$

Obviously, the two transformation matrices are mutually inverse:

$$\mathbf{T}_{BL} \cdot \mathbf{T}_{LB} = \mathbf{T}_{LB} \cdot \mathbf{T}_{BL} = \mathbf{I}, \tag{6.24}$$

Let us now focus, for example, on the mass matrix produced by the abovementioned cubic basis functions.

For the Lagrange polynomials, the mass matrix is given by

$$\mathbf{M}_u = \frac{1}{c^2} \int_0^1 \begin{bmatrix} L_0(\xi)L_0(\xi) & L_0(\xi)L_1(\xi) & L_0(\xi)L_2(\xi) & L_0(\xi)L_3(\xi) \\ L_1(\xi)L_0(\xi) & L_1(\xi)L_1(\xi) & L_1(\xi)L_2(\xi) & L_1(\xi)L_3(\xi) \\ L_2(\xi)L_0(\xi) & L_2(\xi)L_1(\xi) & L_2(\xi)L_2(\xi) & L_2(\xi)L_3(\xi) \\ L_3(\xi)L_0(\xi) & L_3(\xi)L_1(\xi) & L_3(\xi)L_2(\xi) & L_3(\xi)L_3(\xi) \end{bmatrix} L d\xi \tag{6.25}$$

The integrand in Eq. (6.25) can be written as:

$$I = \begin{bmatrix} L_0L_0 & L_0L_1 & L_0L_2 & L_0L_3 \\ L_1L_0 & L_1L_1 & L_1L_2 & L_1L_3 \\ L_2L_0 & L_2L_1 & L_2L_2 & L_2L_3 \\ L_3L_0 & L_3L_1 & L_3L_2 & L_3L_3 \end{bmatrix} \equiv \begin{bmatrix} L_0 \\ L_1 \\ L_2 \\ L_3 \end{bmatrix} \cdot \begin{bmatrix} L_0 & L_1 & L_2 & L_3 \end{bmatrix} \tag{6.26}$$

Substituting the vectors in the right-hand side of Eq. (6.26) by $\mathbf{L} = \mathbf{T}_{BL}\mathbf{B}$ [see Eq. (6.22)], the integrand progressively becomes:

$$I = [\mathbf{T}_{BL}] \begin{bmatrix} B_0 \\ B_1 \\ B_2 \\ B_3 \end{bmatrix} \cdot \begin{bmatrix} B_0 & B_1 & B_2 & B_3 \end{bmatrix} [\mathbf{T}_{BL}]^T$$

$$= [\mathbf{T}_{BL}] \begin{bmatrix} B_0B_0 & B_0B_1 & B_0B_2 & B_0B_3 \\ B_1B_0 & B_1B_1 & B_1B_2 & B_1B_3 \\ B_2B_0 & B_2B_1 & B_2B_2 & B_2B_3 \\ B_3B_0 & B_3B_1 & B_3B_2 & B_3B_3 \end{bmatrix} [\mathbf{T}_{BL}]^T. \tag{6.27}$$

Setting

$$[\mathbf{R}] = [\mathbf{T}_{BL}]^t, \tag{6.28}$$

the substitution of Eq. (6.27) into Eq. (6.25) yields

$$\mathbf{M}_u = \frac{1}{c^2}\int_0^1 [\mathbf{R}]^t \begin{bmatrix} B_0B_0 & B_0B_1 & B_0B_2 & B_0B_3 \\ B_1B_0 & B_1B_1 & B_1B_2 & B_1B_3 \\ B_2B_0 & B_2B_1 & B_2B_2 & B_2B_3 \\ B_3B_0 & B_3B_1 & B_3B_2 & B_3B_3 \end{bmatrix} [\mathbf{R}] L \mathrm{d}\xi$$

$$= [\mathbf{R}]^t \cdot \underbrace{\frac{1}{c^2}\int_0^1 \begin{bmatrix} B_0B_0 & B_0B_1 & B_0B_2 & B_0B_3 \\ B_1B_0 & B_1B_1 & B_1B_2 & B_1B_3 \\ B_2B_0 & B_2B_1 & B_2B_2 & B_2B_3 \\ B_3B_0 & B_3B_1 & B_3B_2 & B_3B_3 \end{bmatrix} L \mathrm{d}\xi}_{\mathbf{M}_a} \cdot [\mathbf{R}]. \tag{6.29}$$

Therefore, Eq. (6.29) is in line with the well-known quadratic form

$$\mathbf{M}_u = [\mathbf{R}]^t \cdot \mathbf{M}_a \cdot [\mathbf{R}] \tag{6.30}$$

Inversely, starting from the inverse of Eq. (6.28), which due to Eq. (6.22) is

$$\mathbf{B} = \mathbf{T}_{LB}\mathbf{L} = [\mathbf{T}_{BL}]^{-1}\mathbf{L},$$

in a similar way we find again a quadratic form due to basis change:

$$\mathbf{M}_a = [\mathbf{R}^*]^t \cdot \mathbf{M}_u \cdot [\mathbf{R}^*], \tag{6.31}$$

where

$$[\mathbf{R}^*] = [\mathbf{T}_{LB}]^t = \left([\mathbf{T}_{BL}]^{-1}\right)^t \equiv \left([\mathbf{T}_{BL}]^t\right)^{-1} \equiv [\mathbf{R}]^{-1}. \tag{6.32}$$

Note Repeating the same procedure for any number of n subdivisions, it can be proven that the mass (**M**) and stiffness matrices (**K**) produced using Lagrange (**L**) and Bézier (**B**) polynomials are mutually depended through a quadratic form. This is due to the basis change, from **L** to **B**, and vice versa.

6.3 Two-Dimensional Problems

6.3.1 General

So far we have learnt that a single macroelement of Coons type (see Chap. 3), as well as of Gordon or Lagrangian type (see Chaps. 4 and 5) may occupy the space of an entire primitive (oblong, circle, ellipse, triangle), of course when the anticipated solution $U(x, y)$ is smooth. The aforementioned macroelements are constructed so as all their boundary *nodes* belong to the real boundary, whereas the internal codes may be automatically generated using Coons interpolation. Of course, when the distribution of the variable $U(x, y)$ within the domain is not smooth, then it becomes necessary to subdivide it (domain decomposition) into a certain number of large CAD-based macroelements.

In this section, we deal with *2D nonrational Bézierian macroelements*, which are based on *control points*. We also compare with classical Lagrangian macroelements, which are based on *nodal points* (see Chap. 4).

The procedure is as follows. We consider the *unit* square (Fig. 6.9a), which is uniformly discretized using n_ξ and n_η subdivisions toward the directions ξ and η, respectively (in the example of Fig. 6.9**a**, $n_\xi = 4$, $n_\eta = 3$). Then we define the images of Coons mapping for the corners A, B, C, and D of the unit square on the actual domain Ω as well as of all rest boundary nodes (i.e., along the sides AB, BC, CD, and DA shown in Fig. 6.9a, b). After the determination of the boundary nodes on the real boundary Γ of the domain Ω, the internal nodes are determined (within Ω) in a second stage using Coons interpolation, which is a mapping between the unit square and the real domain Ω.

Furthermore, at all the abovementioned already known $q_e = (n_x+1)\times(n_y+1)$ *auxiliary* (test) nodal points, (x_k, y_k), $k = 1, \ldots, q_e$, which merely define the geometry as exactly in the Lagrangian elements, the following equations are fulfilled:

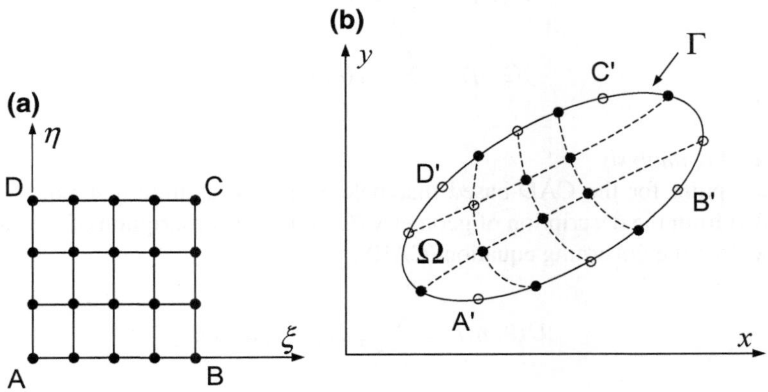

Fig. 6.9 Single biquadratic element (Bézierian or Lagrangian) in reference system

$$\left.\begin{aligned} x_k &= \sum_{m=1}^{q_e} \varphi_m(\xi,\eta)\cdot x_m^C \\ y_k &= \sum_{m=1}^{q_e} \varphi_m(\xi,\eta)\cdot y_m^C \end{aligned}\right\}, \tag{6.33}$$

where x_m^C and y_m^C are the Cartesian coordinates of the control points.

In Eq. (6.33), the shape function $\varphi_m(\xi,\eta)$ that is associated to the mth control point is given by

$$\varphi_m(\xi,\eta) = B_i(\xi)\cdot B_j(\eta), \quad i = 0,\ldots,n_\xi;\ j = 0,\ldots,n_\eta, \tag{6.34}$$

where $B_i(\xi)$ and $B_j(\eta)$ are Bernstein polynomials of degree n_ξ and n_η, respectively.

Then the coordinate vectors $(\mathbf{x}_C, \mathbf{y}_C)$ of the control points involved in Eq. (6.33) are calculated by applying it at all nodal points, $k = 1,\ldots,q_e$, as the solution of the linear equations system

$$\mathbf{x}_{\text{nodal}} = \widetilde{\mathbf{A}}\cdot\mathbf{x}_C \quad \text{and} \quad \mathbf{y}_{\text{nodal}} = \widetilde{\mathbf{A}}\cdot\mathbf{y}_C, \tag{6.35}$$

where the arbitrary element $\tilde{a}_{kl}$ of the abovementioned matrix $\widetilde{\mathbf{A}}$ (of order q_e) refers to the influence of the lth DOF (associated to the *control* point P_l) at the kth *nodal* point and is given as

$$\tilde{a}_{ij} = \varphi_j(\mathbf{x}_i), \quad 1 \le j \le q_e = (n_\xi + 1)(n_\eta + 1) \tag{6.36}$$

Note The aforementioned control points $(x_m^C, y_m^C, m = 1,\ldots,q_e)$ establish an isoparametric mapping for any point $P'(\xi,\eta)$ of the unit square which has its image $P(x,y)$ in the circle given by

$$\left.\begin{aligned} x(\xi,\eta) &= \sum_{m=1}^{q_e} \varphi_m(\xi,\eta)\cdot x_m^C \\ y(\xi,\eta) &= \sum_{m=1}^{q_e} \varphi_m(\xi,\eta)\cdot y_m^C \end{aligned}\right\} \tag{6.37}$$

Engineering analysis

The key point for the CAD-based macroelement is to assume that Eq. (6.37) is extended from the description of geometry (CAD) to the description of the variable $U(x,y)$ into the governing equation (CAE):

$$U(\xi,\eta;t) = \sum_{m=1}^{q_e} \varphi_m(\xi,\eta)\cdot a_m(t), \tag{6.38}$$

where $a_m, m = 1,\ldots,q_e$ denote *generalized coefficients* to be determined.

The application of Galerkin method to the governing PDE [say Eq. (6.39)] leads to the well-known stiffness and mass matrices. For example, in case of the acoustic equation

$$\frac{1}{c^2}\frac{\partial^2 U}{\partial t^2} - \nabla^2 U = 0, \tag{6.39}$$

its matrix formulation becomes:

$$[\mathbf{M}]\{\ddot{\mathbf{a}}(t)\} + [\mathbf{K}]\{\mathbf{a}(t)\} = \{\mathbf{f}(t)\}, \tag{6.40}$$

with the elements of the matrices **[M]** and **[K]**, respectively, given by

$$m_{ij} = \frac{1}{c^2}\iint\limits_{\Omega} \varphi_i\, \varphi_j \mathrm{d}\Omega, \quad \text{and} \quad k_{ij} = \iint\limits_{\Omega} \nabla\varphi_i\, \nabla\varphi_j \mathrm{d}\Omega \tag{6.41}$$

Application of the above formulas will be later shown in Exercise 6.4.

Remarks Control points (dominating in the Bézierian macroelement) should not be confused with the nodal points (also called "nodes", dominating in the Lagrangian macroelement). Although they are generally different one another, in the following Sect. 6.3.2 we shall see that they coincide in a *rectangle* (of course, in conjunction with nonrational Bézier interpolation). In other words, for a uniform arrangement of $q_e = (n_x + 1) \times (n_y + 1)$ nodal points (at the intersection of the lines $\xi = \xi_i$ and $\eta = \eta_j$ in the reference square) which may produce a classical macroelement of Lagrangian type, the control points for a nonrational Bézier approximation of the same degree are *coincident* with the aforementioned nodes.

6.3.2 The Rectangle

Excluding the classical bilinear elements, the simplest Bézierian (or Lagrangian) element uses ($n_x \times n_y = 2 \times 2$) subsquares, as shown in Fig. 6.10a. If we are intended to analyze the rectangle A′B′C′D′ (of dimensions $a \times b$, shown in Fig. 6.10b), the latter can be considered as an image of the *unit square* ABCD.

More generally, the easiest way to produce a large macroelement is to uniformly discretize the unit square using n_x and n_y subdivisions, toward the axes x and y, respectively. In the particular case of a rectangle, the relationship between the local and global coordinates is $x = a\xi$ and $y = b\eta$, with $0 \le \xi, \eta \le 1$.

Let us now focus on the corresponding *control points* and the relevant *Bézierian* macroelement. Based on the abovementioned $(n_x + 1) \times (n_y + 1)$ nodal points, we are able to produce the corresponding control points. But since in the 1D problem (Sect. 6.2) we found that, in conjunction with a straight line, the control points uniformly subdivide the interval [0, a] in the x-direction (see, e.g., Eq. (6.101) and

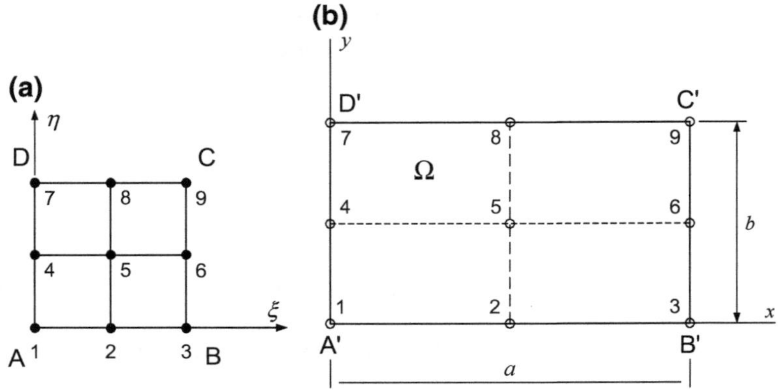

Fig. 6.10 Single biquadratic element (Bézierian or Lagrangian) in **a** the reference, and **b** the global system

its generalization), and the same is valid for the interval [0, b] in the y-direction, it can be easily shown that the same is valid for the tensor product as well. In brief, the control points for a rectangular of arbitrary dimensions $a \times b$ are a *uniform* mesh, which is *identical* with the grid of the uniform nodal points (nodes). From the programming point of view, it becomes necessary to determine only the univariate Bernstein polynomials, $B_i(\xi)$ and $B_j(\eta)$, at the Gauss points for a global quadrature into the reference unit square. The determinant of the Jacobian for the transformation between the local (ξ, η) and the global (x, y) coordinate system is merely the area $A = ab$ of the rectangle. Analytical computations for the stiffness and mass matrices are presented in Exercises 6.4 and 6.5. At this point, it is adequate to remember that the set of all numerical operations involved in a Bézierian macroelement is an analogue of those in a corresponding macroelement of Lagrangian type, with the same number of nodal or control points, respectively. In the next section, it will be shown that these two macroelements, Lagrangian and Bézierian are numerically mathematically *equivalent* to one another.

Remark For a rectangle, both the Lagrangian and Bézierian macroelements share the *same* nodal points or control points (equal number and identical position). Nevertheless, the classical macroelement of Lagrangian type refers to nodal values U_i, whereas the Bézierian macroelement refers to generalized coefficients a_i ($i = 1, \ldots, q_e$, where q_e is the total number of degrees of freedom in the macroelement, in potential problems with one DOF per node). This difference will be better understood through Exercise 6.4.

Exercise 6.4 (Biquadratic element [Bézierian versus Lagrangian]) *Consider the two-dimensional 9-node (quadratic) rectangular element shown in Fig. 6.10. For the particular case, it is rectangular of dimensions $a \times b$, show that the mass matrix in the classical Lagrange formulation is:*

$$\mathbf{M}_u = \frac{ab}{900c^2}\begin{bmatrix} \mathbf{16} & 8 & -4 & 8 & 4 & -2 & -4 & -2 & 1 \\ 8 & \mathbf{64} & 8 & 4 & 32 & 4 & -2 & -16 & -2 \\ -4 & 8 & \mathbf{16} & -2 & 4 & 8 & 1 & -2 & -4 \\ 8 & 4 & -2 & \mathbf{64} & 32 & -16 & 8 & 4 & -2 \\ 4 & 32 & 4 & 32 & \mathbf{256} & 32 & 4 & 32 & 4 \\ -2 & 4 & 8 & -16 & 32 & \mathbf{64} & -2 & 4 & 8 \\ -4 & -2 & 1 & 8 & 4 & -2 & \mathbf{16} & 8 & -4 \\ -2 & -16 & -2 & 4 & 32 & 4 & 8 & \mathbf{64} & 8 \\ 1 & -2 & -4 & -2 & 4 & 8 & -4 & 8 & \mathbf{16} \end{bmatrix} \tag{6.42}$$

Then, show that for the Bézierian formulation the mass matrix is:

$$\mathbf{M}_a = \frac{ab}{900c^2}\begin{bmatrix} \mathbf{36} & 18 & 6 & 18 & 9 & 3 & 6 & 3 & 1 \\ 18 & \mathbf{24} & 18 & 9 & 12 & 9 & 3 & 4 & 3 \\ 6 & 18 & \mathbf{36} & 3 & 9 & 18 & 1 & 3 & 6 \\ 18 & 9 & 3 & \mathbf{24} & 12 & 4 & 18 & 9 & 3 \\ 9 & 12 & 9 & 12 & \mathbf{16} & 12 & 9 & 12 & 9 \\ 3 & 9 & 18 & 4 & 12 & \mathbf{24} & 3 & 9 & 18 \\ 6 & 3 & 1 & 18 & 9 & 3 & \mathbf{36} & 18 & 6 \\ 3 & 4 & 3 & 9 & 12 & 9 & 18 & \mathbf{24} & 18 \\ 1 & 3 & 6 & 3 & 9 & 18 & 6 & 18 & \mathbf{36} \end{bmatrix} \tag{6.43}$$

One may observe that:

(i) *While in Lagrange formulation the mass matrix includes both positive and negative terms, in Bézierian formulation all terms are positive.*

(ii) *The sum of all terms in any of the two expressions, Eq. (6.42) or Eq. (6.43), is equal to the area $A = ab$ of the rectangle (divided by c^2).*

Repeat the calculations for the stiffness matrices **K** *as well.*

Hint The univariate basis functions are given by Eqs. (6.6) and (6.8) for the Lagrange and Bézier polynomials, respectively. If $\varphi_i(x)$, $i = 0, 1, 2$ are the three basis functions in the corresponding formulation for the x-direction [and similarly $\varphi_j(y)$, $j = 0, 1, 2$ for the y-direction], the elements of the stiffness (**K**) and mass (**M**) matrices, already shown in Eq. (6.41), are given by:

$$k_{ij} = \int_0^1\int_0^1 \left(\frac{\partial\varphi_i}{\partial\xi}\frac{\partial\varphi_j}{\partial\xi} + \frac{\partial\varphi_i}{\partial\eta}\frac{\partial\varphi_j}{\partial\eta}\right)|\det J|\,\mathrm{d}\xi\,\mathrm{d}\eta \tag{6.44}$$

and

$$m_{ij} = \frac{1}{c^2} \int_0^1 \int_0^1 \varphi_i \varphi_j |\det J| \mathrm{d}\xi \mathrm{d}\eta \tag{6.45}$$

In both integrals, the determinant of the Jacobian matrix is equal to the area ($|\det J| = ab$).

Applying Eq. (6.45) the first time for Lagrange and the second for Bézier polynomials, the derivation of Eqs. (6.42) and (6.43), respectively, is straightforward.

Finally, concerning the stiffness matrices they are found as follows:

(i) *Lagrangian element*:

$$\mathbf{K}_u = \frac{ab}{90} \begin{bmatrix} \mathbf{56} & -18 & -3 & -18 & -32 & 10 & -3 & 10 & -2 \\ -18 & \mathbf{176} & -18 & -32 & -96 & -32 & 10 & 0 & 10 \\ -3 & -18 & \mathbf{56} & 10 & -32 & -18 & -2 & 10 & -3 \\ -18 & -32 & 10 & \mathbf{176} & -96 & 0 & -18 & -32 & 10 \\ -32 & -96 & -32 & -96 & \mathbf{512} & -96 & -32 & -96 & -32 \\ 10 & -32 & -18 & 0 & -96 & \mathbf{176} & 10 & -32 & -18 \\ -3 & 10 & -2 & -18 & -32 & 10 & \mathbf{56} & -18 & -3 \\ 10 & 0 & 10 & -32 & -96 & -32 & -18 & \mathbf{176} & -18 \\ -2 & 10 & -3 & 10 & -32 & -18 & -3 & -18 & \mathbf{56} \end{bmatrix} \tag{6.46}$$

(ii) *Bézierian element*:

$$\mathbf{K}_a = \frac{ab}{45c^2} \begin{bmatrix} \mathbf{24} & 0 & -4 & 0 & -6 & -4 & -4 & -4 & -2 \\ 0 & \mathbf{20} & 0 & -6 & 2 & -6 & -4 & -2 & -4 \\ -4 & 0 & \mathbf{24} & -4 & -6 & 0 & -2 & -4 & -4 \\ 0 & -6 & -4 & \mathbf{20} & 2 & -2 & 0 & -6 & -4 \\ -6 & 2 & -6 & 2 & \mathbf{16} & 2 & -6 & 2 & -6 \\ -4 & -6 & 0 & -2 & 2 & \mathbf{20} & -4 & -6 & 0 \\ -4 & -4 & -2 & 0 & -6 & -4 & \mathbf{24} & 0 & -4 \\ -4 & -2 & -4 & -6 & 2 & -6 & 0 & \mathbf{20} & 0 \\ -2 & -4 & -4 & -4 & -6 & 0 & -4 & 0 & \mathbf{24} \end{bmatrix} \tag{6.47}$$

6.3.3 The Circle

In contrast to the rectangles, when the boundary is *curved* the control points of the Bézierian element do *not* generally belong to it, whereas the rest control points do not generally coincide with the internal nodes. As an example, below let us consider the construction of a single nonrational Bézierian macroelement for an entire circle.

The procedure is as follows. We consider the uniformly discretized *unit* square (see Fig. 6.11a) and then we define the images of Coons mapping for the corners A, B, C, and D of the unit square on the actual circular domain (of arbitrary radius R). This ideal shape dictates that these four points should be *uniformly* distributed along the circumference of the circle, as shown in Fig. 6.11b. For example, if we divide the unit square into ($n_x \times n_y = 4 \times 4$) subsquares, as shown in Fig. 6.11a, each of the four boundary edges (AB, BC, CD, and DA) are uniformly divided into four equal intervals, thus producing 16 equal arcs along the circumference, as shown in Fig. 6.11b. After the determination of the boundary nodes, the internal nodes are determined using Coons interpolation, which is a mapping between the unit square and the circle, both of them shown in Fig. 6.11.

Following the procedure mentioned in Sect. 6.3.1, the application of Eq. (6.33) to all 25 nodal points shown in Fig. 6.11b and the further solution of the system given by Eq. (6.35) produce the coordinates ($\mathbf{x}_C, \mathbf{y}_C$) of the 25 control points which are also shown therein (by a cross +). One may observe that only a few of the control nodes (those at the four corners A, B, C, and D, plus the center of the circle) coincide with the nodal points of the uniform mesh. Not only that but, interestingly, the closest to the boundary (circumference) control points lie *outside* the circle.

A shortcoming of nonrational Bézierian elements

It was previously mentioned that all the boundary nodes of the unit cube (e.g., Fig. 6.11a) were mapped on the boundary of the real domain (e.g., the circumference of Fig. 6.11b). Based on these boundary nodes as well as on the rest internal auxiliary nodes, the control points were determined using Eq. (6.33) and its matrix form is given finally by Eq. (6.35). The aforementioned control points ($x_m^C, y_m^C, m = 1, \ldots, q_e$) establish an isoparametric mapping for any point $P'(\xi, \eta)$ of the unit square which has its image $P(x, y)$ in the circle given by

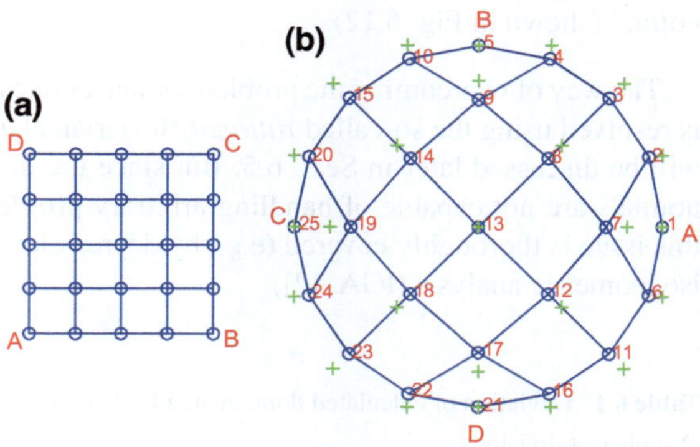

Fig. 6.11 Nodal points (∘) and control points (+) for a nonrational 16-node element in **a** the unit square and **b** the circular domain

$$\left.\begin{aligned} x(\xi,\eta) &= \sum_{m=1}^{q_e} \varphi_m(\xi,\eta)\cdot x_m^C \\ y(\xi,\eta) &= \sum_{m=1}^{q_e} \varphi_m(\xi,\eta)\cdot y_m^C \end{aligned}\right\} \tag{6.48}$$

The shortcoming of using Eq. (6.48) is that it can *not* ensure that an entire side of the unit square (e.g., AB) will be fully mapped along the circumference of the circle. In fact, for a quartic approximation (polynomial degree $n_x = n_y = 4$), only the two endpoints plus the three intermediate points will lie exactly on the circumference (they were taken there by construction) whereas all the other images in the isoparametric mapping described by Eq. (6.48) will not. Consequently, using the formula

$$A = \iint_{A_c} |\det J| \mathrm{d}\xi \mathrm{d}\eta, \tag{6.49}$$

the calculated area will somehow deviate from the exact value. This shortcoming is certainly reduced when using more nodal points along the boundary, as Table 6.1 shows.

Using Eq. (6.33) or Eq. (6.48) at the two ends plus the midpoint M of the arc AB, the calculated control points are shown in Fig. 6.12. One may observe that the intermediate control point P_1 lies between the circumference and the intersection of the two tangents at the ends of the circular arc. Moreover, the deviation between the interpolated and the exact boundary can be distinguished.

Note The use of Lagrange interpolation based on the polynomials of Eq. (6.6) results in the same (identical) approximation with that when using nonrational Bézier polynomial (shown in Fig. 6.12).

The way of overcoming the problem of inaccurate approximation of the boundary is resolved using the so-called *rational* Bézierian elements of second degree, which will be discussed later in Sect. 6.5. But since the involved quadratic rational polynomials are not capable of handling arbitrary profiles of the unknown variable U, this issue is thoroughly covered (e.g., by degree elevation) using the NURBS-based isogeometric analysis (IGA, [2]).

Table 6.1 Deviation of calculated domain area (in %) for a circular single Bézierian element

Number of divisions along the side AB	2	3	4
Error of calculated area (in %)	−1.1785	+0.1574	+0.0072

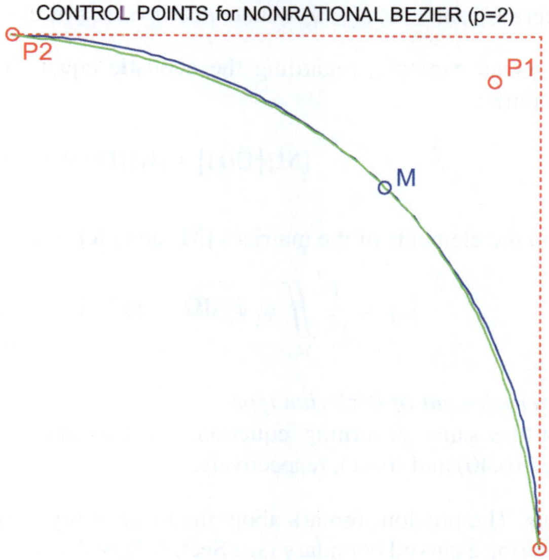

Fig. 6.12 Control points for the quadrant of a circle when nonrational Bézier polynomials are used

6.4 Equivalency Between Lagrange and Nonrational Bézier Elements: 2D Case

6.4.1 General

It will be shown below that the equivalency between the Lagrange and Bézier polynomials which has been already shown that occurs in 1D problems is valid in two-dimensional problems as well. In other words, the classical elements of Lagrangian type are equivalent with the Bézierian elements that were introduced in Sect. 6.3.

Macroelement of Lagrangian type
It is clarified that the classical isoparametric finite element of Lagrangian type interpolates in the same manner, both the geometry (x_m, y_m are the *nodal* coordinates):

$$\left.\begin{aligned} x(\xi,\eta) &= \sum_{m=1}^{q_e} \hat{\varphi}_m(\xi,\eta)\cdot x_m \\ y(\xi,\eta) &= \sum_{m=1}^{q_e} \hat{\varphi}_m(\xi,\eta)\cdot y_m \end{aligned}\right\} \tag{6.48}$$

and the variable

$$U(\xi,\eta;t) = \sum_{m=1}^{q_e} \hat{\varphi}_m(\xi,\eta)\cdot U_m(t), \tag{6.50}$$

where $U_m, m = 1, \ldots, q_e$ denote *nodal values* of the variable U to be determined.

As an example, regarding the acoustic equation, its matrix formulation now becomes:

$$[\widehat{\mathbf{M}}]\left\{\ddot{\mathbf{U}}(t)\right\} + [\widehat{\mathbf{K}}]\{\mathbf{U}(t)\} = \left\{\hat{\mathbf{f}}(t)\right\}, \tag{6.51}$$

with the elements of the matrices $[\widehat{\mathbf{M}}]$ and $[\widehat{\mathbf{K}}]$ are, respectively, given by

$$\hat{m}_{ij} = \frac{1}{c^2} \iint_{\Omega} \hat{\varphi}_i \, \hat{\varphi}_j \, \mathrm{d}\Omega, \quad \text{and} \quad \hat{k}_{ij} = \iint_{\Omega} \nabla\hat{\varphi}_i \cdot \nabla\hat{\varphi}_j \, \mathrm{d}\Omega \tag{6.52}$$

Macroelement of Bézierian type
For the same governing equation, the two above equations are substituted by Eqs. (6.40) and (6.41), respectively.

Note The previous remark about the incapability of the Bézierian elements to fully describe a curved boundary (see Sect. 6.3) is also valid for the Lagrangian elements. For example, Table 6.1 is valid for the Lagrangian elements as well.

Before we proceed with the 2D case, the reader is encouraged to go through Exercise 6.4 to refresh the comparison of two types of macroelements (Lagrangian and Bézierian) in the 1D analogue.

6.4.2 Demonstration of Equivalency on a Rectangular Element

As clearly was shown for the 1D problem, a similar equivalence of the results (e.g., calculated eigenvalues) will be shown for the 2D problem as well. In order to simplify the procedure, we choose the analogue case of the 3-node (quadratic) acoustic pipe of the previous section, where now the mesh consists of either 3×3 control or nodal points as shown in Fig. 6.9.

In general, we distinguish two cases. In the first case, the element is rectangular, while in the second it is not.

If the actual domain is a *rectangle* of dimensions $a \times b$, then the control and the nodal points are *identical* (in this particular case). Starting with the Bézierian macroelement of nine parameters, we shall show that it is equivalent with the classical 9-node Lagrangian element shown in Fig. 6.9.

The starting point is Eq. (6.9), which for simplicity is written by omitting the second subscript referring to the polynomial degree $n = 2$. Therefore, for both directions (x and y) we have:

$$\underbrace{\begin{bmatrix} L_0(x) \\ L_1(x) \\ L_2(x) \end{bmatrix}}_{\mathbf{L}_x} = \underbrace{\begin{bmatrix} 1 & -1/2 & 0 \\ 0 & 2 & 0 \\ 0 & -1/2 & 1 \end{bmatrix}}_{\mathbf{T}} \cdot \underbrace{\begin{bmatrix} B_0(x) \\ B_1(x) \\ B_2(x) \end{bmatrix}}_{\mathbf{B}_x},$$

$$\underbrace{\begin{bmatrix} L_0(y) \\ L_1(y) \\ L_2(y) \end{bmatrix}}_{\mathbf{L}_y} = \underbrace{\begin{bmatrix} 1 & -1/2 & 0 \\ 0 & 2 & 0 \\ 0 & -1/2 & 1 \end{bmatrix}}_{\mathbf{T}} \cdot \underbrace{\begin{bmatrix} B_0(y) \\ B_1(y) \\ B_2(y) \end{bmatrix}}_{\mathbf{B}_y} \tag{6.9}$$

For the sake of generality, the transformation matrix $\mathbf{T}$ involved in Eq. (6.9) is written in full form as

$$\mathbf{T} = \begin{bmatrix} t_{00} & t_{01} & t_{02} \\ t_{10} & t_{11} & t_{12} \\ t_{20} & t_{21} & t_{22} \end{bmatrix} \tag{6.53}$$

Now, it is our purpose to express the column vector $\mathbf{n}_L$ (of dimensions 9×1) which includes the nine tensor-product Lagrange shape functions (associate to the nine nodes shown in Fig. 6.9a) in terms of the column vector $\mathbf{n}_B$ (9×1) that includes tensor products of Bézier basis functions. Both of them are defined by Eq. (6.54):

$$\mathbf{n}_L \triangleq \left\{ \begin{array}{c} L_0(x)\cdot L_0(y) \\ L_1(x)\cdot L_0(y) \\ L_2(x)\cdot L_0(y) \\ \hline L_0(x)\cdot L_1(y) \\ L_1(x)\cdot L_1(y) \\ L_2(x)\cdot L_1(y) \\ \hline L_0(x)\cdot L_2(y) \\ L_1(x)\cdot L_2(y) \\ L_2(x)\cdot L_2(y) \end{array} \right\}, \quad \mathbf{n}_B \triangleq \left\{ \begin{array}{c} B_0(x)\cdot B_0(y) \\ B_1(x)\cdot B_0(y) \\ B_2(x)\cdot B_0(y) \\ \hline B_0(x)\cdot B_1(y) \\ B_1(x)\cdot B_1(y) \\ B_2(x)\cdot B_1(y) \\ \hline B_0(x)\cdot B_2(y) \\ B_1(x)\cdot B_2(y) \\ B_2(x)\cdot B_2(y) \end{array} \right\} \tag{6.54}$$

To this purpose, we start from the upper part of Eq. (6.9a) and multiply both parts with $L_0(y)$, thus receiving:

$$\begin{bmatrix} L_0(x) \\ L_1(x) \\ L_2(x) \end{bmatrix} L_0(y) = \begin{bmatrix} t_{00} & t_{01} & t_{02} \\ t_{10} & t_{11} & t_{12} \\ t_{20} & t_{21} & t_{22} \end{bmatrix} \cdot \begin{bmatrix} B_0(x) \\ B_1(x) \\ B_2(x) \end{bmatrix} L_0(y) \tag{6.55}$$

Then, the term $L_0(y)$ in the right part of Eq. (6.55) is replaced by the first term of the second system of Eq. (6.9a), that is:

$$L_0(y) = \begin{bmatrix} t_{00} & t_{01} & t_{02} \end{bmatrix} \cdot \begin{bmatrix} B_0(y) \\ B_1(y) \\ B_2(y) \end{bmatrix}. \tag{6.56}$$

As a result, substituting Eq. (6.56) into Eq. (6.55), one obtains:

$$\begin{bmatrix} L_0(x)L_0(y) \\ L_1(x)L_0(y) \\ L_2(x)L_0(y) \end{bmatrix} = \begin{bmatrix} t_{00} & t_{01} & t_{02} \\ t_{10} & t_{11} & t_{12} \\ t_{20} & t_{21} & t_{22} \end{bmatrix} \begin{bmatrix} B_0(x) \\ B_1(x) \\ B_2(x) \end{bmatrix} \cdot \begin{bmatrix} t_{00} & t_{01} & t_{02} \end{bmatrix} \cdot \begin{bmatrix} B_0(y) \\ B_1(y) \\ B_2(y) \end{bmatrix}. \tag{6.57}$$

Manipulation in the right-hand side of Eq. (6.57) leads to

$$\begin{bmatrix} L_0(x)L_0(y) \\ L_1(x)L_0(y) \\ L_2(x)L_0(y) \end{bmatrix} = \left[\begin{array}{ccc|ccc|ccc} t_{00}t_{00} & t_{01}t_{00} & t_{02}t_{00} & t_{00}t_{01} & t_{01}t_{01} & t_{02}t_{01} & t_{00}t_{02} & t_{01}t_{02} & t_{02}t_{02} \\ t_{10}t_{00} & t_{11}t_{00} & t_{12}t_{00} & t_{10}t_{01} & t_{11}t_{01} & t_{12}t_{01} & t_{10}t_{02} & t_{11}t_{02} & t_{12}t_{02} \\ t_{20}t_{00} & t_{21}t_{00} & t_{22}t_{00} & t_{20}t_{01} & t_{21}t_{01} & t_{22}t_{01} & t_{20}t_{02} & t_{21}t_{02} & t_{22}t_{02} \end{array}\right] \cdot \left\{\begin{array}{c} B_0(x)\cdot B_0(y) \\ B_1(x)\cdot B_0(y) \\ B_2(x)\cdot B_0(y) \\ \hline B_0(x)\cdot B_1(y) \\ B_1(x)\cdot B_1(y) \\ B_2(x)\cdot B_1(y) \\ \hline B_0(x)\cdot B_2(y) \\ B_1(x)\cdot B_2(y) \\ B_2(x)\cdot B_2(y) \end{array}\right\}. \tag{6.58}$$

Repeating the same as Eq. (6.58) but now for the second and third triplet of the terms in the column vector $\mathbf{n}_L$ of Eq. (6.54), it is easily found that

$$\mathbf{n}_L = [\mathbf{R}] \cdot \mathbf{n}_B \tag{6.59}$$

with the entire transformation matrix $\mathbf{R}$ to be given in terms of the elements of matrix $\mathbf{T}$ as

$$\mathbf{R} = \left[\begin{array}{ccc|ccc|ccc} t_{00}t_{00} & t_{01}t_{00} & t_{02}t_{00} & t_{00}t_{01} & t_{01}t_{01} & t_{02}t_{01} & t_{00}t_{02} & t_{01}t_{02} & t_{02}t_{02} \\ t_{10}t_{00} & t_{11}t_{00} & t_{12}t_{00} & t_{10}t_{01} & t_{11}t_{01} & t_{12}t_{01} & t_{10}t_{02} & t_{11}t_{02} & t_{12}t_{02} \\ t_{20}t_{00} & t_{21}t_{00} & t_{22}t_{00} & t_{20}t_{01} & t_{21}t_{01} & t_{22}t_{01} & t_{20}t_{02} & t_{21}t_{02} & t_{22}t_{02} \\ \hline t_{00}t_{10} & t_{01}t_{10} & t_{02}t_{10} & t_{00}t_{11} & t_{01}t_{11} & t_{02}t_{11} & t_{00}t_{12} & t_{01}t_{12} & t_{02}t_{12} \\ t_{10}t_{10} & t_{11}t_{10} & t_{12}t_{10} & t_{10}t_{11} & t_{11}t_{11} & t_{12}t_{11} & t_{10}t_{12} & t_{11}t_{12} & t_{12}t_{12} \\ t_{20}t_{10} & t_{21}t_{10} & t_{22}t_{10} & t_{20}t_{11} & t_{21}t_{11} & t_{22}t_{11} & t_{20}t_{12} & t_{21}t_{12} & t_{22}t_{12} \\ \hline t_{00}t_{20} & t_{01}t_{20} & t_{02}t_{20} & t_{00}t_{21} & t_{01}t_{21} & t_{02}t_{21} & t_{00}t_{22} & t_{01}t_{22} & t_{02}t_{22} \\ t_{10}t_{20} & t_{11}t_{20} & t_{12}t_{20} & t_{10}t_{21} & t_{11}t_{21} & t_{12}t_{21} & t_{10}t_{22} & t_{11}t_{22} & t_{12}t_{22} \\ t_{20}t_{20} & t_{21}t_{20} & t_{22}t_{20} & t_{20}t_{21} & t_{21}t_{21} & t_{22}t_{21} & t_{20}t_{22} & t_{21}t_{22} & t_{22}t_{22} \end{array}\right] \tag{6.60}$$

In a more handy form, $\mathbf{R}$ is written as

$$\mathbf{R} = \left[\begin{array}{c|c|c} \mathbf{T}\cdot t_{00} & \mathbf{T}\cdot t_{01} & \mathbf{T}\cdot t_{02} \\ \hline \mathbf{T}\cdot t_{10} & \mathbf{T}\cdot t_{11} & \mathbf{T}\cdot t_{12} \\ \hline \mathbf{T}\cdot t_{20} & \mathbf{T}\cdot t_{21} & \mathbf{T}\cdot t_{22} \end{array}\right] \tag{6.61}$$

Therefore, the linear dependency $\mathbf{n}_L = \mathbf{R}\mathbf{n}_B$ in Eq. (6.59) between the two basis vectors imposes a usual quadratic transformation in the mass matrix since

$$\begin{aligned} \mathbf{M}_L &= \frac{1}{c^2}\int\limits_A \mathbf{n}_L(\mathbf{n}_L)^T \mathrm{d}A = \frac{1}{c^2}\int\limits_A (\mathbf{R}\mathbf{n}_B)(\mathbf{R}\mathbf{n}_B)^T \mathrm{d}A \\ &= \mathbf{R}\left[\frac{1}{c^2}\int\limits_A (\mathbf{n}_B)(\mathbf{n}_E)^T \mathrm{d}A\right]\mathbf{R}^T \\ &= \mathbf{R}\cdot\mathbf{M}_a\cdot\mathbf{R}^T \end{aligned} \tag{6.62}$$

Equation (6.62) clearly shows that the mass matrix in terms of Lagrange polynomials ($\mathbf{M}_L$) is a quadratic transformation of the mass matrix ($\mathbf{M}_a$) based on Bezier polynomials of the same degree. In other words, if the functional set of Lagrange polynomials serves as a *local* system then the set of Bézier polynomials serves as a *global* system, and vice versa.

Obviously, since Eq. (6.9a) is also applicable for the partial derivatives of the shape functions, it is concluded that the transformation shown in Eq. (6.62) also occurs for the *stiffness* matrices as well. As a result, either the static solution or the eigenvalue solution using either of the formulations will be *identical*.

6.4.3 Demonstration of Equivalency on a Nonrectangular Element

In contrast, when the domain is *not* rectangular, the determination of the control points becomes necessary in order to calculate the Jacobian at the Gauss points in the procedure of numerical integration. Looking for a fair comparison between a uniform mesh of a single Lagrangian macroelement and a relevant Bézierian one, a logical procedure is to proceed as follows. The reference square ABCD is divided into a uniform mesh using n_ξ and n_η subdivisions along the ξ- and η-directions, respectively. Based on the n_ξ subdivisions made by the control points P_i along the side AB (ξ-direction in the reference square), we define the Bézier polynomials

$$B_{i,n_\xi}(\xi), \quad i = 0, 1, \ldots, n_\xi \tag{6.63a}$$

Similarly, based on the n_η subdivisions along the side BC (η-direction in the reference square), we define the Bézier polynomials

$$B_{j,n_\eta}(\eta), \quad j = 0, 1, \ldots, n_\eta \tag{6.63b}$$

The tensor-product shape functions $\varphi_k(\xi, \eta)$ associated to the kth control point, at the intersection of the ith column and jth row, are produced through

$$\varphi_k(\xi, \eta) = B_{i,n_\xi}(\xi) \cdot B_{j,n_\eta}(\eta), \quad i = 0, 1, \ldots, n_\xi; j = 0, 1, \ldots, n_\eta \tag{6.64}$$

Applying Coons interpolation, all the

$$q_e = (n_\xi + 1)(n_\eta + 1) \tag{6.65}$$

nodes of the reference square are mapped onto the same number of q_e nodal points on the actual domain Ω. This is a standard procedure which is fully controlled by the choice of the boundary points. In this way, the nodal points of the *Lagrangian macroelement* are obtained. Furthermore, at all the aforementioned already known nodal points (x_k, y_k) of the Lagrangian macroelement, the following equations are fulfilled:

$$x_k = \sum_{j=1}^{q_e} \varphi_j(\xi_i, \eta_j) \cdot x_k^C \text{ and } y_k = \sum_{j=1}^{q_e} \varphi_j(\xi_i, \eta_j) \cdot y_k^C, \quad k = 1, \ldots, q_e \tag{6.66}$$

Obviously, the control points that are produced solving the equations system formed by Eq. (6.66) do *not* generally coincide with the nodal points, of course excepting the case of a uniformly meshed rectangle (see Sect. 6.4.2). Therefore, in general, the Jacobian

$$\mathbf{J} = \begin{bmatrix} \partial x/\partial\xi \ \partial y/\partial\xi \\ \partial x/\partial\eta \ \partial y/\partial\eta \end{bmatrix} \tag{6.67}$$

is calculated in accordance with the particular formulation as follows:

Lagrangian macroelement:

$$\left.\begin{aligned} \frac{\partial x}{\partial \xi} &= \sum_{k=1}^{q_e} \frac{\partial \varphi_k(\xi,\eta)}{\partial \xi} \cdot x_k, \quad \frac{\partial y}{\partial \xi} = \sum_{k=1}^{q_e} \frac{\partial \varphi_k(\xi,\eta)}{\partial \xi} \cdot y_k \\ \frac{\partial x}{\partial \eta} &= \sum_{k=1}^{q_e} \frac{\partial \varphi_k(\xi,\eta)}{\partial \eta} \cdot x_k, \quad \frac{\partial y}{\partial \eta} = \sum_{k=1}^{q_e} \frac{\partial \varphi_k(\xi,\eta)}{\partial \eta} \cdot y_k \end{aligned}\right\} \tag{6.68a}$$

Bézierian macroelement:

$$\left.\begin{aligned} \frac{\partial x}{\partial \xi} &= \sum_{k=1}^{q_e} \frac{\partial \varphi_k(\xi,\eta)}{\partial \xi} \cdot x_k^C, \quad \frac{\partial y}{\partial \xi} = \sum_{k=1}^{q_e} \frac{\partial \varphi_k(\xi,\eta)}{\partial \xi} \cdot y_k^C \\ \frac{\partial x}{\partial \eta} &= \sum_{k=1}^{q_e} \frac{\partial \varphi_k(\xi,\eta)}{\partial \eta} \cdot x_k^C, \quad \frac{\partial y}{\partial \eta} = \sum_{k=1}^{q_e} \frac{\partial \varphi_k(\xi,\eta)}{\partial \eta} \cdot y_k^C \end{aligned}\right\} \tag{6.68b}$$

The key point so as the two Jacobians are equal to one another is when Lagrange interpolation coincides with Bezierian one, at all points of the domain. The given conditions are that in each direction we have fulfilled the series expansion at so many points as exactly the degree of the polynomial. This ensures that for a given η-value a coincidence will occur in both types of polynomials. Similarly, for a given ξ-value a coincidence will occur in both types of polynomials. Since these two polynomial interpolations are identical, their Jacobians will be identical as well:

$$\mathrm{d}A = |\det J_L| \mathrm{d}\xi \mathrm{d}\eta = |\det J_B| \mathrm{d}\xi \mathrm{d}\eta \tag{6.69}$$

Therefore, in the general case the mass matrix in both formulations will be:

$$\mathbf{M}_L = \frac{1}{c^2} \int_0^1 \int_0^1 \mathbf{n}_L (\mathbf{n}_L)^T |\det J_L| \mathrm{d}\xi \mathrm{d}\eta \tag{6.70a}$$

and

$$\mathbf{M}_B = \frac{1}{c^2} \int_0^1 \int_0^1 \mathbf{n}_B (\mathbf{n}_B)^T |\det J_B| \mathrm{d}\xi \mathrm{d}\eta \tag{6.70b}$$

Moreover, due to Eq. (6.59), Eq. (6.70a) becomes

$$\begin{aligned} \mathbf{M}_L &= \frac{1}{c^2} \int_A (\mathbf{R}\mathbf{n}_B)(\mathbf{R}\mathbf{n}_B)^T \mathrm{d}A = \mathbf{R} \left[\frac{1}{c^2} \int_A (\mathbf{n}_B)(\mathbf{n}_B)^T \mathrm{d}A \right] \mathbf{R}^T \\ &= \mathbf{R} \cdot \mathbf{M}_B \cdot \mathbf{R}^T \end{aligned} \tag{6.71}$$

Thus, again, a transformation (in quadratic form) between the two matrices occurs.

Exercise 6.5 (Sixteen nodes bicubic elements) *With respect to the unit square shown in Fig. 6.13:*

- *Let us consider a classical tensor-product finite element of Lagrange family, of 16 nodes.*
- *Let us also consider a tensor-product nonrational Bézier element of 16 control points.*

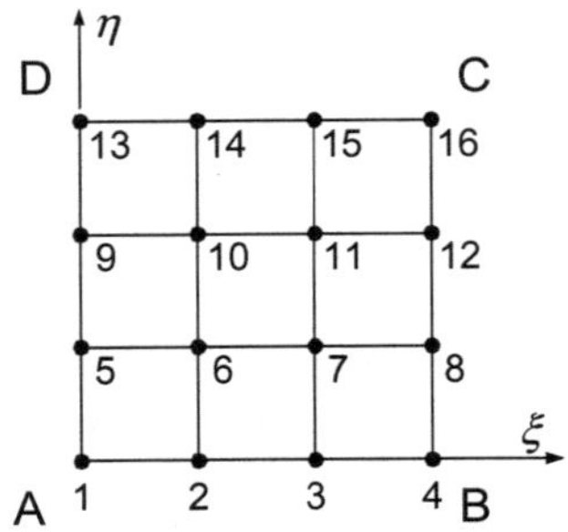

Fig. 6.13 Uniform mesh of 3×3 cells

Questions:

- *Do the control points coincide with the nodal points in the reference square?*
- *What is the relationship between the abovementioned macroelements?*

Answer Since in each direction the control points coincide with the nodal points, the same will be valid for the tensor product as well.

Concerning the matrices, following the same procedure as that of Exercise 6.4, we shall find that again in this 2D problems one matrix is produced by the other, using a proper transformation matrix **T** (here of dimensions 16×16). The reader may find it.

Therefore:

- Despite the fact the basis/shape functions are different from one another, the series expansion leads to the *same* numerical results.
- Stiffness and mass matrices based on the transformation matrices are as follows: $\widetilde{\mathbf{K}} = \mathbf{T}^t\mathbf{K}\mathbf{T}, \quad \widetilde{\mathbf{M}} = \mathbf{T}^t\mathbf{M}\mathbf{T}$

The interested reader can find many solved examples in Provatidis [12].

6.5 Relationship Between Higher-Order Elements (Lagrange, Nonrational Bézier) and P-Methods

At this point of the book, the reader has become aware of higher-order Lagrange and nonrational Bezier macroelements. On the other point of view, he/she may have heard about the so-called p-methods [16]. Although these are two quite different approaches in finite element analysis, they are strongly interrelated from the point of view on a common set of basis functions, which consists of the monomials x^n. This subject is discussed in the next sections.

6.5.1 A Summary of the P-Method

The complete theory is found in the book of the founders [16]. Here we extract modified only the absolutely necessary knowledge so as to be able to continue our discussion (comparison of the *p*-method with Lagrange and nonrational Bézier interpolation).

6.5.1.1 Legendre Polynomials

Legendre polynomials, $P_n(x)$, are the solutions of the *Legendre ordinary differential equation* for $n = 0, 1, 2, \ldots$:

$$(1 - x^2)y'' - 2xy' + n(n+1)y = 0, -1 \le x \le 1. \tag{6.72}$$

The first eight Legendre polynomials, which play a practical role in finite element analysis, are:

$$P_0(x) = 1 \tag{6.73a}$$

$$P_1(x) = x \tag{6.73b}$$

$$P_2(x) = \frac{1}{2}(3x^2 - 1) \tag{6.73c}$$

$$P_3(x) = \frac{1}{2}(5x^3 - 3x) \tag{6.73d}$$

$$P_4(x) = \frac{1}{8}(35x^4 - 30x^2 + 3) \tag{6.73e}$$

$$P_5(x) = \frac{1}{8}(63x^5 - 70x^3 + 15x) \tag{6.73f}$$

$$P_6(x) = \frac{1}{16}(231x^6 - 315x^4 + 105x^2 - 5) \tag{6.73g}$$

$$P_7(x) = \frac{1}{16}(429x^7 - 693x^5 + 315x^3 - 35x). \tag{6.73h}$$

Legendre polynomials fulfill the following condition of orthogonality:

$$\int_{-1}^{+1} P_i(x)P_j(x)\mathrm{d}x = \begin{cases} \frac{2}{2i+1} & \gamma\iota\alpha\; i = j \\ 0 & \gamma\iota\alpha\; i \neq j \end{cases} \tag{6.74}$$

It is remarkable that all roots of Legendre polynomials belong to the interval $-1 < x < 1$. The n roots of $P_n(x)$ are the well-known values of x_i involved in the n-points Gauss integration (quadrature):

$$\int_{-1}^{+1} f(x)\mathrm{d}x \approx \sum_{i=1}^{n} w_i f(x_i) \tag{6.75}$$

6.5.1.2 Shape Functions

In the one-dimensional finite element analysis, the domain Ω is decomposed into elements defined by corresponding nodal points. Following the notation by Szabó and Babuška [16], a 1D mesh is denoted by Δ:

$$\Delta : 0 = x_1 < x_2 < \cdots < x_{M(\Delta)} < x_{M(\Delta)+1} = l \tag{6.76}$$

where $M(\Delta)$ is the number of finite elements, and obviously, $M(\Delta)+1$ is the number of nodes. The kth element is denoted by Ω_k:

$$\Omega_k = \{x | x_k < x < x_{k+1}\}. \tag{6.77}$$

Each finite element is mapped to the standard (or reference) element, where the normalized coordinates are:

$$\Omega_{\mathrm{st}} = \{\xi | -1 < \xi < 1\} \tag{6.78}$$

The reference element is mapped to the kth element as follows:

$$x = Q_k(\xi) = \frac{1-\xi}{2} x_k + \frac{1+\xi}{2} x_{k+1}, \xi \in \Omega_{\mathrm{st}} \tag{6.79}$$

In each finite element, i, we define the degree p_i of the polynomial interpolation. Therefore, the vector that includes the corresponding degrees of all elements involved is:

$$\mathbf{p} = \left\{p_1, p_2, \ldots, p_{M(\Delta)}\right\} \tag{6.80}$$

Within each finite element of degree p (order $k = p + 1$), the p-method uses the following shape functions:

$$N_1(\xi) = \frac{1-\xi}{2}, N_2(\xi) = \frac{1+\xi}{2}, N_i(\xi) = \phi_{i-1}(\xi), \quad i = 3, 4, \ldots, p+1, \tag{6.81}$$

where ϕ_j are defined in terms of the Legendre polynomials P_{j-1} by:

$$\phi_j(\xi) = \sqrt{\frac{2j-1}{2}} \int_{-1}^{\xi} P_{j-1}(t)\mathrm{d}t, \quad j = 2, 3, \ldots \tag{6.82}$$

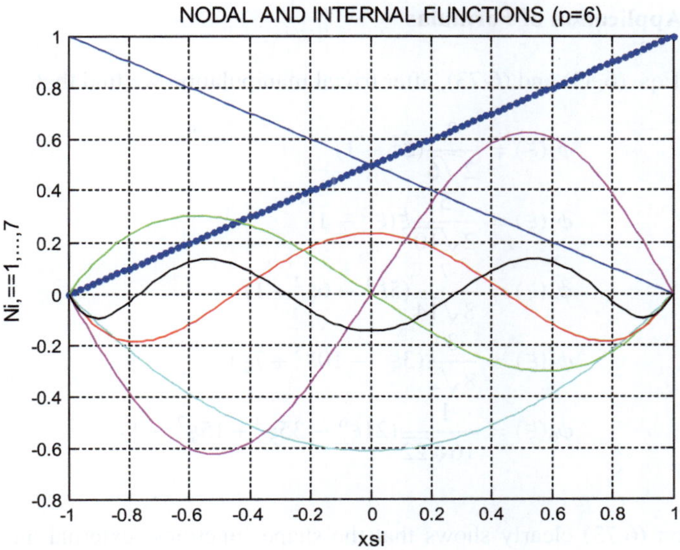

Fig. 6.14 Shape functions in the p-method (for $p = 6$)

The shape functions N_1, N_2 in Eq. (6.81) are called *nodal shape functions* or *external shape functions* or *external modes*, and they are the well-known linear (hat) functions associated to the ends of a usual finite element (see Fig. 2.4, also shown in Fig. 6.14). Moreover, the basis functions $N_i,\ i = 3, 4, \ldots$ are called *internal shape functions* or *internal modes*, or, sometimes, *bubble modes* (here we follow the definitions by Szabó and Babuška [16]).

It is not difficult to prove that Eq. (6.82) can be written as a difference of two Legendre polynomials as follows:

$$\phi_j(\xi) = \frac{1}{\sqrt{2(2j-1)}}\big[P_j(\xi) - P_{j-2}(\xi)\big]. \tag{6.83}$$

Equation (6.83) implies that the function $\phi_j(\xi)$ is a polynomial of jth degree. From the above definition of ϕ_j and the orthogonal property of Legendre polynomials (cf. Equation (6.74)), we have:

$$\phi_j(-1) = \phi_j(+1) = 0, \quad j = 2, 3, \ldots \tag{6.84}$$

and:

$$\int_{-1}^{+1} \frac{d\phi_i}{d\xi}\frac{d\phi_i}{d\xi}d\xi = \begin{cases} 1 \text{ για } i = j \\ 0 \text{ για } i \neq j \end{cases} \tag{6.85}$$

6.5.1.3 Application of Formulas

Based on Eqs. (6.83) and (6.73), after trivial manipulation we find that:

$$\begin{aligned}
\phi_2(\xi) &= \frac{3}{2\sqrt{6}}(\xi^2 - 1) \\
\phi_3(\xi) &= \frac{5}{2\sqrt{10}}\xi(\xi^2 - 1) \\
\phi_4(\xi) &= \frac{7}{8\sqrt{14}}(5\xi^4 - 6\xi^2 + 1) \\
\phi_5(\xi) &= \frac{3}{8\sqrt{2}}(3\xi^5 - 10\xi^3 + 7\xi) \\
\phi_6(\xi) &= \frac{11}{16\sqrt{22}}(21\xi^6 - 35\xi^4 + 15\xi^2 - 1), \qquad (6.86)
\end{aligned}$$

and so on.

Equation (6.73) clearly shows that the shape functions (external and internal) involved in the p-method consist of monomials. Since the maximum useful polynomial degree is practically $p = 6$, Eq. (6.81) shows that the most important shape functions are $N_1, \ldots, N_7 \equiv \phi_6$. In other words, the shape functions in the p-method are monomials that vary between 1 to ξ^6.

For the particular case of sextic interpolation ($p = 6$), for each element the p-method requires $p + 1 = 7$ shape functions, which are illustrated in Fig. 6.14. One may observe the nodal shape functions, $N_1(\xi) = (1 - \xi)/2$, $N_2(\xi) = (1 + \xi)/2$, associated to the ends, as well as the five internal ones (bubble functions) which equal to the functions $\phi_2, \phi_3, \phi_4, \phi_5$, and ϕ_6 included in Eq. (6.86). In contrast to Lagrange and Bernstein polynomials, the bubble functions are not all symmetric with respect to the middle of the interval (i.e., $N_4 = \phi_3$ and $N_6 = \phi_5$ include odd powers of ξ and therefore are antisymmetric).

6.5.2 *Application of P-Method to One-Dimensional Differential Equations of Second Order*

Time-dependent differential equations of second order may be:

(i) Axial deformation of a bar is described by:

$$\left(AEu'\right)' + f_x - (A\rho)\ddot{u} = 0 \qquad (6.87)$$

where A is the cross-sectional area, E is the elastic modulus, ρ is the mass density, f_x is the body force, and $u(x, t)$ is the axial displacement at the position x at the time t.

(ii) Transient heat transfer is described by:

$$(ku')' + q_x - (c\rho)\dot{u} = 0 \tag{6.88}$$

where k is the conductivity, c is the heat capacity, ρ is the mass density, q_x is the heat source per unit volume, and $u(x, t)$ is the temperature.

In both cases above, we assume constant coefficients within each finite element. Although there is no mathematical difference dealing with either of (6.87) and (6.88), in the sense that the produced matrices differ only in one factor, for the sake of briefness we prefer to work below with the first case (i.e., axially vibrating bar).

6.5.2.1 Stiffness Matrix

At any position x within the kth element ($x \in \Omega_k$), we have that $u = u(x) = u(Q_k(\xi))$, which can be written as:

$$u = \sum_{i=1}^{p_k+1} a_i^{(k)} N_i(\xi) \tag{6.89}$$

Each element of the stiffness matrix of a bar is written as:

$$k_{ij}^{(k)} = \int_0^{l_k} \frac{\mathrm{d}N_i}{\mathrm{d}x}(AE)_k \frac{\mathrm{d}N_j}{\mathrm{d}x}\mathrm{d}x = \frac{2}{l_k}\int_{-1}^{+1} (AE)_k \frac{\mathrm{d}N_i}{\mathrm{d}\xi}\frac{\mathrm{d}N_j}{\mathrm{d}\xi}\mathrm{d}\xi \tag{6.90}$$

Considering Eq. (6.85), the final stiffness matrix of the kth element is given by:

$$\left[K^{(k)}\right] = \frac{2(AE)_k}{l_k}\left[\begin{array}{cc|cccc} 1/2 & -1/2 & 0 & 0 & \cdots & 0 \\ -1/2 & 1/2 & 0 & 0 & \cdots & 0 \\ \hline 0 & 0 & 1 & 0 & \cdots & 0 \\ 0 & 0 & 0 & 1 & \cdots & 0 \\ \vdots & \vdots & \vdots & \vdots & \cdots & \vdots \\ 0 & 0 & 0 & 0 & \cdots & 1 \end{array}\right] \tag{6.91}$$

One may observe that:

(1) The upper left part of the stiffness matrix in Eq. (6.91), of dimensions 2×2, is *exactly* what we know from the basics of the finite element method for a truss element.
(2) The terms of the stiffness matrix that correspond to the internal shape functions are *decoupled* from those associated with the external shape functions.

6.5.2.2 Mass Matrix

Similarly, each element of the mass matrix of a bar (truss) element is written as

$$m_{ij}^{(k)} = \int_0^{l_k} N_i (A\rho)_k N_j \, \mathrm{d}x = \frac{l_k}{2} \int_{-1}^{+1} (A\rho)_k N_i N_j \, \mathrm{d}\xi \tag{6.92}$$

Setting the constant $(A\rho)_k = c_k$, then for the polynomial degree $p_k = 5$ we lead to the following mass matrix [generally of dimensions $(p_k + 1) \times (p_k + 1)$]:

$$[M^{(k)}] = \frac{c_k l_k}{2} \begin{bmatrix} 2/3 & 1/3 & -1/\sqrt{6} & 1/(3\sqrt{10}) & 0 & 0 \\ & 2/3 & -1/\sqrt{6} & -1/(3\sqrt{10}) & 0 & 0 \\ & & 2/5 & 0 & -1/(5\sqrt{21}) & 0 \\ & & & 2/21 & 0 & -1/(7\sqrt{45}) \\ & \text{(Symmetric)} & & & 2/45 & 0 \\ & & & & & 2/77 \end{bmatrix} \tag{6.93}$$

More generally, for $p_k \geq 2$ a simple expression may be derived for the diagonal terms. Assuming that c_k is a constant, we have:

$$\begin{aligned} m_{ii}^{(k)} &= \frac{c_k l_k}{2} \int_{-1}^{+1} \phi_{i-1}^2 \mathrm{d}\xi = \frac{c_k l_k}{2} \frac{1}{2(2i-3)} \int_{-1}^{+1} (P_{i-1}(\xi) - P_{i-3}(\xi))^2 \mathrm{d}\xi \\ &= \frac{c_k l_k}{2} \frac{1}{(2i-3)(2i-5)}, \quad i \geq 3 \end{aligned} \tag{6.94}$$

in which the orthogonality property of Legendre functions (Eq. (6.85)) was taken into consideration. In a similar way, it is not difficult to show that all the nondiagonal terms vanish for $i \geq 3$, except:

$$m_{i,i+2}^{(k)} = m_{i+2,i}^{(k)} = \frac{c_k l_k}{2} \int_{-1}^{+1} \phi_{i-1} \phi_{i+1} \mathrm{d}\xi = -\frac{c_k l_k}{2} \frac{1}{(2i-1)\sqrt{(2i-3)(2i+1)}}, \quad i \geq 3. \tag{6.95}$$

Note Regarding a quadratic element in conjunction with (i) Lagrange [cf. Eq. (6.13)] or (ii) Bernstein [cf. Eq. (6.14)] polynomials, in Exercise 6.1 we had found different stiffness and mass matrices but the eigenvalues for both cases were identical, a finding that was attributed to a basis change. The study below will be completed by examining the performance of the p-method compared to the aforementioned Lagrange and Bézier polynomials (see Exercise 6.6).

Exercise 6.6 (The quadratic p-element) *For the one-dimensional wave propagation problem in the interval $[-1, 1]$, compare the stiffness and mass matrices of the p-method with those of the abovementioned Lagrange and Bézier polynomials. Are the obtained eigenvalues once more the same?*

Answer Applying Eqs. (6.90) and (6.92), in conjunction with the shape functions

$$N_1 = \frac{1-\xi}{2}, \quad N_2 = \frac{1+\xi}{2}, \quad N_3 = \frac{3}{2\sqrt{6}}(\xi^2 - 1), \quad -1 \le \xi \le 1, \tag{6.96}$$

it is easily found that for quadratic approximation ($n = 2$) the corresponding stiffness and mass matrices become:

$$\mathbf{K}_P = \frac{1}{3l}\begin{bmatrix} 3 & 0 & -3 \\ 0 & 6 & 0 \\ -3 & 0 & 3 \end{bmatrix}, \quad \mathbf{M}_P = \frac{l}{30\,c^2}\begin{bmatrix} 10 & -15/\sqrt{6} & 5 \\ -15/\sqrt{6} & 6 & -15/\sqrt{6} \\ 5 & -15/\sqrt{6} & 10 \end{bmatrix}. \tag{6.97}$$

A numerical solution has revealed once more identical eigenvalues with those obtained using either Lagrange or Bézier polynomials (see [11]). The reason for this coincidence is as follows.

In terms of Lagrange polynomials, the interpolation is as follows:

$$\begin{aligned} u_L(\xi) &= \left(\frac{\xi^2}{2} - \frac{\xi}{2}\right)u_1 + (1-\xi^2)u_2 + \left(\frac{\xi^2}{2} + \frac{\xi}{2}\right)u_3 \\ &\equiv \xi^2\left(\frac{u_1}{2} - u_2 + \frac{u_3}{2}\right) + \xi\left(-\frac{u_1}{2} + \frac{u_3}{2}\right) + u_2 \end{aligned} \tag{6.98}$$

In terms of the p-method, the interpolation is:

$$\begin{aligned} u_P(\xi) &= \left(\frac{1-\xi}{2}\right)u_1 + \left(\frac{1+\xi}{2}\right)u_3 + \frac{3}{2\sqrt{6}}(\xi^2 - 1)a_3 \\ &\equiv \xi^2\left(\frac{3}{2\sqrt{6}}a_3\right) + \xi\left(-\frac{u_1}{2} + \frac{u_3}{2}\right) + \left(\frac{u_1}{2} + \frac{u_3}{2} - \frac{3}{2\sqrt{6}}a_3\right) \end{aligned} \tag{6.99}$$

Setting the two abovementioned polynomials to be identical, i.e., $u_L(\xi) \equiv u_P(\xi)$ in ξ, the corresponding factors of ξ^2, ξ and the constants must be identical, whence (ignoring two apparent identities) the only useful linear dependency is

$$u_2 = \tfrac{1}{2}u_1 - \frac{3}{2\sqrt{6}}a_3 + \tfrac{1}{2}u_3 \tag{6.100}$$

Since the Lagrangian-based matrix of Eq. (6.13) refers to the vector $\{u_1, u_2, u_3\}^T$ whereas the p-method based in Eq. (6.97) refers to $\{a_1, a_3, a_3\}^T \equiv \{u_1, u_3, a_3\}^T$, the transformation matrix **[T]** which relates two DEF vectors through the formula $\{u_1, u_2, u_3\}^T = [\mathbf{T}]\{a_1, a_3, a_3\}^T$ is

$$[\mathbf{T}] = \begin{bmatrix} 1 & 0 & 0 \\ \frac{1}{2} & -\frac{3}{2\sqrt{6}} & \frac{1}{2} \\ 0 & 0 & 1 \end{bmatrix} \tag{6.101}$$

Therefore, the change of basis from Lagrange polynomials (matrix $[\mathbf{K}_L] \equiv [\mathbf{K}_u]$) to those of the p-method would give the stiffness matrix

$$\begin{aligned}
[\widetilde{\mathbf{K}}_P] &= [\mathbf{T}]^T[\mathbf{K}_L][\mathbf{T}] \\
&= \begin{bmatrix} 1 & \frac{1}{2} & 0 \\ 0 & -\frac{3}{2\sqrt{6}} & 0 \\ 0 & \frac{1}{2} & 1 \end{bmatrix} \cdot \underbrace{\frac{1}{3l}\begin{bmatrix} 7 & -8 & 1 \\ -8 & 16 & -8 \\ 1 & -8 & 7 \end{bmatrix}}_{\text{LAGRANGE}} \cdot \begin{bmatrix} 1 & 0 & 0 \\ \frac{1}{2} & -\frac{3}{2\sqrt{6}} & \frac{1}{2} \\ 0 & 0 & 1 \end{bmatrix} \\
&= \frac{1}{3l}\begin{bmatrix} 3 & 0 & -3 \\ 0 & 6 & 0 \\ -3 & 0 & 3 \end{bmatrix} \equiv \mathbf{K}_P.
\end{aligned} \tag{6.102}$$

Therefore, applying the matrix transformation we obtained the same stiffness matrix as that previously calculated in Eq. (6.97). Similarly, the same conclusion is valid for the mass matrix as well. ■

Exercise 6.7 (The cubic element) *Derive analytical expressions for the stiffness and mass matrices using (i) Lagrange polynomials, (ii) Bernstein polynomials, and (iii) the p-method. Show that all these three give the same eigenvalues for any boundary conditions.*

Hint Based on the above equations, the matrices are

1. Lagrange polynomials:

$$\mathbf{K}_L = \frac{1}{40L}\begin{bmatrix} 148 & -189 & 54 & -13 \\ -189 & 432 & -297 & 54 \\ 54 & -297 & 432 & -189 \\ -13 & 54 & -189 & 148 \end{bmatrix},$$

$$\mathrm{M}_L = \frac{L}{1680}\begin{bmatrix} 128 & 99 & -36 & 19 \\ 99 & 648 & -81 & -36 \\ -36 & -81 & 648 & 99 \\ 19 & -36 & 99 & 128 \end{bmatrix} \tag{6.103}$$

2. Bernstein polynomials:

$$\mathbf{K}_a = \frac{3}{10L}\begin{bmatrix} 6 & -3 & -2 & -1 \\ -3 & 4 & 1 & -2 \\ -2 & 1 & 4 & -3 \\ -1 & -2 & -3 & 6 \end{bmatrix}, \quad \mathrm{M}_a = \frac{L}{420}\begin{bmatrix} 60 & 30 & 12 & 3 \\ 30 & 36 & 27 & 12 \\ 12 & 27 & 36 & 30 \\ 3 & 12 & 30 & 60 \end{bmatrix} \tag{6.104}$$

3. Higher-order p-method:

$$\mathbf{K}_P = \frac{2}{L}\begin{bmatrix} 1/2 & 0 & 0 & -1/2 \\ 0 & 1 & 0 & 0 \\ 0 & 0 & 1 & 0 \\ -1/2 & 0 & 0 & 1/2 \end{bmatrix}, \quad \mathrm{M}_L = \frac{L}{2}\begin{bmatrix} 2/3 & -1/\sqrt{6} & 1/(3\sqrt{10}) & 1/3 \\ -1/\sqrt{6} & 2/5 & 0 & -1/\sqrt{6} \\ 1/(3\sqrt{10}) & 0 & 2/21 & -1/(3\sqrt{10}) \\ 1/3 & -1/\sqrt{6} & -1/(3\sqrt{10}) & 2/3 \end{bmatrix} \tag{6.105}$$

One can notice that in the case of the p-method, although the mass matrix is symmetric, no symmetry exists along the diagonal terms of the mass matrix which correspond to internal modes (i.e., $\frac{2}{5} \neq \frac{2}{21}$). This fact is attributed to the fact that the cubic Lagrange and Bernstein polynomials are symmetric with respect to the middle point of the length at $\xi = 0$, while the same does not hold for the p-method.

It is left to the reader to determine the new transformation matrix **[T]**. Further comparisons with Chebyshev polynomials may be found in Provatidis [11].

6.6 Rational Bézier Elements

6.6.1 Summary of Nonrational Elements

The shortcoming of using control points in conjunction with *nonrational* Bézier polynomials was earlier shown in Table 6.1. Generally, the boundary of the reference square is not entirely mapped along the real boundary, so a deviation appears and the entire image is found *inside* the ideal circle apart from the boundary nodes (see Fig. 6.12). The maximum deviation occurs particularly when applying nonrational Bernstein polynomials of second degree. This case is entirely equivalent to using parabolic Lagrangian elements. This incapability is obvious from background knowledge in analytical geometry, because a circular part (which belongs to an *ellipse*, in the sense that a circle is an ellipse of equal diameters) cannot be accurately represented by three boundary points that determine a *parabolic* part (due to the quadratic Lagrange polynomials).

Therefore, if a circle is modeled using a 9-node element, of either Lagrangian or nonrational Bézierian type, the deviation of the calculated area is -1.8% in both of them (we showed earlier that these are equivalent). Again, the control points associated to this element are shown in Fig. 6.15. We draw the attention of the reader that the points Pij are shown in the unit square (Fig. 6.15a) truly in conjunction with Lagrangian polynomials and only "schematically" for Bézier polynomials, in the latter case to *remind* the tensor product involved. Of course, these would be the control points if the domain were a square or a rectangle. But now, the corner points P00, P20, P22, and P02 are mapped to the nodes A, B, C, and D on the circumference of the circle (Fig. 6.15b). Also, P11 of Fig. 6.15a is mapped to the center of the circle. In contrast, the intermediate points P10, P21, P12, and P01 of the unit square are actually mapped to the midpoints of the circular arcs (points 2, 6, 8, and 4 of Fig. 6.15b). Clearly, the control points P10, P21, P12, and P01 that are outside the circle (Fig. 6.15b) have *no* relationship with those in the middle of the sides in the unit square, but have the *same* role in the tensor product.

For the completeness of this text, we repeat that the full representation of the approximate circle is given either in terms of Lagrange polynomials (see Eq. (6.6)):

$$\mathbf{x}(\xi,\eta)=\sum_{i=0}^{2}\sum_{j=0}^{2}[L_{i,2}(\xi)L_{j,2}(\eta)]\cdot\mathbf{x}_{ij} \tag{6.106a}$$

or in terms of Bézierian polynomials (see Eq. (6.8)):

$$\mathbf{x}(\xi,\eta)=\sum_{i=0}^{2}\sum_{j=0}^{2}[B_{i,2}(\xi)B_{j,2}(\eta)]\cdot\mathbf{x}_{ij}^{P} \tag{6.106b}$$

In Eq. (6.106a), $\mathbf{x}_{ij}$ denotes the coordinates of the *nodal* point (1, ..., 9) in the circle that is an *image* of the point P_{ij} in the unit (reference) square (shown in Fig. 6.15b;

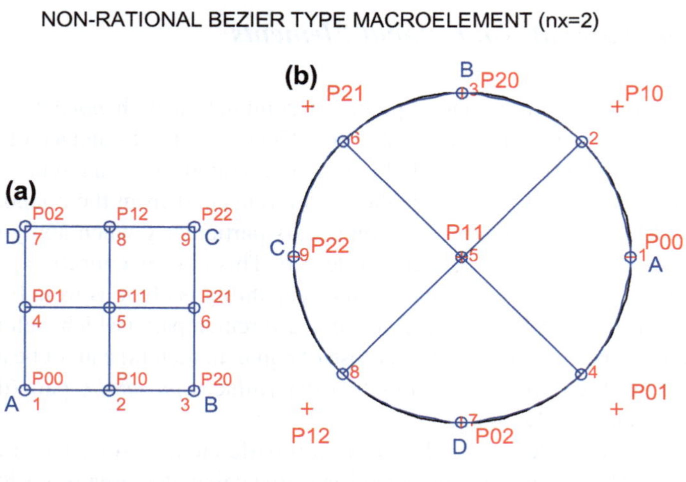

Fig. 6.15 Control points for a *nonrational* 9-node element in **a** the unit square and **b** the circular domain (the length of P11-P10 is approximately $\cong 1.29R$)

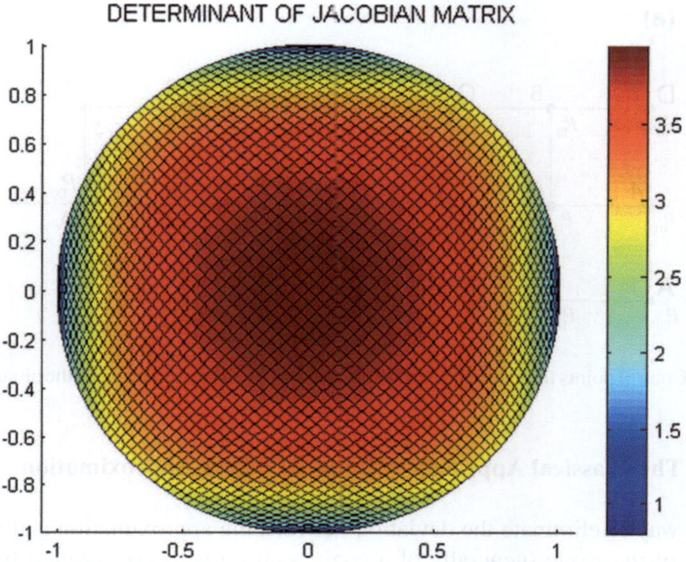

Fig. 6.16 Calculated determinant of the Jacobian matrix for a nonrational 9-node element of circular shape

also in Fig. 6.17a). Finally, $\mathbf{x}_{ij}^P$ in Eq. (6.106b) refers to the coordinates of the *control* points $P_{ij}(i, j = 0, 1, 2)$ shown in Fig. 6.15b.

In order to make clear the difference with the subsequent rational elements, we report that the external control points P10, P21, P12, and P01 in Fig. 6.15b lie on the bisectors of the x- and y-axes, and their abscissae are equal to approximately $x = \pm 0.9142R$, which is smaller than the value $x = \pm R$ (in absolute value). Also, the distribution of the determinant of the Jacobian matrix is shown in Fig. 6.16.

6.6.2 *Rational Bézier*

6.6.2.1 General

In the aforementioned case of a classical quadratic Lagrangian element, the midpoints (P_{10}, P_{21}, P_{12} and P_{01}) of the four edges in the unit square (see Fig. 6.17a) should be taken at the midpoints of the circular arcs (AB, BC, CD, and DA); the latter are shown as white-colored filled circles in Fig. 6.17b (see points 2, 4, 6, and 8). Again, the application of either Eq. (6.106a) or Eq. (6.106b) caused a circle of smaller area.

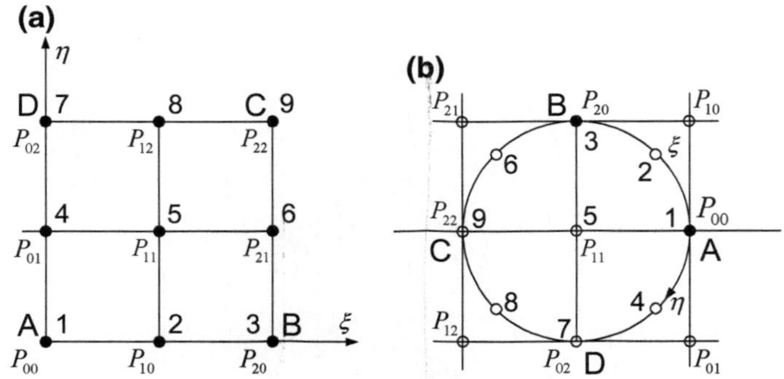

Fig. 6.17 Control points for a rational 9-node element in **a** the unit square and **b** the circular domain

6.6.2.2 The Classical Approach in Rational Bézier Approximation

A drastic way to eliminate the deviation between the approximation and the actual boundary of the circle (generally of a conic section) is to use *rational* basis functions in conjunction with a proper *shift* of the control points (P_{10}, P_{21}, P_{12} and P_{01}) outwards (compared to those found in Fig. 6.15b) and particularly at the intersections of the tangents to the circle at the points (A, B, C, and D). Since one-quarter of the circle requires three control points (e.g., the circular arc AB in Fig. 6.17b is accurately represented using P_{00}, P_{10} and P_{20}), the quadratic Bernstein polynomials are capable of accurately approximating the circumference, providing that proper weighting factors are used. For details, the reader is referred to Chap. 2 or to a more specialized handbook on CAD (e.g., [7]; Piegl and Tiller [10]).

As an example, the entire arc AB (90 degrees central angle) of Fig. 6.17b is accurately represented using the formula

$$\mathbf{x}(\xi) = \sum_{i=0}^{2} R_{i0}(\xi)\,\mathbf{x}_{i0}, \tag{6.107}$$

where $\mathbf{x}_{ij}$ are the Cartesian coordinates of the control points P_{00}, P_{10}, and P_{20}, respectively, whereas the univariate rational shape functions are given as:

$$R_{i0}(\xi) = \frac{w_{i0} B_i(\xi)}{\sum_{k=0}^{2} w_{k0} B_k(\xi)}, \quad i = 0, 1, 2 \tag{6.108}$$

Omitting the second subscript that refers to the isoline $\eta = 0$, in Eq. (6.108) the choice $w_0 = w_2 = 1$ and $w_1 = \frac{\sqrt{2}}{2}$ leads to rational shape functions that appear in symmetry with respect to the midpoint of the circular arc (say AB). The involved quadratic Bernstein polynomials are given by Eq. (6.3).

Having made clear the representation of the geometry in one curvilinear direction (e.g., toward the ξ-direction, along AB), we can now generalize the representation of each point $P(\xi, \eta)$ of the entire circle. To this purpose, the *classical* approach is to use nine control points to describe it (both the boundary and the interior). Then, the four control points (P_{10}, P_{21}, P_{12} and P_{01}) should be found at the intersections of the tangents at four boundary points (A, B, C, and D) that uniformly divide the circumference (every 90 degrees, as shown in Fig. 6.17b), and the tensor product is a reconstruction of Eq. (6.106b) into:

$$\mathbf{x}(\xi, \eta) = \sum_{i=0}^{2} \sum_{j=0}^{2} R_{ij}(\xi, \eta) \cdot \mathbf{x}_{ij}^{P}, \tag{6.109}$$

with the rational shape functions given as:

$$R_{ij}(\xi, \eta) = \frac{w_{ij} B_i(\xi) B_j(\eta)}{\sum_{k=0}^{2} \sum_{l=0}^{2} w_{kl} B_k(\xi) B_l(\eta)}, \quad i, j = 0, 1, 2 \tag{6.110}$$

The basis functions in Eq. (6.110) are the Bernstein polynomials of second degree, which are given by Eq. (6.3). For the choice $w_0 = w_2 = 1$ and $w_1 = \frac{\sqrt{2}}{2}$, the weighting factors in the tensor product are given by $w_{ij} = w_i w_j$. The distribution of the Jacobian is shown in Fig. 6.18. Despite the accurate representation of the boundary and the accurate calculation of the domain area, no significant change in the distribution of the Jacobian is observed.

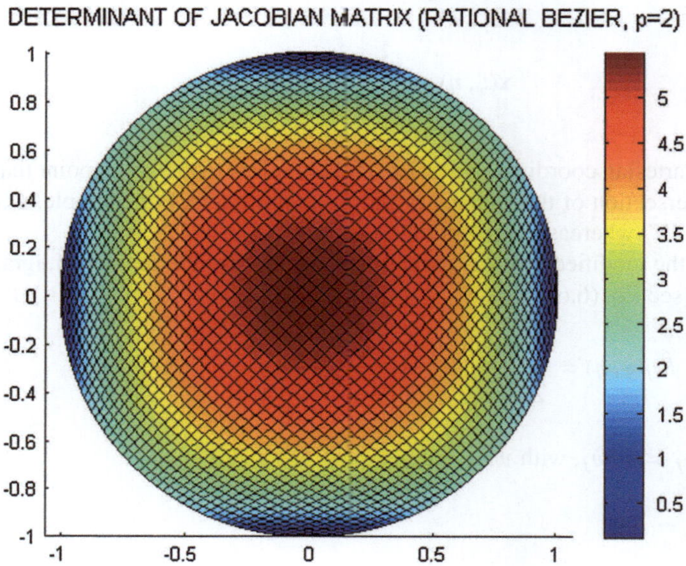

Fig. 6.18 Calculated determinant of the Jacobian matrix for a rational 9-node element of circular shape

Note Although we refer to the term "tensor product," the rational expression of Eq. (6.110) appears different from the previous tensor products (found in Lagrangian and Bézierian elements). This lies in the fact that a denominator, which is a sum of nine terms, should be applied to the numerator. In other words, the term "tensor product" is influenced mostly by the tensor product that actually occurs in the numerator.

6.6.2.3 A "Novel" Approach in Rational Bézier Approximation

Another possibility is to deal directly with the nodal points (1, 2, …, 9) in Fig. 6.17b. This "novel" approach is not found in classical textbooks, but has come as a need to apply the boundary conditions in a direct way (on nodal points) as happens in the classical finite element method (for a detailed discussion on NURBS, see [8, 9, 15]).

In brief, it is possible to shift the intermediate control point P_{10} from the above-mentioned *intersection* of the tangents at A and B to the real *midpoint* of the arc (P_0P_2 $\equiv$ AB). Actually, this modification can be achieved by modifying the weighting factors to the updated value $\hat{w}_1 = \frac{1}{2} + \frac{\sqrt{2}}{4}$ and in conjunction with classical quadratic *Lagrange* polynomials (see Eq. (6.6)) instead of *Bernstein* ones (see Chap. 2, Exercise 2.14). Again, the circular arc AB in Fig. 6.17b may be described either in terms of the control points P_{00}, P_{10}, and P_{20} (classical approach), or in terms of the ends ($1 \equiv P_{00}$, $3 \equiv P_{20}$) plus the midpoint "2" of this arc ("novel" approach). Extending this concept to the rest boundary arcs as well (BC, CD, and DA), the useful points are now the nodal points (1, …, 9 in Fig. 6.17b), and the tensor-product formula becomes:

$$\mathbf{x}(\xi, \eta) = \sum_{i=0}^{2}\sum_{j=0}^{2} \hat{R}_{ij}(\xi, \eta) \cdot \mathbf{x}_{ij}. \tag{6.111}$$

The Cartesian coordinate $\mathbf{x}_{ij}$ in Eq. (6.111) refers to the nodal point that is found at the intersection of the isolines ξ_i and η_j, respectively. For example, $\mathbf{x}_{00}$ refers to the node "1", whereas $\mathbf{x}_{22}$ refers to the node "9".

Also, the modified rational basis functions are given in terms of Lagrange polynomials (see Eq. (6.6)) by:

$$\hat{R}_{ij}(\xi, \eta) = \frac{\hat{w}_{ij} L_i(\xi) L_j(\eta)}{\sum_{k=0}^{2}\sum_{l=0}^{2} \hat{w}_{kl} L_k(\xi) L_l(\eta)}, \quad i, j = 0, 1, 2, \tag{6.112}$$

where $\hat{w}_{ij} = \hat{w}_i \hat{w}_j$, with $\hat{w}_0 = 1$, $\hat{w}_1 = \frac{1}{2} + \frac{\sqrt{2}}{4}$, and $\hat{w}_2 = 1$.

6.7 Unsolved Exercises

Exercise 6.8 *The reader is encouraged to develop a simple computer program that uniformly divides the circumference of a unit circle (in, say, 32 segments) and compares the computations* (x, y) *using (i) Eq. (6.109) and (ii) Eq. (6.111) for arbitrary pairs of normalized* (ξ, η) *values. Not only the two sets should be identical but when either of* (ξ, η) *is equal to zero or unit, the obtained point* $P(x, y)$ *should lie on the ideal circumference (i.e., will fill the condition:* $x^2 + y^2 = 1$*).*

Exercise 6.9 *The reader is also encouraged to develop another computer program that deals with an ellipse of semi-diameters equal to* $a = 2.0$ *and* $b = 1.0$*. Again, compare the computations* (x, y) *using (i) Eq. (6.109) and (ii) Eq. (6.111) for arbitrary pairs of normalized* (ξ, η) *values. Attention should be paid so as the midpoints (2, 6, 8, and 4) are chosen at the intersection of diagonals of the circumscribed rectangle with the boundary of the ellipse (consult Chap. 2, Exercise 2.14).*

For a further discussion, the reader is referred to Provatidis [13].

6.8 Bézier Triangles

6.8.1 General

We return to the nonrational Bernstein–Bézier interpolation but now in *triangular* patches. The topic of interpolation within a triangular patch has been studied with detail in several books ([7], pp. 303–330) and papers. The idea is due to de Casteljau [4] which remained unknown until its discovery by W. Boehm around 1975 (see [7], p. 303). A thorough review focused on CAGD including triangles has been reported by Barnhill [1], whereas two surveys on the field of triangular Bézier patches are due to Farin [6] and de Boor [3]. In the context of computational mechanics, the reader may consult an interesting paper by Zieniuk and Szerszen [17].

In brief, the de Casteljau algorithm can be used as a generalization of that used for curves (see Sect. 6.1 of this chapter). Instead of joining homologous points along the generator, now we join homologous points inside smaller triangles of which the base is initially the boundary of the triangle **ABC**. The procedure continues with a set of fewer triangles and progresses until only one triangle remains at which the algorithm gives the final point and stops. The three steps required for the cubic case are shown in Fig. 6.19, where the final point $1'''$ is determined (Fig. 6.19c).

The control net is not associated to a generator, as happens in one-dimensional interpolation, but is now of a triangle structure that consists of vertices, as shown in Fig. 6.20. More details are provided below.

The arrangement of the involved monomials in the triangle structure is somehow relevant to that previously known in Pascal's triangle but there are also some differences as follows. For a particular number of n subdivisions along each side of the

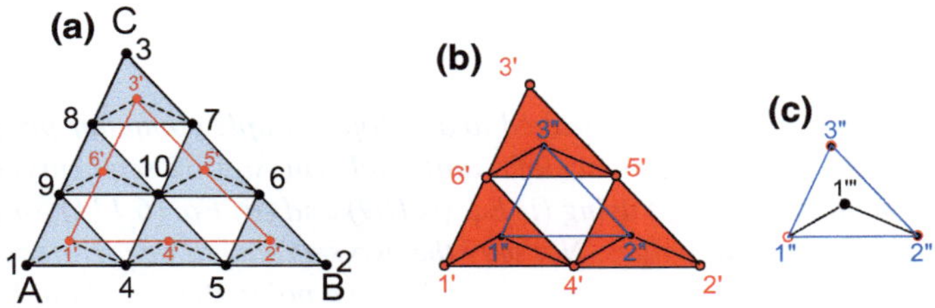

Fig. 6.19 Iterative de Casteljau algorithm inside a triangle: **a** first step, **b** second step, and **c** third step

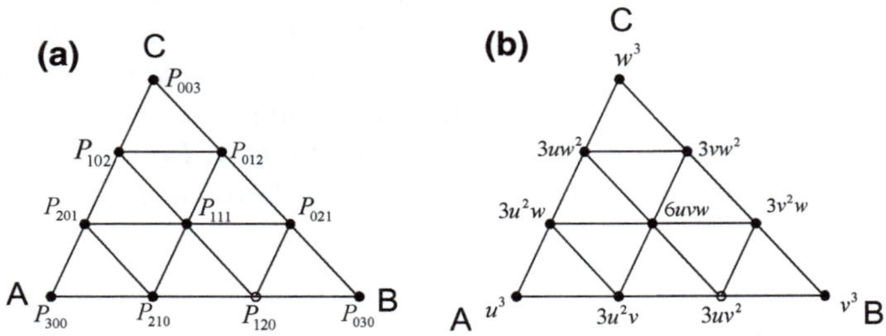

Fig. 6.20 Triangle structure for the cubic case: **a** net of control points, **b** associated monomials

triangle, which determines the polynomial degree p along a side ($p = n$), the control net consists of $\frac{1}{2}(n+1)(n+2)$ vertices, and each vertex is given by

$$B_{ijk}(u, v, w) = \frac{n!}{i!\,j!\,k!} u^i v^j w^k, \quad i + j + k = n. \tag{6.113}$$

Each of the abovementioned basis functions $B_{ijk}(u, v, w)$ is associated to a point, and concretely the point P_{ijk}, of the control net.

The above remark is the 2D analogue of Bézier curves. Again, a Bézier curve requires Bernstein polynomials $B_{i,n}(\xi)$ [described by Eq. (6.2)] and a polygon line $P_0 P_1 \ldots P_n$, whereas a Bézier triangular patch requires extended Bernstein polynomials $B_{ijk,n}(u, v, w)$ or simply B_{ijk} [described by Eq. (6.113)] and a net of *control points* properly arranged into the triangle (see Fig. 6.20).

6.8.2 Comparison and Contradiction of Bézierian with Traditional Triangular Elements

The barycentric coordinates of this chapter are those used in Chap. 5, that is:

$$u = \xi_1, \; v = \xi_2, \; w = \xi_3. \tag{6.114}$$

Let us take the simplest case of a *cubic* element ($n = 3$). Based on Fig. 6.20b, and considering Eq. (6.113) for $n = 3$, the interpolation within the entire triangle will be:

$$\begin{aligned} U(u, v, w) = {} & B_{300} \cdot a_{300} + B_{210} \cdot a_{210} + B_{120} \cdot a_{120} + B_{030} \cdot a_{030} \\ & + B_{201} \cdot B_{201} + B_{102} \cdot a_{102} + B_{003} \cdot a_{003} \\ & + B_{012} \cdot a_{012} + B_{021} \cdot a_{021} + B_{111} \cdot a_{111}. \end{aligned} \tag{6.115a}$$

or, in more details:

$$\begin{aligned} U(u, v, w) = {} & u^3 \cdot a_{300} + 3u^2v \cdot a_{210} + 3uv^2 \cdot a_{120} + v^3 \cdot a_{030} \\ & + 3u^2w \cdot a_{201} + 3uw^2 \cdot a_{102} + w^3 \cdot a_{003} \\ & + 3vw^2 \cdot a_{012} + 3v^2w \cdot a_{021} + uvw \cdot a_{111}. \end{aligned} \tag{6.115b}$$

It is noted that the terms u^3, v^3, and w^3 are associated to the vertices **A**, **B**, and **C**, respectively (the bold style stands for the position vector). Also, the parameters a_{ijk} with $i + j + k = n$ are generalized coefficients (sometimes called generalized "coordinates"); only three of them have a physical meaning [see Eq. (6.115c)].

For the particular case, we interpolate along the side **AB** of the triangle, since $w = 0$ we have $u + v = 1$. Then, Eq. (6 115b) reduces to only the first four terms and therefore (by virtue of $v = 1 - u$) it becomes:

$$U_{ab}(u, v, 0) = u^3 \cdot a_{300} + 3u^2(1-u) \cdot a_{210} + 3u(1-u)^2 \cdot a_{120} + (1-u)^3 \cdot a_{030} \tag{6.116}$$

In other words, Eq. (6.116) takes the form of the univariate Bernstein–Bézier interpolation (cf. Equation (6.4) in which ξ is replaced by u), where a_{300} and a_{030} are the nodal values at the vertices **A** and **B**, respectively, whereas a_{210} and a_{120} are generalized coefficients. Similar observations may be made for the rest two sides (**BC** and **CA**) of the triangle **ABC**.

Finally, for the last term ($uvw \cdot a_{111}$) involved in Eq. (6.115b), we make the following remarks. In order to fulfill the condition of affine invariance (cyclic symmetry among the vertices and relevant area coordinates), this terms should be associated to the center of gravity located at $u = v = w = \frac{1}{3}$, where it takes the value $\frac{1}{27}a_{111}$. On the other point of view, according to Eq. (5.37), the cardinal shape function associated to the node at the centroid is $N_{10} = 27\xi_1\xi_2\xi_3 \equiv 27uvw$ and obviously is equal to the unity at this node. Setting the values $u = v = w = \frac{1}{3}$, in conjunction with the fact that:

$$a_{300} = U_a, \quad a_{030} = U_b, \quad a_{003} = U_c, \tag{6.115c}$$

Equation (6.115b) becomes:

$$\begin{aligned} U_{\text{centroid}} = & \tfrac{1}{27} \cdot (U_a + U_b + U_c) \\ & + \tfrac{3}{27} \cdot (a_{210} + a_{120} + a_{201} + a_{102} + a_{012} + a_{021}) \\ & + \tfrac{1}{27} \cdot a_{111}. \end{aligned} \tag{6.115d}$$

Therefore, the value at the centroid is influenced equally by the three vertices and equally from the six intermediate points (two per side). Practically, the parameter a_{111} has not an obvious physical meaning.

Obviously, if Eq. (6.115b) is fulfilled at the ten nodal points of the control net in the element (note they are identical with the nodes in the classical element shown in Fig. 5.3c for $m = 3$), a linear relationship is established between the nodal values U_i and the generalized coefficients a_{ijk}. In other words, for the cubic interpolation the Bernstein–Bézier formulation is equivalent with that given by Eq. (5.37) in conjunction with classical triangular elements.

Exercise 6.10 *Start with the parametric expression of the geometry, according to which the position of any point* $\mathbf{x} = [x, y]^T$ *inside a triangle with straight sides is given by:*

$$\begin{aligned} \mathbf{x}(u, v, w) = & u^3 \cdot \mathbf{x}_{300} + 3u^2 v \cdot \mathbf{x}_{210} + 3uv^2 \cdot \mathbf{x}_{120} + v^3 \cdot \mathbf{x}_{030} \\ & + 3u^2 w \cdot \mathbf{x}_{201} + 3uw^2 \cdot \mathbf{x}_{102} + w^3 \cdot \mathbf{x}_{003} \\ & + 3vw^2 \cdot \mathbf{x}_{012} + 3v^2 w \cdot \mathbf{x}_{021} + uvw \cdot \mathbf{x}_{111}. \end{aligned} \tag{6.117}$$

(i) *Show that when the triplet* $(u, v, w) = \left(\frac{1}{3}, \frac{1}{3}, \frac{1}{3}\right)$ *is substituted into Eq. (6.117), the center of mass of this triangle,* $\mathbf{x}_G = \frac{1}{3}(\mathbf{x}_{300} + \mathbf{x}_{030} + \mathbf{x}_{003})$*, is derived.*

(ii) *For the same triangle (with straight sides), show that for every triplet* (u, v, w) *the resultant point is* $\mathbf{x}(u, v, w) = u \cdot \mathbf{x}_{300} + v \cdot \mathbf{x}_{030} + w \cdot \mathbf{x}_{003}$.

Hint We repeatedly use the law of linear interpolation, according to which when a point P divides the straight segment 12 into a ratio $\lambda = \left\|\overrightarrow{(1P)}\right\| / \left\|\overrightarrow{(12)}\right\|$, its position is given by $\mathbf{r}_P = (1 - \lambda)\mathbf{r}_1 + \lambda \mathbf{r}_2$.

This example shows that the ten points of the net actually operate like *control points*.

Analogous observations and remarks may be made for the *quartic* element ($n = 4$) described by Eq. (5.38), and so on.

Exercise 6.11 *Using the abovementioned cubic Bernstein–Bézier triangle (generally shown in Fig. 6.20), develop a computer program and determine the eigenvalues of a rectangular isosceles triangle of side a, under Neumann boundary conditions. Compare the numerical results with those previously obtained and shown in Table 5.5, where the classical 10-node Lagrangian triangle has been used. It is reminded that the shape functions of the latter are given by Eq. (5.4c). What do you observe?*

Hint Both numerical results will be found to be *identical*. This is justified due to the fact that the Bernstein–Bézier series of ten terms:

$$U(u, v, w) = \sum_{0}^{3} \sum_{0}^{i} \sum_{0}^{3} \frac{n!}{i!\,j!\,k!} u^i v^j w^k \alpha_{ijk},$$

includes monomials of the form $u^i v^j w^k$, with $i + j + k = n$ and $u + v + w = 1$. In an analogous way, the classical triangular element is a complete expansion of terms $\xi_1^p \xi_2^q \xi_3^r$, etc., in many of their combinations. The essential difference lies in the fact that the classical elements refer to nodal values, U_i, whereas Bernstein–Bézier triangles refer to nodeless coefficients α_{ijk}.

Furthermore, for each side of the triangle (say AB where $w = 0$ and $k = 0$), the function $U(u, v, w)$ becomes univariate, and as has been previously shown in this chapter (see Sect. 6.2), a Bézier curve is equivalent with the Lagrange polynomial representation. Therefore, along the boundary of the triangular patch, the equivalence between Lagrange polynomials and Bernstein–Bézier representation is evident. It only remains to show that the basis function associated to the control point at the centroid ($B_{111} = 6uvw$) is equivalent with the shape function $N_{10} = 27\xi_1\xi_2\xi_3$ that appears in Eq. (5.39) of Chap. 5. Obviously, they only differ by a factor of $N_{10}/B_{111} = 4.5$.

In general, there is a linear relationship between the Bernstein–Bézier coefficients $\{\alpha\}$ and the nodal values $\{\mathbf{U}_{\text{nodal}}\}$, in the form:

$$\{\alpha\} = [\mathbf{T}]\{\mathbf{U}_{\text{nodal}}\}$$

For the particular case for the cubic triangle, the transformation matrix **[T]** is found to be given by:

$$[\mathbf{T}] = \begin{bmatrix} 1 & 0 & 0 & 0 & 0 & 0 & 0 & 0 & 0 & 0 \\ -\frac{5}{6} & 3 & -\frac{3}{2} & \frac{1}{3} & 0 & 0 & 0 & 0 & 0 & 0 \\ \frac{5}{6} & -\frac{3}{2} & 3 & -\frac{5}{6} & 0 & 0 & 0 & 0 & 0 & 0 \\ 0 & 0 & 0 & 1 & 0 & 0 & 0 & 0 & 0 & 0 \\ -\frac{5}{6} & 0 & 0 & 0 & 3 & 0 & 0 & -\frac{3}{2} & 0 & \frac{1}{3} \\ \frac{1}{3} & -\frac{3}{4} & -\frac{3}{4} & \frac{1}{3} & -\frac{3}{4} & \frac{9}{2} & -\frac{3}{4} & -\frac{3}{4} & -\frac{3}{4} & \frac{1}{3} \\ 0 & 0 & 0 & -\frac{5}{6} & 0 & 0 & 3 & 0 & -\frac{3}{2} & \frac{1}{3} \\ \frac{1}{3} & 0 & 0 & 0 & -\frac{3}{2} & 0 & 0 & 3 & 0 & -\frac{5}{6} \\ 0 & 0 & 0 & \frac{1}{3} & 0 & 0 & -\frac{3}{2} & 0 & 3 & -\frac{5}{6} \\ 0 & 0 & 0 & 0 & 0 & 0 & 0 & 0 & 0 & 1 \end{bmatrix}$$

Therefore, considering that the basis of Bernstein–Bézier is (for example) the *local* one, whereas the basis of the classical cubic element is the *global* one (or vice versa), the obvious relationships between their stiffness and mass matrices will be:

$$\mathbf{K}_{\text{classic}} = \mathbf{T}^t \mathbf{K}_{\text{Bezier}} \mathbf{T} \quad \text{and} \quad \mathbf{M}_{\text{classic}} = \mathbf{T}^t \mathbf{M}_{\text{Bezier}} \mathbf{T}$$

The consequence of the above quadratic forms is that they eventually lead to the same values of eigenvalues. Obviously, the same occurs for Poisson's equation and the elastostatic (stress equilibrium equation) problems as well.

6.9 Recapitulation

It was shown that Lagrange and nonrational Bernstein–Bézier polynomials share the same functional space. Therefore, the stiffness and mass matrices (generally **A**) may be transformed from one system to another using the well-known local-to-global matrix transformation ($\mathbf{T}^t\mathbf{AT}$), where **T** is the rotation matrix. The transformation matrix **T** can be easily calculated, for example, using the theorem that concerns to equal polynomials of the same degree. Moreover, it was shown that in one-dimensional problems, the classical p-method spans the same functional space with that spanned by Lagrange and Bézier polynomials. Therefore, static problems as well as eigenvalue 1D analysis lead to the same numerical result in all these three formulations (Lagrange polynomials, Bernstein polynomials, and Legendre polynomials in the p-method). Since a lot of 2D problems were solved in Chap. 4 using tensor-product elements of Lagrangian type, identical results will be found when solved using tensor-product nonrational Bézierian elements as well. Furthermore, the identity between classical high-order elements and Bernstein–Bézier triangular macroelements was shown, provided that they share the same number of nodes or control points, respectively. Finally, the rational Bézier approximation was fully discussed and was opposed to the nonrational formulations (both for quadratic approximation).

Appendix 1: An Algebraic Proof of de Casteljau Algorithm

Theorem *Given a sequence of control points $P_0 P_1 \ldots P_{n-1} P_n$, when each segment $P_k P_{k+1}$ is divided according to a constant ratio ξ by the new point $P_k^{(1)}$, and this is repeated progressively until, after n steps, a single point $\mathbf{P}(\xi) \equiv \mathbf{P}_0^{(n)}$ remains, it holds*

$$\mathbf{P}(\xi) = \sum_{i=0}^{n} \binom{n}{i} (1-\xi)^{n-i} \xi^i \mathbf{P}_i \tag{6.118}$$

Below, the following notation is used. The initial set of control points is denoted by $P_i^{(0)}$, $i = 0, \ldots, n$. The produced points at the end of the first step are declared as $P_i^{(1)}$ with $i = 0, \ldots, n-1$ (one point less), and generally those at the end of the *kth step as $P_i^{(k)}$ with $i = 0, \ldots, n-k$, and so on. The procedure continues until the final point $P_0^{(n)}$ is produced.*

Proof The proof will be given by mathematical induction. First it will be shown that (6.118) is valid for $n = 2$ points ($P_0 P_1 P_2$). Then, assuming it holds for n points, we shall show it also holds for $n + 1$ ones.

For $n = 2$, given the polygon line $P_0 P_1 P_2$, and setting $P_0^{(0)} = P_0$, $P_1^{(0)} = P_1$, $P_2^{(0)} = P_2$, at the end of the first step we have:

$$\begin{aligned} P_0^{(1)} &= (1-\xi)P_0^{(0)} + \xi\, P_1^{(0)} \\ P_1^{(1)} &= (1-\xi)P_1^{(0)} + \xi\, P_2^{(0)} \end{aligned} \tag{6.119}$$

Then, connecting the points $P_0^{(1)}$ and $P_1^{(1)}$, the final point will be:

$$P_0^{(2)} = (1-\xi)P_0^{(1)} + \xi\, P_1^{(1)} \tag{6.120}$$

Substituting (6.119) into (6.120), one obtains

$$P_0^{(2)} = (1-\xi)\Big[(1-\xi)P_0^{(0)} + \xi\, P_1^{(0)}\Big] + \xi\Big[(1-\xi)P_1^{(0)} + \xi\, P_2^{(0)}\Big],$$

and after the reduction (of second and third term) the final expression becomes

$$P_0^{(2)} = (1-\xi)^2 P_0^{(0)} + 2(1-\xi)\xi\, P_1^{(0)} + \xi^2 P_2^{(0)} \tag{6.121}$$

Omitting the superscripts $^{(0)}$, (6.121) takes the form:

$$P_0^{(2)} = \sum_{i=0}^{2} \binom{2}{i} (1-\xi)^{2-i} \xi^i P_i, \tag{6.122}$$

which is the proof of the theorem for $n = 2$. One may observe in (6.121) that the sequence of points can be taken either as $P_0 P_1 P_2$ (with respect to ξ) or in its inverse direction, as $P_2 P_1 P_0$ (with respect to $\xi' = 1 - \xi$). From the geometrical point of view, this fact was anticipated, since the procedure is not influenced where we start dividing from.

Let us now assume that (6.118) is valid for a sequence of any sequence of n points. Selecting the polygon line $P_0 P_1 \ldots P_{n-1}$ with respect to ξ after $(n-1)$ steps (6.118) will give the point:

$$\mathbf{P}_0^{(n-1)} = \sum_{i=0}^{n-1} \binom{n-1}{i} (1-\xi)^{n-1-i} \xi^i \mathbf{P}_i \tag{6.123}$$

Similarly, selecting the n points for the inverse sequence $P_n P_{n-1} \ldots P_1$ with respect to ξ' (see Fig. 6.21) after $(n-1)$ steps (6.118) will offer the point:

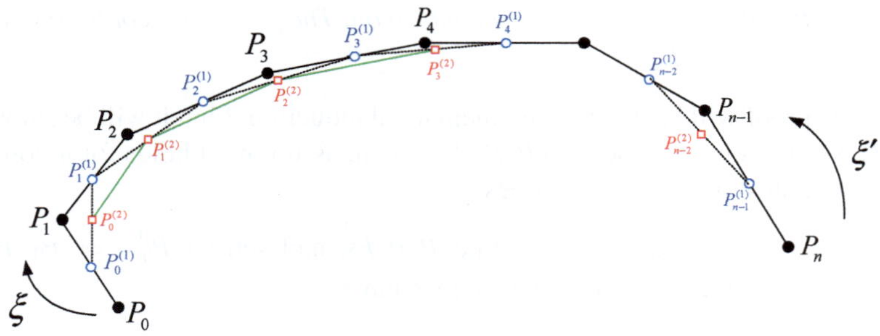

Fig. 6.21 De Casteljau algorithm

$$\mathbf{P}_0'^{(n-1)} = \sum_{i=0}^{n-1} \binom{n-1}{i} (1-\xi')^{n-1-i} \xi'^i \mathbf{P}_i' \tag{6.124}$$

Based on (6.123) and (6.124), which both are valid for n points, we shall show that it also holds for $(n+1)$ points, i.e., for the entire line $P_0 P_1 \dots P_{n-1} P_n$.

Actually, the final point for the entire generator $P_0 P_1 \dots P_{n-1} P_n$ is produced by linear interpolation along the abovementioned segment $\mathbf{P}_0^{(n-1)} \mathbf{P}_0'^{(n-1)}$ which gives

$$\mathbf{P}_0^{(n)} = (1-\xi)\mathbf{P}_0^{(n-1)} + \xi \mathbf{P}_0'^{(n-1)} \tag{6.125}$$

Substituting (6.123) and (6.124) into (6.125), one receives

$$\mathbf{P}_0^{(n)} = (1-\xi) \cdot \sum_{i=0}^{n-1} \binom{n-1}{i} (1-\xi)^{n-1-i} \xi^i \mathbf{P}_i + \xi \cdot \sum_{i=0}^{n-1} \binom{n-1}{i} (1-\xi')^{n-1-i} \xi'^i \mathbf{P}_i' \tag{6.126}$$

Further manipulation of (6.126) in conjunction with the relationship $\xi' = 1 - \xi$ gives

$$\mathbf{P}_0^{(n)} = \sum_{i=0}^{n-1} \binom{n-1}{i} (1-\xi)^{n-i} \xi^i \mathbf{P}_i + \sum_{i=0}^{n-1} \binom{n-1}{i} (1-\xi)^i \xi^{n-i} \mathbf{P}_i' \tag{6.127}$$

Concerning the last variable in (6.127), we can easily notice that

$$\mathbf{P}_i' = \mathbf{P}_{n-i}, \quad i = 0, \dots, n-1 \tag{6.128}$$

Therefore, introducing a change of variable in the second term

$$i' = n - i, \tag{6.129}$$

by virtue of (6.128) and (6.129), (6.127) becomes

$$\mathbf{P}_0^{(n)} = \sum_{i=0}^{n-1} \binom{n-1}{i} (1-\xi)^{n-i} \xi^i \mathbf{P}_i + \sum_{i'=1}^{n} \binom{n-1}{n-i'} (1-\xi)^{n-i'} \xi^{i'} \mathbf{P}_{i'} \quad (6.130)$$

Since i' in the second sum of (6.130) is merely a counter, we can replace it by i, and therefore it becomes:

$$\mathbf{P}_0^{(n)} = \sum_{i=0}^{n-1} \binom{n-1}{i} (1-\xi)^{n-i} \xi^i \mathbf{P}_i + \sum_{i=1}^{n} \binom{n-1}{n-i} (1-\xi)^{n-i} \xi^i \mathbf{P}_i \quad (6.131)$$

But since the two sums involved in (6.131) do not have exactly the same values of the index i, we perform as follows. The first sum in the right-hand side of (6.131) is split in the single term $(1-\xi)^n \mathbf{P}_0$ for $i = 0$, plus another sum for the rest $i = 1, \ldots, n-1$ terms. Also, the second sum in the right-hand side of (6.131) is split in a sum for the terms $i = 1, \ldots, n-1$ plus the remaining term $\xi^n \mathbf{P}_n$ for $i = n$. After these operations, the two sums refer to the same values $i = 1, \ldots, n-1$ and therefore (6.131) is written as

$$\mathbf{P}_0^{(n)} = (1-\xi)^n \mathbf{P}_0 + \sum_{i=1}^{n-1} \left[\binom{n-1}{i} + \binom{n-1}{n-i} \right] (1-\xi)^{n-i} \xi^i \mathbf{P}_i + \xi^n \mathbf{P}_n \quad (6.132)$$

One can easily prove the identity:

$$\binom{n-1}{i} + \binom{n-1}{n-i} \equiv \binom{n}{i} \quad (6.133)$$

Substituting (6.133) in (6.132), and then collecting all terms in a single sum, we can write:

$$\mathbf{P}_0^{(n)} = \sum_{i=0}^{n} \binom{n}{i} (1-\xi)^{n-i} \xi^i \mathbf{P}_i \quad (6.134)$$

Equation (6.134) completes the proof. ■

Additional remarks:

(i) For a given value of the ratio ξ, the entire de Casteljau procedure requires n steps until the final point $\mathbf{P}_0^{(n)}(\xi)$ is obtained. This is the only point of the entire procedure that belongs to the Bézier curve $\mathbf{C}(\xi)$.

(ii) The introduction of the additional point P_n influences n points, those of which the sum of the lower and upper index is equal to n, as follows: $P_{n-1}^{(1)}, P_{n-2}^{(2)}, \ldots, P_1^{(n-1)}, P_0^{(n)}$.

Demonstration for $n = 4$

For the sake of clearness, we additionally present the case of five control points $P_0 P_1 P_2 P_3 P_4$, which corresponds to a generator made of $n = 4$ straight segments

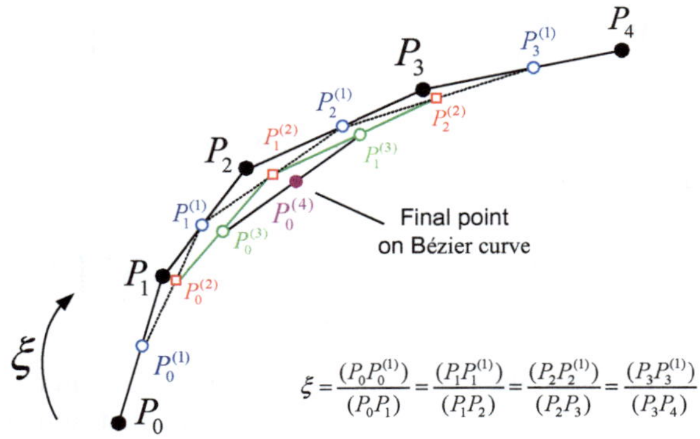

Fig. 6.22 De Casteljau algorithm for quartic polynomial ($n = 4$)

(Fig. 6.22). Starting from the initial state of the line $P_0P_1P_2P_3P_4$, it will be shown that after $n = 4$ steps a point of the corresponding Bézier curve is obtained.

The required steps are as follows:

Step #1: At the end of this step, we have:

$$\left.\begin{aligned} \mathbf{P}_0^{(1)} &= (1-\xi)\mathbf{P}_0 + \xi\mathbf{P}_1, \quad \mathbf{P}_1^{(1)} = (1-\xi)\mathbf{P}_1 + \xi\mathbf{P}_2 \\ \mathbf{P}_2^{(1)} &= (1-\xi)\mathbf{P}_2 + \xi\mathbf{P}_3, \quad \mathbf{P}_3^{(1)} = (1-\xi)\mathbf{P}_3 + \xi\mathbf{P}_4 \end{aligned}\right\} \tag{6.135}$$

Step #2: At the end of this step, we have:

$$\left.\begin{aligned} \mathbf{P}_0^{(2)} &= (1-\xi)\mathbf{P}_0^{(1)} + \xi\mathbf{P}_1^{(1)}, \quad \mathbf{P}_1^{(2)} = (1-\xi)\mathbf{P}_1^{(1)} + \xi\mathbf{P}_2^{(1)} \\ \mathbf{P}_2^{(2)} &= (1-\xi)\mathbf{P}_2^{(1)} + \xi\mathbf{P}_3^{(1)} \end{aligned}\right\} \tag{6.136}$$

Step #3: At the end of this step, we have:

$$\mathbf{P}_0^{(3)} = (1-\xi)\mathbf{P}_0^{(2)} + \xi\mathbf{P}_1^{(2)}, \quad \mathbf{P}_1^{(3)} = (1-\xi)\mathbf{P}_1^{(2)} + \xi\mathbf{P}_2^{(2)} \tag{6.137}$$

Step #4: At the end of this step, we have:

$$\mathbf{P}_0^{(4)} = (1-\xi)\mathbf{P}_0^{(3)} + \xi\mathbf{P}_1^{(3)}. \tag{6.138}$$

First we substitute Eq. (6.135) into (6.136), second the derived expression into (6.137), and third the new expression into (6.138). Therefore, after reduction of common terms the point $\mathbf{P}_0^{(4)}$ is expressed in terms of $\mathbf{P}_0, \mathbf{P}_1, \mathbf{P}_2, \mathbf{P}_3$, and $\mathbf{P}_4$ as follows:

$$\mathbf{P}_0^{(4)} = (1-\xi)^4\mathbf{P}_0 + 4(1-\xi)^3\xi\,\mathbf{P}_1 + 6(1-\xi)^2\xi^2\,\mathbf{P}_2 + 4(1-\xi)\xi^3\,\mathbf{P}_3 + \xi^4\mathbf{P}_4. \tag{6.139}$$

Obviously, Eq. (6.139) is the anticipated expression:

$$\mathbf{P}_0^{(4)} = \sum_{i=0}^{4} \frac{4!}{i!(4-i)!}(1-\xi)^{n-i}\xi^i \mathbf{P}_i. \tag{6.140}$$

Appendix 2: Position of Control Points in Nonrational Univariate Bézier Interpolation Along a Straight Segment

Theorem 6.1 *When the univariate domain* $u \in [0, 1]$ *is a straight segment AB, all the control points lie along AB and uniformly subdivide it.*

Proof First we consider the well-known binomial theorem (e.g., [14]), according to which for a positive integer n and two real variables, X and Y:

$$(X+Y)^n = \sum_{i=0}^{n} \binom{n}{i} X^{n-i} Y^i = \sum_{i=0}^{n} \binom{n}{i} X^i Y^{n-i} \tag{6.141}$$

Taking $X = u$ and $Y = 1 - X$, the first and the third terms of Eq. (6.141) give

$$\sum_{i=0}^{n} \binom{n}{i} u^i (1-u)^{n-i} = 1 \tag{6.142}$$

Equation (6.142) is the well-known "partition of unity" property (otherwise the "rigid body" property) since the function that appears into the symbol of summation is the nonrational Bézier (Bernstein) polynomial $B_{i,n}(u) = \binom{n}{i} u^i (1-u)^{n-i}$ [cf. its definition in Eq. (6.2)].

The determination of the $(n + 1)$ control points may come from the fulfillment of the condition

$$x(u) = \sum_{i=0}^{n} B_{i,n}(u) \cdot x_i^C, \tag{6.143}$$

at $(n + 1)$ arbitrary test positions along the straight segment AB. It can be proven that if the test points are discrete, their choice does *not* influence the control points. However, the most reasonable choice is to subdivide the straight segment AB into n equal subsegments, at the test locations given by the vector

$$\{\mathbf{x}_{\text{test}}\} = \{0, 1/n, 2/n, \ldots, k/n, \ldots, (n-1)/n, 1\} \tag{6.144}$$

One may observe that the endpoints A$(u = 0)$ and B$(u = 1)$ are included as well. The application of Eq. (6.143) to all the locations included in the test vector of Eq.

(6.144) implies:

$$\begin{bmatrix} B_{0,n}(0) & B_{1,n}(0) & \cdots & B_{k,n}(0) & \cdots & B_{n,n}(0) \\ B_{0,n}(\frac{1}{n}) & B_{1,n}(\frac{1}{n}) & \cdots & B_{k,n}(\frac{1}{n}) & \cdots & B_{n,n}(\frac{1}{n}) \\ \vdots & \vdots & \vdots & \vdots & \vdots & \vdots \\ \hline B_{0,n}(\frac{k}{n}) & B_{1,n}(\frac{k}{n}) & \cdots & B_{k,n}(\frac{k}{n}) & \cdots & B_{n,n}(\frac{k}{n}) \\ \hline \vdots & \vdots & \vdots & \vdots & \vdots & \vdots \\ B_{0,n}(1) & B_{1,n}(1) & \cdots & B_{k,n}(1) & \cdots & B_{n,n}(1) \end{bmatrix} \begin{Bmatrix} x_0^C \\ x_1^C \\ \vdots \\ \hline x_k^C \\ \hline \vdots \\ x_n^C \end{Bmatrix} = \begin{Bmatrix} 0 \\ \frac{1}{n} \\ \vdots \\ \hline \frac{k}{n} \\ \hline \vdots \\ 1 \end{Bmatrix} \tag{6.145}$$

Equation (6.145) can be shortly written as

$$[\mathbf{A}]\{\mathbf{x}_C\} = \{\mathbf{x}_{\text{Test}}\} \tag{6.146}$$

By inverting the matrix **[A]**, the coordinate vector of the control points is numerically determined as $\{\mathbf{x}_C\} = [\mathbf{A}]^{-1}\{\mathbf{x}_{\text{Test}}\}$. For the purpose of this Appendix, it is sufficient to prove that the vector $\{\mathbf{x}_C\}$ is equal to $\{\mathbf{x}_C\} = \{0, \frac{1}{n}, \frac{2}{n}, \ldots, \frac{n-1}{n}, 1\}^t$, which here happens to be equal to the chosen particular vector $\{\mathbf{x}_{\text{Test}}\}$. Equivalently, it will be rigorously proven that the vector $\{\mathbf{x}_{\text{Test}}\}$ is a solution of the linear equations system given by Eq. (6.146). More specifically, it will be shown that the product of the arbitrary kth row by the vector $\{\mathbf{x}_{\text{Test}}\}$ implies the term k/n at the right-hand side.

Actually, the abovementioned product is written as

$$\begin{aligned} P(u) &= \sum_{i=0}^{n} B_{i,n}(u) x_i^C \\ &= \sum_{i=0}^{n} \binom{n}{i} u^i (1-u)^{n-i} \frac{i}{n} \end{aligned} \tag{6.147}$$

Also, it can be easily validated that

$$\binom{n}{i}\frac{i}{n} = \binom{n-1}{i-1} \tag{6.148}$$

Substituting Eq. (6.148) into Eq. (6.147), the latter becomes

$$P(u) = \sum_{i=0}^{n} \binom{n-1}{i-1} u^i (1-u)^{n-i} \tag{6.149}$$

The term $P(u)$ in Eq. (6.149) consists of $(n+1)$ terms, of which the first vanishes ($i = 0 \therefore u = i/n = 0$). Considering the transformations

$$n^* = n - 1 \quad \text{and} \quad i^* = i - 1, \tag{6.150}$$

Equation (6.149) is written as

$$P(u) = \sum_{i^*=0}^{n^*} \binom{n^*}{i^*} u^{i^*+1}(1-u)^{n^*-i^*}$$
$$= u \cdot \sum_{i^*=0}^{n^*} \binom{n^*}{i^*} u^{i^*}(1-u)^{n^*-i^*}$$
$$= u \cdot \sum_{i^*=0}^{n^*} B_{i^*,n^*}(u) \tag{6.151}$$

Finally, taking into consideration the partition of unity (Eq. (6.142)), Eq. (6.151) implies the anticipated general expression

$$P(u) = u \tag{6.152}$$

In a different way, for the particular kth row of the matrix **[A]** at which the argument of the Bernstein polynomials is $u = k/n$,

$$P(k/n) = k/n, \tag{6.153}$$

which actually is the right-hand side term in Eq. (6.145). This finding completes the proof of this theorem. ■

Appendix 3: Bernstein–Bézier Triangles

I. Net of control points

A Bernstein–Bézier triangular patch of polynomial degree n, in the three-dimensional space (curvilinear), is based on a net of $\frac{(n+1)(n+2)}{2}$ control points, which are arranged along isolines with constant area coordinates (i.e., "parallel" to the sides of the triangle). For a perfectly uniform net, the area coordinates of these control points will be $u = 0, \frac{1}{n}, \ldots, \frac{n-1}{n}, 1$, $v = 0, \frac{1}{n}, \ldots, \frac{n-1}{n}, 1$, and $w = 0, \frac{1}{n}, \ldots, \frac{n-1}{n}, 1$. The zeros correspond to the sides and the units to the vertices.

The particular case of $n = 3$ is shown in Figs. 6.19 and 6.20. Each of the ten involved control points corresponds to a triplet (u_i, v_j, w_k), with $i, j, k = 0, \ldots, n$. For example, the node 4 is the third one on the side represented by the vector $\overrightarrow{BA}$ toward the u-direction (where the sequence of points along $\overrightarrow{BA}$ is $u_0 = 0, u_1 = \frac{1}{3}, u_2 = \frac{2}{3}, u_3 = 1$); thus, it holds the position $i = 2$ (the index of the third term u_2). Moreover, it holds the second position in the v-direction (i.e., $j = 1$) and the first one in the w-direction (i.e., $k = 0$). In brief, the aforementioned point "4" is described by the triplet of indexes $(i, j, k) = (2, 1, 0)$ or shortly is denoted by $\mathbf{P}_{ijk} = \mathbf{P}_{210}$ (Fig. 6.20a).

The net of the abovementioned control points P_{ijk} is further triangulated. This means that these points are properly connected so as to form n^2 flat subtriangles.

Note Instead of using two indexes as generally used in tensor products, here three indexes are necessary.

II. **Construction of curvilinear patch**

The creation of a curvilinear triangular Bézier patch in the 3D space is very similar to the construction of the Bézier curve. First we note that, for any number of n divisions per side of the triangular patch, there are $3n$ boundary control points and an equal number of $3n$ subtriangles of which the bases belong to the boundary (shown in gray color, Fig. 6.19a). The total number of subtriangles is n^2. The position of a point P in the triangular patch is determined by the triplet of area coordinates (u, v, w) with reference to the outer sides of the patch (AB, BC, and CA). Similarly, for a subtriangle $A'B'C'$ the position of another point P' is again determined by a triplet of *local* area coordinates (u', v', w').

In the de Casteljau construction, we select a constant triplet (u, v, w) for the initial (ABC) as well as for all the subsequent triangles involved, in the same way we had chosen a constant value of ξ for the construction of the Bézier curve. The whole procedure is performed in n steps by connecting homologous points into the continuously produced new triangles until a unique point remains (as happened in the Bézier curve). The procedure is shown with an example for $n = 3$ (see Fig. 6.19).

In the first step (Fig. 6.19a):

$$\begin{aligned} P_1^{(1)} &= uP_1^{(0)} + vP_4^{(0)} + wP_9^{(0)}, \quad & P_2^{(1)} &= uP_5^{(0)} + vP_2^{(0)} + wP_6^{(0)}, \\ P_3^{(1)} &= uP_8^{(0)} + vP_7^{(0)} + wP_3^{(0)}, \quad & P_4^{(1)} &= uP_4^{(0)} + vP_5^{(0)} + wP_{10}^{(0)}, \\ P_2^{(1)} &= uP_{10}^{(0)} + vP_6^{(0)} + wP_7^{(0)}, \quad & P_6^{(1)} &= uP_9^{(0)} + vP_{10}^{(0)} + wP_8^{(0)} \end{aligned} \tag{6.154}$$

where $P_i^{(0)}$, $i = 1, \ldots, 10$ are the control points $P_1 \ldots P_{10}$ in the initial net.

In the second step (Fig. 6.19b), the new homologous points will be:

$$\begin{aligned} P_1^{(2)} &= uP_1^{(1)} + vP_4^{(1)} + wP_6^{(1)}, \quad P_2^{(2)} = uP_4^{(1)} + vP_2^{(1)} + wP_5^{(1)}, \\ P_3^{(2)} &= uP_6^{(1)} + vP_5^{(1)} + wP_3^{(1)} \end{aligned} \tag{6.155}$$

Combining the above relationships (6.154) and (6.155), after manipulation one receives (the upper index $^{(0)}$ is omitted):

$$\begin{aligned} P_1^{(2)} &= u^2P_1 + v^2P_5 + w^2P_8 + 2(uvP_4 + uwP_9 + vwP_{10}) \\ P_2^{(2)} &= u^2P_4 + v^2P_2 + w^2P_7 + 2(uvP_5 + uwP_{10} + vwP_6) \\ P_3^{(2)} &= u^2P_9 + v^2P_6 + w^2P_3 + 2(uvP_{10} + uwP_8 + vw{\cdot}P_7) \end{aligned} \tag{6.156}$$

In the third step (Fig. 6.19c), the final point, $P(u, v, w) = P_1^{(3)}$, is determined as follows:

$$P_1^{(3)} = uP_1^{(2)} + vP_2^{(2)} + wP_3^{(2)} \tag{6.157}$$

Substituting (6.156) into (6.157), after manipulation one eventually obtains:

$$\begin{aligned} P(u, v, w) = & u^3 P_1 + v^3 P_2 + w^3 P_3 \\ & + 3u^2 v\, P_4 + 3uv^2 P_5 + 3v^2 w\, P_6 + 3w^2 v\, P_7 \\ & + 3uw^2 P_8 + 3u^2 w\, P_9 + 3uvw\, P_{10} \end{aligned} \tag{6.158}$$

Equation (6.158) is written in short form by

$$\mathbf{P}(u, v, w) = \sum_{\substack{i,j,k=0 \\ i+j+k=1}}^{3} \frac{n!}{i!\,j!\,k!} u^i v^j w^k \mathbf{P}_{ijk} \tag{6.159}$$

Equation (6.159) is the proof for $n = 3$ and can be easily extended for any other value n. For a rigorous proof, the reader may apply mathematical induction.

References

1. Barnhill R (1985) Surfaces in computer aided geometric design: a survey with new results. Comput Aided Geom Des 2:1–17
2. Cottrell JA, Hughes TJR, Bazilevs Y (2009) Isogeometric analysis: towards integration of CAD and FEA. Wiley, Chichester
3. De Boor C (1987) B-form basics. In: Farin G (ed) Geometric modeling: algorithms and new trends. SIAM, Philadelphia, pp 131–148
4. De Casteljau P (1959) Courbes à poles. National Industrial Property Institute (INPI), France
5. de De Casteljau PF (1999) De Casteljau's autobiography: my life at Citroën. Comput Aided Geom Des 16:583–586
6. Farin G (1986) Triangular Bernstein-Bézier patches. Comput Aided Geom Des 3(2):83–128
7. Farin G (1990) Curves and surfaces for computer aided geometric design: a practical guide. Academic Press, Boston
8. Liu B, Xing Y, Wang Z, Lu X, Sun H (2017) Non-uniform rational Lagrange functions and its applications to isogeometric analysis of in-plane and flexural vibration of thin plates. Comput Methods Appl Mech Eng 321:173–208
9. Natekar D, Zhang X, Subbarayan G (2004) Constructive solid analysis: a hierarchical, geometry-based meshless analysis procedure for integrated design and analysis. Comput Aided Des 36(5):473–486
10. Piegl L, Tiller W (1997) The NURBS book. Springer, New York
11. Provatidis CG (2011) Equivalent finite element formulations for the calculation of eigenvalues using higher-order polynomials. Appl Math 1(1):13–23. https://doi.org/10.5923/j.am.20110101.02. Available online at: http://article.sapub.org/10.5923.j.am.20110101.02.html

12. Provatidis CG (2014) Bézier versus Lagrange polynomials-based finite element analysis of 2-D potential problems. Adv Eng Softw 73:22–34
13. Provatidis CG (2018) Engineering analysis with CAD-based macroelements. Arch Appl Mech 88:121–140
14. Ralston A, Rabinowitz P (2001) A first course in numerical analysis. Dover Publications, Mineola, New York
15. Schillinger D, Ruthala PK, Nguyen LH (2016) Lagrange extraction and projection for NURBS basis functions: a direct link between isogeometric and standard nodal finite element formulations. Int J Numer Meth Eng 108(6):515–534
16. Szabó B, Babuška I (1991) Finite element analysis. Wiley, New York
17. Zieniuk E, Szerszen K (2013) Triangular Bézier surface patches in modeling shape of boundary geometry for potential problems in 3D. Eng Comput 29:517–527

Chapter 7
Interpolation Using B-Splines and Relevant Macroelements

Abstract In this chapter, we present the theoretical background and discuss the numerical performance of CAD-macroelements in which the solution is approximated with B-splines. Univariate as well as two-dimensional tensor-product interpolation will be considered. Not only the usual Curry–Schoenberg B-splines normalized to provide a partition of unity, but also "reduced cardinal B-splines" are studied (to fully explain some older papers of the CAD/CAE group at NTUA since 1989). Numerical experiments of this chapter restrict to the Galerkin–Ritz formulation and refer to domains or structures of simple 2D primitive shapes (rectangles, circles, and ellipses). The numerical analysis is performed using a single macroelement only, without domain decomposition. The results are also compared with the FEM solution of the same mesh density.

Keywords B-splines · Truncated power series · Reduced cardinal B-splines
De Boor formulation · MATLAB · Tensor product · Solved problems

Nomenclature

ξ	Normalized coordinate ($\xi = x/L$), with $\xi \in [0, 1]$ and $x \in [0, L]$
n	Number of breakpoint spans in the interval $\xi \in [0, 1]$
$(n + 1)$	Number of breakpoints
p	Degree of polynomial
$(n + p)$	Number of coefficients in truncated power series formulation [for $C^{(p-1)}$-continuity]
ϕ_i, U_i	Shape functions and nodal displacements, respectively, based on $(n + 3)$ test points
$\bar{\phi}_i, \bar{U}_i$	Reduced shape functions and nodal displacements, respectively, associated to the breakpoints only. Based on the *first* derivative (slope) at the two ends
$\hat{\phi}_i, \hat{U}_i$	Reduced shape functions and nodal displacements, respectively, associated to the breakpoints only. Based on the *second* derivative (curvature) at the two ends

C. G. Provatidis, *Precursors of Isogeometric Analysis*, Solid Mechanics and Its Applications 256, https://doi.org/10.1007/978-3-030-03889-2_7

$(n_p + 1)$ Number of control points $(P_0, P_1, \ldots, P_{n_p})$
$N_{i,p}(\xi)$ De Boor basis functions, $i = 0, \ldots, n + p - 1$

7.1 One-Dimensional Problems

7.1.1 General

In Chap. 2, the basics of spline interpolation (using B-splines) were fully explained for one-dimensional problems. In a sense, B-splines are the splines of minimal support for a given degree. We have seen that for a nondecreasing sequence of breakpoints $(\xi_0, \xi_1, \xi_2 \ldots \xi_{n-2}, \xi_{n-1}, \xi_n)$, there are two basic formulations as follows.

I. *Truncated power series formulation*
This is the prior to Schoenberg formulation to which he gave the name "B-spline" [27]. The approximate solution $\hat{U}(\xi)$ is set in the form of a series expansion using a complete polynomial of degree p plus certain truncated monomials $< x - x_i >^q$ within the breakpoint spans: $(\xi_1, \xi_n), \ldots, (\xi_{n-1}, \xi_n)$. The power q depends on the continuity adopted. For the particular case of C^{p-1}-continuity (multiplicity of inner breakpoints: $\lambda = 1$), the full expression is given by:

$$U(\xi) = a_0 + a_1\xi + a_2\xi^2 + \cdots + a_p\xi^p + \sum_{i=1}^{n-1} b_i \langle \xi - \xi_i \rangle^p \tag{7.1}$$

Between any two successive breakpoints, the approximate solution is a polynomial of degree p; thus, we have a *piecewise* polynomial interpolation. While the number of breakpoints is $(n+1)$, in case of C^{p-1}-continuity the number of unknown coefficients in Eq. (7.1) is $(n + p)$. For example, if the interval [0, 1] is divided into four equal segments (i.e., $n = 4$ with five breakpoints) and the polynomial degree is taken $p = 3$, the number of coefficients will be seven (because $n + p = 4 + 3 = 7$).

II. *Curry–Schoenberg (de Boor) formulation*
Curry–Schoenberg [6] proposed the series expansion in a different (equivalent) but much more convenient set of noncardinal basis functions $N_{i,p}(\xi)$, $i = 0, \ldots, n+p-1$, which have some important advantages. De Boor [8] developed a robust and efficient numerical procedure to calculate them.

Among others, we can distinguish the following properties:

(i) The first half of the bell-shaped functions $N_{i,p}$ is *symmetric* (mirrors) of the rest half ones, with respect to the middle $\left(\xi = \frac{1}{2}\right)$ of the interval [0, 1].
(ii) The functions $N_{i,p}$ have the property of *local support*. Each of them is multiplied by a different unknown coefficient α_i and is associated to the *control point* P_i.

(iii) Given the breakpoints, the knot vector $\mathbf{U} = \{u_0, u_1, \ldots, u_m\}$ is easily computed, and then the abovementioned local support is as follows: $N_{i,p}(\xi) = 0$ if ξ is *outside* the interval $[u_i, u_{i+p+1})$. Moreover, concerning the intervals where the first derivative $\left(N'_{i,p}\right)$ vanishes, the same property can be applied to the formula:

$$N'_{i,p} = \frac{p}{u_{i+p} - u_i} N_{i,p-1} - \frac{p}{u_{i+p+1} - u_{i+1}} N_{i+1,p-1}$$

(iv) The functions $N_{i,p}$ have the *partition of unity* property (rigid-body), i.e., $\sum_{i=0}^{n+p-1} N_{i,p}(\xi) \equiv 1$.
(v) The basis functions associated to the ends of the interval are equal to the unity: $N_{0,p}(0) = 1$ and $N_{p+n-1,p}(1) = 1$.
(vi) The basis functions can be efficiently calculated by simple recursions; thus, no matrix inversion is required.

7.1.2 *Truncated Power Series Formulation: Full and Reduced Cardinal B-Spline Shape Functions*

Full formulation
Most finite element researchers are familiar with the interpolation which is based on the $(q + 1)$ *nodal* values U_i at the nodal points of the element:

$$\hat{U}(\xi) = \sum_{i=0}^{q} \phi_i(\xi) \cdot U_i, \tag{7.2}$$

where the shape functions fulfill the cardinality conditions:

$$\phi_i(\xi_j) = \delta_{ij} = \begin{cases} 1, \ i = j \\ 0, \ i \neq j \end{cases} \tag{7.3}$$

The question that here arises is whether B-splines can be adjusted to the abovementioned scheme, i.e., to determine the *cardinal* $\phi_i(\xi)$ shape functions, and the answer is "yes" under certain modifications.

To be more specific, let us divide the interval [0, 1] into n segments, for example, putting the breakpoints at uniform positions:

$$\xi_0 = 0, \quad \xi_1 = \tfrac{1}{n}, \ldots \xi_{n-1} = \tfrac{n-1}{n}, \quad \xi_n = 1. \tag{7.4}$$

For the sake of convenience, let us also write Eq. (7.1) in the shorter form:

$$U(\xi) = \sum_{i=0}^{q} \tilde{B}_i(\xi) \cdot \alpha_i, \tag{7.5}$$

where

$$q = p + n - 1, \tag{7.6}$$

while the basis functions form a row vector $\tilde{\mathbf{B}} = [\tilde{B}_0, \ldots, \tilde{B}_{n+p}]$ including the following elements:

$$\left.\begin{array}{l} \tilde{B}_0(\xi) = 1,\ \tilde{B}_1(\xi) = \xi, \ldots, \tilde{B}_p(\xi) = \xi^p, \\ \tilde{B}_{p+1}(\xi) = \langle \xi - \xi_1 \rangle^p, \ldots, \tilde{B}_{p+n}(\xi) = \langle \xi - \xi_{n-1} \rangle^p. \end{array}\right\} \tag{7.7}$$

Now, it is obvious that the coefficients α_i, $i = 0, \ldots, q$ in Eq. (7.5) stand for both sets of coefficients (a_i and b_i) that appear in Eq. (7.1).

Let us now write Eq. (7.5) to all the (n + 1) breakpoints which are included in Eq. (7.4), thus obtaining the equations system:

$$\begin{bmatrix} U(\xi_0) \\ \vdots \\ U(\xi_n) \end{bmatrix} = \begin{bmatrix} \tilde{B}_0(\xi_0) & \ldots & \tilde{B}_q(\xi_0) \\ \vdots & \ldots & \vdots \\ \tilde{B}_0(\xi_n) & \ldots & \tilde{B}_q(\xi_n) \end{bmatrix} \cdot \begin{bmatrix} \alpha_0 \\ \vdots \\ \alpha_q \end{bmatrix}, \tag{7.8}$$

Since the matrix in the right part of Eq. (7.8) is *nonsquare*, with dimensions $(n+1) \times (q+1) \equiv (n+1) \times (n+p)$, we have to write Eq. (7.5) at $(p-1)$ *additional* (test) points within the interval (0, 1) since the endpoints have been already used. Then the total number of test points will become (n + p), as required.

The extension of Eq. (7.8) at the abovementioned (n + p) test points is written in matrix form as follows:

$$\mathbf{d} = \mathbf{A}\boldsymbol{\alpha}, \tag{7.9}$$

where the vector containing the degrees of freedom is:

$$\mathbf{d} = [\underbrace{U(\xi_0), \ldots, U(\xi_n)}_{\text{breakpoints}}, \underbrace{d_{n+1} \ldots, d_{n+p-1}}_{\text{extra points}} \equiv d_q]^{\mathrm{T}}, \tag{7.10}$$

while the vector containing the coefficients is:

$$\boldsymbol{\alpha} = [\alpha_0, \ldots, \alpha_q]^{\mathrm{T}}. \tag{7.11}$$

Alternatively, for a straight segment the test points could be chosen to be quite different than the breakpoints, for example, at equidistant positions, thus defining a vector of different degrees of freedom:

$$\mathbf{d} = [U_0, U_1, \ldots, U_q]^{\mathrm{T}}, \tag{7.12}$$

In any case, solving Eq. (7.9) in $\boldsymbol{\alpha}$ and then substituting the result ($\boldsymbol{\alpha} = \mathbf{A}^{-1}\mathbf{d}$) into Eq. (7.5), one progressively obtains:

$$U(\xi) = \left[\tilde{B}_0(\xi), \ldots, \tilde{B}_q(\xi)\right] \cdot \begin{bmatrix} \alpha_0 \\ \vdots \\ \alpha_q \end{bmatrix} = \underbrace{\left(\left[\tilde{B}_0(\xi), \ldots, \tilde{B}_q(\xi)\right] \cdot \mathbf{A}^{-1}\right)}_{\text{Shape Functions}} \cdot \mathbf{d},$$

and, finally:

$$U(\xi) = \left[\phi_0(\xi), \ldots, \phi_q(\xi)\right] \cdot \mathbf{d} \tag{7.13}$$

The result described by Eq. (7.13) is here described by the term "*full*" cardinal B-splines (shape) functions, because the *entire* matrix $\mathbf{A}^{-1}$ is utilized.

Example 7.1 (Cubic B-splines in several normalizations) Let us consider the interval [0, 1], which is uniformly divided into $n = 3$ breakpoint spans, using four breakpoints (1–4) as shown below the line in Fig. 7.1a, b. We assume cubic B-splines ($p = 3$).

For this configuration, two sets of test points were studied as follows:

(i) The first set of test points refers to the six equidistant points shown in the upper part of Fig. 7.1a. The vector of DOF refers to equidistant test points and is $\mathbf{d} = [U(0), U(\frac{1}{5}), U(\frac{2}{5}), U(\frac{3}{5}), U(\frac{4}{5}), U(1)]^{\mathrm{T}}$. The corresponding shape functions are shown in Fig. 7.2a.

(ii) The second set of test points refers to all the breakpoints $\xi = 0, \frac{1}{3}, \frac{2}{3}, 1$, plus two extra test points at $\xi = \frac{1}{9}$ and $\xi = \frac{8}{9}$. The reader may later show that all

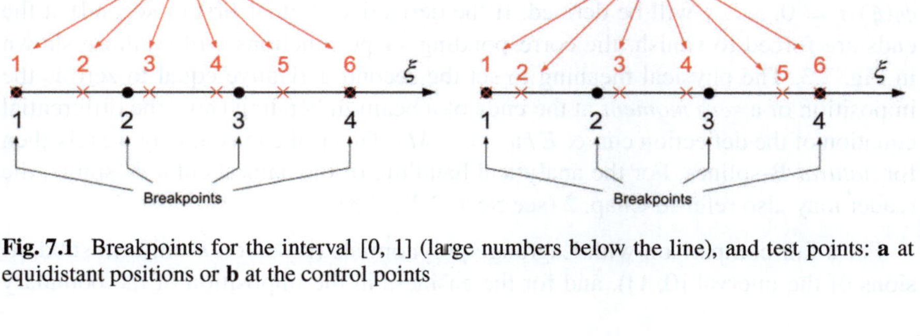

Fig. 7.1 Breakpoints for the interval [0, 1] (large numbers below the line), and test points: **a** at equidistant positions or **b** at the control points

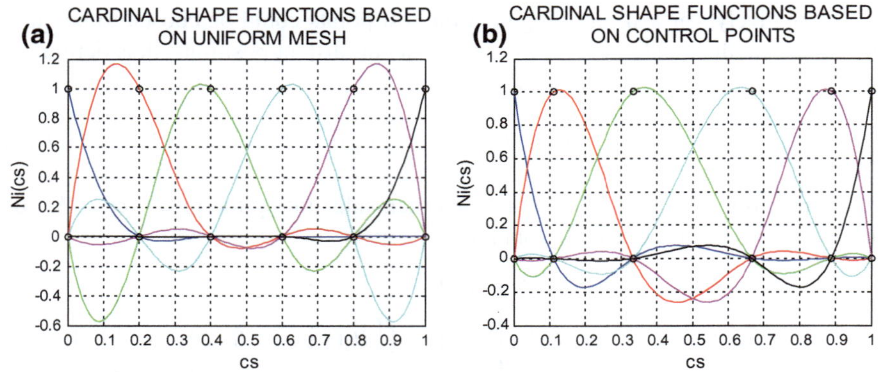

Fig. 7.2 Full cubic B-spline cardinal shape functions for $n=3$ breakpoint spans (four breakpoints), based on six test points located at **a** uniform positions and **b** the control points $(0, \frac{1}{9}, \frac{1}{3}, \frac{2}{3}, \frac{8}{9}, 1)$

these six test points are the control points (in the sense of Curry–Schoenberg splines, also adopted by de Boor). Now, the vector of DOF at the control points is $\mathbf{d} = [U(0), U(\frac{1}{9}), U(\frac{1}{3}), U(\frac{2}{3}), U(\frac{8}{9}), U(1)]^{\mathrm{T}}$. The corresponding shape functions are shown in Fig. 7.2b.

One may observe (in Fig. 7.2) that although the condition of *cardinality* [cf. Eq. (7.3)] is fulfilled at *all test points*, the values of the oscillating shape functions exceed the unity at neighboring places, as happens with the Lagrange polynomials as well (see Fig. 6.5). Again, although we started with four breakpoints, the number of coefficients involved in Eq. (7.1) is six, while the shape of these six shape functions highly depends on the position of the test points (at the circles of the ξ-axis where they always vanish). Both sets are mathematically *equivalent* one another, because they utilize the *same* basis functions [given by Eq. (7.7)]. Merely, the same vector-row $\tilde{\mathbf{B}} = [\tilde{B}_0, \ldots, \tilde{B}_{n+p}]$ is multiplied by the columns of a different matrix $\mathbf{A}^{-1}$.

Reduced formulation
For simplicity, let us remain in the case of cubic B-splines. Instead of using two extra test points than the number of breakpoints (as happened above), we can utilize the first or the second derivatives of the series expansion shown in Eq. (7.5)—completed by Eq. (7.7)—at the two ends. Then, for each choice a different set of shape functions $\phi_i(\xi), i = 0, \ldots, q$ will be derived. If the derivatives (either first or second) at the ends are forced to vanish, the corresponding shape functions will be those shown in Fig. 7.3. The physical meaning to set the second derivative equal to zero is the imposition of a *zero moment* at the ends of a beam in bending (note the differential equation of the deflection curve: $EIw'' = -M$). This is the reason why we talk then for *natural* B-splines. For the analytical handling of the natural cubic B-spline, the reader may also refer to Chap. 2 (see Sect. 2.3.7.3.5).

For a fair comparison with Lagrange polynomials (based on n uniform subdivisions of the interval [0, 1]), and for the easiness in the imposition of the boundary

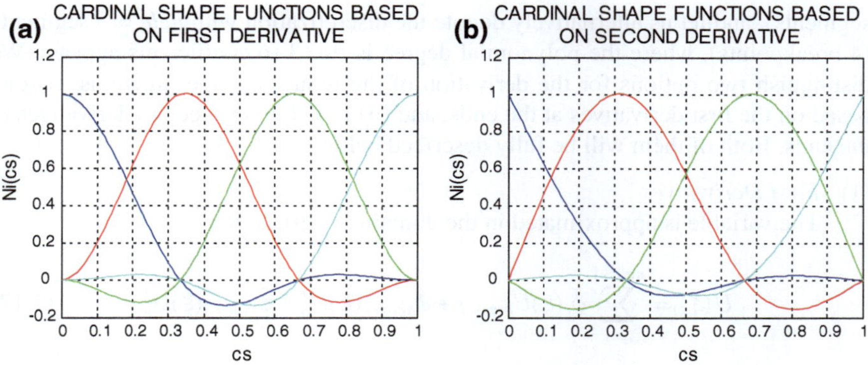

Fig. 7.3 Reduced cubic B-spline shape functions, for ($n = 3$) breakpoint spans, based on vanishing **a** the first and **b** the second derivative (*natural* B-splines) at the ends

conditions, it is our intension to use a reduced vector of the degrees of freedom, strictly related to the $(n + 1)$ *breakpoints* only:

$$\tilde{\mathbf{d}} = [U_0, \ldots, U_n]^{\mathrm{T}}. \tag{7.14}$$

Before we reach this goal, we note that if the *second* derivative is obtained at the two ends, we actually work with a hypothetical vector of $(n + p) \equiv (n + 3)$ elements:

$$\hat{\mathbf{d}} = [U_0, \ldots, U_n, U_0'', U_n'']^{\mathrm{T}}. \tag{7.15}$$

In this way, the $(p - 1)$ missing equations are completed. The total set of the $(n+3)$ shape functions is derived in a very similar way, and the so-obtained shape functions are again full. In the general case, the determination of the second derivatives, U_0'' and U_n'', is part of the problem.

However, if the two curvatures at the ends are forced (a priori) to vanish ($d_{n+1} = U_0'' = 0$ and $d_{n+2} = U_n'' = 0$), then the last two columns in the matrix $\mathbf{A}^{-1}$ are de facto ignored (are multiplied by the two curvatures of zero value). In this case, we receive the *reduced* shape functions. A typical graph for the same breakpoints as those in Figs. 7.1 and 7.2 ($n = 3$, $p = 3$) are now shown in Fig. 7.3b.

The above procedure may be repeated for the vector that includes the *first* derivatives at the two ends:

$$\bar{\mathbf{d}} = [U_0, \ldots, U_n, U_0', U_n']^{\mathrm{T}}, \tag{7.16}$$

and now, the associated set of reduced shape functions is shown in Fig. 7.3a.

Let us now see more closely the above produced reduced but cardinal shape functions, which do *not* exceed the unity. In order to avoid confusion with indexes (here we start from 1 and finish to *nele* + 1, where *nele* is the number of breakpoint segments), and let us alternatively denote the discretization with $nele = 3$ segments (4 breakpoints), where the polynomial degree is $p = 3$ (6 coefficients appear). We distinguish two options for the derivation of the reduced shape functions: (i) one based on the first derivatives at the ends, and (ii) based on the second derivatives at the ends. Both of them will be fully described below.

(1) *First Derivatives*
The variable is approximated in the domain $\xi \in [0, 1]$ by:

$$U(\xi) = \sum_{i=1}^{nele+1} \bar{\phi}_i(\xi) U_{\text{break},i} + \bar{\phi}_{nele+2}(\xi) U_0' + \bar{\phi}_{nele+3}(\xi) U_n' \tag{7.17}$$

with

$$\begin{aligned}
\bar{\phi}_1(\xi) &= 1 - 19.8\xi^2 + 32.4\xi^3 - 48.6\langle \xi - \tfrac{1}{3} \rangle^3 + 21.6\langle \xi - \tfrac{2}{3} \rangle^3 \\
\bar{\phi}_2(\xi) &= \quad 25.2\xi^2 - 48.6\xi^3 + 86.4\langle \xi - \tfrac{1}{3} \rangle^3 - 59.4\langle \xi - \tfrac{2}{3} \rangle^3 \\
\bar{\phi}_3(\xi) &= \quad -7.2\xi^2 + 21.6\xi^3 - 59.4\langle \xi - \tfrac{1}{3} \rangle^3 + 86.4\langle \xi - \tfrac{2}{3} \rangle^3 \\
\bar{\phi}_4(\xi) &= \quad 1.8\xi^2 - 5.4\xi^3 + 21.6\langle \xi - \tfrac{1}{3} \rangle^3 - 48.6\langle \xi - \tfrac{2}{3} \rangle^3
\end{aligned}$$

$$\begin{aligned}
\bar{\phi}_5(\xi) &= 1.0\xi - 5.2\xi^2 + 6.6\xi^3 - 8.4\langle \xi - \tfrac{1}{3} \rangle^3 + 2.4\langle \xi - \tfrac{2}{3} \rangle^3 \\
\bar{\phi}_6(\xi) &= -0.2\xi^2 + 0.6\xi^3 - 2.4\langle \xi - \tfrac{1}{3} \rangle^3 + 8.4\langle \xi - \tfrac{2}{3} \rangle^3
\end{aligned}$$

One may easily validate that:

$$\bar{\phi}_1(\xi) + \bar{\phi}_2(\xi) + \bar{\phi}_3(\xi) + \bar{\phi}_4(\xi) \equiv 1, \quad \forall \xi \in [0, 1] \tag{7.18}$$

Also, one may observe that:

(i) In the first four shape functions, "*the term* ξ *is missing*"; thus, the terminology "reduced" basis is justified.
(ii) None of the four shape functions $\bar{\phi}_1(\xi), \ldots, \bar{\phi}_4(\xi)$ exceeds the unity.

Again, the first four shape functions (above reduced set) are shown in Fig. 7.3a, whereas the rest two $(\bar{\phi}_5(\xi)$ and $\bar{\phi}_6(\xi))$ in Fig. 7.4a.

(2) *Second Derivatives*
The variable is approximated by:

$$U(x) = \sum_{i=1}^{nele+1} \hat{\phi}_i(\xi) U_{\text{break},i} + \hat{\phi}_{nele+2}(\xi) U_0'' + \hat{\phi}_{nele+3}(\xi) U_n'' \tag{7.19}$$

with

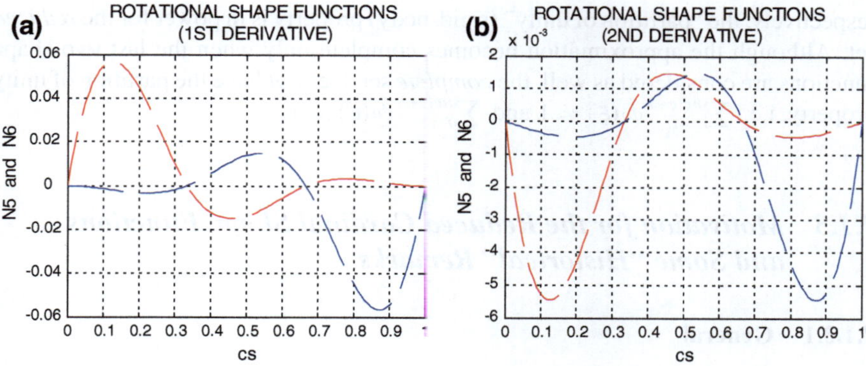

Fig. 7.4 Shape functions associated to rotational DOF at the endpoints, i.e., based on **a** first derivatives (slopes) and **b** second derivatives (curvatures)

$$\left|\begin{aligned}
\hat{\phi}_1(\xi) &= 1 - 3.8\xi + 7.2\xi^3 - 16.2\langle \xi - \tfrac{1}{3}\rangle^3 + 10.8\langle \xi - \tfrac{2}{3}\rangle^3 \\
\hat{\phi}_2(\xi) &= \quad 4.8\xi - 16.2\xi^3 + 43.2\langle \xi - \tfrac{1}{3}\rangle^3 - 37.8\langle \xi - \tfrac{2}{3}\rangle^3 \\
\hat{\phi}_3(\xi) &= \quad -1.2\xi + 10.8\xi^3 - 37.8\langle \xi - \tfrac{1}{3}\rangle^3 + 43.2\langle \xi - \tfrac{2}{3}\rangle^3 \\
\hat{\phi}_4(\xi) &= \quad 0.2\xi - 1.8\xi^3 + 10.8\langle \xi - \tfrac{1}{3}\rangle^3 - 16.2\langle \xi - \tfrac{2}{3}\rangle^3
\end{aligned}\right.$$

$$\begin{aligned}
\hat{\phi}_5(\xi) &= -0.0963\xi + 0.5\xi^2 - 0.63333\xi^3 + 0.8\langle \xi - \tfrac{1}{3}\rangle^3 - 0.2\langle \xi - \tfrac{2}{3}\rangle^3 \\
\hat{\phi}_6(\xi) &= -0.0037\xi \qquad + 0.0333\xi^3 - 0.2\langle \xi - \tfrac{1}{3}\rangle^3 + 0.8\langle \xi - \tfrac{2}{3}\rangle^3
\end{aligned}$$

One may easily validate that:

$$\hat{\phi}_1(\xi) + \hat{\phi}_2(\xi) + \hat{\phi}_3(\xi) + \hat{\phi}_4(\xi) \equiv 1, \quad \forall \xi \in [0, 1] \tag{7.20}$$

Also, one may observe that:

(i) In the first four shape functions, "*the term* ξ^2 *is missing*"; thus, the terminology "reduced" basis is justified. Essentially, the assumption of zero curvatures (second derivatives) at the two ends means that the piecewise bending of an interpolating beam suffers *vanishing bending moments* (but not vanishing shear forces) at its two ends. The latter condition seems to be "natural," and therefore, we talk about "natural B-splines."

(ii) None of the four shape functions $\hat{\phi}_1(\xi), \ldots, \hat{\phi}_4(\xi)$ exceeds the unity.

(iii) Again, the first four shape functions (above reduced set) are shown in Fig. 7.3b, whereas the rest two ($\hat{\phi}_5(\xi)$ and $\hat{\phi}_6(\xi)$) in Fig. 7.4b.

Conclusion: Considering either of the above two expressions [Eq. (7.17) or (7.19)] in conjunction with vanishing endpoints rotational DOF, (U'_0, U'_1) or (U''_0, U''_1), respectively, the "partition of unity" (rigid-body) property is in power for the *reduced* set. Although the approximation becomes complete only when the last two shape functions are considered as well, the *complete* set does *not* have the partition of unity property, i.e., $\sum_{i=1}^{nele+3} \bar{\phi}_i(\xi) \neq 1$ and $\sum_{i=1}^{nele+3} \hat{\phi}_i(\xi) \neq 1$.

7.1.3 Motivation for the Reduced Cardinal Shape Functions and Some "Historical" Remarks

7.1.3.1 General

The remarks of this subsection are mostly a self-criticism of our past work, which was part of the activities within the CAD/CAE Group of NTUA (Department of Mechanical Engineering).

The abovementioned procedure of creating several univariate B-splines was not an end in itself. Clearly, in our particular engagement toward the development of Gordon–Coons macroelements (see Chaps. 3 and 4), given the *nele* ($=n$) subdivisions along a side of the quadrilateral patch (say AB where $\eta = 0$), we had necessarily to deal with the univariate function $U(\xi)$ into the interval [0, 1]. And our primary purpose was to replace Lagrange polynomials with the more stable cubic B-splines.

In more details, given a number of nodal points at which the function $U(\xi)$ is known, i.e., in pairs $(\xi_0, U_0), \ldots, (\xi_n, U_n)$, B-splines based interpolation is preferred because it is widely accepted that the Lagrange polynomials may fail, due to numerical oscillations. This issue has been clearly shown through the *Runge example* ([9], p. 17). In our particular case of Gordon–Coons macroelements, we wished to replace Lagrange polynomials with cubic B-splines. Due to the fact that the number of coefficients in cubic B-spline is $(n+3)$, which is by two higher than the number of $(n+1)$ breakpoints, raises the question how this replacement will be *fairly* done.

The first thought was to use the same number of parameters for both the Lagrange polynomials and the B-spline approximation. In other words, since the aforementioned *nele* ($=n$) subdivisions determine $(n + 1)$ nodes, a fair replacement would require $(n + 1)$ breakpoints. This choice, however, is unfair for the B-splines, since we do not exhaust all their capabilities in performing high accuracy.

The dilemma (second thought) was whether it was better to implement all $(n+3)$ involved shape functions at $(n + 3)$ uniformly distributed trial points along the true boundary, but it was found to be rather unfair for Lagrange polynomials, particularly in case where the boundary is curved (e.g., along a circular arc) where the B-splines a priori would better since they would use more test points along the curve.

Anyhow, the urgent problem we had in 1984 was *not* to implement the best B-spline formulation but rather to study the performance of Coons–Gordon macroelements in engineering analysis (see details in Chaps. 3 and 4).

7.1.3.2 Past Activities and Publications

In order to justify the reason of the aforementioned B-spline formulations, I ought to give some explanations about the origin of the relevant research.

In early 1980s and systematically since the summer 1984, the CAD/CAE group within the Department of Mechanical Engineering at NTUA implemented the idea of generating large (CAD-based) finite elements which would occupy an entire Coons patch. At that time, we were all fully aware about the "old-fashioned" truncated power series splines (before Schoenberg) as well as about the most "modern" Curry–Schoenberg (adopted also by de Boor) splines. Both of them were described in the book of Böhmer [3] in which FORTRAN codes were attached.

Before we start developing our computer code, we had read about a subclass of splines called *natural splines*, already studied by Greville [13]. Among several options, these splines may be produced considering zero derivatives at the ends of the interval (see also Chap. 2 of this book). Another reference to a natural spline is found in the book of Faux and Pratt [12]. Also, the term *cardinal* splines is met in the book of Ahlberg et al. [1]. These works were closer to our comprehension of cardinal shape functions, local, or global, than those of Curry–Schoenberg [see details in de Boor [9], revised in 2001] who was referring to "control points," noncardinal basis functions, and generalized coefficients (not having a clear physical meaning, except that it was the value at another point in the neighborhood of the associated control point). Particularly, the imposition of the boundary conditions was a fear.

In 1981, we were already working with Taylor finite element code FEAP [29], which we had extended with time-integration (explicit, implicit) and eigenvalue (Bathe's subspace iteration) subroutines. Later, probably in 1984, FEAP [29] was translated in FORTRAN 77 and was migrated first on a Stride and then on a MicroVax-II workstation (4 MB RAM) where it remained until 1989. At the same time, FEAP was modified so as to be capable of coupling finite elements (FEM) with boundary element (BEM) macroelements. Research was conducted simultaneously toward FEM and BEM, either independently or on their coupling.

Since a true (usually complex) structure has to be divided into a certain number of large CAD-based (e.g., Coons–Gordon) macroelements, and probably we had to perform FEM/BEM coupling inside the same computer codes, we preferred to deal with physical nodal values U_i instead of the nodeless coefficients α_i involved in Curry–Schoenberg (de Boor) formulation. This decision was taken for the following reasons. First, in this way the FEAP environment is practically preserved and modifications are kept in the lowest level affecting some dimension or common FORTRAN commands. Second, the novel Coons–Gordon macroelements are fairly compared either with conventional finite elements or with boundary elements, at the *same* nodal points.

Therefore, in a period where the best practice was to comply with a standard academic finite element program, like Taylor FEAP (see [29]), it was decided to comply with cardinal (i.e., of [0, 1]-type) and also *bounded* shape functions, without oscillations and without exceeding the unity at all, as those described above (in Fig. 7.3b). The reviewing character of this book justifies the term "*reduced set*", which is used in this chapter for the first time. In conclusion, in the beginning of 1980s the choice of the *reduced* B-spline cardinal shape functions was found to be the most appropriate conservative draft solution (remedy) to extend a typical finite element code such as Taylor FEAP (in our case).

Based on the abovementioned reduced (*natural*) cubic B-spline cardinal shape functions (in conjunction with $U_0'' = U_n'' = 0$), a number of papers were released much later [19, 20, 22, 24], among others in refereed journals and many conference proceedings; the complete list is given in Chap. 14. In these, as well as other papers of our group [25] till Provatidis [26], whenever the term "*cubic B-splines*" or "*cardinal cubic B-splines*" is met, we mean the abovementioned reduced natural and cardinal cubic B-splines, in conjunction with zero second derivatives at the ends $\left(U_0'' = U_n'' = 0\right)$.

7.1.4 De Boor Univariate Formulation

De Boor [8] [see Fig. 7.5], and at the same time Cox [5], proposed a formulation of B-splines which *does* have a compact support. A further advantage of de Boor formulation is the efficient and stable computation that does not require any matrix inversion (as those of matrix $\mathbf{A}^{-1}$ mentioned in the previous subsections) but only recursions. These recursions start from trivial step functions, and then they continue with ramps (linear), quadratic, and so on until the relevant polynomial degree p is reached. Details and FORTRAN 77 codes may be found in de Boor (1978) and later in the second edition of his popular book [9]. Later his codes were embedded in MATLAB® through the famous `spcol` function among others.

Fig. 7.5 Professor Carl-Wilhelm Reinhold de Boor in 2004 (his Courtesy)

In order to be consistent and make easier the communication with the isogeometric (IGA) community which has adopted rather the book of Piegl and Tiller [21] as a standard, henceforth we shall use the expression

$$U(\xi) = \sum_{i=0}^{n_p} N_{i,p}(\xi) \cdot a_i, \tag{7.21}$$

where the symbol $N_{i,p}(\xi)$ refers to the basis functions (`spcol` function in MATLAB®), with n_p denoting the number of control points.

Control points

The control points were developed in Citroën and patented by de Casteljau ([10,11]). Also, they were independently developed by Bézier (before 1970) who was working for many years in Renault. Not surprisingly, in an older book on splines (such as [1]) the term "control point" is *not* included in its index. Concerning B-splines, Curry–Schoenberg [6] do not need and do not refer to the control points at all. The introduction of the control points in the definition and mathematical treatment of B-splines can be found much later into the pioneering book of de Boor (1978), particularly in its revised version [9].

In the design phase of a component, the control points of its geometric model are given and can be further used for the analysis, in an iterative cycle. If however the control points are unknown but the geometry is fully prescribed, we can work as follows. First we determine the number n_p of coefficients that contribute in the series expansion given by Eq. (7.21). Then we easily construct a uniform mesh of also n_p test points (so many as exactly the number of the aforementioned coefficients). For each of these test points, we write

$$\mathbf{x}(\xi) = \sum_{i=0}^{n_p} N_{i,p}(\xi) \cdot \mathbf{x}_{Pi}, \tag{7.22}$$

where $\mathbf{x} = [x, y]^{\mathrm{T}}$ is the vector of the Cartesian coordinates. In this way, a linear system of $(n_p + 1)$ equations is constructed. The solution of this system, once for the x- and then for the y-components, determines the desired coordinates $\mathbf{x}_{Pi}$ of the control points.

It should be clarified that in the above procedure, "the position of the control points depends on the selection of the test points." If the curve is a straight-line segment, then the associated control points belong to this segment. Furthermore, concerning not straight curves, except the parabola (which is accurately represented by a polynomial of second degree), in all other cases a deviation generally appears between the B-spline interpolation, expressed by Eq. (7.22), and the exact curve. By construction, the only positions where no deviation appears are the aforementioned trial points because they have been chosen so as to belong to the true curve.

For the sake of brevity, we do not repeat issues which the reader may find in Chap. 2, or study on the prototype [9]. Nevertheless, in Fig. 7.6 we present the same case

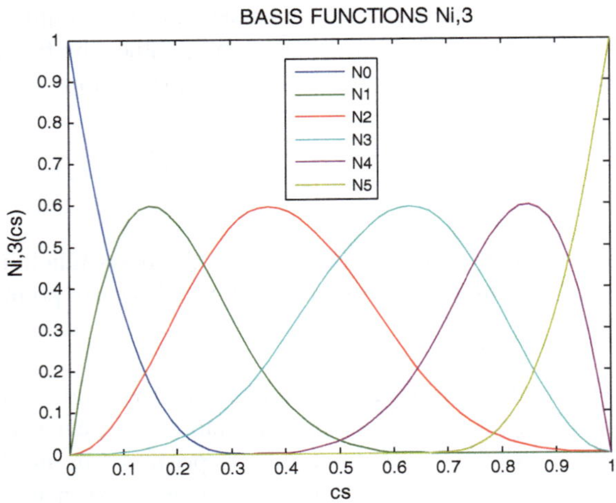

Fig. 7.6 Cubic B-splines, $N_{i,3}$, for ($n = 3$) breakpoint spans based on Curry–Schoenberg (de Boor) formulation

with that in Fig. 7.1, where again now the number of breakpoint spans is $n = 3$ and the polynomial degree is $p = 3$. One may observe that at the left end ($\xi = 0$):

(i) The function N_0, which is associated to the first control points (P_0), equals to the unity: $N_0(0) = 1$, whereas all the other functions vanish: $N_i(0) = 0, i = 1, \ldots, n_p$.
(ii) Except N_0' and N_1', all the other derivatives (i.e., of the functions N_2 to N_5 vanish at the end $\xi = 0$. This is very important because boundary conditions of Neumann type lead to the conditions $N_0'(0) \cdot \alpha_0 + N_1'(0) \cdot \alpha_1 = 0$, which is very useful as we shall see later, particularly in collocation methods.

A short parenthesis
A rigorous explanation can be provided when the following rules (properties) are taken into consideration:

For a given number of n_{ele} (say uniform) subdivisions in the interval (say [0, 1]), ($n_{ele} + 1$) breakpoints are produced at the positions: $\{0, \frac{1}{n_{ele}}, \ldots, \frac{n_{ele}-1}{n_{ele}}, 1\}$. The latter set includes ($n_{ele} - 1$) inner breakpoints plus two at the ends (0, 1) of the interval (boundary). For a given polynomial degree, p, and single multiplicity, the following knot vector is produced:

$$\mathbf{U}_{\text{knot}} = \left\{ \underbrace{0, \ldots, 0}_{(p+1)\,\text{times}}, \frac{1}{n_{ele}}, \ldots, \frac{n_{ele}-1}{n_{ele}}, \underbrace{1, \ldots, 1}_{(p+1)\,\text{times}} \right\} = \{u_0, u_1, \ldots, u_m\} \tag{7.23}$$

Since $N_{i,p}(\xi) = 0$ if ξ is *outside* the interval $[u_i, u_{i+p+1})$, and due to the fact that the first derivative $\left(N_{i,p}'\right)$ is given by the formula:

$$N_{i,p}' = \frac{p}{u_{i+p} - u_i} N_{i,p-1} - \frac{p}{u_{i+p+1} - u_{i+1}} N_{i+1,p-1},$$

the reader may easily conclude the abovementioned claims concerning the reason why $N_i(0) = 0, i = 1, \ldots, n_p$, as well as why except N_0' and N_1', all the other derivatives vanish at $\xi = 0$. This is the end of the parenthesis. ■

The MATLAB® command sequence to determine the knot vector, knots, is:

```
%...DEFINE THE 'KNOT-SEQUENCE' (KNOTS):
nele=3;                                    %Number of ELEments
nodes=nele+1;                              %Number of nodes (breakpoints)
a=1;                                       %Length of interval [0,a]
p=3;                                       %Polynomial degree
C=1;                                       %Multiplicity of inner knots
knots=augknt(linspace(0,a,nodes),p+1,C);   %knot sequence.
```

Based on the abovementioned knot vector, $\mathbf{U}_{\text{knot}}$, the number of control points $\left(P_0, \ldots, P_{n_p}\right)$ is determined according to the formula:

$$m = n_p + p + 1 \tag{7.24}$$

Let us now proceed with the computation of the functions $N_{i,p}(\xi)$. Before we call the function spcol that determines the basis functions $N_{i,p}(\xi)$ and their derivatives $N_{i,p}^{(k)}(\xi)$ [up to the order $(p-1), k = 1, \ldots, p-1$], we have to define an arbitrary set of q points, tau:

$$\mathbf{t} = \{t_1, t_2, \ldots, t_q\}, \tag{7.25}$$

where we wish to calculate all of them. For example, if we want to refine the mesh of points where the calculations will be found, using ndiv=8 subintervals per knot span, then the total number of smallest intervals will be ndiv*nele and the total number of points will be q=ndiv*nele+1.

The execution of the MATLAB® function spcol will give the matrix colmat, which includes $(n+1)$ columns (i.e., so many as exactly the number of control points) and q*nodes rows. In more detail, the set of the first p rows of the matrix colmat includes the basis function $N_{0,p}(\xi)$ and its derivatives $N_{0,p}', N_{0,p}'', \ldots, N_{0,p}^{(p-1)}$. The second set of the next p rows includes the basis function $N_{1,p}(\xi)$ and its derivatives $N_{1,p}', N_{1,p}'', \ldots, N_{1,p}^{(p-1)}$, and so on. Moreover, the first column refers to the control point P_0, the second column to the P_1, and so on up to P_n. One can store the basis functions and the first derivatives in the vectors Basis1 and Dbasis1, respectively, using the following commands.

```
%...DEFINE THE 'tau' VECTOR WHERE THE VALUES NEED TO BE CALCULATED:
    ndiv=8;                          %Number of divisions per breakpoint span
    q=ndiv*nele+1;                   %Number points to calculate the basis functions
    tau(1:q)=linspace(0,a,q);
%...DEFINE THE 'colmat1' MATRIX:
    colmat1=spcol(knots,p+1,brk2knt(tau,p)); %Nip and (p-1) derivatives.
    [i1,i2]=size(colmat1);
    ncontrolpoints=i2;               %Number of control points.
    Basis1 =colmat1(1:p:i1,:);       %Vector including B-splines Basis, Ni,p.
    Dbasis1=colmat1(2:p:i1,:);       %Vector of B-splines First Derivatives.
%...PLOT THE BASIS FUNCTIONS
    plot(tau,Basis1)
    xlabel('t')
    ylabel('Ni,p(t)')
    title('Ni,p')
%...END OF PROGRAM ----------------
```

Finally, it may be useful to determine the position of the *control* points. This can be accomplished by applying Eq. (7.22) to an arbitrary set of $(n + 1)$ *trial* points that belong to the interval [0, 1]. Therefore, selecting these trial points at the test position involved in the trial vector:

$$\bar{\mathbf{t}} = \{\bar{t}_1, \bar{t}_2, \ldots, \bar{t}_{n+1}\} = \left\{0, \tfrac{1}{n}, \ldots, \tfrac{n-1}{n}, 1\right\}, \tag{7.26}$$

a linear equations system can be constructed in the following form:

$$[\mathbf{A}] \cdot \left\{\mathbf{x}_p\right\} = \bar{\mathbf{t}}, \tag{7.27}$$

with

$$[\mathbf{A}]_{ij} = N_{i,p}(t_j) \tag{7.28}$$

The inversion of the matrix $[A]$ in Eq. (7.27) leads to the numerical determination of the control points. The procedure is shown below in terms of MATLAB® commands, where the vector `tau_bar` corresponds to Eq. (7.26).

```
%---DEFINE THE CONTROL POINTS:
    tau_bar=linspace(0,1,ncontrolpoints);
    colmat_bar=spcol(knots,p+1,brk2knt(tau_bar,1)); %Nip, without derivatives.
    [i1,i2]=size(colmat_bar);
    ncontrolpoints=i2;                    %Number of control points.
    Basis_bar =colmat_bar(1:1:i1,:);      %Vector including B-splines Basis, Ni,p.
%---EQUATIONS SYSTEM:
    A=Basis_bar(1:ncontrolpoints,:);
    xcontrol = A\tau_bar(1:ncontrolpoints)';
```

The results of the vector `xcontrol` are shown below. The reader may observe that in addition to the given breakpoints, {0, 0.3333, 0.6667, 1}, two additional control points appear, at the positions $\xi = \frac{1}{9} \cong 0.1111$ and $\xi = \frac{8}{9} \cong 0.8888$. The latter divide the two extreme intervals in ratio 1:3, with the smallest distance from the endpoints.

```
xcontrol =
        0
   0.111111111111111
   0.333333333333333
   0.666666666666667
   0.888888888888889
   1.000000000000000
>>
```

After the above introduction, it is now time to continue with the application of B-spline formulations in the Galerkin–Ritz (finite element) solution of a typical engineering problem that refers to the numerical extraction of the eigenvalues (free vibrations) in an axially vibrating elastic bar. All the abovementioned B-spline formulations will be discussed.

Example 7.2 (Axial vibrations of a bar fixed at one end) Determine the numerical eigenvalues (natural frequencies) of an elastic bar of length L, modulus of elasticity E and mass density ρ, under the boundary conditions shown in Fig. 7.7 (one fixed end).

The exact eigenvector of the axial eigenmotion is given by:

$$u_i(x) = A_i \sin \frac{i\pi x}{2L}, \quad i = 1, 2, \dots . \tag{7.29}$$

Therefore, the analytical eigenvalues are given by

$$\omega_i^2 = (2i-1)^2\pi^2/4L^2(E/\rho), \quad i = 1, 2, \dots . \tag{7.30}$$

Several formulations of *cubic* B-splines (degree $p = 3$) are tested and compared with Lagrange polynomials of the maximum degree which is allowed by the same discretization, as well as conventional linear finite elements for the same mesh (Table 7.1). In all cases, the model depends on the sequence of $(n+1)$ uniform points: $x_0 = 0, x_1 = \frac{L}{n}, \dots, x_n = L$, which contribute either as breakpoints or as nodal points.

Seven models were tested as follows:

MODEL-1: *Reduced* natural cubic B-splines ($p = 3$), of cardinal ([0, 1]-) type, based on the abovementioned $(n + 1)$ breakpoints. The series expansion includes $(n + 1)$ shape functions, similar to those shown in Fig. 7.3a.
MODEL-2: Lagrange polynomials of n-th degree based on the abovementioned $(n+1)$ nodes. The series expansion includes $(n + 1)$ shape functions.

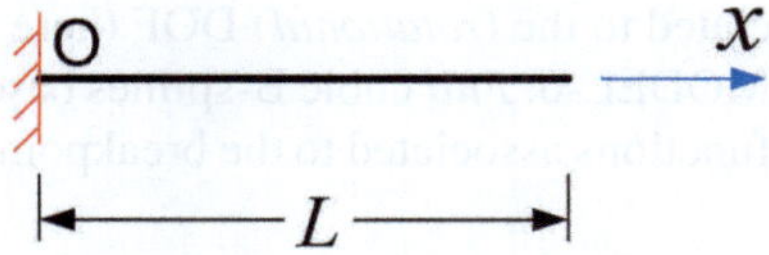

Fig. 7.7 Clamped beam in axial free vibrations

Table 7.1 The seven models tested for $(n + 1)$ breakpoints

Model	Approximation	Number of equations (free DOF) *after* the imposition of the BCs
Model-1 (Cardinal natural B-splines: $p = 3$)	$U(\xi) = \sum_{i=1}^{n+1} {}^{(1)}N_i(\xi) \cdot {}^{(1)}U_i$	n
Model-2 (Lagrange polynomials: $p = n$)	$U(\xi) = \sum_{i=1}^{n+1} {}^{(1)}L_i(\xi) \cdot {}^{(1)}U_i$	n
Model-3 (Uniform subintervals: $p = 3$)	$U(\xi) = \sum_{i=1}^{n+3} {}^{(2)}N_i(\xi) \cdot {}^{(2)}U_i$	$(n + 2)$
Model-4 (Control points: breakpoints plus two extra points: $p = 3$)	$U(\xi) = \sum_{i=1}^{n+3} {}^{(3)}N_i(\xi) \cdot {}^{(3)}U_i$	$(n + 2)$
Model-5 (Breakpoints plus two extra slopes at endpoints: $p = 3$)	$U(\xi) = \sum_{i=1}^{n+3} {}^{(4)}N_i(\xi) \cdot {}^{(4)}U_i$	$(n + 2)$
Model-6 (Breakpoints plus two extra curvatures at endpoints: $p = 3$)	$U(\xi) = \sum_{i=1}^{n+3} {}^{(5)}N_i(\xi) \cdot {}^{(5)}U_i$	$(n + 2)$
Model-7 (de Boor B-splines: $p = 3$)	$U(\xi) = \sum_{i=1}^{n+3} N_{i,3}(\xi) \cdot \alpha_i$	$(n + 2)$

MODEL-3: *Full* cubic B-splines based on the abovementioned $(n + 1)$ breakpoints, using test points that divide the interval [0, L] into $(n + 2)$ equal subintervals. A representative case is that of Fig. 7.2a which however corresponds to the specific case $n = 3$.

MODEL-4: *Full* cubic B-splines based on the abovementioned $(n + 1)$ breakpoints, using test points that coincide with the control points (breakpoints plus to extra points in one-third of the distance from the endpoints to the nearest internal breakpoint). A representative case is that of Fig. 7.2b which however corresponds to the specific case $n = 3$.

MODEL-5: *Full* cubic B-splines based on the abovementioned $(n+1)$ reduced shape functions associated to the breakpoints, plus the two additional shape functions associated to the (*rotational*) DOF (here the *first* derivatives or slopes) at the two ends.

MODEL-6: *Full* cubic B-splines based on the abovementioned $(n+1)$ reduced shape functions associated to the breakpoints, plus the two additional shape functions associated to the (*rotational*) DOF (here the *second* derivatives or curvatures) at the two ends.

MODEL-7: De Boor cubic B-splines based on the abovementioned $(n + 1)$ breakpoints, for polynomial degree $p = 3$ and multiplicity $\lambda = 1$ (C^2-continuous), using the `spcol` function in MATLAB.

In summary, Model-1 and Model-2 deal with $(n + 1)$ shape functions. Model-3 to Model-6 deal with $(n + 3)$ shape functions. Finally, Model-7 deals with $(n + 3)$ basis functions $N_{i,3}$. All these shape or basis functions exist before the boundary conditions are imposed. In all cases, after the imposition of the BCs one degree of freedom is deleted and a concise comparison between the above seven models is shown in Table 7.1. The upper left superscript refers to the particular set of shape functions as well as to the particular set (Model-1 to Model-6) of nodal values U_i (DOF).

For the first six models (Model-1, Model-2, Model-3, and Model-4), the equation of the free vibrations is:

$$[\mathbf{M}_u]\{\ddot{\mathbf{U}}\} + [\mathbf{K}_u]\{\mathbf{U}\} = \{0\}, \tag{7.31}$$

whereas for the seventh one (Model-5), it is:

$$[\mathbf{M}_\alpha]\{\ddot{\boldsymbol{\alpha}}\} + [\mathbf{K}_\alpha]\{\boldsymbol{\alpha}\} = \{0\}, \tag{7.32}$$

Model-5 deals with the vector $\bar{\mathbf{d}} = [U_0, \ldots, U_n, U_0', U_n']^{\mathrm{T}}$, whereas Model-6 deals with $\hat{\mathbf{d}} = [U_0, \ldots, U_n, U_0'', U_n'']^{\mathrm{T}}$ [previously shown in Eq. (7.15)]. In both these models, due to the fact that $\sum_{i=1}^{nele+3} \bar{\phi}_i(\xi) \neq 1$ and $\sum_{i=1}^{nele+3} \hat{\phi}_i(\xi) \neq 1$, the sum of the elements in the corresponding mass matrices slightly differs than the true mass of the rod. The higher the number of knot spans the smaller is the difference.

Numerical integration

For all models, each element of the mass matrix is given by $m_{ij} = \int_0^L N_i(\rho A)N_j \mathrm{d}x$, which means that the number of integration points depends on the polynomial degree of the integrand $(N_i N_j)$. In more details:

(i) In case of cubic B-splines, either of cardinal natural or of usual form (de Boor type), the polynomial degree of the integrand equals to 6 $\left(x^3 x^3 = x^6\right)$; thus, the accurate numerical integration requires four Gauss points *per span* $[x_k, x_{k+1}]$ (totally $4n$ integration points for all n spans in the sequence of breakpoints $x_0 = 0, x_1 = \frac{L}{n}, \ldots, x_n = L$). It is clarified that for these $(n + 1)$ breakpoints, which means $(n + 3)$ coefficients due to $p = 3$, after the imposition of the boundary conditions, the cardinal splines (initially with $(n + 1)$ DOF) deal with n equations (derivatives at the ends are missing), whereas full de Boor splines (initially with $(n + 3)$ coefficients) will refer to $(n + 2)$ equations.

(ii) In contrast, using the abovementioned $(n+1)$ nodal points in conjunction with $(n + 1)$ Lagrange polynomials of degree n, the integrand $(N_i N_j)$ is of degree

$2n$ and the number of Gauss points n_g is the minimum value that fulfills the condition $2n_g - 1 \geq 2n$. In the latter case, the Gauss points refer to the *entire* length L of the bar.

Results. Considering a sequence of ten uniform spans ($n = 10$), with ($x_0 = 0, x_1 = \frac{L}{10}, \ldots, x_{10} = L$), the calculated results are illustrated in Table 7.2. The findings are commented as follows:

(1) The reported degrees of freedom (DOF) in Table 7.2 refer to the number of equations after the imposition of the boundary conditions at the fixed point. One may observe that de Boor formulation (12 DOF) is competitive to Lagrange polynomials (10 DOF), and slightly outperforms compared to the reduced cardinal B-splines (10 DOF). The latter are generally worse than Lagrange polynomials (of 10th degree), whereas all methods are far superior to the usual FEM solution (10 DOF).

(2) Interestingly, all full truncated power series models, i.e., Model-3 to Model-6, lead to the *same* numerical results as those using the de Boor formulation.

General observations

This example offers the chance for some additional general remarks.

Remark-1 Truncated power series natural cubic B-spline interpolation in conjunction with the two additional (*rotational*) DOF at the two ends led to results that were found to be *identical* with those produced by the Curry–Schoenberg (de Boor) formulation. The reason for the aforementioned coincidence is due to the fact that the de Boor basis functions are linear combinations of the monomials (complete or truncated) involved in the truncated power series formula. In other words, a transformation matrix is always hidden between these two functional sets.

As an illustration for the above statement, let us consider the *simpler* case of cubic B-splines over two equal spans (i.e., three breakpoints: $n = 2$) in the unit

Table 7.2 Calculated eigenvalues of axial vibrations for a bar fixed at one end ($n = 10$ spans)

Mode (i)	Exact (ω_i^2)	Errors (in %) of calculated eigenvalues			
		Model-1 Cardinal (cubic) 10 DOF	Models-3,4,5,6,7 De Boor (cubic) 12 DOF	Model-2 Lagrange (10th degree) 10 DOF	FEM (linear) 10 DOF
1	2.46740e+00	0.01	0.00	0.00	0.21
2	2.22066e+01	0.06	0.00	0.00	1.86
3	6.16850e+01	0.19	0.00	0.00	5.24
4	1.20903e+02	0.41	0.01	0.00	10.42
5	1.99859e+02	0.81	0.05	0.09	17.44
6	2.98556e+02	1.60	0.23	1.34	26.06
7	4.16991e+02	3.27	0.83	8.36	35.32

length, as shown in Fig. 7.8. In this case, the control points $P_0P_1P_2P_3P_4$ are at {0, 1/6, ½, 5/6, 1}. Using de Boor formula [function spcol in MATLAB®, to which the functions $N_{i,p}(x)$ in the book of Piegl and Tiller [21] have been fully harmonized (i.e., are the same)], the reader may validate that the plot of the basis functions $N_{i,3}(x)$, $i = 0, 1, 2, 3, 4$ fulfills the conditions:

$$\begin{Bmatrix} N_{0,3} \\ N_{1,3} \\ N_{2,3} \\ N_{3,3} \\ N_{4,3} \end{Bmatrix} = \begin{bmatrix} 1 & -6 & 12 & -8 & 8 \\ 0 & 6 & -18 & 14 & -16 \\ 0 & 0 & 6 & -8 & 16 \\ 0 & 0 & 0 & 2 & -16 \\ 0 & 0 & 0 & 0 & 8 \end{bmatrix} \begin{Bmatrix} 1 \\ x \\ x^2 \\ x^3 \\ \left\langle x - \frac{1}{2} \right\rangle_+^3 \end{Bmatrix} \tag{7.33}$$

Also, again for cubic B-splines using three uniform segments (i.e., four nodes: $n = 3$) in the unit length, the control points $P_0P_1P_2P_3P_4P_5$ are at {0, 1/9, 1/3, 2/3, 8/9, 1} (see Fig. 7.1b). Then it can be numerically validated that the relationships between the de Boor and the Schoenberg basis functions are:

$$\begin{Bmatrix} N_{0,3} \\ N_{1,3} \\ N_{2,3} \\ N_{3,3} \\ N_{4,3} \\ N_{5,3} \end{Bmatrix} = \begin{bmatrix} 1 & -9 & 27 & -27 & 27 & 0 \\ 0 & 9 & -40.5 & 47.25 & -54 & 6.75 \\ 0 & 0 & 13.5 & -24.75 & 40.5 & -20.25 \\ 0 & 0 & 0 & 4.5 & -20.25 & 40.5 \\ 0 & 0 & 0 & 0 & 6.75 & -54 \\ 0 & 0 & 0 & 0 & 0 & 27 \end{bmatrix} \begin{Bmatrix} 1 \\ x \\ x^2 \\ x^3 \\ \left\langle x - \frac{1}{3} \right\rangle_+^3 \\ \left\langle x - \frac{2}{3} \right\rangle_+^3 \end{Bmatrix} \tag{7.34}$$

One may observe that the sum of all terms in each column equals to zero, except the first one in which it equals to the unity.

Remark-2 Lagrangian polynomials $L_{i,n}$ are a particular class (the simplest one) of B-splines of degree n. Lagrangian polynomials are also combinations of Bern-

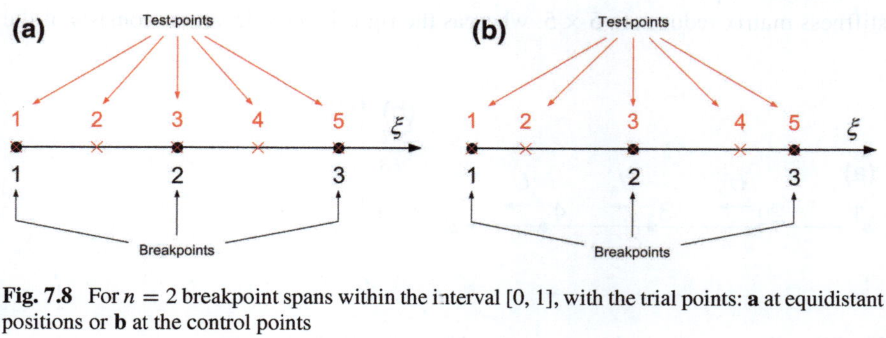

Fig. 7.8 For $n = 2$ breakpoint spans within the interval [0, 1], with the trial points: **a** at equidistant positions or **b** at the control points

stein–Bézier polynomials and vice versa. The latter use $(n+1)$ control points and no inner breakpoints.

Example 7.3 (Bar in tension due to an intermediate force) Let us now consider the same bar as that in Example 7.2, fixed again at the left end ($x = 0$), on which a concentrated force of $\bar{F} = 1000N$ (like a Dirac delta function) is applied to the point $x = 2L/3$. The exact solution is:

$$U(x) = \begin{cases} \frac{F}{EA}x, & 0 \le x \le \frac{2L}{3} \\ \frac{F}{EA}\frac{2L}{3}, & \frac{2L}{3} \le x \le L, \end{cases} \tag{7.35}$$

and its graph is shown (as a ramp) in Fig. 7.9.

This problem is solved using cubic B-splines as well as Lagrange polynomials of third degree. In all cases, we divide the interval $[0, L]$ into three equal segments ($n = 3$), thus introducing four breakpoints at $\{0, \frac{1}{3}, \frac{2}{3}, 1\}$ (see Fig. 7.9a).

(1) *De Boor model*. In case we use a de Boor approximation for the entire interval $[0, L]$, six control points are introduced at $\{0, \frac{1}{9}, \frac{1}{3}, \frac{2}{3}, \frac{8}{9}, 1\}$. Thus, the stiffness matrix is of dimensions 6×6, whereas the force vector of 6×1 given by:

$$\mathbf{F}_\alpha^{\mathrm{T}} = \int_0^L [N_{0,3}(x), N_{1,3}(x), N_{2,3}(x), N_{3,3}(x), N_{4,3}(x), N_{5,3}(x)]^{\mathrm{T}} \cdot \left[1000\delta(x - \tfrac{2}{3})\right] \mathrm{d}x,$$

or

$$\mathbf{F}_\alpha^{\mathrm{T}} = 1000 \cdot [N_{0,3}(\tfrac{2}{3}), N_{1,3}(\tfrac{2}{3}), N_{2,3}(\tfrac{2}{3}), N_{3,3}(\tfrac{2}{3}), N_{4,3}(\tfrac{2}{3}), N_{5,3}(\tfrac{2}{3})]^{\mathrm{T}}$$

and finally as:

$$\mathbf{F}_\alpha^{\mathrm{T}} = 1000 \cdot [0, 0, \tfrac{1}{6}, \tfrac{7}{12}, \tfrac{1}{4}, 0]^{\mathrm{T}} \tag{7.36}$$

After the imposition of the Dirichlet boundary condition, $U(0) = 0$, the active stiffness matrix reduces to 5×5, whereas the right-hand-side vector consists of the

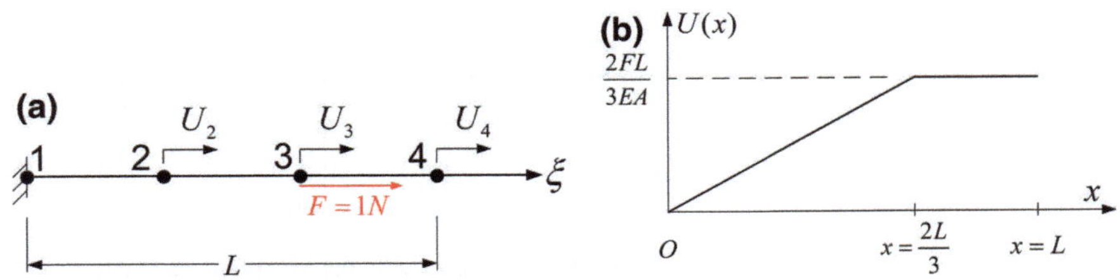

Fig. 7.9 **a** Bar in tension under a concentrated force, **b** axial displacement

last five elements of the abovementioned vector **F**. The linear equations' system is first solved in $\boldsymbol{\alpha}$ and then, using Eq. (7.21), the nodal values U_2, U_3 and U_4 are determined. Improvement of the results at the second and third breakpoint is possible, by changing the multiplicity of inner knots.

(2) *Natural (reduced) B-splines*. Since the shape functions are cardinal and the degrees of freedom are associated to the breakpoints shown in Fig. 7.9a, the force vector is calculated immediately by

$$\mathbf{F}_u^{\mathrm{T}} = 1000 \cdot [0, 0, 1, 0]^{\mathrm{T}} \tag{7.37}$$

After the imposition of the Dirichlet boundary condition, $U(0) = 0$, the active stiffness matrix reduces to 3×3, whereas the right-hand-side vector consists of the last three elements of the abovementioned vector **F**.

(3) *Lagrange polynomials*. The force vector is identical with that of natural (reduced) splines, according to Eq. (7.37). After the imposition of the BCs, the active stiffness matrix reduces to 3×3 as well.

For the above three models, the obtained numerical solution is compared with the exact one at the breakpoints, as shown in Table 7.3.

In this particular example, the findings of Table 7.3 suggest:

(1) De Boor (noncardinal, bell-shaped) and full cardinal based on truncated power series form of cubic B-splines (irrespectively to the choice of test points, uniform, or at control points) led to identical results. In the de Boor formulation, we need to calculate the force components at many places according to Eq. (7.36). In truncated power series formulation (using cardinal shape functions), we need

Table 7.3 Calculated axial displacements of a bar in tension

Model	Error (in %) of calculated displacement				Number of equations (after imposition of BCs)
	At breakpoint shown in Fig. 7.9				
	1	2	3	4	
De Boor (or Schoenberg)	–	−1.55	−3.28	0.00	5
Natural B-splines (second derivative)	–	−0.37	−4.03	0.00	3
Lagrange polynomials	–	2.88	−7.00	0.00	3
Exact	0.0000	0.3333	0.6667	0.6667	

to put the given force at the concrete nodal point only. The overall quality of the numerical solution is the best among the three models.

(2) Natural cubic B-spline formulation and higher-order Lagrange polynomials use less equations than the de Boor and truncated power series formulation, for the same breakpoints. Moreover, they treat a concentrated force in the same way, according to Eq. (7.37). The quality of the numerical solution is very good.

(3) All methods determine the same displacement at the free end.

It should become clear that for this particular problem, which is a *ramp* function, any B-spline approximation has a rather low accuracy because of the inherent C^2-continuity that pushes to erroneous values in the interior.

7.2 Two-Dimensional Problems

The theory of this section concerns a smooth quadrilateral patch ABCD, which can be mapped to a unit square in the $\xi\eta$-plane. Concerning B-spline finite elements in engineering analysis, a rather forgotten paper is due to Aristodemo [2], and selected later works are due to Brown et al. [4], as well as Kagan and Fischer [18], among others. A detailed monograph, including a literature survey, was published by Höllig [15, 16]. In the two latter monographs, it seems that the only relevant paper in B-spline finite element analysis is due to Höllig et al. [14]. In our opinion, the literature survey in the previous works is incomplete and may be justified by the fact (the therein literature shows that) a patent had been filled in 2002. Some of the older works (before 2001) may be found in Chap. 14 of the book in hands.

7.2.1 The Truncated Power Series (Older Than Schoenberg) Approach

7.2.1.1 General

A mathematical summary for generalized splines in two dimensions may be found in Ahlberg et al. [1], Chap. VIII.

In principle, the truncated power series formulation is applicable to two directions, ξ and η. Extending Eqs. (7.5) and (7.7), one can interpolate at every pair (ξ, η) by the tensor product

$$U(\xi, \eta) = \sum_{i=0}^{n_{p,x}} \sum_{j=0}^{n_{p,y}} \tilde{B}_{i,n_{p,x}}(\xi) \cdot \tilde{B}_{j,n_{p,y}}(\eta) \cdot a_{ij}, \tag{7.38}$$

where $(1 + n_{p,x})$ and $(1 + n_{p,y})$ are the number of coefficients in each direction.

Researchers started working with cubic B-splines and later continued with higher polynomial degrees. Considering n_x and n_y spans in the x- and y-direction, respectively, the complete tensor-product cubic B-splines require $(n_x + 3) \times (n_y + 3)$ generalized coefficients associated to the 2D basis functions given by:

$$\tilde{N}_{ij}(\xi, \eta) = \tilde{B}_{i,n_{p,x}}(\xi) \cdot \tilde{B}_{j,n_{p,y}}(\eta). \tag{7.39}$$

Obviously, all models presented in Sect. 7.1 (regarding 1D problems) are applicable to 2D problems as well. Between these versions, one can ignore the rotational DOF and consider only those associated to the $(n_x + 1) \times (n_y + 1)$ breakpoints. However, the completeness of the approximate solution is not ensured. Below we present only the particular case of natural cubic B-splines, which is a reduced basis that needs a smaller number of parameters than the full truncated power series formulation.

7.2.1.2 Tensor Product of Natural (Reduced) B-Splines

In this case, the meaning of the control points is not needed. Instead, the usual nodal points (at the position of the well-known breakpoints) are used, very similarly to the case of Lagrangian-type macroelements. The computational mesh consists of $(n_\xi + 1) \times (n_\eta + 1)$ nodes, which are produced by dividing the ξ- and η-direction in n_ξ and n_η subdivisions, respectively (the case $n_\xi = n_\eta = 4$ is shown in Fig. 7.10). The nodes along the opposite sides AB and DC are located at the positions included as elements of the vector $\boldsymbol{\xi} = \{\xi_0, \xi_1, \ldots, \xi_{n_\xi}\}$. Similarly, the nodes along the opposite sides BC and AD are located at the positions included as elements of the vector $\boldsymbol{\eta} = \{\eta_0, \eta_1, \ldots, \eta_{n_\eta}\}$. In general, the aforementioned $(n_\xi + 1) \times (n_\eta + 1)$ nodes $\mathbf{P}_{ij}, i = 0, \ldots, n_\xi, j = 0, \ldots, n_\eta$ are located at the intersections (ξ_i, η_j) made of the elements included in the vectors $\boldsymbol{\xi}$ and $\boldsymbol{\eta}$.

Based on the idea of *isoparametric* elements, we assume that the geometry is described by:

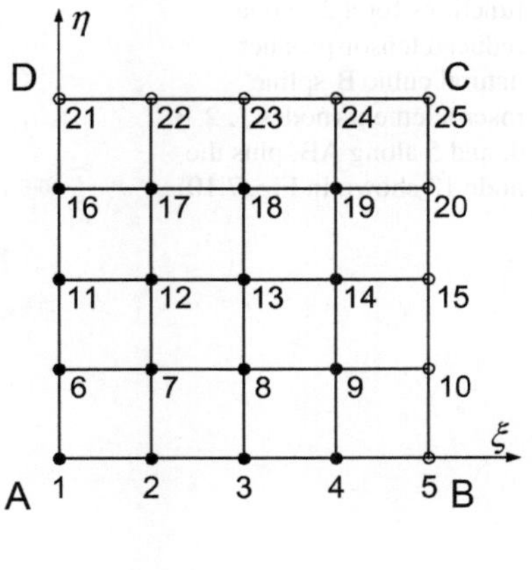

Fig. 7.10 Tensor-product cubic B-spline macroelement

$$\mathbf{x}(\xi,\eta)=\sum_{i=0}^{n_\xi}\sum_{j=0}^{n_\eta}\left[\hat{\phi}_i(\xi)\cdot\hat{\phi}_j(\eta)\right]\cdot\mathbf{x}_{Pij},\quad i=1,\ldots,n_\xi,\quad j=1,\ldots,n_\eta. \tag{7.40}$$

and also the variable of the problem is approximated by:

$$U(\xi,\eta)=\sum_{i=0}^{n_\xi}\sum_{j=0}^{n_\eta}\left[\hat{\phi}_i(\xi)\cdot\hat{\phi}_j(\eta)\right]\cdot U_{ij},\quad i=1,\ldots,n_\xi,\quad j=1,\ldots,n_\eta. \tag{7.41}$$

The advantage of this approach is that we refer directly to the nodal value U_{ij} (as happens in the conventional FEM analysis) and not to arbitrary nodeless coefficients α_{ij} similar to those mentioned in the previous subsection (1D Curry–Schoenberg formulation implemented by the de Boor software).

The derivation of the product $\hat{\phi}_i(\xi)\cdot\hat{\phi}_j(\eta)$ is achieved extending the 1D methodology using now tensor-product natural (reduced) B-splines like those discussed in Sect. 7.1. The selected breakpoints form a uniform mesh of $n_\xi\times n_\eta$ subdivisions, and these also operate as the usual nodal points of the finite element mesh. In this way, in addition to the $2(n_\xi+n_\eta)$ boundary nodes, $(n_\xi-1)\times(n_\eta-1)$ internal nodes appear as well. Again, the computational mesh is the same as in the usual finite element analysis (*local* approximation) but now a *global* approximation dominates. Again, the proposed natural cubic B-spline macroelement refers to positions at which the nodal values of U_{ij} are *directly* calculated.

As an example, for the 25-node macroelement of Fig. 7.10, six relevant shape functions (associated to the nodal points 1–5 along the side AB plus the center 13) are cardinal as shown in Fig. 7.11. One may observe that the maximum value of each shape function equal always to the unity but also negative values may appear. Nevertheless, the unity is the upper limit of the bounded shape functions $\hat{\phi}_i(\xi)\cdot\hat{\phi}_j(\eta)$, and this happens because (as we saw in Sect. 7.1) each of the factor is bounded by the unity.

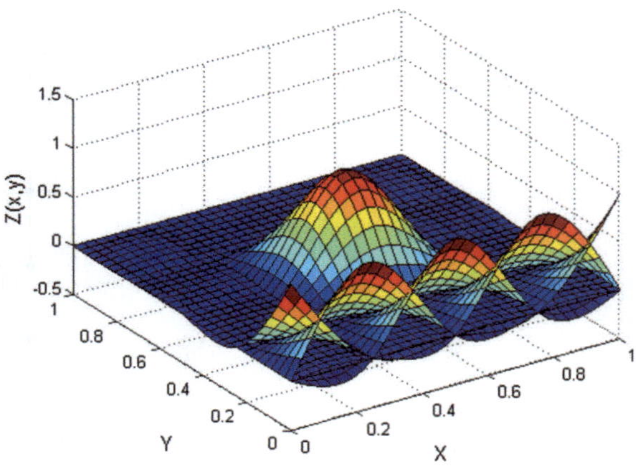

Fig. 7.11 Cardinal shape functions for a 25-node reduced tensor-product natural cubic B-spline macroelement (nodes 1, 2, 3, 4, and 5 along AB, plus the node 13 shown in Fig. 7.10)

An advantage of this type of interpolation is that the peak values of the shape functions equal to unity, whereas no matter how many breakpoints exist in any direction (no numerical oscillation appears as in Lagrange polynomials). A disadvantage is that *natural* B-spline has a vanishing *second* derivative at the ends; thus, it is against the accurate representation of the particular function $U(x) = x^3$ of which the second derivative at $x = L$ is not properly considered (reduced basis). But since in eigenvalue problems the traction free part is taken equal to zero, the method is ideal as will be demonstrated in the subsequent examples.

7.2.2 Curry–Schoenberg (de Boor) Tensor-Product B-Splines

Since the full truncated power series formulation (Model-1, and Model-3 to Model-6) of Example 7.2 are equivalent to Curry–Schoenberg (de Boor) formulation in which normalized *bell-shaped* basis functions appear, it is a loss of time to repeat them in the general 2D problem. Therefore, henceforth we only deal with the pure de Boor formulation using his software.

7.2.2.1 Determination of Control Points

In the case of the dominating de Boor [8] formulation, in order to determine the possibly unknown control points of the patch ABCD, a uniform mesh (as much as possible) of $(n_{p,x} + 1) \times (n_{p,y} + 1)$ *test points* on its real surface is generated. Then, for each test point we apply the formula

$$\mathbf{x}(\xi, \eta) = \sum_{i=0}^{n_{p,x}} \sum_{j=0}^{n_{p,y}} \left[N_{i,p_x}(\xi) \cdot N_{j,p_y}(\eta)\right] \cdot \mathbf{x}_{Pij}, \quad i = 1, \ldots, n_x, \quad j = 1, \ldots, n_y. \tag{7.42}$$

Control points

From the solution of the linear equations system coming from Eq. (7.42), we can calculate the coordinates of the control points for an entire 2D patch.

Global approximation of the variable

Here, the physical problem variable $U(\xi, \eta)$ refers to nodeless coefficients a_{ij} as follows:

$$U(\xi, \eta) = \sum_{i=0}^{n_{p,x}} \sum_{j=0}^{n_{p,y}} \left[N_{i,p_x}(\xi) \cdot N_{j,p_y}(\eta)\right] \cdot \alpha_{ij}, \quad i = 1, \ldots, n_x, \quad j = 1, \ldots, n_y. \tag{7.43}$$

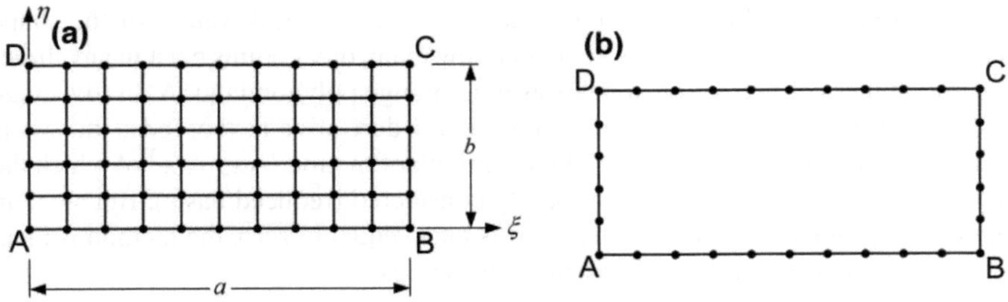

Fig. 7.12 Rectangular acoustic cavity for a single **a** tensor product and **b** Coons macroelement

It is clarified that the coefficient α_{ij} in Eq. (7.43) is not the value U_{ij} at the nodal point P_{ij}, but sometimes it corresponds to another unknown point in its neighborhood. In practice, the determination of the coefficients α_{ij} is of first priority, whereas the nodal values U_{ij} are calculated afterward.

Example 7.4 (Rectangular acoustic cavity) Consider an acoustic cavity of dimensions $a \times b = 2.5 \times 1.1$ with hard walls (infinite resistance or zero impedance), which means that it is subject to Neumann boundary conditions ($\partial p/\partial n = 0$). The advantage of this example is that its exact solution is analytically known in closed form (see Example 3.3 of Chap. 3).

Using a uniform mesh of breakpoints according to Fig. 7.12a ($n_\xi \times n_\eta = 10 \times 5$ subdivisions), the obtained tensor-product results are shown in Table 7.4. In summary, the findings and the comments are as follows:

Table 7.4 Calculated eigenvalues of an acoustic cavity using tensor-product approximations (10 × 5 subdivisions according to Fig. 7.12a)

Mode	Exact ω^2	Errors (in %) of calculated Eigenvalues			
		FEM	Natural cubic B-splines 2 × 2 GPTs	Lagrange polynomials 12 × 12 GPTs	De Boor cubic polynomials
		66 DOF	66 DOF	66 DOF	104 DOF
1	0.00	–	–	–	–
2	1.57914	0.83	0.04	0.00	0.00
3	6.31655	3.33	0.15	0.00	0.00
4	8.15670	3.33	0.31	0.00	0.00
5	9.73583	2.92	0.27	0.00	0.00
6	14.2122	5.23	0.29	0.00	0.00
7	14.4732	5.66	0.24	0.00	0.00
8	22.3689	6.04	0.30	0.00	0.00
9	25.2662	13.70	0.37	0.00	0.02
10	32.6268	13.70	1.48	0.73	0.02

(1) The model with natural tensor-product B-splines has the same number of DOF with the macroelement of Lagrangian type (66 DOF) as well as with the uniform mesh of 4-node finite elements (66 DOF). In all these three cases, the position of the breakpoints and the nodal points is the same. Nevertheless, the accuracy highly depends on each model.

2) The natural cubic tensor-product B-splines are of much better quality than the FEM solution based on bilinear elements.

3) The Lagrangian macroelement is exceptionally accurate, but it necessarily uses different polynomial degrees per direction, i.e., $p_x = 10$ and $p_y = 5$, according to the mesh.

4) The de Boor formulation leads to the same accuracy with the Lagrangian element (it outperforms only in the tenth mode), but it uses 104 DOF due to the number of control points: $n_{p,x} \times n_{p,y} = (10+3) \times (5+3) = 104$.

(5) Obviously, the same numerical result with the Lagrangian macroelement may be eventually derived through the tensor-product de Boor formulation, provided Bernstein–Bézier polynomials are used. This is accomplished by applying `spcol` MATLAB function in conjunction with the knot vector $\mathbf{U}_{\text{knot},x} = \{\underbrace{0,\ldots,0}_{p_x+1},\underbrace{a,\ldots a}_{p_x+1}\}$ in the x-direction, and $\mathbf{U}_{\text{knot},y} = \{\underbrace{0,\ldots,0}_{p_y+1},\underbrace{b,\ldots b}_{p_y+1}\}$ in the y-direction. It is apparent that within this context, the uniform mesh of 66 points shown Fig. 7.12a will also refer to uniformly distributed control points of the tensor-product Bézierian element (eventually equivalent to the Lagrangian element, through a linear transformation of the basis functions: For details, the reader is referred to Chap. 6).

For the completeness of the presentation, in Table 7.5 we also show the results obtained by a single Coons macroelement (30 DOF) using only the boundary nodes as shown in Fig. 7.12b. One may observe that the natural B-spline (as trial functions along each side of the rectangle) solution is of the same quality with the tensor-product natural B-spline solution shown in Table 7.4, and outperforms in comparison with the FEM solution. Moreover, the Lagrange polynomials (as trial functions along each side of the rectangle) are exceptionally accurate and competitive to the tensor product solution (shown in Table 7.4), except some higher modes.

Example 7.5 (Circular acoustic cavity) A circular acoustic cavity of unit radius is analyzed, again under Neumann boundary conditions ($\partial p/\partial n = 0$). The problem has been previously defined in Chap. 3. The quality of convergence using de Boor cubic B-splines is shown in Table 7.6.

Choosing the case of $n_\xi = n_\eta = 8$ subdivisions per side of the unit square (Fig. 7.13), the overall results are shown in Table 7.7. In addition of the tensor product methods of this chapter, for comparison purposes, boundary-only Coons macroelements are also included in the right side of the table.

Further comparative results for six subdivisions per side are shown in Table 7.8.

In this particular example, the boundary-only Coons macroelement does not converge to the accurate value and is not so accurate as previously happened with the

Table 7.5 Calculated eigenvalues of a rectangular acoustic cavity using COONS interpolation in conjunction with two different trial functions (10×5 subdivisions on the boundary according to Fig. 7.12b)

Mode	Exact ω^2	Errors (in %) of calculated eigenvalues		
		FEM	Trial functions	
			Natural cubic B-splines 2×2 GPTs	Lagrange polynomials 12×12 GPTs
		66 DOF	30 DOF	30 DOF
1	0.00	–	–	–
2	1.57914	0.83	0.04	0.00
3	6.31655	3.33	0.15	0.00
4	8.15670	3.33	0.31	0.00
5	9.73583	2.92	0.56	0.28
6	14.2122	5.23	0.29	0.00
7	14.4732	5.66	12.23	12.16
8	22.3689	6.04	7.96	7.78
9	25.2662	13.70	0.37	0.00
10	32.6268	13.70	1.48	0.73

Table 7.6 Calculated eigenvalues of a circular acoustic cavity of radius 1.0 m under Neumann boundary conditions, using various subdivisions in conjunction with de Boor cubic B-splines ($p = 3$)

Mode	m	Exact ω^2	Error (in %) of calculated eigenvalues			
			Number of circumferential subdivisions ($4n$)			
			4	8	16	32
1	0	0.00	–	–	–	–
2	1	3.3900	0.09	0.29	0.00	0.00
3	1	3.3900	0.09	0.29	0.00	0.00
4	2	9.3284	−1.13	0.18	0.06	0.00
5	2	9.3284	50.37	2.71	0.09	0.00
6	0	14.6820	68.85	0.50	0.15	0.00
7	3	17.6500	63.78	9.68	0.76	0.00
8	3	17.6500	63.78	9.68	0.76	0.00
9	4	28.2764	67.23	9.80	0.45	0.02
10	4	28.2764	90.50	88.13	5.17	0.03
Number of equations			16	25	49	121

Results are shown as percentage errors for "tensor-product splines" approximation using Galerkin–Ritz formulation with multiplicity $\lambda = 1$

Table 7.7 Calculated eigenvalues of a circular acoustic cavity of radius 1.0 m under Neumann boundary conditions, using $n_\xi = n_\eta = 8$ subdivisions per one-quarter of the circumference (i.e., 32 circumferential subdivisions)

Mode	Exact ω^2	Error (in %) of calculated eigenvalues						
		Tensor-product methods (boundary and domain discretization)				COONS (boundary-only discretization)		
		Natural cubic B-splines	Lagrange	De Boor	FEM	Natural B-splines	Lagrange	Piecewise-linear
1	0.00	–	–	–	–	–	–	–
2	3.3900	0.07	0.00	0.00	1.57	4.15	4.12	5.32
3	3.3900	0.07	0.00	0.00	1.57	4.15	4.12	5.32
4	9.3284	0.01	0.00	0.00	3.42	3.74	3.83	7.82
5	9.3284	0.18	0.00	0.00	4.74	11.55	11.43	13.40
6	14.6820	0.04	0.00	0.00	5.78	22.86	23.10	31.84
7	17.6500	0.12	0.02	0.00	7.98	11.57	11.81	18.54
8	17.6500	0.12	0.02	0.00	7.98	11.57	11.81	18.54
9	28.2764	-0.04	0.06	0.02	11.50	11.84	13.08	21.73
10	28.2764	0.25	0.24	0.03	11.50	27.03	29.10	42.64
Area deviation (in%)		−4.71e−2	8.91e−8	4.00e−7	−6.41e−1	−4.70e−2	2.14e−7	−6.41e−1
Number of equations		81	81	121	81	32	32	32

Several macroelements of tensor product and Coons type are reported

Table 7.8 Calculated eigenvalues of a circular acoustic cavity of radius 1.0 m under Neumann boundary conditions, using $n_\xi = n_\eta = 6$ subdivisions per one-quarter of the circumference (i.e., 32 circumferential subdivisions)

Mode	Exact ω^2	Error (in %) of calculated eigenvalues						
		Tensor-product methods (boundary and domain discretization)				COONS (boundary-only discretization)		
		Natural B-splines	Lagrange	De Boor	FEM	Natural B-splines	Lagrange	Piecewise-linear
1	0.00	–	–	–	–	–	–	–
2	3.3900	0.15	0.00	0.00	2.78	4.25	4.12	6.24
3	3.3900	0.15	0.00	0.00	2.78	4.25	4.12	6.24
4	9.3284	−0.03	0.02	0.00	5.99	3.84	3.87	10.95
5	9.3284	0.38	0.05	0.01	8.42	11.89	11.70	14.84
6	14.6820	−0.04	0.08	0.02	10.34	23.52	23.53	39.14
7	17.6500	0.15	0.72	0.04	13.98	11.95	13.77	23.79
8	17.6500	0.15	0.72	0.04	13.98	11.95	13.77	23.79
9	28.2764	−0.36	1.36	0.18	20.19	12.35	23.93	28.92
10	28.2764	−0.36	5.33	0.22	20.19	29.29	44.84	54.20
Area deviation (in%)		−1.12e-01	−3.05e-05	−1.15e-05	−1.14e+00	−1.12e−01	−2.98e−05	−1.14e+00
Number of equations		49	49	81	49	24	24	24

Several macroelements of tensor product and Coons type are reported

Table 7.9 Accuracy of alternative macroelements for the elliptical bar in torsion

	ERROR (in %) of calculated torsional rigidity and area			
	Natural B-splines (cubic)	Lagrange polynomials (8th degree)	De Boor B-splines (cubic)	FEM (4-node)
Torsional rigidity (J)	−2.0866e−01	4.0211e−4	3.8839e−3	−3.33
Area (A)	−1.0396e−01	7.6380e−03	5.1031e−03	−0.76
Number of equations	49	48	81	49

rectangular cavity. However, this quality is better than that obtained using the competitive DR/BEM dual reciprocity boundary element method (see [23], p. 51, Table 7.9 therein). Almost the same poor results were found for the Coons macroelement using either higher-order Lagrange polynomials or natural cubic B-splines. These findings are an evidence that a single boundary-only Coons macroelement is not capable of accurately solving the eigenvalue problem of a circular cavity under Neumann BCs.

In conclusion, the above two example in acoustics (rectangular and circular cavities) show the high quality of the tensor-product numerical solution (calculated eigenvalues) for all methods. In brief:

(1) Despite the fact that the tensor-product natural cubic B-splines are not complete they achieve an acceptable level of numerical accuracy.
(2) Tensor-product Lagrange polynomials perform well up to the tenth degree, without any numerical oscillations.
(3) Tensor-product de Boor cubic B-splines are the most superior, but they need more equations for the same mesh of breakpoints. The interested reader may

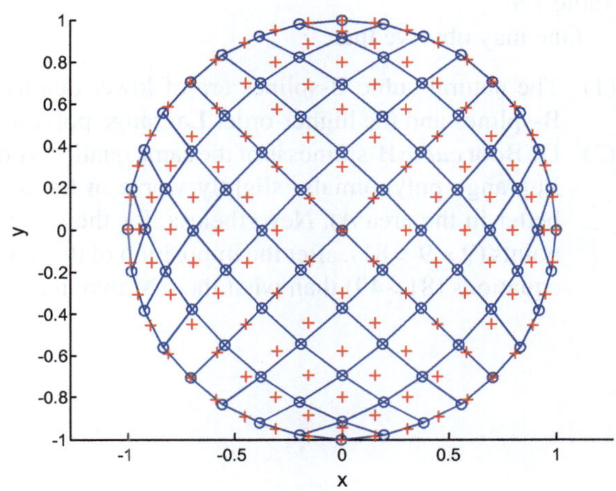

Fig. 7.13 Breakpoints and control points for a circular cavity idealized with a single tensor-product macroelement with $n_\xi = n_\eta = 8$ (81 breakpoints: blue circle, and 121 control points: red plus)

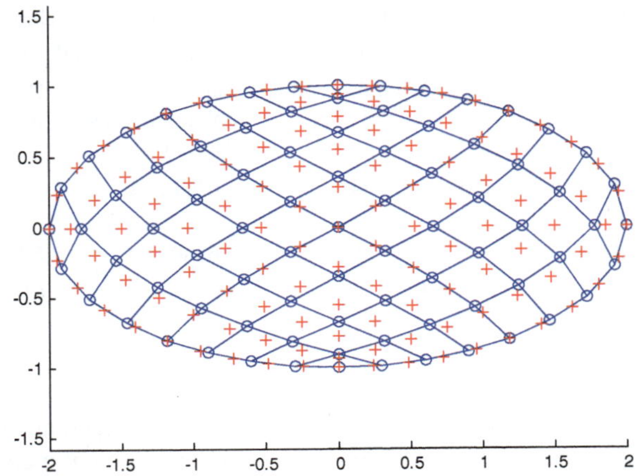

Fig. 7.14 Breakpoints (blue circle) and control points (red plus) for an ellipse idealized with a single tensor-product macroelement with $n_\xi = n_\eta = 8$

continue with a detailed comparison using error graphs (error versus number of equations), as well as higher polynomial degrees ($p > 3$).

Example 7.6 (Torsion of rod with elliptic cross section) A rod with elliptic cross section with semi-diameters $a \times b = 2 \times 1$ is subjected to torsion. The governing equation is $\nabla^2 U = -2$ and the boundary conditions are of Dirichlet type, i.e., $U = 0$. The torsional rigidity is given by the integral $\int_A 2U \mathrm{d}x\mathrm{d}y$ over the area $A = \pi ab$ of the ellipse, and its accurate value is given by

$$J = G\frac{\pi a^3 b^3}{(a^2 + b^2)} \tag{7.44}$$

For a uniform mesh with eight subdivisions per side of the single macroelement that occupies the entire ellipse (see Fig. 7.14), comparative results are shown in Table 7.9.

One may observe that:

(1) The natural cubic B-splines are of lower quality than both the de Boor cubic B-splines and the higher-order Lagrange polynomials.
(2) De Boor *cubic* B-splines is of the same quality with the *higher-order* (8th degree) Lagrange polynomials (slightly worse in the torsional rigidity J and slightly better in the area A). Nevertheless, for the same number of breakpoints/nodal points ($9 \times 9 = 81$), after the imposition of the boundary conditions, it uses more equations ($81 > 49$) than what the first two macroelements do.

Example 7.7 (Thick-walled cylindrical section subject to internal pressure)
Consider a thick-walled hollow cylinder of inner radius R_1, outer radius R_2 and height $H = 2$, with uniform pressure P prescribed on the inner surface. The ends of the cylinder are constrained to move in the radial direction only, i.e., $\varepsilon_{zz} = 0$. Therefore, the analysis of the section becomes a plane strain problem of the known analytical solution:

$$u_r = C_1 r + \frac{C_2}{r}, \quad \sigma_r = C_3 - \frac{C_4}{r^2}, \quad \sigma_\vartheta = C_3 + \frac{C_4}{r^2} \tag{7.45}$$

The traction boundary conditions at $r = R_1$ and R_2 imply immediately the constants C_3 and C_4. Then the two normal stresses σ_r and σ_ϑ are considered in Hooke's law. From this system, the rest two constants C_1 and C_2 are determined. As a result, the four constants are given by:

$$\left.\begin{array}{ll} C_1 = C_3 \frac{(1+\nu)(1-2\nu)}{E} \cong 0.1733, & C_2 = C_4 \frac{(1+\nu)}{E} = 15.6 \\ C_3 = \frac{1}{R_2^2} \cdot C_4 = \frac{1}{3}, & C_4 = \frac{P}{\left(\frac{1}{R_1^2} - \frac{1}{R_2^2}\right)} = 12.0 \end{array}\right\} \tag{7.46}$$

The cross section of the thick cylinder is a rectangle of dimensions $(R_2 - R_1) \times H = 3 \times 2$. The height H is the z-direction is constantly divided into four equal segments ($n_z = 4$), whereas the radial part $(R_2 - R_1)$ in a number that varies from $n_r = 6$ to $n_r = 20$. The particular case of $n_r = 6$ is shown in Fig. 7.15.

It is noted that the breakpoints in the tensor-product macroelement are the only nodal points used in natural B-splines and Lagrange polynomials. Moreover, these breakpoints are also some of the control points participating in de Boor formulation, where additional ones are shown in Fig. 7.15 by the symbol ($\times$) and lie at one-third of the extreme segments measured from the corresponding endpoints.

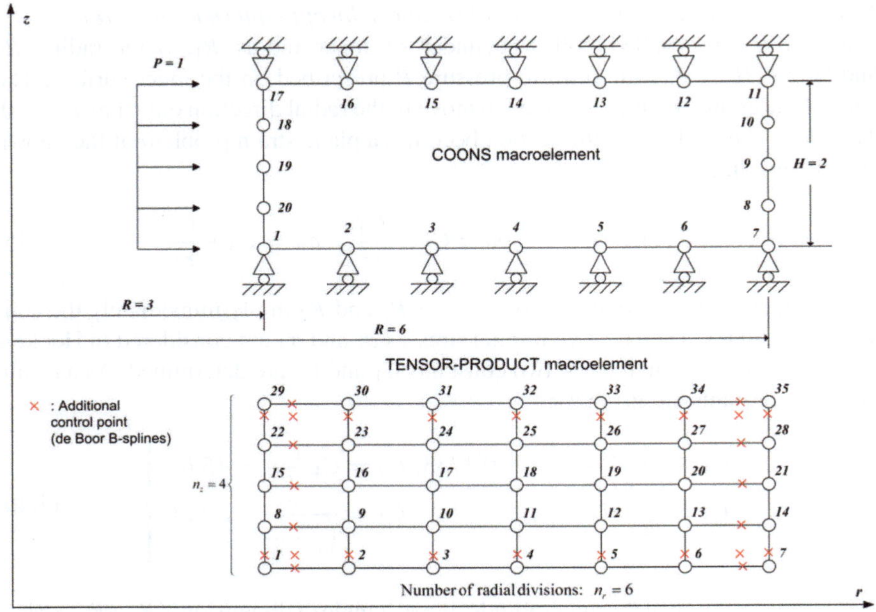

Fig. 7.15 Discretization of a rectangular cross section of a thick cylinder (here $n_r = 6$ and $n_z = 4$ subdivisions formed by the breakpoints)

The accuracy of the results is studied in terms of the normalized L^2-norms:

$$\left.\begin{aligned} &\text{Displacement: } L_u(\text{in}\ \%) = \frac{\sqrt{\int_\Omega (\mathbf{u}_{\text{calculated}} - \mathbf{u}_{\text{exact}})^2 d\Omega}}{\sqrt{\int_\Omega (\mathbf{u}_{\text{exact}})^2 d\Omega}} \times 100, \\ &\text{Stress: } \qquad L_\sigma(\text{in}\ \%) = \frac{\sqrt{\int_\Omega (\boldsymbol{\sigma}_{\text{calculated}} - \boldsymbol{\sigma}_{\text{exact}})^2 d\Omega}}{\sqrt{\int_\Omega (\boldsymbol{\sigma}_{\text{exact}})^2 d\Omega}} \times 100. \end{aligned}\right\} \tag{7.47}$$

For several radial subdivisions ($6 \le n_r \le 20$), the results are tabulated in Table 7.10 for L_u (in %) and Table 7.11 for L_σ(in %).

7.3 Discussion

The numerical results of the previous subsection show a lot of interesting facts and relevant conclusions may be done.

First of all, it was clearly shown that the *boundary-only Coons* interpolation does not always converge to the accurate solution, but it works well in some test cases in which the solution includes the legs of Pascal's triangle with a surplus of two layers

Table 7.10 Quality of overall numerical solution for the thick cylinder (in normalized L^2-norm L_u according to Eq. (7.47)) using several macroelements

Number of radial subdivisions (n_r)	Normalized L^2-norm of calculated displacement L_u (in %)						
	Tensor-product macroelement (boundary and domain discretization)				COONS macroelement (boundary-only discretization)		
	Natural B-splines	Lagrange	De Boor	FEM	Natural B-splines	Lagrange	Piecewise-linear
$n_r = 6$ (NEQ) →	7.09384e−02	5.43425e−04	4.15491e−04	1.92993e−01	7.46881e−02	5.43268e−04	2.04312e−01
	(56)	(56)	(108)	(56)	(26)	(26)	(26)
$n_r = 8$ (NEQ) →	3.47192e−02	1.54943e−05	1.41248e−04	1.09759e−01	3.63878e−02	1.54910e−05	1.16246e−01
	(72)	(72)	(132)	(72)	(30)	(30)	(30)
$n_r = 10$ (NEQ) →	1.98648e−02	4.47278e−07	6.07311e−05	7.06090e−02	2.07394e−02	4.47207e−07	7.47960e−02
	(88)	(88)	(156)	(88)	(34)	(34)	(34)
$n_r = 15$ (NEQ) →	7.17994e−03	2.63004e−07	1.28426e−05	3.15435e−02	7.44978e−03	2.96654e−07	3.34199e−02
	(128)	(128)	(216)	(128)	(44)	(44)	(44)
$n_r = 20$ (NEQ) →	3.48657e−03	3.74206e−04	4.20525e−06	1.77753e−02	3.60522e−03	8.52182e−03	1.88339e−02
	(168)	(168)	276	(168)	(54)	(54)	(54)

NEQ number of equations (after the imposition of the boundary conditions)

Table 7.11 Quality of overall numerical solution for the thick cylinder [in normalized L^2-norm L_σ, in %, according to Eq. (7.47)] using several macroelements

Number of radial subdivisions (n_r)	Normalized L^2-norm of calculated stress L_σ (in %)						
	Tensor-product methods (boundary and domain discretization)			FEM	COONS (boundary-only discretization)		
	Natural cubic B-splines	Lagrange	De Boor		Natural cubic B-splines	Lagrange	Piecewise-linear
6	3.06038e+00	2.56483e−02	3.78584e−02	9.99947e+00	3.37307e+00	2.56524e−02	9.56957e+00
8	1.99473e+00	9.73616e−04	1.66448e−02	7.54444e+00	2.20408e+00	9.73730e−04	7.21059e+00
10	1.42501e+00	3.50964e−05	8.77018e−03	6.05249e+00	1.57874e+00	3.50997e−05	5.78104e+00
15	7.71031e−01	3.32607e−07	2.71290e−03	4.04629e+00	8.58298e−01	6.39600e−07	3.86239e+00
20	4.98560e−01	4.52993e−04	1.17166e−03	3.03771e+00	5.56562e−01	9.87661e−03	2.89901e+00

of terms. Overall, from the practical engineering point of view, one could claim that it can be taken as a first-order approximation (with less than 5% error overall).

Second, it was shown that the *natural cardinal cubic B-splines* are not bad, they work even for very high numbers of breakpoints (e.g., 20 segments in Table 7.11), but converge slowly than the Lagrange polynomials.

Third, *Lagrange polynomials* do *not easily* lead to numerical oscillations as it has been widely written in books. With all respect to previous recognized researchers, in a wider variety of engineering problems than those presented in this short chapter, the overall numerical experience of the author has shown that they can safely use:

- Up to the degree $p = 10$ or $p = 12$ in potential problems, and
- Up to the degree $p = 8$ in elasticity problems.

In the particular results of Table 7.11, one may observe that Lagrange polynomials led to monotonic convergence up to $p = 15$ (L_σ = 3.33e−07%) and then started to give a little higher errors at $p = 20$ (L_σ = 4.53e−04%).

Fourth, *de Boor cubic B-splines* are probably the most superior in the sense that they never diverged till now. The only "disadvantage" is that they need more complicated programming since the analyst has to work with coefficients α_i and then to change to nodal values U_i. Again, the equivalency of the de Boor cubic B-splines with the abovementioned natural cubic B-splines is achieved when the two extra rotational degrees of freedom associated to the curvatures at the endpoints (in each direction) are considered.

De Boor B-splines have not been here studied for higher polynomial degrees $p > 3$, since this fact has been extensively done by others (see [16]) as well as by the successor isogeometric community that works with NURBS. In general, cubic B-splines are of lower accuracy, but in all cases one has to decide where to limit the degree p and in how many subregions one has to decompose the domain. For example, the closely related famous p-method uses (Legendre) polynomials up to the sixth (or seventh) degree (see Appendix in Szabó and Babuška [28]) and this could be taken as a practical "rule of thumb" for the domain decomposition.

7.4 Recapitulation

The use of cubic splines in the numerical solution of initial- and boundary value problems in ordinary differential equations, as well as for the approximate solution of integral equations, was known since the 1960s. On this issue, the interested reader may consult Ahlberg et al. [1]. In the late 1960s, it had been well understood that spline functions could also be used for the solution of nonlinear boundary value and eigenvalue problems as well [7, 17]. Later, the splines of higher polynomial degree were implemented. A thorough representation of the topic is out the scope of this book, but a complete summary can be found in the textbook of de Boor [9]. Additional documentation may be found in the book of Höllig [16].

In this chapter, we focused on two tensor-product B-spline versions as follows. The first version is somehow "heretic" since it is not complete (the terms/DOF related to the two derivatives at the ends have been left out). Nevertheless, it works particularly well when vanishing the second derivative at the endpoints (*natural* B-splines); it has been used since 1989 in many older papers of our research group. It is based on truncated power series formulation in conjunction with cubic polynomials from which upper *bounded* shape functions (cardinal B-splines) can be determined. These functions are associated to the breakpoints involved, one-to-one, for both the geometry $\mathbf{x} = [x, y]^{\mathrm{T}}$ and the variable $U(x, y)$ (as happens to the isoparametric elements in the classical finite element method). Therefore, no difficulty appears concerning the imposition of the boundary conditions (no need for penalty methods) and no need to transform from the nodeless coefficients α_i to selected point values U_i. Moreover, it was clearly shown that if the two second derivatives at the ends are included as additional degrees of freedom, then the reduced-plus formulation becomes identical with that of the de Boor B-splines.

The second version refers to de Boor B-splines, under the dominating symbol $N_{i,p}$ which is found in the papers and the popular book of Piegl and Tiller [21]. This method is bound with the so-called *control points* which are auxiliary quantities related to the geometry of the domain, but they also influence the approximate solution $U(x, y)$ through the associated generalized coefficients α_i. The readers should be aware that after the determination of the coefficients α_i, they should again use the basis functions $N_{i,p}$, this time at selective points where they wish to determine the value of the variable U.

It is of great importance to understand that truncated power series formulation leads to fully populated matrices, whereas Curry–Schoenberg (de Boor) formulation is characterized by local support and gives *bell-shaped* functions. Although both formulations are mathematically equivalent, there is no ambiguity that de Boor formulation is by far superior because it is extremely efficient and stable, as it is based on recursions; thus, no matrix inversion is required. Moreover, the produced basis functions are nonnegative and the first half of them is a *mirror* of the rest half with respect to the middle of the interval. Obviously, a similar mirroring occurs with the Lagrange polynomials as well.

This chapter dealt with B-splines of degree p with multiplicity equal to one, which means $C^{(p-1)}$-continuity. Nevertheless, it is important to keep in mind that de Boor formulation can easily deal with any lower continuities, just by affecting the known vector.

The personal opinion of the author is that despite it is widely accepted that Lagrange polynomials do not perform well (as also discussed in Chap. 2), the numerical results of this chapter show that this opinion is not absolutely true. In potential problems, one can safely use polynomials of tenth degree and sometimes may be still higher (not suggested). In elasticity problems, the rule of eighth degree may be adopted. Not to forget that in the p-method, the involved bubble functions are usually of sixth degree [28] but this choice is probably for the sake of conservatism. Of course, one can use macroelements up to the sixth degree and decompose the domain in some of them.

References

1. Ahlberg JH, Nilson EN, Walsh JL (1967) The theory of splines and their applications. Academic Press, New York
2. Aristodemo M (1985) A high-continuity finite element model for two-dimensional elastic problems. Comput Struct 21(5):987–993
3. Böhmer K (1974) Spline-Funktionen, B. G. Teubner-Studienbücher-Mathematik, Stuttgart
4. Brown JM, Bloor MIG, Bloor MS, Wilson MJ (1998) The accuracy of B-spline finite element approximations to PDE. Comput Methods Appl Mech Eng 158:221–234
5. Cox MG (1972) The numerical evaluation of B-splines. J Inst Math Its Appl 10:134–149
6. Curry HB, Schoenberg IJ (1966) On Pólya frequency functions IV: the fundamental spline functions and their limits. J Anal Math 17(1):71–107
7. De Boor C (1966) The method of projections as applied to the numerical solutions of two-point boundary value problems using cubic splines, Ph.D. thesis, University of Michigan. ftp://ftp.cs.wisc.edu/Approx/
8. De Boor C (1972) On calculating with B-splines. J Approx Theory 6:50–62
9. De Boor C (2001) A practical guide to splines (revised edition). Springer, New York; also: First edition (1978)
10. De Casteljau P (1959) Courbes à poles. National Industrial Property Institute (INPI, France
11. De Casteljau P (1999) De Casteljau's autobiography: my life at Citroën. Comput Aided Geom Des 16:583–586
12. Faux ID, Pratt MJ (1979) Computational geometry for design and manufacture. Ellis Horwood, Chichester
13. Greville TNE (1969) Introduction to spline functions. In: Greville TNE (ed) Theory and applications of spline functions. Academic Press, New York, pp 1–35
14. Höllig K, Reif U, Wipper J (2001) Weighted extended B-spline approximation of Dirichlet problems. SIAM J Numer Anal 39:442–462
15. Höllig K (2002) Finite element approximation with splines. In: Farin G, Hoschek J, Kim MS (eds) Handbook of computer aided geometric design, North-Holland, Amsterdam, Chapter 11, pp 283–307
16. Höllig K (2003) Finite element methods with B-splines. SIAM, Philadelphia
17. Jerome JW, Varga RS (1969) Generalizations of spline functions and applications to nonlinear boundary value and eigenvalue problems. In: Greville TNE (ed) Theory and applications of spline functions. Academic Press, New York, pp 103–155
18. Kagan P, Fischer A (2000) Integrated mechanical based CAE systems using B-spline based finite elements. Comput Aided Des 32(8–9):539–552
19. Kanarachos A, Deriziotis D (1989) On the solution of Laplace and wave propagation problems using 'C-elements'. Finite Elem Anal Des 5:97–109
20. Kanarachos A, Provatidis C, Deriziotis D, Foteas N (1999) A new approach of the FEM analysis of two-dimensional elastic structures using global (Coons's) interpolation functions. In: Wunderlich J (ed) CD proceedings first european conference on computational mechanics, München-Germany, Aug–Sept, 1999
21. Piegl L, Tiller W (1997) The NURBS book, 2nd edn. Springer, Berlin
22. Provatidis C, Kanarachos A (2001) Performance of a macro-FEM approach using global interpolation (Coons') functions in axisymmetric potential problems. Comput Struct 79(19):1769–1779
23. Provatidis CG (2004) On DR/BEM for eigenvalue analysis of 2-D acoustics. Comput Mech 35:41–53
24. Provatidis CG (2005) Three-dimensional Coons macroelements in Laplace and acoustic problems. Comput Struct 83(4–5):1572–1583
25. Provatidis CG (2012) Two-dimensional elastostatic analysis using Coons-Gordon interpolation. Meccanica 47(4):951–967
26. Provatidis CG (2018) Engineering analysis with CAD-based macroelements. Arch Appl Mech 88:121–140

27. Schoenberg IJ (1946) Contributions to the problem of approximation of equidistant data by analytic functions. Q Appl Math 4:45–99
28. Szabó B, Babuška I (1991) Finite element analysis. Wiley, New York
29. Zienkiewicz OC (1977) The finite element method. McGraw-Hill, London

Chapter 8
Rational Elements (BEZIER, NURBS)

Abstract This chapter deals with single macroelements in which the approximation of the variable U is based mostly on rational Bézier and less on nonuniform rational B-splines (NURBS). Since univariate rational Bernstein–Bézier polynomials is a special case of univariate NURBS, it becomes obvious that tensor-product rational Bézier is also a specific case of tensor-product NURBS. The major significance of rational elements is that they accurately represent the geometry of conics (circles, ellipses, parabolas, and hyperbolas). In an instructive way, we focus on the analysis of a circular cavity using a single tensor-product macroelement. It is shown that a single quadratic Bézier macroelement, although is capable of accurately representing the entire circle, it leads to a numerical solution of low quality (slightly worse than the classical nine-node finite element of Lagrangian type). In both cases, this is due to its insufficiency to approximate the eigensolutions (e.g., the eigenvectors in dynamics). Nevertheless, after a sufficient degree elevation which maintains the shape of the circle, it is shown that the higher-order Bézier converges to the exact solution. The presentation continues with a very short summary on the NURBS-based dominating IGA, and the reader is advised for further study.

Keywords Rational Bézier · Circular macroelement · MATLAB Freeware software · Degree elevation · Circular acoustic cavity · NURBS issues

8.1 Introduction

In Chap. 2, we met the definitions of rational Bézier and nonuniform rational B-spline (NURBS). We learnt that the term "rational" refers to the "ratio" that expresses the basis functions. In the case of rational Bézier, there are no inner breakpoints while in the case of NURBS there are inner breakpoints of a desired multiplicity. In both cases (rational Bézier and NURBS), control points exist and the formulation follows the Curry–Schoenberg [7] B-spline definitions that can be easily implemented according to the procedure due to de Boor [8]. In other words, we deal with basis functions associated to nodeless coefficients and not nodal values of the variable U itself. In the

C. G. Provatidis, *Precursors of Isogeometric Analysis*, Solid Mechanics and Its Applications 256, https://doi.org/10.1007/978-3-030-03889-2_8

former case (i.e., rational Bézier), fully occupied matrices are built, whereas in the latter banded matrices are produced due to the compact support property of NURBS.

It was previously shown that Bernstein–Bézier polynomials $B_{i,n}$ and the bell-shaped normalized Curry–Schoenberg basis functions $N_{i,p}$ (without inner break-points/knots), of the same degree (i.e., $p = n$), coincide one another when referring to the same control points which determine a certain curve. Consequently, similar coincidences appear in tensor-product surfaces and volume blocks as well.

Curve
Based on the above remark, a curve is generally described as

$$\boldsymbol{C}(\xi) = \sum_{i=0}^{n} R_{i,p}(\xi)\boldsymbol{P}_i, \quad 0 \leq \xi \leq 1, \tag{8.1}$$

where the involved univariate rational basis functions are given by

$$R_{i,p}(\xi) = \frac{N_{i,p}(\xi)w_i}{\sum_{j=0}^{n} N_{j,p}(\xi)w_j} \tag{8.2}$$

The first derivative of Eq. (8.2) is:

$$\frac{\mathrm{d}}{\mathrm{d}\xi}R_{i,p}(\xi) = w_i \frac{\sum_{i=0}^{n} N_{i,p}(\xi)N'_{i,p}(\xi) - \sum_{i=0}^{n} N'_{i,p}(\xi)N_{i,p}(\xi)}{\left(\sum_{i=0}^{n} N_{i,p}(\xi)\right)^2} \tag{8.3}$$

Setting the denominator of Eq. (8.3) in the form:

$$W(\xi) = \sum_{i=0}^{n} N_{i,p}(\xi)w_i, \tag{8.4}$$

Equation (8.3) takes the following short expression:

$$\frac{\mathrm{d}}{\mathrm{d}\xi}R_{i,p}(\xi) = w_i \frac{W(\xi)N'_{i,p}(\xi) - W'(\xi)N_{i,p}(\xi)}{(W(\xi))^2}, \tag{8.5}$$

with

$$W'(\xi) = \sum_{i=0}^{n} N'_{i,p}(\xi)w_i \tag{8.6}$$

For a given point $P(x, y)$ along a curve which is described either by rational Bézier or by rational B-splines (shortly NURBS) through Eqs. (8.1) and (8.2), the value of the relevant parameter ξ is not always obvious and generally requires the utilization of an iterative scheme such as Newton-Raphson. Even for an circular arc, which is the simplest possible case of a curvilinear curve, the correspondence between ξ and the polar angle θ (measured from its left end) is complicated, as we shall see later in this chapter. This fact justifies the term "nonuniform" in the two initials of NURBS.

Rational Bézier (quadratic, i.e., $p = 2$)
Between several options, we introduce one specific case for the description of a circle, which is the closest to the classical macroelements we met so far.

Let us consider a circle of center K and unit radius, of which the boundary (i.e., the circumference) is divided into four equal parts using the points A, B, C, and D. Obviously, the position of the first point (i.e., A) can be everywhere along the boundary, and it is considered as the origin of the parametric system $O\xi\eta$. Given the point A, the next "corner" point B is determined in the counterclockwise direction, and then the ξ-axis coincides with the arc AB, whereas the η-axis coincides with the arc AD (as shown in Fig. 8.1).

In general, each pair (ξ, η) of the parametric space (unit square $[0, 1] \times [0, 1]$) has an image $P(x, y)$ into the circle. When only boundary data are available, Coons interpolation is a convenient formula to establish the mapping $M : (\xi, \eta) \rightarrow (x, y)$, as follows:

$$\begin{aligned}\begin{bmatrix} x(\xi,\eta) \\ y(\xi,\eta) \end{bmatrix} &= (1-\xi)\begin{bmatrix} \cos\left(\frac{\pi}{2}\eta\right) \\ -\sin\left(\frac{\pi}{2}\eta\right) \end{bmatrix} + \xi\begin{bmatrix} -\sin\left(\frac{\pi}{2}\eta\right) \\ \cos\left(\frac{\pi}{2}\eta\right) \end{bmatrix} \\ &\quad + (1-\eta)\begin{bmatrix} \cos\left(\frac{\pi}{2}\xi\right) \\ \sin\left(\frac{\pi}{2}\xi\right) \end{bmatrix} + \eta\begin{bmatrix} -\sin\left(\frac{\pi}{2}\xi\right) \\ -\cos\left(\frac{\pi}{2}\xi\right) \end{bmatrix} \\ &\quad - (1-\xi)(1-\eta)\begin{bmatrix} 1 \\ 0 \end{bmatrix} - \xi(1-\eta)\begin{bmatrix} 0 \\ 1 \end{bmatrix} - \xi\eta\begin{bmatrix} -1 \\ 0 \end{bmatrix} - (1-\xi)\eta\begin{bmatrix} 0 \\ -1 \end{bmatrix}\end{aligned} \tag{8.7}$$

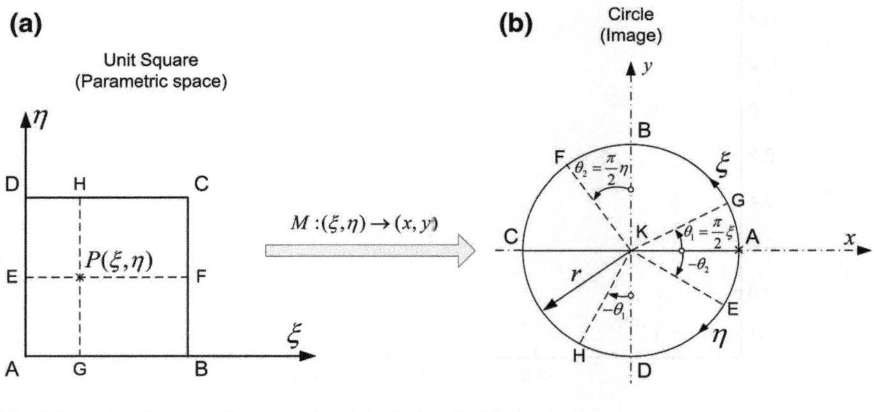

Fig. 8.1 **a** A unit square is mapped to **b** a circle of radius r

Obviously, if the radius of the circle is $r \neq 1$, then the right-hand side of Eq. (8.7) has to be multiplied by r.

Equation (8.7) can be explained in terms of Chap. 3, where two projections are written, the first along EF and the second along GH (Fig. 8.1). In more detail, there are two unique points (G, and H) along the opposite arcs AB and CD having the same given normalized coordinate ξ, and two points (F and E) along the arcs BC and DA having the same given parameter η, as shown in Fig. 8.1. The application of Coons interpolation easily produces the isolines ξ = constant and η = constant, of which the intersection is a point $P(\xi, \eta)$. In such as case, the position of a boundary point, say G, is easily determined by demanding $(AG)/(AB) = \xi$, which means that the polar angle of the position vector KG is $\theta = \sphericalangle(Kx, KG) = \frac{\pi}{2}\xi$ (henceforth, the polar angle θ_1 shown in Fig. 8.1a is permanently replaced by θ). In the classical Gordon–Coons elements with internal nodes, the latter are easily determined by the Coons formula (Eq. 8.7) as they lie at the intersections of isolines with ξ = constant and η = constant.

Let us now return to *rational* elements and consider again the quadratic Bézier macroelement as shown in Fig. 8.2 (we met it earlier in Chap. 6) for the same circle. Every side of the curvilinear quadrilateral $ABCD$, say AB, lies into a region that is determined by the polar angle $0 \leq \theta \leq \frac{\pi}{2}$ and is described by the parameter $\xi \in [0, 1]$. Clearly, the values $\xi = 0$ and $\xi = 1$ correspond to the endpoints A and B, respectively. Nevertheless, except of the ends and the middle of the arc AB, the abovementioned formula $\theta = \frac{\pi}{2}\xi$ is no more applicable. This fact will be explained below.

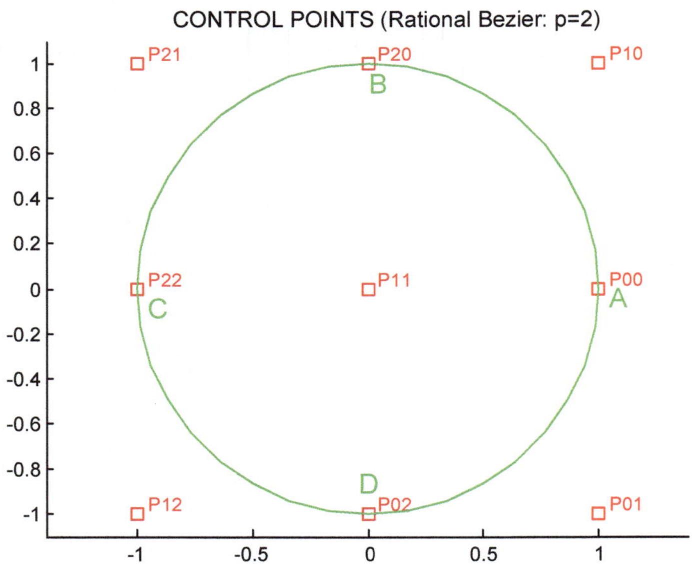

Fig. 8.2 Control points for the accurate representation of a unit circle using quadratic *rational* Bézier basis functions

Let us now focus on the sector *KAB* of the circle (Fig. 8.2). The control points of the arch *AB* are the endpoints *A* and *B* as well as the intersection P_{10} of the two tangents at *A* and *B*; thus, the coordinates of the control points are $P_{00}(1, 0)$, $P_{10}(1, 1)$ and $P_{20}(0, 1)$. Therefore, we can consider equal weights for the ends such as $w_0 = w_2 = 1$ (for $P_{00} \equiv A$ and $P_{20} \equiv B$) and a suitable weight for the middle control point P_{10} (at the corner) such as $w_1 = \frac{1}{\sqrt{2}}$. The triangle $AP_{10}B$ is the *control triangle* of the arc *AB* and is isosceles ($AP_{10} = BP_{10}$), whereas the weight w_1 equals to the cosine of the angle between the chord joining the endpoints and one of the legs of the control triangle [19]. Alternatively, the weights w_0 and w_2 at the ends may take any value (of course, provided $w_0' = w_2'$) and the correspondent w_1' must be equal to $w_1' = (w_0'w_2'/2)^{1/2}$. The latter comes from the condition of *shape invariance*, which requires a constant ratio w_0w_2/w_1^2, which here equals to 2.

In order to demonstrate that generally a nonuniform relationship exists between the polar angle and the variable ξ, i.e., $\theta \neq \frac{\pi}{2}\xi$, two different cases are studied as follows. The first case concerns the subdivision of the interval $\xi \in [0, 1]$ into six (unequal) subintervals, taking a constant step equal to $\Delta\xi = \frac{1}{8}$. The second case concerns the *uniform* subdivision of the arc *AB* into eight *equal* subarcs (constant variation of the polar angle, $\Delta\theta = \frac{\pi}{16}$). In this second case, the produced nine points along *AB* correspond to polar angles $\theta = 0, \frac{\pi}{16}, \frac{2\pi}{16}, \frac{3\pi}{16}, \frac{4\pi}{16}, \frac{5\pi}{16}, \frac{6\pi}{16}, \frac{7\pi}{16}$, and $\frac{\pi}{2}$, and therefore their Cartesian coordinates will be $x = r\cos\theta$ and $y = r\sin\theta$. In contrast, in the first case, for each value of the parameter ξ among $\xi = 0, \frac{1}{8}, \frac{2}{8}, \frac{3}{8}, \frac{4}{8}, \frac{5}{8}, \frac{6}{8}, \frac{7}{8}, 1$, Eq. (8.1) suggests that the Cartesian coordinates of a point $P(x, y)$ along the exact circular arc *AB* should be related to ξ by the formula:

$$\begin{bmatrix} x(\xi) \\ y(\xi) \end{bmatrix} = \frac{1\cdot(1-\xi)^2\cdot\begin{bmatrix}1\\0\end{bmatrix} + \frac{1}{\sqrt{2}}2(1-\xi)\xi\cdot\begin{bmatrix}1\\1\end{bmatrix} + 1\cdot\xi^2\cdot\begin{bmatrix}0\\1\end{bmatrix}}{1\cdot(1-\xi)^2 + \frac{1}{\sqrt{2}}2(1-\xi)\xi + 1\cdot\xi^2} \tag{8.8a}$$

For both cases, the obtained Cartesian coordinates are given in Table 8.1. One may observe that those pairs (x, y) at the *ends* and the *middle* of the interval are equal one another, while the rest are not. In other words, generally the normalized parameter differs from the polar angle θ (i.e., $\xi \neq \frac{\theta}{(\pi/2)}$), except of the ends and the middle. Inversely, if Eq. (8.1) is applied to the ends and the middle of a 90° arc in which the weights $w_0 = w_2 = 1$ and $w_1 = \frac{1}{\sqrt{2}}$ are given, the solution of the linear equations system (3 equations) will lead to the accurate position of the three control points [$P_{00} = (1, 0)$, $P_{10} = (1, 1)$ and $P_{20} = (0, 1)$].

Exercise 8.1 (Control points of nonrational circular arc)
Use a variation of Eq. (8.8) to determine the control points for the nonrational representation of the arc *AB* (90°).

Solution In Fig. 6.15 (Chap. 6), we mentioned that the control point P_{10}^* which is found at $x = y \cong 0.9142$, but not a rigorous proof was given. By virtue of Eq. (8.8a) in conjunction with equal weights (nonrational approximation), the determination of the control points is realized using the following formula:

Table 8.1 Cartesian coordinates using nonuniform rational and uniform formulations

Point	Based on quadratic rational Bézier [$p = 2$ in Eq. (8.1)], using a constant step $\Delta\xi = \frac{1}{8}$		Based on *uniform* divisions of the polar angle: $\theta = 0, \frac{\pi}{16}, \ldots, \frac{7\pi}{16}, \frac{\pi}{2}$ ($x = \cos\theta$, $y = \sin\theta$)		Standard condition
	x	y	x	y	$x^2 + y^2$
1	1.00000e+00	0.00000e+00	1.00000e+00	0.00000e+00	1
2	9.83305e−01	1.81963e−01	9.80785e−01	1.95090e−01	1
3	9.29788e−01	3.68095e−01	9.23880e−01	3.82683e−01	1
4	8.36996e−01	5.47210e−01	8.31470e−01	5.55570e−01	1
5	7.07107e−01	7.07107e−01	7.07107e−01	7.07107e−01	1
6	5.47210e−01	8.36996e−01	5.55570e−01	8.31470e−01	1
7	3.68095e−01	9.29788e−01	3.82683e−01	9.23880e−01	1
8	1.81963e−01	9.83305e−01	1.95090e−01	9.80785e−01	1
9	0.00000e+00	1.00000e+00	0.00000e+00	1.00000e+00	1

$$\begin{bmatrix} x(\xi) \\ y(\xi) \end{bmatrix} = \frac{1 \cdot (1-\xi)^2 \cdot \begin{bmatrix} x_{P_0} \\ y_{P_0} \end{bmatrix} + 1 \cdot 2(1-\xi)\xi \cdot \begin{bmatrix} 1 \\ 1 \end{bmatrix} + 1 \cdot \xi^2 \cdot \begin{bmatrix} 0 \\ 1 \end{bmatrix}}{1 \cdot (1-\xi)^2 + 1 \cdot 2(1-\xi)\xi + 1 \cdot \xi^2}, \tag{8.8b}$$

for three discrete values such as $\xi = 0, \frac{1}{2}, 1$. Since the denominator is identically equal to the unity, Eq. (8.8b) takes the nonrational form:

$$\begin{bmatrix} x(\xi) \\ y(\xi) \end{bmatrix} = (1-\xi)^2 \cdot \begin{bmatrix} x_{P_0} \\ y_{P_0} \end{bmatrix} + 2(1-\xi)\xi \cdot \begin{bmatrix} x_{P_1} \\ y_{P_1} \end{bmatrix} + \xi^2 \cdot \begin{bmatrix} x_{P_2} \\ y_{P_2} \end{bmatrix}, \tag{8.8c}$$

For $\xi = 0$, the corresponding point is the begin $A(1,0)$ which belongs to the circular arc AB, and therefore Eq. (8.8c) implies ($x_{P_0} = 1$, $y_{P_0} = 0$).

Similarly, for $\xi = 1$, the corresponding point is the end $B(0,1)$ which belongs to the circular arc AB, and therefore Eq. (8.8c) implies ($x_{P_2} = 0$, $y_{P_2} = 1$).

Finally, for $\xi = \frac{1}{2}$, the corresponding point is the middle point $M\left(\frac{\sqrt{2}}{2}, \frac{\sqrt{2}}{2}\right)$ of the circular arc AB, and therefore Eq. (8.8c) implies:

$$\begin{bmatrix} \frac{\sqrt{2}}{2} \\ \frac{\sqrt{2}}{2} \end{bmatrix} = \left(1 - \frac{1}{2}\right)^2 \cdot \begin{bmatrix} 1 \\ 0 \end{bmatrix} + 2\left(1 - \frac{1}{2}\right)\frac{1}{2} \cdot \begin{bmatrix} x_{P_1} \\ y_{P_1} \end{bmatrix} + \left(\frac{1}{2}\right)^2 \cdot \begin{bmatrix} 0 \\ 1 \end{bmatrix}, \tag{8.8d}$$

The obvious solution of the equations system (8.8d) is:

$$x_{P_1} = y_{P_1} = \sqrt{2} - \frac{1}{2} \cong 0.9142. \tag{8.8e}$$

Remark It is noted that using the abovementioned found three control points $\left[P_0(1,0), P_1^*\left(\sqrt{2}-\frac{1}{2}, \sqrt{2}-\frac{1}{2}\right) \text{ and } P_0(0,1)\right]$, the produced curve is a perfect parabola that *coincides* with the curve that passes through the endpoints and the middle and is interpolated using three quadratic Lagrange polynomials. ■

8.2 The 9-Node Rational Bernstein–Bézier Element

Let us now proceed with the formulation of the nine-node *rational* element. Before we are able to propose the tensor-product procedure, first we study the 9-node Lagrange element.

The 9-node Lagrange element is characterized by eight nodes along the boundary of the circle plus one node at its center (Fig. 8.3). Following Provatidis [24], the entire circle is consider as a Coons patch *ABCD*, where the nodes (1, 2, and 3) belong to the isoline $\eta = 0$, the nodes (4, 5, and 6) lie on the isoline $\eta = \frac{1}{2}$, and finally the nodes (7, 8, and 9) belong to the isoline $\eta = 1$. In the conventional macroelement of Lagrange type, each node is associated to a tensor product of the corresponding Lagrange polynomials ([35], p. 155). For the purposes of this book, this is called "Lagrangian macroelement".

In Chap. 6, we saw that the abovementioned Lagrangian macroelement is equivalent to the so-called "Bézierian macroelement". For the current case of a circle, due to the result of Exercise 8.1, for each edge of the Coons patch (say *AB* with a midpoint *M*), the sequence of the successive nodes (*A*, *M*, and *B*, i.e., the points 1, 2, and 3 in Fig. 8.3) in conjunction with three quadratic Lagrange polynomials is equivalent with the three control points (P_{00}, P_{10}, and P_{20} shown in Fig. 8.4) in conjunction with quadratic Bernstein–Bézier polynomials.

Continuing with the rest three quarters of the boundary (edges *BC*, *CD*, and *DA*), eight control points are derived for the boundary (four of them coincide with *A*, *B*,

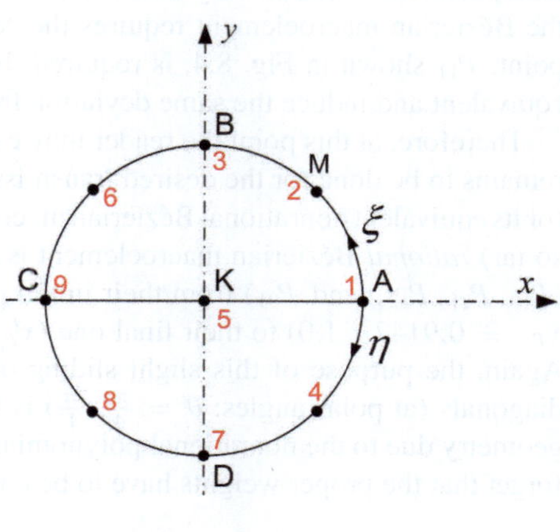

Fig. 8.3 Nine-node Lagrange element for the entire circle

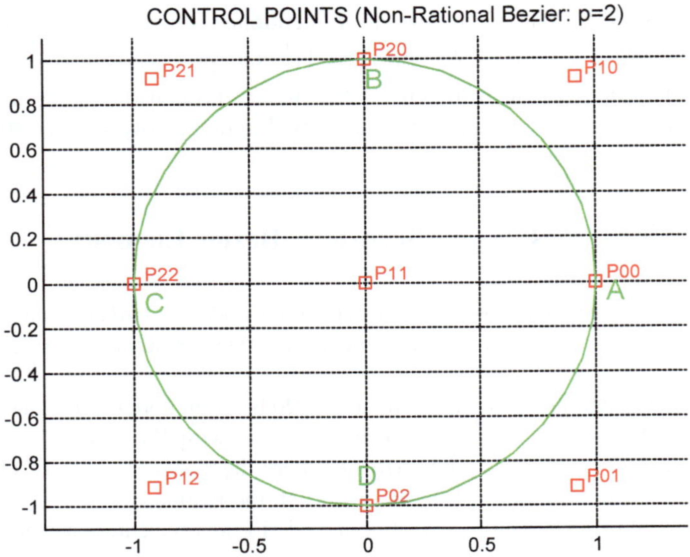

Fig. 8.4 Control points for the approximate representation of a unit circle using quadratic *nonrational* Bézier basis functions

C, and D), whereas the remaining four (P_{10}, P_{21}, P_{12} and P_{01}) are symmetrically arranged along lines that pass through the center and form angles of 45 and 135° with respect to the x-axis. These eight control points, in conjunction with the corresponding quadratic Bernstein–Bézier polynomials, may are used to represent the boundary of the circle with the same accuracy as the quadratic Lagrange polynomials. Clearly, in both cases each quarter (say AB) is represented by a parabolic arc.

In order to establish the mapping between the *interior* of the parametric unit square ([0, 1] × [0, 1]) and the interior of the circle $[(x, y) : x^2 + y^2 = 1]$, the Lagrangian macroelement requires only one extra node at the center of the circle. Similarly, the Bézierian macroelement requires the center as well, to which an extra control point, P_{11} shown in Fig. 8.4, is required. Both representations are mathematically equivalent and induce the same deviation from the true boundary of the circle.

Therefore, at this point the reader may easily understand that the only action that remains to be done for the desired transmission from the Lagrangian macroelement (or its equivalent nonrational Bézierian macroelement) toward the novel (for this book so far) *rational* Bézierian macroelement is a simple *shift* of the four control points (P_{10}, P_{21}, P_{12}, and P_{01}) from their initial position (shown in Fig. 8.4, where, e.g., $x_{P_{10}} \cong 0.9142 < 1.0$) to their final one ($x'_{P_{10}} = 1.0$, shown previously in Fig. 8.2). Again, the purpose of this slight sliding of the four control points along the two diagonals (at polar angles: $\theta = \frac{\pi}{4}, \frac{3\pi}{4}$) is to settle the minor violation of the true geometry due to the nonrational polynomials (either Lagrange or Bernstein). Not to forget that the proper weights have to be applied.

The above discussion is a natural explanation why the tensor product can be extended from the Lagrange polynomials (elements) to the tensor product of rational Bernstein–Bézier ones.

Rational Bézier (cubic, i.e., $p = 3$)
Although rational quadratic Bézier is adequate to accurately represent a circle, sometimes it is necessary to fit a cubic rational Bézier polynomial "*without changing the curve's shape.*" Increasing the degree of a Bézier curve *without* changing its shape is referred to as *degree elevation*. If we remain in the case of the circular arc AB of 90°, the degree elevation by-one increases the number of control points from the previous three in Fig. 8.2 to those four ones (P_{00}, P_{10}, P_{20} and P_{30}) shown in Fig. 8.5. Applying Eq. (8.1) in conjunction with Eq. (8.2) and replacing $N_{i,p}$ with the well-known Bernstein polynomials, $B_{i,n} = \frac{n!}{i!(n-i)!}(1-\xi)^{n-i}\xi^i$, after a tedious algebra process based on the theory of polynomials, the condition $x^2 + y^2 - 1 \equiv 0$ leads to the following control points (each column corresponds to one control point, posed as a label above the columns of the following matrix):

$$\begin{bmatrix} X \\ Y \\ Z \\ W \end{bmatrix} = \begin{bmatrix} P_0 & P_1 & P_2 & P_3 \\ 1 & 1 & a & 1 \\ 0 & a & 1 & 1 \\ 0 & 0 & 0 & 0 \\ 1 & w_1 & w_2 & 1 \end{bmatrix}, \tag{8.9}$$

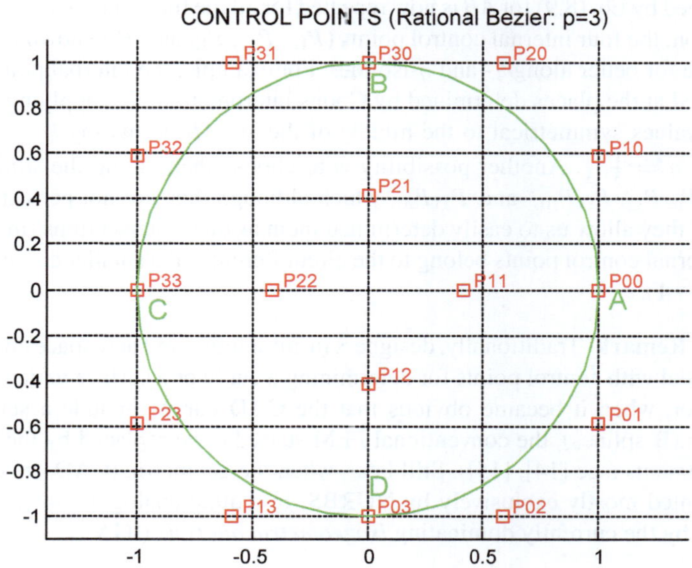

Fig. 8.5 Control points after the degree elevation, from $p=2$ to $p=3$

where

$$a = 2 - \sqrt{2} \quad \text{and} \quad w_1 = w_2 = \frac{1 + \sqrt{2}}{3} \tag{8.10}$$

Surface patch
Extending the concept from 1D to the 2D space, a surface patch is described by

$$\boldsymbol{S}(\xi, \eta) = \frac{\sum_{i=0}^{n_{p,x}} \sum_{j=0}^{n_{p,y}} \left[w_{ij} N_{i,p_x}(\xi) \cdot N_{j,p_y}(\eta) \right] \boldsymbol{P}_{ij}}{\sum_{i=0}^{n_{p,x}} \sum_{j=0}^{n_{p,y}} w_{ij} N_{i,p_x}(\xi) N_{j,p_y}(\eta)}, \quad 0 \leq \xi, \ \eta \leq 1. \tag{8.11}$$

One may observe in Eq. (8.11) that the tensor product appears mainly in the nominator (as also happened in Lagrange and Bernstein polynomials as well as in tensor B-splines), whereas the denominator is a common factor including all basis functions involved in this patch.

In the previous chapters, we saw that an entire circle may be considered as a single CAD patch *ABCD* made of four equal sides (circular arcs of 90°). As an example, if one consider the quadratic rational Bézier for each side of the quadrilateral patch, it is possible to deal with the tensor product of nine basis functions associated to the nine control points shown in Fig. 8.2. Merely, in addition to the eight control points related to the boundary of the circle, there is one more at its center (P_{11}).

If the above concept is applied to a cubic rational Bézier, the accurate description of the entire circle requires now sixteen control points of which four belong to its interior (symmetrically arranged along the diagonals *BD* and *AC*). The pattern which is described by Eq. (8.9) for *AB* is now repeated for all the four quadrants of the circle. In addition, the four internal control points (P_{11}, P_{21}, P_{12} and P_{22} shown in Fig. 8.4) should lie for better along ξ- and η-isolines. For example, the internal points can be considered at the places determined by Coons interpolation, i.e., applying Eq. (8.7) for two values symmetrical to the middle of the arc *AB*, let us say $\xi = \frac{1}{3}, \frac{2}{3}$, and similarly $\eta = \frac{1}{3}, \frac{2}{3}$. Another possibility is to choose them along the straight lines $P_{01}P_{31}$, $P_{02}P_{32}$, $P_{13}P_{10}$, and $P_{23}P_{20}$, which although they are not perfect isolines; however, they allow us to easily determined them as their intersections (in this case, these internal control points belong to the circumference of a smaller circle of radius $b = \sqrt{2} - 1$).

General Remark Traditionally, designers in automotive and aerospace engineering used to deal with control points for determining a curve or a surface under construction. Later, when it became obvious that the CAD curves include a set of basis functions (B-splines), the conventional FEM started to be replaced by the B-Spline macroelements (see [14], [15]). Still later, when the geometric (CAD) models were implemented mostly exclusively by NURBS, in a quite analogous way, FEM was replaced by the currently dominating *isogeometric analysis* (IGA).

8.3 Numerical Implementation

8.3.1 Curry–Schoenberg (de Boor) Implementation

There are many software platforms where NURBS have been implemented. For example, de Boor programmed himself in FORTRAN [10], and then his codes were transferred into the MATLAB environment [11] through approximately forty (40) functions (initial version):

(i) `aptknt, augknt, aveknt, bkbrk, brk2knt, bspligui, bspline, chbpnt, csape, csapi, csaps, cscvn, fn2fm, fnbrk, fncmb, fnder, fndir, fnint, fnjmp, fnplt, fnrfn, fntlr, fnval, getcurve, knt2brk, knt2mlt, newknt optknt.`
(ii) `ppmak, rpmak, rsmak, slvblk, sorted, spap2, spapi, spaps, spcol, spcrv, splinetool, splpp, sprpp, spmak.`

In addition to the MATLAB help command, the reader may use the initial Spline Toolbox User's Guide-version 3 by Carl de Boor [11], as well as the most updated of today (see, e.g., Curve Fitting Toolbox™ for R2018a). Usually, the polynomial degree, p, is replaced by the order, k, where by definition $k \triangleq p + 1$.

Regarding the application of B-splines and NURBS in computational methods (Galerkin–Ritz, collocation), some of the most important among the above functions are shown in Table 8.2.

Example 8.2 (A MATLAB subroutine for univariate NURBS) Develop a MATLAB function for the estimation of the rational basis functions and their derivatives.

Answer For a given knot vector `knots`, given order k, given vector `tau` of the sites $\xi_i, i = 1, \ldots,$ `length(tau)` where we wish to calculate the NURBS interpolation, and given the weighting factors weights, the values of the rational basis functions, $R_{i,p}(\xi)$ and its first derivatives $\frac{\mathrm{d}}{\mathrm{d}\xi} R_{i,p}(\xi)$, are produced by the following in house MATLAB function "`nurbscol`".

Below we use the variable `nderiv` that refers to the number of terms we consider for each rational basis function. In more detail, it takes the value "1" when we wish only the value of the basis function $R_{i,p}$, the value "2" when we additionally wish to calculate its first derivative, the value "3" we on top of that we wish the second derivative, and so on. Here we selected the case `nderiv` = 2 because (in the Galerkin–Ritz formulation of second-order PDEs) we are interested only up to the first derivative (involved in the stiffness matrix). Moreover, in plate bending analysis the second derivatives are needed as well. Higher-order derivatives are also involved in collocation methods (see Chap. 11 of this book).

Based on the MATLAB function spcol (in Spline Toolbox) which refers to the B-splines $N_{i,p}(\xi)$, one can easily understand the following subroutine that calculates

Table 8.2 Most essential MATLAB functions for B-spline CAE applications

	Command	Action
1.	`augknt(knots,k,mults)`	This MATLAB function returns a nondecreasing and *augmented* knot sequence that has the first and last knot with exact multiplicity `k`. Moreover, it returns the multiplicity `mults` to the rest elements. Therefore, the command: `augmenknots = augknt(knots,k,k-(m+1))`, where knots are the sequence of *breakpoints*, and C^m is the desired continuity [achieved when selecting `mults = k-(m+1)`], eventually gives the augmented knot sequence
2.	`spcol(knots,k,tau)`	This MATLAB function returns the matrix, with `length(tau)` rows and `length(knots)-k` columns, i.e., the number of control points, whose (*i*,*j*)th entry is $\frac{d^m N_j}{d\xi^m}$(tau(*i*)). This is the value at `tau`(*i*) of the m(*i*)th derivative of the *j*th B-spline ($N_{j,p}$) of order `k (=p+1)` for the knot sequence knots. Here, `tau` is a sequence of sites, assumed to be nondecreasing. Therefore, the basis function N_1 and its derivatives for the first block of rows, and so on until the N_{n_p} basis function and its derivatives, with n_p representing the number `length(knots)-k`, i.e., the number of control points
3.	`spmak(knots,coefs)`	Given the "`knots`" is the knot sequence and "`coefs`" which are the coefficients involved in a B-spline series, this command constructs the ppf $f(x) = \sum_{i=0}^{f-1} \lambda_i B_i(x)$ Therefore, the command: $y =$ spmak(augmenknots,coefs); gives the required approximation function

(continued)

Table 8.2 (continued)

	Command	Action
4.	f = **fnval(y,t);**	For a given vector **t** of points that belong to the interval, and for the interpolation function *y*, this command calculates the value of *y* at the points of the interval *t*
5.	ftrue = **feval(@realsol,t);**	If *t* is a vector of points and realsol is a function of variable *x*, then the command ftrue = **feval(@realsol,t)**; Calculates its value at the points of the interval *t*

the rational basis functions $R_{i,p}(\xi)$ given by Eq. (8.2), as well as its first derivative over ξ, $R'_{i,p}(\xi)$, given by Eqs. (8.5) and (8.6).

```
function[nurbs_bf,nurbs_dbf] = nurbscol(knots,k,tau,weights)

%---PROGRAM P8-1

% © 2016: Prof. C. Provatidis, and Dipl.-Ing. Aris Papangelakis, NTUA

%-------------------------------------------------------------------------------
% NOMENCLATURE:
%...Input variables:
% knots   : knot row-vector
% k       : order (= degree of piecewise polynomial + 1)
% tau     : row-vector including the points where the basis functions are to be calculated
% weights: row-vector including the weights that correspond to control points
%...Output variables:
% nurbs_bf : matrix including the basis functions Nj,p relevant to control points Pj
% nurbs_dbf: matrix including the first derivative of basis functions Nj,p
%-------------------------------------------------------------------------------
nderiv = 2;     %less than 'k-1' (number of derivatives of Ni,p + 1)
bsplines_basis_functions = spcol(knots,k,brk2knt(tau,nderiv));
bspline_bf = bsplines_basis_functions(1:nderiv:end,:); % (basis functions)

bspline_dbf = bsplines_basis_functions(2:nderiv:end,:);% (Derivatives of Basis Functions)
npoints = length(tau); %number of points (sites) where values are required
ncntrl  = size(bsplines_basis_functions,2); %number of control points.

% Calculation of rational NURBS basis functions:
for i = 1:npoints
    for j = 1:ncntrl
        nurbs_bf(i,j) = bspline_bf(i,j)*weights(j)/sum(bspline_bf(i,:).*weights);
    end
end

% Calculation of first derivatives of rational NURBS basis functions:
for i = 1:npoints
    for j = 1:ncntrl
        nurbs_dbf(i,j) = weights(j)*(sum(bspline_bf(i,:).*weights)*bspline_dbf(i,j)-
                         sum(bspline_dbf(i,:).*weights)*bspline_bf(i,j))/...
                         (sum(bspline_bf(i,:).*weights))^2;

    end
end
%%%%%%%%%%%%%%%%%%%%%%%%%%%%%%%%%%%%%%%%%%%%%%%%%%%%%%%%%%%%%%%%%%%%%%%%%%%%%%%%
```

Example 8.3 (Application example using "`nurbscol`"*)* Let us now divide the domain [0,1] into four equal segments. Considering a cubic NURBS ($p = 3$), the knot vector will be: $\boldsymbol{U} = \left[0, 0, 0, 0, \frac{1}{4}, \frac{3}{4}, \frac{3}{4}, 1, 1, 1, 1\right]$. Since there will be seven control points, a possible vector of weights may be $\boldsymbol{W} = [1, 1, 1, 3, 1, 1, 1]$. If the reader writes a simple MATLAB code, he/she will find the graphs shown in Fig. 8.6.

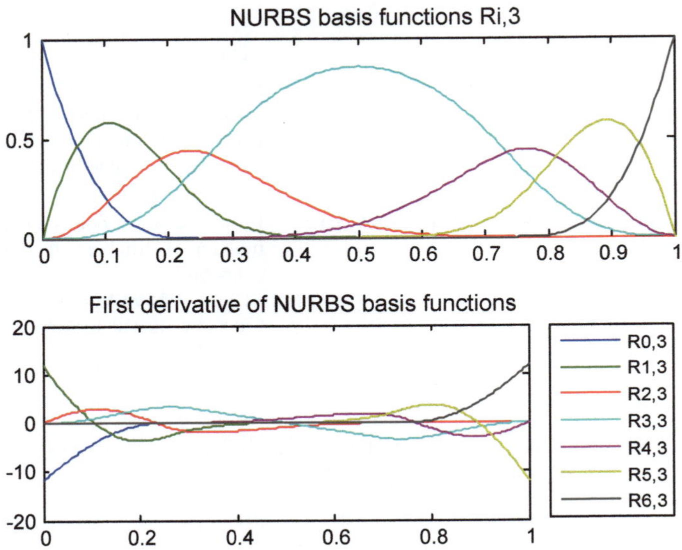

Fig. 8.6 Rational basis functions (NURBS) and their first derivatives in the interval [0,1]

8.3.2 Freeware Software

A particular increase of computer codes was noted after the publication of the popular NURBS book by Piegl and Tiller [20], which is full of pseudocodes that can be easily rewritten by the interested reader to any computer language. Within this context, the so-called *NurbsToolbox* was developed by Mark Spink and was released in the year 2000. Currently, it is not embodied in the official MATLAB but in the period this book was published the *m*-files had been stored at the following Web site address:

https://www.mathworks.com/matlabcentral/fileexchange/26390-nurbs-toolbox-by-d-m-spink

Later, the maintenance of the above collection was completed by a larger team (Mark Spink, Daniel Claxton, Carlo de Falco, Rafael Vazquez), which has added some new functions and has corrected some mistakes. The latest version can be downloaded from the website (of the open-source code OCTAVE):

https://octave.sourceforge.io/nurbs/index.html

One critical point is the operation of the function `nrbmak` (coming from NuRBs and MAKe). This function processes the data of the initial geometry and creates the `nurbs` structure. The peculiarity of this function is that it does not read the Cartesian

Table 8.3 Most essential NURBS Toolbox functions

Treatment of geometry (NURBS structure)	
nrbmak	Constructs the NURBS structure given the control points and the knots
nrbderiv	Constructs the first and second derivative representation of a NURBS curve, surface, or volume
nrbdegelev	Elevates the polynomial degree of the NURBS curve or surface
nrbkntins	Insert a single knot or multiple knots into a NURBS curve or surface
Plot of NURBS	
nrbplot	Plot a NURBS curve or surface, or the boundary of a NURBS volume
nrbkntplot	Plot a NURBS entity with the knots subdivision
nrbctrlplot	Plot a NURBS entity along with its control points
Numerical evaluation of geometry in the cartesian space	
nrbeval	Evaluate a NURBS at parametric points
nrbdeval	Evaluation of the derivative and second derivatives of a NURBS curve, surface, or volume

coordinates of the actual control points themselves but the "projective" ones. In other words, if we want to define a control point of coordinates (X, Y, Z) and weight W, we have to input the following matrix:

$$\begin{bmatrix} WX \\ WY \\ WZ \\ W \end{bmatrix}.$$

The most useful functions are shown in Table 8.3.

8.4 Numerical Implementation of Rational Bézier and NURBS in CAE Applications

8.4.1 Rational Bézier

It was clearly shown that even using quadratic interpolation, the nine control points shown in Fig. 8.2 are a sufficient set to fully describe the entire boundary (circumference) of a circle. This fully controlled case is a good example to test the quality of domain integration. In more details, although the rational Bézier basis functions operate in the entire circular domain, the unit square (i.e., the "master" or "parent" element to which the entire circle is mapped) in a first case is considered as one integration cell, and in a second case is divided into four equal integration cells (arranged in a 2×2 configuration). In both cases, a standard $n_g \times n_g$ integration per cell is implemented. From the results in Table 8.4, it becomes apparent that the conventional Gaussian quadrature does not ensure the most accurate results one would anticipate for a method that fully represents the geometry along the boundary. Nevertheless, in both cases one may observe a monotonic convergence towards the accurate value of the circle's area. Based on this finding, one could say that, whatever is gained in terms of boundary accuracy is "lost" in terms of high computational effort for the domain integration.

The numerical results of Table 8.4 are easily explained since the determinant of the Jacobian is *not* a perfect polynomial (in fact it is of rational character) and therefore the usual quadrature scheme of n_g Gauss points for a polynomial up to the degree $\tilde{p} = (2n_g - 1)$ is not easily applicable. Based on this simple example, one may anticipate that a similar shortcoming will appear for NURBS as well.

Example 8.4 (Eigenvalues of circular acoustic cavity) This problem concerns a circle of unit radius under Neumann boundary conditions, which have been previously solved (in Chap. 7) using B-splines and other methods. The novel point here is to

Table 8.4 Convergence of the calculated area of a circle modeled with nine control points

Gaussian quadrature per cell ($n_g \times n_g$)	Calculated area error (in %)	
	One cell	Four cells
2×2	−7.6448	−0.0806
3×3	0.6043	−0.0027
4×4	−3.9990e−02	1.1180e−04
5×5	2.3841e−03	−2.4438e−06
6×6	−1.3261e−04	3.9136e−08
7×7	7.0202e−06	−5.2276e−10
8×8	−3.5818e−07	5.5412e−12
9×9	1.7759e−08	−1.1309e−13

Table 8.5 Calculated eigenvalues for a circular acoustic cavity using a single 9-node macroelement of rational Bézier type and conventional Lagrangian type

Exact eigenvalues	Errors (in %) of calculated eigenvalues					
	Quadratic ($p=2$)		Cubic ($p=3$)		Quartic ($p=4$)	
	Rational Bézier	Lagrangian	Rational Bézier	Lagrangian	Rational Bézier	Lagrangian
$\omega_1^2 = 0$	–	–	–	–	–	–
$\omega_2^2 =$ 3.3900	13.89	13.75	0.39	0.43	0.09	0.28
$\omega_3^2 =$ 3.3900	13.89	13.75	0.39	0.43	0.09	0.28
$\omega_4^2 =$ 9.3284	47.15	30.76	2.31	2.11	0.64	1.13
$\omega_5^2 =$ 9.3284	172.92	135.99	36.44	47.16	1.69	2.87
$\omega_6^2 =$ 14.6820	74.08	50.60	58.40	72.16	1.62	2.59
Error of calculated area (in %)	0.00	−1.17848	0.00	0.157448	0.00	0.007191
Number of equations	9	9	16	16	25	25

apply an approximation of the variable $U(x, y)$ using rational Bézier functions and comment on the quality of the results in conjunction with the parameterization and the location of the internal control points.

Concerning the *quadratic* rational Bézier, there is no ambiguity that the internal control point P_{11} should be at the center of the circle, thus making a tensor product of rational Bernstein polynomials using totally 3×3 control points.

The numerical results for this 9-DOF rational macroelement are compared with those obtained using a standard nine-node Lagrange element applicable to the entire circle. Surprisingly, Table 8.5 ($p=2$, second and third column) depicts that for this coarse representation the quality of the rational Bézier macroelement is not better than the standard (old-fashioned) nine-node Lagrange element (for the first two nonzero eigenvalues: 13.89 > 13.75%). Using an extremely high Gaussian quadrature scheme (20×20 integration points in the entire circle), the only superiority of the Bézier macroelement is that it accurately catches the area of the circle, whereas the Lagrangian element appears an error of approximately 1.2%.

Furthermore, using a single *cubic* rational Bézier macroelement of 16 control points (see Fig. 8.5), which—as was earlier mentioned—is produced by a degree elevation of the abovementioned 9-DOF rational Bézier macroelement, the new results are again shown in Table 8.5. For this particular macroelement, the quality of the 16-

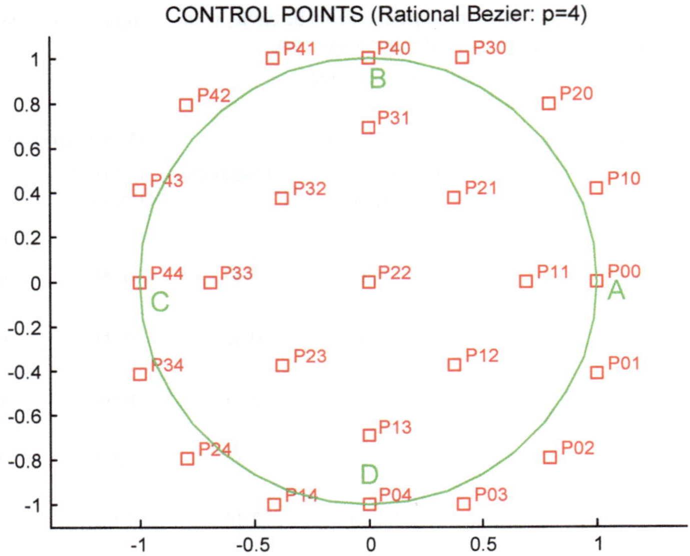

Fig. 8.7 Control points after the degree elevation, from $p = 3$ to $p = 4$

DOF rational Bézier macroelement is again of similar quality (slightly better) than that of the classical 16-DOF macroelement of Lagrangian type which now deviates from the exact area by 0.16% only. For example, for the first two nonzero eigenvalues, the numerical error is $0.39 < 0.43\%$.

Moreover, applying degree elevation for one more time, the 25-DOF *quartic* rational Bézier macroelement (shown in Fig. 8.7), appears a still better behavior as shown in Table 8.5. A further degree elevation at $p = 5$ produces still better results, where the Lagrangian-type element is still competitive (appears errors of 0.01% for the first two and 0.05% for the next eigenvalue).

Remark on the degree elevation It is worthy to mention (to the younger readers) that degree elevation is known at least since early 1980s [21] and has kept the researchers busy about how to represent the full circle [3]. For more details, the reader is referred to Piegl and Tiller ([20], p. 200).

Remarks on the internal control points (parameterization) Concerning the cubic rational Bézier macroelement which has been produced from the degree elevation of the quadratic element along each of the four arcs (*AB*, *BC*, *CD*, and *DA*) of the quadrilateral, there is *no* doubt about the position of the control points associated to the *boundary*. To this purpose, the reader may use either the function `nrbdegelev` of the NURBS Toolbox for one-quarter of the boundary or derive them analytically. In contrast, the proper location of the four internal control points (arranged in a 2×2 symmetric scheme around the center of the circle), which are necessary to complete the tensor product of totally 4×4 control points, give us a trouble.

Clearly, in case of the analogous tensor-product Lagrange macroelement (of 16 nodes) the twelve boundary nodes would be uniformly distributed along the circumference, whereas the internal nodes could be easily constructed along the four isolines $\xi = \frac{1}{3}, \frac{2}{3}$ and $\eta = \frac{1}{3}, \frac{2}{3}$ which note that *pass through* the boundary nodes, using the simple Coons interpolation which is based on the boundary nodes only [see Eq. (8.7)]. The difficulty here with the cubic rational Bézier macroelement under examination is that the control point P_{10} does not belong to the boundary (it lies outside the circle), and therefore, it is under question where to pass and how to construct the isolines (note that in the tensor product the unknown internal control points participate as well).

To make this point more clear, the straight line that connects P_{10} with the center K of the circle intersects the circumference at the point Q (see Fig. 8.8) with parametric coordinate $\xi_Q \cong 0.34495$, which differs from the ratio $(AQ)/(AB)$ of the arc lengths. Furthermore, if we connect the corresponding control points P_{10} and P_{13}, this line forms 45° with the x-axis and intersects the circumference at another point R with parameter $\xi_R \cong 0.31926$. Obviously, for each value $\xi_I (< \frac{1}{2})$, which can be easily written in the form $\xi_I = 0.5 - \xi_a$, the parameters (ξ, η) of the four internal control points will be: $P_{11}\big(\frac{1}{2} - \xi_a, \frac{1}{2} - \xi_a\big)$, $P_{21}\big(\frac{1}{2} + \xi_a, \frac{1}{2} - \xi_a\big)$, $P_{12}\big(\frac{1}{2} - \xi_a, \frac{1}{2} + \xi_a\big)$ and $P_{22}\big(\frac{1}{2} + \xi_a, \frac{1}{2} + \xi_a\big)$. Therefore, eventually the problem is to select the proper value of the parameter ξ_I on which the control points P_{11} and P_{12} lie. In case the points P_{10}, P_{11}, P_{12} and P_{13} are collinear, the reader may validate that the four internal control points (see Fig. 8.8) lie on a circle of radius $b = \sqrt{2} - 1$, thus $\xi_a \cong 0.1388$ $(\xi_I = 0.361177)$. For this particular choice, the obtained eigenvalues for the circular acoustic cavity lead to slightly improved results than those shown in Table 8.5 under the label cubic Bézier.

In principle, one could start with an initial mesh of internal control points produced by Eq. (8.7), for example, for a cubic Bézerian element ($p = 3$) could start with the isolines $\xi = \frac{1}{3}, \frac{2}{3}$ and $\eta = \frac{1}{3}, \frac{2}{3}$ based on boundary data only and then perform a few iterations (until convergence within a given tolerance) using Eq. (8.1) in conjunction with the previously found values at internal control points. Such a relaxation scheme may lead to a fully controlled and reproducible net of inner control points.

Similar remarks may be made for the quartic rational Bézier macroelement (25-DOF, degree $p = 4$) as well. In order to offer an easy solution that can be easily repeated by the reader, the internal points were determined using Coons interpolation using Eq. (8.7).

8.4.2 NURBS-Based Isogeometric Analysis (IGA)

Obviously, the rational Bézier patch which we treated in the previous section is known since 1970s, that is well before the logo NURBS was even known. With this, we want to point out that we can deal with rational Bézier macroelements in a direct way, using rational Bernstein polynomials $B_{i,n}$ only, without resorting to concepts such as knot vector and subsequent basis functions $N_{i,p}$ that require a short but special education. On the other point of view, rational Bézier is a rational B-spline (NURBS) *without* having inner knots; thus, it is a special case (subset) of NURBS. Therefore, in general lines both sets of basis functions are anticipated to have similar advantages or to suffer from similar shortcomings. In brief, one of the basic differences between them is that rational Bézier does not have any compact support, whereas NURBS does have. Moreover, the advantage of NURBS is that the knot vector offers the capability to fully control the continuity of the approximate solution or/and the geometry, locally and globally. For an overview until 2009, the reader is referred to Cottrell et al. [5]. The NURBS-based isogeometric analysis was founded by Hughes et al. [12].

If we remain to the case of the circular cavity, which is a typical case that requires the use of *rational* elements, it is worthy to mention that some early IGA papers referred to several alternative parameterizations of a circle. For example, Cottrell et al. ([6], p. 5280) analyzed a clamped circular plate using "eight elements within a single patch". Cohen et al. ([4], p. 343) presented three ways to represent a disk of radius one. Among them, the third one is probably that used by us in Sect. 8.4.1 of the book in hands, of course in conjunction with rational Bézier polynomials (not advanced NURBS). It is perhaps of interest to mention that the idea of using a single CAD-based (Coons or transfinite) macroelement for modeling the entire circle had been proposed before the presentation of IGA, initially for acoustics using boundary-only formulation ([22], p. 51) and at the same time for Poisson problems using one

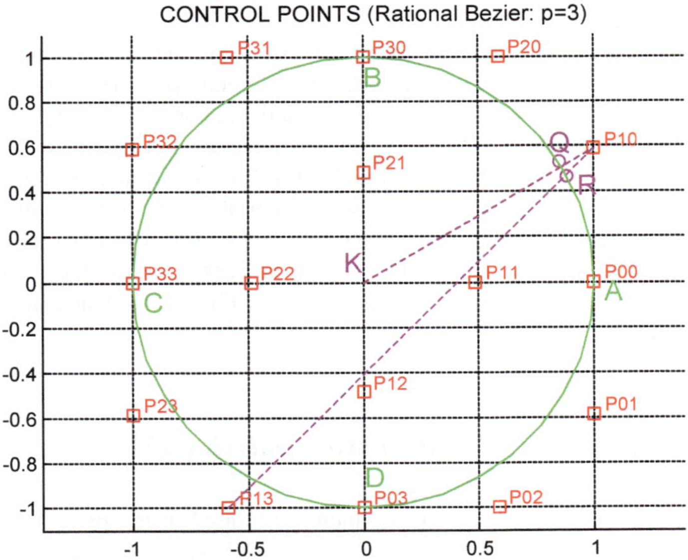

Fig. 8.8 A possible scenario for alternative control points inside a circle approximated with the cubic rational Bézier curve

internal point only in conjunction with Lagrange polynomials [24]. Later, transfinite interpolation was applied again for 2D acoustics but now using as many internal points as we like ([25], p. 490).

Two different parameterizations of a circular plate have been studied by Schmidt et al. ([29], pp. 320–322). The passage from the initial geometry description of a disk to the computational domain has been showed by Vuong et al. ([30], pp. 653–654). In a similar way as happened with the Gordon–Coons macroelement (see, for instance, [23]), isogeometric analysis requires more elaborate integration schemes for its computation [17]. Efficient quadrature schemes have been reported by many investigators [1, 13, 27, 28], among others.

The significance of the geometric parameterization is an old topic that has been studied initially within the context of the Coons interpolation [16] as well as occasionally in many papers of the author (e.g., Provatidis [26], pp. 345–347). In the framework of NURBS-based IGA, selected reports are [31–33] and [34] among others.

Regarding software tools for research in isogeometric analysis, two interesting papers describe GeoPDEs [9] and ISOGAT [30], whereas an overview and further computer implementation aspects were reported by Nguyen et al. [18].

Since this book deals with the *Precursors* of IGA, it is not our aim to deal or comment on the successive NURBS-based IGA. The reader who is interested in this topic may consult the current book of IGA founders [5], and also a more recent one by Buffa and Sangalli [2] which includes T-splines as well. In any case, self-programming is the best guide and offers the maximum elucidation.

8.5 Recapitulation

In Chap. 4, it was shown that older CAD-based macroelements (Coons, Gordon) are capable of successfully handling curved patches such as a circle. For an acoustic circular cavity, a single Coons-type macroelement (for the entire circle) in which the nodal points are along the boundary only leads to a numerical solution of low quality (converges with an error 4.1 % for the first two nonzero eigenvalues). In contrast, a single Gordon-type transfinite element gives improved and rapidly converging results, with a top of accuracy when this element becomes a conventional tensor-product Lagrangian element. Despite the fact that the latter element does not accurately represent the geometry, an increase of the nodes (arranged in a $n \times n$ uniform scheme in the parametric unit square) rapidly reduces the deviation from the accurate area and also rapidly converges to the accurate solution (eigenvalues and eigenmodes). It is worthy to point out that, although the dominating view that Lagrangian elements suffer from numerical oscillations is not false, excellent numerical results are generally obtained for relatively small number of unknowns (degrees of freedom) well before this instability is active. For example, using eight to ten divisions per side of the unit square the numerical solution is usually excellent and we have observed no instability so far (particularly in potential problems).

Before we move from the classical transfinite elements, of which the crest is the tensor-product Lagrangian element, we have to remind that the latter are mathematically equivalent to the *nonrational* Bézierian elements because they share the same functional set of monomials $x^i y^j$ (see details in Chap. 6). Therefore, the transmission from the classical world of Lagrangian macroelements to the "modern" world of rational ones passes through the transmission from *nonrational to rational* Bézierian elements. In an instructive way, in this chapter we showed that a quadratic nonrational Bézierian element has nine control points (weights equal to the unity) among which five points coincide with those of the rational element, whereas the rest four control points lie along the diagonals of the circumscribed square at $x, y \cong \pm 0.9142$. Therefore, the conceptual transmission from the nonrational to the rational Bézierian element is the rather slight shift of the four aforementioned control points from the positions $x, y \cong \pm 0.9142$ to $x, y = \pm 1$, provided the proper weights are used.

For the case of the acoustic cavity that we dealt in this chapter, it was found that the quadratic rational Bézierian element did not perform better than the corresponding nonrational one (or its equivalent quadratic Lagrangian element). Nevertheless, degree elevation led to excellent results, clearly better but not dramatically more accurate than those obtained applying the nonrational Bézierian element. It was also shown that the classical Gaussian scheme requires many integration points or a dedicated quadrature scheme. Another interesting point for further study is the proper choice of the internal nodes/control points (affecting the parameterization of the circle), which plays a certain role in the accuracy of the numerical solution. Overall, there is no ambiguity about the usefulness of CAD-based macroelements due to the easiness they provide in the coupling of FEA with CAD models. In this chapter, we obtained numerical results for only nonrational and rational Bézier elements (these curves are found in older CAD systems such as UNISURF, EUCLID_IS). For the particular case of a circular cavity which is analyzed using a single macroelement, it is still a matter of discussion whether it is worthy to apply rational instead of classical nonrational Bézierian elements. In other words, after so many tedious modifications, the profit over the classical Lagrangian elements (equivalent to nonrational Bézierian elements) is not that tremendous. Therefore, in primitive shapes, a single classical Lagrangian macroelement (or its equivalent nonrational Bézierian macroelemeent) is still valuable. Of course, a fair numerical comparison had to include NURBS as well but this issue is outside the scope of the book in hands.

Of course, in real structures the analyst is obliged to decompose the structure into smaller substructures (using tensor-product macroelements), where coupling (mortar) difficulties along the interface of neighboring macroelements should be resolved. This topic is discussed in Chap. 13 of this book.

References

1. Auricchio F, Calabrò F, Hughes TJR, Reali A, Sangalli G (2012) A simple algorithm for obtaining nearly optimal quadrature rules for NURBS-based isogeometric analysis. Comput Methods Appl Mech Eng 249–252:15–27

2. Buffa A, Sangalli G (eds) (2016) Isogeometric analysis: a new paradigm in the numerical approximation of PDEs. Springer International Publishing, Switzerland
3. Chou JJ (1995) Higher order Bézier circles. Comput Aided Des 27(4):303–309
4. Cohen E, Martin T, Kirby RM, Lyche T, Riesenfeld RF (2010) Analysis-aware modeling: understanding quality considerations in modeling for isogeometric analysis. Comput Methods Appl Mech Eng 199:334–356
5. Cottrell J, Hughes TJR, Bazilevs Y (2009) Isogeometric analysis: toward integration of CAD and FEA. Wiley, New York
6. Cottrell JA, Reali A, Bazilevs Y, Hughes TJR (2006) Isogeometric analysis of structural vibrations. Comput Methods Appl Mech Eng 195:5257–5296
7. Curry HB, Schoenberg IJ (1966) On Pólya frequency functions IV: the fundamental spline functions and their limits. J Anal Math 17(1):71–107
8. De Boor C (1972) On calculating with B-splines. J Approximation Theor 6:50–62
9. De Falco C, Reali A, Vázque R (2011) GeoPDEs: a research tool for isogeometric analysis of PDEs. Adv Eng Softw 42:1020–1034
10. De Boor C (2001) A practical guide to splines, revised edition. Springer, New York, Berlin
11. De Boor C (2000) Spline toolbox for use in MATLAB, user's guide, version 3. The MathWorks Inc
12. Hughes TJR, Cottrell JA, Bazilevs Y (2005) Isogeometric analysis: CAD, finite elements, NURBS, exact geometry and mesh refinement. Comput Methods Appl Mech Eng 194(39–41):4135–4195
13. Hughes TJR, Reali A, Sangalli G (2010) Efficient quadrature for NURBS-based isogeometric analysis. Comput Methods Appl Mech Eng 199:301–313
14. Höllig K (2003) Finite element methods with B-splines. SIAM, Philadelphia
15. Höllig K (2002) Finite element approximation with splines. In: Farin G, Hoschek J, Kim MS (eds) Handbook of computer aided geometric design. North-Holland, Amsterdam, Chapter 11, pp 283–307
16. Kanarachos A, Röper O (1979) Rechnerunterstützte netzgenerierung mit hilfe der Coonsschen abbildung. VDI-Z 121:297–303
17. Karatarakis A, Karakitsios P, Papadrakakis M (2014) GPU accelerated computation of the isogeometric analysis stiffness matrix. Comput Methods Appl Mech Eng 269:334–355
18. Nguyen VP, Anitescu C, Bordas SPA, Rabczuk T (2015) Isogeometric analysis: an overview and computer implementation aspects. Math Comput Simul 117:89–116
19. Piegl L, Tiller W (1989) Circles: a menagerie of rational B-splines circles. IEEE Comput Appl 9(5):48–56
20. Piegl L, Tiller W (1997) The NURBS book. Springer, Berlin
21. Prautzsch H (1984) Degree elevation of B-spline curves. Comput Aided Geom Des 1(1):193–198
22. Provatidis CG (2004) On DR/BEM for eigenvalue analysis of 2-D acoustics. Comput Mech 35:41–53
23. Provatidis CG (2004) Coons-patch macroelements in two-dimensional eigenvalue and scalar wave propagation problems. Comput Struct 82(4–5):383–395
24. Provatidis CG (2004) Solution of two-dimensional Poisson problems in quadrilateral domains using transfinite Coons interpolation. Commun Numer Methods Eng 20(7):521–533
25. Provatidis CG (2009) Eigenanalysis of two-dimensional acoustic cavities using transfinite interpolation. J Algorithms Comput Technol 3(4):477–502
26. Provatidis CG (2006) Coons-patch macroelements in two-dimensional parabolic problems. Appl Math Model 30(4):319–351
27. Rypl D, Patzák B (2012) Study of computational efficiency of numerical quadrature schemes in the isogeometric analysis. In: Proceedings 18th international conference engineering mechanics 2012, Svratka, Czech Republic, 14–17 May 2012, Paper #304, pp 1135–1143
28. Schillinger D, Hossain SJ, Hughes TJR (2014) Reduced Bézier element quadrature rules for quadratic and cubic splines in isogeometric analysis. Comput Methods Appl Mech Eng 277:1–45

29. Schmidt R, Kiendl J, Bletzinger K-U, Wüchner R (2010) Realization of an integrated structural design process: analysis-suitable geometric modeling and isogeometric analysis. Comput Vis Sci 13:315–330
30. Vuong A-V, Heinrich Ch, Simeon B (2010) ISOGAT: a 2D tutorial MATLAB code for isogeometric analysis. Comput Aided Geom Des 27:644–655
31. Xu G, Mourrain B, Duvigneau R, Galligo A (2011) Parameterization of computational domain in isogeometric analysis: methods and comparison. Comput Methods Appl Mech Eng 200:2021–2031
32. Xu G, Mourrain B, Duvigneau R, Galligo A (2013) Analysis-suitable volume parameterization of multi-block computational domain in isogeometric applications. Comput Aided Des 45:395–404
33. Xu G, Mourrain B, Duvigneau R, Galligo A (2013) Optimal analysis-aware parameterization of computational domain in 3D isogeometric analysis. Comput Aided Des 45:812–821
34. Xu G, Mourrain B, Galligo A, Rabczuk T (2014) High-quality construction of analysis-suitable trivariate NURBS solids by reparameterization methods. Comput Mech 54:1303–1313
35. Zienkiewicz OC (1977) The finite element method, 3rd edn. McGraw-Hill, London

Chapter 9
Plate Bending Macroelements

Abstract This chapter deals with plate bending analysis applying several CAD-based interpolations. First the performance of boundary-only Coons interpolation is studied; it will be shown that its simplest form coincides with the well-known BFS element of mid-1960s. Then Gordon interpolation is used (i.e., internal nodes are inserted) in order to improve the accuracy of the numerical solution; it will be shown that the Hermite tensor-product element is a special case. The applicability of Bernstein–Bézier interpolation, as a substitute of Lagrange and Hermite polynomials, is discussed in detail. Also, the use of B-splines is examined and it is clearly shown that the barriers are broken when a control points-based tensor product is applied to curvilinear domains. Numerical examples include rectangular and circular thin plates which are solved using a single macroelement.

Keywords Plate bending · Blending functions · C^1-continuity Coons formula
BFS element · Transfinite · Tensor-product Hermites · Eigenvalues
Bernstein–Bézier · Beam bending · B-splines · Rational macroelement
Test cases

9.1 General Introduction

Plate bending analysis requires the use of deflections, slopes and sometimes of the curvature. All these parameters have been independently studied in the context of CAD (e.g. Farin [9]) and the FEM ([2, 12, 16], among others) as well. Below we shall show that some ideas can be transferred from CAD to FEM in a similar way they were transferred in potential and plane elasticity problems (in previous chapters of this book).

We start with the first recipe proposed in CAD, and particularly the C^1-version of *Coons* interpolation formula [7]. The latter formula deals with numerical data along the four edges of the patch ABCD as usual. In addition to the previously mentioned cases in this book which are concerned with the basic variable $w(x, y)$ along the boundary, it is possible to deal with the *derivatives* $w_x = \partial w/\partial x$ and $w_y = \partial w/\partial y$

C. G. Provatidis, *Precursors of Isogeometric Analysis*, Solid Mechanics and Its Applications 256, https://doi.org/10.1007/978-3-030-03889-2_9

as well. Based on this particular C^1-version of Coons interpolation, we can extract the classical Bogner-Fox-Schmit (BFS) plate bending element [4]. For a *fourth* time in this book, this finding validates the rule that the generalized Coons (and Barnhill) interpolation formulas are a powerful "vehicle" for producing all the classical finite elements. Concerning large Coons macroelements with degrees of freedom along the boundary only, a great number of numerical examples for static analysis have been reported by Provatidis and Angelidis [22].

Later in this book, issues of *transfinite* plate macroelements (with internal points into the *Gordon*-like macroelement) will be discussed. The extreme transfinite case of *Hermitian* tensor-product elements is fully studied, and many numerical results are presented to demonstrate its superiority over small size classical BFS [4] and MZC [15, 27] rectangular finite elements. An early report on the performance of tensor-product Hermitian elements in eigenvalue problems is due to Provatidis [20]. It is pointed out that the tensor-product Hermite elements have been criticized that they "must be restricted to consist of *rectangles*" [6].

Concerning the applicability of *Bernstein–Bézier* macroelements in plate bending, we begin our discussion with beam bending, which (the latter) is an instructive analogue that substantially elucidates the former topic. Hermite polynomials are compared with Bernstein polynomials and some equivalencies are discussed. It will be shown that Hermite polynomials that can be used for the global approximation of the deflection, are closely related with proper Bernstein ones.

Finally, interpolation using B-splines is studied for beam bending analysis, whereas later the tensor-product *B-splines* interpolation is discussed for rectangle and curvilinear plates. For the case of beam bending, it will be shown that cubic B-splines in conjunction with C^1-continuity (multiplicity of inner knots equal to 2) is equivalent with the traditional piecewise Hermite polynomials of degree three that may be found in all finite element texts (see, Provatidis and Isidorou [23]).

For the case of plate bending, in contrast to the cases of Coons- and Gordon-based transfinite macroelements that were restricted to *rectangles*, the use of knots and control points "breaks the barriers" and make *the same idea* to survive in curvilinear plates as well. Again, although the opinion according to which tensor-product Hermite polynomials "must be restricted to consist of rectangles" [6] is correct, however, through the use of cubic polynomials in conjunction with double knots the former opinion is abolished. Now, the position of the quartet $(w, \partial w/\partial x, \partial w/\partial y, \partial^2 w/\partial x \partial y)$ is taken over by a large number of degrees of freedom (DOF) associated to many control points. The performance of these macroelements will be presented for square and circular macroelements (simply supported and clamped).

9.2 Coons Interpolation Using Derivatives

9.2.1 CAD Applications

In case of plate bending where C^1-continuity is required between two adjacent surface patches, the relevant theory is included again into the famous report by the late Professor Coons [7] at MIT that was published twice in the years 1964–1967. Below we shall see that, using this new formula, we can build up large CAD-based macroelements useful for thin plate bending and shell analysis. The basic difference between the Coons macroelement and the most known plate elements is that in the former each nodal point is associated, not only to the flexure and the two slopes, but also the second mixed derivative $(\partial^2 w/\partial x \partial y)$.

We consider again a unit square patch $A'B'C'D'$ on the parametric $\xi\eta$-plane ($0 \leq \xi, \eta \leq 1$) which is mapped to the actual quadrilateral domain $ABCD$ of the xy-plane, as shown in Fig. 9.1.

In order to ensure C^1-continuity, the first partial derivatives with respect to ξ and η along the four edges of the patch should be given:

$$\frac{\partial w(0,\eta)}{\partial \xi}, \frac{\partial w(1,\eta)}{\partial \xi}, \frac{\partial w(\xi,0)}{\partial \eta}, \frac{\partial w(\xi,1)}{\partial \eta}$$

as well as the second mixed derivatives at the corners $A(0,0)$, $B(1,0)$, $C(1,1)$, and $D(0,1)$ of the unit reference square:

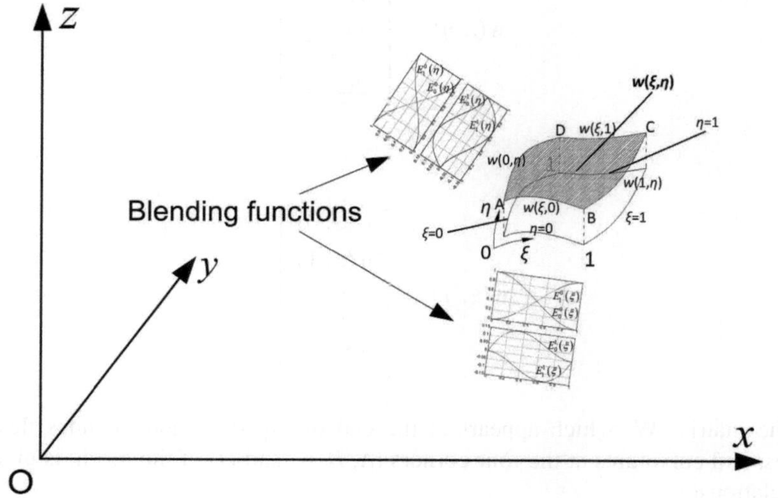

Fig. 9.1 Coons macroelement for the analysis of thin plate bending

$$\frac{\partial^2 w(0,0)}{\partial\xi\partial\eta}, \frac{\partial^2 w(1,0)}{\partial\xi\partial\eta}, \frac{\partial^2 w(1,1)}{\partial\xi\partial\eta}, \frac{\partial^2 w(0,1)}{\partial\xi\partial\eta}.$$

With given the values of the variable $\mathbf{w}$ (deflections and first derivatives) along the four edges of the boundary ($\mathbf{w}(\xi, 0), \mathbf{w}(\xi, 1), \mathbf{w}(0, \eta), \mathbf{w}(1, \eta)$) of the patch, according to Coons interpolation (for C^1-continuity) at the point $P(\xi, \eta)$ of its interior $\Omega = ABCD$, the corresponding value is approximated by:

$$\mathbf{w}(\xi,\eta) = \mathbf{E}^{\mathrm{T}}(\xi)\mathbf{w}(i,\eta) + \mathbf{E}^{T}(\eta)\mathbf{w}(\xi,j) - \mathbf{E}^{T}(\xi)\mathbf{W}(i,j)\mathbf{E}(\eta), \tag{9.1}$$

where the corresponding blending functions are the Hermite polynomials of third degree (well known from beam bending FEM analysis) given by the vector:

$$\mathbf{E}(s) = \begin{bmatrix} E_0^0(s) \\ E_1^0(s) \\ E_0^1(s) \\ E_1^1(s) \end{bmatrix} = \begin{bmatrix} 1 - 3s^2 + 2s^3 \\ 3s^2 - 2s^3 \\ s - 2s^2 + s^3 \\ -s^2 + s^3 \end{bmatrix}, \quad s = \xi, \eta. \tag{9.2}$$

Equation (9.1) is called "Coons interpolation formula with C^1-continuity".

Also, the vectors including the flexure and the slopes along the boundary are given as:

$$\mathbf{w}(i,\eta) = \begin{bmatrix} w(0,\eta) \\ w(1,\eta) \\ \frac{\partial w(0,\eta)}{\partial\xi} \\ \frac{\partial w(1,\eta)}{\partial\xi} \end{bmatrix} \tag{9.3}$$

and

$$\mathbf{w}(\xi,j) = \begin{bmatrix} w(\xi,0) \\ w(\xi,1) \\ \frac{\partial w(\xi,0)}{\partial\eta} \\ \frac{\partial w(\xi,1)}{\partial\eta} \end{bmatrix}. \tag{9.4}$$

The matrix, $\mathbf{W}$, which appears at the end of Eq. (9.1) and includes flexures, slopes, and curvatures at the four corners (A, B, C, and D) of the patch, is given by the relationship:

$$\mathbf{W}(i, j) = \begin{bmatrix} w_{00} & w_{01} & w_{00,\eta} & w_{01,\eta} \\ w_{10} & w_{11} & w_{10,\eta} & w_{11,\eta} \\ w_{00,\xi} & w_{01,\xi} & w_{00,\xi\eta} & w_{01,\xi\eta} \\ w_{10,\xi} & w_{11,\xi} & w_{10,\xi\eta} & w_{11,\xi\eta} \end{bmatrix}, \tag{9.5}$$

with $w_{kl} = w(k, l)$ and $w_{kl,s} = \partial w(k, l)/\partial s$, $s = \xi, \eta$.

The index 'i' (or 'k') takes the values 0 or 1 and refers to the parameter ξ. Similarly, the index 'j' (or 'l') takes the values 0 or 1 and refers to the parameter η. Also, in Eq. (9.2), the upper index '0' denotes flexure only, whereas '1' denotes slopes. Furthermore, the subscript '0' denotes the left and the right side of the patch *ABCD* which are vertical to the corresponding axes ξ or η. For example, if $s = \xi$, then the blending function $E_0^0(\xi)$ affects the flexures along the edge *DA*, whereas $E_1^0(\xi)$ affects the flexures along the edge *BC*. Correspondingly, $E_0^1(\xi)$ affects the slopes $\partial w/\partial\eta$ along the edge *DA*, whereas $E_1^1(\xi)$ affects the slopes $\partial w/\partial\eta$ along the edge *BC*.

9.2.2 Development of Coons Macroelements for Plate Bending (C^1-Continuity)

Definition of trial functions (Hermite polynomials)

Suppose that one out of the four edges of the Coons patch *ABCD* (say *AB*) consists of n nodes, that is of $(n-1)$ intervals. Then, the flexure $w(s_i)$ and the slope $w'(s_i)$, $i = 1, \ldots, n$ along it may be interpolated in terms of the nodal values, based on the well-known formula of numerical analysis (see also Chap. 2):

$$w(s) = \sum_{i=1}^{n} w(s_i)H_i(s) + \sum_{i=1}^{n} w'(s_i)\overline{H}_i(s), \tag{9.6}$$

where

$$H_i(s) = \left[1 - 2(s - s_i)L_i'(s_i)\right][L_i(s)]^2, \tag{9.7}$$

and

$$\overline{H}_i(s) = (s - s_i)[L_i(s)]^2, \tag{9.8}$$

represent Hermite polynomials, which in turn are expressed in terms of Lagrange polynomials $L_i(s)$:

$$L_i(s) = \prod_{j=1, j\neq i}^{n} \frac{s - s_j}{s_i - s_j} \tag{9.9}$$

and $L_i'(s)$ is the first derivative with respect to the parameter s (=ξ or η).

The cardinality properties of Hermite polynomials (which participate in our problem either as blending or as trial functions) are:

$$\left.\begin{aligned} H_j(s_i) &= \delta_i^j,\ H_j'(s_i) = 0 \\ \overline{H}_j(s_i) &= 0,\ \overline{H}_j'(s_i) = \delta_i^j \end{aligned}\right\} \tag{9.10}$$

where δ_i^j is Kronecker delta.

Discretization of the Coons macroelement

Considering now that q_1, q_2, q_3, and q_4 is the number of nodes along the four boundary edges AB, BC, CD, and DA, the application of (9.6) on each side of the quadrilateral Coons patch gives:

$$\begin{aligned} w(\xi,0) &= \sum_{i=1}^{q_1} w(\xi_i,0)H_i(\xi) + \sum_{i=1}^{q_1} w'(\xi_i,0)\overline{H}_i(\xi) \\ w(1,\eta) &= \sum_{i=1}^{q_2} w(1,\eta_i)H_i(\eta) + \sum_{i=1}^{q_2} w'(1,\eta_i)\overline{H}_i(\eta) \\ w(\xi,1) &= \sum_{i=1}^{q_3} w(\xi_i,1)H_i(\xi) + \sum_{i=1}^{q_3} w'(\xi_i,1)\overline{H}_i(\xi) \\ w(0,\eta) &= \sum_{i=1}^{q_4} w(0,\eta_i)H_i(\eta) + \sum_{i=1}^{q_4} w'(0,\eta_i)\overline{H}_i(\eta) \end{aligned} \tag{9.11}$$

Therefore, substituting (9.11) into (9.3) and (9.4), the vectors of generalized displacement become:

$$\mathbf{w}(\xi,j) = \begin{bmatrix} w(\xi,0) \\ w(\xi,1) \\ \frac{\partial w(\xi,0)}{\partial \eta} \\ \frac{\partial w(\xi,1)}{\partial \eta} \end{bmatrix} = \begin{bmatrix} \sum_{i=1}^{q_1} w(\xi_i,0)H_i(\xi) + \sum_{i=1}^{q_1} w_\xi(\xi_i,0)\overline{H}_i(\xi) \\ \sum_{i=1}^{q_3} w(\xi_i,1)H_i(\xi) + \sum_{i=1}^{q_3} w_\eta(\xi_i,1)\overline{H}_i(\xi) \\ \sum_{i=1}^{q_1} w_\eta(\xi_i,0)H_i(\xi) + \sum_{i=1}^{q_1} w_{\eta\xi}(\xi_i,0)\overline{H}_i(\xi) \\ \sum_{i=1}^{q_3} w_\eta(\xi_i,1)H_i(\xi) + \sum_{i=1}^{q_3} w_{\eta\xi}(\xi_i,1)\overline{H}_i(\xi) \end{bmatrix} \tag{9.12}$$

and

$$\mathbf{w}(i,\eta)=\begin{bmatrix} w(0,\eta) \\ w(1,\eta) \\ \frac{\partial w(0,\eta)}{\partial \xi} \\ \frac{\partial w(1,\eta)}{\partial \xi} \end{bmatrix}=\begin{bmatrix} \sum_{i=1}^{q_4} w(0,\eta_i)H_i(\eta)+\sum_{i=1}^{q_4} w_\eta(0,\eta_i)\overline{H}_i(\eta) \\ \sum_{i=1}^{q_2} w(1,\eta_i)H_i(\eta)+\sum_{i=1}^{q_2} w_\eta(1,\eta_i)\overline{H}_i(\eta) \\ \sum_{i=1}^{q_4} w_\xi(0,\eta_i)H_i(\eta)+\sum_{i=1}^{q_4} w_{\xi\eta}(0,\eta_i)\overline{H}_i(\eta) \\ \sum_{i=1}^{q_2} w_\xi(1,\eta_i)H_i(\eta)+\sum_{i=1}^{q_2} w_{\xi\eta}(1,\eta_i)\overline{H}_i(\eta) \end{bmatrix} \tag{9.13}$$

As usual, the flexure $\mathbf{w}(\xi,\eta)$ in the interior of Coons macroelement is expressed as a superposition of the degrees of freedom, that is

$$\mathbf{w}(\xi,\eta)=\sum_{i=1}^{q_{tot}} N_i(\xi,\eta)\mathbf{d}_i, \tag{9.14}$$

where q_{tot} is the total number of the boundary nodes obviously given by:

$$q_{tot}=q_1+q_2+q_3+q_4-4. \tag{9.15}$$

In Eq. (9.14), the generalized nodal vector $\mathbf{d}_i$ is:

$$\mathbf{d}_i=\begin{bmatrix} w_i & \frac{\partial w_i}{\partial \xi} & \frac{\partial w_i}{\partial \eta} & \frac{\partial^2 w_i}{\partial \xi \partial \eta} \end{bmatrix}^T, \tag{9.16}$$

and refers to boundary only degrees of freedom at the ith nodal point.

Combining Eqs. (9.3), (9.4), (9.5), and equating (9.1) with (9.14), after manipulation the global shape functions are classified in three categories and are determined as follows:

A. **Shape functions along the edges *AB* and *CD***, excluding the corner nodes, i.e., at the nodes $i=(2,\ \ldots,\ q_1-1)$ and $(q_1+q_2,\ \ldots,\ q_1+q_2+q_3-3)$, which are placed along $w(\xi,j),\ j=0,1$, are:

$$N_i(\xi,\eta)=\begin{bmatrix} N_1(\xi,\eta) \\ N_2(\xi,\eta) \\ N_3(\xi,\eta) \\ N_4(\xi,\eta) \end{bmatrix}=\begin{bmatrix} E_j^0(\eta)H_i(\xi) \\ E_j^0(\eta)\overline{H}_i(\xi) \\ E_j^1(\eta)H_i(\xi) \\ E_j^1(\eta)\overline{H}_i(\xi) \end{bmatrix} \tag{9.17}$$

B. **Shape functions along the edges *BC* and *DA***, excluding the corners nodes, i.e., at the nodes $i=(q_1+1,\ \ldots,\ q_1+q_2-2), (q_1+q_2+q_3-1), \ldots, (q_1+q_2+q_3+q_4-4)$, which are placed along $w(j,\eta),\ j=0,1$, are:

$$N_i(\xi,\eta)=\begin{bmatrix}N_1(\xi,\eta)\\N_2(\xi,\eta)\\N_3(\xi,\eta)\\N_4(\xi,\eta)\end{bmatrix}=\begin{bmatrix}E_j^0(\xi)H_i(\eta)\\E_j^1(\xi)H_i(\eta)\\E_j^0(\xi)\overline{H}_i(\eta)\\E_j^1(\xi)\overline{H}_i(\eta)\end{bmatrix}\tag{9.18}$$

C. **Shape functions at corner nodes**

For each corner node of the quadrilateral $ABCD$, we have:
Node $\boldsymbol{A}$, located in the corner at the position $(\xi_A,\ \eta_A)=(0,\ 0)$:

$$\mathbf{N}_1(\xi,\eta)=\begin{bmatrix}N_1(\xi,\eta)\\N_2(\xi,\eta)\\N_3(\xi,\eta)\\N_4(\xi,\eta)\end{bmatrix}=\begin{bmatrix}E_0^0(\xi)H_1(\eta)+E_0^0(\eta)H_1(\xi)-E_0^0(\xi)E_0^0(\eta)\\E_0^1(\xi)H_1(\eta)+E_0^0(\eta)\overline{H}_1(\xi)-E_0^1(\xi)E_0^0(\eta)\\E_0^0(\xi)\overline{H}_1(\eta)+E_0^1(\eta)H_1(\xi)-E_0^0(\xi)E_0^1(\eta)\\E_0^1(\xi)\overline{H}_1(\eta)+E_0^1(\eta)\overline{H}_1(\xi)-E_0^1(\xi)E_0^1(\eta)\end{bmatrix}\tag{9.19a}$$

Node $\boldsymbol{B}$, located in the corner at the position $(\xi_B,\ \eta_B)=(1,\ 0)$:

$$\mathbf{N}_{q_1}(\xi,\eta)=\begin{bmatrix}N_1(\xi,\eta)\\N_2(\xi,\eta)\\N_3(\xi,\eta)\\N_4(\xi,\eta)\end{bmatrix}=\begin{bmatrix}E_1^0(\xi)H_{q_1}(\eta)+E_0^0(\eta)H_{q_1}(\xi)-E_1^0(\xi)E_0^0(\eta)\\E_1^1(\xi)H_{q_1}(\eta)+E_0^0(\eta)\overline{H}_{q_1}(\xi)-E_1^1(\xi)E_0^0(\eta)\\E_1^0(\xi)\overline{H}_{q_1}(\eta)+E_0^1(\eta)H_{q_1}(\xi)-E_1^0(\xi)E_0^1(\eta)\\E_1^1(\xi)\overline{H}_{q_1}(\eta)+E_0^1(\eta)\overline{H}_{q_1}(\xi)-E_1^1(\xi)E_0^1(\eta)\end{bmatrix}\tag{9.19b}$$

Node $\boldsymbol{C}$, located in the corner at the position $(\xi_C,\ \eta_C)=(1,\ 1)$:

$$\mathbf{N}_{q_1+q_2-1}(\xi,\eta)=\begin{bmatrix}N_1(\xi,\eta)\\N_2(\xi,\eta)\\N_3(\xi,\eta)\\N_4(\xi,\eta)\end{bmatrix}$$

$$=\begin{bmatrix}E_1^0(\xi)H_{q_1+q_2-1}(\eta)+E_1^0(\eta)H_{q_1+q_2-1}(\xi)-E_1^0(\xi)E_1^0(\eta)\\E_1^1(\xi)H_{q_1+q_2-1}(\eta)+E_1^0(\eta)\overline{H}_{q_1+q_2-1}(\xi)-E_1^1(\xi)E_1^0(\eta)\\E_1^0(\xi)\overline{H}_{q_1+q_2-1}(\eta)+E_1^1(\eta)H_{q_1+q_2-1}(\xi)-E_1^0(\xi)E_1^1(\eta)\\E_1^1(\xi)\overline{H}_{q_1+q_2-1}(\eta)+E_1^1(\eta)\overline{H}_{q_1+q_2-1}(\xi)-E_1^1(\xi)E_1^1(\eta)\end{bmatrix}\tag{9.19c}$$

Node *D*, located in the corner at the position $(\xi_D,\ \eta_D) = (0,\ 1)$:

$$\mathbf{N}_{q_1+q_2+q_3-2}(\xi,\eta) = \begin{bmatrix} N_1(\xi,\eta) \\ N_2(\xi,\eta) \\ N_3(\xi,\eta) \\ N_4(\xi,\eta) \end{bmatrix}$$

$$= \begin{bmatrix} E_0^0(\xi)H_{q_1+q_2+q_3-2}(\eta) + E_1^0(\eta)H_{q_1+q_2+q_3-2}(\xi) - E_0^0(\xi)E_1^0(\eta) \\ E_0^1(\xi)H_{q_1+q_2+q_3-2}(\eta) + E_1^0(\eta)\overline{H}_{q_1+q_2+q_3-2}(\xi) - E_0^1(\xi)E_1^0(\eta) \\ E_0^0(\xi)\overline{H}_{q_1+q_2+q_3-2}(\eta) + E_1^1(\eta)H_{q_1+q_2+q_3-2}(\xi) - E_0^0(\xi)E_1^1(\eta) \\ E_0^1(\xi)\overline{H}_{q_1+q_2+q_3-2}(\eta) + E_1^1(\eta)\overline{H}_{q_1+q_2+q_3-2}(\xi) - E_0^1(\xi)E_1^1(\eta) \end{bmatrix} \tag{9.19d}$$

The above relationships can be easily programmed in any computer language and have been successfully implemented in the solution of many test cases of plate bending (see, Provatidis [21], Angelidis [1], Provatidis and Angelidis [22]).

Remark It is worth mentioning that the application of Eqs. (9.19a–d) in conjunction with only the corner nodes (*A*, *B*, *C*, and *D*), leads to the well-known *classical BFS element* [4]. This is left as an exercise to the reader, while details may be found in Provatidis [21] as well as in the Appendix of Provatidis and Angelidis [22].

9.3 Transfinite Plate Macroelement

9.3.1 General Considerations

Transfinite plate macroelement is an extension of Coons macroelement where the vector **E** of the blending functions is not restricted to the third degree (cf. Eq. (9.2) that corresponds to the configuration of Fig. 9.2a), but it is now increased in rows by twice the number of inter-boundaries. A special family of transfinite elements is to arrange the internal nodes in a structured scheme of $n \times m$ points, whereas the number of the boundary nodes along a side, say *AB*, may be different than the number of internal ones [i.e., $(n + 2)$ in the ξ-direction]. But even if the number of internal nodes equals to that of the intermediate boundary ones (i.e., n), the position of the former nodes may generally be different than that of the latter ones, as for example is shown in Fig. 9.2b.

Obviously, the n internal points along the ξ-direction define $2(n + 2)$ blending functions. Half of them refer to the translational DOF (flexures) and are denoted by $E_0^0(\xi), E_1^0(\xi), \ \ldots, \ E_n^0(\xi), E_{n+1}^0(\xi)$. The rest half refer to the rotational DOF (slopes) and are denoted by $E_0^1(\xi), E_1^1(\xi), \ \ldots, \ E_n^1(\xi), E_{n+1}^1(\xi)$. Clearly, the subscript refers

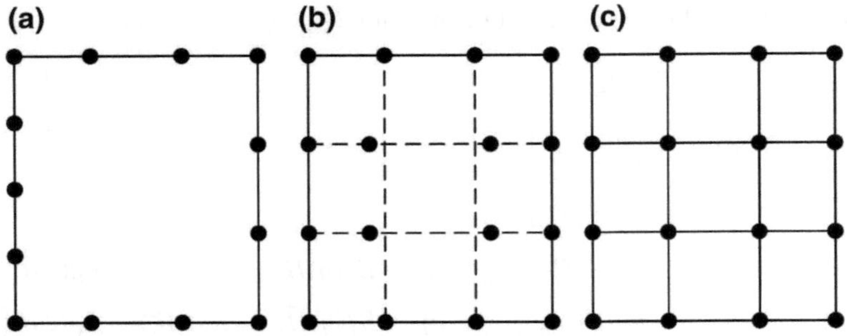

Fig. 9.2 **a** Coons macroelement and generalized transfinite plate elements with **b** arbitrary internal nodes and **c** structured nodes (Hermitian element)

to the position, whereas the superscript to the type of the DOF ('0' for translation, '1' for rotation).

Given these blending functions the relevant vector **E** is formed and, again, the Gordon–Coons (*transfinite*) interpolation preserves its form and is given again by the same expression as in Eq. (9.1). However, now the matrix $\mathbf{W}(i, j)$ is of order $2n$.

After the above modifications, Eqs. (9.17)–(9.19d) are still valid with the proper blending functions.

Moreover:

Shape functions at the internal nodes with ascending numbers. $k = (q_{B,\text{tot}} + 1), \ \ldots, \ (q_{B,\text{tot}} + nm)$, are:

$$N_k(\xi, \eta) = \begin{bmatrix} N_1(\xi, \eta) \\ N_2(\xi, \eta) \\ N_3(\xi, \eta) \\ N_4(\xi, \eta) \end{bmatrix}$$

$$= \begin{bmatrix} E_j^0(\eta) E_i^0(\xi) \\ E_j^0(\eta) E_i^1(\xi) \\ E_j^1(\eta) E_i^0(\xi) \\ E_j^1(\eta) E_i^1(\xi) \end{bmatrix}, \quad i = 1, \ldots, n; \ j = 1, \ldots, m \text{ (local numbering)} \tag{9.20}$$

9.3.2 *Hermitian Macroelement (Tensor-Product Hermites)*

In this case, we consider that the opposite sides of the patch have an equal number of nodes ($q_1 = q_3 = n + 2$, $q_2 = q_4 = m + 2$), and particularly at the *same* relative position (along ξ_i =constant, and η_j =constant). Also, the internal nodes have also the *same* relative position as the boundary ones, as shown in Fig. 9.2c. Under these circumstances, it can be easily shown that Eq. (9.1) degenerates to the following tensor product:

$$\mathbf{w}(\xi,\eta) = \mathbf{E}^{\mathrm{T}}(\xi)\mathbf{W}(i,j)\mathbf{E}(\eta) \tag{9.21}$$

in which the blending functions *coincide* with those of the Hermite interpolants along the sides.

Then all shape functions are given by the same formula which is that of Eq. (9.20), but now the latter is valid for all the nodal points $k = 1, \ldots, (n+2)(m+2)$.

Remarks (1) Obviously, the simplest element of this class is the previously mentioned BFS element [4].

(2) These Hermitian tensor-product elements form the *analogue* of the elements of Lagrangian type. The former are applicable to plate bending, whereas the latter to potential and plane elasticity problems.

9.3.3 Arbitrary Transfinite Interpolation

This is the arbitrary case, which generally corresponds to a situation *in-between* the boundary-only Coons (Sect. 9.2) and the tensor-product Hermite (Sect. 9.3.2) formulation. This means that not only the internal nodes may be shifted with respect to the boundary ones, but they can be also less than those determined by the boundary discretization. For the completeness of this text, it should be mentioned that the transfinite interpolation can be applied also to more internal nodes than those used in tensor-product Hermite type; however, in such a case no significant improvement of the accuracy is usually achieved.

9.4 Derivation of Stiffness and Mass Matrices

9.4.1 Governing Equation

The standard elastodynamic equation of motion (second Newton's law) for a thin plate undertaking bending vibrations is described by

$$D\nabla^4 w(x,y;t) + \rho h \frac{\partial^2 w(x,y;t)}{\partial t^2} = f_z(x,y;t), \tag{9.22}$$

where $w(x, y; t)$ denotes the time-dependent deflection, $f_z(x, y; t)$ the normal pressure, $D = Eh^3/(12(1-\nu^2))$ the flexural rigidity, E is the modulus of elasticity, ν is Poisson's ratio, h is the thickness of the plate, ρ is the mass density, and t is the time.

Using the standard Galerkin–Ritz formulation, the matrix formulation becomes

$$[\mathbf{M}]\{\ddot{\mathbf{d}}\} + [\mathbf{K}]\{\mathbf{d}\} = \{\mathbf{r}\} + \{\mathbf{f}\} \tag{9.23}$$

where the elements of the stiffness matrix $[\mathbf{K}]$ are given in terms of the second derivatives of the global shape functions by:

$$k_{ij} = \int_{\Omega} D \cdot \left[\frac{\partial^2 N_i}{\partial x^2} \left(\frac{\partial^2 N_j}{\partial x^2} + \nu \frac{\partial^2 N_j}{\partial y^2} \right) + \frac{\partial^2 N_i}{\partial y^2} \left(\nu \frac{\partial^2 N_j}{\partial x^2} + \frac{\partial^2 N_j}{\partial y^2} \right) \right.$$
$$\left. +2(1-\nu) \cdot \frac{\partial^2 N_i}{\partial x \partial y} \frac{\partial^2 N_j}{\partial x \partial y} \right] \cdot \mathrm{d}\Omega \tag{9.24}$$

and these of mass matrix $[\mathbf{M}]$ are expressed in products of the global shape functions themselves by:

$$m_{ij} = \rho h \int_{\Omega} N_i N_j \mathrm{d}\Omega, \tag{9.25}$$

with $\{\mathbf{r}\}$ and $\{\mathbf{f}\}$ denoting the vector of the internal and external load, respectively ($i, j = 1, \ldots, q_{\text{tot}}$). In the above expressions, the domain Ω denotes the area of the macroelement ($\mathrm{d}\Omega = \mathrm{d}x\mathrm{d}y$), while ν is the Poisson's ratio. Obviously, both matrices $[\mathbf{M}]$ and $[\mathbf{K}]$ are symmetric and fully populated.

9.4.2 Estimation of the Stiffness and Mass Matrices

For a given discretization of the boundary, the maximum number of *nodes per direction,* is $n_\xi = \max(q_1, q_3)$ and $n_\eta = \max(q_2, q_4)$ for the directions ξ and η, respectively. By virtue of Eq. (9.17), the shape functions associated to a node along the side AB includes a Hermitian polynomial of $(2n_\xi - 1)$ th degree in ξ, while along the side BC it becomes of $(2n_\eta - 1)$th degree in η. Obviously, the integrands in the mass matrix, which are products in the form $N_i N_j$, will be polynomials of degree $2(2n_\xi - 1)$ in ξ, whereas for the stiffness matrix (where the maximum degree is controlled by the first derivative shown into the mixed derivative) the integrand will be a polynomial of $4(n_\xi - 1)$ in ξ.

Recalling that n_g Gauss points are required for the accurate integration of a polynomial of $(2n_g - 1)$ degree, and taking the ceiling for each separate case, it is easily concluded that:

(i) The accurate estimation of the *stiffness* matrix requires $n_{\text{gauss}} = 2n_\xi - 1$ integration points, in the ξ direction.
(ii) The accurate estimation of the *mass* matrix requires $n_{\text{gauss}} = 2n_\xi$ integration points, in the ξ direction.

Overall, it is generally sufficient to use $2n_\xi \times 2n_\eta$ Gauss points for the accurate calculation of all elements in both matrices.

9.4.3 Numerical Results Using Coons–Gordon Macroelements

Within the context of the abovementioned older CAD-based macroelements, two test cases will be studied below. Both of them refer to a square plate of dimensions $a \times a$, with $a = 1$ m. The first concerns a static analysis, whereas the second an eigenvalue problem. A single boundary-only Coons is compared with a Hermitian macroelement, and both of them with the assemblage of classical BFS and MZC finite elements for the same mesh density.

Example 9.1 (Simply supported plate subject to central concentrated load) This example is solved either within the entire domain $a \times a$ ($ABCD$: $0 \leq x, y, \leq a$, shown in Fig. 9.3a, c) or in one-fourth of that $a/2 \times a/2$ (hatched square $A'B'C'D'$: $0 \leq x', y', \leq a/2$, shown in Fig. 9.3b, d). Material properties were chosen as follows: $E{=}2 \times 10^{11}$ N/m^2, $\nu{=}0.3$, $\rho{=}7800$ kg/m^3, while the thickness of the plate was taken equal to $h{=}0.01$ m.

Below, the error (in %) of either the calculated deflection (w) or the eigenvalue (ω^2), is defined as follows:

$$err(\%) = \frac{w_{\text{calculated}} - w_{\text{exact}}}{w_{\text{exact}}} \times 100 \text{ or } \frac{\omega^2_{\text{calculated}} - \omega^2_{\text{exact}}}{\omega^2_{\text{exact}}} \times 100 \tag{9.26}$$

It is clarified that, for given subdivisions per direction (e.g., $n_x = 4$ in Fig. 9.3a, and $n_x = 2$ in Fig. 9.3b), the single Coons macroelement is formed by $2(n_x{+}n_y)$ nodes that lie along the boundary only, whereas the corresponding Hermite macroelement is formed by $(n_x+1)(n_y+1)$ nodes in total. For this convention, the numerical results obtained with both types of CAD-based macroelements, i.e.:

(i) boundary-only *Coons* and
(ii) tensor-product (transfinite) *Hermite*,

are shown in Table 9.1.

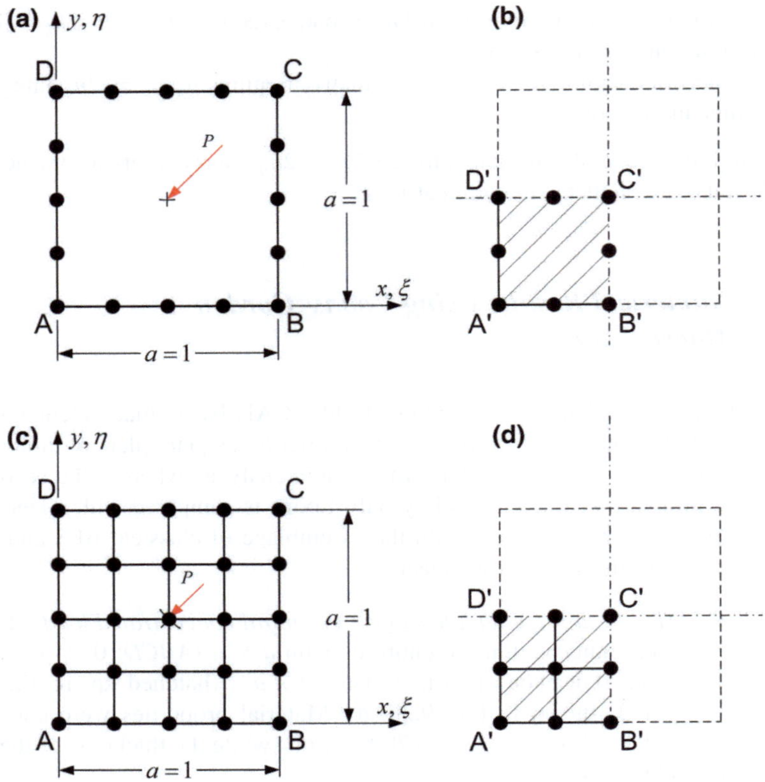

Fig. 9.3 Square plate subjected to the central concentrated load *P*, which is solved using a single boundary-only *Coons* macroelement in **a** the entire domain and **b** the one-fourth, as well as a single *Hermite* macroelement in **c** the entire domain and **d** the one-fourth of it

One may observe in the upper part of Table 9.1 that:

(i) The Coons macroelement that occupies the entire plate model (indicated in the table as "Entire") converges very slowly, whereas
(ii) The one-fourth Coons model (indicated in the table as "1/4") converges faster but to a false numerical solution (with error about 1%).

Obviously, although the Coons solution can be accepted for good practical engineering purposes, the abovementioned findings lead to the need of using internal nodes as well. Therefore, utilizing internal nodes in the form of a uniform mesh in accordance with the uniformly discretized boundary (shown in Fig. 9.3c, d), the performance of the relevant single macroelement of Hermite type (tabulated in the lower part of Table 9.1) improves and a rapid convergence appears (entire model: $0.3023 < 1.8446$, and $0.0692 < 0.9962$, as shown in Table 9.1).

Table 9.1 Maximum deflection at the center of the simply supported plate (expressed by error in %) using a single Coons macroelement (as in Fig. 9.3a, b) or a single Hermite macroelement (as in Fig. 9.3c, d)

	Error (in %)					
	$n_x = 1$	$n_x = 2$	$n_x = 4$	$n_x = 6$	$n_x = 8$	$n_x = 10$
Model **Coons Element**: Boundary nodes only						
Entire	−31.1202	−10.3151	−4.6841	−4.0070	−3.8455	−1.8446
1/4	−4.5005	−1.0426	−0.9964	−0.9964	−0.9964	−0.9962
Model **Hermite Element**: Boundary and internal nodes						
Entire	−31.1202	−9.7007	−2.3752	−1.0230	−0.5643	−0.3023
1/4	−33.5257	−0.3940	−0.02781	−8.5946e−04	0.0045	0.0692

Note It should be clarified that in both above cases (i.e., Coons and Hermite macroelements), for a specific number of subdivisions n_x, the "Entire" model uses a Hermite polynomial of $(2n_x + 1)$ degree for the entire length a, whereas the one-fourth ("1/4") model uses the same degree for its half. Note that, in contrast to the small size conventional elements that repeat the same local interpolation between any two successive nodes, the *entire* model is not equivalent to the *one-fourth* model

Moreover, single Coons and Hermite macroelements, both based on the "one-fourth" of the domain, are compared to the classical BFS and MZC elements (see, Fig. 9.4). One may observe the permanent error of the Coons macroelement (about 1%) and the excellent convergence of the Hermite one which outperforms.

Remark The one-fourth Hermite model appears the fact that the error slightly increases when the discretization exceeds the critical number of seven divisions per side (Hermite polynomials of fifteenth degree). This case is also met in plane elasticity problems and may support the claim of some researchers who prefer spline interpolation instead of Lagrange polynomials. Nevertheless, one could oppose the argument that the accuracy is sufficiently excellent for $n_\xi = 6$.

Example 9.2 (Eigenvalues of a free–free square plate) This example is solved in the entire domain $a \times a$ $(ABCD)$ using a single macroelements (of Coons and Hermite type). The numerical solution is also compared with conventional finite elements (BFS and MZC) of the same mesh density.

According to Blevins [3] and Leissa [14], the exact solution is given by:

$$f_{ij}[\mathrm{Hz}] = \frac{\lambda_{ij}^2}{2\pi a^2}\left[\frac{Eh^2}{12\rho(1-\nu^2)}\right]^{1/2}; \quad i = 1,\ 2,\ 3,\ \ldots; \quad j = 1,\ 2,\ 3,\ \ldots \tag{9.27}$$

where i and j are the numbers of half-waves in mode shape along horizontal and vertical axis, respectively, and λ_{ij}^2 is a parameter that depends on the ratio (a/b) of the length over the width of the plate. According to Leissa [14], in the particular case

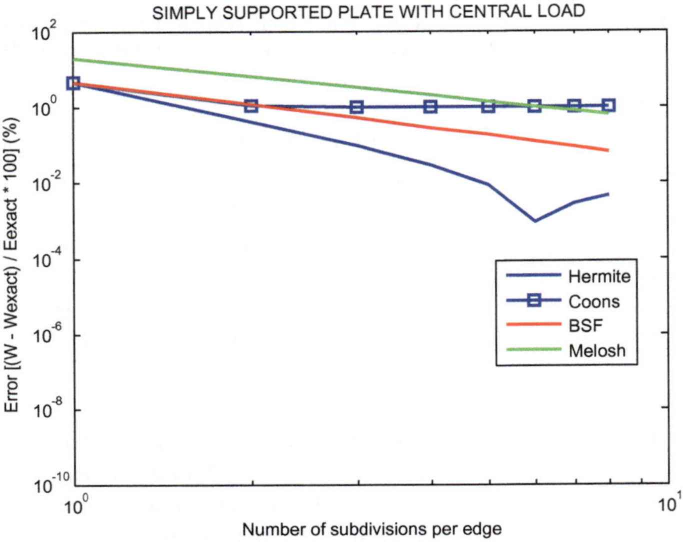

Fig. 9.4 Convergence quality of the CAD-based macroelements (Coons, Hermite) versus older classical elements (BFS, MZC)

of the unit square ($a=1$ m) the approximate values are given below the values of λ_{ij}^2 are given in Table 9.2. At the same table, we also present the correcting factors according to the corrected formula proposed by Gorman [10] (valid for a square plate with Poisson's ratio of 0.3); therefore, the exact values used in this paper are illustrated at the bottom of Table 9.2. It is noted that the first three modes $(i, j)=(1, 1)$, $(1, 2)$, and $(2, 1)$ are rigid body modes, and therefore are not included.

The results of the (boundary-only) Coons macroelement are shown in Table 9.3. One may observe the rapid convergence toward *almost* the exact values, despite the fact that no internal nodes exist. Nevertheless, a small error remains even for a fine

Table 9.2 Scaling for the determination of the exact normalized eigenvalues [λ_{ij}^2, which appear in Eq. (9.27)] for the free-free unit square plate ($E=2\times10^{11}$ N/m^2, $\nu=0.30$, $a=1$ m, $h=0.01$ m)

Eigenvalues	Mode sequence			
	1	2	3	4
(i, j)	(2,2)	(1,3)	(3,1)	(3,2)
Approximate λ_{ij}^2 Blevins [3]	13.49	19.79	24.43	35.02
Correcting factor Gorman [10]	3.367/3.372	4.899/4.947	6.068/6.108	8.700/8.756
Exact λ_{ij}^2 (three digits)	13.470	19.598	24.270	34.796

Table 9.3 Calculated eigenvalues for a square plate under free–free boundary conditions, using a single Coons macroelement

Exact eigenvalue ω^2	Error (in %)				
	$n_x = 1$	$n_x = 2$	$n_x = 4$	$n_x = 8$	$n_x = 10$
42603.74	2.84	−0.01	−0.02	−0.02	−0.02
90185.23	31.22	1.31	−0.01	−0.01	−0.01
138309.74	58.90	2.33	0.15	0.15	0.15
284296.84	27.14	2.91	0.17	0.17	0.17

discretization. Of course, due to the two-digit accuracy given by Leissa/Blevins, and the three digits in the given correcting factors, the exact values should be somehow different (probably slightly smaller) than the really exact value, a fact that justifies the minor negative errors that appear to the Coons (and later to the Hermite) macroelements.

For the completeness of this study, not only the convergence quality of a single Hermite macroelement is presented in Fig. 9.5, but its accuracy is also compared with conventional finite elements of the same mesh density as well as with the abovementioned single Coons macroelement. Note that in the particular case of the Hermite macroelement, the position as well as the number of nodes is the same as in the assemblage of conventional elements. It is noted that the first eigenvalue shown in Table 9.3 is the fourth one, since the first three equal to zero (*rigid-body*) modes.

Remarks

(1) Concerning the boundary-only Coons macroelement, we have considered *Hermite polynomials* as trial functions along the entire of any of the four sides in the rectangular patch *ABCD*. Nevertheless, in principle, the aforementioned Hermitian polynomials could be replaced by another suitable functional set (such as Bernstein–Bézier, B-splines or NURBS). In addition, one could also choose piecewise cubic Hermite interpolation between any two successive nodes (in both directions). However, this matter is not so crucial to interrupt the continuation of the study.
(2) Concerning the tensor-product Hermitian macroelement, it should become clear that so far we have considered global *Hermite polynomials* as trial functions in both directions. In principle, these could be replaced by any reasonable alternative as above was mentioned.
(3) Within the context of a Hermitian macroelement, interestingly, if the abovementioned piecewise cubic Hermite interpolation is combined with a modification of the blending functions in order they are piecewise cubic Hermite polynomials as well, then the obtained macroelement is equivalent to an assemblage of $(n_\xi \times n_\eta)$ classical BSF elements. Similar findings (however with respect to Lagrangian macroelements) have been previously reported by Provatidis ([19], p. 956, among others).

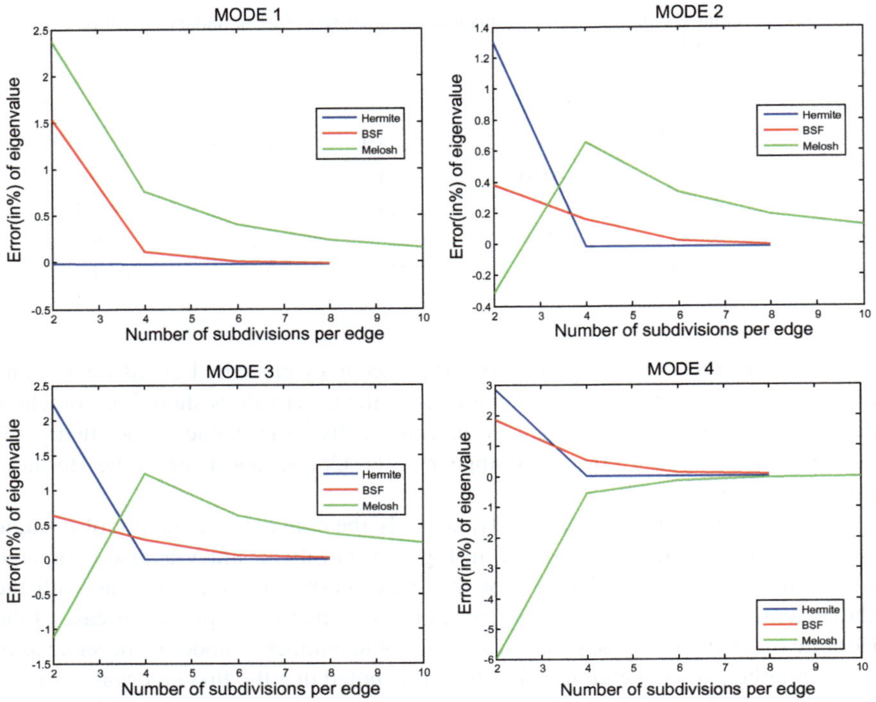

Fig. 9.5 Convergence quality of the calculated eigenvalues for the first four nonrigid modes (from Provatidis [17] with permission)

9.5 Bernstein–Bézier Macroelements Applied to Curvilinear Plates

9.5.1 A General Note

So far in this chapter, it was shown that the classical BFS rectangle plate element may be extended at any arbitrary noded macroelement in which the trial functions along any side of the rectangular patch *ABCD* are Hermite polynomials. Overall, a single Coons macroelement exhibited a performance that varies between good and very good, which of course becomes excellent when a tensor-product Hermitian macroelement was used.

The most crucial issue with these CAD-based macroelements is the difficulty of Coons interpolation to deal with curvilinear patches (again, this is an older general remark by Carey and Oden [6]), an important topic that will be studied below to some extent. While this difficulty remains still an open problem for the boundary-only Coons macroelement, later we shall give the receipt to replace the Hermitian macroelement by an equivalent formulation that permits dealing with curvilinear plates as well.

Before we go in details, as a *pilot study* it is very instructive to deal with the one-dimensional *analogue* of the plate bending, which is the *beam* bending analysis. In both cases, in addition to the deflection the slope plays a significant role.

9.5.2 The 1D Analogue

From the beginning, it is important to make some remarks. First, for a given number of nodal segments along one direction, say n_ξ in the ξ-direction, the Lagrangian polynomial which is involved in Eq. (9.7) is of degree n_ξ and therefore the Hermite functions (H_i and $\bar{H}_i$) will be of degree $(2n_\xi + 1)$. The 1D analogue of a plate is the beam in bending (two DOF per node: w and $\partial w/\partial x$). A beam of length L can be uniformly divided into n_ξ nodal segments (i.e., $n_\xi + 1$ nodal points), which define $(n_\xi + 1)$ Hermite polynomials of type H_i (cf. Eq. (9.7)) which refer to the nodal deflections, and another $(n_\xi + 1)$ Hermite polynomials of type $\bar{H}_i$ (cf. Eq. (9.8)) which refer to the nodal rotations. In the traditional macroelement approach (global approximation), the two DOF which are associated to each nodal point lead to a stiffness matrix of dimensions $[2(n_\xi + 1)] \times [2(n_\xi + 1)]$ and the same holds for the mass matrix. Then imposing the boundary conditions, either the static or the eigenvalue problem is solved as usual.

Bernstein–Bézier

The abovementioned procedure in the 1D problem can be solved in an *equivalent* way as follows. Instead of using the $(n_\xi + 1)$ *nodal* points with two DOF per node, we start our analysis considering a single interval for the entire domain $[0, L]$ and a polynomial interpolation of degree $p = 2n_\xi + 1$. But as has been previously discussed in Chap. 6, the aforementioned polynomial is equivalent to the use of Bernstein–Bézier representation of the solution $w(x)$ of the same degree, i.e., $p = 2n_\xi + 1$. The latter is accomplished using $n_{\text{cntrl}} = 2n_\xi + 2$ control points, i.e., $P_0, \ldots, P_{n_{\text{cntrl}}-1}$, which are uniformly distributed in the interval $[0, L]$. It is reminded that the relevant knot vector is $\mathbf{U} = \{\underbrace{0, \ldots, 0}_{p+1}, \underbrace{1, \ldots, 1}_{p+1}\}$ (i.e., none inner knots).

Using the basis functions (i.e., the Bernstein–Bézier polynomials) that are associated to the above control points, we can express the approximate solution, $w(x)$, thus replacing the Hermite polynomials in an equivalent way. For example, the well-known cubic Hermite polynomials that are involved in the trivial beam bending finite element of length L (found in all FEM textbooks):

$$\left.\begin{array}{ll} N_1(x) = \left[1 - 3\left(\frac{x}{L}\right)^2 + 2\left(\frac{x}{L}\right)^3\right], & N_2(x) = \left[\left(\frac{x}{L}\right) - 2\left(\frac{x}{L}\right)^2 + \left(\frac{x}{L}\right)^3\right]L \\ N_3(x) = \left[3\left(\frac{x}{L}\right)^2 - 2\left(\frac{x}{L}\right)^3\right], & N_4(x) = \left[-\left(\frac{x}{L}\right)^2 + \left(\frac{x}{L}\right)^3\right]L \end{array}\right\} \quad (9.28)$$

are mathematical *equivalent* (through a proper transformation matrix **T**, see Chap. 6) with the cubic Bernstein–Bézier polynomials:

$$\left.\begin{aligned} B_{0,3}(x) &= (1-x/L)^3, \quad B_{1,3}(x) = 3(1-x/L)^2(x/L) \\ B_{2,3}(x) &= 3(1-x/L)(x/L)^2, \quad B_{3,3}(x) = (x/L)^3 \end{aligned}\right\} \tag{9.29}$$

In other words, considering four control points, $(P_0,\ P_1,\ P_2,\ P_3)$, which divide the interval $[0, L]$ into three equal segments, we have:

$$\begin{aligned} w(x) &= N_1(x)\cdot w_1 + N_2(x)\cdot w_1' + N_3(x)\cdot w_2 + N_4(x)\cdot w_2' \\ &= B_{0,3}(x)\cdot a_0 + B_{1,3}(x)\cdot a_1 + B_{2,3}(x)\cdot a_2 + B_{3,3}(x)\cdot a_3 \end{aligned} \tag{9.30}$$

B-splines

Obviously, the Bernstein–Bézier polynomials, which were used in the interval $[0, L]$ on the entire beam, are characterized by C^{p-1}-continuity. Using some breakpoints in addition to the endpoints of the beam, and considering a certain multiplicity for inner knots, one has full control on the approximate solution $w(x)$. In more details, the degree p may be now lower than the number of knot spans, and one may consider a multiplicity of inner knots different than one. For example, choosing double inner knots the continuity becomes C^{p-2}, choosing triple inner knots it becomes C^{p-3}, and so on.

In the particular case of cubic splines ($p = 3$), taking the multiplicity equal to *two* (i.e., double inner knots), the numerical solution will be C^1-continuous, a fact that is consistent with the C^1-continuity involved in the piecewise Hermite approximation (also of third degree). Therefore, if a beam of length L is uniformly divided into n finite (beam) elements (Hermite cubics), the model is mathematically equivalent with that in which n knot spans are introduced over the interval $[0, L]$, and each of the $(n-1)$ inner knots is taken to be double, while the polynomial degree is taken as $p = 3$. This issue is clearly shown in Fig. 9.6, where the B-spline basis functions are compared with the Hermite polynomials. The entire interval [0, 1] has been uniformly divided into four subintervals. Focusing on the second subinterval, i.e., [0.25, 0.50], only four out of the seven basis functions are nonzero. Despite the fact, these four nonzero functions differ than the local Hermite polynomials in which Eq. (9.28) is applied for the interval [0.25, 0.50], both sets are linearly depended through a transformation matrix, as it has been extensively discussed in Chap. 6. This issue is elucidated by Example 9.3, where, for the sake of simplicity in manual operations, the interval $[\frac{3}{4}, 1]$ was chosen instead of [0.25, 0.50].

Example 9.3 (Equivalency between cubic doubly knotted B-splines and Hermite polynomials) Show, in a numerical way, that for the knot vector $\mathbf{U} = \{0,\ 0,\ 0,\ 0,\ \frac{1}{4},\ \frac{1}{4},\ \frac{2}{4},\ \frac{2}{4},\ \frac{3}{4},\ \frac{3}{4},\ 1,\ 1,\ 1,\ 1\}$, the basis functions in the subinterval $[\frac{3}{4}, 1]$ are equivalent with the cubic Hermite polynomials that refer to the same interval.

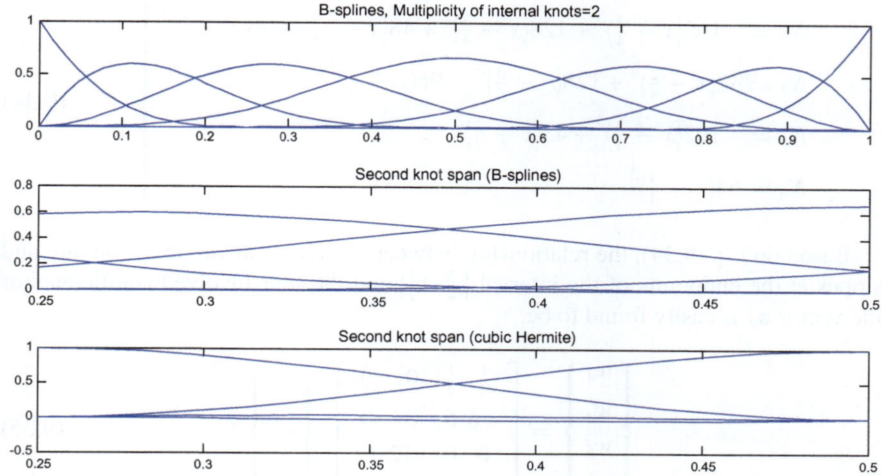

Fig. 9.6 B-splines for the entire domain [0,1] as well as for the subinterval [0.25,0.50]

Solution The given knot vector gives ten control points and ten basis functions, $N_{0,3}, \ \ldots, \ N_{9,3}$. By virtue of the property of compact support, within the interval $[\frac{3}{4}, 1]$ the only nonzero basis functions are $N_6, \ N_7, \ N_8$, and N_9. Due to the double multiplicity, as discussed in Chap. 2, the flexural displacement is approximated by the formula:

$$\begin{aligned} w(x) = &\left[1, x, x^2, x^3, < x - \tfrac{1}{4} >^3, < x - \tfrac{2}{4} >^3, \right. \\ &\left. < x - \tfrac{3}{4} >^3 < x - \tfrac{1}{4} >^2, < x - \tfrac{2}{4} >^2, < x - \tfrac{3}{4} >^2 \right] \cdot \mathbf{a}_s \end{aligned} \tag{9.31}$$

where $\mathbf{a}_s$ is a column vector of dimensions 10×1.

Implementing Eq. (9.31) at ten test points (preferably, uniformly distributed) within the interval [0, 1], one obtains, $\mathbf{W}_{\text{test}} = \mathbf{A}_s \mathbf{a}_s$, whence

$$\mathbf{a}_s = (\mathbf{A}_s)^{-1} \mathbf{W}_{\text{test}} \tag{9.32}$$

On the other point of view, implementing the MATLAB function `spcol` on a `tau` vector that is formed by the abovementioned ten uniformly distributed test points in the interval [0, 1], one may determine the matrix $\mathbf{A}_d$ (of dimensions 10×10) that fulfills the relationship

$$\mathbf{W}_{\text{test}} = \mathbf{A}_d \mathbf{a}_d. \tag{9.33}$$

Eliminating $\mathbf{W}_{\text{test}}$ between Eqs. (9.33) and (9.32), and further substituting the vector $\mathbf{a}_s$ into Eq. (9.31), the global shape functions are eventually found. Between them, the four useful ones are given by the following analytical expressions:

$$\left.\begin{aligned} N_6 &= -160\left\langle x-\tfrac{2}{4}\right\rangle^3 + 128\left\langle x-\tfrac{3}{4}\right\rangle^3 + 48\left\langle x-\tfrac{2}{4}\right\rangle^2 + 96\left\langle x-\tfrac{3}{4}\right\rangle^2 \\ N_7 &= 32\left\langle x-\tfrac{2}{4}\right\rangle^3 + 128\left\langle x-\tfrac{3}{4}\right\rangle^3 - 96\left\langle x-\tfrac{3}{4}\right\rangle^2 \\ N_8 &= -192\left\langle x-\tfrac{3}{4}\right\rangle^3 + 48\left\langle x-\tfrac{3}{4}\right\rangle^2 \\ N_9 &= 64\left\langle x-\tfrac{3}{4}\right\rangle^3 \end{aligned}\right\} \tag{9.34}$$

Based on Eq. (9.34), the relationship between the classical variables (flexure and slopes at the endpoints of the interval $\left[\frac{3}{4}, 1\right]$) and the four involved coefficients of the vector $\mathbf{a}_d$ is easily found to be:

$$\begin{Bmatrix} w_4 \\ w'_4 \\ w_5 \\ w'_5 \end{Bmatrix} = \underbrace{\begin{bmatrix} \frac{1}{2} & \frac{1}{2} & 0 & 0 \\ -6 & 6 & 0 & 0 \\ 0 & 0 & -12 & 12 \\ 0 & 0 & 0 & 1 \end{bmatrix}}_{[\mathbf{T}]} \cdot \begin{Bmatrix} a_6 \\ a_7 \\ a_8 \\ a_9 \end{Bmatrix} \tag{9.35}$$

Right multiplying of the row vector $[N_6, N_7, N_8, N_9]$ by the inverse of the transformation matrix $[\mathbf{T}]$ that appears in Eq. (9.35), the obtained four shape functions are eventually found to be coincident with the Hermite polynomials, where in Eq. (9.28) the length L is replaced by the length of the subinterval under consideration $\left(\left[\frac{3}{4}, 1\right]\right)$, i.e., $L = \frac{1}{4}$. ■

In conclusion, not only the cubic Bernstein–Bézier (that was previously shown in the first part of this subsection) but here also the cubic B-splines in conjunction with double inner knots give an equivalent functional space with the classical piecewise Hermite polynomials.

9.5.3 Imposition of Boundary Conditions in Beam Bending Analysis

The operation of the Bernstein–Bézier elements is easily demonstrated in beam bending, which is an analogue of plate bending. The flexural displacement is expanded into the following series

$$w(\xi) = B_{0,n}(\xi) \cdot a_0 + B_{1,n}(\xi) \cdot a_1 + \cdots + B_{n,n}(\xi) \cdot a_n, \tag{9.36}$$

with

$$B_{i,n}(\xi) = \frac{n!}{i!\,(n-i)!}(1-\xi)^{n-i}\xi^i \tag{9.37}$$

The slope of the beam is produced by taking the first derivative of Eq. (9.36) with respect to ξ:

$$w'(\xi) = B'_{0,n}(\xi) \cdot a_0 + B'_{1,n}(\xi) \cdot a_1 + \cdots + B'_{n,n}(\xi) \cdot a_n, \tag{9.38}$$

where one can easily validate that at the two ends:

$$\left.\begin{array}{ll} B_{0,n}(0) = 1,\ B'_{0,n}(0) = -n; & B_{1,n}(0) = 0,\ B'_{1,n}(0) = 1 \\ B_{n,n}(1) = 1,\ B'_{n,n}(1) = n; & B_{n-1,n}(1) = 0,\ B'_{n-1,n}(1) = -1 \end{array}\right\} \tag{9.39}$$

Therefore, if the beam is clamped at $\xi = 0$, then by virtue of Eqs. (9.39) and (9.36) in conjunction with the boundary condition $w(0) = 0$ gives $a_0 = 0$. Also, Eq. (9.38) in conjunction with the condition $w'(0) = 0$ implies $a_1 = 0$. In other words, since the first two coefficients vanish, it becomes obvious that the first two rows and columns in the stiffness matrix should be deleted (recall that $\mathbf{Ka} = \mathbf{f}$).

Earlier in this chapter, we saw that cubic finite elements are equivalent with cubic Bezier. This finding may be generalized. In other words, similarly to what occurred with the Lagrangian polynomials in Chap. 6, the high order Hermite approximation is *equivalent* with the Bernstein–Bézier expansion of the approximate flexure. One may observe that for a nodal sequence $\xi_0, \ldots, \xi_n$, the Hermite polynomials, H_i and $\bar{H}_i$, given by Eqs. (9.7) and (9.8), are both of degree $(2n + 1)$. For example, when $n = 1$ the classical Hermite polynomials are all of third degree [cf. Eqs. (9.2) and (9.28)]. Again, the maximum polynomial degree ($p = 2n + 1$) is higher than that determined by the number of nodal points ($n + 1$). In order to make both functional sets, i.e., Hermite polynomials and monomials, to be equivalent, it is sufficient to increase the number of segments.

Therefore, a beam idealized with ($n + 1$) nodal points according to the sequence $\xi_0, \ldots, \xi_n$, can be treated in two equivalent ways as follows:

(i) Two degrees of freedom ($w, \partial w/\partial x$) may be assigned to each nodal point, thus receiving $2(n + 1)$ DOF, in total. The deflections, which correspond to the odd DOF (1, 3, …, $2n + 1$) are associated to the global shape function H_i given by Eq. (9.7). The slopes, which correspond to the even DOF (2, 4, …, $2n + 2$) are associated to the global shape function $\overline{H}_i$ given by Eq. (9.8). The stiffness matrix [**K**] as well as the mass matrix [**M**] will be of dimensions $[2(n + 1)] \times [2(n + 1)]$.

(ii) Within the context of Bernstein–Bézier approximation, in order to achieve the same results with the abovementioned traditional Hermite expansion, we only need to use a series of Bernstein polynomials of degree $p = 2n+1$. This choice implies a stiffness matrix $[\overline{\mathbf{K}}]$ of dimensions $(p+1)\times(p+1) = (2n+1)\times(2n+1)$, which is the same with that in the abovementioned traditional formulation.

The theoretical explanation for the abovementioned coincidence between the Hermitian and the Bézierian macroelement is very easy and can be based on the findings of Chap. 6. An extra numerical validation is given through Example 9.4.

Example 9.4 (Eigenvalues of a clamped beam) Calculate the first eigenvalues of a clamped beam with given data (E, A, I, L), which is allowed to undertake transverse vibrations in the vertical plane only. The exact eigenvalues may be found in many references such as Blevins [3] or in the textbook of Weaver et al. [25].

Choosing, for example, $n_x = 4$ subintervals along the entire beam:

- The classical FEM model consists of $n_x = 4$ cubic finite elements, with totally $(n_x + 1) = 5$ nodes and 10 DOF.
- The single Hermite macroelement consists of five nodes, with totally 10 DOF. Each of the ten involved Hermite polynomials is of degree $p = (2n_x + 1) = 9$.
- The single nonrational Bézier macroelement consists of 10 control points, with totally ten DOF. Each of the involved Bézier polynomials is of degree $p = 9$.

In all the abovementioned three models, two out of the ten DOF are restricted due to the fixed end of the clamped beam; thus, the produced stiffness and mass matrices are all of dimension 8×8, which means that eight eigenvalues are eventually calculated.

The numerical results are shown in Table 9.4, where one may observe:

(i) The coincidence between the "old-fashioned" high order Hermite macroelement and the CAD-based "modern" Bézierian one.
(ii) The superiority of the global approximation (either Hermitian or Bézierian) over the local one (FEM solution).

Comprehension questions

Consider the Bernstein–Bézier formulation (of degree p) of a beam in bending.

(1) If the beam is simply supported at its both ends ($x = 0, x = L$), which rows and columns of the stiffness and mass matrices should be deleted?
(2) If the beam is clamped at its both ends ($x = 0, x = L$), which rows and columns of the matrices (**K**, **M**) should be deleted?

Table 9.4 Bending eigenvalues of a clamped beam

Exact eigenvalues ω^2	Errors (in %)		
	Hermitian macroelement	Bézierian macroelement	FEM solution
12.3622	0.00	0.00	0.01
485.5188	0.00	0.00	0.23
3806.5663	0.00	0.00	1.55
14,617.2733	0.36	0.36	2.93
39,943.8318	1.24	1.24	30.30
Number of equations (after imposing the B.C.s)	8	8	8

Answer In the first case, we delete the first and the $(p + 1)$th rows and columns, which refer to the flexural degree of freedom. In the second case, we delete again the same (first and $(p + 1)$th), but in addition we also should delete the second and the pth, which refer to the restricted rotational DOF (slopes at the two ends).

The reader is encouraged to implement the above test cases in an in-house code.

9.5.4 General Plate Analysis

Plate bending (static and dynamic) analysis is merely an extension of what we mentioned in the above beam bending analysis. The outer boundary of the control points refers to the deflections, whereas the immediately inner layer refers to the slopes. Thus, the deletion of rows and columns is quite similar with the beam bending analogue (discussed in Sect. 9.5.3).

In contrast to the beam bending analysis where the Bernstein–Bézier formulation does not add something new (it rather causes confusion to a beginner), in plate bending analysis it offers the advantage of dealing with *curvilinear* shapes. Therefore, one may use the tensor product of Bernstein polynomials:

$$w(\xi, \eta) = \sum_{i=0}^{n} \sum_{j=0}^{m} B_{i,n}(\xi) \cdot B_{j,m}(\eta) \cdot a_{ij}, \tag{9.40}$$

Based on the isoparametric concept, the geometry is also described by

$$\mathbf{x}(\xi, \eta) = \sum_{i=0}^{n} \sum_{j=0}^{m} \underbrace{B_{i,n}(\xi) \cdot B_{j,m}(\eta)}_{\overline{N}(\xi,\eta)} \cdot \mathbf{x}_{ij}, \tag{9.41}$$

where $\mathbf{x}(\xi, \eta) = [x, y]^{\mathrm{T}}$ are the Cartesian coordinates, whereas the global shape functions are:

$$\left.\begin{aligned} \overline{N}_k(\xi, \eta) = B_{i,n}(\xi) \cdot B_{j,m}(\eta), \quad i = 0, \ldots, n; \quad j = 0, \ldots, m \\ \text{with} \quad k = 1, \ldots, (n+1)(m+1) \end{aligned}\right\} \tag{9.42}$$

Using again the standard Galerkin–Ritz formulation, Eq. (9.23) is replaced by

$$[\mathbf{M}]\{\ddot{\mathbf{a}}\} + [\mathbf{K}]\{\mathbf{a}\} = \{\mathbf{r}\} + \{\mathbf{f}\} \tag{9.43}$$

where the elements of the stiffness matrix $[\mathbf{K}]$ are given in terms of the derivatives of the global tensor-product shape functions by:

$$k_{ij} = \int_\Omega D\left[\frac{\partial^2 \overline{N}_i}{\partial x^2}\left(\frac{\partial^2 \overline{N}_j}{\partial x^2} + \nu\frac{\partial^2 \overline{N}_j}{\partial y^2}\right) + \frac{\partial^2 \overline{N}_i}{\partial y^2}\left(\nu\frac{\partial^2 \overline{N}_j}{\partial x^2} + \frac{\partial^2 \overline{N}_j}{\partial y^2}\right)\right.$$
$$\left. +2(1-\nu)\frac{\partial^2 \overline{N}_i}{\partial x \partial y}\frac{\partial^2 \overline{N}_j}{\partial x \partial y}\mathrm{d}x\mathrm{d}y\right] \tag{9.44}$$

Similarly, the mass matrix [**M**] is expressed in products of the global shape functions themselves by:

$$m_{ij} = \rho h \int_\Omega \overline{N}_i \overline{N}_j \mathrm{d}x\mathrm{d}y \tag{9.45}$$

with $\{\mathbf{r}\}$ and $\{\mathbf{f}\}$ denoting the vector of the internal and external load, respectively.

Note that the meaning of the indexes (i, j) on the matrix elements k_{ij} and m_{ij} in Eqs. (9.44) and (9.45) is different than what it was in Eqs. (9.24) and (9.25).

The advantage of using Eq. (9.41) is that now we can calculate the second order derivatives in the global system in terms of those in the local one as follows:

$$\frac{\partial^2 \overline{N}}{\partial x^2} = \frac{\partial^2 \overline{N}}{\partial \xi^2}\left(\frac{\partial \xi}{\partial x}\right)^2 + 2\frac{\partial^2 \overline{N}}{\partial \xi\, \partial \eta}\frac{\partial \xi}{\partial x}\frac{\partial \eta}{\partial x} + \frac{\partial^2 \overline{N}}{\partial \eta^2}\left(\frac{\partial \eta}{\partial x}\right)^2 + \frac{\partial \overline{N}}{\partial \xi}\frac{\partial^2 \xi}{\partial x^2} + \frac{\partial \overline{N}}{\partial \eta}\frac{\partial^2 \eta}{\partial x^2} \tag{9.46}$$

$$\frac{\partial^2 \overline{N}}{\partial x\, \partial y} = \frac{\partial^2 \overline{N}}{\partial \xi^2}\frac{\partial \xi}{\partial x}\frac{\partial \xi}{\partial y} + \frac{\partial^2 \overline{N}}{\partial \xi\, \partial \eta}\left(\frac{\partial \xi}{\partial x}\frac{\partial \eta}{\partial y} + \frac{\partial \xi}{\partial y}\frac{\partial \eta}{\partial x}\right) + \frac{\partial^2 \overline{N}}{\partial \eta^2}\frac{\partial \eta}{\partial x}\frac{\partial \eta}{\partial y}$$
$$+ \frac{\partial \overline{N}}{\partial \xi}\frac{\partial^2 \xi}{\partial x\, \partial y} + \frac{\partial \overline{N}}{\partial \eta}\frac{\partial^2 \eta}{\partial x\, \partial y} \tag{9.47}$$

$$\frac{\partial^2 \overline{N}}{\partial y^2} = \frac{\partial^2 \overline{N}}{\partial \xi^2}\left(\frac{\partial \xi}{\partial y}\right)^2 + 2\frac{\partial^2 \overline{N}}{\partial \xi\, \partial \eta}\frac{\partial \xi}{\partial y}\frac{\partial \eta}{\partial y} + \frac{\partial^2 \overline{N}}{\partial \eta^2}\left(\frac{\partial \eta}{\partial y}\right)^2$$
$$+ \frac{\partial \overline{N}}{\partial \xi}\frac{\partial^2 \xi}{\partial y^2} + \frac{\partial \overline{N}}{\partial \eta}\frac{\partial^2 \eta}{\partial y^2} \tag{9.48}$$

Again, the traditional tensor-product Hermite element has *not* the capability to treat plates of nonrectangular shape, whereas the Bernstein–Bézier does have. But the latter elements have still the restriction that they are of large degree (nth and mth in each direction accordingly) and thus $C^{(n-1)}$ and $C^{(m-1)}$ continuous, respectively.

9.6 B-Splines Macroelements

These elements expand the abovementioned Bernstein–Bézier plate macroelements in that they can handle inner breakpoints as well, thus one can "play" with the multiplicity of the inner knots. A relevant one-dimensional case was studied earlier in Example 9.3.

The simplest case is to consider cubic polynomials in conjunction with inner knots of multiplicity equal to two ($\lambda = 2$). For a rectangular plate, this choice is mathematically *equivalent* to the use of the classical BFS elements (Bogner et al. [4]). This is so because the produced basis functions $\tilde{N}$ are a linear transformation of those in the classical cubic Hermite polynomials (see Example 9.3, as well as in [23]). In other words, for this particular choice, although one may believe that he/she has implemented a global approximation, in reality the true approximation is local.

Moreover, the analyst may implement any number (n) of spans, in conjunction with any polynomial degree (p) and any multiplicity (λ) at the ($n - 1$) inner knots. Then, the procedure described in Sect. 6.4.2 is repeated, with the basic difference that Eq. (9.40) is replaced by:

$$w(\xi, \eta) = \sum_{i=0}^{n_p} \sum_{j=0}^{m_p} N_{i,p_x}(\xi) \cdot N_{j,p_y}(\eta) \cdot a_{ij}, \tag{9.49}$$

where p_x and p_y are the polynomial degrees in the x- and y-directions, respectively. The univariate B-splines functions, $N_{i,p_x}(\xi)$ and $N_{j,p_y}(\eta)$, refer to n_p and m_p control points, respectively, which are dependent on the polynomial degrees (p_x, p_y) and the multiplicities (λ_x, λ_y), according to the general theory of B-splines (see, Chap. 2).

The reader is encouraged to write computer codes and implement this formulation to rectangular and circular plates, for both static and dynamic analyses.

Example 9.5 (Simply supported circular plate under a concentrated central load) This example is concerned with circular simply supported (SS) plate of radius $R = 1$, at the center of which a concentrated transverse load P is exerted (on the vertical plane). The analytical exact (analytical) solution of this problem may be found in several textbooks such as in Ugural [24]:

$$w_{\max} = \kappa \frac{Pa^2}{16\pi D}, \text{ where } \kappa = \begin{cases} 1, & \text{clamped} \\ (3+\nu)/(1+\nu), & \text{simply supported} \end{cases} \tag{9.50}$$

Following the standard procedure for circular domains, the boundary is divided into four equal parts where the corners of the unit reference square $ABCD$ are located. For a certain polynomial degree p, within the context of the *nonrational* Bézier formulation, the $(p+1) \times (p+1)$ control points are easily determined using an equal number of test points of the domain (boundary plus Coons interpolation, as we saw earlier in Chap. 6; shown by a set of crosses in Fig. 9.7) to which Eq. (9.41) is applied.

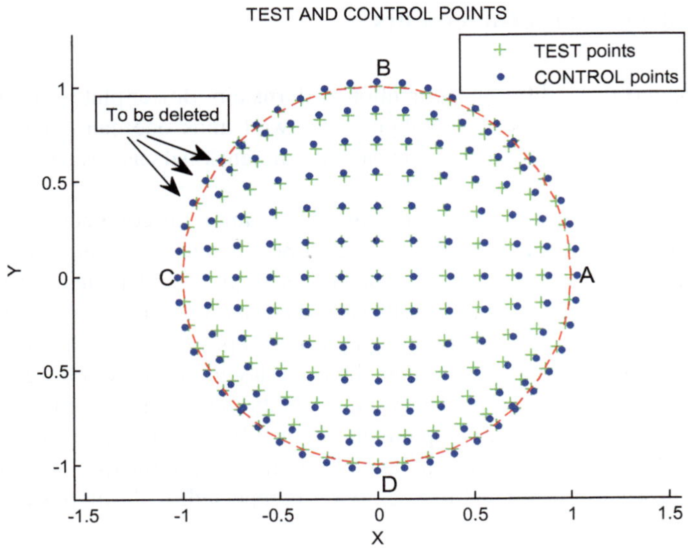

Fig. 9.7 Control points related to the corner points (A, B, C, and D) of a circular simply supported plate (here the polynomial degree of the bézierian element is $p = 12$)

The stiffness matrix is calculated by substituting Eqs. (9.46)–(9.48) into Eq. (9.44). The simply supported boundary conditions are easily imposed, by deleting the rows and columns that are related to the control points that are very close to the boundary (in fact they are outside the circle), as shown by the dark points in Fig. 9.7.

The results are shown in Fig. 9.8 in terms of the ratio:

$$(\rho = w_{\text{calculated}}/w_{\text{exact}})_{\text{centre}} \tag{9.51}$$

where $w_{\text{calculated}}$ is the calculated deflection at the center and w_{exact} is the corresponding analytical value.

One may observe a good convergence toward the unity (i.e., that $w_{\text{calculated}} \rightarrow w_{\text{exact}}$ at the center of the plate).

With regards to Fig. 9.8, it is noted that:

(1) In the Bézierian macroelement, it was found that odd polynomial degrees lead to almost the same result as the previous even degree. For example, for $p = 7$ the ratio of the deflections at the center, $(\rho = w_{\text{calculated}}/w_{\text{exact}})_{\text{centre}}$, is 0.9465, whereas for $p = 6$ is was equal to 0.9466 (no improvement was achieved). Nevertheless, all findings are included in Fig. 9.8. If only the *even* values of p had been considered, the graph would be quite *smooth*.

(2) In the tensor B-spline macroelement, where permanently the degree was $p = 3$ and the multiplicity of inner knots was equal to 2, odd numbers of knot spans lead

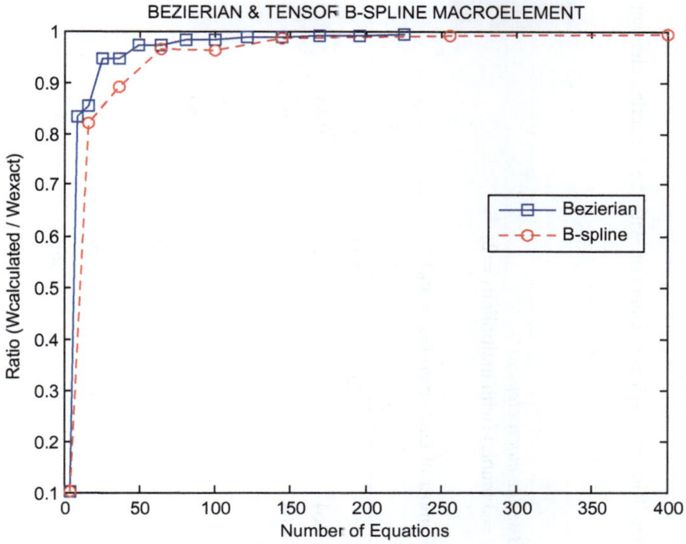

Fig. 9.8 Convergence quality of calculated deflection at the center of a circular plate

to somehow lower accuracy than even ones. This may be probably attributed to the fact that in the first case there is not any control point exactly at the center of the circle. Nevertheless, all findings (even the not-in-favor ones) are included in Fig. 9.8, where the convergence is characterized by some numerical oscillations. If only the *even* values of $n_\xi = n_\eta$ had been considered, the graph would be quite *smooth*.

Example 9.6 (Natural vibrations of circular clamped plate) A circular clamped plate of radius $R = 1$ was analyzed and the proposed CAD-based macroelement Galerkin–Ritz method was successfully compared with the exact solution [26].

Similarly to the previous Example 9.5, the parameterization is the simplest one, where the edges of the unit square *ABCD* have been located at the boundary which is divided into four equal parts. In the single nonrational Bézierian macroelement, we use one knot span ($n_\xi = n_\eta = 1$) for each entire side (say *AB*) of the quadrilateral *ABCD*. Moreover, when we use more than one knot spans ($n_\xi = n_\eta > 1$), the formulation is a pure single B-spline macroelement.

First, in Table 9.5 we study the performance of a single Bézierian macroelement in comparison with the older cubic B-spline (single macroelement as well). In both cases, the convergence is mostly monotonic. One may also observe, for example, that for a high polynomial degree such as $p = 10$, the Bézierian macroelement (with 49 DOF) is more accurate than the cubic in conjunction with five knot spans along one-fourth of the boundary (say side AB) [with 64 DOF]. It is noted that the number of equations, shown at the bottom of Table 9.5, is the number of free DOF that comes out after the imposition of the boundary conditions.

Table 9.5 Clamped circular plate of unit radius. Calculated natural frequencies using the Galerkin–Ritz approach in conjunction with the Bézierian element and the cubic B-spline macroelement (multiplicity = 2)

Mode	Exact parameter λ_{ij}^2 [cf. Eq. (9.27)]	Errors in % of the calculated natural frequencies							
		Bezierian macroelement Number of knot spans ($n_\xi = n_\eta = 1$), (Multiplicity is not applicable)				*B-spline macroelement* $p = 3$ (cubic) (with multiplicity = 2)			
		Polynomial degree (p)				Number of knot spans ($n_\xi = n_\eta$)			
		5	6	7	10	2	3	4	5
1	10.2158	6.59	0.14	0.14	0.00	1.05	1.25	0.31	0.13
2	21.2604	14.44	7.38	0.54	0.01	14.64	1.00	1.21	0.43
3	21.2604	14.44	7.38	0.54	0.01	14.64	1.00	1.21	0.43
4	34.8770	30.09	15.22	1.50	0.05	11.20	1.99	0.73	0.67
5	34.8770	–	30.13	15.22	0.06	–	14.77	2.62	1.71
6	39.7711	–	14.51	14.51	0.06	–	10.99	0.41	1.59
7	51.0306	–	58.49	22.62	0.67	–	14.15	4.46	1.89
8	51.0306	–	58.49	22.62	0.67	–	14.15	4.46	1.89
9	60.8287	–	126.99	24.10	2.19	–	24.20	8.76	0.83
Number of equations		4	9	16	49	4	16	36	64

Furthermore, Table 9.6 presents typical numerical results when using a single macroelement based on B-splines of a high polynomial degree ($p=7$) in conjunction with various knot spans of multiplicity of inner knots. Again, the number of equations (after the imposition of the boundary conditions) is shown at the bottom of Table 9.6. One may observe that for the same number of equations, the use of double inner knots leads to a somehow better accuracy

Finally, in order to get a better idea about the excellent convergence quality of the abovementioned nonrational Bézierian and B-splines macroelements (shown in Tables 9.5 and 9.6), conventional finite elements for plate bending analysis (in the form of a structured mesh using three different element types from ANSYS library) were also used. As clearly is shown in Table 9.7, the convergence rate of the latter is very slow and the error of the first natural eigenfrequency is *considerably greater* than that of the single Macroelement method (Tables 9.5 and 9.6). Regarding the rest nine modes, it is noted that the error of the eigenfrequency is practically maintained at the same level with the first one.

Remark Based on the same global approximations using nonrational Bézier or nonrational B-splines, Ritz–Galerkin approach may be replaced by the Global Collocation Method [18]. Details are also given in Chap. 11.

Table 9.6 Clamped circular plate of unit radius

Mode	Exact parameter λ_{ij}^2 [cf. Equation (9.27)]	Errors in % of the calculated natural frequencies					
		Multiplicity = 1			Multiplicity = 2		
		Number of knot spans ($n_\xi = n_\eta$)			Number of knot spans ($n_\xi = n_\eta$)		
		2	3	4	2	3	4
1	10.2158	0.00	0.00	0.00	0.00	0.00	0.00
2	21.2604	0.36	0.01	0.03	0.00	0.00	0.00
3	21.2604	0.36	0.01	0.03	0.00	0.00	0.00
4	34.8770	0.60	0.02	0.01	0.00	0.00	0.00
5	34.8770	1.50	1.50	0.13	0.60	0.03	0.00
6	39.7711	0.72	1.63	0.01	0.72	0.04	0.00
7	51.0306	5.22	2.07	0.91	0.91	0.05	0.00
8	51.0306	5.22	2.07	0.91	0.91	0.05	0.00
9	60.8287	19.39	2.02	3.27	0.75	0.03	0.01
Number of equations		25	36	49	36	64	100

Calculated natural frequencies using the Galerkin-Ritz approach, in conjunction with a single B-spline macroelement of seventh degree ($p = 7$), for two different multiplicities of inner knots (1 and 2)

Table 9.7 Calculated fundamental natural frequency of a clamped circular plate of unit radius, using three alternative ANSYS elements

Exact parameter λ_{ij}^2 [cf. Eq. (9.27)]	ANSYS element	Errors in % of the calculated fundamental natural frequency		
		Number of uniform boundary segments per quarter of the circle ($n_\xi = n_\eta$)		
		16	32	64
10.2158	SHELL43	4.13	3.57	3.43
	SHELL63	3.14	3.34	3.39
	SHELL181	4.97	3.77	3.47
Number of unrestrained nodal points		225	961	3969
Number of equations (free bending DOFs)		675	2883	11,907

9.7 Rational Bézier and Rational B-Splines Macroelements

Since a single tensor-product B-spline macroelement performs well in beam bending and (in the particular parameterization of Examples 9.5 and 9.6) in circular plate bending analysis, it is anticipated that also rational plate elements will perform well. After a circle or another part of a conical section has been modeled using the proper control points and weights (in the same with the abovementioned or a different parameterization), and after a suitable parametric space (ξ, η) has been defined, automatic *knot insertion* and *degree elevation* (alone or in combination) assist the isogeometric analysis. For further details, the interested reader may consult Cottrell et al. [8] and papers therein. A further study that includes the use of T-splines may be the book edited by Buffa and Sangalli [5], as well as recent journal papers.

Unsolved exercises

(1) Apply Coons theory and determine the closed-form analytical shape functions for an eight-noded plate bending element of which all nodes belong to the boundary.

(2) Extend the classical 4-node BFS element from four nodes to five ones, by adding one internal node at the center of the rectangle ABCD of dimensions $a \times b$. Note that the new blending functions will be Hermite polynomials which obviously should vanish at the position of the internal node, i.e., at $x = \frac{a}{2}$ and $y = \frac{b}{2}$. Since these Hermite polynomials operate over two spans $\left(\left[0, \frac{L}{2}\right] \text{ and } \left[\frac{L}{2}, L\right]\right)$ of total length L (either a or b), they should be of fifth degree.

(3) Repeat the static calculations of Example 9.5 (using nonrational B-splines) imposing the boundary conditions of a clamped plate.

(4) Repeat the static calculations of Example 9.5 considering a uniform load, p_0. In this case, the exact value of the deflection at the center is $w_{\max} = \frac{p_0 a^4}{64D} \frac{5+\nu}{1+\nu}$.

9.8 Recapitulation

Plate bending analysis is a topic of major importance in engineering. The purpose of this chapter was to make clear that all older CAD interpolation formulas (i.e., Coons, Gordon, Bézier, and B-splines), which are precursors of NURBS, are capable of constructing large plate bending macroelements, for static and dynamic analyses. For the sake of brevity, Barnhill's patches with C^1-continuity as well as examples related to triangular patches have been omitted. Based on these numerical results, since NURBS is chronologically the fifth important formula in CAD theory, it is reasonable that also the NURBS-based IGA should work perfectly as initially was reported by Hughes et al. [13]. The reader may develop his/her own computer codes to discover the magnitude of the improvement caused by the accurate representation of circles through rational expressions for both the geometric model and the deflection.

In case of quadrilaterals, it was shown that the classical BFS element (Bogner et al. [4]) is the first element in the family of Coons (boundary-only) plate elements. This particular element happens to be also the simplest element in the family of the transfinite (Gordon) elements, in the sense that it is a tensor-product Hermite element as well.

For any number of subdivisions ($n_x \times n_y$) in both directions, between the boundary-only (from one side) and the tensor-product Hermite macroelements (from the other side), there are many alternative arrangements of internal nodes, and all of them can be treated by the transfinite formula initially proposed by Coons [7] and later by Bill Gordon's team at General Motors.

Moreover, it was shown that the older Hermitian tensor-product macroelement is equivalent to a tensor-product Bézierian one, for the same polynomial degree. This finding is very similar to the previously recognized equivalency between the Lagrangian and Bézierian elements, which we met earlier in Chap. 6.

Furthermore, one of the most interesting issues is as follows. In an instructive way, it was shown that the two older CAD-based methods which are based on control points, i.e. Bernstein–Bézier and B-splines [11] *break the barriers* and make the tensor-product elements applicable to *curvilinear* domains as well. Specific details concerning the imposition of the boundary conditions were shown initially through the example of a clamped beam. Later the discussion was extended to the static as well as the natural frequency analysis of circular plates, either simply supported or clamped, along the entire boundary.

References

1. Angelidis D (2005) Development of Coons macroelements for static thin plate-bending analysis, Diploma Work. National Technical University of Athens, School of Mechanical Engineering, July 2005
2. Blaauwendraad J (2010) Plates and FEM: surprises and pitfalls, solid mechanics and its applications (Book 171). Springer, Berlin

3. Blevins RD (1979) Formulas for natural frequency and mode shape. Van Nostrand, New York
4. Bogner FK, Fox RL, Schmit LA Jr (1965) The generation of inter-element compatible stiffness and mass matrices by use of interpolation formulas. In: Proceedings of 1st conference on matrix method in structural mechanics, Ohio, pp 397–443 (Oct 1965)
5. Buffa A, Sangalli G (eds) (2016) IsoGeometric analysis: a new paradigm in the numerical approximation of PDEs. Springer International Publishing, Switzerland
6. Carey GF, Oden JT (1983) Finite elements: a second course, vol II. Prentice-Hall, Englewood Cliffs, New Jersey, p 68
7. Coons SA (1964) Surfaces for computer aided design of space form, Project MAC, MIT (1964), revised for MAC-TR-41 (1967). Springfield, VA 22161, USA: Available as AD 663 504 from the National Technical Information Service (CFSTI), Sills Building, 5285 Port Royal Road. Available online: http://publications.csail.mit.edu/lcs/pubs/pdf/MIT-LCS-TR-041.pdf
8. Cottrell J, Hughes TJR, Bazilevs Y (2009) Isogeometric analysis: toward integration of CAD and FEA. Wiley, New York
9. Farin G (1990) Curves and surfaces for computer aided geometric design, 2nd edn. Academic Press, New York
10. Gorman DJ (1978) Free vibration analysis of the completely free rectangular plate by the method of superposition. J Sound Vib 57(3):437–447
11. Höllig K (2003) Finite element methods with B-Splines. SIAM, Philadelphia
12. Hrabok MM, Hrudey TM (1984) A review and catalogue of plate bending finite elements. Comput Struct 19(3):479–495
13. Hughes TJR, Cottrell JA, Bazilevs Y (2005) Isogeometric analysis: CAD, finite elements, NURBS, exact geometry and mesh refinement. Comput Methods Appl Mech Eng 194:4135–4195
14. Leissa AW (1973) The free vibration of rectangular plates. J Sound Vib 31(2):257–293
15. Melosh RJ (1965) A stiffness matrix for the analysis of thin plates in bending. J Aeronaut Sci 28(1):34–42
16. Oñate E (2013) Structural analysis with the finite element method. linear statics: volume 2: beams, plates and shells. Lecture Notes on Numerical methods in engineering and sciences, vol 2. Springer, Berlin
17. Provatidis CG (2018) Engineering analysis with CAD-based macroelements. Arch Appl Mech 88(1–2):121–140
18. Provatidis CG (2017) B-splines collocation for plate bending eigenanalysis. J Mech Mater Struct 12(4):353–371
19. Provatidis CG (2012) Two-dimensional elastostatic analysis using Coons-Gordon interpolation. Meccanica 47(4):951–967
20. Provatidis CG (2008) Plate bending analysis using transfinite interpolation. In: Talaslidis D, Manolis G (eds) Proceedings 6th GRACM international congress on computational mechanics. Thessaloniki, Greece
21. Provatidis CG (2003) Coons-patch macroelements for boundary-value problems governed by fourth-order partial differential equation. In: Lipitakis EA (ed) Proceedings of the 'sixth hellenic-european conference on computer mathematics and its applications' (HERCMA 2003), Sept 25–27, 2003, Athens, Greece, pp 634–643, Lea Publications. ISBN: 960-87275-1-0
22. Provatidis CG, Angelidis D (2014) Performance of Coons' macroelements in plate bending analysis. Int J Comput Methods Eng Sci Mech 15:110–125
23. Provatidis CG, Isidorou SK (2012) Solution of one-dimensional hyperbolic problems using cubic B-Splines collocation. Int J Comput Sci Appl (IJCSA) 1(1):12–18
24. Ugural AC (1981) Stresses in plates and shells. McGraw-Hill, New York
25. Weaver W Jr, Timoshenko SP, Young DH (1990) Vibration problems in engineering, 5th edn. Wiley-Interscience, New York
26. Zhou ZH, Wong KW, Xu XS, Leung AYT (2011) Natural vibration of circular and annular thin plates by Hamiltonian approach. J Sound Vib 330(5):1005–1017
27. Zienkiewicz OC, Cheung YK (1964) The finite element method for analysis of elastic isotropic and orthotropic slabs. Proc Inst Civ Eng 28:471–488

Chapter 10
Three-Dimensional Macroelements

Abstract This chapter deals with three-dimensional macroelements (large solid bricks) based on several CAD-based interpolation formulas. First, three alternative expressions are derived for the boundary-only Coons interpolation; the first of them is complete, whereas the next two cover particular cases. It will be shown that the classical eight-node trilinear and the twenty-node triquadratic solid elements are the simplest ones of the Coons family. Second, Gordon interpolation in conjunction with internal nodes is fully explained, and it is shown that the classical 27-node tensor-product Lagrangian element is the simplest element of this class. Based on the two aforementioned CAD interpolations (i) simplified edge-only Coons macroelements will be developed, which are hierarchically improved (ii) through enhanced boundary-only formulation (based on the entire faces) as well as (iii) full tensor-product Coons–Gordon macroelements. Nonrational Bézier elements as well as elements based on B-splines will be presented. For the sake of brevity, numerical results reduce to the acoustic analysis of rectangular and spherical cavities, where all formulations are compared to the closed-form exact solution and are thoroughly criticized. The reader is also referred to previously published numerical examples that include potential (Laplace and Poisson's equations) and elasticity problems in bodies of cuboidal (parallelepiped with six rectangular faces), cylindrical and spherical shape, using a single macroelement.

Keywords 3D Coons interpolation · C-element · *p*-method · Transfinite Bernstein–Bézier element · B-spline solid element · Rational bézier · Cuboid Sphere

10.1 General

Having the reader been deeply introduced in the whole topic from the material of the previous chapters, this chapter merely applies *all* previous interpolations (Coons, Gordon, Bezier, B-splines, rational Bézier) in 3D primitives such as rectangles and spheres. For the particular case of Coons interpolation [4], the three alternative ver-

C. G. Provatidis, *Precursors of Isogeometric Analysis*, Solid Mechanics and Its Applications 256, https://doi.org/10.1007/978-3-030-03889-2_10

sions of the 3D Coons interpolations (based on three hierarchies of the hexahedral: 8 corners, 12 edges, 6 surfaces) are discussed in Sect. 10.2. Later, in subsequent subsections other interpolation formulas will follow.

10.2 Three-Dimensional (3D) Coons Interpolations

10.2.1 General Formulation of Coons Macroelements

The standard 3D Coons interpolation was introduced as a small part in the MIT Report of 1967 ([4], also updated in 1967), and here, it refers to the hexahedral ABCDEFGH shown in Fig. 10.1.

Three different expressions for the interpolation within the volume block ABCDE-FGH will be presented below.

Expression C1: The most general expression The most complete formula is the one proposed by Steven Coons himself [4, p. 41]:

$$\mathbf{x}(\xi, \eta, \zeta) = S\{\mathbf{x}\}(\xi, \eta, \zeta) - \widehat{E}\{\mathbf{x}\}(\xi, \eta, \zeta) + C\{\mathbf{x}\}(\xi, \eta, \zeta) \tag{10.1}$$

with the three quantities given by:

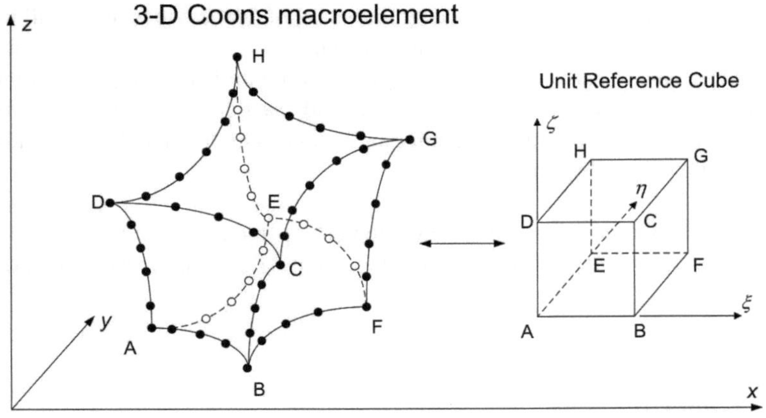

Fig. 10.1 Definition of a three-dimensional macroelement (*ABCDEFGH*) in the form of a curvilinear hexahedral

$$\begin{aligned} S\{\mathbf{x}\} &= \left(P_\xi + P_\eta + P_\zeta\right)\{\mathbf{x}\} && : \textbf{Surfaces} \\ \widehat{E}\{\mathbf{x}\} &= \left(P_{\xi\eta} + P_{\eta\zeta} + P_{\zeta\xi}\right)\{\mathbf{x}\} && : \textbf{Edges} \\ C\{\mathbf{x}\} &= P_{\xi\eta\zeta}\{\mathbf{x}\} && : \textbf{Corners} \end{aligned} \tag{10.2}$$

whereas the involved projections are (repeated index implies a summation over all possible values of the index):

$$\left.\begin{aligned} P_\zeta\{\mathbf{x}\} &= \mathbf{x}(\xi, \eta, \zeta_k) \cdot E_k(\zeta) \\ P_{\eta\zeta}\{\mathbf{x}\} &= P_\eta P_\zeta\{\mathbf{x}\} = \mathbf{x}\left(\xi, \eta_j, \zeta_k\right) \cdot E_j(\eta) \cdot E_k(\zeta) \\ P_{\xi\zeta}\{\mathbf{x}\} &= P_\xi P_\zeta\{\mathbf{x}\} = \mathbf{x}(\xi_i, \eta, \zeta_k) \cdot E_i(\xi) \cdot E_k(\zeta) \\ P_{\xi\eta\zeta}\{\mathbf{x}\} &= P_\xi P_\eta P_\zeta\{\mathbf{x}\} = \mathbf{x}\left(\xi_i, \eta_j, \zeta_k\right) \cdot E_i(\xi) \cdot E_j(\eta) \cdot E_k(\zeta) \end{aligned}\right\} \tag{10.3}$$

The expression of Eq. (10.1) covers the most general case, where based on Fig. 10.1, the data are given:

(i) Over the 6 surfaces of the volume block ("*S*" in Eq. (10.1) stands for surface).
(ii) Along the 12 edges of the volume block ("*E*" stands for edges).
(iii) At the 8 corners of the volume block ("*C*" stands for corners).

Clearly, let us consider a point $P(\xi, \eta, \zeta)$ inside the volume block *ABCDEFGH* shown in Fig. 10.2. Let us also consider that the line, which passes through *P* and is parallel to the ξ-axis, intersects the planes $\xi = 0$ and $\xi = 1$ at the points *I* and *J*, respectively. Also, consider that the line through *P* and parallel to the η-axis, intersects the planes $\eta = 0$ and $\eta = 1$ at the points *K* and *L*, respectively. Finally, the line through P and parallel to the ζ-axis intersects the planes $\zeta = 0$ and $\zeta = 1$ at the points *M* and *N*, respectively. Under these definitions, the entire projection *S* (cf. Eq. (10.2)) is the sum of the linear interpolation of the value at the point $P(\xi, \eta, \zeta)$ in terms of the values at the points I and J (projection P_ξ), plus the interpolation between the points *K* and *L* (projection P_η), and eventually plus the interpolation between the points *M* and *N* (projection P_ζ).

If, temporarily, one replaces the Cartesian coordinate with a physical magnitude $U(\xi, \eta, \zeta)$ and considers that its value is a constant, U_0, the abovementioned sum *S* would lead to the triple of its value ($3U_0$); thus, it is incorrect. Due to this fact, in order to maintain the values of the variable over the six surfaces and along the twelve edges, the corrective term ($-\widehat{E} + C$) was introduced.

After Coons [4] presented Eq. (10.1) for the description of the geometry $\mathbf{x}(\xi, \eta, \zeta)$, as well as of a variable $U(\xi, \eta, \zeta)$, Cook [5] used this complete expression for 3D mesh generation of a structured net of nodes within a volume block *ABCDEFGH*, using conventional 8-node brick elements He also implemented variations, similar to those shown below.

Expression C2: Using surfaces and corners only Having fixed the data along the 12 edges of the volume block, and expressing the values of each point on every

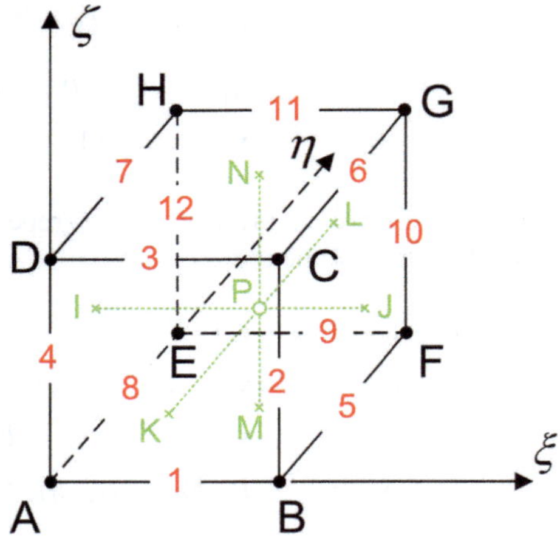

Fig. 10.2 Brick (solid) element in terms of Coons interpolation

surface in terms of the boundary data of the relevant patch (six patches in total), it can be easily proven that for every triplet (ξ, η, ζ), the interpolating quantity is given by (see also in [2]):

$$\mathbf{x}(\xi, \eta, \zeta) = 1/2S\{\mathbf{x}\} - 1/2C\{\mathbf{x}\}, \tag{10.4a}$$

which, in more details, is written as

$$\begin{aligned}\mathbf{x}(\xi, \eta, \zeta) = \tfrac{1}{2} \cdot [&(1-\eta)\mathbf{x}(\xi, 0, \zeta) + \eta\mathbf{x}(\xi, 1, \zeta) \\ &+ (1-\xi)\mathbf{x}(0, \eta, \zeta) + \xi\mathbf{x}(1, \eta, \zeta) \\ &+ (1-\zeta)\mathbf{x}(\xi, \eta, 0) + \zeta\mathbf{x}(\xi, \eta, 1)] \\ &- \tfrac{1}{2} \cdot [(1-\xi)(1-\eta)(1-\zeta)\mathbf{x}(0,0,0) + (1-\xi)(1-\eta)\zeta\mathbf{x}(0,0,1) \\ &+ (1-\xi)\eta(1-\zeta)\mathbf{x}(0,1,0) + (1-\xi)\eta\zeta\mathbf{x}(0,1,1) \\ &+ \xi(1-\eta)(1-\zeta)\mathbf{x}(1,0,0) + \xi(1-\eta)\zeta\mathbf{x}(1,0,1) \\ &+ \xi\eta(1-\zeta)\mathbf{x}(1,1,0) + \xi\eta\zeta\mathbf{x}(1,1,1)]\end{aligned} \tag{10.4b}$$

Expression C3: Using edges and corners only Assuming that each of the six surfaces (S) can be interpolated in terms of the four surrounding edges of the relevant quadrilateral patch, they are easily eliminated, thus leading to the following third expression:

$$\mathbf{x}(\xi, \eta, \zeta) = \widehat{E}\{\mathbf{x}\} - 2C\{\mathbf{x}\}, \tag{10.5a}$$

which, in more details, is written as

$$
\begin{aligned}
\mathbf{x}(\xi,\eta,\zeta) = {} & E_0(\eta)E_0(\zeta)\mathbf{x}(\xi,0,0) + E_1(\xi)E_0(\eta)\mathbf{x}(1,0,\zeta) \\
& + E_0(\eta)E_1(\zeta)\mathbf{x}(\xi,0,1) + E_0(\xi)E_0(\eta)\mathbf{x}(0,0,\zeta) \\
& + E_1(\xi)E_0(\zeta)\mathbf{x}(1,\eta,0) + E_1(\xi)E_1(\zeta)\mathbf{x}(1,\eta,1) \\
& + E_0(\xi)E_1(\zeta)\mathbf{x}(0,\eta,1) + E_0(\xi)E_0(\zeta)\mathbf{x}(0,\eta,0) \\
& + E_1(\eta)E_0(\zeta)\mathbf{x}(\xi,1,0) + E_1(\xi)E_1(\eta)\mathbf{x}(1,1,\zeta) \\
& + E_1(\eta)E_1(\zeta)\mathbf{x}(\xi,1,1) + E_0(\xi)E_1(\eta)\mathbf{x}(0,1,\zeta) \\
& - 2[E_0(\xi)E_0(\eta)E_0(\zeta)\mathbf{x}(0,0,0) + E_1(\xi)E_0(\eta)E_0(\zeta)\mathbf{x}(1,0,0) \\
& + E_1(\xi)E_0(\eta)E_1(\zeta)\mathbf{x}(1,0,1) + E_0(\xi)E_0(\eta)E_1(\zeta)\mathbf{x}(0,0,1) \\
& + E_0(\xi)E_1(\eta)E_0(\zeta)\mathbf{x}(0,1,0) + E_1(\xi)E_1(\eta)E_0(\zeta)\mathbf{x}(1,1,0) \\
& + E_1(\xi)E_1(\eta)E_1(\zeta)\mathbf{x}(1,1,1) + E_0(\xi)E_1(\eta)E_1(\zeta)\mathbf{x}(0,1,1)]
\end{aligned} \tag{10.5b}
$$

Equation (10.5a) has been the vehicle to develop the first category of arbitrary noded 3D Coons macroelements with nodal points arranged along the *twelve edges* only. An example is provided in the next subsection, while a lot of numerical examples may be found in Provatidis [11–13].

10.2.2 The Eight-Node (Solid) Isoparametric Element Under the Prism of Coons Interpolation

Let us consider the simplest case where the volume block becomes a unit cube (see Fig. 10.2), so each of the twelve edges is a straight line along which the variable varies linearly.

The starting point to develop global shape functions is to apply the idea of isoparametric element by extending Coons interpolation formula [4] from geometry to the interpolation of the scalar potential U or the displacement $\mathbf{u}$. Therefore, with $(0 \le \xi, \eta, \zeta \le 1)$, the term which is related to the twelve edges is written as:

$$
\begin{aligned}
\widehat{E}\{U\} = \left(P_{\xi\eta} + P_{\eta\zeta} + P_{\zeta\xi}\right)\{U\} = {} & (1-\eta)(1-\zeta)U_{AB} + \xi(1-\eta)U_{BC} \\
& + (1-\eta)\zeta U_{CD} + (1-\xi)(1-\eta)U_{DA} \\
& + \xi(1-\zeta)U_{BF} + \xi\zeta U_{CG} + (1-\xi)\zeta U_{DH} \\
& + (1-\xi)(1-\zeta)U_{AE} + \eta(1-\zeta)U_{EF} + \xi\eta U_{FG} \\
& + \eta\zeta U_{GH} + (1-\xi)\eta U_{EH}
\end{aligned} \tag{10.6}
$$

Also, the correcting term which is related to the corners is:

$$
\begin{aligned}
C(\xi,\eta,\zeta) = {} & (1-\xi)(1-\eta)(1-\zeta)U_A \\
& + \xi(1-\eta)(1-\zeta)U_B + \xi(1-\eta)\zeta U_C \\
& + (1-\xi)(1-\eta)\zeta U_D + (1-\xi)\eta(1-\zeta)U_E \\
& + \xi\eta(1-\zeta)U_F + \xi\eta\zeta U_G + (1-\xi)\eta\zeta U_H
\end{aligned} \tag{10.7}
$$

The expressions of the quantities $\widehat{E}$ and C in Eqs. (10.6) and (10.7), respectively, are general and can be applied for any set of trial functions along the twelve edges of the macroelement. In the particular case of this element in which linear interpolation is implemented along the twelve edges, for the particular numbering of the edges according to Fig. 10.2, we have:

$$\begin{aligned}
&\text{Edge \#1:}\ \ U_{AB} \equiv U(\xi,0,0) = (1-\xi)U_A + \xi U_B\\
&\text{Edge \#2:}\ \ U_{BC} \equiv U(1,0,\zeta) = (1-\zeta)U_B + \zeta U_C\\
&\text{Edge \#3:}\ \ U_{CD} \equiv U(\xi,0,1) = (1-\xi)U_C + \xi U_D\\
&\text{Edge \#4:}\ \ U_{AD} \equiv U(0,0,\zeta) = (1-\zeta)U_A + \zeta U_D\\
&\text{Edge \#5:}\ \ U_{BF} \equiv U(1,\eta,0) = (1-\eta)U_B + \eta U_F\\
&\text{Edge \#6:}\ \ U_{CG} \equiv U(1,\eta,1) = (1-\eta)U_C + \eta U_G\\
&\text{Edge \#7:}\ \ U_{DH} \equiv U(0,\eta,1) = (1-\eta)U_D + \eta U_H\\
&\text{Edge \#8:}\ \ U_{AE} \equiv U(0,\eta,0) = (1-\eta)U_A + \eta U_E\\
&\text{Edge \#9:}\ \ U_{EF} \equiv U(\xi,1,0) = (1-\xi)U_E + \xi U_F\\
&\text{Edge \#10:}\ U_{FG} \equiv U(1,1,\zeta) = (1-\zeta)U_F + \zeta U_G\\
&\text{Edge \#11:}\ U_{GH} \equiv U(\xi,1,1) = (1-\xi)U_H + \xi U_G\\
&\text{Edge \#12:}\ U_{EH} \equiv U(0,1,\zeta) = (1-\zeta)U_E + \zeta U_H
\end{aligned} \tag{10.8}$$

Then, substituting Eqs. (10.6) and (10.7) in conjunction with the linear trial functions Eq. (10.8) into Eq. (10.5a), after manipulation the following expression is derived:

$$\begin{aligned}
U(\xi,\eta,\zeta) = {} & U_A\cdot(1-\xi)(1-\eta)(1-\zeta) + U_B\cdot\xi(1-\eta)(1-\zeta)\\
& + U_C\cdot\xi(1-\eta)\zeta + U_D\cdot(1-\xi)(1-\eta)\zeta\\
& + U_E\cdot(1-\xi)\eta(1-\zeta) + U_F\cdot\xi\eta(1-\zeta) + U_G\cdot\xi\eta\zeta\\
& + U_H\cdot(1-\xi)\eta(1-\zeta).
\end{aligned} \tag{10.9}$$

One may observe that Eq. (10.9) is identical with that of the classical eight-node hexahedral element (brick) (see "*Linear*" element (8 nodes), in: [19, p. 169]). Therefore, as previously happened in the 2D analogue case, the simplest element produced by the 3D Coons interpolation is the well-known brick element. This issue has been previously mentioned by Provatidis [11].

Exercise 10.1 (Twenty-node (solid) isoparametric element)
Derive the shape functions of the twenty-node classical *quadratic* solid element shown in Fig. 10.3, using Coons interpolation.

Hint: Eq. (10.8) is replaced by quadratic interpolation along the 12 edges of the superblock ABCDEFGH. For example, along the edge *AB* we have (see Fig. 10.3):

$$U(\xi,0,0) = L_0^{AB}(\xi)\cdot U_1 + L_{1/2}^{AB}(\xi)\cdot U_9 + L_1^{AB}(\xi)\cdot U_2, \tag{10.10}$$

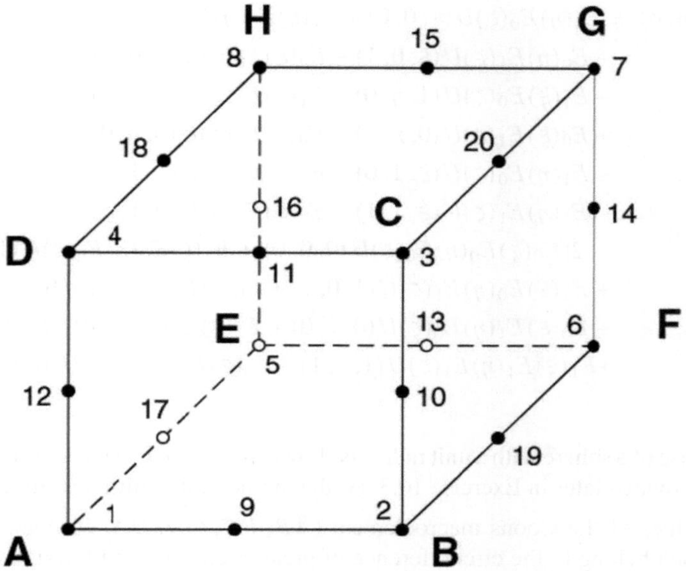

Fig. 10.3 Quadratic hexahedral 20-node finite element of serendipity family

with

$$L_0^{AB}(\xi) = (2\xi - 1)(\xi - 1), \quad L_{1/2}^{AB}(\xi) = 4\xi(1 - \xi), \quad L_1^{AB}(\xi) = \xi(2\xi - 1) \tag{10.11}$$

After similar expressions with Eq. (10.11) have been written for all the rest eleven edges, these are substituted first into Eqs. (10.6) and (10.7) and finally into Eq. (10.5a). After manipulation, the final expression includes all the shape functions which are identical with those of the classical triquadratic element of Serendipity family (see "*Quadratic*" element (20 nodes), in: [19, pp. 169–170]).

10.3 Closed-Form Shape Functions of "C1-Element"

By the term "C1-element," here we refer to the case described by Eq. (10.5b), in which only *edges* and *corners* participate (it should not be confused with the C^1-continuity). Below, this element will be also called "*edge-only*" Coons macroelement. Thus for this isoparametric element, we can write:

$$
\begin{aligned}
U(\xi,\eta,\zeta) = {} & E_0(\eta)E_0(\zeta)U(\xi,0,0) + E_1(\xi)E_0(\eta)U(1,0,\zeta)\\
& + E_0(\eta)E_1(\zeta)U(\xi,0,1) + E_0(\xi)E_0(\eta)U(0,0,\zeta)\\
& + E_1(\xi)E_0(\zeta)U(1,\eta,0) + E_1(\xi)E_1(\zeta)U(1,\eta,1)\\
& + E_0(\xi)E_1(\zeta)U(0,\eta,1) + E_0(\xi)E_0(\zeta)U(0,\eta,0)\\
& + E_1(\eta)E_0(\zeta)U(\xi,1,0) + E_1(\xi)E_1(\eta)U(1,1,\zeta)\\
& + E_1(\eta)E_1(\zeta)U(\xi,1,1) + E_0(\xi)E_1(\eta)U(0,1,\zeta)\\
& - 2[E_0(\xi)E_0(\eta)E_0(\zeta)U(0,0,0) + E_1(\xi)E_0(\eta)E_0(\zeta)U(1,0,0)\\
& + E_1(\xi)E_0(\eta)E_1(\zeta)U(1,0,1) + E_0(\xi)E_0(\eta)E_1(\zeta)U(0,0,1)\\
& + E_0(\xi)E_1(\eta)E_0(\zeta)U(0,1,0) + E_1(\xi)E_1(\eta)E_0(\zeta)U(1,1,0)\\
& + E_1(\xi)E_1(\eta)E_1(\zeta)U(1,1,1) + E_0(\xi)E_1(\eta)E_1(\zeta)U(0,1,1)]
\end{aligned}
\tag{10.12}
$$

The case of a sphere with a unit radius is shown in Fig. 10.4. Details of construction will be provided later in Exercise 10.3. At the moment, it is sufficient to say that the twelve edges of the Coons macroelement ($\widehat{AB}$, $\widehat{BC}$, $\widehat{CD}$, $\widehat{DA}$, etc.) are *spherical arcs*, which belong to the circumference of great circles passing through the center of the sphere.

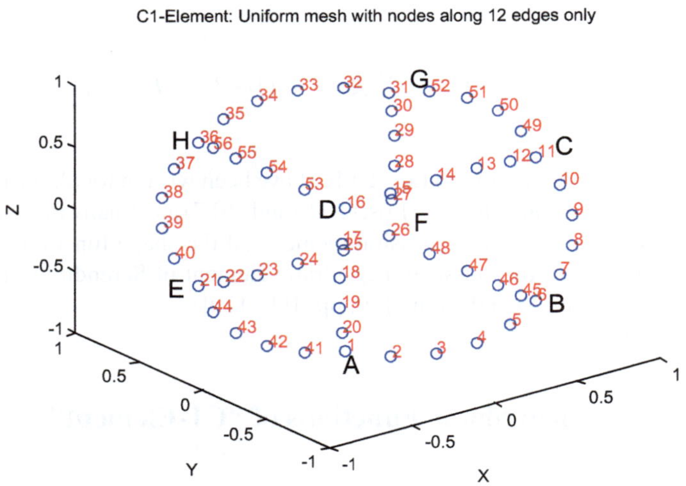

Fig. 10.4 C1-element (56 nodes) for a sphere with $n_s = 5$ subdivisions per edge

10.3.1 Univariate Interpolation

Let us assume that the edges *AB*, *BC*, *CD*, *DA*, *EF*, *FG*, *GH*, *HE*, *AE*, *DH*, *BF*, and *CG* include $q_1, \ldots, q_{12}$ nodes, respectively. Then, the total number of nodes along the 12 edges of the 3D Coons macroelement becomes:

$$q_e = \sum_{i=1}^{12} q_i - 16 \tag{10.13}$$

Each of the *univariate* functions along the 12 abovementioned edges included in Eq. (10.12), for example, $U(\xi, 0, 0)$ and $U(1, 0, \zeta)$, should be further approximated using a conventional interpolation such as piecewise-linear, natural cubic B-splines, Lagrange polynomials. Therefore, one can write:

$$\text{Edge } AB: \quad U(\xi, 0, 0) = \sum_{j=1}^{q_1} B_j^{AB}(\xi) \cdot U(\xi_j, 0, 0) \tag{10.14a}$$

$$\text{Edge } BC: \quad U(1, 0, \zeta) = \sum_{j=1}^{q_2} B_j^{BC}(\zeta) \cdot U(1, 0, \zeta_j) \tag{10.14b}$$

and similarly for the rest 10 edges.

10.3.2 Trivariate Interpolation

By substituting Eq. (10.14a, b, etc) into Eq. (10.12), it is easily found that the problem variable $U(\xi, \eta, \zeta)$ inside the reference cube (macroelement) is approximated as follows:

$$U(\xi, \eta, \zeta; t) = \sum_{j=1}^{q_e} N_j(\xi, \eta, \zeta) \cdot U_j(t), \tag{10.15}$$

with $N_j(\xi, \eta, \zeta)$ denoting *trivariate* cardinal global shape functions and $U_j(t)$, $j = 1, \ldots, q_e$ the (generally) time-dependent nodal variables at the *j*th nodal point along the 12 edges of the macroelement. A suitable re-arrangement of the degrees of freedom produces the global shape functions in terms univariate and blending functions. Typical expressions for a corner node (e.g., corner *A*) and one between two corners along an edge (e.g., edge *AB*) are given below:

(i) Corner (vertex) node A:

$$N_A(\xi,\eta,\zeta) = B_1^{AB}(\xi)E_0(\eta)E_0(\zeta) + B_{q_4}^{DA}(\zeta)E_0(\xi)E_0(\eta) + B_1^{AE}(\eta)E_0(\xi)E_0(\zeta) - 2E_0(\xi)E_0(\eta)E_0(\zeta) \tag{10.16}$$

(ii) Interior nodes to A:

$$N_j(\xi,\eta,\zeta) = B_j^{AB}(\xi)E_0(\eta)E_0(\zeta), \quad 2 \le j \le q_1 - 1 \tag{10.17}$$

The other corner- and edge-dependent shape functions are analogous.

10.3.3 General Remarks

(a) In Eqs. (10.16) and (10.17), $B_j^m(w)$ denote cardinal *univariate* functions of $w(=\xi, \eta$ or ζ, accordingly), while the global shape functions $N_j(\xi,\eta,\zeta)$ are also cardinal but *trivariate* functions. The aforementioned superscript $m = 1, 2, \ldots, 12$ corresponds to the edges *AB*, *BC*, *CD*, *DA*, *EF*, *FG*, *GH*, *HE*, *AE*, *DH*, *BF*, and *CG*, respectively.

(b) It should be clarified that it is not generally necessary to preserve the same interpolation along each of the 12 edges.

(c) The overall above formulation described by Eq. (10.12) restricts to box-like structures (hexahedra) provided that each of the six faces (surfaces) can be approximated with adequate accuracy on the basis of the four corresponding (surrounding) edges.

(d) In principle, *pentahedrals* can be easily treated by degenerating one whole face to a single point. Also, there are alternative dedicated CAGD expressions for this case.

(e) Moreover, if in the aforementioned pentahedral an edge of the intact face is degenerating to a single point, then a *tetrahedral* is received.

10.3.4 Values at Internal Points

After the boundary values have been calculated, the variable U at every internal point (x, y, z) can be calculated using Eq. (10.12) or Eq. (10.15). For nonrectangular domains, the relationship between Cartesian and normalized coordinates is not straightforward and may be necessary to apply a Newton–Raphson algorithm or a similar procedure on Eq. (10.15) so that to determine the triplet (ξ, η, ζ). Also, internal fluxes (or strains) can be easily derived by differentiating the global shape functions involved in Eq. (10.12) or Eq. (10.15).

10.3.5 Remarks on the Completeness of Coons "C1" Macroelements

Let us consider the particular case in which the basis functions B_j^m are Lagrangian polynomials. For simplicity purposes only, let us also consider that the parallel edges of the reference cube possess the same number of subdivisions, i.e., $(q_1 = q_3 = q_9 = q_{11})$, $(q_2 = q_4 = q_{10} = q_{12})$ and $(q_5 = q_6 = q_7 = q_8)$. Then, the univariate function $B_j^{AB}(\xi)$ involved in Eq. (10.14a) is a complete polynomial of $(q_1 - 1)$th degree in ξ, and therefore, (after the substitution) the first term in Eq. (10.12), i.e., $B^{AB}(\xi)E_0(\eta)E_0(\zeta)$ includes the following functional set:

$$\begin{aligned}&\{1,\xi,\xi^2,\ldots,\xi^{q_1-1}\},\{1,\xi,\xi^2,\ldots,\xi^{q_1-1}\}\eta,\{1,\xi,\xi^2,\ldots,\xi^{q_1-1}\}\zeta,\\&\text{and } \{1,\xi,\xi^2,\ldots,\xi^{q_1-1}\}\eta\zeta\end{aligned} \tag{10.18a}$$

In a similar way, the remaining monomials included in Eq. (10.14b) are (by cyclic permutation):

$$\begin{aligned}&\{1,\eta,\eta^2,\ldots,\eta^{q_5-1}\},\{1,\eta,\eta^2,\ldots,\eta^{q_5-1}\}\zeta,\{1,\eta,\eta^2,\ldots,\eta^{q_5-1}\}\xi,\\&\text{and } \{1,\eta,\eta^2,\ldots,\eta^{q_5-1}\}\zeta\xi\end{aligned} \tag{10.18b}$$

$$\begin{aligned}&\{1,\zeta,\zeta^2,\ldots,\zeta^{q_2-1}\},\{1,\zeta,\zeta^2,\ldots,\zeta^{q_2-1}\}\xi,\{1,\zeta,\zeta^2,\ldots,\zeta^{q_2-1}\}\eta,\\&\text{and } \{1,\zeta,\zeta^2,\ldots,\zeta^{q_2-1}\}\xi\eta\end{aligned} \tag{10.18c}$$

From this analysis, it becomes obvious that the abovementioned 3D Coons macroelement is *incomplete* and it is essentially a *Serendipity*-type element of an arbitrary number of boundary nodes (again along the 12 edges of the superbrick).

In order to get full polynomials, there are at least two procedures to follow:

- The first procedure refers to the subdivision of the domain into a certain number of the abovementioned incomplete macroelements, preferably of convex shape, with the nodes arranged along the 12 edges only.
- The second procedure refers to the addition of more nodes in the Coons macroelement. In this case, the addition can be performed on two levels. The *first* level concerns the introduction of boundary nodes over (some of) the six surfaces. The *second* level concerns the introduction of internal nodes inside the volume of the domain. A general methodology is presented in Sect. 10.5.

10.3.6 Relationships of "C-Elements" with Higher-Order p-Methods

In this section, the generalized 3D Coons–Gordon element is compared with the p-methods (of polynomial degree p). As a reference, we consider the book by Szabó and Babuška [18, pp. 239–242] where details can be found. A detailed comparison of a 3D Coons (*edge-only*) macroelement with nodal points $q_i = p, i = 1, \ldots, 12$, for which Lagrange polynomials ($B_j, j = 1, \ldots, p+1$) are used, may be found in Provatidis [13, pp. 118–119].

Here, we go one step further, discussing the complete Lagrangian macroelement. If, for convenience, we assume the same polynomial degree toward each direction, this element includes $(p+1)^3$ monomials in the form $\xi^i \eta^j \zeta^k$ (with $0 \le i, j, k \le p$), i.e., all those that can be produced by the tensor product of the vectors: $\{1, \xi, \ldots, \xi^p\} \times \{1, \eta, \ldots, \eta^p\} \times \{1, \zeta, \ldots, \zeta^p\}$.

In contrast, higher-order p-methods divide the basis functions in four categories as follows: (i) nodal nodes at corners, (ii) edge modes, (iii) face modes, and (iv) internal nodes. In more details:

(1) Higher-order p-methods use eight nodal (***corner***) shape functions that are the same as those in eight-node h-version computer programs ("linear" 8-node brick element). This leads to the functional set $\{1, \xi, \eta, \zeta, \xi\eta, \eta\zeta, \zeta\xi, \xi\eta\zeta\}$, which appears in both formulations (see, for example, Eq. 10.9).

(2) Higher-order p-methods deal with $12(p-1)$ ***edge*** modes of which the shape functions are given, e.g., by $N_{i-1}^{AB}(\xi, \eta, \zeta)|_{p\text{-method}} = \phi_i(\xi)(1-\eta)(1-\zeta), i = 2, 3, \ldots, p$ (local numbering). Obviously, when only the 12 edges of the macroelement are used, the blending functions are linear; thus, the edge-only Coons element coincides with the p-method. In contrast, the tensor-product Lagrangian element has more terms since the function $\phi_i(\xi)$ is multiplied by higher-order monomials (the blending functions are polynomials of pth degree as well).

(3) Moreover, higher-order p-methods deal with $3(p-2)(p-3)$ ***face*** modes ($p \ge 4$) for which the shape functions are given, for example, over the face *ABCD*, by $N_m^{ABCD}(\xi, \eta, \zeta)|_{p\text{-method}} = \phi_i(\xi)\phi_j(\zeta)(1-\eta), i = 2, 3, \ldots, p-2; i + j = 4, 5, \ldots, p$ (local numbering). Regarding Coons–Gordon element, we have to notice that:

   a. Each face includes $(p-1)^2$ nodal points, which somehow differs than $3(p-2)(p-3)$.
   b. The blending term $(1-\eta)$ in the product $\phi_i(\xi)\phi_j(\zeta)(1-\eta)$ is of lower degree in η, compared to the $\phi_k(\eta)$ of pth degree which is allowed in the Lagrangian element.

(4) Higher-order p-methods deal with $(p-3)(p-4)(p-5)/6$ ***internal*** nodes ($p \ge 6$) for which the shape functions are given by $N_m^{(0)}(\xi, \eta, \zeta)\big|_{p\text{-method}} = \phi_i(\xi)\phi_j(\eta)\phi_k(\zeta)$, where the subscript $m = m(i, j, k)$ and $i, j = 2, 3, \ldots, p-4; i + j + k = 6, 7, \ldots, p$. For example, if $p = 6$, then the only term related to internal node is $\phi_2(\xi)\phi_2(\eta)\phi_2(\zeta)$ which is proportional to the bubble function $(1-\xi^2)(1-\eta^2)(1-\zeta^2)$.

From the above discussion, it comes out that higher-order terms such as $\xi^6\eta^6\zeta^6$ do not appear in the higher-order p-method, at least as it is described in [18, pp. 239–242].

10.4 Transfinite Hexahedral Coons–Gordon Macroelements

The expression (10.5) is dedicated to superbrick elements of the *Serendipity* family, in which nodal points are distributed along the twelve edges including the eight corners.

Moreover:

- When additional nodes are inserted over the surfaces (boundary) then Eq. (10.4a) is to be used.
- When internal nodes are inserted then the general Eq. (10.1) is to be used.

Equation (10.1) enhances Eq. (10.5a) and constitutes the most general Coons–Gordon interpolation formula allowing C^0-continuity properties to three-dimensional space with an arbitrary number of inter-boundaries and arbitrary information level on any boundary. Furthermore, Eq. (10.1) is a "vehicle" to derive the global shape functions of a Coons–Gordon macroelement in a systematic way. Notice that not only an element of the *Serendipity* but also of the *Lagrangian* family, or any other finite element "in-between" them (with less or even more face or/and internal nodes), can be described on the basis of implementing Eq. (10.1). In the particular case of Lagrangian-type elements, it can be easily proven that Eq. (10.1) degenerates to the well-known *tensor-product* $U(\xi, \eta, \zeta) = P_{\xi\eta\zeta}\{U\}$, in which the blending functions E_i coincide with the (trial) basis functions B_i (univariate Lagrange polynomials in ξ, η and ζ).

Exercise 10.2 [Twenty-six-node-enhanced Coons brick element ("C2-element")] For the 26-node-enhanced Coons element shown in Fig. 10.5 (having a different philosophy of node numbering), find the shape functions. It is noted than this element comes from the well-known 27-node solid element of Lagrange family (with nodes in arrangement $3 \times 3 \times 3$), in which the internal nodes (at the centroid) is eliminated.

Solution: The full Coons formula, Eq. (10.1), is:

$$U(\xi, \eta, \zeta) = S - E + C \tag{i}$$

where the three projections are written as:

$$S = (1-\xi)U_I + \xi U_J + (1-\eta)U_K + \eta U_L + (1-\zeta)U_M + \zeta U_N \tag{ii}$$

also,

$$\begin{aligned} E &= (1-\xi)(1-\eta)U_{AD} + \xi(1-\eta)U_{BC} + \xi\eta U_{FG} + (1-\xi)\eta U_{EH} \\ &\quad + (1-\eta)(1-\zeta)U_{AB} + \eta(1-\zeta)U_{EF} + \eta\zeta U_{HG} + (1-\eta)\zeta U_{DC} \\ &\quad + (1-\xi)(1-\zeta)U_{AE} + \xi(1-\zeta)U_{BF} + \xi\zeta U_{CG} + (1-\xi)\zeta U_{DH} \end{aligned} \tag{iii}$$

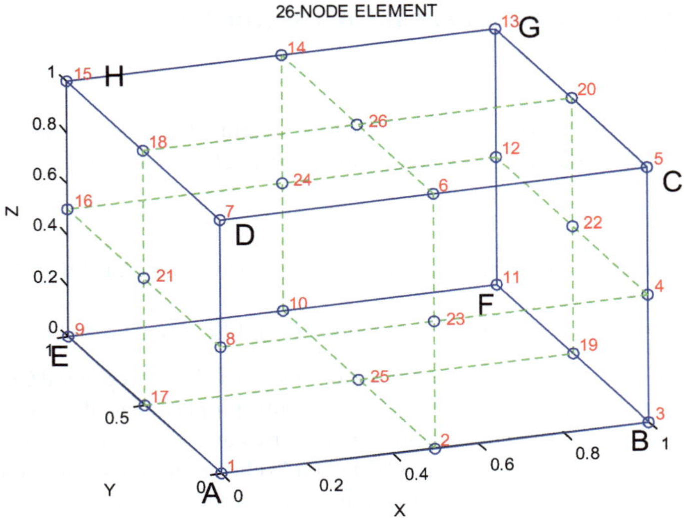

Fig. 10.5 Transfinite hexahedral 26-node Coons element (enhanced)

and

$$\begin{aligned} C = &(1-\xi)(1-\eta)(1-\zeta)\cdot U_A + \xi(1-\eta)(1-\zeta)\cdot U_B \\ &+ \xi(1-\eta)\zeta\cdot U_C + (1-\xi)(1-\eta)\zeta\cdot U_C \\ &+ (1-\xi)\eta(1-\zeta)\cdot U_E + \xi\eta(1-\zeta)\cdot U_F \\ &+ \xi\eta\zeta\cdot U_G + (1-\xi)\eta\zeta\cdot U_H \end{aligned} \tag{iv}$$

The six values $U_I, U_J, U_K, U_L, U_M, U_N$ (not shown here but only in the previous Fig. 10.2), involved in (ii), are the projections of the point $P(\xi, \eta, \zeta)$ onto the faces *AEHD*, *BFGC*, *ABCD*, *EFGH*, *ABFE*, *DCGH*, respectively. Obviously, the first two are normal to the ξ-axis at $\xi = 0$ and $\xi = 1$, the next two are normal to the η-axis (at $\eta = 0$ and $\eta = 1$), while the last two are normal to the ζ-axis (at $\zeta = 0$ and $\zeta = 1$). Therefore, since U_I belongs to the face *AEHD*, it can be interpolated in terms of a tensor product of quadratic Lagrange polynomials that are restricted to this face, as follows:

$$\begin{aligned} U_I = &L_1(\eta)L_1(\zeta)\cdot U_1 + L_2(\eta)L_1(\zeta)\cdot U_{17} + L_3(\eta)L_1(\zeta)\cdot U_9 \\ &+ L_1(\eta)L_2(\zeta)\cdot U_8 + L_2(\eta)L_2(\zeta)\cdot U_{21} + L_3(\eta)L_2(\zeta)\cdot U_{16} \\ &+ L_1(\eta)L_3(\zeta)\cdot U_7 + L_2(\eta)L_3(\zeta)\cdot U_{18} + L_3(\eta)L_3(\zeta)\cdot U_{15} \end{aligned} \tag{v}$$

It is reminded that the abovementioned quadratic Lagrange polynomials in the s-direction (where $s = \xi, \eta, \zeta$) are given by:

$$L_1(s) = (2s-1)(s-1),\ L_2(s) = 4s(1-s),\ L_3(s) = s(2s-1). \quad \text{(vi)}$$

Similarly, all rest five projections are interpolated in the same way, thus forming five additional equations like Eq. (v).
Moreover, the twelve univariate functions such as U_{AD} along the relevant edges can be interpolated in terms of Lagrange polynomials associated to nodal values. For example,

$$U_{AD} = L_1(\zeta)U_1 + L_2(\zeta)U_8 + L_3(\zeta)U_7, \quad \text{(vii)}$$

and still eleven similar equations may be derived.
Substituting Eq. (v) and the similar five ones into Eq. (ii), also Eq. (vii) and the similar eleven ones into Eq. (iii), the projections S, E, C are expressed in terms of the twenty-six nodal values, $U_i, i = 1, \ldots, 26$. Finally, all these three projections S, E and C are substituted into Eq. (i), and eventually, the shape functions are analytically obtained in closed form.
Typical functions for a corner node ($1 \equiv A$), a midpoint node along an edge (node 2 along AB), and a face node (node 21 along the face $AEHD$) are:

$$\begin{aligned}
N_1(\xi,\eta,\zeta) &= (1-\xi)L_1(\eta)L_1(\zeta) + (1-\eta)L_1(\xi)L_1(\zeta) + (1-\zeta)L_1(\xi)L_1(\eta) \\
&\quad - (1-\xi)(1-\eta)L_1(\zeta) - (1-\eta)(1-\zeta)L_1(\xi) \\
&\quad - (1-\xi)(1-\zeta)L_1(\eta) - (1-\xi)(1-\eta)(1-\zeta). \\
N_2(\xi,\eta,\zeta) &= L_2(\xi)\cdot[(1-\eta)L_1(\zeta) + (1-\zeta)L_1(\eta) - (1-\eta)(1-\zeta)] \\
N_{21}(\xi,\eta,\zeta) &= (1-\xi)L_2(\eta)L_2(\zeta).
\end{aligned} \quad \text{(viii)}$$

Substituting Eq. (vi) into Eq. (viii), the closed-form analytical expressions of the global shape functions are:

$$\begin{aligned}
N_1(\xi,\eta,\zeta) &= -(\xi-1)(\eta-1)(\zeta-1)(4\xi\eta - 2\eta - 2\zeta - 2\xi + 4\xi\zeta + 4\eta\zeta + 1), \\
N_2(\xi,\eta,\zeta) &= 4\xi(\xi-1)(\eta-1)(\zeta-1)(2\eta + 2\zeta - 1), \\
N_{21}(\xi,\eta,\zeta) &= -16\eta\zeta(\xi-1)(\eta-1)(\zeta-1),
\end{aligned}$$

etc.
After a careful inspection, Eq. (viii) elucidates the automatic procedure for the arbitrary noded enhanced Coons macroelement. Based on this representative pattern, a computer code was developed using (n_x, n_y, n_z) subdivisions per edge. The only restriction is that all "parallel" edges should have the same number of subdivisions (e.g., the edges AE, BF, CG, and DH have the same n_y). Numerical results will be reported for a vibrating sphere, later in this chapter.

Remark One may observe in Eq. (viii) that:

(1) The shape functions associated to the eight corner nodes (A, B, C, D, E, F, G, H) of the enhanced Coons macroelement are influenced by all three projections,

i.e., S, E and C, and therefore, they include *seven* terms. In contrast, Eq. (10.16) depicts that the edge-only Coons "C-element" includes four terms only.

2) The shape functions associated to intermediate nodes along any edge are influenced only by the projections S and E, and therefore, they consist of *three* terms. In contrast, Eq. (10.17) depicts that the edge-only Coons "C-element" includes one term only.
3) The shape functions associated to nodes in the interior of a face include only one product, which is linearly projected in the direction to which the current face is normal. This case does not appear in the edge-only "C-element."
4) Given the triplet of subdivisions, (n_x, n_y, n_z), the edge-only "C-element" includes $4(n_x + n_y + n_z - 1)$ nodes, while the enhanced Coons macroelement requires additional $2\big[(n_x - 1)(n_y - 1) + (n_y - 1)(n_z - 1) + (n_z - 1)(n_x - 1)\big]$ face nodes.
5) The tensor-product (Gordon) macroelement includes $\big[(n_x + 1)(n_y + 1)(n_z + 1)\big]$ nodes in total.

10.5 Other Tensor-Product CAD-Based Macroelements

10.5.1 Trivariate Bernstein–Bézier Elements

From Chap. 5, we have learnt that *nonrational* Bernstein–Bézier polynomials are related to the Lagrange polynomials through a linear transformation and vice versa. This fact was tested in detail for 2D problems [14] but obviously is valid for three-dimensional problems as well.

Extending these ideas, we have the right to claim that tensor-product Lagrange elements:

$$\hat{N}(\xi, \eta, \zeta) = L_{i,n}(\xi) \cdot L_{j,m}(\eta) \cdot L_{k,p}(\zeta), \tag{10.19}$$

are equivalent with tensor-product (nonrational) Bernstein–Bézier elements:

$$N(\xi, \eta, \zeta) = B_{i,n}(\xi) \cdot B_{j,m}(\eta) \cdot B_{k,p}(\zeta) \tag{10.20}$$

In the former elements, the global shape functions $\hat{N}(\xi, \eta, \zeta)$ are associated to *nodal values*, U_i, while in the latter elements the variable is globally approximated using generalized *coefficients*, $a_i, i = 0, 1, \ldots, n$ associated to the control points.

Although the numerical results in both formulations are identical, the only advantage of using nonrational Bernstein–Bézier elements is due to their flexibility in design thanks to the involved control points. In both functional sets (Lagrange, nonrational Bernstein–Bézier), the number of degrees of freedom (DOF) is the same.

One of the major advantages of using Bernstein–Bézier formula is its capability of dealing with tetrahedral domains as well. This procedure is quite similar to that applied to a triangular domain (see Chap. 6), and is performed using the de Casteljau algorithm to Bernstein–Bézier tetrahedral. One of the recent works on this topic is due to Ainsworth et al. [1], among others.

10.5.2 Trivariate B-Spline Solid Elements

Now, we can handle and achieve local control using tensor product of univariate B-splines:

$$\tilde{N}(\xi, \eta, \zeta) = N_{i,p_\xi}(\xi) \cdot N_{j,p_\eta}(\eta) \cdot N_{k,p_\zeta}(\zeta), \tag{10.21}$$

where p_ξ, p_η and p_ζ are the polynomial degrees in the three directions of the unit cube (reference or master or parent, element).

The functions $N_{i,p_\xi}(\xi)$ are those mentioned as B-splines in the book of Piegl and Tiller [10], which can be easily calculated by de Boor function "`spcol`" in MATLAB®.

In all cases, we use a set of control points which highly depends on both (i) the number of spans, (ii) the degree of piecewise polynomials, and (iii) the multiplicity of inner knots.

The most primitive approximation is to use cubic B-splines in conjunction either single ($\lambda = 1$, for C^p-continuity) or double ($\lambda = 2$, for $C^{(p-1)}$-continuity) multiplicity of inner knots. For a rectangular 3D domain, the latter is equivalent with the use of local Hermite polynomials between two successive breakpoints (in each direction). A higher polynomial degree generally leads to better results.

10.5.3 Trivariate NURBS-Based Solid Elements

Replacing each of the three factors that appear in Eq. (10.21) with rational functions, the NURBS-based isogeometric macroelement is derived. However, in this book we reduce to the rational Bézier only (see Sect. 10.7).

Below we shall present several numerical results using a single macroelement for the entire domain. Since the aforementioned macroelement is applicable to hexahedral domains, sometimes we may call it by the term "*superbrick*".

10.6 Numerical Examples

A variety of numerical examples in 3D elasticity and 3D potential problems on primitive solids (hexahedral, cylinder, sphere) may be found in four papers by Provatidis [11–13, 15]. In the latter works, engineering analysis includes steady-state uniaxial heat conduction in rectangular and cylindrical walls, a thick-walled spherical part, a compound sphere, a rectangular acoustic cavity, elastic analysis (tension, bending, and torsion of rectangular, cylindrical, and elliptic beams) , a thick cylinder under internal pressure, extraction of eigenfrequencies in (a rectangular beam, a spherical cavity), transient analysis (wave propagation problem) in rectangular structures, as well as Poisson's equation into a sphere (electrostatics), among others.

Nevertheless, a systemic comparison between several CAD-based approaches has not been reported so far. Here, for the first time, three variations of a single edge-only ("C") Coons macroelement (extreme/reduced family) is compared with a single tensor-product macroelement (complete one) based on Lagrange polynomials, Bernstein–Bézier, natural cubic B-splines and de Boor B-splines, as well as with an assemblage of conventional finite elements of the same mesh density.

For saving power, we fully analyze only two typical test cases, which can be considered as a *bed test* to judge the accuracy of several methodologies in 3D analysis. Again, the numerical examples of this subsection are concerned with a *single* macroelement.

Exercise 10.3 (Hexahedral rectangular acoustic cavity)
The sound propagation into acoustic cavities is described by the following partial differential equation:

$$\frac{1}{c^2}\frac{\partial^2 p}{\partial t^2} - \nabla^2 p = 0 \tag{10.22}$$

where p is the acoustic pressure, c is the wave speed, and t is the time.

As usual, the boundary conditions are either of Dirichlet type ($p = 0$) or of Neumann type ($\partial p/\partial n = 0$). The latter case occurs when the boundary (walls) of the cavity are entirely hard (infinite impedance). Walls with acoustic impedance different than infinity are out the scope of this book.

In the macroelement approach, as well as in the conventional FEM solution, for the eigenvalue problem the Galerkin procedure on Eq. (10.12) leads to the following matrix formulation:

$$\mathbf{M}\ddot{\mathbf{p}} + \mathbf{K}\mathbf{p} = \mathbf{0}, \tag{10.23}$$

where $\mathbf{M}$, $\mathbf{K}$ are the mass and stiffness matrices, respectively. Given the set of global shape (or basis) functions ϕ_j, $j = 1, \ldots, n$, the elements of these matrices are given by:

$$[\mathbf{M}]_{ij} = \frac{1}{c^2}\int\limits_V \phi_i\phi_j \mathrm{d}V \quad \text{and} \quad [\mathbf{K}]_{ij} = \int\limits_V \nabla\phi_i\nabla\phi_j \mathrm{d}V \tag{10.24}$$

As a first step to demonstrate the performance of 3D CAD-based macroelements, we determine the eigenvalues of a rectangular acoustic cavity of dimensions $L_x \times L_y \times L_z = 2.5 \times 1.1 \times 1.0$ m, with reference wave velocity $c = 1.0$ m/s. The walls of the cavity are considered to be hard ($\partial p/\partial n = 0$); i.e., we have boundary conditions of Neumann type. This means that, in Eq. (10.13), *no* rows or columns of the stiffness and mass matrices (**K**, M) have to be deleted.

For this problem, the analytical eigenvalues are given by:

$$\omega_{mnl}^2 = \pi c^2 \left[\left(\frac{m}{L_x} \right)^2 + \left(\frac{n}{L_y} \right)^2 + \left(\frac{l}{L_z} \right)^2 \right], \quad m, n, l = 0, 1, \ldots, \tag{10.25}$$

The twelve edges are uniformly divided into $10 \times 5 \times 5$ segments, thus leading to:

- 76 nodes along the twelve edges (see Fig. 10.6).
- 396 nodes all together (edges, surfaces, interior) (see Fig. 10.7).

Therefore:

- The single *edge-only "C" (Coons)* element consists of 76 nodes.
- The *tensor-product* (Gordon) macroelement consists of the abovementioned 396 nodes.
- In addition, a *Bernstein–Bézierian* macroelement (with the same 396 uniformly distributed nodes, but now in the role of *control* points),
- as well as a B-spline macroelement will be considered.
- Moreover, for comparison purposes, an assemblage of 250 eight-node trilinear finite elements (bricks, again based on the same 396 nodes) was considered.

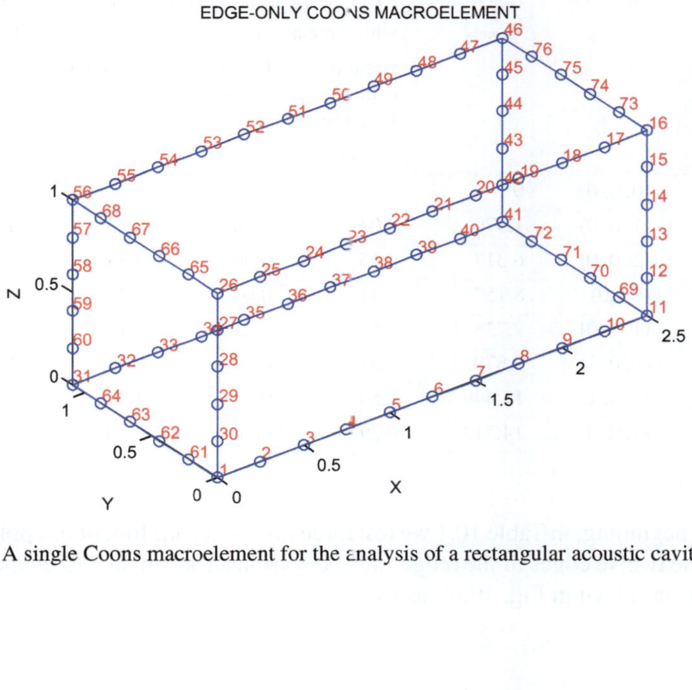

Fig. 10.6 A single Coons macroelement for the analysis of a rectangular acoustic cavity

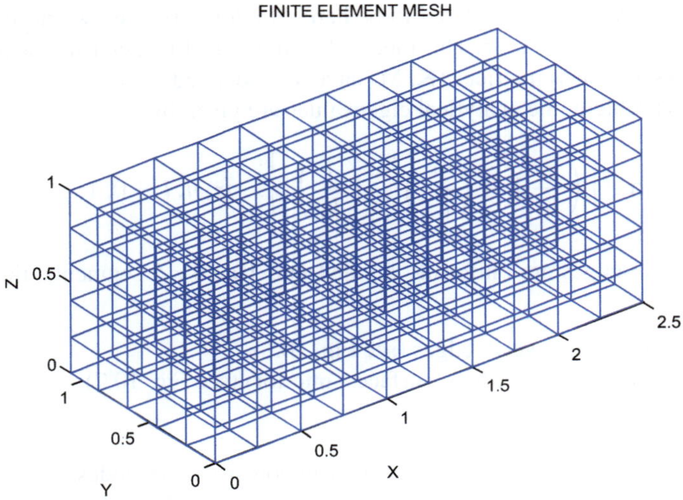

Fig. 10.7 Assemblage of 250 usual eight-node brick elements (396 nodes) for the analysis of a rectangular acoustic cavity. The same mesh is applied to a single tensor-product Lagrangian and a single tensor-product Bézierian macroelement as well

Table 10.1 Eigenvalues of a rectangular acoustic cavity of dimensions $L_x \times L_y \times L_z = 2.5 \times 1.1 \times 1.0$ m with wave velocity $c = 1.0$ m/s, using a single edge-only COONS macroelement (76 nodes) of several trial functions, and conventional finite elements

Ascending number	Mode (m, n, l)	Exact Eigenvalue (ω^2_{mnl})	Errors in %			
			COONS macroelement (Fig. 10.6) with three alternative trial functions			FEM (Fig. 10.7)
			Natural cubic B-splines	Lagrange polynomials	Piecewise-linear	396 nodes
			(76 nodes)			
1	(0, 0, 0)	0	–	–	–	–
2	(1, 0, 0)	1.579	0.04	0.00	0.83	0.83
3	(2, 0, 0)	6.317	0.15	0.00	3.33	3.33
4	(0,1,0)	8.157	0.31	0.00	3.33	3.33
5	(1, 1, 0)	9.736	0.56	0.28	3.17	2.92
6	(0, 0, 1)	9.870	0.31	0.00	3.33	3.33
7	(1, 0, 1)	11.449	0.55	0.27	3.23	2.99
8	(3, 0, 0)	14.212	0.29	0.00	7.60	5.23

In the beginning, in Table 10.1 we test three different *trial* functions applied along each of the twelve edges of the (edge-only) Coons macroelement ("C-element," with 76 nodes, as shown in Fig. 10.6), as follows:

Table 10.2 Eigenvalues of a rectangular acoustic cavity of dimensions $L_x \times L_y \times L_z = 2.5 \times 1.1 \times 1.0$ m with wave velocity $c = 1.0$ m/s, using several tensor-product CAD-based macroelements

Mode	Exact eigen-value	Errors in %					
		Edge-only (natural cubic B-spline)	Tensor-product				FEM
			Natural cubic B-spline	Lagrange	Bernstein	De Boor B-spline	
(0, 0, 0)	0	–	–	–	–	–	–
(1, 0, 0)	1.579	0.04	−0.00	0.00	0.00	0.00	0.83
(2, 0, 0)	6.317	0.15	−0.03	0.00	0.00	0.00	3.33
(0, 1, 0)	8.157	0.31	−0.03	0.00	0.00	0.00	3.33
(1, 1, 0)	9.736	0.56	-0.03	0.00	0.00	0.00	2.92
(0, 0, 1)	9.870	0.31	−0.03	0.00	0.00	0.00	3.33
(1,0,1)	11.449	0.55	−0.03	0.00	0.00	0.00	2.99
(3, 0, 0)	14.212	0.29	−0.16	0.00	0.00	0.01	5.23
Number of equations		76	396	396	396	396	396

(i) Natural cardinal cubic B-splines
(ii) Lagrange polynomials
(iii) Piecewise-linear interpolation.

One may observe in Table 10.1 that the edge-only 3D-Coons macroelement has an overall excellent behavior. In more details, the model of "Lagrange polynomials" outperforms. The model of "natural cubic B-splines" follows. Also, the model based on "piecewise-linear" trial functions comes third in performance. Nevertheless, even for the poorer model of the "piecewise-linear" interpolation (along each of the twelve edges), one may also observe that the set of its first three nonzero calculated eigenvalues is identical with that obtained using the conventional FEM solution. In other words, the Coons macroelement based on 76 DOF is generally overall superior to the FEM solution (396 DOF).

Although there is not much remaining to improve (with respect to the excellent solution based on Lagrange polynomials), for the completion of this study in Table 10.2 we extend our calculations by implementing tensor-product CAD-based macroelements, as follows:

- Tensor product of natural cardinal cubic B-splines, based on $n_x = 10, n_y = 5$, and $n_z = 5$ uniform subdivisions of the edges (396 DOF).
- Tensor product Lagrangian polynomials [transfinite-Gordon, i.e., $P_{\xi\eta\zeta}$ according to Eq. (10.3)], based on $n_x = 10$, $n_y = 5$, and $n_z = 5$ uniform subdivisions of the edges (396 DOF).
- Tensor product nonrational Bernstein–Bézier polynomials of degree $p_x = 10$, $p_y = 5$, and $p_z = 5$, respectively (396 DOF).

- Tensor-product cubic $(p_x = p_y = p_z = 3)$ de Boor B-splines, based on $n_x = 8, n_y = 3$, and $n_z = 3$ uniform knot spans of the edges (396 DOF).

One may observe in Table 10.2 that:

- The edge-only Coons element (76 DOF) based on natural cubic B-spline trial functions (third column in Table 10.2) substantially improves when tensor product of the same trial functions is used (fourth column of the same table).
- Lagrange and Bernstein–Bézier polynomials lead to *identical* results (up to 15th decimal point) and appear an error less than 0.005%. It is noted that only at the eighteenth mode, the error exceeds the value 0.005 and becomes equal to 0.73 percent.
- Cubic de Boor B-splines use two more control points than the number of breakpoints in each direction (two endpoints plus inner knots). Therefore, in order to have a fair comparison, Table 10.2 shows the case of $n_x = 8, n_y = 3$, and $n_z = 3$ uniform knot spans. This choice, in conjunction with cubic piecewise polynomials, gives a total number of 396 DOF, which is the *same* as that of the other tensor product macroelements.

Exercise 10.4 (Spherical acoustic cavity)
This problem has been previously successfully solved using tensor-product Lagrange polynomials [15], but is still of major interest because it offers the framework for a further detailed discussion. Here, for the first time, most of the possible CAD-based formulations will be applied and compared.

In the four first formulations of a single CAD-based macroelement (Coons, Gordon, Bézier, B-spline), the parameterization of the sphere is formed primarily by an inscribed cube ABCDEFGH (according to its definition in Fig. 10.1), of side $a = 2R/\sqrt{3}$. The center K of the sphere is initially connected with the eight corners of this cube. Moreover, each of the twelve edges of the aforementioned cube (say AB), plus the center K, forms a triangle of which the plane intersects the boundary of the sphere at a major arc $(\overset{\frown}{AB})$, part of a great circle. The so-produced twelve major arcs (at the intersections of the great planar circles passing through and the straight edges of the inscribed cube) when organized in quartets such as ($\overset{\frown}{\mathrm{AB}}$, $\overset{\frown}{\mathrm{BC}}$, $\overset{\frown}{\mathrm{CD}}$ and $\overset{\frown}{\mathrm{DA}}$) form the boundary of one spherical patch (cap), out of the six ones. Similarly, the rest five spherical caps are created. Selecting the vertex A as the origin of the curvilinear reference system, $\xi\eta\zeta$, in which the arcs $\overset{\frown}{AB}$, $\overset{\frown}{AE}$, and $\overset{\frown}{AD}$ are the ξ-, η- and ζ-axes, respectively, each point $P(\xi, \eta, \zeta)$ of the parametric space has an image $P(x, y, z)$ in the Cartesian space $Oxyz$, respectively. Each of the so-produced twelve edges is discretized into a relatively large number of uniform subdivisions, which is best to be the same for all three directions. The parametric axes $\xi\eta\zeta$ are shown in Fig. 10.8.

I. *Edge-only Coons macroelement*

Considering n_s spans per edge (i.e., $n_\xi = n_\eta = n_\zeta = n_s$), the edge-only Coons macroelement consists of $n_b = 4(3n_s - 1)$ nodal points along the twelve edges. The particular case of $n_s = 5$ is shown in Fig. 10.4. Repeating the same procedure for a larger number of subdivisions, say $n_s = 9$ (as shown in Fig. 10.9), the 3D Coons interpolation fails to accurately represent the boundary of the sphere, thus leading to an *underestimation* of the sphere's volume by about 21%. Moreover, Table 10.3 shows that in all three alternative models of various trial functions (natural cubic B-splines, Lagrange polynomials, piecewise-linear) a relatively large error appears, which is of the same order with that obtained by the assemblage of the brick elements that are produced in terms of the 3D Coons mesh generation scheme, using only the twelve edges.

The above findings suggest that (in the case of a sphere) in a hierarchical order:

(i) As a *first* step of improvement, each of the spherical caps (say *ABCD*) has to be represented with *more* boundary points, which should be accurately put on the boundary of the sphere, i.e., in the interior of the relevant patch ABCD.
(ii) As a *second* step of improvement, additional internal nodes should be inserted into the sphere.

II. *Enhanced boundary-only Coons macroelement*

Step-1: Regarding the *first* step, we start from the fact that each of the four edges of a spherical cap (say *ABCD*) has been previously subdivided into smaller arcs of equal length. The further procedure is generally applicable to all six cups, but it is easier to explain it through an example. Therefore, let us consider the cap *ABCD*, of which the edge *AB* has been previously (uniformly) divided into n_s segments by the points $P_1, \ldots, P_{n_s-1}$, where the opposite edge *CD* similarly by the points $Q_1, \ldots, Q_{n_s-1}$. Considering the existing $(n_s - 1)$ great circles that all pass through the center *K* and

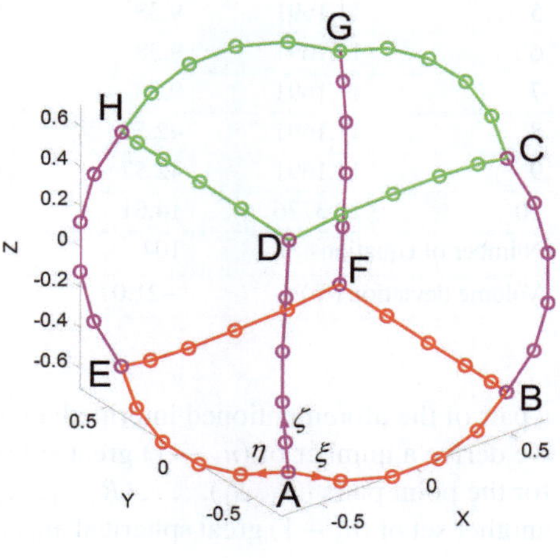

Fig. 10.8 Parameterization of the sphere

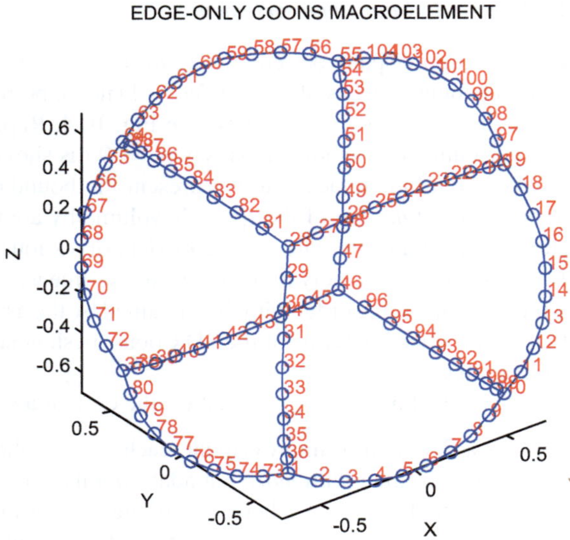

Fig. 10.9 Edge-only Coons macroelement using $n_s = 9$ subdivisions per edge

Table 10.3 Calculated eigenvalues of a spherical cavity (of unit radius) using a single 3D edge-only Coons macroelement in conjunction with $n_s = 9$ subdivisions per edge

Mode	Exact ω_i^2	ERROR (in %) of calculated eigenvalues ω_i^2			
		Edge-only COONS macroelement			FEM
		Natural cubic B-splines	Lagrange polynomials	Piecewise-linear	
1	0	–	–	–	–
2	4.3332	16.50	16.45	17.75	15.66
3	4.3332	16.50	16.45	17.75	15.66
4	4.3332	16.50	16.45	17.75	15.66
5	11.1691	9.38	9.37	10.59	3.99
6	11.1691	9.38	9.37	10.59	3.99
7	11.1691	9.38	9.37	10.59	3.99
8	11.1691	42.57	42.49	47.79	42.47
9	11.1691	42.57	42.49	47.79	42.47
10	20.3776	14.61	14.65	15.69	1.20
Number of equations		104	104	104	1000
Volume deviation (%)		−21.07	−21.03	−21.69	−21.69

a pair of the aforementioned intermediate points, i.e., $(P_1, Q_1), \ldots, (P_{n_s-1}, Q_{n_s-1})$, we derive a number of $(n_s - 1)$ great spherical arcs. Repeating the same procedure for the point pairs $(R_1, S_1), \ldots, (R_{n_s-1}, S_{n_s-1})$ along the opposite sides BC and DA, another set of $(n_s - 1)$ great spherical arcs is derived. Then, the intersections of these

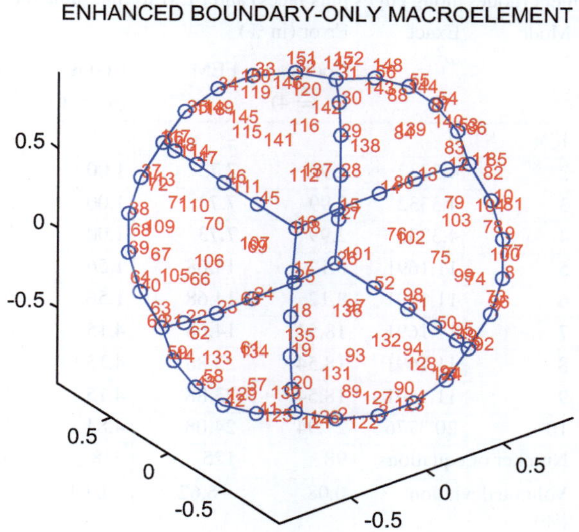

Fig. 10.10 Enhanced Coons macroelement of 152 boundary nodes, using $n_s = 5$ subdivisions per edge

two sets of perpendicular isolines eventually lead to the construction of $(n_s - 1)^2$ nodal points, by which the spherical cup *ABCD* is enriched. The same procedure is repeated for the rest five cups as well.

Based on the abovementioned Step-1, the enhanced boundary-only Coons macroelement comprises of totally $n_b = 4(3n_s - 1) + 6(n_s - 1)^2$ boundary nodes. For $n_s = 5$, the produced enhanced macroelement of 152 nodes is shown in Fig. 10.10. Using this type of single macroelement, varying n_s the results is shown in Table 10.4.

The results shown in Table 10.4 suggest that the enhanced Coons–Gordon macroelement, which is built by nodal points along the twelve edges plus the six spherical caps, is substantially improved compared with the previous edge-only Coons macroelement. Although the error for the first mode is half of that of the FEM solution (for the same mesh density but with boundary discretization only), the convergence rate is rather slow, a fact that dictates the necessity of internal nodes.

III. *Tensor-product Gordon macroelement*

Step-2: This is the final stage that permits the introduction of additional nodes inside the sphere. Based on the boundary nodes, the internal points are derived using 3D Coons interpolation formula [either Eq. (10.1) or Eq. (10.4b)]. Eventually, the tensor product Coons macroelement consists of totally $(n_s + 1)^3$ nodal points.

Clearly, all 216 nodes in the tensor-product macroelement shown in Fig. 10.11 are arranged along the isolines $\xi, \eta, \zeta = 0, \frac{1}{5}, \frac{2}{5}, \frac{3}{5}, \frac{5}{5}, 1$. Essentially, this is a classical element of Lagrange family, well known since the 1960s but not deeply studied so far.

Table 10.4 Calculated eigenvalues of a spherical cavity using a single enhanced Coons macroelement (nodes along edges plus faces) and FEM for the same mesh density

Mode	Exact ω_i^2	Error (in %)					
		Coons	FEM	Coons	FEM	Coons	FEM
		($n_s = 4$)		($n_s = 6$)		($n_s = 8$)	
1	0	–	–	–	–	–	–
2	4.3332	2.99	7.73	1.00	3.51	0.92	1.99
3	4.3332	2.99	7.73	1.00	3.51	0.92	1.99
4	4.3332	2.99	7.73	1.00	3.51	0.92	1.99
5	11.1691	8.12	14.68	1.56	6.79	1.09	3.96
6	11.1691	8.12	14.68	1.56	6.79	1.09	3.96
7	11.1691	18.54	14.68	4.15	6.79	3.21	3.96
8	11.1691	18.54	23.88	4.15	11.38	3.21	6.43
9	11.1691	18.54	23.88	4.15	11.38	3.21	6.43
10	20.3776	27.94	24.08	4.84	10.98	2.83	6.41
Number of equations		98	125	218	343	386	729
Volume deviation (%)		0.08	−6.62	−0.01	−3.01	0.00	−1.71

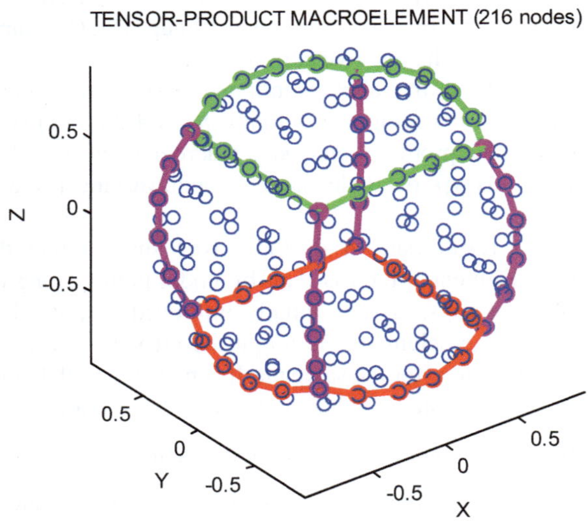

Fig. 10.11 Tensor-product macroelement (216 nodes) for the entire sphere

Preserving the same position and the same number of the abovementioned nodes, a finite element mesh with $5^3 = 125$ brick elements and $6^3 = 216$ nodes is automatically produced as shown in Fig. 10.12.

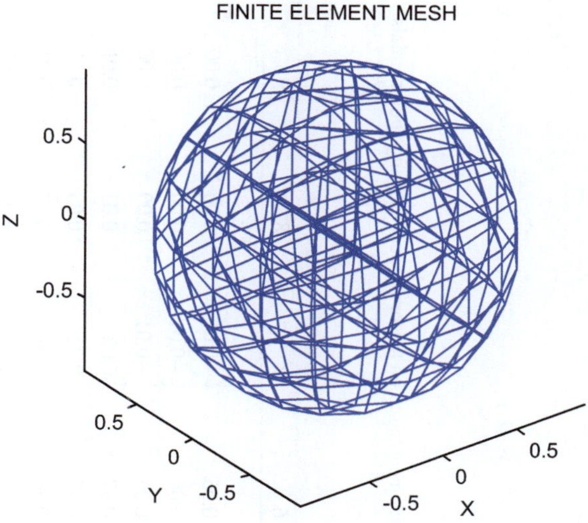

Fig. 10.12 Finite element mesh (216 nodes, 125 brick elements) produced through 3D Coons interpolation using $n_s = 5$ subdivisions per edge

In fact, using the same parameterization in conjunction with a tensor product of Lagrange polynomials, the results in Table 10.5 show a drastically improved numerical solution. Interestingly, although the idea of the isoparametric element is known since 1960s and can be found in every FEM textbook (e.g., [19, p. 171]), to the best of our knowledge so far it was never applied to a sphere as a single element.

Note Transfinite elements, with a different number of internal nodal points than those of the abovementioned $(n_s - 1)^3$ ones, may be easily constructed. Nevertheless, in this particular case study (of the sphere) there is no need to present this choice.

IV. *Other macroelements*

Let us now study some other CAD-based tensor products, for which the results are as follows:

(1) The global approximation of the solution using a single macroelement of tensor product nonrational Bernstein–Bézier polynomials leads to *identical* results with those obtained using the Lagrange polynomials.
(2) The tensor-product natural B-splines are of better quality than the FEM solution but of lower quality than the Lagrange polynomials. Representative results are shown in Table 10.6.
(3) A single macroelement for the entire spherical cavity using the (complete) de Boor B-splines leads to excellent results, as shown in Table 10.7. The results depend on the number of knot spans and the polynomial degree.

In summary, all older CAD-based methods (Gordon, Bernstein–Bézier, B-splines), which are the precursors of the NURBS-based IGA, demonstrate an exceptional accuracy. This happens despite the fact that the boundary of the sphere is

Table 10.5 Calculated eigenvalues of a spherical cavity using a single (Gordon–Coons) macroelement based on tensor-product Lagrange polynomials

Mode	Exact ω_i^2	Error (in %)							
		Number of subdivisions per side of each cap (n_s)							
		2	3	4	5	6	7	8	9
1	0	–	–	–	–	–	–	–	–
2	4.3332	9.79	1.90	1.14	0.17	0.06	−0.01	0.00	0.00
3	4.3332	9.79	1.90	1.14	0.17	0.06	−0.01	0.00	0.00
4	4.3332	9.79	1.90	1.14	0.17	0.06	−0.01	0.00	0.00
5	11.1691	24.85	7.24	5.45	0.70	0.37	0.02	0.03	0.01
6	11.1691	24.85	7.24	5.45	0.70	0.37	0.02	0.03	0.01
7	11.1691	99.67	7.24	8.04	0.70	0.53	0.02	0.03	0.01
8	11.1691	99.67	106.90	8.04	5.03	0.53	0.38	0.03	0.02
9	11.1691	99.67	106.90	8.04	5.03	0.53	0.38	0.03	0.02
10	20.3776	45.81	21.41	9.86	1.56	0.21	0.10	−0.82	-0.82
Number of equations		27	64	125	216	343	512	729	1000
Volume deviation (%)		−2.6737	1.3531	0.0779	−0.0806	−0.0057	0.0077	5.6782e−04	−8.8310e−04

Table 10.6 Calculated eigenvalues of a spherical cavity using a single macroelement based on tensor-product natural cardinal cubic B-splines

Mode	Exact ω_i^2	Error (in %)							
		Number of subdivisions per side of each cap (n_s)							
		3	4	5	6	7	8	9	9 FEM
1	0	–	–	–	–	–	–	–	–
2	4.3332	31.02	7.86ff	3.91	2.35	1.62	1.17	0.89	1.58
3	4.3332	31.02	7.86	3.91	2.35	1.62	1.17	0.89	1.58
4	4.3332	37.39	8.69	4.08	2.40	1.63	1.18	0.90	1.58
5	11.1691	44.54	10.47	3.25	2.01	1.39	1.02	0.79	3.17
6	11.1691	44.54	10.47	3.27	2.03	1.50	1.02	0.79	3.17
7	11.1691	50.20	10.48	3.27	2.03	1.50	1.02	0.79	3.17
8	11.1691	85.87	12.70	6.00	2.81	1.87	1.36	1.05	5.09
9	11.1691	85.87	12.70	6.00	2.81	1.87	1.36	1.05	5.09
10	20.3776	57.61	10.98	−0.15	−0.51	−0.37	−0.40	−0.44	4.88
Number of equations		64	125	216	343	512	729	1000	1000

Table 10.7 Calculated eigenvalues of a spherical cavity using a single macroelement based on tensor-product *de Boor* B-splines, for several combinations of n_s knot spans per edge (side) of the hexahedral and polynomial degree p

Mode	Exact ω_i^2	Error (in %)							
		Number of subdivisions per side of each cap (n_s)							
		$p = 3$			$p = 4$		$p = 5$	$p = 6$	$p = 7$
		$n_s = 2$	$n_s = 3$	$n_s = 4$	$n_s = 2$	$n_s = 3$	$n_s = 2$	$n_s = 2$	$n_s = 2$
1	0	–	–	–	–	–	–	–	–
2	4.3332	1.10	0.07	0.05	0.09	0.09	0.07	−0.01	−0.00
3	4.3332	1.10	0.07	0.05	0.09	0.09	0.07	−0.01	−0.00
4	4.3332	1.10	0.07	0.05	0.09	0.09	0.07	−0.01	−0.00
5	11.1691	1.82	0.11	0.08	0.30	0.05	0.15	0.01	0.01
6	11.1691	1.82	0.11	0.08	0.30	0.05	0.15	0.01	0.01
7	11.1691	7.63	0.11	0.51	0.30	0.66	0.53	0.01	0.03
8	11.1691	7.63	5.28	0.51	4.93	0.66	0.53	0.38	0.03
9	11.1691	7.63	5.28	0.51	4.93	0.66	0.53	0.38	0.03
10	20.3776	3.18	0.09	−0.88	0.59	−0.60	−0.31	0.05	−0.86
Number of equations		125	216	343	216	343	343	512	729

slightly violated by all of them (only NURBS accurately represents a sphere). To our opinion, this excellent behavior is due to the *global* character of the approximate solution, independently on the particular choice of the CAD formula. The only exception to this tendency is the poor performance of the "edge-only" Coons interpolation, which is incapable of approximating a sphere.

The skilled reader is encouraged to perform his/her comparisons with IGA on this crucial test case. This task is obviously out the scope of this book, and any not deliberate bad operation could offend the latter method.

10.7 Rational Bézier

This subsection deals with the particular case of accurately representing a sphere tile or sphere cap, which is one of the six ones faces that are determined by the great circles passing through the edges of the abovementioned inscribed cube. The case of accurately modeling a circular arc and then a spherical sector is produced by rotation about an axis is not considered at all.

A question here arises, whether it is possible to accurately represent a spherical cap (say the southern one, *ABFE*) by a rational biquadratic Bézier surface which requires nine control points and corresponding weights. Of course, the four control points should coincide with the given vertices *A*, *B*, *F*, and *E* (with weights equal to unity), whereas the next four control points should be at the intersections of the tangents to the great circles for each arc, i.e., *AB*, *BF*, *FE*, and *EA*. The latter can be easily found at the positions $x, y = \pm 3R/(2\sqrt{3})$ and $z = -3R/(2\sqrt{3})$, and the corresponding weight will be the cosine of the half angle of $\measuredangle(AKB)$, which equals to $\sqrt{\frac{2}{3}}$. After some manual work, the reader may ascertain that the condition that the south pole (0, 0, −1) belongs to the spherical patch leads to a linear dependency between the vertical position $z_{P_{11}}$ of the middle control point P_{11} (if it exists) and its corresponding weight w_{11}. If the value w_{11} is taken as the square of those at P_{10} and P_{21} (in the tensor-product sense) , it would be $(\sqrt{\frac{2}{3}})^2 = \frac{2}{3}$ and this would require that $z_{P_{11}} \cong -1.9621$. Unfortunately, the aforementioned combination $\left(z_{P_{11}} \cong -1.9621, w_{11} = \frac{2}{3}\right)$ does *not* accurately represent the spherical cap at all other points. Moreover, a parametric analysis, in which the weight w_{11} is varied to other values, showed that there is not a value that achieves the condition $x^2+y^2+z^2 = R^2$ to all points $P(\xi, \eta)$ of the spherical tile ABFE. This finding does not allow us to repeat the comparison between a tensor-product 27-node (Lagrangian-type) element and the rational Bézier analogous for the entire sphere. To make this point more clear, it is reminded that it was possible for a 9-node Lagrangian element and rational Bézier analogue of nine control points (see Chap. 8, Table 8.5, for polynomial degree $p = 2$).

Nevertheless, one of the earliest works in the accurate representation of a spherical cap (one-sixth of the spherical surface) is due to Cobb [3] and offers a set of 25 control points and weights in a configuration of 5 rows by 5 columns, as shown in Fig. 10.13. The latter coefficients may be also found in MATLAB®, and the reader may be assisted by the following computer code, which is just a validation that all pairs $(u, v) \equiv (\xi, \eta)$ of the parametric patch fulfill the condition $x^2 + y^2 + z^2 = R^2$.

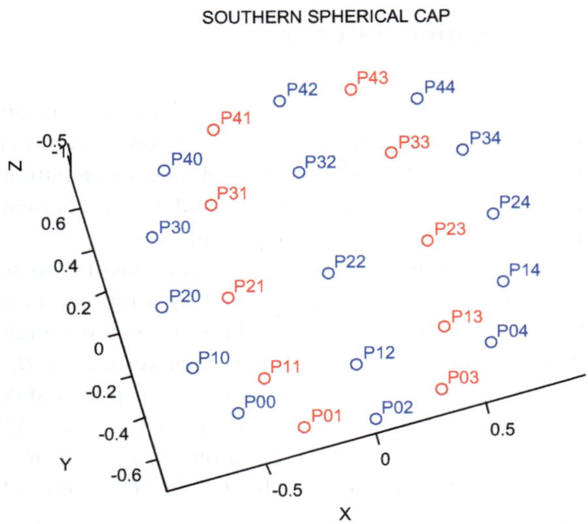

Fig. 10.13 Control points on the southern cap

```
%-------------------------------------------------------------------------------
% This code validates the control points and the weights for the bi-quartic
% rational Bezier approximation, as mentioned in the paper of Cobb(1988),
% Table 4, p.11.
%-------------------------------------------------------------------------------
clear all; clc;
%-------------------------------------------------------------------------------
%% Define the southern spherical cap:
%-------------------------------------------------------------------------------
southcap=rsmak('southcap');
%-Optional: Store (and if you like print) the coefficients:
all_coefficients=southcap.coefs;
%% Build the matrices of projective coordinates:
    for rows=1:5
        WX(rows,1:5)=southcap.coefs(1,:,rows);
        WY(rows,1:5)=southcap.coefs(2,:,rows);
        WZ(rows,1:5)=southcap.coefs(3,:,rows);
        W(rows,1:5)=southcap.coefs(4,:,rows);
    end
%% Calculate the Cartesian coordinates of control points:
    X=WX./W; Y=WY./W; Z=WZ./W;
%% Divide the parametric patch in "seg" divisions per direction.
%... For each pair (u,v), calculate the term x^2+y^2+z^2
nseg=4;
du=1/nseg; dv=1/nseg;
nc=0;
for j=1:nseg+1
    v=(j-1)*dv;
    for i=1:nseg+1
        u=(i-1)*du;
        nc=nc+1;
        Basis=[(1-v)^4;4*(1-v)^3*v;6*(1-v)^2*v^2;4*(1-v)*v^3;v^4]*...
                [(1-u)^4,4*(1-u)^3*u,6*(1-u)^2*u^2,4*(1-u)*u^3,u^4];
        nominatorX=dot(reshape(WX,1,25),reshape(Basis,1,25));
        nominatorY=dot(reshape(WY,1,25),reshape(Basis,1,25));
        nominatorZ=dot(reshape(WZ,1,25),reshape(Basis,1,25));
        denominator=dot(reshape(W,1,25),reshape(Basis,1,25));
        x(nc)=nominatorX/denominator;
        y(nc)=nominatorY/denominator;
        z(nc)=nominatorZ/denominator;
        Rsq(nc)=x(nc)^2+y(nc)^2+z(nc)^2;
        fprintf('i=%3i  j=%3i   Rsquared=%12.5e\n',i,j,Rsq(nc));
    end
end
```

For the purposes if this subsection, a computer program for single bi-quartic (i.e., of fourth degree) rational Bézier macroelement was developed. The axis origin was taken at the corner *A* of the hexahedral ABCDEFGH, while the $\xi-$, $\eta-$ and $\zeta-$ axes were taken along the curves (arcs of great circles) *AB*, *AE,* and *AD*, respectively. Based on this parameterization (previously studied by Provatidis [15] in Poisson problems as well), the rational shape functions are here active in the entire sphere and the approximate solution is a *rational tensor product*. In other words, this three-dimensional situation is quite analogous with the circular cavity we studied at the end of Chap. 8 (9-node rational Bézierian element).

In more details, the southern cap was rotated around the *x*-axis by ninety degrees three times successively, and two more times around the *y*-axis until the entire sphere it coved by six similar caps. In this way, 125 control points in total are produced, as shown in Fig. 10.14.

The results are shown in Table 10.8, where one may observe the questionable (only due to the underestimation of the lowest eigenvalues) outperformance of the

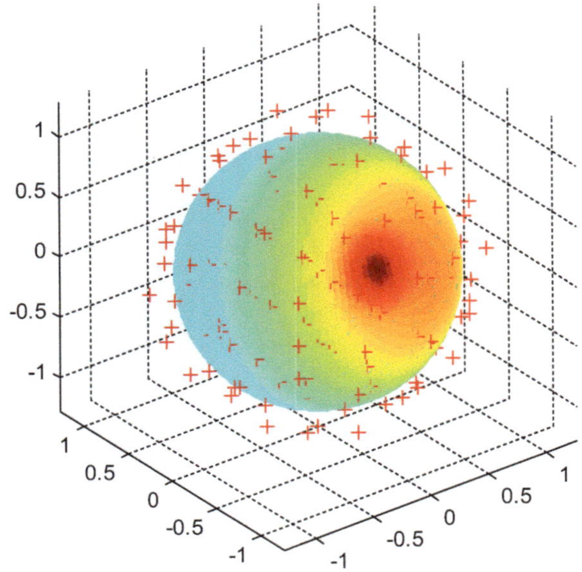

Fig. 10.14 Control points for a 125-node rational Bézier macroelement

rational element in comparison with the 125-node tensor-product Lagrangian. It is reminded that the latter is (numerically and mathematically) equivalent with the *non-rational* Bézier element. Nevertheless, as Table 10.5 clearly shows the tensor-product Lagrangian (equivalently, nonrational Bézierian) element drastically improves (the error for the first calculated eigenvalues reduces from 1.14 to *only* 0.17%) when one more node per side is introduced, and it is a matter of further study to see how better the rational elements are in terms of equal numbers of DOF in both models. Not to forget that the numerical solution generally depends on the parameterization of the sphere. Table 10.8 was obtained using the same internal points, which are produced by Coons interpolation based on uniform division of the spherical caps.

The main lesson learned from this example is that all these CAD-based macroelements (based on a *global* approximation) are superior to the FEM solution (based on a *local* approximation).

10.8 Discussion

Similarly to the 2D problems where a global CAD-based approximation was applicable to the entire surface patch ABCD (Chaps. 3 and 4), in 3D problems all the corresponding CAD formulas are applicable to the volume block ABCDEFGH (superbrick) as well. While in 2D problems a C-element required the use of the boundary

Table 10.8 Eigenvalues for a coarse 125-node bi-quartic rational Bézier macroelement applied to a sphere of unit radius

Mode	Exact ω_i^2	Error (in %)		
		Bi-quartic rational Bezier	Bi-quartic Lagrange (nonrational Bezier)	FEM (64 bricks)
1	0	–	–	–
2	4.3332	−0.47	1.14	7.73
3	4.3332	−0.47	1.14	7.73
4	4.3332	0.26	1.14	7.73
5	11.1691	−0.14	5.45	14.68
6	11.1691	0.53	5.45	14.68
7	11.1691	1.19	8.04	14.68
8	11.1691	1.19	8.04	23.88
9	11.1691	1.25	8.04	23.88
10	20.3776	1.38	9.86	24.08
Number of equations		125		

(Chap. 3), in 3D problems its analogue can be reduced to the set of the twelve edges in the superbrick.

The abovementioned twelve edges can work satisfactorily and have led to successful numerical results in a variety of problems found in literature [11–13, 16]. Nevertheless, if the boundary is intensively curved, the edge-only Coons interpolation cannot accurately "catch" it; thus, we need to use additional data over the Coons patches, i.e., to apply a type of Gordon interpolation on the boundary.

But, despite the fact that the boundary has been adequately approximated, the exact solution is generally still poor and needs a set of internal points in order to be accurately interpolated. In this case, we need a certain higher-order interpolation (blending) between the opposite faces of the superbrick. The latter is easily accomplished using a tensor product in three directions, of which all three curvilinear axes $(\xi\eta\zeta)$ belong to the boundary.

The reader may wonder why we focused all numerical results in acoustics. This was done because we believe that acoustics is a fine "test bed" to compare alternative and competitive methods. Of course the CAD-based macroelements which we presented in this chapter are equally accurate in elastodynamics (see [11]). But since, according to the "Helmholtz decomposition", the displacement vector can be written in terms of *two* potential functions in the form $\vec{\mathbf{u}} = grad\Phi + curl\vec{\Psi}$, where both of them fulfill a different wave propagation equation (i.e., four acoustic equations in total, see [6], as well as [7]), we believe that it is better to start any dynamic research with acoustics, where closed-formed analytical solutions exist, and then move to elastodynamics.

In addition, the reader should be cautious about the size of the elastic macroelement for the following reason. Any CAD-based (FEM-like) macroelement approach is based on the assumption that the displacement component (say u toward the x-axis) is interpolated in terms of the u values (nodal or generalized coefficients). In contrast, horizontal displacements at the boundary do not impose only horizontal displacements elsewhere but also toward the two remaining directions as well. Obviously, the latter shortcoming is not met in acoustics where the acoustic pressure P is a scalar quantity; the same holds for all other potential problems as well.

10.9 A Theoretical Note on the Generalization of Coons Interpolation

One may observe that the three-dimensional version of Coons interpolation (Eq. 10.1) includes three terms, whereas the two-dimensional Coons interpolation includes only two. In other words, the *dimensionality* of the problem determines the number of terms involved in Coons interpolation. Below we generalize Coons theory.

Having introduced the one-, two-, and the three-dimensional operators in Eq. (10.3), one may observe that:

I. The 3D Coons formula may be written as (cf. Eq. 10.1):

$$\begin{aligned} U(\xi,\eta,\zeta) = &-(-1)^1\cdot(P_\xi + P_\eta + P_\zeta)\{U\} \\ &-(-1)^2\cdot(P_{\xi\eta} + P_{\xi\zeta} + P_{\eta\zeta})\{U\} \\ &-(-1)^3\cdot P_{\xi\eta\zeta}\{U\} \end{aligned} \tag{10.26}$$

II. The 2D Coons formula may be written as (cf. Eq. 3.1):

$$\begin{aligned} U(\xi,\eta) = &-(-1)^1\cdot(P_\xi + P_\eta)\{U\} \\ &-(-1)^2\cdot P_{\xi\eta}\{U\} \end{aligned} \tag{10.27}$$

Equation (10.26) has been formulated in such a way that it can be easily generalized for a n-dimensional space Ω. If, for example, Ω is a four-dimensional space with coordinates ξ, η, ζ, t, then Eq. (10.26) is replaced by:

$$\begin{aligned} U(\xi,\eta,\zeta,t) = &-(-1)^1\cdot(P_\xi + P_\eta + P_\zeta + P_t)\{U\} \\ &-(-1)^2\cdot(P_{\xi\eta} + P_{\xi\zeta} + P_{\xi t} + P_{\eta\zeta} + P_{\eta t} + P_{\zeta t})\{U\} \\ &-(-1)^3\cdot(P_{\xi\eta\zeta} + P_{\xi\eta t} + P_{\eta\zeta t} + P_{\xi\zeta t})\{U\} \\ &-(-1)^4\cdot P_{\xi\eta\zeta t}\{U\} \end{aligned} \tag{10.28}$$

The above idea immerged in the Diploma Work by our student D. Grekas (1989). It was further enhanced and was presented in May 31, 1994, during a one-day Research

Seminar entitled "*Computer-Aided Geometric Design: From Theory to Practice,*" co-organized by P. D. Kaklis and N. S. Sapidis, at NTUA, Zografou Campus, to honor the first visit of Professor Gerald Farin, of Arizona State University, to NTUA's Computer-Aided Geometric Design Group (Department of Naval Engineering). The complete text may be found in the proceedings volume [8].

10.10 Recapitulation

It was shown that all four basic CAD interpolation formulas (Coons, Gordon, Bézier, B-splines) lead to corresponding 3D single macroelements.

In more details, the use of 3D Coons interpolation in a hexahedral volume block considering nodal points along the twelve edges only, gives the well-known elements of the Serendipity family (when univariate Lagrange polynomials are used). However, Coons element is more flexible than the aforementioned Serendipity element, because (in addition to Lagrange polynomials) one may use any other univariate approximation (such as piecewise linear, piecewise quadratic, cardinal natural cubic B-splines, Bernstein–Bézier, B-splines, NURBS) along the twelve edges of the superbrick. Therefore, a "mixed" formulation Coons elements in which the twelve edges are modeled using Bernstein–Bézier polynomials is possible, and the produced results will be identical with those obtained using Lagrange polynomials. This issue has been also showed in the 2D analogue (Sect. 3.8).

Of major importance and originality was the consideration of a sphere by all six faces in the Coons superbrick (enhanced boundary model). This consideration is an intermediate configuration between the extreme Serendipity and the tensor-product Lagrangian (or similar) element.

Moreover, the use of additional internal nodes located at the same position as the boundary ones gives the classical family of Lagrangian-type elements. Gordon's transfinite elements sweep the entire spectrum between the Serendipity (poor) and Lagrangian (rich) elements. In principle, natural cardinal cubic B-splines are easily applicable, but as also happened in 2D problems these are of questionable value, and this is left as an exercise for the reader to form his/her opinion.

On the other bank of the river, one can deal with control points and build tensor products based on Bernstein–Bézier polynomials and nonrational B-splines (piecewise polynomials). Although these elements offer a higher flexibility in handling curvilinear boundaries, sometimes they may suffer from the accurate imposition of boundary conditions, a topic on which several progress has been recently remarked [9, 17]. Anyway, the central message we want to pass is that old-fashioned CAD-based macroelements work satisfactorily despite the small violation of the actual volume as, for example, happens in the sphere. In general, the high accuracy of the CAD-based macroelements is the *global* character of the approximation in their interior. This explains the excellent accuracy in the calculation of the eigenvalues, due to the adequately accurate approximation of the smooth eigenmodes through any of the CAGD interpolation formulas.

References

1. Ainsworth M, Davydov O, Schumaker L (2016) Bernstein-Bézier finite elements on tetrahedral–hexahedral–pyramidal partitions. Comput Methods Appl Mech Eng 304:140–170
2. Beer G, Watson JO (2002) Introduction to finite and boundary element methods for engineers (Chap. 11). Wiley, Chichester, pp 357–377
3. Cobb JE (1988) Tilling the sphere with rational Bezier patches. UUCS-88-009, University of Utah
4. Coons SA (1964) Surfaces for computer aided design of space form. Project MAC, MIT (1964), revised for MAC-TR-41 (1967). Springfield, VA 22161, USA. Available online: http://publications.csail.mit.edu/lcs/pubs/pdf/MIT-LCS-TR-041.pdf
5. Cook WA (1974) Body oriented (natural) co-ordinates for generating three-dimensional meshes. Int J Numer Meth Eng 8:27–43
6. Eringen AC, Suhubi ES (1974) Elastodynamics, vols I & II. Academic Press, New York and London
7. Graff KF (1975) Wave motion in elastic solids, Oxford University Press. Also: Dover (1991)
8. Kanarachos A, Grekas D, Provatidis C (1995) Generalized formulation of Coons' interpolation. In: Kaklis P, Sapidis N (eds) Computer aided geometric design from theory to practice, 1995, NTUA, Chap. 7, pp 65–76 (a one-day research seminar to honour the first visit of Prof. G. Farin from the Arizona University to the National Technical University of Ahens, Greece). ISBN 960-254-068-0
9. Liu B, Xing Y, Wang Z, Lu X, Sun H (2017) Non-uniform rational Lagrange functions and its applications to isogeometric analysis of in-plane and flexural vibration of thin plates. Comput Methods Appl Mech Eng 321:173–208
10. Piegl L, Tiller W (1997) The NURBS book. Springer, Berlin
11. Provatidis CG (2005) Analysis of box-like structures using 3-D Coons' interpolation. Commun Numer Methods Eng 21:443–456
12. Provatidis CG (2005) Three-dimensional Coons macroelements in Laplace and acoustic problems. Comput Struct 83:1572–1583
13. Provatidis CG (2006) Three-dimensional Coons macroelements: application to eigenvalue and scalar wave propagation problems. Int J Numer Meth Eng 65:111–134
14. Provatidis CG (2014) Bézier versus Lagrange polynomials-based finite element analysis of 2-D potential problems. Adv Eng Softw 73:22–34
15. Provatidis CG (2018) Engineering analysis with CAD-based macroelements. Arch Appl Mech 88:121–140
16. Provatidis CG, Vossou CG, Theodorou EG (2006) On the CAD/CAE integration using Coons interpolation. In: Tsahalis D (ed) Proceedings 2nd international conference "from scientific computing to computational engineering", Athens, Greece, 5–8 July 2006
17. Schillinger D, Ruthala PK, Nguyen LH (2016) Lagrange extraction and projection for NURBS basis functions: a direct link between isogeometric and standard nodal finite element formulations. Int J Numer Meth Eng 108(6):515–534
18. Szabó B, Babuška I (1991) Finite element analysis. Wiley, New York
19. Zienkiewicz OC (1977) The finite element method. McGraw-Hill, London

Chapter 11
Global Collocation Using Macroelements

Abstract The CAD-based global approximation of the approximate solution within a patch (or a volume block) leads to large matrices, and therefore, high computer effort is required. The involved matrices may be either fully populated (as happens when using Lagrange and Bernstein–Bézier polynomials) or partially populated (thanks to the compact support of B-splines and NURBS). This fact is the motivation for preserving the global basis functions but replacing the Galerkin–Ritz with a collocation method which is here called the "global collocation method." In the latter method, each element of the large matrices can be calculated without performing domain integration, since only a substitution of the basis functions into the partial differential operator is needed. Nevertheless, the numerical solution is highly influenced by the location of the so-called collocation points, and this is an open topic for research. Through a number of test problems, it will be shown that the global collocation method performs equally well in 1D, 2D, and 3D, static and dynamic, problems. Numerical results are presented for a broad spectrum of test problems, such as eigenvalue analysis of rods and beams, transient thermal analysis, plane elasticity, rectangular and circular acoustic cavities and plates.

Keywords Orthogonal collocation · MATLAB · Piecewise Hermite Lagrange polynomials · Bernstein-Bézier · B-splines Neumann boundary condition · Lumped mass · Transient heat analysis Wave propagation · Acoustics · Elastodynamics · Plate bending

11.1 Introduction

The use of collocation methods in the numerical solution of partial differential equations has been an issue of intensive research in the past [7, 17, 20, 27, 32, 41, 67, 68] and continues till today. For many years, the research was concerned with ordinary differential equations (ODE), dealing with functions $U(x)$ in one variable x, in which the collocation method applies very well, while in 1960s it had shown a development particularly in the field of chemical engineering [1, 26, 27]. Another branch

C. G. Provatidis, *Precursors of Isogeometric Analysis*, Solid Mechanics and Its Applications 256, https://doi.org/10.1007/978-3-030-03889-2_11

of applicability is the environmental engineering, where similar equations such as diffusion, mass transport, energy equilibrium, and generally particular expressions in fluid mechanics dominate (e.g., [7], and papers therein).

The collocation method is capable of solving any kind of problems, elliptic, hyperbolic and parabolic, linear and nonlinear, and its mathematical foundation as a method of approximation solution of a general boundary value problem is found in the famous book of Collatz [15]. Other useful reference books are de Boor [18] and Ascher et al. [3]. Also, a detailed collection of research activities until the year 1989 are found in the paper of Fairweather and Meade [23].

Generally speaking, the conventional collocation method focuses on the local interpolation of the variable U of the problem, whereas a few attempts are concerned with its global interpolation in two-dimensional problems [28, 31, 71].

As mentioned above, in its initial form the collocation method is met in combination with trial functions of low degree, such as cubic Hermite polynomials (i.e., of third degree) [9], but we should point out that already in 1966 the collocation was used in conjunction with splines (spline collocation), of which a special case is B-spline collocation [16]. Transission from low degree to global approximation has been carried out in the context of Trefftz [42] and radial base functions [13, 14, 21, 43–45, 72]. Other interesting works are [34, 37, 40].

We remind that in 1970s the global approximation had been successfully implemented in conjunction with the standard Galerkin–Ritz method, either using the Gordon–Coons interpolation [24, 29, 30, 10] or the so-called p-methods (summarized in the book of Szabó and Babuška [69]). Interestingly, upon finishing the writing of this book, it was found that Cavendish [11] (who also cooperated with Bill Gordon at General Motors Laboratory) conducted his PhD thesis on collocation methods, while later he published a joint paper on the L_∞ convergence for both the collocation and Galerkin approaches [12]. It seems that those people at Mathematics Department of General Motors Research Laboratories (Bill Gordon, James Cavendish) and Department of Mathematics at the University of Pittsburgh (Charles Hall) had understood a lot of things on the value of using the same basis for both the geometry and the physical field and therefore could be considered as the forefathers of the CAD-based approach, which starts by Coons–Gordon, continues with tensor-product Bernstein–Bézier and B-splines, and currently ends with the IGA approach.

As was demonstrated in previous chapters, Coons-based "C-elements" (or "CPM": Coons Patch Macroelements) were applied initially to 2D potential Kanarachos and Deriziotis [37] and later to elasticity problems Kanarachos et al. [38], while a systematic study on Gordon-Coons-based approximation in 2D and 3D problems was continued by Provatidis [47–50]. In all these works, the conventional Galerkin–Ritz procedure was implemented, with the result to obtain stiffness and mass matrices that require time-consuming *domain integration*. Although the CPM method works perfectly (under conditions) and can handle patches with a complexity up to a "Π"-shape (see [47, 50]); however, it has the shortcoming of *fully populated matrices*, particularly when Lagrange polynomials are used. Already in 2004 (by instinct as well as due to previous reviewers' comments on research papers submitted since 1999), it had been well understood by the author of this book that a *global collocation method* (GCM) was necessary to be implemented to reduce the computational cost of the global approximation in CAD-based macroelements (see [50, p. 6704] and literature therein).

The need to reduce the abovementioned high computational time which is inherent in the Galerkin-Ritz based "C-element", combined with the need to preserve the CAD-based global interpolation within the macroelement (whatever it is, i.e., Coons, Gordon, etc), created a competitive global collocation method. The first pilot study showed a very good behavior in one-dimensional problems of dynamic analysis of elastic bars and showed the need to use collocation points that do not coincide with internal nodes of the mesh, because then some unrealistic complex eigenvalues may be obtained [51]. Later, the research expanded to other interpolation functions (Lagrange, Bézier, monomials) and to several sites for the collocation points (roots of Legendre, i.e., Gauss points, roots of Chebyshev polynomials, etc.) [33, 52, 58, 53]. A portion of research concerned with wave propagation problems [54].

Concerning two-dimensional problems, the first application of the global collocation in conjunction with tensor-product Lagrange polynomials (or generally, the global Gordon–Coons interpolation) gave encouraging results for potential problems (Poisson's equation) in rectangles [55] and curvilinear domains [56]. At the same time, the method was applied to plane elasticity problems [57, 65] and plate bending eigenanalysis [66]. Extensive investigation was performed in eigenfrequency extraction problems of acoustic cavities using B-spline collocation [61].

In order to achieve a greater reduction in the CPU time required to estimate the matrices, further attempts have been made by shifting the mesh nodes at the position of the Gauss points and then a nodal collocation was applied, thus resulting in a mass matrix equal to the *unitary* matrix for which no computation is required [53, 60, 63].

Later, the collocation approach was adopted by the IGA as well, thus creating the terminology "*isogeometric collocation.*" Despite the initial reservations (see [4, pp. 2104–2105]), the method was improved and continues to be developed till today ([2, 19, 22, 46], among others).

Most papers that have appeared are concerned with Dirichlet-type boundary conditions, while the shortcoming of the method lies on the safe treatment of Neumann type ones (for a penalty-like method, see, e.g., [5]). Closely related is the proper procedure for the coupling between subregions, where the selection of some collocation along the interfaces in combination with the Neumann boundary conditions (penalty-like method) has been proposed. Research in this area continues to grow at high speed, and this chapter is only a brief attempt to capture the most important research efforts to date.

For the benefit of the reader, we start exhaustively dealing with the 1D problem and then continue with 2D and 3D applications where CAD-based macroelements are applicable.

11.2 General Formulation

We consider the ordinary differential equation (ODE):

$$L\{U(x)\} = b(x), \tag{11.1}$$

where L is a linear operator which describes a physical phenomenon (e.g., stress equilibrium, heat transfer, wave propagation, mechanics of fluids, etc.).

Similar to what happens in most computational methods, here also the collocation method is based on the approximation of the unknown function U of the problem in a series expansion of the form:

$$U(x) = \sum_{j=1}^{n} f_j(x) \cdot a_j. \tag{11.2}$$

In Eq. (11.2), the $f_j(x)$ are known basis functions and a_j are the coefficients that have to be determined so as to properly fulfill the conditions of the problem (ODE and BCs).

Sometimes, the above coefficients a_j may be of arbitrary nature, particularly when f_j is a Bernstein–Bézier, B-splines or a NURBS function. Some other times they may represent either the nodal values U_j of the variable U (in conjunction with Lagrange polynomials) or a mixture of nodal values and their derivatives (U_j, U_j') (in conjunction with Hermite polynomials).

If the problem domain is the interval [0, L], the determination of the coefficients a_j depends on:

(i) The fulfillment of the BCs, e.g.,

$$U(0) = u_0 \quad \text{and} \quad U(L) = u_L, \tag{11.3}$$

as well as on

(ii) The fulfillment of the ODE at a sufficiently large number of points:

$$L\{U(x_i)\} = b(x_i), \quad i = 1, \ldots, n \tag{11.4a}$$

Substituting Eq. (11.2) into Eq. (11.4a) and considering that the operator L is linear, one obtains:

$$\sum_{i=1}^{n} L\{f(x_i)\} \cdot a_i = b(x_i), \quad i = 1, \ldots, n, \tag{11.4b}$$

The set of Eqs. (11.3) and (11.4b) comprises a system of linear equations, of which the solution is capable to give us the numerical value of the unknown coefficients a_j which are involved in Eq. (11.2).

11.3 Polynomial Interpolation

Sometimes, if this is possible, the functions $f_j(x)$ in Eq. (11.2) are taken so as to satisfy per se the BCs of first kind (Dirichlet type). For example, if the domain of definition is $x \in [0, 1]$ and the BCs are $U(0) = 0$ and $U(1) = 0$, then the function $U(x)$ could be set in the form of a product of the two monomials x and $(x - 1)$. Therefore, if the exact solution corresponds to a Taylor's expansion with a degree larger than 2, then the trial solution could be set in the form:

$$U(x) = x(x - 1)g(x), \tag{11.5}$$

where $g(x)$ is a proper (e.g., polynomial) function.

Example 11.1 (A Simple Elliptic Problem Using Polynomials) The ordinary differential equation:

$$U''(x) - x^2 = 0, \quad \text{defined in the interval } [0, 1] \tag{11.6}$$

with boundary conditions $U(0) = U(1) = 0$, has the theoretical solution $U(x) = \frac{1}{12}x(x-1)(x^2+x+1)$. Therefore, in this case we have $g(x) = \frac{1}{12}(x^2+x+1)$. Given that the BCs are satisfied per se due to the term $x(x - 1)$ in Eq. (11.2), the determination of the three coefficients of the polynomial $g(x)$ requires the fulfillment of the ODE at three positions. If we select these positions at the three Gauss points of which the normalized coordinates are $\xi = 0, \pm\sqrt{\frac{3}{5}}$, that is at the Cartesian coordinates $x = (1 + \xi)/2$, then due to

$$L(x) \equiv U''(x) - x^2 = 2g + 2(2x-1)g' + x(x-1)g'' - x^2,$$

with

$$g(x) = a_0 + a_1 x + a_2 x^2, \quad g'(x) = a_1 + 2a_2 x, \quad g''(x) = 2a_2,$$

Equation (11.6) becomes:

$$\begin{aligned} L(x) \equiv U''(x) - x^2 &= 2(a_0 + a_1 x + a_2 x^2) + 2(2x-1)(a_1 + 2a_2 x) + x(x-1)(2a_2) \\ &= 2a_0 + (-2+6x)a_1 + (-6x+12x^2)a_2 - x^2, \end{aligned} \quad (11.7a)$$

Since $L(U)=0$, Eq. (11.7a) is finally written as

$$2a_0 + (-2+6x)a_1 + (-6x+12x^2)a_2 = x^2. \quad (11.7b)$$

Applying Eq. (11.7b) for the three collocation points $x = x_1$, x_2, and x_3, the following equations system is obtained:

$$\begin{bmatrix} 2 & (-2+6x_1) & (-6x_1+12x_1^2) \\ 2 & (-2+6x_2) & (-6x_2+12x_2^2) \\ 2 & (-2+6x_3) & (-6x_3+12x_3^2) \end{bmatrix} \begin{Bmatrix} a_0 \\ a_1 \\ a_2 \end{Bmatrix} = \begin{Bmatrix} x_1^2 \\ x_2^2 \\ x_3^2 \end{Bmatrix}. \quad (11.7c)$$

Setting the values of the Gauss points, i.e., $x_1 = \frac{1}{2}(1-\sqrt{3/5})$, $x_2 = \frac{1}{2}$ and $x_3 = \frac{1}{2}(1+\sqrt{3/5})$, after manipulation the solution of the system of Eqs. (11.7c) is found to be: $a_0 = a_1 = a_2 = \frac{1}{12}$, that is exactly the anticipated exact solution of the problem.

Remark 1 The above approach was found to be exactly the same as the theoretical one, and this is because the exact solution of the problem is a perfect polynomial. If, however, it was not a polynomial, we usually need more terms in its series expansion (11.2), but again there would be a calculation error that generally would decrease as the number of terms in the series increases.

Remark 2 Due to the purely polynomial expression of the analytical solution, the same result would have ended even if we had received the collocation points at equidistant positions. We urge the reader to verify this claim for the points: $x_1 = 1/3$, $x_2 = 0$, and $x_3 = 2/3$, replacing them into Eq. (11.7c).

Remark 3 Instead of the above laborious process of analytically calculated derivatives and the subsequent isolation of the polynomial coefficients $a_0, \ldots, a_n$, it is preferable to work with equivalent *Lagrange polynomials*, $f_j(x) = L_{j,n}(x)$ according to Eq. (11.2). But then it is necessary to have a subroutine that provides the output of the Lagrange polynomials and their derivatives at the collocation points.

Remark 4 Later we shall see that the abovementioned polynomials may be replaced by *Bernstein–Bézier* polynomials $f_j(x) = B_{j,n}(x)$, as well as by *B-splines* $f_j(x) = N_{j,p}(x)$ or even by NURBS $f_j(x) = R_{j,p}(x)$, where $f_j(x)$ is the basis function involved in Eq. (11.2).

Example 11.2 (Collocation Method Using Cubic Hermite Polynomials) For the same differential Eq. (11.6), we will now consider that the interval [0, 1] is subdivided into four equal subintervals. At each of these intervals, of length l and with local edges "1" and "2" (as if they were like finite elements), we accept that the function $u(x)$ is locally interpolated by means of four cubic Hermite polynomials, which are exactly the same with those we encountered in beam bending, that is:

$$U(x) = N_1(x)U_1 + N_2(x)U_1' + N_3(x)U_2 + N_4(x)U_2' \tag{11.8a}$$

with

$$\begin{aligned} N_1(x) &= \left[1 - 3(x/l)^2 + 2(x/l)^3\right] \\ N_2(x) &= \left[(x/l) - 2(x/l)^2 + (x/l)^3\right]l \\ N_3(x) &= \left[3(x/l)^2 - 2(x/l)^3\right] \\ N_4(x) &= \left[-(x/l)^2 + (x/l)^3\right]l \end{aligned} \tag{11.9a}$$

Then, the second derivative of the function $u(x)$ is written in the form:

$$U''(x) = N_1''(x)U_1 + N_2''(x)U_1' + N_3''(x)U_2 + N_4''(x)U_2' \tag{11.8b}$$

The first derivatives of the shape functions are:

$$\begin{aligned} N_1'(x) &= -\frac{6}{l^2}x + \frac{6}{l^3}x^2, \quad N_2'(x) = 1 - \frac{4}{l}x + \frac{3}{l^2}x^2 \\ N_3'(x) &= \frac{6}{l^2}x - \frac{6}{l^3}x^2, \quad N_4'(x) = -\frac{2}{l}x + \frac{3}{l^2}x^2 \end{aligned} \tag{11.9b}$$

The second derivatives of the shape functions are:

$$N_1''(x) = -\frac{6}{l^2} + \frac{12}{l^3}x, \quad N_2''(x) = -\frac{4}{l} + \frac{6}{l^2}x$$
$$N_3''(x) = \frac{6}{l^2} - \frac{12}{l^3}x, \quad N_4''(x) = -\frac{2}{l} + \frac{6}{l^2}x \tag{11.9c}$$

Remark 1 It should be clarified that the variable x, which is involved in the second derivatives of Eq. (11.9c), is measured relative to the left node of the space. That is, if we refer to the ith interval, with absolute coordinates $[x_i, x_{i+1}]$, then x of Eqs. (11.9a–11.9c) refers to the difference $(x - x_i)$.

Substituting Eqs. (11.8a) and (11.8b) in the initial ODE (11.6), in combination with (11.9a) and (11.9c), we receive:

$$U''(x) - x^2$$
$$= \boxed{\left(-\tfrac{6}{l^2} + \tfrac{12}{l^3}x\right)u_1 + \left(-\tfrac{4}{l} + \tfrac{6}{l^2}x\right)u_1' + \left(\tfrac{6}{l^2} - \tfrac{12}{l^3}x\right)u_2 + \left(-\tfrac{2}{l} + \tfrac{6}{l^2}x\right)u_2' - x^2 = 0} \tag{11.10}$$

The subdivision of the domain [0, 1] into four equal segments implies the creation of five nodes at the locations $x = 0, 0.25, 0.50, 0.75, 1$, at which nodes ten variables correspond (two per node) that are (u_1, u_1'), (u_2, u_2'), (u_3, u_3'), (u_4, u_4'), (u_5, u_5').

If we impose the two BCs, $u_1 = u(0)$ and $u_5 = u(1) = 0$, then out of the ten only eight unknowns remain, the following ones: (u_1'), (u_2, u_2'), (u_3, u_3'), (u_4, u_4'), (u_5'). This means that we have to satisfy the ODE at eight collocation points. In the current case, if we choose 2 such points per subinterval, then for the 4 intervals the number of collocation points will be $2 \times 4 = 8$, which is exactly the number we need.

Following the usual way [17], the abovementioned two collocation points are taken at the positions that correspond to the corresponding Gauss points, that is those with normalized coordinates $\xi = \pm\sqrt{3}/3$. From the fulfillment of the ODE, we obtain a system of 8 equations with 8 unknowns, from the solution of which we take the nodal values of Table 11.1, and all of them coincide with the analytical solution, $u_{\text{exact}}(x) = \frac{1}{12}x(x-1)(x^2+x+1)$.

The MATLAB code that solves the abovementioned problem is given in the Program P11.1 that follows.

Program P11.1: Collocation code for piecewise Hermite polynomials

```
%%%%%%%%%%%%%%%%%%%%%%%%%%%%%%%%%%%%%%%%%%%%%%%%%%%%%%%%%%%%%%%%%%%%%%%%%%%%%%
% Program for the solution of the ODE U''-f(x) = 0,
%         using HERMITE COLLOCATION                                          %
% The function f(x) is determined by the variable ichoice, as follows:      %
% ichoice = 1: f(x)=1,   u(x)=x^2/2,                    u(0)=0, u(1)=1/2    %
% ichoice = 2: f(x)=x,   u(x)=x^3/6,                    u(0)=0, u(1)=1/6    %
% ichoice = 3: f(x)=x^2, u(x)=1/12*x*(x-1)*(x^2+x+1), u(0)=0, u(1)=0      %
%%%%%%%%%%%%%%%%%%%%%%%%%%%%%%%%%%%%%%%%%%%%%%%%%%%%%%%%%%%%%%%%%%%%%%%%%%%%%%
    clear all; clc;
%----------------------------------------
U0=0;
ichoice=3;
if(ichoice == 1)
    UL=1/2; %CHOICE-1
elseif(ichoice==2)
    UL=1/6; %CHOICE-2
else
    UL=0;    %CHOICE-3
end
%----------------------------------------
L=1;
nele=4;
nodes=nele+1;
dx=L/nele;
gi = [-sqrt(3)/3, sqrt(3)/3];    %normalized position of Gauss point
nc=0;
%Loop over all segments
for iel=1:nele
    xm=(iel-1)*dx+dx/2;          %middle of interval No. "iel"
    i1=iel; i2=i1+1;             %nodes of the interval "iel"
    xstart=(iel-1)*dx;           %starting node in the interval
    for i=1:2
        xcol=xm+(dx/2)*gi(i);    %position of collocation point
        s=xcol-xstart;           %relative distance from the left node
        nc=nc+1;                 %Ascending number of row in matrix [A]
        der2N1=-6/dx^2+12/dx^3*s;
        der2N2=-4/dx+6/dx^2*s;
        der2N3=6/dx^2-12/dx^3*s;
        der2N4=-2/dx+6/dx^2*s;
        A(nc,2*i1-1) = der2N1;
        A(nc,2*i1  ) = der2N2;
        A(nc,2*i2-1) = der2N3;
        A(nc,2*i2  ) = der2N4;
%---Computation of the right hand side, according to the case:
        if(ichoice==1)
            rhs(nc)=1;       %CHOICE-1
        elseif(ichoice==2)
            rhs(nc)=xcol;    %CHOICE-2
        else
            rhs(nc)=xcol^2; %CHOICE-3
        end
    end
end
%Imposition of Boundary Conditions:
    rhs = rhs' - A(:,1)*U0 - A(:,2*nodes-1)*UL;
%Separation between known and unknown terms:
    fixed_dofs=[1,2*nodes-1];
    free_dofs=[2:2*nele,2*nodes];
    solv = A(:,free_dofs)\rhs
%Store the nodal values of u(x) and u'(x) in the vector {U}:
    U=[];
    U(fixed_dofs)=0; U(1)=U0;U(2*nodes-1)=UL;
    U(free_dofs)=solv;
%Computation of the exact solution and comparison with the calculated values
for i=1:nodes
    x=(i-1)*dx;
    %------------------------------------
    if(ichoice==1)
    %CHOICE-1:
        Uexact=x^2/2;
        UderExact=x;
    elseif(ichoice==2)
    %CHOICE-2:
        Uexact=x^3/6;
        UderExact=3*x^2/6;
    else
        Uexact=1/12*x*(x-1)*(x^2+x+1);
        UderExact=x^3/3-1/12;
    end
    %------------------------------------
    fprintf('%5i  %12.5e  %12.5e  %12.5e  %12.5e\n',i,U(2*i-1),U(2*i),...
                                                         Uexact,UderExact);
end
%%%%%%%%%%%%%%%%%% END-OF-PROGRAM %%%%%%%%%%%%%%%%%%%%%%%%%%%%%%%%%%%%%%%%%%%%%
```

Table 11.1 Solution of ODE given by Eq. (11.6), using 4 equal subintervals and collocation

Nodal point	u	u'
1	0 (given data)	−8.33333e−002
2	−2.05078e−002	−7.81250e−002
3	−3.64583e−002	−4.16667e−002
4	−3.61328e−002	5.72917e−002
5	0 (given data)	2.50000e−001

Note 1

It should be mentioned that according to the selection of the variable `ichoice` in Program P11.1, not only the particular ODE $u''(x) - x^2 = 0$ described by Eq. (11.6), but totally the following three cases are included:

$$\text{ichoice} = 1:\quad u'' - 1 = 0 \quad \therefore\ u_{\text{exact}}(x) = x^2/2$$
$$\text{ichoice} = 2:\quad u'' - x = 0 \quad \therefore\ u_{\text{exact}}(x) = x^3/6$$
$$\text{ichoice} = 3:\quad u'' - x^2 = 0 \quad \therefore\ u_{\text{exact}}(x) = \tfrac{1}{12}x(x-1)(x^2+x+1)$$

Note 2
For the two abovementioned cases (`ichoice` =1, 2), the Hermite interpolation coincides with the accurate solution because its cubic polynomials are of the same degree as the analytical solution. In addition, even in the case (`ichoice` =3) where $f(x) = x^2$, and the exact solution is of fourth degree (higher than the Hermites), again the nodal values are accurately calculated (deviation appears between the nodes). We encourage the reader to construct a graph.

Note 3
Using Hermite polynomials (in 1D problems), we treat on an equal basis, both the variable u of the problem and its first derivative u'.

11.4 Collocation Types

Usually, the position of the collocation points influences the numerical solution, particularly when the accurate solution is not a perfect polynomial. According to their positions, we can distinguish the following alternative schemes:

- **Nodal collocation**. The problem domain $[x_1, x_n]$ is divided into a number of $(n-1)$ intervals, uniformly or not, and then the ODE is fulfilled at the $(n-2)$ internal points $(x_2, \ldots, x_{n-1})$ of the nodal sequence, $x_1, x_2, \ldots, x_{n-1}, x_n$. This method works adequately well but the relevant numerical solution is not the best one.

- **Orthogonal collocation**. Instead of a uniform arrangement of collocation points, it is preferable to choose particular positions that correspond to the roots of characteristic polynomials such as Legendre polynomials (Gauss points) or Chebyshev polynomials [6, 8]. Moreover, Greville and Demko points have been proposed [18].
- **Least-squares collocation**. Sometimes it may be useful to fulfill the ODE at a greater number of points than the number of unknown coefficients which are involved in Eq. (11.2). Then, since we have more equations than the number of unknowns, the standard treatment is to apply the least-squares method. Therefore, the initial system $\mathbf{AX} = \mathbf{b}$, with $\mathbf{A}$ (a matrix of order $m \times n$, with $m > n$) and $\mathbf{b}$ (a vector of order $m \times 1$), is transformed into $(\mathbf{A}^T\mathbf{A})\mathbf{X} = \mathbf{A}^T\mathbf{b}$, with $\mathbf{A}^T\mathbf{A}$ (of dimensions $n \times n$) and $\mathbf{A}^T\mathbf{b}$ (of dimensions $n \times 1$).

In case of a Taylor series, the optimum position of collocation points is rather that of Gauss points for the entire problem domain $[0, L]$. Nevertheless, it is not bad to choose the roots of Chebyshev polynomials, of first and second kinds in the same interval (i.e., $[0, L]$). For a detailed study, the reader is referred to Provatidis [52–54].

Example 11.3 (Steady-State Conduction in a Thick Cylindrical Wall) Consider a thick cylinder in which the heat is steady and flows only in the radial direction. The temperature along the inner radius $R_1 = 1$ is $T_1 = 1000$ °C, whereas along the outer radius $R_2 = 32$ is $T_2 = 0$ °C. Considering cylindrical coordinates, (r, θ, z), the governing Laplace equation is: $\nabla^2 T = \frac{\partial^2 T}{\partial r^2} + \frac{1}{r}\frac{\partial T}{\partial r} + \frac{1}{r^2}\frac{\partial^2 T}{\partial \theta^2} + \frac{\partial^2 T}{\partial z^2} = 0$. For the particular description of the problem where the variation in the directions θ and z is ignored, the governing ODE takes the following simplified form:

$$\frac{\partial^2 T}{\partial r^2} + \frac{1}{r}\frac{\partial T}{\partial r} = 0 \tag{i}$$

The analytical temperature distribution is given by:

$$T(r) = T_1 + \frac{(T_2 - T_1)}{\ln(R_2/R_1)} \ln(r/R_1). \tag{ii}$$

This example has been previously solved elsewhere using Lagrange polynomials among others [47, 56]. Therefore, for the purpose of variety, here results will be presented in terms of Bernstein–Bézier polynomials. The interval $[R_1, R_2] = [1, 32]$ is divided into one knot span (a single macroelement) and solution is presented for several polynomial degrees, p. The collocation points were chosen at the position of the Gaussian points, and for this choice the average error of the calculated temperature is shown in Fig. 11.1 (in usual and double logarithmic graphs), where monotonic convergence may be observed.

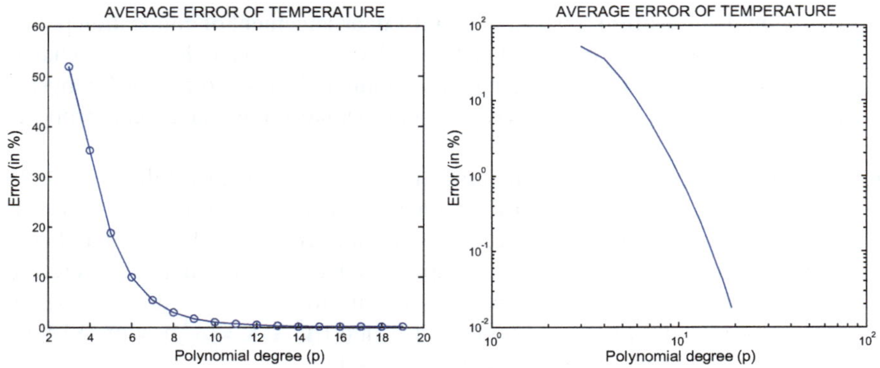

Fig. 11.1 Convergence of the calculated temperature (normalized L^2 error norm)

11.5 Parabolic Problems

In addition to the elliptic problems as those in the previous three examples, the collocation method is capable of solving time-dependent problems as well, in which the variable is $u = u(x,\ t)$. One category of these problems appears in transient heat transfer phenomena, which are governed by the following partial differential equation:

$$\rho c \frac{\partial u}{\partial t} - k\nabla^2 u = 0 \tag{11.11}$$

In one dimension, the above PDE is written as:

$$\frac{\partial u}{\partial t} - \alpha \cdot \frac{\partial^2 u}{\partial x^2} = 0, \quad \text{where } \alpha = \frac{k}{\rho c} \tag{11.12}$$

Fulfilling Eq. (11.12) at a number of n_{col} collocation points, a system of n_{col} equations with n unknowns (cf. Eq. 11.2) is derived. In the aforementioned system, we have to consider the two boundary conditions. Therefore, a proper choice regarding the number of collocation points is $n_{col} = n-2$. Then, before the boundary conditions are imposed we have two nonsquare matrices, **C** and **K**, each of dimensions $(n-2) \times n$, that are involved in the following system:

$$\mathbf{C} \cdot \dot{\mathbf{u}} + \mathbf{K} \cdot \mathbf{u} = \mathbf{0} \tag{11.13}$$

After the BCs are imposed, we derive two square matrices, **C** and **K**, each of dimensions $(n-2) \times (n-2)$. The right-hand side is produced by transferring terms from the left-hand side, according to the type of BCs. Therefore, in case of Dirichlet-type BCs for both ends ($u(0) = u_1,\ \ u(L) = u_n$), the RHS consists first from the terms

$b_i = -k_{i1}u_1 - k_{in}u_n$, and second by those terms that are produced by expressing the term $\dot{\mathbf{u}}$ in a finite-difference scheme with respect to the time.

For example, if we define a constant time step Δt and choose the trapezoidal rule as a difference scheme at an intermediate point of any arbitrary time interval (t_n, t_{n+1}), that is $\mathbf{u}_{n+\beta} = (1-\beta)\mathbf{u}_n + \beta\mathbf{u}_{n+1}$ and $\dot{\mathbf{u}}_{n+\beta} = (\mathbf{u}_{n+1} - \mathbf{u}_n)/\Delta t$, then we conclude again in Eq. (11.37) which constitutes a recursive formula for determining the temperatures:

$$\left(\frac{1}{\Delta t}\mathbf{C} + \beta\mathbf{K}\right) \cdot \mathbf{u}_{n+1} = \left(\frac{1}{\Delta t}\mathbf{C} - (1-\beta)\mathbf{K}\right) \cdot \mathbf{u}_n \qquad (11.14)$$

In the general case, the above matrices **C** and **K** are nonsquare, whereas the vectors $\mathbf{u}_n$ and $\mathbf{u}_{n+1}$ are concerned with all degrees of freedom including the boundary ones. In the sequence, we will see that after the imposition of the BCs, the left-hand matrix is replaced by a square one and then the equations system is solved for the unknowns.

Without particular reduction of the generality, let us consider the one-dimensional problem, and particularly that of only Dirichlet type (given temperatures at both ends: u_0 and u_L which form the boundary value vector $\mathbf{u}_b = [u_0, u_L]^T$). Then the vector $\mathbf{u}_{n+1} = [\mathbf{u}_i, \mathbf{u}_b]^T$ consists of the temperature values of the internal nodes $\mathbf{u}_i$, and the aforementioned boundary values $\mathbf{u}_b$. An analogous partition is applied to the vector $\mathbf{u}_n$.

In accordance with the partition of the temperature vector $\mathbf{u}_{n+1} = [\mathbf{u}_i, \mathbf{u}_b]^T$, a consistent partition is also applied to the matrices, where Eq. (11.11) becomes:

$$\left[\mathbf{A}_{ii} \middle| \mathbf{A}_{ib}\right] \left[\frac{\mathbf{u}_i}{\mathbf{u}_b}\right]_{n+1} = \left[\mathbf{B}_{ii} \middle| \mathbf{B}_{ib}\right] \left[\frac{\mathbf{u}_i}{\mathbf{u}_b}\right]_n, \qquad (11.15a)$$

where the matrices are of order $\mathbf{A}_{ii}[n_c \times (n_c + 2)]$, $\mathbf{A}_{ib}[n_c \times 2]$, $\mathbf{B}_{ii}[n_c \times (n_c + 2)]$, and $\mathbf{B}_{ib}[n_c \times 2]$. Consequently, the number of columns that appear in the matrices is increased by two against the number of their rows.

Carrying on the operations with the given Dirichlet BCs in the left-hand side and then transferring the result in the right-hand side (changing the sign), Eq. (11.15a) becomes:

$$\mathbf{A}_{ii}(\mathbf{u}_i)_{n+1} = \left[\mathbf{B}_{ii} \middle| \mathbf{B}_{ib}\right] \cdot \left[\frac{\mathbf{u}_i}{\mathbf{u}_b}\right]_n - \mathbf{A}_{ib} \cdot (\mathbf{u}_b)_{n+1} \qquad (11.15b)$$

Now the matrix $\mathbf{A}_{ii}$ is *square*, of dimensions $(n_c \times n_c)$, and therefore Eq. (11.15b) can be easily solved in the unknown vector $(\mathbf{u}_i)_{n+1}$. It is noted that the right-hand side consists of data that refer to the previous time instance (subscript n) but also to the current BCs (subscript $n+1$).

At this point, and before we go on with the solution algorithm, we should clarify the meaning:

(i) of the initial given temperature distribution $\mathbf{u}_0$ (including the boundary) and

(ii) of the boundary conditions $\mathbf{u}_b$.

It is important to mention that between the initial and the permanent boundary conditions continuity may be or not. For example, a rod of length L (defined in the interval [0, L]) may be characterized, e.g., by a uniform constant temperature $u_0 =$ 100 °C at every point of the rod, and suddenly both ends of this rod are immersed into a bath of constant temperature $u_b = 0$ °C. In this case, we have a discontinuity of the temperature at the two ends (boundary) of the rod, from 100 °C (for $t < 0$) to 0 °C (for $t \geq 0$).

There is no sense to consider Eq. (11.11) at the time instance $t = 0$, because it is satisfied as an identity. For example, if the vector $\mathbf{u}_n$ has a constant value for all degrees of freedom, then the product $\mathbf{K}\mathbf{u}_n$ will lead to a null result (the matrix $\mathbf{K}$ is singular, the sum of rows equals to zero). A similar zeroing occurs for the product $\mathbf{K}\mathbf{u}_{n+1}$ as well. Therefore after the simplification of the matrix $\mathbf{C}$ in both parts, the result of Eq. (11.14) will be the obvious relationship $\mathbf{u}_{n+1} = \mathbf{u}_n$. Therefore, a differentiation appears in $\mathbf{u}_{n+1}$ only when concrete boundary values $\mathbf{u}_b$ are imposed, either of constant value or even of time-dependent form [that is $\mathbf{u}_b = \mathbf{u}_b(t)$].

Applying Eq. (11.15b) at the end of the first time step, $t = \Delta t$, the first vector in the right-hand side, that is the $\mathbf{u}_n = [\mathbf{u}_i, \mathbf{u}_b]_n^{\mathrm{T}}$, will have the *initial* value $\mathbf{u}_0$, including the two boundary values that correspond to the two ends: u_0 and u_{L} (for $t = 0$). Also, the second vector that is involved in Eq. (11.15b), generally the $(\mathbf{u}_b)_{n+1}$, is concerned with the boundary conditions that dominate at the time instance $t = \Delta t$. Consequently, the result of the first step, including the boundary conditions as well, is the total vector $\mathbf{u}_1$.

Similarly, applying Eq. (11.15b) at the end of the second time step, $t = 2\Delta t$, the first vector of the right part, $\mathbf{u}_n = [\mathbf{u}_i, \mathbf{u}_b]_n^{\mathrm{T}}$, will have the total value $\mathbf{u}_1$ which was calculated above (including the two boundary values: u_0 and u_{L}, for $t = \Delta t$). Also, the second vector involved, $(\mathbf{u}_b)_{n+1}$, is concerned with the (probably updated) boundary values that dominate at the time instance $t = 2\Delta t$, and so on.

Example 11.4 (Transient Heat Analysis with Sinusoidal Initial Condition) The problem is defined in a metallic rod of unit length. Initially, the temperature distribution along the entire rod is of sinusoidal form:

$$u(x; t = 0) = \sin \pi x, \quad 0 \leq x \leq 1. \tag{i}$$

Suddenly, the rod is sinked into a liquid of permanent zero temperature, and therefore the boundary conditions become:

$$u(0, t) = u(1, t) = 0 \tag{ii}$$

For this example, there is not any discontinuity between initial and boundary conditions at the ends $x = 0, 1$, because in both states (initial and final) the potential remain equal to zero. The analytical solution for this problem is easily shown to be

$$u(x;t) = \mathrm{e}^{-\pi^2 t} \sin \pi x, \tag{iii}$$

and obviously behaves well for all values of x and t.

Using a set of eleven Lagrange polynomials associated to ten uniform subdivisions, i.e., using polynomials of tenth degree with a constant time step $\Delta t = 1 \times 10^{-3}$ s, the results are excellent as shown in Fig. 11.2. Identical results were obtained in the following cases:

(i) When nonrational Bernstein–Bézier basis functions of the same degree are used.
(ii) When the Lagrange polynomials are associated to the Gauss points, thus leading to a lumped mass matrix equal to the identity matrix.

11.6 Hyperbolic Problems

11.6.1 General

The hyperbolic problems are concerned with wave propagation, of which a special case is the scalar wave (in terms of field potential U) given by:

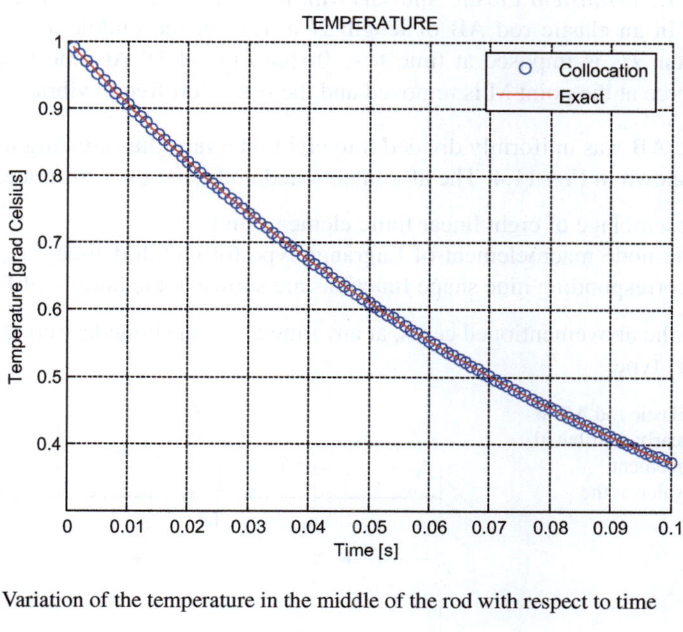

Fig. 11.2 Variation of the temperature in the middle of the rod with respect to time

$$\left(\frac{1}{c^2}\right)\frac{\partial^2 U}{\partial t^2} - \nabla^2 U = 0 \tag{11.16a}$$

In one dimension, the above equation becomes:

$$\left(\frac{1}{c^2}\right)\frac{\partial^2 U}{\partial t^2} - \frac{\partial^2 U}{\partial x^2} = 0. \tag{11.16b}$$

Similarly, the elastodynamic equation of motion for a bar under axial loading is:

$$\frac{\partial}{\partial x}\left(AE\frac{\partial U}{\partial x}\right) + b - (A\rho)\frac{\partial^2 U}{\partial t^2} = 0, \tag{11.17}$$

where $U = U(x,t)$ is the axial displacement, x is the Cartesian coordinate, t is the time, A is the cross-sectional area, E is the modulus of elasticity, ρ is the mass density, and $b = b(x,t)$ are the body forces in the x-direction.

Dividing by the constant term (AE) and considering the body forces equal to zero $(b = 0)$, Eq. (11.17) takes exactly the form of Eq. (11.16b) with a wave speed given by $c = \sqrt{E/\rho}$.

As already known from the FEM theory, Eq. (11.16b) and its equivalent Eq. (11.17) can be numerically solved either in the time domain or in the frequency domain. Below we present an example in the time domain (transient analysis).

Example 11.5 (Transient Elastic Analysis with Initial Displacement) The problem is defined in an elastic rod AB of length $2l$ at rest, at the middle M of which a displacement U_0 is imposed at time $t < 0$ (see Fig. 11.3). At time $t = 0_+$, the external force at the point M is removed and the rod is left free to vibrate.

The rod AB was uniformly divided into eight intervals, thus creating nine nodal points as shown in Fig. 11.4. The aforementioned nodes are used to create:

(i) An assemblage of eight linear finite elements and
(ii) A nine-node macroelement of Lagrange type (of eighth degree, i.e., $p = 8$). The corresponding nine shape functions are shown at the bottom of Fig. 11.4.

In both the abovementioned cases, at any time $t \geq 0$ the boundary conditions are of Dirichlet type:

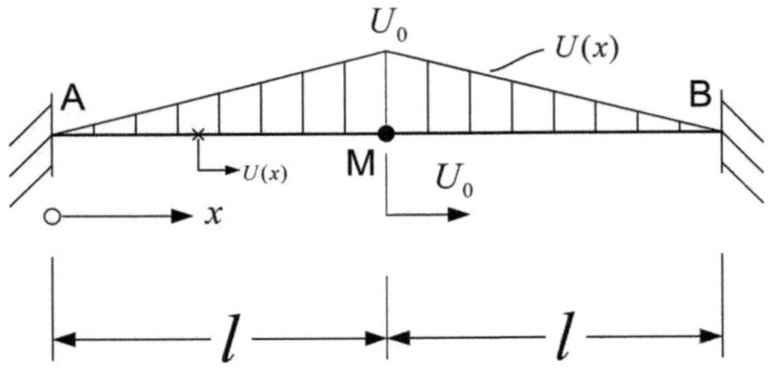

Fig. 11.3 Elastic rod AB at rest with linearly distributed initial displacement (maximum value at the middle M)

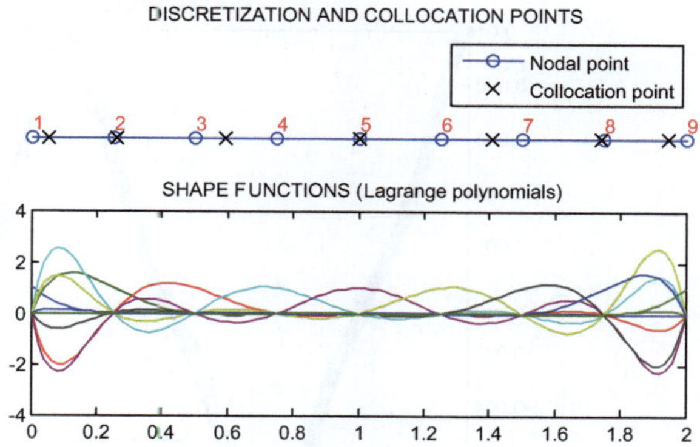

Fig. 11.4 Discretization and shape functions for the elastic rod

$$x = 0: \quad U_A = U(0, t) = 0$$
$$x = 2l: \quad U_B = U(2l, t) = 0 \tag{a}$$

Therefore, the equations system will eventually include *seven* unknown degrees of freedom.

Considering a constant cross-sectional area along the entire rod AB, the variable A involved in Eq. (11.17) is eliminated, and the latter ordinary differential equation (ODE) has to be satisfied at *seven* collocation points (as many as the number of the unknown DOF). Between several possible sets, below we present the case where the latter are images of the roots of Legendre polynomials, i.e., they are taken at the position of the 7-point Gaussian quadrature.

The seven ODEs produced by the collocation at the abovementioned seven points give a system of seven ODEs in the form:

$$\mathbf{M\ddot{u}} + \mathbf{Ku} = 0 \tag{b}$$

with

$$m_{ij} = -\rho L_{i+1}(x_j) \quad \text{and} \quad k_{ij} = EL''_{i+1}(x_j) \;\; i, j = 1, \ldots 7 \tag{c}$$

The functions $L_i(x)$ in the relationship (c) represent the Lagrange polynomials shown at the bottom of Fig. 11.4.

The differential system of seven equations (displacements) can be easily solved using the Runge–Kutta method (function `ode45` in MATLAB®) by doubling the number of unknowns (seven displacements plus seven velocities). The obtained numerical solution is shown in Fig. 11.5, where one may observe a good accordance between the collocation method as well as the FEM solution. Both of them are very close to the exact solution.

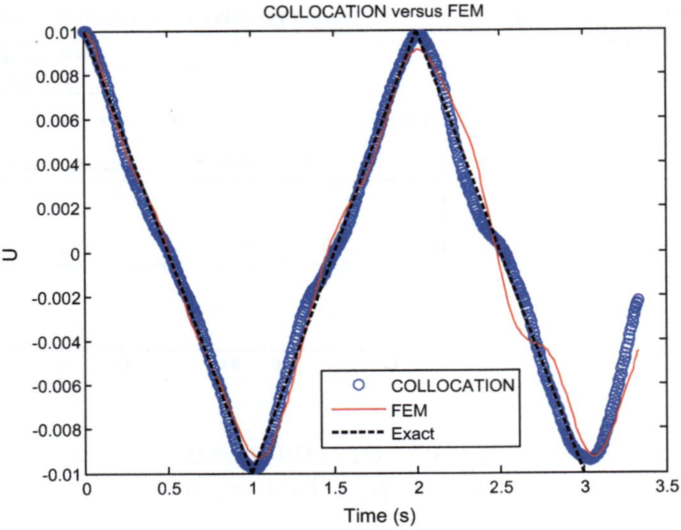

Fig. 11.5 Displacement history at the middle M of the rod

11.6.2 Eigenvalue and General Transient Analysis

Given the fact that a characteristic example in wave propagation is the case of a clamped beam under a sudden loading (of temporal Heaviside shape) at the free end, it is imperative to deal with the way the Neumann-type BCs are imposed in practice.

Throughout the following text, the spatial derivatives are denoted by the symbol prime ($'$).

In the absence of body forces ($b = 0$), and assuming constant parameters A, E, and ρ, as well as considering harmonic conditions at resonance, that is $U(x, t) = \overline{U}(x) \sin \omega t$, Eq. (11.16b) takes the well-known form of *Helmholtz equation*:

$$\overline{U}''(x) + \lambda \overline{U}(x) = 0, \quad x \in [0, L], \tag{11.18}$$

where $\lambda = (\omega\rho/E)^2$ is the eigenvalue.

In this problem, three alternative types of boundary values are possible as follows (Fig. 11.6):

(i) Both ends are fixed:

$$\overline{U}(0) = 0, \quad \overline{U}(L) = 0 \tag{11.19a}$$

(ii) One end is fixed, the other free:

$$\overline{U}(0) = 0, \quad \overline{U}'(L) = 0 \tag{11.19b}$$

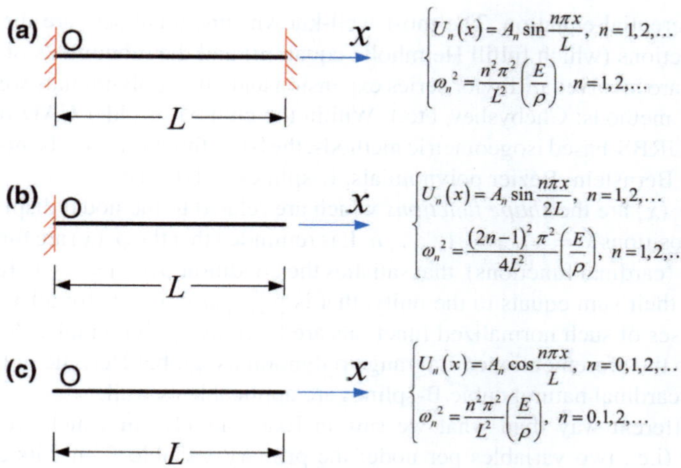

Fig. 11.6 Types of boundary conditions and exact modes for an elastic rod

(iii) Both ends are free (free-free):

$$\overline{U}'(0) = 0, \quad \overline{U}'(L) = 0 \tag{11.19c}$$

For the sake of brevity, henceforth the "bar" in $\overline{U}$ will be suppressed.

Notice The hyperbolic problems can be solved either using small size interpolation in a number of subintervals (see Examples 11.1 and 11.2) or in a global way (see Examples 11.3–11.5). In the sequence, we deal only with the case of global interpolation, as it is much closer to the modern isogeometric methods.

Therefore, the solution of Eq. (11.18) can be expressed as a series expansion using two alternative ways:

$$U(x) = \sum_{j=1}^{n} f_j(x) a_j, \tag{11.20a}$$

or

$$U(x) = \sum_{j=1}^{n} \phi_j(x) U_j. \tag{11.20b}$$

The $f_j(x)$, $j = 1, \ldots, n$, in Eq. (11.20a) are the *basis functions* and a_j are generalized coefficients that refer to n nodeless parameters. The basis functions may be generally selected among well-known linear independent and complete functional sets. Sometimes such a functional set is the corresponding solution of a homogeneous differential equation. The most well-known functional sets are the trigonometric functions (which fulfill Helmholtz equation) and the monomials of the form x^j (which are involved in Taylor series expansion and other polynomials well known in spectral methods: Chebyshev, etc.). Within the context of older CAD-based and modern NURBS-based isogeometric methods, the basis functions may be nonrational or rational Bernstein–Bézier polynomials, B-splines or NURBS.

Also, $\phi_j(x)$ are the *shape functions* which are related to the nodal displacements U_j at the positions $x = x_j$, $j = 1, \ldots, n$. It is reminded that the $\phi_j(x)$ are functions of [0, 1]-type (cardinal functions), that satisfies the condition $\phi_j(x_i) = \delta_{ij}$ (=Kronecker delta) and their sum equals to the unity, that is $\sum_{j=1}^{n} \phi_j(x) \equiv 1$, for all $x \in [0, L]$. Typical cases of such normalized functions are Lagrange polynomials. However, in addition to the aforementioned Lagrange polynomials, global Hermite polynomials as well as cardinal natural cubic B-splines are applicable as well.

In a different way than what we saw in Example 11.2 in which we had C^1-continuity (i.e., two variables per node: the primary variable U and its derivative $\mathrm{d}U/\mathrm{d}x$, in conjunction with *local* collocation points between two successive breakpoints), the global collocation method (Provatidis [52, 53]) demands the direct fulfillment of the PDE (11.18) at some *global* (collocation) points into the interval $\tilde{x}_i \in [0, L]$, $i = 1, \ldots, n_{\mathrm{eq}}$, whence:

$$U''(\tilde{x}_i) + \lambda U(\tilde{x}_i) = 0, \quad i = 1, \ldots, n_{\mathrm{eq}} \tag{11.21}$$

In principle, the abovementioned collocation points $\tilde{x}_i$ could coincide even with the nodes on which Lagrange polynomials are defined. However, based on previous experience ([51, 56] and papers therein)—in elliptic problems—the collocation points can be safely taken at the location of Gauss points. In addition, the roots of Chebyshev polynomials, of first and second kind, have given very good results (see [53]). Also, considering the findings of Chap. 6 where the nonrational Bernstein–Bézier was found to be a change of basis of the Lagrange polynomials, it is easy to understand that the nonrational Bézier-based collocation is equivalent to the Lagrange polynomial based one. The latter has been early reported by Provatidis [52].

For both formulations, either Eq. (11.20a) [CAD-based] or Eq. (11.20b) [Lagrange and Hermite polynomial based], the numerical procedure is the same. In more details, in a problem of purely Dirichlet type, the nodal values at the ends are known and also the coefficients associated to them are also known ($a_1 = U_1$ and $a_n = U_n$). Therefore, after the fulfillment of the ODE at the collocation points, in the produced matrix of equations system, we have only to arrange in the right-hand side those columns related to the two ends (i.e., the first and the last column). This philosophy may be found in classical textbooks [15], Finlayson [27].

The imposition of the Neumann-type BCs is quite different than the case of Dirichlet one. The first papers on this topic were published by Provatidis [51–53, and later], where the key ingredient (novel point) was to impose the Neumann BC through elimination of the corresponding boundary DOF (in contrast to the FEM practice where the Neumann DOF are not eliminated). As an alternative to this tedious task, although we had suspected that the boundary condition had to be imposed and the PDE to be satisfied at the same boundary point (of a single body) or the interface between the two cooperating bodies in equilibrium (see [53, p. 919]), we failed to find a better solution. This is due to the fact that, even the sign (addition or subtraction) of redundant equations plays a significant role.

Later, researchers from the isogeometric analysis group [5] proposed a penalty-like method, which includes both the fulfillment of the PDE and the Neumann-type boundary conditions as well. Concerning Lagrange polynomials, a shortcoming of the latter (appearance of complex eigenvalues in higher modes) has been commented in the Appendix of Provatidis [65].

Below we start with the former method of eliminating the Neumann DOF, inspired in 2004 by Provatidis [50, p. 6704, and papers therein] and eventually published a few years later [51, etc.].

11.6.3 Formulation Using Basis Functions

Substituting Eq. (11.20a) into Eq. (11.21), one obtains:

$$\sum_{j=1}^{n}\left[f_j''(\tilde{x}_i)+\lambda f_j(\tilde{x}_i)\right]a_j=0,\quad i=1,\ldots,n_{\mathrm{eq}} \tag{11.22}$$

We distinguish the following two cases.

11.6.3.1 Both Ends Fixed (Dirichlet Type)

Similarly to Example 11.5, the boundary conditions are imposed substituting Eq. (11.19b) into Eq. (11.20a), thus leading to:

$$\begin{aligned} U(0)&=\sum_{j=1}^{n} f_j(0)a_j=0\\ U(L)&=\sum_{j=1}^{n} f_j(L)a_j=0 \end{aligned} \tag{11.23}$$

Equations (11.23) impose two relationships between the n unknown generalized coordinates. Therefore, we will need ($n_{eq}=n-2$) equations of the form (11.22) so as to derive a system of n algebraic equations, equal to the number of unknowns ($a_i, i=1,\ldots,n$). The derived homogeneous algebraic system, of dimensions $n\times n$, has the form (see [52]):

$$\begin{bmatrix} \underbrace{\mathbf{K}_{11}}_{2\times 2} & \underbrace{\mathbf{K}_{12}}_{2\times(n-2)} \\ \underbrace{\mathbf{K}_{21}}_{(n-2)\times 2} & \underbrace{\mathbf{K}_{22}}_{(n-2)\times(n-2)} \end{bmatrix} \cdot \begin{bmatrix} \underbrace{\mathbf{a}_1}_{2\times 1} \\ \underbrace{\mathbf{a}_2}_{(n-1)\times 1} \end{bmatrix} - \lambda \begin{bmatrix} \underbrace{\mathbf{0}}_{2\times 2} & \underbrace{\mathbf{0}}_{2\times(n-2)} \\ \underbrace{\mathbf{M}_{21}}_{(n-2)\times 2} & \underbrace{\mathbf{M}_{22}}_{(n-2)\times(n-2)} \end{bmatrix} = \begin{bmatrix} \underbrace{\mathbf{0}}_{2\times 1} \\ \underbrace{\mathbf{0}}_{(n-2)\times 1} \end{bmatrix} \tag{11.24}$$

In Eq. (11.24), the vectors $\mathbf{a}_1$ and $\mathbf{a}_2$ may be generalized coordinates, for example, $\mathbf{a}_1 = [a_1, a_2]^{\mathrm{T}}$ and $\mathbf{a}_2 = [a_3, \ldots, a_n]^{\mathrm{T}}$, but this choice is not restrictive.

Obviously, the elements of the matrices $\mathbf{K}_{11}$ and $\mathbf{K}_{12}$, which are produced from Eq. (11.23), i.e., the boundary conditions of Dirichlet type, are given by:

$$(k)_{ij} = \begin{cases} f_j(0), & \text{if } i = 1 \\ f_j(L), & \text{if } i = 2 \end{cases} \tag{11.25a}$$

Moreover, the elements of the rest matrices ($\mathbf{K}_{21}$, $\mathbf{K}_{22}$, $\mathbf{M}_{21}$ and $\mathbf{M}_{22}$) are given by:

$$k_{ij} = f_j''(\tilde{x}_i) \quad \text{and} \quad m_{ij} = -f_j(\tilde{x}_i), \quad \text{for } i = 1, \ldots, n-2; \quad j = 1, \ldots, n, \tag{11.25b}$$

Equation (11.24) splits into the two following subsystems:

$$\mathbf{K}_{11}\mathbf{a}_1 + \mathbf{K}_{12}\mathbf{a}_2 = \mathbf{0} \tag{11.26}$$

$$(\mathbf{K}_{21} - \lambda\mathbf{M}_{21})\mathbf{a}_1 + (\mathbf{K}_{22} - \lambda\mathbf{M}_{22})\mathbf{a}_2 = \mathbf{0} \tag{11.27}$$

From Eq. (11.26), the vector $\mathbf{a}_1$ is eliminated as follows:

$$\mathbf{a}_1 = -(\mathbf{K}_{11})^{-1}\mathbf{K}_{12}\mathbf{a}_2, \tag{11.28}$$

and after substitution in Eq. (11.27), one obtains:

$$(\mathbf{K} - \lambda\mathbf{M})\mathbf{a}_2 = \mathbf{0}, \tag{11.29}$$

where

$$\mathbf{K} = \mathbf{K}_{22} - \mathbf{K}_{21}(\mathbf{K}_{11})^{-1}\mathbf{K}_{12} \quad \text{and} \quad \mathbf{M} = \mathbf{M}_{22} - \mathbf{M}_{21}(\mathbf{K}_{11})^{-1}\mathbf{K}_{12} \tag{11.30}$$

Based on the two above matrices, $\mathbf{K}$ and $\mathbf{M}$, the $(n-2)$ eigenvectors may be easily calculated by finding the roots of the characteristic polynomial in λ, that is, produced by the determinant of the usual matrix ($\det\|\mathbf{K} - \lambda\mathbf{M}\| = 0$). Due to the nonsymmetric formulation, the QR algorithm could be implemented. In the MATLAB® environment, the usual "`eig`" function is applicable.

Remark The above procedure is very general and covers all possible cases of basis functions such as monomials of any kind, trigonometric functions. However, in the particular case of CAD-based basis functions, such as B-splines and NURBS, the fact that $a_1 = U_1$ and $a_n = U_n$ simply implies $\mathbf{K}_{11} = \begin{pmatrix} 1 & 0 \\ 0 & 1 \end{pmatrix}$ is the identity matrix, and $\mathbf{K}_{12}$ vanishes. Therefore, in terms of Eq. (11.30), only the matrix $\mathbf{K}_{22}$ participates in the final matrix $\mathbf{K}$. Again, this happens because the first and last columns of the initial matrix (i.e., the submatrix $\mathbf{K}_{21}$) are deleted.

11.6.3.2 One End Fixed, the Other Free

Substituting Eqs. (11.19a) and (11.19b) (for $x = 0$ and $x = L$, respectively) into Eq. (11.20a), one obtains:

$$U(0) = \sum_{j=1}^{n} f_j(0)a_j = 0$$

$$U'(L) = \sum_{j=1}^{n} f_j'(L)a_j = 0 \tag{11.31}$$

Obviously, Eq. (11.31) imposes new dependencies between the n generalized coefficients, and therefore the second row of the tables $\mathbf{K}_{11}(2 \times 2)$ and $\mathbf{K}_{12}(2 \times 2)$ will be modified. From this point and further, the numerical procedure is similar to that of the former Sect. 11.6.1.

11.6.4 Shape Function Formulation

The above treatment of arbitrary base functions (Eq. 11.20a) requires a special treatment (Eqs. 11.24–11.30) in order to incorporate the boundary conditions by proper elimination. The alternative use of shape functions described by Eq. (11.20b) (instead of base functions) is related to an implicit (closed formed) expression for the matrix inversion, as will then be explained below.

Clearly, whatever the type boundary conditions at the two ends, the governing equation [stress equilibrium (11.21)] is fulfilled at all the $(n-2)$ internal (collocation) points $\tilde{x}_i$, $i = 1, \ldots, (n-2)$, so as Eq. (11.22) is substituted by:

$$\sum_{j=1}^{n}\left[\phi_j''(\tilde{x}_i)+\lambda\phi_j(\tilde{x}_i)\right]U_j=0;\quad i=1,\ldots,(n-2) \tag{11.32}$$

When Eq. (11.32) is applied to all (n – 2) points, this leads to a matrix system (see Eq. 11.33), which is similar to that of Eq. (11.27) but with $\hat{k}_{ij}=\phi_j''(\tilde{x}_i)$ and $\hat{m}_{ij}=-\phi_j(\tilde{x}_i)$:

$$\Bigg(\underbrace{\widehat{\mathbf{K}}_{22}}_{(n-2)\times(n-2)}-\lambda\underbrace{\widehat{\mathbf{M}}_{22}}_{(n-2)\times(n-2)}\Bigg)\underbrace{\mathbf{U}_2}_{(n-2)\times 1}=-(\mathbf{K}_{21}-\lambda\mathbf{M}_{21})\mathbf{U}_1 \tag{11.33a}$$

According to the kind of boundary conditions, two alternative formulations are taken as follows.

11.6.4.1 Both Ends' Fixed

Since vanishing of displacements at the ends implies $\mathbf{U}_1=\mathbf{0}$, the $(n-2)$ eigenvalues are calculated through the following formula

$$\Bigg(\underbrace{\widehat{\mathbf{K}}_{22}}_{(n-2)\times(n-2)}-\lambda\underbrace{\widehat{\mathbf{M}}_{22}}_{(n-2)\times(n-2)}\Bigg)\underbrace{\mathbf{U}_2}_{(n-2)\times 1}=\mathbf{0} \tag{11.33b}$$

11.6.4.2 One End Fixed, the Other Free

In addition to Eq. (11.13), one more equation is written at the traction-free end (nth node):

$$U_n'=\sum_{j=1}^{n}\phi_j'(L)U_j=0, \tag{11.34}$$

Based on Eq. (11.34), since $U_1=0$, the nth degree of freedom is eliminated as follows:

$$U_n=-\sum_{j=2}^{n-1}\phi_j'(L)U_j/\phi_n'(L) \tag{11.35}$$

Now, substituting Eq. (11.35) into Eq. (11.32), we obtain:

$$\sum_{j=2}^{n-1}\left(\phi_j''(x_i)-\phi_n''(x_i)\frac{\phi_j'(L)}{\phi_n'(L)}\right)U_j$$

$$-\lambda\sum_{j=2}^{n-1}\left(-\phi_j(x_i)+\phi_n(x_i)\frac{\phi_j'(L)}{\phi_n'(L)}\right)U_j=0,\quad i=2,\ldots,(n-1) \qquad (11.36)$$

Writing Eq. (11.36) for all the $(n-2)$ internal nodes, we derive again the well-known expression:

$$(\overline{\mathbf{K}}-\lambda\overline{\mathbf{M}})\overline{\mathbf{U}}_2=\mathbf{0},\quad \overline{\mathbf{U}}_2=\left[U_2,\ldots,U_{n-1}\right]^{\mathrm{T}}, \qquad (11.37\text{a})$$

in which the elements of the stiffness and mass matrices are given by:

$$\bar{k}_{ij}=\phi_{j+1}''(\tilde{x}_{i+1})-\phi_n''(\tilde{x}_{i+1})\frac{\phi_{j+1}'(L)}{\phi_n'(L)},\quad i,j=1,\ldots,(n-2)$$

$$\bar{m}_{ij}=-\phi_{j+1}(\tilde{x}_{i+1})+\phi_n(\tilde{x}_{i+1})\frac{\phi_{j+1}'(L)}{\phi_n'(L)} \qquad (11.38\text{a})$$

Remark Interestingly, Eq. (11.37a) does not lead to $(n-1)$ eigenvalues, as happened in the conventional finite element method, but in $(n-2)$ ones.

11.6.5 Lumped Mass Formulation

Lumped mass is very important in the speed up of time-integration procedure. Therefore, many commercial FEM codes use it as a default option.

In principle, lumped mass could be achieved using nodal collocation, i.e., taking the collocation points exactly at the position of the nodal points involved in 1D, 2D, or 3D tensor-product Lagrange polynomials. Nevertheless, it has been definitely shown that using nodal collocation, either complex eigenvalues appear [51, 52] or the quality of the solution is poor [56].

Following the famous proverb "*If the mountain will not come to Muhammad, then Muhammad must go to the mountain*" (Francis Bacon, in Essays, 1625), we have worked as follows. We have reversed the position of the nodal nodes by putting them at the position of the collocation points (i.e., at images of the roots of Legendre polynomials). Then nodal collocation is performed. Interestingly, no complex eigenvalues appear. Not only that but the (new nodal collocation based) numerical solution is identical with that in which the nodal points are uniformly arranged and the collocation points are taken at those of the Gaussian quadrature.

In more details, no mass matrix is needed to be calculated, as it coincides with the unitary matrix ($\mathbf{M}_{\text{lumped}}=\mathbf{I}$). Thus, the eigenvalues of the problem are only those of the relevant stiffness matrix, i.e., $\det(\mathbf{K}-\lambda\mathbf{I})=0$.

For example, in the latter case of one fixed node the final formulation is:

$$(\overline{\mathbf{K}} - \lambda \mathbf{I})\overline{\mathbf{U}}_2 = \mathbf{0}, \quad \overline{\mathbf{U}}_2 = \left[U_2, \ldots, U_{n-1}\right]^{\mathrm{T}} \tag{11.37b}$$

The abovementioned phenomenon was first remarked by Provatidis [53] and followed by Provatidis [60, 63, 65]. Therein theoretical explanations, based on basis change may be found.

In addition to the eigenvalue problem, it should be noted that the lumped mass method has been successfully applied to the transient wave propagation analysis as well [54, 55].

Example 11.6 (Axial Natural Vibrations of a Beam) A beam of uniform length ($L = 1$), with material properties (E, ρ) and constant cross-sectional area (A) equal to the unity, is fixed at the position $x = 0$. Both ends of the beam are considered to be fixed. An axial displacement is imposed and then is left to vibrate as a bar.

Using Lagrange polynomials and fulfilling the governing equation at the Gauss points, the results shown in Table 11.2 are derived. The errors of each eigenvalue are given in percent deviation (%) from the exact solution which is given as:

$$\omega_{i,\text{exact}}^2 = \frac{i^2\pi^2}{L^2} \cdot \frac{E}{\rho}, \quad i = 1, 2, \ldots \tag{11.38b}$$

Remark 1 We notice that as the number of nodes increases (i.e., the polynomial degree), the global collocation method leads to more accurate numerical results.

Remark 2 Whatever polynomial form we use (power series of monomials x^k, Lagrange, Bernstein, Chebyshev), the numerical results of Table 11.2 remain the same.

Remark 3 If the nodal points are put at the Gauss points, and these are also taken as collocation points, the results of Table 11.2 do not change.

Remark 4 For further development of this issue, using other types of BCs, as well as using other types of polynomials (Lagrange, Bernstein, and Chebyshev), as well as several other positions of the collocation points, the reader is referred to [53, 54].

11.7 Boundary Conditions of Neumann Type

11.7.1 One-Dimensional Problems

The way we imposed the boundary conditions of Neumann type so far in the above text has probably the disadvantage that even in a problem without any support (free-free problem), the number of final equations has to be reduced (by two) with respect to that of classical finite elements, while in FEM no (row or column) deletion is performed.

Table 11.2 Calculated eigenvalues using global collocation and polynomials of nth degree

Mode	Accurate ω^2 (s^{-2})	Errors (in %) Degree of Lagrange polynomial (n)									
		2	3	4	5	6	7	8	9	10	11
1	9.870	18.94	21.59	−0.98	0.06	0.00	0.00	0.00	0.00	0.00	0.00
2	39.478	–	−8.81	51.98	−4.24	0.73	−0.05	0.00	0.00	0.00	0.00
3	88.826	–	–	10.58	91.53	−7.62	2.70	−0.36	0.01	0.00	0.00
4	157.914	–	–	–	40.71	140.79	−9.47	6.33	−1.20	0.22	−0.03
5	246.740	–	–	–	–	81.60	199.47	-9.16	11.77	-2.61	0.69
6	355.306	–	–	–	–	–	132.36	267.18	−6.40	19.00	−4.26
7	483.611	–	–	–	–	–	–	192.28	343.64	−1.11	27.96
8	631.655	–	–	–	–	–	–	–	260.95	454.54	6.53

The partial differential equation is satisfied at $(n-1)$ Gauss points

From the other point of view, in order to preserve the traction given node, we have to impose a Neumann-type BC and at the same time to satisfy the PDE at the same boundary node. The most apparent way is to add or subtract the aforementioned two equations, probably applying a proper factor depending on the ratio (E/ρ) so as the stiffness and mass terms become comparable (see later in Eq. 11.41). While relevant researchers claim that this is a way that resolves the difficulty (e.g., [5]), our personal research [65, p. 72] shows that, at least in the case of Lagrange polynomials, the aforementioned alternative may lead even to fictitious complex values at medium modes.

Within this context, if we consider a bar of length L which is supported at its one end ($x = 0$), whereas the other end ($x = L$) is allowed for performing axial-free vibration, then the exact eigenvalues are given by:

$$\omega_j^2 = \frac{(2j-1)^2\pi^2}{4L^2}\left(\frac{E}{\rho}\right), \quad j = 1, 2, \ldots \tag{11.39}$$

Considering the nodes $1, 2, \ldots, n$ on which the Lagrange polynomials are supported, the abovementioned bar is uniformly subdivided into $(n - 1)$ subintervals of n nodal points. The corresponding Lagrange polynomials $L_j(x)$, of degree $p = (n - 1)$, interpolate the function of the axial displacement $U(x)$ in a global way within the interval $[0, L]$.

Extending what we have already talked about, we present two separate approaches of Neumann BCs, which both depend on the same position of collocation points defined by the $(n - 2)$ Gauss points in the interior of the elastic bar. Moreover, the second method uses one additional collocation point at the free end.

Approach 1: Elimination of Neumann-type DOF The *first* approach is the one we mentioned earlier [51, 53] and consists of eliminating one of the n nodal degrees of freedom (in addition to the fixed end, $u_1 = 0$), for example that of the free node, u_n. As a result, the produced mass and stiffness matrices become of dimensions $(n - 2) \times (n - 2)$, and therefore, we can calculate until the $(n - 2)$th eigenvalue.

After the elimination, the stiffness matrix is given by the formula:

$$k_{ij} = -L_j''(x_i) + \frac{L_j'(L)}{L_n'(L)} L_n''(x_i), \quad i = 2, \ldots, n-1; \; j = 1, \ldots, n-1, \tag{11.40a}$$

whereas the mass matrix becomes:

$$m_{ij} = L_j(x_i) - \frac{L_j'(L)}{L_n'(L)} L_n(x_i), \quad i = 2, \ldots, n-1; \; j = 1, \ldots, n-1, \tag{11.40b}$$

Approach 2: Adding the PDE with the Neumann-type boundary condition The *second* approach was developed five years later by Auricchio et al. [5] in the context of 2D and 3D problems, but is explained here in full detail for the case of a one-dimensional bar. Therefore, stress equilibrium (the governing PDE) and the Neumann BC at the free end are added in parts (in fact this is a kind of condensation without using a least-squares scheme but only implementing a Lagrange multiplier of unit value), so as the following equation is obtained:

$$-\underbrace{\left(EAu'' - \rho A\ddot{u}\right)}_{\text{ODE}} + \underbrace{\frac{EA}{L}u'}_{\text{BC}} = 0, \quad x = L \tag{11.41}$$

Setting unit values ($E = 1, A = 1, L = 1$) so as to avoid any possible scaling factors, for each interval point we have:

$$k_{ij} = -L_j''(x_i), \quad i = 2, \ldots, n-1; \ \ j = 1, \ldots, n, \tag{11.42}$$

while for the free end:

$$k_{nj} = -L_j''(x_n) + L_j'(x_n), \quad j = 1, \ldots, n. \tag{11.43}$$

In both cases of the second approach (i.e., either in conjunction with Eq. (11.42) or with Eq. (11.43)), the mass matrix includes elements given by:

$$m_{ij} = L_j(x_i), \quad i = 2, \ldots, n; \ \ j = 1, \ldots, n, \tag{11.44}$$

Based on the above considerations, for $n = 11$ (i.e., Lagrange polynomials of degree $p = 10$, implemented through ten equidistant segments), the obtained results are provided into Table 11.3. There one may observe that the approach of eliminating the u_n DOF which corresponds to the free end (Neumann-type BC) leads to very small errors (up to the 6th mode), while the approach of fulfilling the PDE and at the same time the Neumann BC fails after the fourth mode as it leads to fictitious complex eigenvalues. This study was reduced in the case of Lagrange polynomials only. Similar findings were observed for either smaller or larger values of the variable n.

The reader may repeat the above two alternative procedures for a bar under free-free boundary conditions, for which the exact solution is shown in Fig. 11.6c.

11.7.2 Coupling Two Adjacent Macroelements

The value of a computational method is judged upon its applicability to domain decomposition. So far, the first approach of the previous subsection [51–53] has been used to assemble two curvilinear domains for the solution of the 2D potential flow past a cylinder [56, p. 550] as well as the collinear impact between two elastic bars [53], p. 913).

In brief, in the kth subregion there are boundary nodes under Dirichlet ($\mathbf{U}_D$ at n_D nodes) and Neumann ($\mathbf{U}_N$ at n_N nodes) boundary conditions, as well as internal

Table 11.3 (One-dimensional bar fixed at one end) Calculated eigenvalues obtained either by elimination or without elimination of the DOF at the free end, using Lagrange polynomials

Mode	Exact eigenvalues ω^2 (s^{-2})	Applying elimination of the DOF at the free end (Eqs. 11.40a, 11.40b)		Without elimination of the DOF at the free end (Eqs. 11.42–11.44)	
		Calculated eigenvalue	Error (%)	Calculated eigenvalue	Error (%)
1	2.46740110	2.46740110	0.00	2.46740109	0.00
2	22.20660990	22.20660990	0.00	22.20446026	0.01
3	61.68502751	61.68503257	0.00	61.88334349	0.32
4	120.90265391	120.90382191	0.00	139.28 − 27.40*i*	Complex
5	199.85948912	199.80516704	0.03	139.28 + 27.40*i*	Complex
6	298.55553313	297.58782188	0.32	193.82535792	–
7	416.99078595	452.95004744	8.62	−95.74 − 381.27*i*	Complex
8	555.16524756	908.59061476	63.66	−95.74 + 381.27*i*	Complex
9	713.07891798	3420.83645043	379.73	444.98811642	–
10	890.73179720	–	–	3418.56102041	–

ones ($\mathbf{U}_I$ at n_I nodes). The interface between two adjacent macroelements is always considered to be of Neumann type. Collocating the Neumann BCs and also the PDE, after arrangement of the Dirichlet DOF to the RHS, the produced algebraic system becomes:

$$^{(k)}\begin{bmatrix} \mathbf{K}_{NN} & \mathbf{K}_{NI} \\ \mathbf{K}_2^N & \mathbf{K}_2^I \end{bmatrix} \cdot \begin{bmatrix} \mathbf{U}_N \\ \mathbf{U}_I \end{bmatrix} = {}^{(k)}\begin{bmatrix} \mathbf{q}_N - \mathbf{K}_{ND}\mathbf{U}_D \\ -\mathbf{K}_2^D\mathbf{U}_D \end{bmatrix} \quad (11.45)$$

This system consists of $(n_N + n_I)$ equations (Neumann plus internal DOF). If the problem domain is occupied by a single macroelement, Eq. (11.45) contains also $(n_N + n_I)$ unknowns and can be directly solved. However, in the general case of adjacent subregions, the vector $\mathbf{q}_N$ will contain interface fluxes as well. The coupling (assembling) procedure is as follows.

For each macroelement, the matrix appearing on the left side of Eq. (11.45) is written. The assembly of these matrices is performed by imposing compatibility and equilibrium conditions at the interface nodes, similar to the conventional FEM procedure. Thus, the unknown flux along the interface is temporarily eliminated. The produced equation system includes only the unknown Neumann DOF $\mathbf{U}_{N,\text{assembly}}$ (boundary and interface) plus the internal ones ($\mathbf{U}_{I,\text{assembly}}$). After the aforementioned potential (or displacement) vectors have been calculated, using the first set of equations in each macroelement the interface fluxes $\mathbf{q}_N$ can be calculated.

The consequence of the elimination of all nodal DOF under Neumann-type BCs leads to the appearance of only the internal nodes in the final equations system [53].

In order to describe details for the alternative approach, let us consider a rod AB, fixed at A and free at B, with M denoting its middle point (Fig. 11.7). The macroelement (I), i.e., AM, is discretized using n_1 nodes (two of them are its ends A and M and the rest are in the interior), thus making possible the generation of a series of Lagrange polynomials, ${}^{(1)}\phi_j(x)$, $j = 1, \ldots, n_1$, each of degree $(n_1 - 1)$. Similarly, the rodMB is discretized using n_2 nodes (two of them are its ends$M$ and$B$); again, this choice is related to another series of Lagrange polynomials, ${}^{(2)}\phi_j(x)$, $j = 1, \ldots, n_2$, each of degree $(n_2 - 1)$. Therefore, the total number of geometric nodes in the entire structure becomes $(n_1 + n_2 - 1)$. The special case in which $n_1 = 5$ and $n_2 = 5$, i.e., with nine nodes $(n_1 + n_2 - 1 = 9)$ in total, is shown in Fig. 11.7.

With respect to Fig. 11.7, in the first approach node 1 is deleted due to Dirichlet boundary conditions, node 5 is eliminated due to force balance at the junction$M$ $(x = l)$, and node 9 is also eliminated due to Neumann BC (traction free). Therefore, the vector of the primary (master) DOF becomes $\mathbf{U}_m = [U_2, U_3, U_4, U_6, U_7, U_8]^{\mathrm{T}}$, the vector $\mathbf{U}_s = [U_5, U_9]^{\mathrm{T}}$ of the eliminated (slave) DOF is related to the primary $\mathbf{U}_m$ through a transformation matrix $\mathbf{B}$ as follows:

$$\mathbf{U}_s = \mathbf{B}\mathbf{U}_m, \tag{11.46}$$

and the final equations of motion become:

$$\mathbf{M}\ddot{\mathbf{U}}_m + (\mathbf{K}_{ms}\mathbf{B} - \mathbf{K}_{mm})\mathbf{U}_m = \mathbf{0}. \tag{11.47}$$

Again, in the macroelement (I) the ODE is fulfilled at three collocation points (e.g., those in 3-point Gauss quadrature); alternatively, the three nodes (2, 3, 4) could be shifted to the position of Gaussian points and then nodal collocation (lumped mass) may be applied. Similarly, in the macroelement (II) the ODE is fulfilled also at three collocation points with the same remarks as in macroelement (I). This approach leads to a dynamic system of six equations.

In contrast, in the second approach one additional equation of motion is written for the node at the junction (node 5) and another at the traction-free end (node 9),

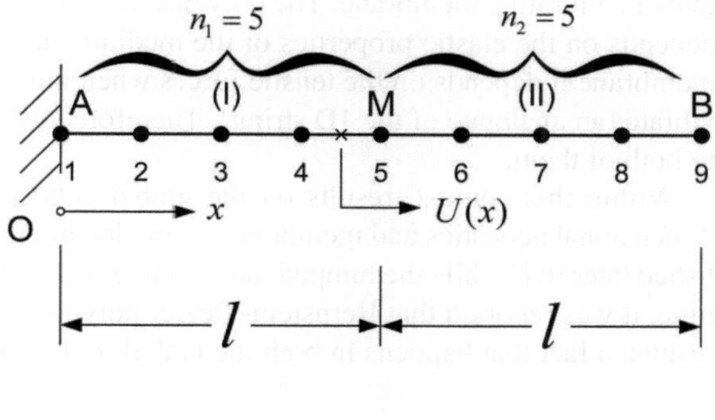

Fig. 11.7 Assembly of two macroelements: (I) and (II)

at which also Eq. (11.41) is applied. In this way, the dynamic system includes eight equations, as also happened in the FEM solution.

11.7.3 Two-Dimensional Potential Problems

The basis functions which we tested so far are based on transfinite interpolation or tensor products made of Lagrange or Bernstein–Bézier polynomials ([52–56, 57]) as well as B-splines ([58], [59]). Relevant research under the label "NURBS-based isogeometric collocation" (Aurichio et al. [4, 5], etc.) seems to meet similar problems and continues till today.

We give some comparative results on the truly 2D problem which is treated using a single tensor-product macroelement.

Example 11.7 (Thermal Analysis in a Rectangle Plate) Let us consider a rectangular plate of dimensions $L \times H = 2 \times 1$, with Dirichlet boundary conditions: $U = 100\cos(\pi x/L)$ at the top side and $U = 0$ elsewhere (Fig. 11.8). Due to the symmetry with respect to the y-axis, only the half right part is discretized so that the domain becomes a square of unit edge. Each side of the unit square is uniformly divided into n segments; thus, $(n - 1)$ nodes of Neumann type appear along the boundary. Using a mesh of eight segments per side (81 nodes in total), the normalized error norm was found equal to $L_u = 8.6 \times 10^{-6}(\%)$. It is noted that the same result was found either putting the collocation points at the images of Legendre roots for the uniform mesh shown in Fig. 11.9 or by using a nonuniform mesh with the nodal points (white circles) at the roots of the Legendre points and the collocation points on them. A further detailed parametric analysis for a variable number of internal nodes (more or less than those on the boundary per direction) may be found in Provatidis [57, p. 547].

11.7.4 2D Acoustics and Membranes

It is well known that the form of the partial differential of 2D acoustics is identical to that of a vibrating membrane. The only physical difference is that sound propagation depends on the elastic properties of the medium and the temperature, whereas in a membrane it depends on the tensile forces when induced at the time it is left free to vibrate (an analogue of the 1D string). Therefore, every new method can be applied to both of them.

Within this context, results on the global collocation method applied to two-dimensional acoustics and membranes were documented by [25] but they were published later [64] while the lumped mass analogue was also reported [60]. At the same time, it was reported that Bernstein–Bézier polynomials lead to the same numerical results, a fact that happens in both the Galerkin–Ritz [62], while for the global col-

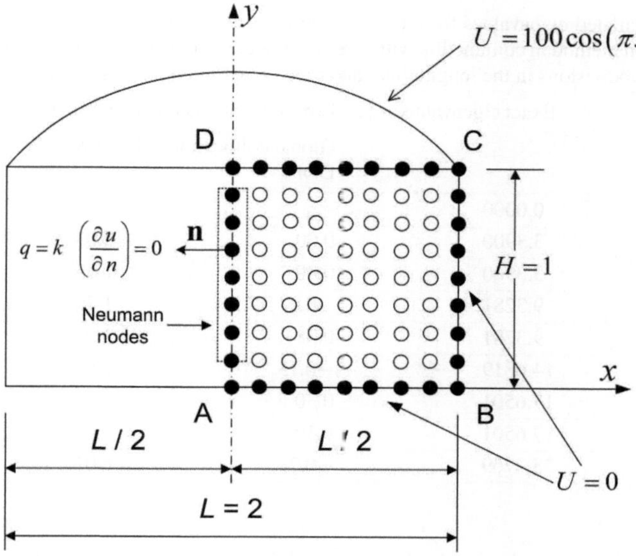

Fig. 11.8 Square domain for the analysis of a rectangular plate under Dirichlet and Neumann boundary conditions

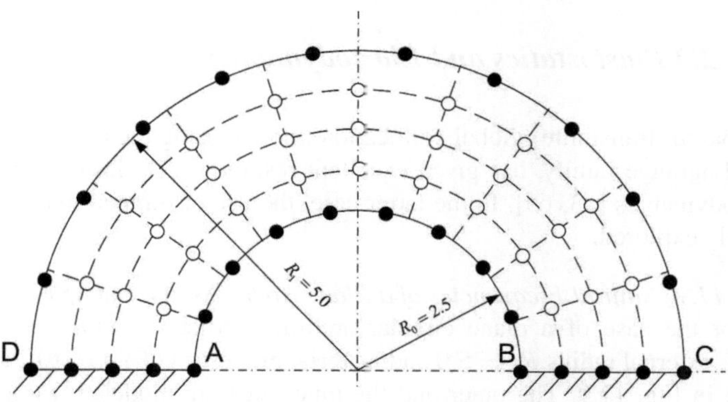

Fig. 11.9 Plane arch using a single transfinite element with (side *AB*: $n_x = 9$, side *BC*: $n_y = 4$, side *CD*: $n'_x = 11$, and side *DA*: $n'_y = 4$)

location approach it had been reported much earlier [52, 53]. Also, the performance of B-splines in acoustics was presented by Provatidis [61].

Example 11.8 (Circular Acoustic Cavity with Hard Walls) Here, in Table 11.4, we compare the global collocation method with the FEM solution in one circular acoustic cavity of unit radius with hard walls. More results may be found in the above-mentioned papers.

Table 11.4 Calculated eigenvalues for a circular cavity of unit radius with hard walls, using the global collocation method in conjunction with tensor-product Lagrange polynomials for $n_x \times n_y = 8 \times 8$ uniform subdivisions in the longitudinal and transverse directions, respectively

Mode	Exact eigenvalues ω_{mn}^2	Errors of calculated eigenvalues in %	
		Global collocation (91 DOF)	FEM (127 DOF)
1	0.0000	–	–
2	3.3900	0.00	3.03
3	3.3900	0.00	1.35
4	9.3281	−0.04	1.79
5	9.3281	0.08	2.27
6	14.6819	−0.15	2.64
7	17.6501	0.30	3.64
8	17.6501	0.30	4.01
9	28.2769	−0.71	6.01

Similar results were obtained when the nodal points were shifted at the position of the Gauss points.

11.7.5 2D Elastostatics and Elastodynamics

Gordon-based (transfinite) global collocation, mostly using the classical macroelement of Lagrange family, has given excellent results in 2D elastostatics [57] and 2D elastodynamics [63, 65]. In the latter case, the use of lumped masses has been extensively explored.

Example 11.9 (Natural Frequencies of a Plane Arch) As an example, we present results for the case of a plane circular arch-like structure, with internal radius $R_0 = 2.5$, external radius $R_1 = 5.0$ and material properties ($E/\rho = 10^4$, $\nu = 0.25$), as shown in Fig. 11.9. The outer and the inner arcs are modeled using 11 and 9 subdivisions, respectively. This allows for using an averaged uniform net of 9×3 internal nodes.

Applying Approach 1, all the boundary DOFs (black circles in Fig. 11.9) are deleted or eliminated, thus only the internal nodes (white circles) remain. The results are shown in Table 11.5.

One may observe that although the transfinite collocation method is less accurate than the corresponding (same global shape functions) Galerkin–Ritz, it is much more accurate than the FEM solution for the same mesh density.

Table 11.5 Calculated eigenvalues using the transfinite global collocation method, which is compared with the corresponding Galerkin–Ritz and the FEM solution for the same mesh density

Exact eigenvalues ω_{mn}^2	Errors of calculated eigenvalues (in %)		
	Global collocation	Galerkin–Ritz	FEM
198.06	−0.14	0.23	14.95
441.60	0.32	0.04	9.32
1062.68	0.26	−0.01	10.89
1274.21	2.81	0.22	18.47
2518.66	6.76	0.99	16.67
2670.55	1.43	0.11	22.29

11.7.6 Plate Bending

The partial differential equation of motion for a plate in bending is given by:

$$D\nabla^4 w(x, y, t) + \rho h \frac{\partial^2 w(x, y, t)}{\partial t^2} = f(x, y, t), \tag{11.48}$$

where D is the flexural rigidity, ρ is the mass density, h is the plate thickness, and f is the loading toward the z-direction.

The use of B-splines goes back to the period of Antes (1974) and Höllig (2003). Following them, Provatidis [66] has expanded the deflection into a B-spline tensor product of the form

$$w(x, y; t) = \sum_{i=1}^{n} \sum_{j=1}^{m} N_{i,p_n}(x) \cdot N_{j,p_m}(y) \cdot a_{ij}(t), \quad i = 1, \ldots, n; \quad j = 1, \ldots, m, \tag{11.49}$$

where $N_{i,p_n}(x)$ and $N_{j,p_m}(y)$ are B-splines in the x- and y-directions of p_nth and p_mth degree, respectively (with $p_n < n$ and $p_m < m$). Obviously, NURBS can be easily applied as well.

In the sequence, a certain number of n_c collocation points (x_c, y_c) are chosen, at which the governing equation is satisfied as follows:

$$D \cdot \nabla^4 \left[\sum_{i=1}^{n} \sum_{j=1}^{m} N_{i,p_n}(x_c) \cdot N_{j,p_m}(y_c) \right] \cdot a_{ij}(t) + \rho h \cdot \sum_{i=1}^{n} \sum_{j=1}^{m} N_{i,p_n}(x_c) \cdot N_{j,p_m}(y_c) \cdot \ddot{a}_{ij}(t) = f(t). \tag{11.50}$$

Setting

$$k_{c,ij} = D \cdot \nabla^4 \left[N_{i,p_n}(x_c) \cdot N_{j,p_m}(y_c) \right] \tag{11.51}$$

and

$$m_{c,ij} = \rho h \cdot \left[N_{i,p_n}(x_c) \cdot N_{j,p_m}(y_c) \right], \tag{11.52}$$

Equation (11.50) takes the following matrix form:

$$[\mathbf{M}] \cdot \{\ddot{\mathbf{a}}(t)\} + [\mathbf{K}] \cdot \{\mathbf{a}(t)\} = \mathbf{f}(t) \tag{11.53}$$

In this way, mass and stiffness matrices of dimensions $n_c \times (nm)$ are produced.

11.7.6.1 Arbitrary Curvilinear Plate

Since the numerical value and the partial derivatives of the basis functions $N_{i,p_n}(\xi)$ and $N_{j,p_m}(\eta)$ are known in the unit square ($0 \le \xi, \eta \le 1$), it becomes necessary to transform the biharmonic operators involved in Eq. (11.51), from the aforementioned local to the global Cartesian coordinate system (the subscript *i* or *j* has been dropped):

$$\nabla^4 N(x, y) = \frac{\partial^4 N(x, y)}{\partial x^4} + 2\frac{\partial^4 N(x, y)}{\partial x^2\, \partial y^2} + \frac{\partial^4 N(x, y)}{\partial y^4}, \tag{11.54}$$

This is accomplished first by calculating the Jacobian matrix $\mathbf{J} = \begin{bmatrix} \partial x/\partial\xi & \partial y/\partial\xi \\ \partial x/\partial\eta & \partial y/\partial\eta \end{bmatrix}$ at each collocation point P($x(\xi, \eta)$, $y(\xi, \eta)$). Then, by numerical inversion, the four terms of its inverse, $\mathbf{J}^{-1}$, are found:

$$\mathbf{J}^{-1} = \begin{bmatrix} \partial\xi/\partial x & \partial\eta/\partial x \\ \partial\xi/\partial y & \partial\eta/\partial y \end{bmatrix} \tag{11.55}$$

Second, every term that appears in (11.54) is processed by successively applying the chain rule. The procedure has been described in detail elsewhere, where the final result for the term $\partial^4 N/\partial x^4$ (first in Eq. 11.54) can be found in Provatidis [66], p. 369).

Moreover, the derivative $\partial^4 N/\partial y^4$ is given by:

$$\begin{aligned} \frac{\partial^4 N}{\partial y^4} &= \frac{\partial^4 N}{\partial \xi^4} \cdot \left(\frac{\partial \xi}{\partial y}\right)^4 + \frac{\partial^4 N}{\partial \eta^4} \cdot \left(\frac{\partial \eta}{\partial y}\right)^4 \\ &\quad + \frac{\partial^4 N}{\partial \xi^3 \partial \eta} \cdot \left[4\left(\frac{\partial \xi}{\partial y}\right)^3 \left(\frac{\partial \eta}{\partial y}\right)\right] + \frac{\partial^4 N}{\partial \xi^2 \partial \eta^2} \cdot \left[6\left(\frac{\partial \xi}{\partial y}\right)^2 \left(\frac{\partial \eta}{\partial y}\right)^2\right] \end{aligned}$$

$$
\begin{aligned}
&+\frac{\partial^4 N}{\partial \xi\, \partial \eta^3}\cdot\bigg[4\left(\frac{\partial \xi}{\partial y}\right)\left(\frac{\partial \eta}{\partial y}\right)^3\bigg]+\frac{\partial^3 N}{\partial \xi^3}\cdot\bigg[6\left(\frac{\partial \xi}{\partial y}\right)^2\left(\frac{\partial^2 \xi}{\partial y^2}\right)\bigg]\\
&+\frac{\partial^3 N}{\partial \eta^3}\cdot\bigg[6\left(\frac{\partial \eta}{\partial y}\right)^2\left(\frac{\partial^2 \eta}{\partial y^2}\right)\bigg]+\frac{\partial^3 N}{\partial \xi^2 \partial \eta}\cdot\bigg[12\left(\frac{\partial \xi}{\partial y}\right)\left(\frac{\partial \eta}{\partial y}\right)\left(\frac{\partial^2 \xi}{\partial y^2}\right)\\
&+6\left(\frac{\partial \xi}{\partial y}\right)^2\left(\frac{\partial^2 \eta}{\partial y^2}\right)\bigg]+\frac{\partial^3 N}{\partial \xi\, \partial \eta^2}\cdot\bigg[12\left(\frac{\partial \xi}{\partial y}\right)\left(\frac{\partial \eta}{\partial y}\right)\left(\frac{\partial^2 \eta}{\partial y^2}\right)+6\left(\frac{\partial \eta}{\partial y}\right)^2\left(\frac{\partial^2 \xi}{\partial y^2}\right)\bigg]\\
&+\frac{\partial^2 N}{\partial \xi^2}\cdot\bigg[4\left(\frac{\partial \xi}{\partial y}\right)\left(\frac{\partial^3 \xi}{\partial y^3}\right)+3\left(\frac{\partial^2 \xi}{\partial y^2}\right)^2\bigg]+\frac{\partial^2 N}{\partial \eta^2}\cdot\bigg[4\left(\frac{\partial \eta}{\partial y}\right)\left(\frac{\partial^3 \eta}{\partial y^3}\right)\\
&+3\left(\frac{\partial^2 \eta}{\partial y^2}\right)^2\bigg]+\frac{\partial^2 N}{\partial \xi\, \partial \eta}\cdot\bigg[4\left(\frac{\partial \xi}{\partial y}\right)\left(\frac{\partial^3 \eta}{\partial y^3}\right)+4\left(\frac{\partial \eta}{\partial y}\right)\left(\frac{\partial^3 \xi}{\partial y^3}\right)\\
&+6\left(\frac{\partial^2 \xi}{\partial y^2}\right)\left(\frac{\partial^2 \eta}{\partial y^2}\right)\bigg]+\frac{\partial N}{\partial \xi}\cdot\left(\frac{\partial^4 \xi}{\partial y^4}\right)+\frac{\partial N}{\partial \eta}\cdot\left(\frac{\partial^4 \eta}{\partial y^4}\right)
\end{aligned}
\tag{11.56}
$$

One may observe in Eq. (11.56) that in places where a Cartesian coordinate exists, this is only the variable y.

Finally, the mixed derivative $\partial^4 N/\partial x^2 \partial y^2$ is the most lengthy and is given by:

$$
\begin{aligned}
\frac{\partial^4 N}{\partial x^2 \partial y^2}=&\frac{\partial^4 N}{\partial \xi^4}\cdot\bigg[\left(\frac{\partial \xi}{\partial x}\right)^2\left(\frac{\partial \xi}{\partial y}\right)^2\bigg]+\frac{\partial^4 N}{\partial \eta^4}\cdot\bigg[\left(\frac{\partial \eta}{\partial x}\right)^2\left(\frac{\partial \eta}{\partial y}\right)^2\bigg]\\
&+\frac{\partial^4 N}{\partial \xi^3 \partial \eta}\cdot 2\bigg[\frac{\partial \xi}{\partial x}\left(\frac{\partial \xi}{\partial y}\right)^2\frac{\partial \eta}{\partial x}+\frac{\partial \xi}{\partial y}\left(\frac{\partial \xi}{\partial x}\right)^2\frac{\partial \eta}{\partial y}\bigg]+\frac{\partial^4 N}{\partial \xi\, \partial \eta^3}\\
&\cdot 2\bigg[\frac{\partial \xi}{\partial x}\left(\frac{\partial \eta}{\partial y}\right)^2\frac{\partial \eta}{\partial x}+\frac{\partial \xi}{\partial y}\left(\frac{\partial \eta}{\partial x}\right)^2\frac{\partial \eta}{\partial y}\bigg]+\frac{\partial^4 N}{\partial \xi^2 \partial \eta^2}\cdot\bigg[\left(\frac{\partial \xi}{\partial x}\right)^2\left(\frac{\partial \eta}{\partial y}\right)^2\\
&+\left(\frac{\partial \xi}{\partial y}\right)^2\left(\frac{\partial \eta}{\partial x}\right)^2+4\frac{\partial \xi}{\partial x}\frac{\partial \xi}{\partial y}\frac{\partial \eta}{\partial x}\frac{\partial \eta}{\partial y}\bigg]+\frac{\partial^3 N}{\partial \xi^3}\cdot\bigg[\frac{\partial^2 \xi}{\partial x^2}\left(\frac{\partial \xi}{\partial y}\right)^2+\frac{\partial^2 \xi}{\partial y^2}\left(\frac{\partial \xi}{\partial x}\right)^2\\
&+4\frac{\partial^2 \xi}{\partial \xi \partial y}\left(\frac{\partial \xi}{\partial x}\right)^2\frac{\partial \eta}{\partial y}+4\frac{\partial^2 \xi}{\partial \eta \partial y}\frac{\partial \xi}{\partial x}\frac{\partial \eta}{\partial y}\frac{\partial \eta}{\partial x}\bigg]+\frac{\partial^3 N}{\partial \xi^2 \partial \eta}\cdot\bigg[\frac{\partial^2 \eta}{\partial x^2}\left(\frac{\partial \xi}{\partial y}\right)^2\\
&+\frac{\partial^2 \eta}{\partial y^2}\left(\frac{\partial \xi}{\partial x}\right)^2+4\frac{\partial^2 \eta}{\partial \xi \partial y}\left(\frac{\partial \xi}{\partial x}\right)^2\frac{\partial \xi}{\partial y}+4\frac{\partial^2 \xi}{\partial \xi \partial y}\left(\frac{\partial \xi}{\partial x}\right)^2\frac{\partial \eta}{\partial y}\\
&+4\frac{\partial^2 \xi}{\partial \eta \partial y}\left(\frac{\partial \eta}{\partial x}\right)^2\frac{\partial \xi}{\partial y}+2\frac{\partial^2 \xi}{\partial x^2}\frac{\partial \xi}{\partial y}\frac{\partial \eta}{\partial y}+2\frac{\partial^2 \xi}{\partial y^2}\frac{\partial \xi}{\partial x}\frac{\partial \eta}{\partial x}+4\frac{\partial^2 \xi}{\partial \xi \partial y}\frac{\partial \xi}{\partial x}\frac{\partial \xi}{\partial y}\frac{\partial \eta}{\partial x}\\
&+4\frac{\partial^2 \eta}{\partial \eta \partial y}\frac{\partial \xi}{\partial x}\frac{\partial \xi}{\partial y}\frac{\partial \eta}{\partial x}+4\frac{\partial^2 \xi}{\partial \eta \partial y}\frac{\partial \xi}{\partial x}\frac{\partial \eta}{\partial x}\frac{\partial \eta}{\partial y}\bigg]+\frac{\partial^3 N}{\partial \eta^3}\cdot\bigg[\frac{\partial^2 \eta}{\partial x^2}\left(\frac{\partial \eta}{\partial y}\right)^2
\end{aligned}
$$

$$+\frac{\partial^2\eta}{\partial y^2}\left(\frac{\partial\eta}{\partial x}\right)^2+4\frac{\partial^2\eta}{\partial\eta\partial y}\left(\frac{\partial\eta}{\partial x}\right)^2\frac{\partial\eta}{\partial y}+4\frac{\partial^2\eta}{\partial\xi\partial y}\frac{\partial\xi}{\partial x}\frac{\partial\eta}{\partial x}\frac{\partial\eta}{\partial y}\Bigg]+\frac{\partial^3 N}{\partial\xi\,\partial\eta^2}$$

$$\cdot\Bigg[\frac{\partial^2\xi}{\partial x^2}\left(\frac{\partial\eta}{\partial y}\right)^2+\frac{\partial^2\xi}{\partial y^2}\left(\frac{\partial\eta}{\partial x}\right)^2+4\frac{\partial^2\eta}{\partial\xi\partial y}\left(\frac{\partial\xi}{\partial x}\right)^2\frac{\partial\eta}{\partial y}+4\frac{\partial^2\eta}{\partial\eta\partial y}\left(\frac{\partial\eta}{\partial x}\right)^2\frac{\partial\xi}{\partial y}$$

$$+4\frac{\partial^2\xi}{\partial\eta\partial y}\left(\frac{\partial\eta}{\partial x}\right)^2\frac{\partial\eta}{\partial y}+2\frac{\partial^2\eta}{\partial x^2}\frac{\partial\xi}{\partial y}\frac{\partial\eta}{\partial y}+2\frac{\partial^2\eta}{\partial y^2}\frac{\partial\xi}{\partial x}\frac{\partial\eta}{\partial x}+4\frac{\partial^2\eta}{\partial\xi\partial y}\frac{\partial\xi}{\partial x}\frac{\partial\xi}{\partial y}\frac{\partial\eta}{\partial x}$$

$$+4\frac{\partial^2\xi}{\partial\xi\partial y}\frac{\partial\xi}{\partial x}\frac{\partial\eta}{\partial x}\frac{\partial\eta}{\partial y}+4\frac{\partial^2\eta}{\partial\eta\partial y}\frac{\partial\xi}{\partial x}\frac{\partial\eta}{\partial x}\frac{\partial\eta}{\partial y}\Bigg]+\frac{\partial^2 N}{\partial\xi^2}\cdot\Bigg[2\left(\frac{\partial^2\xi}{\partial\xi\partial y}\right)^2\left(\frac{\partial\xi}{\partial x}\right)^2$$

$$+2\left(\frac{\partial^2\xi}{\partial\eta\partial y}\right)^2\left(\frac{\partial\eta}{\partial x}\right)^2+\frac{\partial^2\xi}{\partial x^2}\frac{\partial^2\xi}{\partial y^2}$$

$$+2\frac{\partial^3\xi}{\partial\xi\partial y^2}\left(\frac{\partial\xi}{\partial x}\right)^2+2\frac{\partial^3\xi}{\partial\xi^2\partial y}\left(\frac{\partial\xi}{\partial x}\right)^2\frac{\partial\xi}{\partial y}+2\frac{\partial^3\xi}{\partial\eta^2\partial y}\left(\frac{\partial\eta}{\partial x}\right)^2\frac{\partial\xi}{\partial y}$$

$$+2\frac{\partial^2\xi}{\partial x^2}\frac{\partial^2\xi}{\partial\xi\partial y}\frac{\partial\xi}{\partial y}+2\frac{\partial^2\eta}{\partial x^2}\frac{\partial^2\xi}{\partial\eta\partial y}\frac{\partial\xi}{\partial y}+2\frac{\partial^3\xi}{\partial\eta\partial y^2}\frac{\partial\xi}{\partial x}\frac{\partial\eta}{\partial x}$$

$$+4\frac{\partial^2\xi}{\partial\xi\partial y}\frac{\partial^2\xi}{\partial\eta\partial y}\frac{\partial\xi}{\partial x}\frac{\partial\eta}{\partial x}+4\frac{\partial^3\xi}{\partial\xi\partial\eta\partial y}\frac{\partial\xi}{\partial x}\frac{\partial\xi}{\partial y}\frac{\partial\eta}{\partial x}\Bigg]$$

$$+\frac{\partial^2 N}{\partial\eta^2}\cdot\Bigg[2\left(\frac{\partial^2\eta}{\partial\xi\partial y}\right)^2\left(\frac{\partial\xi}{\partial x}\right)^2+2\left(\frac{\partial^2\eta}{\partial\eta\partial y}\right)^2\left(\frac{\partial\eta}{\partial x}\right)^2+\frac{\partial^2\eta}{\partial x^2}\frac{\partial^2\eta}{\partial y^2}$$

$$+2\frac{\partial^3\eta}{\partial\eta\partial y^2}\left(\frac{\partial\eta}{\partial x}\right)^2+2\frac{\partial^3\eta}{\partial\xi^2\partial y}\left(\frac{\partial\xi}{\partial x}\right)^2\frac{\partial\eta}{\partial y}+2\frac{\partial^3\eta}{\partial\eta^2\partial y}\left(\frac{\partial\eta}{\partial x}\right)^2\frac{\partial\eta}{\partial y}$$

$$+2\frac{\partial^2\xi}{\partial x^2}\frac{\partial^2\eta}{\partial\xi\partial y}\frac{\partial\eta}{\partial y}+2\frac{\partial^2\eta}{\partial x^2}\frac{\partial^2\eta}{\partial\eta\partial y}\frac{\partial\eta}{\partial y}+2\frac{\partial^3\eta}{\partial\xi\partial y^2}\frac{\partial\xi}{\partial x}\frac{\partial\eta}{\partial x}+4\frac{\partial^2\eta}{\partial\xi\partial y}\frac{\partial^2\eta}{\partial\eta\partial y}\frac{\partial\xi}{\partial x}\frac{\partial\eta}{\partial x}$$

$$+4\frac{\partial^3\eta}{\partial\xi\partial\eta\partial y}\frac{\partial\xi}{\partial x}\frac{\partial\eta}{\partial x}\frac{\partial\eta}{\partial y}\Bigg]+\frac{\partial^2 N}{\partial\xi\,\partial\eta}\cdot\Bigg[\frac{\partial^2\xi}{\partial x^2}\frac{\partial^2\eta}{\partial y^2}+\frac{\partial^2\xi}{\partial y^2}\frac{\partial^2\eta}{\partial x^2}+2\frac{\partial^3\eta}{\partial\xi\partial y^2}\left(\frac{\partial\xi}{\partial x}\right)^2$$

$$+2\frac{\partial^3\xi}{\partial\eta\partial y^2}\left(\frac{\partial\eta}{\partial x}\right)^2+4\frac{\partial^2\xi}{\partial\xi\partial y}\frac{\partial^2\eta}{\partial\xi\partial y}\left(\frac{\partial\xi}{\partial x}\right)^2+4\frac{\partial^2\xi}{\partial\eta\partial y}\frac{\partial^2\eta}{\partial\eta\partial y}\left(\frac{\partial\eta}{\partial x}\right)^2$$

$$+2\frac{\partial^3\eta}{\partial\xi^2\partial y}\left(\frac{\partial\xi}{\partial x}\right)^2\frac{\partial\xi}{\partial y}+2\frac{\partial^3\xi}{\partial\xi^2\partial y}\left(\frac{\partial\xi}{\partial x}\right)^2\frac{\partial\eta}{\partial y}+2\frac{\partial^3\eta}{\partial\eta^2\partial y}\left(\frac{\partial\eta}{\partial x}\right)^2\frac{\partial\xi}{\partial y}$$

$$+2\frac{\partial^3\xi}{\partial\eta^2\partial y}\left(\frac{\partial\eta}{\partial x}\right)^2\frac{\partial\eta}{\partial y}+2\frac{\partial^2\xi}{\partial x^2}\frac{\partial^2\eta}{\partial\xi\partial y}\frac{\partial\xi}{\partial y}+2\frac{\partial^2\eta}{\partial x^2}\frac{\partial^2\eta}{\partial\eta\partial y}\frac{\partial\xi}{\partial y}+2\frac{\partial^2\xi}{\partial x^2}\frac{\partial^2\xi}{\partial\xi\partial y}\frac{\partial\eta}{\partial y}$$

$$+2\frac{\partial^2\eta}{\partial x^2}\frac{\partial^2\xi}{\partial\eta\partial y}\frac{\partial\eta}{\partial y}+2\frac{\partial^3\xi}{\partial\xi\partial y^2}\frac{\partial\xi}{\partial x}\frac{\partial\eta}{\partial x}+2\frac{\partial^3\eta}{\partial\eta\partial y^2}\frac{\partial\xi}{\partial x}\frac{\partial\eta}{\partial x}+4\frac{\partial^2\xi}{\partial\xi\partial y}\frac{\partial^2\eta}{\partial\eta\partial y}\frac{\partial\xi}{\partial x}\frac{\partial\eta}{\partial x}$$

$$+4\frac{\partial^2\xi}{\partial\eta\partial y}\frac{\partial^2\eta}{\partial\xi\partial y}\frac{\partial\xi}{\partial x}\frac{\partial\eta}{\partial x}+4\frac{\partial^3\eta}{\partial\xi\partial\eta\partial y}\frac{\partial\xi}{\partial x}\frac{\partial\xi}{\partial y}\frac{\partial\eta}{\partial x}+4\frac{\partial^3\xi}{\partial\xi\partial\eta\partial y}\frac{\partial\xi}{\partial x}\frac{\partial\eta}{\partial x}\frac{\partial\eta}{\partial y}\Bigg]$$

$$+\frac{\partial N}{\partial \xi}\cdot\left(\frac{\partial^4 \xi}{\partial x^2 \partial y^2}\right)+\frac{\partial N}{\partial \eta}\cdot\left(\frac{\partial^4 \eta}{\partial x^2 \partial y^2}\right) \tag{11.57}$$

From the educational point of view, it is instructive to start working with the beam bending analysis, which is very close to the plate analysis (both are governed by fourth-order ordinary or partial differential equation, respectively).

11.7.6.2 Rectangular Plate

In a rectangular plate of dimensions $a \times b$, the relationship between the Cartesian and the natural coordinates ($0 \leq \xi, \eta \leq 1$) is:

$$x = a\xi \quad \text{and} \quad y = b\eta \tag{11.58}$$

Therefore, in the above case the biharmonic operator (Eq. 11.54) becomes:

$$\nabla^4 N(x,y) = \frac{1}{a^4}\frac{\partial^4 N(\xi,\eta)}{\partial \xi^4} + \frac{2}{a^2 b^2}\frac{\partial^4 N(\xi,\eta)}{\partial \xi^2 \partial \eta^2} + \frac{1}{b^4}\frac{\partial^4 N(\xi,\eta)}{\partial \eta^4}. \tag{11.59}$$

Therefore, the treatment of rectangular plates is rather a trivial task, in the sense that the tedious Eqs. (11.56) and (11.57) are almost useless.

Therefore, for the sake of brevity below we study only the case of a circular plate.

Example 11.10 (Uniformly Loaded Circular Plate) A circular plate of radius a is subjected to a uniform normal load p_0. The exact solution depends on the type of boundary conditions, Clamped (CL) or simply supported (SS) and is given by (e.g., [70], pp. 33–34):

For a circular plate with clamped edge:

$$w_{\max} = \frac{p_0 a^4}{64D} \tag{11.60a}$$

For a circular plate with simply supported edge:

$$w_{\max} = \frac{p_0 a^4}{64D}\frac{5+\nu}{1+\nu} \tag{11.60b}$$

As usual, we parameterize the circle considering that its circumference corresponds to the boundary of a unit square *ABCD*. The corner points *A*, *B*, *C*, and D obviously belong to the boundary of the circle and split it into four equal parts. The position of the point *A* can be arbitrary chosen, but then the position of the rest three is determined.

Using $n_s = 4$ segments per side of the abovementioned square (i.e., five breakpoints along each side, including the ends), as well as the polynomial degree, say $p = 7$, and assuming C^3-continuity (see [65]), we have a multiplicity of four knots per breakpoint. In order to construct reasonable control points, we create uniform test points (so many as the number of the unknown control points) as we have explained in previous chapters. The obtained control points are shown in Fig. 11.10 where they are represented by a cross.

In the case of a clamped plate, the degrees of freedom associated to the two outer layers have to be deleted. The case of a simply supported plate is easier, since only the DOF associated to the outer layer has to be deleted.

Numerical experience has shown that the minimum suggested polynomial degree is $p_{\min} = 5$. For $n_s = 4$ breakpoint spans, the influence of the polynomial degree in the maximum calculated deflection is shown in Table 11.6.

For further details and applications on the extraction of eigenvalues, the reader is suggested to consult [66].

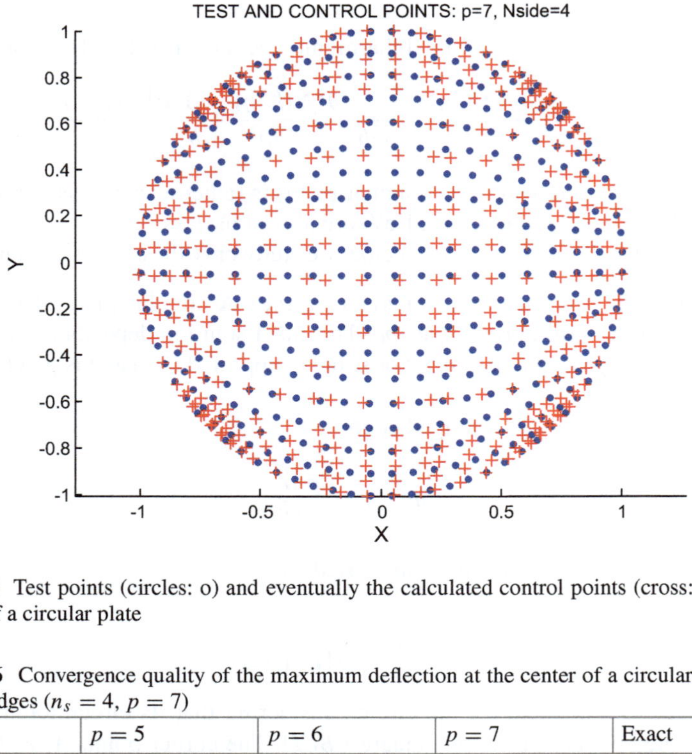

Fig. 11.10 Test points (circles: o) and eventually the calculated control points (cross: +) for the analysis of a circular plate

Table 11.6 Convergence quality of the maximum deflection at the center of a circular plate with clamped edges ($n_s = 4$, $p = 7$)

	$p = 5$	$p = 6$	$p = 7$	Exact
Deflection →	8.52324e−07	8.52569e−07	8.52605e−07	8.5312e−07
Error (in %) →	−0.0933	−0.0646	−0.0604	

11.7.7 3D Problems

The application of the global collocation method to 3D problems is a straightforward extension of the abovementioned 2D ones. Therefore, in general, the approximate solution is written as:

$$U(x, y, z; t) = \sum_{i=1}^{n}\sum_{j=1}^{m}\sum_{k=1}^{p}\left[N_i(x) \cdot N_j(y) \cdot N_k(z)\right] \cdot a_{ijk}(t),$$
$$\text{with } i = 1, \ldots, n; \quad j = 1, \ldots, m; \quad k = 1, \ldots, p. \tag{11.61}$$

The interested reader may easily extend his/her 2D codes to 3D continuum.

11.8 Approximation Using B-Splines

As we have repeatedly mentioned, in addition to the polynomials, today there are other basis functions such as B-splines, cf which a particular case is Bézier (Bernstein) polynomials and NURBS. The use of B-splines in the solution of ordinary differential equations was promoted by the PhD thesis of de Boor [16] and established the method we call today as *B-spline collocation* (particular case of the spline collocation).

The first methods of spline collocation used to implement piecewise cubic polynomials and the relevant method were called "cubic spline collocation." Considering that in the decomposition of the problem domain [0, L] into N subintervals, the B-spline expansion (with C^2-continuity) requires 4 coefficients for the complete cubic polynomial, and another $(N-1)$ coefficients for the truncated monomials (see Chap. 6), thus in the cubic B-spline approximation $N-1+4=N+3$ coefficients are needed in total.

Example. Therefore, if the interval [0, 1] is divided into four segments ($N=4$), then we need seven coefficients. If we subtract the two boundary conditions, then five unknown coefficients remain to be determined. This means that we have to satisfy the PDE at five positions, whereas the number of intervals is only four. In other words, in contrast to Lagrange polynomials the selection of the collocation points is not obvious.[1] If we demand to treat all subintervals equally, since the location of one point per interval is not adequate, then the minimum number of two points per interval gives $2 \times 4 = 8$ collocation points. Then we have more equations (eight) than the number of unknowns (five), and the result would be to apply a *Least-Squares Scheme* (as shown in subsection 11.4).

[1] On the contrary, in the case of Lagrange polynomials there is no skepticism, since then we have to put the collocation points at the location of the roots of Legendre polynomials (which correspond to 5-point Gauss quadrature).

Multiplicity. In contrast, if we take the multiplicity $\lambda = 2$, then at the existing $(N - 1)$ breakpoints the corresponding variables are $2(N - 1)$, and another 4 (2 per each end of the interval [0, L]), totally $2(N + 1)$ coefficients. If, however, we subtract the two boundary conditions, then only $2N$ unknowns remain. Therefore, taking two collocation points (to satisfy the PDE) in each of the N subintervals, we will have so many equations as exactly the number of the unknown coefficients. This is the solution adopted by de Boor and Swartz [17], in which the collocation points were taken at the position of the Gauss points that correspond to each subinterval.

Isogeometric collocation. In principle, to the best of our understanding this is not an extremely original method, since it extends the former B-spline collocation. However, the way of simultaneous consideration of the PDE and the Neumann BCs is innovative.

In full accordance with the B-spline collocation methods, the isogeometric collocation takes advantage of the particular special places such as the Demko and Greville points. For more information, the interested reader may refer to the classical book of de Boor [18] and also to the recent literature (e.g., Auricchio 2010; [61, 66]).

11.9 Unsolved Exercise

1. Solve the Example 11.2 using a global cubic B-spline polynomial, and multiplicity $\lambda = 2$. Which is the difference of this interpolation with the classical Hermite interpolation?

Hint

Draw the basis functions of the cubic B-spline in a common diagram with cubic Hermite polynomials.

11.10 Discussion

All older CAD-based macroelements, i.e., Coons, Gordon (transfinite), Bernstein–Bézier and B-splines can be treated equally well using the so-called CAD-based collocation method. While the idea appeared for the first time rather in 1993 by others [71], the practical need of this method (at least in the framework of Gordon–Coons macroelements) was understood by the author in 2004 and systematically was published progressively from the year 2005 in Provatidis [47–50, p 6704], [51–55] till today. The motivation of preserving the global CAD-based character of the shape functions and the substitution of the Galerkin–Ritz by a global collocation method came as a reaction to repeated concerns of the reviewers (in many papers submitted between 1999 and 2004) that the Galerkin–Ritz procedure in conjunction with Coons–Gordon interpolation is a time-consuming approach. It is self-evident that all the aforementioned methods, as well as the B-spline collocation (introduced by de Boor [16] and then implemented in 2D and 3D problems by Johnson [35, 36], are precursors of the so-called NURBS-based "isogeometric collocation" method ([4], etc.), which although initially "seemed to offer little theoretical advantage" (see [4], pp. 2104–2105), it continues to survive successfully for almost a decade till today [46]. Of course, the critical point that has to be definitely answered is whether the CAD-based collocation methods can be successfully applied to a decomposed domain.

11.11 Recapitulation

It was shown that the fulfillment of the partial differential equation at so many points as the number of unknown coefficients in the CAD-based series expansion of the approximate numerical solution (minus the number of boundary conditions) produces a linear equations system. When the aforementioned system is solved, the coefficients are found and the approximate solution is known throughout the domain. The quality of the numerical solution highly depends on the choice of the collocation points. Much of the current work is toward the proper way of implementing boundary conditions of Neumann type, whereas the applicability of the global collocation method to decomposed domains keeps the researcher still busy.

References

1. Acrivos A (1958) Combined laminar free-and forced-convection heat transfer in external flows. AIChE 4:285–289
2. Anitescu C, Jia Y, Zhang Y, Rabczuk T (2015) An isogeometric collocation method using superconvergent points. Comput Methods Appl Mech Eng 284:1073–1097
3. Ascher UM, Mattheij RMM, Russell RD (1995) Numerical solutions of boundary value problems for ordinary differential equations, 2nd edn. SIAM, Philadelphia, PA
4. Auricchio F, Beirão da Veiga L, Hughes TGR, Reali A, Sangalli G (2010) Isogeometric collocation methods. Math Models Methods Appl Sci 20(11):2075–2107
5. Auricchio F, Beirão da Veiga L, Hughes TJR, Reali A, Sangalli G (2012) Isogeometric collocation for elastostatics and explicit dynamics. Comput Methods Appl Mech Eng 249–252:2–14
6. Bialecki B, Fairweather G (2001) Orthogonal spline collocation methods for partial differential equations. J Comput Appl Math 128(1–2):55–82
7. Botha JF, Pinder GF (1983) Fundamental concepts in the numerical solution of differential equations. Wiley, New York
8. Carey GF, Finlayson BA (1975) Orthogonal collocation on finite elements. Chem Eng Sci 30:587–596
9. Carey GF, Oden JT (1983) Finite elements: a second course, vol II. Prentice-Hall, Englewoods Cliffs, New Jersey
10. Cavendish JC, Gordon WJ, Hall CA (1976) Ritz-Galerkin approximations in blending function spaces. Numer Math 26:155–178
11. Cavendish JC (1972) Collocation methods for elliptic and parabolic boundary value problems. Ph.D. Dissertation, University of Pittsburgh, USA

12. Cavendish JC, Hall CA (1974) L∞-convergence of collocation and Galerkin approximations to linear two-point parabolic problems. Aequationes Math 11(2–3):230–249
13. Chen JT, Chang MH, Chen KH, Lin SR (2002) The boundary collocation method with meshless concept for acoustic eigenanalysis of two-dimensional cavities using radial basis function. J Sound Vib 257(4):667–711
14. Chen JT, Chang MH, Chen KH, Chen IL (2002) Boundary collocation method for acoustic eigenanalysis of three-dimensional cavities using radial basis function. Comput Mech 29:392–408
15. Collatz L (1960) The Numerical Treatment of Differential Equations. Springer, Berlin
16. De Boor C (1966) The method of projections as applied to the numerical solution of the two point boundary value problems using cubic splines. Dissertation, Univ. Michigan
17. De Boor C, Swartz B (1973) Collocation at Gaussian points. SIAM J Numer Anal 10:582–606
18. De Boor C (2001) A practical guide to splines, revised edition. Springer, New York (former edition: Applied mathematical sciences, vol 27. Springer, New York, 1978)
19. De Lorenzis L, Evans JA, Hughes TJR, Reali A (2015) Isogeometric collocation: Neumann boundary conditions and contact. Comput Methods Appl Mech Eng 284:21–54
20. Diaz J (1977) A collocation-Galerkin method for the two-point boundary-value problem using continuous piecewise polynomial spaces. SIAM J Numer Anal 14(5):844–858
21. Elansari M, Quazar D, Cheng AH-D (2001) Boundary solution of Poisson's equation using radial basis function collocated on Gaussian quadrature nodes. Commun Numer Methods Eng 17:455–464
22. Evans JA, Hiemstra RR, Hughes TJR, Reali A (2018) Explicit higher-order accurate isogeometric collocation methods for structural dynamics. Comput Methods Appl Mech Eng 338:208–240
23. Fairweather G, Meade D (1989) Survey of spline collocation methods for the numerical solution of differential equations. In: Diaz JC (ed) Mathematics for large scale computing. Lecture notes in pure applied mathematics, vol 120. Marcel Dekker, New York, pp 297–341
24. Farin G (1990) Curves and surfaces for computer aided geometric design: a practical guide. Academic Press, Boston
25. Filippatos A (2010) Eigenvalue extraction for acoustic cavities and elastic structures using the global collocation method, MSc—Diploma Work (supervised by Prof. C. G. Provatidis), National Technical University of Athens, Athens, School of Mechanical Engineering
26. Finlayson BA, Scriven LE (1966) The method of weighted residuals—a review. Appl Mech Rev 19:735–748
27. Finlayson BA (1972, 2014) The method of weighted residuals and variational principles. Academic Press, New York (SIAM, Philadelphia, 2014)
28. Frind EO, Pinder GF (1979) A collocation finite element method for potential problems in irregular domains. Int J Numer Methods Eng 14:681–701
29. Gordon WJ (1971) Blending functions methods of bivariate multivariate interpolation and approximation. SIAM J Numer Anal 8:158–177
30. Gordon WJ, Hall CA (1973) Transfinite element methods: blending-function interpolation over arbitrary curved element domains. Numer Math 21:109–129
31. Hayes LJ (1980) An alternative-direction collocation method for finite element approximations on rectangles. Comput Math Appl 6:45–50
32. Houstis E (1978) A collocation method for systems of nonlinear ordinary differential equations. J Math Anal Appl 62:24–37
33. Isidorou SK, Provatidis CG (2009) Comparison of advanced collocation methods for the solution of ordinary differential equations. In: Proceedings 3rd international conference on experiments/process/system modeling/simulation and optimization (3rd IC-EpsMsO), Athens, 8–11 July 2009
34. Jator S, Sinkala Z (2007) A high order B-spline collocation method for linear boundary value problems. Appl Math Comput 191:100–116
35. Johnson RW (2003) Progress on the development of B-spline collocation for the solution of differential model equations: a novel algorithm for adaptive knot insertion. INEEL/CON-02-01471, Preprint, Idaho National Engineering and Environmental Laboratory. Available online from: https://pdfs.semanticscholar.org/fe20/02f2bd49798ad2a73ebc022bde7e6dcc14ae.pdf

36. Johnson RW (2005) Higher order B-spline collocation at the Greville Absissae. Appl Numer Math 52(1):63–75
37. Kanarachos AE, Deriziotis DG (1989) On the solution of Laplace and wave propagation problems using "C-elements". Finite Elem Anal Des 5:97–109
38. Kanarachos A, Provatidis C, Deriziotis D, Foteas N (1999), A new approach of the FEM analysis of two-dimensional elastic structures using global (Coons's) interpolation functions. In: Wunderlich J (ed) CD proceedings 1st european conference on computational mechanics, München-Germany, August–September, 1999
39. Kołodziej JA, Zielinski AP (2009) Boundary collocation techniques and their application in engineering. WIT Press, Ashurst Lodge
40. Kwok WY, Moser RD, Jiménez J (2001) A critical evaluation of the resolution properties of B-spline and compact finite difference methods. J Comput Phys 174(2):510–551
41. Lanczos C (1938) Trigonometric interpolation of empirical and analytical functions. J Math Phys 17:123–199
42. Li ZC, Lu TT, Huang HT, Cheng AHD (2007) Trefftz, collocation, and other boundary methods—a comparison. Numer Methods Partial Differ Equ 23:93–144
43. Liu X, Liu GR, Tai K, Lam KY (2005) Radial point interpolation collocation method (RPICM) for the solution of nonlinear Poisson problems. Comput Mech 36:298–306
44. Liu GR, Wang JG (2002) A point interpolation meshless method based on radial basis functions. Int J Numer Meth Eng 54:1623–1648
45. Liszka TJ, Duarte CA, Tworzydlo WW (1996) Hp-meshless cloud method. Comput Methods Appl Mech Eng 139:263–288
46. Morganti S, Auricchio F, Callari C, De Lorenzis L, Evans JA, Hughes TJR, Reali A (2018) Mixed isogeometric collocation methods. In: 6th European conference on computational mechanics (ECCM6), pp 11–June 15 2018, Glasgow. Available online at: http://congress.cimne.com/eccm_ecfd2018/admin/files/fileabstract/a2210.pdf
47. Provatidis CG (2006) Coons patch macroelements in two-dimensional parabolic problems. Appl Math Model 30(4):319–351
48. Provatidis CG (2006) Free vibration analysis of two-dimensional structures using Coons patch macroelements. Finite Elem Anal Des 42(6):518–531
49. Provatidis CG (2006) Three-dimensional Coons macroelements: application to eigenvalue and scalar wave propagation problems. Int. J. Nume. Methods Eng 65(1):111–134
50. Provatidis CG (2006) Transient elastodynamic analysis of two-dimensional structures using Coons patch macroelements. Int J Solids Struct 43(22–23):6688–6706
51. Provatidis CG (2007) Performance of a Lagrange based global finite element collocation method for eigenvalue structural analysis. In: Proceedings 8th Hellenic society of theoretical and applied mechanics (HSTAM) congress, 12–14 July 2007, Patras, Greece
52. Provatidis CG (2008) Global collocation method for 2-D rectangular domains. J Mech Mater Struct 3(1):185–194
53. Provatidis CG (2008) Free vibration analysis of elastic rods using global collocation. Arch Appl Mech 78(4):241–250
54. Provatidis CG (2008) Time- and frequency-domain analysis using lumped mass global collocation. Arch Appl Mech 78(11):909–920
55. Provatidis CG (2008d) Analysis of one-dimensional wave propagation problems using global collocation. In: Proceedings 6th GRACM international congress on computational mechanics, 19–21 June, 2008, Thessaloniki, Greece
56. Provatidis CG (2009) Integration-free Coons macroelements for the solution of 2-D Poisson problems. Int J Numer Meth Eng 77:536–557
57. Provatidis CG, Ioannou KS (2010) Static analysis of two-dimensional elastic structures using global collocation. Arch Appl Mech 80(4):389–400
58. Provatidis CG, Isidorou SK (2011) B-splines collocation eigenvalue analysis of 1-D problems. In: Proceedings 7th GRACM international congress on computational mechanics, June 30–2 July 2011, Athens, Greece

59. Provatidis CG, Isidorou SK (2012) Solution of one-dimensional hyperbolic problems using cubic B-splines collocation. Int J Comput Sci Appl (IJCSA) 1(1):12–18. Journal Website: http://www.ij-csa.org/, Paper in PDF: http://www.ij-csa.org/Download.aspx?ID=2225
60. Provatidis CG (2014) Lumped mass acoustic and membrane eigenanalysis using the global collocation method. Cogent Eng 1(1):981366. (http://www.tandfonline.com/doi/abs/10.1080/23311916.2014.981366#.VIKnImc-eSo)
61. Provatidis CG (2014) B-splines collocation eigenanalysis of 2D acoustic problems. J Mech Mater Struct 9(3):259–285
62. Provatidis CG (2014) Bézier versus Lagrange polynomials-based finite element analysis of 2-D potential problems. Adv Eng Softw 73:22–34
63. Provatidis C (2015) Lumped mass collocation method for 2D elastodynamic analysis. Int J Adv Eng Technol Comput Sci 2(2):1–15. (http://www.irosss.org/ojs/index.php/IJAETCS/article/view/676/207)
64. Provatidis C, Fillipatos A (2015) Numerical determination of eigenfrequencies in two-dimensional acoustic cavities using a global collocation method. Rom J Acoust Vibr 12(2):111–115
65. Provatidis CG (2017) CAD-based collocation eigenanalysis of 2-D elastic structures. Comput Struct 182:55–73
66. Provatidis CG (2017) B-splines collocation for plate bending eigenanalysis. J Mech Mater Struct 12(4):353–371
67. Ronto NI (1971) Application of the method of collocation to solve boundary value problems. Ukrainskii Matematicheskii Zhurnal 23(3):415–421
68. Russell RD, Shampine LF (1972) A collocation method for boundary value problems. Numer Math 19:1–28
69. Szabó B, Babuška I (1991) Finite element analysis. Wiley, New York
70. Ugural AC (1981) Stresses in plates and shells. McGraw-Hill, New York
71. Van Blerk JJ, Botha JF (1993) Numerical solution of partial differential equations on curved domains by collocation. Numer Methods Partial Diff Equ 9:357–371
72. Zhang X, Song KZ, Lu MW, Liu X (2003) Meshless methods based on collocation with radial basis function. Comput Mech 30:396–409

Chapter 12
The Boundary Element Method Using CAD-Based Macroelements

Abstract This chapter deals with the numerical solution of three-dimensional boundary value problems using the Boundary Element Method (BEM) in conjunction with CAD-based macroelements. In more details, the boundary is discretized into a certain number of CAD-based patches (Coons, Gordon, Bernstein–Bézier, B-splines, NURBS, Barnhill, etc.), where both the geometry $\mathbf{x}(\xi, \eta)$ and the variable $U(x, y, z)$ are interpolated through the same CAGD formula. Each of the aforementioned patches is a single isoparametric (or isogeometric) macroelement to which a global approximation of the variable U is applied. The theory is accompanied with numerical results in elasticity problems and acoustics.

Keywords Integral equation · Betti-Maxwell · Direct BEM
Coons/transfinite patch · Brebbia–Nardini · Elasticity · Sound radiation

12.1 Introduction

In the standard Boundary Element Method (BEM) , the boundary Γ of the domain Ω (structure, solid, or field) is divided into a number of segments which are called "boundary elements." In two-dimensional problems, each boundary element is a line segment (straight or curved) on which the variable U (potential or displacement), its generalized derivative (flux, or traction) and the geometry (x, y) are all interpolated through piecewise constant, linear, or quadratic shape functions (isoparametric elements). For a detailed presentation, the reader is referred to Brebbia and Dominguez [3]. It is worthy to mention that BEM is not used only for analysis but it is offered for shape optimization as well ([26], among others). The substitution of the aforementioned (lower-order) elements by higher-order ones was proposed in the early 1990s using B-splines [5–7, 33, 35] and later using NURBS [13, 27–30, 32, 36]. For an overview of the isogeometric (IGA) BEM in 3D potential problems, the interested reader may consult the paper by Gong and Dong [10]. Isogeometric boundary element analysis using T-splines has been published by Ginnis et al. [9].

C. G. Provatidis, *Precursors of Isogeometric Analysis*, Solid Mechanics and Its Applications 256, https://doi.org/10.1007/978-3-030-03889-2_12

At the same time, the paper by Kanarachos and Deriziotis [11] was released from our CAD/CAE Group at NTUA, and the author of this book realized the idea to discretize the boundary of a three-dimensional structure into a small number of large CAD-based macroelements (e.g., Coons patches), provided the variation of the variable U is adequately *smooth*, and perform boundary element analysis. As a first step toward this direction, large Coons-patch macroelements were used to the BEM analysis of structures, and for the sake of convenience, primitive shapes (cubes, rectangular parallelepipeds, cylindrical segments, etc.) were chosen for numerical tests. A part of this research has been documented in the manuscript of two diploma theses under his supervision, dealing with stress analysis [34] and sound radiation problems [14]. Nevertheless, the pending defense of a Ph.D. thesis (officially started in January 1985) and relevant IP claims on the use of CAD (Coons)-based isoparametric elements was the reason that a very small part of the aforementioned studies was published and this happened ten years later [17, 18, 19, 23, 24].

Since the beginning of our involvement in computer methods, we were concerned to clarify which approach is better, the boundary element or the single CAD-based macroelement approach [16, 19, 22].

According to the abstract of a relevant ASME-Greek paper by Provatidis [17], "…Since the involved geometrical entities can be the absolutely necessary quantities that build-up the CAD-model, *the proposed method seems to 'marry' CAD with CAE*." A similar comment is cited at the end of the "Discussion" section of the aforementioned paper as well as of Provatidis [18], p. 409. The author still remembers the enthusiastic reaction of some young attendees in these conferences when he told them that the proposed BEM macroelement method does not restrict to the (old-fashioned) Coons interpolation only but it is easily extendable to any other CAD-based interpolation formulas such as Bézier, B-splines, etc.

As was mentioned above, the idea is to divide the arbitrary boundary of the domain (or the structural component) into a number of CAD-based surface patches. Each patch is described using the minimum allowable number of parameters. If the size of each patch is relatively small and the variation of the variable U is adequately smooth, we can assume that within each patch the quantity U can be globally approximated. This idea was supported by the fact that since the graph of a scalar variable $U(\xi, \eta)$ is a smooth *surface* it is reasonable to assume that it can be approximated like a CAD surface. Therefore, any known CAD (better term is GAGD)-based interpolation formula is applicable for this approximation within a patch. Consequently, Coons, Gordon, Bézier, B-splines, NURBS, Barnhill, and other types (e.g., Gregory patch) of boundary elements may be developed.

Although the concept is applicable to arbitrary shaped bodies, it is reasonable to start the test on primitive shapes such as rectangular or cylindrical cuboids, spheres. For example, in a cuboid it was initially found that sometimes it is sufficient to arrange the nodal points along the twelve edges only [14, 34]. But as we saw in Chap. 10, intensively curved bodies (such as a sphere) may need the discretization of the interior of the *boundary* patch. The latter is in favor of a Gordon transfinite interpolation using some internal nodes in each CAD patch, or a tensor product of any kind (Bézier, B-splines, NURBS).

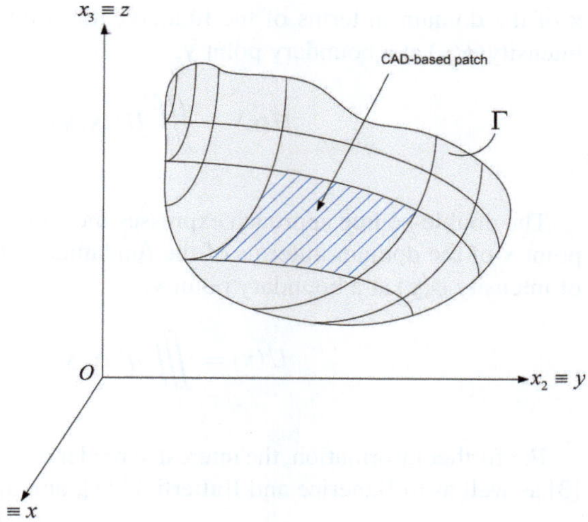

Fig. 12.1 Decomposition of the boundary Γ into CAD-based macroelements

The structure of this chapter is as follows. First, theoretical background knowledge will be given. Full expressions are presented for potential and elasticity problems. Second, the application examples include a cubical cantilever beam in bending, a part of a hollow cylinder of infinite length under internal pressure, and a cubical acoustic cavity in sound radiation. The decomposition of the boundary Γ of a structure into a certain number of CAD-based macroelements is schematically shown in Fig. 12.1 (the nodal points are not shown). It is worthy to mention that, although in the test cases of this chapter the number of patches is restricted to six, the applicability of the "BEM macroelement approach" is quite general.

12.2 Boundary Element Formulations

The Boundary Element Method (BEM) appears in two forms, of which the former is the "Indirect Boundary Element Method" (noted by IBEM) and the latter is the "*Direct Boundary Element Method*" (noted by DBEM).

12.2.1 Indirect Boundary Element Method (IBEM)

The IBEM is also met in two alternative expressions, of which the first is the "single-source" approach whereas the second is the "double-source" (dipoles) approach. The single-source approach expresses the value of the variable U at an arbitrary point

$\mathbf{x}$ of the domain in terms of the fundamental solution $U^*(\mathbf{x}, \mathbf{y})$ due to a source of intensity $\phi(\mathbf{y})$ at a boundary point $\mathbf{y}$.

$$U(\mathbf{x}) = \iiint_{\Gamma} U^*(\mathbf{x}, \mathbf{y}) \cdot \phi(\mathbf{y})\, \mathrm{d}\Gamma \tag{12.1}$$

The double-source approach expresses the value of the variable U at an arbitrary point $\mathbf{x}$ of the domain in terms of the fundamental traction $q^*(\mathbf{x}, \mathbf{y})$ due to a dipole of intensity $\hat{\phi}(\mathbf{y})$ at a boundary point $\mathbf{y}$.

$$U(\mathbf{x}) = \iiint_{\Gamma} q^*(\mathbf{x}, \mathbf{y}) \cdot \hat{\phi}(\mathbf{y})\, \mathrm{d}\Gamma \tag{12.2}$$

For further information, the interested reader may refer to Brebbia and Dominguez [3] as well as to Banerjee and Butterfield [1], among others.

12.2.2 Direct Boundary Element Method (DBEM)

The DBEM is obtained from the weighted Galerkin procedure in which Green's theorem is applied so many times as the order of the PDE. Therefore, Poisson's or a Laplace's equation, which is of the second order, requires two successive applications of Green's theorem. The first integration is performed on the usual PDE (governing equation: $L(U) = 0$), whereas the second one for the unit source or the fundamental solution $[L(u^*) + \Delta^i = 0]$. The aforementioned domain integrals are subtracted, and in this way, the domain terms are eventually eliminated (see details in Sect. 12.3.1). For plate bending which is described by a biharmonic functions (fourth order), we apply two successive Green's theorems for the usual $(\nabla^4 w - p/D = 0)$ and another two for its fundamental solution $(\nabla^4 w + \Delta^i = 0)$. In all cases, and particularly when no distributed sources exist in the interior, the domain terms are eliminated and only a boundary integral equation appears at the end.

In the particular case of problems in elasticity, the *Betti-Maxwell* theorem between the actual and a virtual (fundamental) situation may be applied as well. This was found for the first time by Somigliana [31], at the end of the nineteenth century. It is worthy to mention that, starting from the DBEM, and applying it to an interior and an exterior problem, the IBEM can be derived [2].

12.3 The Most Common Boundary Integral Equations in Engineering

As a special case of the above, particular engineering interest appear in the following equations:

12.3.1 Laplace Equation

Laplace equation describes potential problems of elliptic type such as steady-state heat transfer, velocity field of inviscid fluid, electrostatics field:

$$\nabla^2 U = 0 \tag{12.3}$$

One possible procedure is to start with the weighted Galerkin in which (12.3) is written as

$$\int_{\Omega} u_i^* \nabla^2 U \mathrm{d}\Omega = 0, \quad i = 1, \ldots, n \tag{12.4}$$

where u_i^* is the so-called fundamental solution that fulfills the fundamental equation:

$$\nabla^2 u_i^* = -\Delta^i, \tag{12.5}$$

Obviously, Eq. (12.5) is the same as the initial governing equation which is described by Eq. (12.3), but now a concentrated flux of unit integral (generalized "force") appears (in the form of a Dirac delta function).

Applying Green's identity, Eq. (12.4) becomes:

$$\int_{\Omega} u_i^* \nabla^2 U \, \mathrm{d}\Omega = \int_{\Gamma} u_i^* \nabla U \, \mathrm{d}\Gamma - \int_{\Omega} \nabla u_i^* \nabla U \, \mathrm{d}\Omega, \tag{12.6}$$

and due to Eq. (12.3)

$$0 = \int_{\Gamma} u_i^* \nabla U \, \mathrm{d}\Gamma - \int_{\Omega} \nabla u_i^* \nabla U \mathrm{d}\Omega \tag{12.7}$$

Interchanging the functions u_i^* and U in Eq. (12.6), the adjoint integral becomes

$$\int_{\Omega} U \, \nabla^2 u_i^* \, \mathrm{d}\Omega = \int_{\Gamma} U \, \nabla u_i^* \, \mathrm{d}\Gamma - \int_{\Omega} \nabla U \, \nabla u_i^* \mathrm{d}\Omega, \tag{12.8}$$

and due to (12.5) it further becomes

$$-c_i U_i = \int_{\Gamma} U \, \nabla u_i^* \, \mathrm{d}\Gamma - \int_{\Omega} \nabla U \, \nabla u_i^* \mathrm{d}\Omega, \tag{12.9}$$

where the constant c_i depends on the smoothness of the boundary at the point "i". So, for a smooth boundary point $c_i = 1/2$, in the interior of the domain $c_i = 1$, whereas in the exterior of the domain $c_i = 0$. If the point "i" is a boundary point at a corner where a solid angle α is formed, then $c_i = \alpha/(4\pi)$; this formula obviously covers the case of a smooth boundary where $\alpha = 2\pi$.

Subtracting Eqs. (12.7) and (12.9) in parts, the domain integrals are eliminated; thus, we receive

$$c_i U_i + \int_{\Gamma} (\nabla u_i^*) U \, \mathrm{d}\Gamma = \int_{\Gamma} u_i^* (\nabla U) \, \mathrm{d}\Gamma \tag{12.10}$$

In the three-dimensional problem, it can be shown that the *fundamental solution* of Eq. (12.5) is given by:

$$u_i^* = \frac{1}{4\pi r}, \tag{12.11}$$

where r is the distance between the source point at "i" and the field point. The related fundamental flux is

$$q_i^* = \frac{\partial u_i^*}{\partial \mathbf{n}} = \mathbf{n} \nabla u_i^*. \tag{12.12}$$

Therefore, the final integral equation will be:

$$c_i U_i + \int_{\Gamma} q_i^* U \mathrm{d}\Gamma = \int_{\Gamma} u_i^* q \mathrm{d}\Gamma \tag{12.13}$$

12.3.2 *Poisson's Equation*

If a term of internal sources, $f(x, y)$ or $f(x, y, z)$, appears, the PDE becomes:

$$\nabla^2 u = f,$$

and the final integral expression will be:

$$c_i U_i + \int_{\Gamma} q_i^* U \mathrm{d}\Gamma + \int_{\Omega} u_i^* f \mathrm{d}\Omega = \int_{\Gamma} u_i^* q \mathrm{d}\Gamma \tag{12.14}$$

12.3.3 Helmholtz Equation

Wave propagation, such as sound, is described through the following PDE:

$$\nabla^2 U + k^2 U = 0 \tag{12.15}$$

where $k = \omega/c$ is the wavenumber, ω is the cyclic frequency, and c is the wave propagation velocity (or wave speed).

The fundamental solution of (12.15) fulfills the PDE:

$$\nabla^2 u_i^* + k^2 u_i^* = -\Delta^i, \tag{12.16}$$

and for the three-dimensional problems is given by:

$$u_i^* = \frac{e^{-jkr}}{r} \tag{12.17}$$

Considering again the symbols (that refer to fluxes = particle velocities):

$$q = \frac{\partial U}{\partial n} \quad \text{and} \quad q_i^* = \frac{\partial u_i^*}{\partial n}, \tag{12.18}$$

the final integral equation takes again the same expression as previously, which is

$$c_i U_i + \oiint_{\Gamma} q_i^* U \, \mathrm{d}\Gamma = \oiint_{\Gamma} u_i^* q \, \mathrm{d}\Gamma \tag{12.19}$$

12.3.4 Navier Equation in Elasticity

The complete system of equations that describe the elastostatic behavior of a body is:

Strain–displacement relationships

$$\boldsymbol{\varepsilon} = \mathbf{L}\mathbf{u} \tag{12.20}$$

Stress–strain relationships

$$\boldsymbol{\sigma} = \mathbf{E}\boldsymbol{\varepsilon} \tag{12.21}$$

Stress equilibrium

$$\mathbf{L}^{\mathrm{T}}\boldsymbol{\sigma} + \mathbf{f}^* = \mathbf{0} \tag{12.22}$$

where $\mathbf{L}$ is a well-known matrix operator of order 6×3.

The elimination of the stress vector $\boldsymbol{\sigma}$ leads to the following PDE:

$$\mathbf{D}(\mathbf{u}) + \mathbf{f}^* = \mathbf{0} \text{ with } \mathbf{D} = \mathbf{L}^{\mathrm{T}}\mathbf{E}\mathbf{L}. \tag{12.23}$$

The integral form of Eq. (12.23), in the sense of weighted Galerkin, will be:

$$\int_{\Omega} (\mathbf{u}^*)^{\mathrm{T}} \left\{ \mathbf{D}(\mathbf{u}) + \mathbf{f}^* \right\} \mathrm{d}\Omega = \mathbf{0} \tag{12.24}$$

The application of Green's theorem in the first term of Eq. (12.24) gives:

$$\begin{aligned} \int_{\Omega} (\mathbf{u}^*)^{\mathrm{T}} \mathbf{D}(\mathbf{u}) \mathrm{d}\Omega &= \int_{\Omega} (\mathbf{u}^*)^{\mathrm{T}} \left(\mathbf{L}^{\mathrm{T}} \mathbf{E} \mathbf{L} \right) \mathbf{u} \mathrm{d}\Omega = \int_{\Omega} (\mathbf{u}^*)^{\mathrm{T}} \mathbf{L}^{\mathrm{T}} (\mathbf{E}\mathbf{L}\mathbf{u}) \mathrm{d}\Omega \\ &= \oint (\mathbf{u}^*)^{\mathrm{T}} (\mathbf{E}\mathbf{L}\mathbf{u}) \mathrm{d}\Gamma - \int_{\Omega} (\mathbf{L}\mathbf{u}^*)^{\mathrm{T}} (\mathbf{E}\mathbf{L}\mathbf{u}) \mathrm{d}\Omega \\ &= \oint (\mathbf{u}^*)^{\mathrm{T}} p \mathrm{d}\Gamma - \int_{\Omega} (\varepsilon^*)^{\mathrm{T}} \sigma \mathrm{d}\Omega, \end{aligned} \tag{12.25}$$

Similarly, the application of Green's theorem to the conjugate expression gives:

$$\int_{\Omega} \mathbf{u}^{\mathrm{T}} \mathbf{D}(\mathbf{v}) d\Omega = \oint \mathbf{u}^{\mathrm{T}} \mathbf{p}_{\Gamma}^* \mathrm{d}\Gamma - \int_{\Omega} \boldsymbol{\varepsilon}^{\mathrm{T}} \boldsymbol{\sigma}^* \mathrm{d}\Omega, \tag{12.26}$$

where $\mathbf{p}_{\Gamma}$, $\boldsymbol{\sigma}$, and $\boldsymbol{\varepsilon}$ are the actual tractions, the internal stresses, and the displacements, respectively, whereas $\mathbf{p}_{\Gamma}^*$, $\boldsymbol{\sigma}^*$, and $\boldsymbol{\varepsilon}^*$ are the corresponding quantities that correspond to the fundamental displacement:

$$\left. \begin{aligned} \boldsymbol{\varepsilon}^* &= \mathbf{L}\,\mathbf{u}^* \\ \boldsymbol{\sigma}^* &= \mathbf{E}\,\boldsymbol{\varepsilon}^* \end{aligned} \right\} \tag{12.27}$$

We shall prove that:

$$\int_{\Omega} (\boldsymbol{\varepsilon}^*)^{\mathrm{T}} \boldsymbol{\sigma} \mathrm{d}\Omega = \int_{\Omega} (\boldsymbol{\varepsilon})^{\mathrm{T}} \boldsymbol{\sigma}^* \mathrm{d}\Omega \tag{12.28}$$

Actually, replacing the integrant of Eq. (12.28) with Eq. (12.27), one progressively obtains:

$$\begin{aligned} (\boldsymbol{\varepsilon}^*)^{\mathrm{T}} \cdot \boldsymbol{\sigma} &= (\mathbf{L}\mathbf{u}^*)^{\mathrm{T}} \cdot (\mathbf{E}\mathbf{L}\mathbf{u}) = \left\{ (\mathbf{L}\mathbf{u}^*)^{\mathrm{T}} \cdot (\mathbf{E}\mathbf{L}\mathbf{u}) \right\}^{\mathrm{T}} \quad \text{(because it is scalar quantity)} \\ &= \left\{ (\mathbf{E}\mathbf{L}\mathbf{u})^{\mathrm{T}} \cdot (\mathbf{L}\mathbf{u}^*) \right\} = (\mathbf{L}\mathbf{u})^{\mathrm{T}} \cdot \mathbf{E}.(\mathbf{L}\mathbf{u}^*) = \boldsymbol{\varepsilon}^{\mathrm{T}} \cdot \boldsymbol{\sigma}^* \end{aligned} \tag{12.29}$$

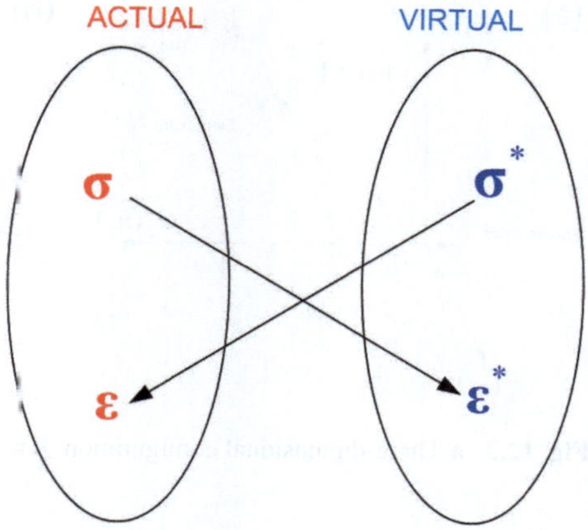

Fig. 12.2 Betti-Maxwell theorem: a crossed relationship between the actual (**σ**, **ε**) and the virtual situation (**σ***, **ε***)

Equation (12.19) is the well-known *Betti-Maxwell theorem*, which is a crossed relationship between the actual (**σ**, **ε**) and the virtual situation (**σ***, **ε***), as schematically is shown in Fig. 12.2.[1]

The fundamental displacements u^*_{lk} fulfill the homogeneous PDE:

$$\mathbf{D}(\mathbf{u}_i^*) + \mathbf{f}^* = \Delta^i, \quad \mathbf{D} = \mathbf{L}^{\mathrm{T}}\mathbf{E}\mathbf{L} \tag{12.30}$$

Combining Eq. (12.25), (12.26), and (12.28), the boundary integral equation becomes:

$$c_i\mathbf{u}_i + \int_{\Gamma} \mathbf{p}^*\mathbf{u}\mathrm{d}\Gamma = \int_{\Gamma} \mathbf{u}^*\mathbf{p}\mathrm{d}\Gamma + \int_{\Omega} \mathbf{u}^*\mathbf{f}^*\mathrm{d}\Gamma. \tag{12.31}$$

The quantities that appear in Eq. (12.31) are as follows:

$\mathbf{u}_i = [u_1, u_2, u_3]^{\mathrm{T}}$ = the displacement vector at the source point "*i*", with components in the directions x_1, x_2, x_3 $(x_1 = x,\ x_2 = y,\ x_3 = z)$.
$\mathbf{u}$ = displacement vector at any point on the boundary Γ.
$\mathbf{p} = [p_1, p_2, p_3]^{\mathrm{T}}$ = tractions at any point on the boundary Γ.
$\mathbf{f}^* = [f_1, f_2, f_3]^{\mathrm{T}}$ = body forces at any point in the interior domain Ω.
$\mathbf{p}^* = \begin{bmatrix} p^*_{11} & p^*_{12} & p^*_{13} \\ p^*_{21} & p^*_{22} & p^*_{23} \\ p^*_{31} & p^*_{32} & p^*_{33} \end{bmatrix}$ = Matrix in which the coefficients p^*_{lk} are forces in the direction "*k*" due to a unit force at the source point "*i*" that acts in the direction "*l*"

[1]It is reminded that Somigliana [31] applied Betti-Maxwell principle in order for the first time to derive integral equations in elasticity. The first application and numerical solution of these equations was published in April 1967 by Rizzo [25] that is 90 years after the theoretical formulation.

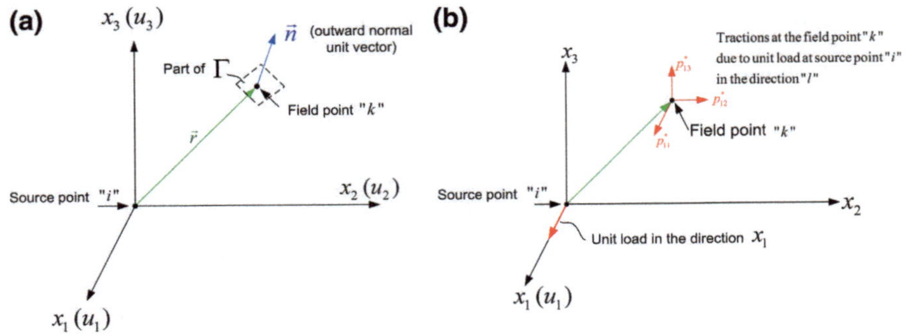

Fig. 12.3 **a** Three-dimensional configuration, **b** unit source load at point "*i*"

$\mathbf{u}^* = \begin{bmatrix} u_{11}^* & u_{12}^* & u_{13}^* \\ u_{21}^* & u_{22}^* & u_{23}^* \\ u_{31}^* & u_{32}^* & u_{33}^* \end{bmatrix}$ = Matrix in which the coefficients u_{lk}^* are the displacements in the direction "*k*" due to a unit force at the source point "*i*" that acts in the direction "*l*"

The fundamental solution and the corresponding traction for a three-dimensional isotropic elastic continuum are calculated using the following relationships (e.g., [3]):

$$u_{lk}^* = \frac{1}{16\pi G(1-\nu)}\frac{1}{r}\left[(3-4\nu)\Delta_{lk} + \frac{\partial r}{\partial x_l}\frac{\partial r}{\partial x_k}\right]$$
$$p_{lk}^* = -\frac{1}{8\pi(1-\nu)r^2}\left[\frac{\partial r}{\partial \mathbf{n}}\left\{(1-2\nu)\Delta_{lk} + 3\frac{\partial r}{\partial x_l}\frac{\partial r}{\partial x_k}\right\} - (1-2\nu)\left\{\frac{\partial r}{\partial x_l}n_k - \frac{\partial r}{\partial x_k}n_l\right\}\right] \tag{12.32}$$

where $\mathbf{n}(n_x, n_y, n_z)$ is the outward unit vector normal to the boundary Γ, Δ_{lk} is Kronecker delta (=1 if $i=j$, =0 if $i \neq j$), and r is the distance between the point of load application and the point under consideration (Fig. 12.3).

12.4 CAD-Based Macroelement Approach

12.4.1 General

Although CAD-based macroelements are applicable to both IBEM and DBEM, for the sake of brevity and without loss of generality below we restrict our discussion only to DBEM.

In general, the boundary of the domain or the structural component is made of curvilinear CAGD patches, either quadrilateral (ABCD) or triangular (ABC) . The traditional procedure is to subdivide these patches into a mesh of adequately small constant, linear, or quadratic elements and then apply the integral equations [cf. Eq. (12.31)] at each of the involved nodal points. Established commercial codes such as BEASY (https://www.beasy.com/) are based on this philosophy.

Alternatively, in the CAD-based macroelement procedure, provided the solution [potential $U(x, y, z)$ or displacement $\mathbf{u}(x, y, z)$] is sufficiently smooth and regular over the entire patch (ABCD or ABC), we apply a certain 2D CAD-based approximation for the entire patch. For example, boundary-only Coons interpolation has been previously applied to elastostatic analysis [17], free vibration analysis [23], acoustic cavities [24], and sound radiation [18, 19]. Although the results using only the boundary of every patch were of acceptable quality, the accuracy improves when using additional internal points inside the contour of each patch.

Details concerning the computation of the normal vector at any point on a parametrically defined patch is given in the Appendix.

12.4.2 *The Procedure of the Boundary Element Method (BEM)*

The boundary Γ of the domain Ω is divided into n_{pa} patches. Each patch may be of the following type:

- *Coons patch* [8], with nodal points only along the contour of each patch.
- *Gordon–Coons patch*, with nodal points along the contour as well as in the interior (transfinite approximation using the **Cij** family that was introduced in Chap. 4). Several trial functions may be used such as:
  - Piecewise-linear
  - Piecewise-quadratic, etc.
  - Natural cubic B-splines (dependent on nodal U-values)
  - Lagrange polynomials.
- Tensor-product patches using one of the following sets:
  - Lagrange polynomials
  - Bernstein–Bézier polynomials (nonrational)
  - Bernstein–Bézier polynomials (rational)
  - B-splines
  - NURBS
  - Etc.

For the sake of brevity, although the proposed theory covers *all* the abovementioned alternatives, below we deal in detail the case of CAGD patched based on the nodal values U_i and not the nodeless coefficients a_i (Bézier, B-splines, NURBS). However, in principle the latter case follows the same rules.

Moreover, since the elasticity problem is more difficult and more general than the potential problem, the theory deals with it. In contrast, the numerical applications refer to both the elasticity and the potential (field) problem.

Therefore, in the case of elasticity problems, the coordinate vector within the l-th CAGD patch, possessing q_l nodes, is interpolated on the basis of the boundaries of the patch as follows:

$$\mathbf{x}(r,s)=\sum_{j=1}^{q_l}N_j(r,s)\mathbf{x}_i^j=\mathbf{N}\,\mathbf{x} \tag{12.33}$$

Since each patch is considered as a large isoparametric element, it holds that:

$$\mathbf{u}(r,s)=\sum_{j=1}^{q_l}N_j(r,s)\mathbf{u}_i^j=\mathbf{N}\,\mathbf{u}$$

$$\mathbf{p}(r,s)=\sum_{j=1}^{q_l}N_j(r,s)\mathbf{p}_i^j=\mathbf{N}\,\mathbf{p} \tag{12.34}$$

I. ***Static analysis***

Substituting Eq. (12.34) in the usual integral equation given by Eq. (12.31), and summarizing over the N_p patches in which the boundary has been divided, one obtains:

$$c^i\mathbf{u}^i+\sum_{ip=1}^{Np}\left\{\oiint_{\Gamma_{ip}}\mathbf{p}_{lk}^{*}\mathbf{N}_k\,\mathrm{d}\Gamma\right\}\mathbf{u}_{ip}=\sum_{ip=1}^{Np}\left\{\oiint_{\Gamma_{ip}}\mathbf{u}_{lk}^{*}\mathbf{N}_k\,\mathrm{d}\Gamma\right\}\mathbf{p}_{ip} \tag{12.35}$$

According to Appendix, in Eq. (12.35) the infinitesimal area $\mathrm{d}\Gamma$ is given by:

$$d\Gamma=|G(r,s)|\mathrm{d}r\,\mathrm{d}s, \tag{12.36}$$

where the Jacobian is calculated as:

$$|G(r,s)|=\left(g_1^2+g_2^2+g_3^2\right)^{1/2}, \tag{12.37}$$

with

$$\left.\begin{aligned}g_1&=\frac{\partial x_2}{\partial r}\frac{\partial x_3}{\partial s}-\frac{\partial x_2}{\partial s}\frac{\partial x_3}{\partial r},\\ g_2&=\frac{\partial x_1}{\partial r}\frac{\partial x_3}{\partial s}-\frac{\partial x_1}{\partial s}\frac{\partial x_3}{\partial r},\\ g_3&=\frac{\partial x_1}{\partial r}\frac{\partial x_2}{\partial s}-\frac{\partial x_1}{\partial s}\frac{\partial x_2}{\partial r}.\end{aligned}\right\} \tag{12.38}$$

Now, for the purposes of the numerical integration only, the patch is divided into $N_r \times N_s$ cells where a second set of normalized coordinates $(-1 \le r', s' \le 1)$ is introduced (see [17]. Therefore, the term $|G(r, s)|dr\, ds$ in Eq. (12.36) is replaced by $\left|G(r, s) \cdot G'(r', s')\right| dr'\, ds'$, which requires a trivial (e.g., 2×2, 3×3, 4×4) Gaussian quadrature. A selective integration scheme can be also implemented according to the selected trial functions.

Substituting Eqs. (12.36–12.38) into Eq. (12.35), the final algebraic system obtains the form:

$$\mathbf{C}\,\mathbf{U} + \sum_{ip=1}^{Np} \widehat{\mathbf{H}}^{ip}\,\mathbf{U}^{ip} = \sum_{ip=1}^{Np} \mathbf{G}^{ip}\,\mathbf{P}^{ip} \tag{12.39}$$

where $\mathbf{U}$ is the displacement vector of all nodes on the boundary of the structure (along the patch edges or inner nodes on the boundary patch as well), $\mathbf{U}^{ip}$ and $\mathbf{P}^{ip}$ are vectors of displacement and traction referring to the *ip*-th patch, and $\mathbf{H}^{ip}$ and $\mathbf{G}^{ip}$ are the well-known nonsymmetric influence matrices [3].

By properly assembling the submatrices in Eq. (12.39), we obtain:

$$\mathbf{H}\,\mathbf{U} = \mathbf{G}\,\mathbf{P} \tag{12.40}$$

If q is the total number of nodes along all edges and the interior of the patches on the *boundary* of the structure, the dimensions of the vectors and matrices in Eq. (12.40) are as follows:

- $\mathbf{U}$: displacement vector $(3q \times 1)$
- $\mathbf{P}$: traction vector $(3\bar{q} \times 1)$
- $\mathbf{H}$: total displacement–influence matrix $(3q \times 3q)$
- $\mathbf{G}$: total traction–influence matrix $(3q \times 3\bar{q})$

The above symbol $\bar{q}$ is larger than q ($\bar{q} = q + \Delta q$) with Δq depending on the number of sharp corners and their multiplicity in traction discontinuity [3].

Furthermore, the numerical procedure is as follows. In the boundary value problem under consideration, a part Γ_1 of the boundary is characterized by boundary conditions of Dirichlet type ($\mathbf{u} = \overline{\mathbf{u}}_1$) whereas in the rest part Γ_2 Neumann conditions ($\mathbf{p} = \overline{\mathbf{p}}_2$) dominate.

Note In potential problems, in the Neumann area of the boundary the given tractions are replaced by the given fluxes (given normal derivative) as shown in Fig. 12.4.

The usual way is to split the displacement and traction vectors into two parts, related to the Dirichlet (subscript "1") and the Neumann (subscript "2") boundary conditions:

$$\mathbf{U} = [\mathbf{U}_1, \mathbf{U}_2]^{\mathrm{T}} \quad \text{and} \quad \mathbf{P} = [\mathbf{P}_1, \mathbf{P}_2]^{\mathrm{T}} \tag{12.41}$$

Therefore, by similarly partitioning the two influence matrices, Eq. (12.40) becomes:

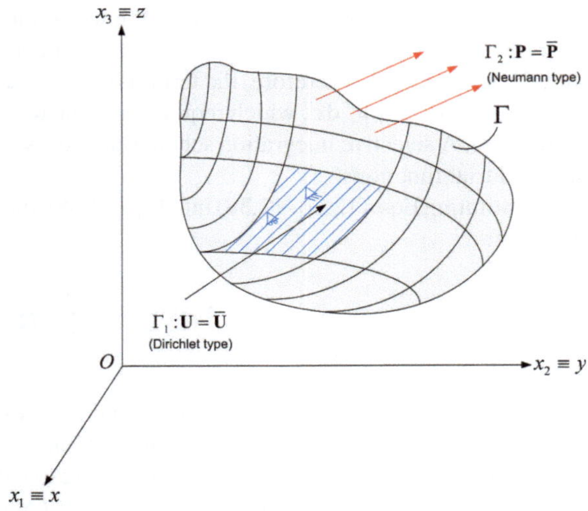

Fig. 12.4 Domain and boundary conditions of two types: Dirichlet (Γ_1) and Neumann (Γ_2)

$$\begin{bmatrix} \mathbf{H}_{11} & \mathbf{H}_{12} \\ \mathbf{H}_{21} & \mathbf{H}_{22} \end{bmatrix} \begin{bmatrix} \overline{\mathbf{U}}_1 \\ \mathbf{U}_2 \end{bmatrix} = \begin{bmatrix} \mathbf{G}_{11} & \mathbf{G}_{12} \\ \mathbf{G}_{21} & \mathbf{G}_{22} \end{bmatrix} \begin{bmatrix} \mathbf{P}_1 \\ \overline{\mathbf{P}}_2 \end{bmatrix} \tag{12.42}$$

Arranging the known terms in the right-hand side, Eq. (12.42) takes the final form:

$$\begin{bmatrix} -\mathbf{G}_{11} & \mathbf{H}_{12} \\ -\mathbf{G}_{21} & \mathbf{H}_{22} \end{bmatrix} \begin{bmatrix} \mathbf{P}_1 \\ \mathbf{U}_2 \end{bmatrix} = \begin{bmatrix} -\mathbf{H}_{11} & \mathbf{G}_{12} \\ -\mathbf{H}_{21} & \mathbf{G}_{22} \end{bmatrix} \begin{bmatrix} \overline{\mathbf{U}}_1 \\ \overline{\mathbf{P}}_2 \end{bmatrix}. \tag{12.43}$$

The solution of the equations system, in Eq. (12.43), gives the unknown vectors $\mathbf{P}_1$ and $\mathbf{U}_2$.

In the sequence, if the displacements are desired at a certain internal point "*i*", Eq. (12.35) is applied for $c^i = 1$ for each component. Also, according to the standard DBEM, the stress components are derived by taking the derivatives of the displacement function that appears in Eq. (12.35) (see details in [3]).

II. ***Dynamic analysis***

Older BEM formulations used to implement either *frequency-* or *time-*dependent fundamental solutions, for which the abovementioned Betti-Maxwell theorem is again applicable. In the former case, the eigenvalue problem is *nonalgebraic*, whereas in the latter case *convolution* integrals are involved. Details may be found in a classical textbook edited by Brebbia [4].

The above static analysis described by Eq. (12.40) can be extended to the solution of dynamic problems as well. Briefly, using a set of radial basis functions which was previously used in 2D elastodynamics [15], a mass matrix may be constructed for the entire structure and it is combined with the static matrices **H** and **G** shown in Eq. (12.40).

This is accomplished starting from the equation of motion (equilibrium):

$$\mathbf{L}^{\mathrm{T}}\boldsymbol{\sigma} + \mathbf{f}^{*} = \rho\ddot{\mathbf{U}}, \tag{12.44}$$

where ρ is the mass density and $\ddot{\mathbf{U}}$ the acceleration.

The weighted Galerkin method is again applied to Eq. (12.44), where now the inertial terms are replaced using a series extension as follows

$$\mathbf{U}(x, y, z; t) = \sum_{j=1}^{q} f_j(x, y, z) \cdot \boldsymbol{\alpha}_j(t). \tag{12.45}$$

Ignoring the body forces ($\mathbf{f}^{*} = \mathbf{0}$), this leads to the following matrix expression

$$\widehat{\mathbf{M}}\ddot{\boldsymbol{\alpha}} + \mathbf{H}\mathbf{U} = \mathbf{G}\mathbf{P}, \tag{12.46}$$

where the terms of the mass matrix are given as a domain integral by

$$\hat{m}_{ij} = \rho \int_{\Omega} \mathbf{U}_i^{*} f_j(x, y, z)\, \mathrm{d}\Omega \tag{12.47}$$

Moreover, considering the basis functions $f_j(x, y, z)$ as body forces, and taking the corresponding dual functions (displacements ψ_j and tractions η_j) through a second Betti-Maxwell relationship, the mass matrix takes its final nice form:

$$\widehat{\mathbf{M}} = \rho(\mathbf{H} \cdot \boldsymbol{\psi} - \mathbf{G} \cdot \boldsymbol{\eta}) \tag{12.48}$$

If Eq. (12.45) is collocated to all nodal points, we derive a relationship between the vector $\mathbf{U}$ of displacements and the vector $\boldsymbol{\alpha}$ of coefficients, $\mathbf{U} = \mathbf{F} \cdot \boldsymbol{\alpha}$, whence:

$$\ddot{\mathbf{U}} = \mathbf{F} \cdot \ddot{\boldsymbol{\alpha}}, \tag{12.49}$$

Substituting Eq. (12.49) into Eq. (12.46), the latter takes its final form:

$$\left(\widehat{\mathbf{M}}\mathbf{F}^{-1}\right)\ddot{\mathbf{U}} + \mathbf{H}\mathbf{U} = \mathbf{G}\mathbf{P} \tag{12.50}$$

Equation (12.50) reminds the usual FEM where the eigenvalue extraction is an algebraic problem. The main difference is that here we need to apply a nonsymmetric algorithm such as QR.

In the past, the performance of Eq. (12.50) has become a matter of debate, particularly in the numerical solution of problems in acoustics [16]. Coons-based eigenanalysis of 3D structures and acoustic cavities has been published by Provatidis and Zafiropoulos [23, 24].

It is noted that the vector of coefficients $\boldsymbol{\alpha}$ involved in Eq. (12.45) is related to inertial degrees of freedom only. Obviously, if a NURBS-based formulation is applied

Table 12.1 Calculated displacement at the loaded surface (in tension)

Figure	Number of nodes	Gauss points per cell			Exact solution
		2 × 2	3 × 3	4 × 4	
12.5a	20	0.939	0.972	0.984	1.0
12.5b	44	0.978	0.989	0.994	1.0
12.5c	68	0.989	0.994	0.997	1.0

to replace Eq. (12.33) and Eq. (12.34), then a second set of B-spline coefficients, **a**, will participate in the solution. Nevertheless, for the sake of brevity and without loss of generality in the same concept, this chapter will restrict in the "old-fashioned" Gordon–Coons patches that refer to nodal **U**-values.

12.5 Numerical Results

Example 12.1 (Cube in Tension) A cube of unit length (side $a = 1$ m) is fixed at its left surface ($x=0$), whereas the opposite face is uniformly loaded in tension ($P_x = 1$). The elastic modulus and Poisson's ratio are $E = 1$, and $\nu = 0$, respectively.

This problem is considered to be composed of six patches, which are the faces of the cube. It was solved for three different uniform meshes of twenty, forty-four, and sixty-eight geometry (displacement) nodes, respectively, as shown in Fig. 12.5. Each case was also solved for three different quadrature schemes, i.e., using 2 × 2, 3 × 3, and 4 × 4 Gaussian points per integration cell. Coons interpolation-based results (using natural cubic B-splines as univariate trial functions along each of the 12 edges) are presented in Table 12.1, where one may observe the convergence toward the exact value.

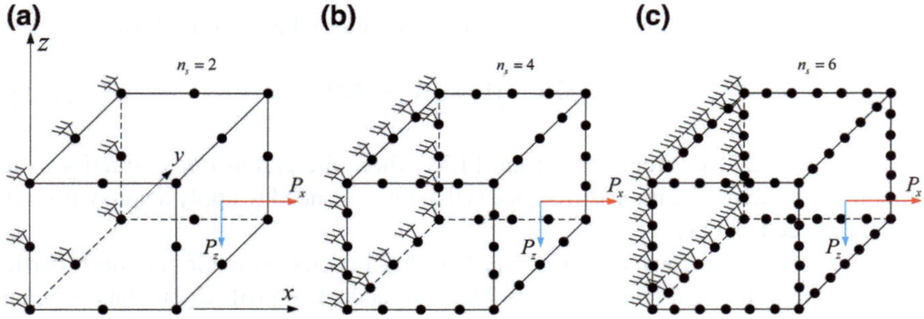

Fig. 12.5 Clamped unit cube discretized using **a** twenty, **b** forty-four, and **c** sixty-eight uniformly distributed nodes

Table 12.2 Calculated displacement at the loaded surface (bending)

Displacement	Gauss points per cell			ALGOR (Inventor)
	2×2	3×3	4×4	
u	9.53	10.06	10.25	10.13
v	0.00	0.00	0.00	0.0
w	20.18	21.14	21.49	21.35

Obviously, due to the zero value of Poisson's ratio, from the theoretical point of view no contraction should occur (axial displacement only, no vertical), and therefore, it is correct to restrict all the degrees of freedom at the left surface of the cube and expect that the nodes of the loaded surface will axially move by $u = \Delta L = \sigma L/E = 1.0$.

It should be noted that even for the coarser mesh of 20 nodes, the axial displacement may be improved. For example, using 6×6 Gauss points the axial displacement improves (from 0.984) to the value 0.993.

Remark 1 For the abovementioned coarse mesh (20 nodes in total), very similar results were found when the interpolation on each face was replaced by the well-known quadratic shape functions of the eight-node element of the *Serendipity family*. It is reminded that the latter elements are coming from boundary-only Coons interpolation as well, but the involved trial functions (along each side) are classical Lagrange polynomials (not natural B-splines). In the latter case, increasing the integration to 9×9 Gauss points per cell the axial displacement was found equal to 0.997, which is an excellent result.

Remark 2 It should be noted that the use of eight-node (quadratic) boundary elements is the standard in all BEM packages. This means that the case of Fig. 12.5a in conjunction with the Serendipity family is identical with conventional BEM. In contrast, the cases of Fig. 12.5b, c have no relation to the usual BEM.

Example 12.2 (Cube in Bending) The same cube of unit length (shown in Fig. 12.5) is fixed at its one surface ($x = 0$) as previously, while the opposite side is now loaded by a uniform shear traction in bending ($P_z = 3.2$). The mechanical properties (elastic modulus and Poisson's ratio) are $E = 1.0$ and $\nu = 0.2$. This problem was solved for a uniform mesh of forty-four geometry (displacement) nodes (Fig. 12.5b) using three different Gaussian quadrature schemes: 2×2, 3×3, and 4×4 integration points per cell. Results are presented in Table 12.2 and are compared with a commercial FEM code (for the same number of subdivisions).

Example 12.3 (Thick Hollow Cylinder Under Internal Pressure) A thick hollow cylinder (of radii $R_i = 10$ mm, $R_e = 20$ mm and height $L = 20$ mm) is subjected to a uniform internal pressure $P = 20$ MPa. The model consists of one-fourth (90°) of the cylinder circumferentially, where the lower and upper sides do not move along the vertical axis of revolution (z) but they only roll on the horizontal plane xy

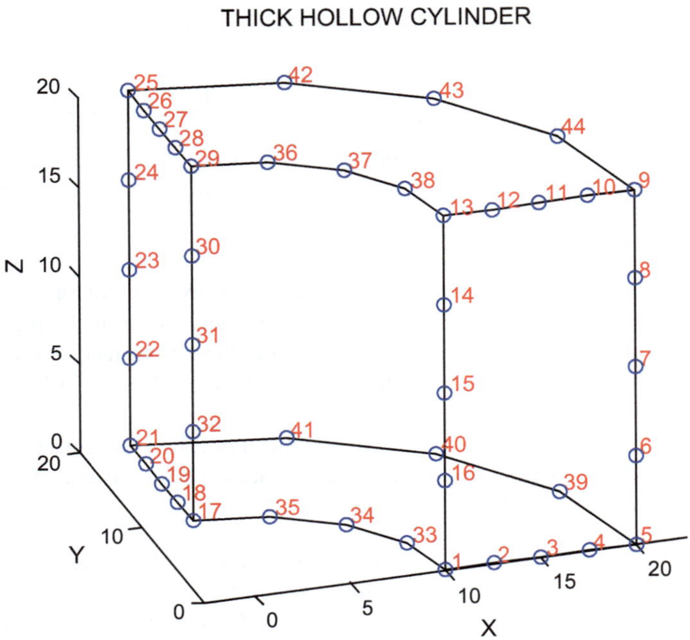

Fig. 12.6 One-fourth of a thick hollow cylinder

Table 12.3 Radial displacement ($u_r \times 10^3$) on the inner boundary

Coons-based BEM solution			Exact
Gauss points			
2 × 2	3 × 3	4 × 4	
1.79	1.83	1.85	1.87

Table 12.4 Circumferential stress (σ_θ) [MPa] on boundary

Coons-based BEM solution			Exact
Gauss points			
2 × 2	3 × 3	4 × 4	
−36.10	−34.58	−34.28	−33.33

(Fig. 12.6). The material is isotropic and linear elastic with mechanical properties ($E = 210{,}000$ MPa, $\nu = 0.30$).

The numerical model consists of forty-four geometry (displacement) nodal points. In more details, it consists of six Coons patches (four flat and two cylindrical) with sixteen nodes per patch. Numerical results (using natural cubic B-splines) are presented at the boundary nodes in Tables 12.3, 12.4, 12.5, and 12.6 and are compared with the analytical solution given as

Table 12.5 Radial displacement ($u_r \times 10^3$) in the interior at $z = L/2 = 10$ mm

Radius (R)	Coons-based BEM solution			Exact
	Gauss points			
	2×2	3×3	4×4	
12.5	1.60	1.56	1.57	1.59
15.0	1.38	1.39	1.41	1.43
17.5	1.31	1.32	1.31	1.33

Table 12.6 Circumferential stress (σ_θ) [MPa] in the interior at $z = L/2 = 10$ mm

Radius (R)	Coons-based BEM solution			Exact
	Gauss points			
	2×2	3×3	4×4	
12.5	24.92	24.30	23.41	23.73
15.0	17.35	18.07	18.26	18.52

$$\sigma_\theta = \frac{a^2 P}{b^2 - a^2}\left(1 + \frac{b^2}{r^2}\right), \quad \sigma_r = \frac{a^2 P}{b^2 - a^2}\left(1 - \frac{b^2}{r^2}\right) \tag{12.51}$$

and

$$E\,\varepsilon_\theta = \sigma_\theta - \nu\sigma_r, \quad \varepsilon_\theta = \frac{u}{r}, \quad u = \frac{r}{E}(\sigma_\theta - \nu\sigma_r) \tag{12.52}$$

Example 12.4 (Rectangular Acoustic Cavity) A rectangular cavity of dimensions $L_x \times L_y \times L_z = 2.5 \times 1.1 \times 1.0$ was divided into 76 nodes using 10, 5, and 5 uniform segments along the aforementioned edges parallel to x-, y-, and z-axis, respectively. Based on the Nardini–Brebbia approach in conjunction with radial basis functions $f_j = C - r$ (with $C = \sqrt{L_x^2 + L_y^2 + L_z^2}$), the calculated natural wavenumbers ($k = \omega/c$) and the corresponding modes are shown in Table 12.7. The exact wavenumbers are given by

$$k_{mnq} = \left[\left(\frac{n}{L_x}\right)^2 + \left(\frac{m}{L_y}\right)^2 + \left(\frac{q}{L_z}\right)^2\right]^{1/2} \tag{12.53}$$

Despite the fact that the proposed (Nardini mass matrix-based) BEM macroelement analysis is less accurate than the FEM, for the same number of 76 nodes along the twelve edges, however for engineering purposes it is acceptable.

Example 12.5 (Rectangular Beam Under Free-Free Boundary Conditions) A rectangular structure of dimensions $L_x \times L_y \times L_z = 3 \times 3 \times 2$ under free-free boundary conditions was considered. Material properties were taken as follows: $E/\rho = 10^4$ and $\nu = 0.30$. The edges were uniformly divided into a variable number

Table 12.7 Calculated eigen wavenumbers for a rectangular cavity of dimensions $2.5 \times 1.1 \times 1.0$ under free-free boundary conditions

Exact eigen wavenumber (k_{mn})	Mode (m, n, q)	Error (in %) of calculated eigenvalues	
		BEM macroelement analysis	FEM (396 nodes, 250 elements)
0.000	(0, 0, 0)	–	–
1.2566	(1, 0, 0)	−0.02	0.41
2.5133	(2, 0, 0)	4.48	1.65
2.8560	(0, 1, 0)	6.40	1.65
3.1202	(1, 1, 0)	4.32	1.45
3.1416	(0, 0, 1)	6.91	1.65
3.3836	(1, 0, 1)	5.08	1.48
3.7699	(3, 0, 0)	8.23	2.58

of (n_x, n_y, n_z) segments, respectively. Wherever comparison is performed, the same number of nodal points exists along the twelve edges of the rectangular.

Since the exact eigenvalues are not analytically known (contrary to the previous acoustic problem) , relative errors were calculated with respect to an adequately fine finite element mesh ($n_x = 15, n_y = 15, n_z = 10$), i.e., 2816 nodes and 2250 elements.

Table 12.8 presents the results for two different discretizations (n_x, n_y, n_z) using both the proposed Coons-BEM macroelement analysis [the constant C in radial basis function $f_j = C - r$ is: $C = (32+32+22)^{1/2} = 22$] and finite elements with the same number of nodes along the edges. Obviously, due to the free-free boundary conditions, the first six eigenvalues equal to zero (rigid body motion modes). One may observe in Table 12.8 that, for a small number of nodes, the proposed Coons-BEM macroelement analysis is more accurate than the FEM solution. As the number of nodes increases, the first nonzero eigenvalue (calculated using the proposed Coons-BEM formulation) is still adequately accurate but the higher ones become slightly less accurate than the FEM solution, however being acceptable for engineering purposes. It is remarkable that the FEM solution overestimates the eigenvalues (monotonically converges from higher values) while the BEM solution does not.

Furthermore, for comparison purposes the same problem was solved using conventional boundary elements (using again the same mass matrix philosophy in conjunction with radial basis functions: $f_j = C - r$). In order to avoid shortcomings associated to the corners, *discontinuous* nine-node Lagrangian elements were chosen (see Fig. 12.7). Each surface of the rectangle was uniformly divided into $2 \times 2 = 4$ boundary elements, thus leading to totally 24 elements (98 geometrical nodes and 216 collocation nodes) . It is noted that the geometric nodes of this choice ensure four divisions per element, i.e., eight subdivisions per edge, which is comparable with the six and nine subdivisions imposed by the contours of each Coons-patch (identical to

Table 12.8 Calculated eigenvalues (ω^2) using Coons-BEM macroelement analysis and finite elements with the same number of nodes (n_x, n_y and n_z segments) along the twelve edges of the rectangle

"EXACT" eigenvalues (FEM: 2816 nodes)	Error (in %) of calculated eigenvalues			
	($n_x = 6, n_y = 6, n_z = 4$)		($n_x = 9, n_y = 9, n_z = 4$)	
	BEM (60 nodes)	FEM (245 nodes, 144 elements)	BEM (84 nodes)	FEM (500 nodes, 324 elements)
2615.8	+1.30	+5.07	−0.93	+2.70
5101.7	−0.41	+7.28	−3.71	+3.01
6632.8	+1.83	+5.11	+2.46	+1.77
7284.4	−3.19	+7.69	−4.46	+3.77

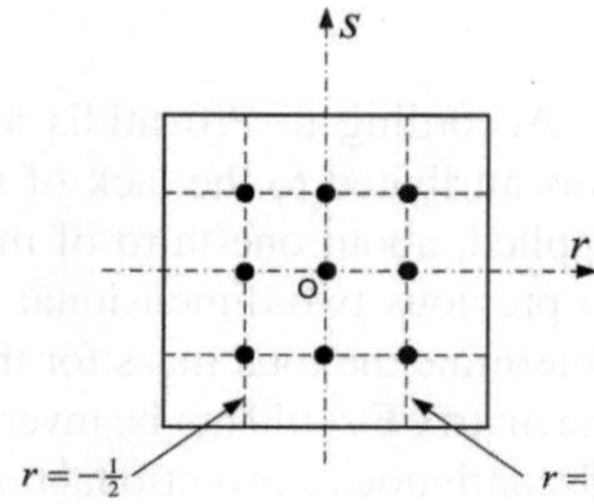

Fig. 12.7 Discontinuous conventional (quadratic) boundary element with the outer collocation points at $r, s = \pm\frac{1}{2}$

Table 12.9 Influence of the position of collocation points in each discontinuous nine-node Lagrangian boundary element ($2 \times 2 = 4$ per surface of the rectangle) on the calculated eigenvalues

Error (in %) of calculated eigenvalues			
$[r, s] = \pm\frac{1}{2}$	$[r, s] = \pm\frac{1}{\sqrt{3}}$	$[r, s] = \pm\frac{3}{4}$	$[r, s] = \pm 0.85$
−11.31	−11.47	−10.99	−10.31
−15.32	−15.35	−14.08	−12.81
−13.44	−15.82	−19.18	−20.67
−18.22	−19.51	−21.47	−22.40

Radial basis functions were considered with the same constant C as in Table 12.8. Mesh consists of 24 boundary elements and 216 collocation nodes

the corresponding surface). A sensitivity analysis was performed with respect to the position of the outer collocation points, which are defined in normalized coordinates within the interval $[r, s] = [-1, 1]$. The relative results are shown in Table 12.9.

It is observed from Table 12.9 that for this coarse mesh, the conventional discontinuous boundary elements are not adequately accurate, as the error in first calculated eigenvalue varies between 10 and 11%. This happens despite the fact that the number of collocation nodes in the conventional BEM is 216: 60 = *3.6 times* greater than the number of nodes participating in the proposed Coons-BEM macroelement.

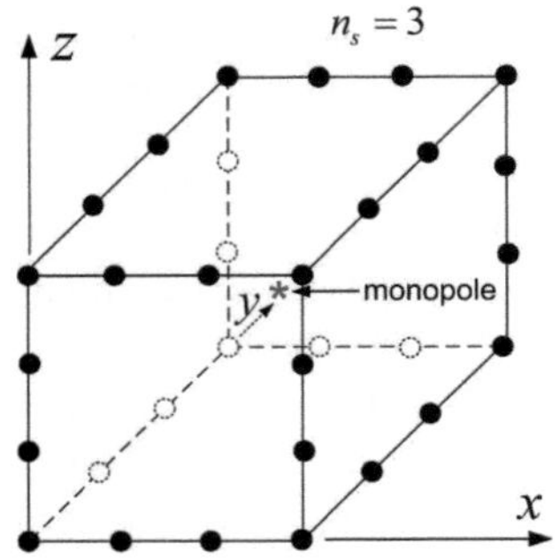

Fig. 12.8 Mesh for the Coons-BEM acoustic analysis

According to Provatidis and Zafiropoulos [23], the reason for this bad result was attributed to the lack of mass. In more details, when the latter technique was applied, about one-third of the total mass was missing, a finding being consistent to previous two-dimensional observations [12, 16]. An attempt was also made to determine the total mass for the simplest case of one boundary element per side but the matrix **F** could not be inverted. In the sequence, we tried to increase the number of discontinuous conventional boundary elements from 24 to 42 (378 collocation nodes, i.e., 1134 degrees of freedom). In this case, the computer effort became extremely high and the mass matrix could not be immediately inverted. Several multipliers were applied (e.g., 3×106) so that the determinant was estimated of the order of 10–70, but nevertheless the quality of the results was not satisfactory because of many complex values found.

Example 12.6 (Radiating Cube [External Problem]) The efficiency of the proposed method will be elucidated by a typical test problem that refers to the sound radiation due to a monopole at the center of a cube with sides of unit length. Throughout the six surfaces of the vibrating cube, the boundary conditions were taken as the exact velocities due to this monopole (see also [21]). The exact solution for the acoustic pressure at a distance r from the center of the cube is given by:

$$p(r) = \frac{a}{r} U_a \frac{j z_0 k a}{1 + jka} e^{-jk(r-a)}, \tag{12.53}$$

where U_a is the normal velocity on the sphere at $r = a$, z_0 is the acoustic impedance ($z_0 = \rho c$), and k is the wavenumber.

The applied mesh consists of six Coons patches (with each side uniformly divided into $n_s = 3$ divisions) with totally 32 nodes, as shown in Fig. 12.8.

For any corner node of the cube, the obtained results are shown in Fig. 12.9.

One may observe in Fig. 12.9 that for small values of the wavenumber $k = \omega/c$, the BEM solution follows the exact solution, in both the real and imaginary components of the acoustic pressure. Probably, the results will somehow improve when the Coons

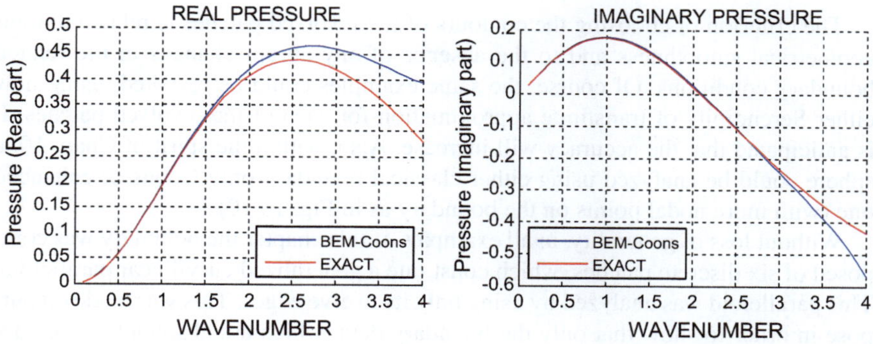

Fig. 12.9 Real and imaginary part of the acoustic pressure at a corner node of the cube

patches will be replaced by tensor-product Lagrange elements or similar transfinite elements (the same holds for tensor-product Bézier, B-splines, etc.). However, a similar deviation will appear to higher wavenumbers and this is due to two main reasons as follows.

First, BEM analysis (similar to FEM analysis) requires a fine mesh so that each wavelength λ is covered by eight elements at least. In our case, one can easily validate that the latter condition implies the upper bound of the wavenumber $k \leq 2\pi/(8\Delta x) \cong 2.36$, and actually, the numerical solution is accurate till this value. Second, BEM suffers from the appearance of *fictitious* eigenvalues, which appear at the eigenvalues of the dual internal problem. This means that since for this unit cube the first eigenvalue of the internal Dirichlet problem is $k_1 = \pi\sqrt{3} \cong 5.44$, a fictitious eigenvalue will appear at its neighborhood for the sound radiation (exterior) problem. In general, the fictitious eigenvalue influences an entire zone (on the left and on the right), which somehow influences the area in Fig. 12.9 as well. Several ways to avoid this shortcoming have been discussed extensively in the literature (e.g., [21]), but this is out of the scope of this book.

12.6 Discussion

A characteristic of the Coons-based Boundary Element Method is that it leads to good results even for a few number of boundary nodes. Moreover, a small increase in accuracy appears when increasing the number of boundary nodes (convergence) and/or the number of integration points per cell. Although the applied numerical integration scheme was a primitive one, it can be easily improved applying a selective integration. This means that we can use less integration points when the field point is far away from the source point. Moreover, additional attention has to be paid on the treatment of the singular integrals.

The criterion of choosing the contours of the involved patches is related to their geometrical smoothness and to the absence of any abrupt changes in the related boundary conditions. Of course, the same examples could be revisited, using now either Serendipity of transfinite approximation for each of the involved patches. It is anticipated that the accuracy will increase. Also, within the spirit of Chap. 10, a sphere could be analyzed using either classical boundary-only Coons or enhanced one (with more nodal points on the boundary as in Fig. 10.10).

Without loss of generality, in all examples of this chapter the boundary was composed of six discrete patches, which constitute a generalized curvilinear paralleloid. This paralleloid was analyzed by using only its twelve edges. This was made on purpose in order to show that only the boundary data, which are absolutely necessary for the development of the CAD (geometry) model, were involved in the analysis. Now, the reader surely understands the reason that in the past (before 2001) we have written that "…in this sense, the proposed method '*marries*' CAD with CAE" [17]. Thanks to a classmate professor of CAD, in January 2003 the word "marriage" was replaced by the correct term "integration" ([20], slightly revised).

12.7 Recapitulation

The basics of the Boundary Element Method (BEM) were exposed in a concise manner. It was shown that after the discretization of the boundary, not only internal but also external problems may be solved. It was shown that static and dynamic analyses of structures may be successfully performed dividing the boundary of the domain into a certain number of large CAD-based patches. In this chapter, most of the analysis concerned with Coons patches in conjunction with cardinal natural cubic B-splines along each of the four edges of each patch. Nevertheless, it is worthy to mention that additional results show that other types of CAD patches such as tensor products lead to still better results.

Appendix: The General Curvilinear Patch

Let us consider a small portion (within a large curvilinear CAGD patch of arbitrary orientation) and let us attach a natural coordinate system, of which the axes (ξ, η) are over the elementary portion whereas the axis ζ is normal to it, as shown in Fig. 12.10.

The relationship between the differentials of the Cartesian coordinates ($\mathrm{d}x$, $\mathrm{d}y$, $\mathrm{d}z$) and the natural ones ($\mathrm{d}\xi$, $\mathrm{d}\eta$, $\mathrm{d}\zeta$), follows the chain rule:

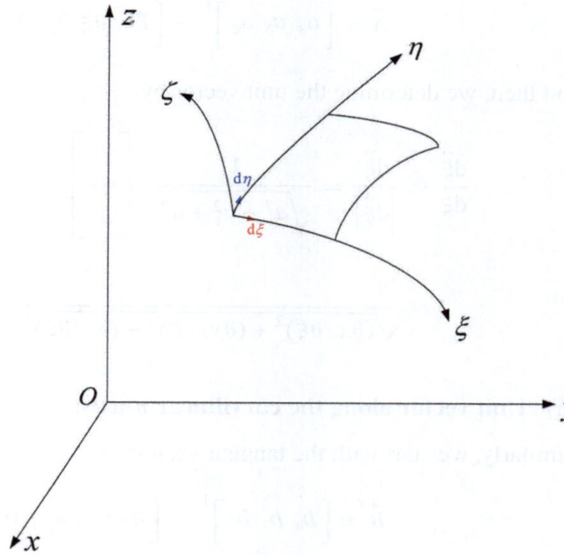

Fig. 12.10 General curvilinear boundary element in the natural coordinate system (ξ, η, ζ)

$$\begin{aligned} dx &= \frac{\partial x}{\partial \xi}d\xi + \frac{\partial x}{\partial \eta}d\eta + \frac{\partial x}{\partial \zeta}d\zeta \\ dy &= \frac{\partial y}{\partial \xi}d\xi + \frac{\partial y}{\partial \eta}d\eta + \frac{\partial y}{\partial \zeta}d\zeta \\ dz &= \frac{\partial z}{\partial \xi}d\xi + \frac{\partial z}{\partial \eta}d\eta + \frac{\partial z}{\partial \zeta}d\zeta \end{aligned} \tag{12.54}$$

In matrix form, Eq. (12.54) is further written as:

$$\begin{Bmatrix} dx \\ dy \\ dz \end{Bmatrix} = \begin{bmatrix} \frac{\partial x}{\partial \xi} & \frac{\partial x}{\partial \eta} & \frac{\partial x}{\partial \zeta} \\ \frac{\partial y}{\partial \xi} & \frac{\partial y}{\partial \eta} & \frac{\partial y}{\partial \zeta} \\ \frac{\partial z}{\partial \xi} & \frac{\partial z}{\partial \eta} & \frac{\partial z}{\partial \zeta} \end{bmatrix} \begin{Bmatrix} d\xi \\ d\eta \\ d\zeta \end{Bmatrix} \tag{12.55}$$

The matrix that appears in the right-hand side of Eq. (12.55) is called the *Jacobian*, and its determinant, det J, is useful for the numerical integration within an elementary volume, according to the formula:

$$dV = |\det J|\, d\xi\, d\eta\, d\zeta \tag{12.56}$$

Other useful quantities are:

(i) **Unit vector along the curvilinear ξ-axis:**

We start with the tangent vector:

$$\vec{A} = \begin{bmatrix} a_x & a_y & a_z \end{bmatrix}^{\mathrm{T}} = \begin{bmatrix} \partial x/\partial\xi & \partial y/\partial\xi & \partial z/\partial\xi \end{bmatrix}^{\mathrm{T}},$$

and then, we determine the unit vector by:

$$\frac{\mathrm{d}\vec{\xi}}{\mathrm{d}\xi} \equiv \frac{\mathrm{d}\vec{\xi}}{\left|\mathrm{d}\vec{\xi}\right|} = \frac{1}{\sqrt{a_x^2 + a_y^2 + a_z^2}} \cdot \begin{bmatrix} a_x \\ a_y \\ a_z \end{bmatrix}$$

$$= \frac{1}{\sqrt{(\partial x/\partial\xi)^2 + (\partial y/\partial\xi)^2 + (\partial z/\partial\xi)^2}} \cdot \begin{bmatrix} \partial x/\partial\xi \\ \partial y/\partial\xi \\ \partial z/\partial\xi \end{bmatrix} \tag{12.57}$$

(ii) **Unit vector along the curvilinear η-axis:**

Similarly, we start with the tangent vector:

$$\vec{B} = \begin{bmatrix} b_x & b_y & b_z \end{bmatrix}^{\mathrm{T}} = \begin{bmatrix} \partial x/\partial\eta & \partial y/\partial\eta & \partial z/\partial\eta \end{bmatrix}^{\mathrm{T}},$$

and then, we determine the unit vector by:

$$\frac{\mathrm{d}\vec{\eta}}{\mathrm{d}\eta} \equiv \frac{\mathrm{d}\vec{\eta}}{|\mathrm{d}\vec{\eta}|} = \frac{1}{\sqrt{b_x^2 + b_y^2 + b_z^2}} \cdot \begin{bmatrix} b_x \\ b_y \\ b_z \end{bmatrix}$$

$$= \frac{1}{\sqrt{(\partial x/\partial\eta)^2 + (\partial y/\partial\eta)^2 + (\partial z/\partial\eta)^2}} \cdot \begin{bmatrix} \partial x/\partial\eta \\ \partial y/\partial\eta \\ \partial z/\partial\eta \end{bmatrix} \tag{12.58}$$

(iii) **Area:**

It is determined as the measure of the cross-product of the abovementioned vectors $\vec{A}$ and $\vec{B}$:

$$\vec{A} \times \vec{B} = \begin{vmatrix} \vec{i} & \vec{j} & \vec{k} \\ a_x & a_y & a_z \\ b_x & b_y & b_z \end{vmatrix}$$

$$= \underbrace{(a_y b_z - a_z b_y)}_{J_1}\vec{i} \underbrace{-(a_x b_z - a_z b_x)}_{J_2}\vec{j} + \underbrace{(a_x b_y - a_y b_x)}_{J_3}\vec{k}$$

$$\equiv J_1\vec{i} + J_2\vec{j} + J_3\vec{k} \tag{12.59}$$

The quantity $J = \sqrt{J_1^2 + J_2^2 + J_3^2} = \left|\vec{A} \times \vec{B}\right|$ equals the area of the boundary element.

(iv) **Normal unit vector**

The three components of the unit normal vector $\vec{n}$ on the boundary element are obviously given by:

$$\begin{aligned} n_1 &= J_1/J \\ n_2 &= J_2/J \\ n_3 &= J_3/J \end{aligned} \tag{12.60}$$

and fulfill the apparent condition:

$$n_1^2 + n_2^2 + n_3^2 = 1 \tag{12.61}$$

References

1. Banerjee PK, Butterfield R (1981) Boundary element methods in engineering science. McGraw-Hill, London
2. Brebbia CA, Butterfield R (1978) Formal equivalence of direct and indirect boundary element methods. Appl Math Model 2:132–134
3. Brebbia CA, Dominguez J (1992) Boundary elements: an introductory course. Computational Mechanics Publications, Mc-Graw Hill Book Company, Southampton
4. Brebbia CA (ed) (1985) Time-dependent and vibration problems. Springer, Berlin
5. Cabral JJSP, Wrobel LC, Brebbia CA (1990) A BEM formulation using B-splines: I-uniform blending functions. Eng Anal Bound Elem 7(3):136–144
6. Cabral JJSP, Wrobel LC, Brebbia CA (1991) A BEM formulation using B-splines: II-multiple knots and non-uniform blending functions. Eng Anal Bound Elem 8(1):51–55
7. Cerrolaza M, Annicchiarico W, Martinez M (2000) Optimization of 2D boundary element models using B-splines and genetic algorithms. Eng Anal Bound Elem 24(5):427–440
8. Coons SA (1964) Surfaces for computer aided design of space form, Project MAC, MIT (1964), revised for MAC-TR-41 (1967), Springfield, VA 22161, USA: Available as AD 663 504 from the National Technical Information Service (CFSTI), Sills Building, 5285 Port Royal Road. Available online: http://publications.csail.mit.edu/lcs/pubs/pdf/MIT-LCS-TR-041.pdf
9. Ginnis AI, Kostas KV, Politis CG, Kaklis PD, Belibassakis KA, Gerostathis ThP, Scott MA, Hughes TJR (2014) Isogeometric boundary-element analysis for the wave-resistance problem using T-splines. Comput Methods Appl Mech Eng 279(1):425–439
10. Gong YP, Dong CY (2017) An isogeometric boundary element method using adaptive integral method for 3D potential problems. J Comput Appl Math 319(1):141–158
11. Kanarachos A, Deriziotis D (1989) On the solution of Laplace and wave propagation problems using 'C-elements'. Finite Elem Anal Des 5:97–109
12. Kanarachos A, Provatidis C (1992) Potential and wave propagation problems using the boundary element method and BEM-subregions. Eng Anal Bound Elem 9:117–124
13. Mallardo V, Ruocco E (2016) A NURBS boundary-only approach in elasticity. Eur J Comput Mech 25(1–2):71–90
14. Mastorakis I (1992) Coons based boundary elements in acoustics (in Greek). Diploma Work, National Technical University of Athens, Department of Mechanical Engineering (supervisor: Dr. C. G. Provatidis)
15. Nardini D, Brebbia CA (1983) A new approach to free vibration analysis using boundary elements. Appl Math Model 7(3):157–162

16. Provatidis CG (1987) On the application of the boundary element method in the analysis of field and dynamic problems. Doctoral Dissertation, National Technical University of Athens, Department of Mechanical Engineering, November 1987, Athens (in Greek). Available online at: https://www.didaktorika.gr/eadd/handle/10442/0616
17. Provatidis CG (2001) Stress analysis of 3D solid structures using large boundary elements derived from 2D-Coons' interpolation. In: Drakatos PA (ed) CD proceedings of ASME—Greek Section, September 17–20, 2001, Patras, Greece (Paper ANG1/P129, pages 1–6). Available online: http://users.ntua.gr/cprovat/yliko/C26.pdf
18. Provatidis CG (2002a) Analysis of three-dimensional sound radiation problems using trimmed patch boundary elements. In: Tsahalis D (ed) CD proceedings 4th GRACM congress on computational mechanics, Patras, 27–29 June, 2002, pp 402–409. Available online: http://users.ntua.gr/cprovat/yliko/C36.pdf
19. Provatidis CG (2002b) A comparative study between Coons-patch macroelements and boundary elements in two-dimensional potential problems. In: 4th GRACM congress on computational mechanics, Patras, 27–29 June, 2002 (CD proceedings)
20. Provatidis CG (2002c) CAD-FEA integration using Coons interpolation. Technical Report, submitted to two Journals. Available online: http://users.ntua.gr/cprovat/yliko/Provatidis_CAD_CAE_Integration.pdf
21. Provatidis C, Zafiropoulos N (2002) On the "interior Helmholtz integral equation formulation" in sound radiation problems. Eng Anal Bound Elem 26:29–40
22. Provatidis CG (2003) Free vibrations of two-dimensional structures using Coons-patch macroelements, FEM and BEM. In: Atluri SN (ed) Proceedings ICCES' 03, Corfu, Greece, 24–29 July 2003
23. Provatidis CG, Zafiropoulos NK (2003a) Free-vibration analysis of three-dimensional solids using Coons-patch Boundary Superelements. In: Aifantis EC (ed) Proceedings fifth European solids mechanics conference, Thessaloniki, Greece, 17–22 August 2003. Extended paper submitted, available online: http://users.ntua.gr/cprovat/yliko/C55.pdf
24. Provatidis CG, Zafiropoulos NK (2003b) Determination of eigenfrequencies in three-dimensional acoustic cavities using Coons-patch Boundary Superelements. In: Aifantis EC (ed) Proceedings fifth European solids mechanics conference, Thessaloniki, Greece, 17–22 August 2003. Extended paper submitted, available online: http://users.ntua.gr/cprovat/yliko/C56.pdf
25. Rizzo FJ (1967) An integral equation approach to boundary value problems of classical elastostatics. Q. J. Appl. Math. 25:83–95
26. Sandgren E, Wu S-J (1988) Shape optimization using the boundary element method with substructuring. Int J Numer Meth Eng 26(9):1913–1924
27. Schramm U, Pilkey WD (1994) Higher order boundary elements for shape optimization using rational *B*-splines. Eng Anal Boundary Elem 14(3):255–266
28. Scott MA, Simpson RN, Evans JA, Lipton S, Bordas SPA, Hughes TJR, Sederberg TW (2013) Isogeometric boundary element analysis using unstructured T-splines. Comput Methods Appl Mech Eng 254:197–221
29. Simpson RN, Scott MA, Taus M, Thomas DC, Lian H (2014) Acoustic isogeometric boundary element analysis. Comput Methods Appl Mech Eng 269(1):265–290
30. Simpson RN, Bordas SPA, Trevelyan J, Rabczuk T (2012) A two-dimensional isogeometric boundary element method for elastostatic analysis. Comput Methods Appl Mech Eng 209–212(1):87–100
31. Somigliana C (1886) "Sopra l' equilibrio di un corpo elastic isotropo", Il Nuovo Ciemento, pp 17–19
32. Takahashi T, Matsumoto T (2012) An application of fast multipole method to isogeometric boundary element method for Laplace equation in two dimensions. Eng Anal Bound Elem 36(12):1766–1775
33. Turco E, Aristodemo M (1998) A three-dimensional B-spline boundary element. Comput Methods Appl Mech Eng 155:119–128

34. Tzanakis Chr (1991) Elastostatic analysis of 3D bodies using large Coons-based boundary elements (in Greek), Diploma Work, National Technical University of Athens, Department of Mechanical Engineering, April 1991 (supervisor: Dr. C. G. Provatidis)
35. Ushatov R, Power H, Rêgo Silva JJ (1994) Uniform bicubic B-splines applied to boundary element formulation for 3-D scalar problems. Eng Anal Bound Elem 13(4):371–381
36. Zhou W, Liu B, Wang Q, Cheng Y, Ma G, Chang X, Chen X (2017) NURBS-enhanced boundary element method based on independent geometry and field approximation for 2D potential problems. Eng Anal Bound Elem 83:158–166

Chapter 13
Domain Decomposition and Other Advanced Issues

Abstract While previous chapters focused on the performance of single macroelements, here we study the case of assembling adjacent CAD-based macroelements in which the computational domain has been decomposed. The discussion starts with Coons–Gordon interpolation using Lagrange polynomials, in which no difficulty appears provided the same nodes are used along an interface being at the same time an entire side in both adjacent patches. Obviously, the aforementioned easiness appears to all tensor-product CAD-based macroelements (Bézier, B-splines). If, however, the interface between two adjacent patches is not an entire edge, then Gordon interpolation has to be extended in a proper way using artificial external nodes. The case of two macroelements that share the same edge but do not have the same number of nodes along it is also studied. A short discussion is devoted to issues such as closed surface patches and local control. Numerical examples refer to potential problems (heat flow and acoustics) and elasticity problems (tension and bending).

Keywords Domain decomposition · Mortar finite element methods
Lagrange multiplier · Master–slave · Mismatching · Artificial nodes
Degeneration · Non-rational Bernstein–Bézier and B-splines · Test cases

13.1 Usual Shortcomings in Subregion Coupling

13.1.1 General Problem

The value of a computational method (such as the method of using CAD-based macroelements) should be judged regarding its capability to be applied in conjunction with the decomposition of the domain. Usually, each subregion is handled separately (e.g., in a parallel processing procedure or even probably using different computational methods for each subregion) and then all components somehow should be synthesized in a whole. Obviously, this happens so even in the case of using conventional small or large finite elements. The general pattern of domain decomposition is

C. G. Provatidis, *Precursors of Isogeometric Analysis*, Solid Mechanics and Its Applications 256, https://doi.org/10.1007/978-3-030-03889-2_13

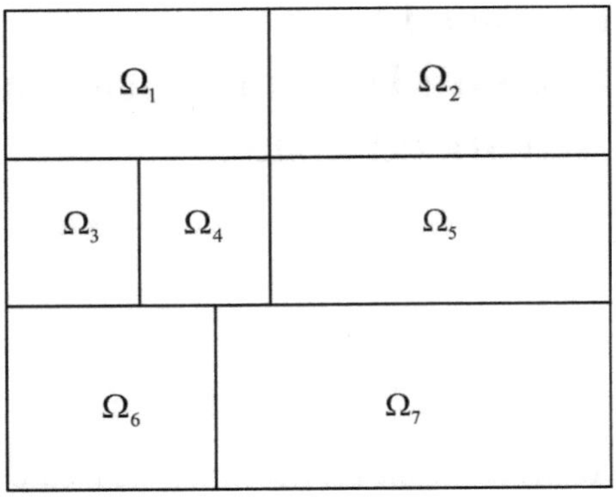

Fig. 13.1 Domain decomposition

schematically shown in Fig. 13.1, where overlapping and non-overlapping interfaces may appear.

Regarding the use of CAD-based *transfinite* elements within a decomposed domain, the earliest work is a rather forgotten paper due to Cavendish et al. [8], which continued later by Cavendish and Hall [9]. It is noted that before the submission of the latter paper, William (Bill) Gordon had already moved from the Mathematics Department, General Motors Research Laboratories (Warren, Michigan, USA) to Drexel University, where he continued his work [13]. Therefore, if the meaning of isogeometric analysis (IGA) is broadened so as to include all CAD methods (in addition to NURBS), the author believes that the aforementioned works are the Precursors of IGA in conjunction with domain decomposition.

Moreover, concerning the pure NURBS-based *isogeometric* analysis, two interesting works are due to Hesch and Betsch [14] as well as Apostolatos et al. [3], and papers therein.

In general, the simplest case to connect two different CAD patches is when they share the *same* interpolation along their interface. For example, compatibility and equilibrium conditions are easily imposed between the patches ABCD and BEFC (shown in Fig. 13.2a) due to their common interface BC. Clearly, whatever is the common approximation along the side BC, both the compatibility and equilibrium conditions can be easily imposed and ensured. Within this context, Coons–Gordon, Bézier, and B-splines macroelements that share the same side BC can be easily and automatically coupled, provided that the same number (and position) of nodes or coefficients is used along this interface.

In contrast, if the side BC is shorter or larger than its neighboring interface side (see Fig. 13.2b where BC is shorter than HE), some difficulties may arise. We say "may" because in the particular case piecewise linear interpolation along the sides of the Coons–Gordon patch has been chosen, the compatibility is ensured per se, for the cases shown in Figs. 13.2, 13.3a, b (shorter or larger). A similar conclusion is produced when using piecewise quadratic or cubic elements, and so on irrespectively to the relationship between the sides BC and EH at the interface (see Fig. 13.3c, d).

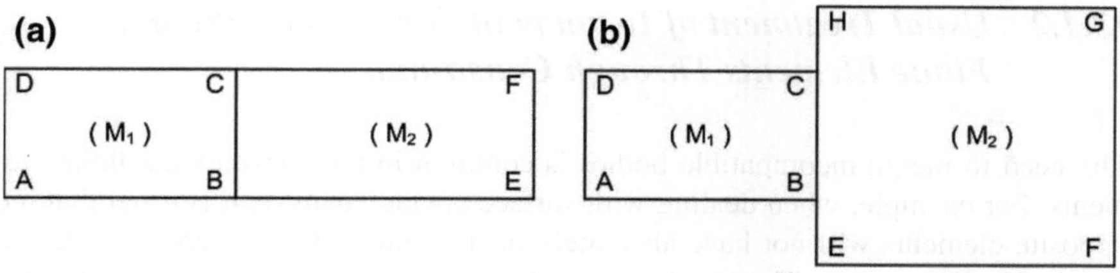

Fig. 13.2 The patches ABCD and CBEF share the same edge (BC)

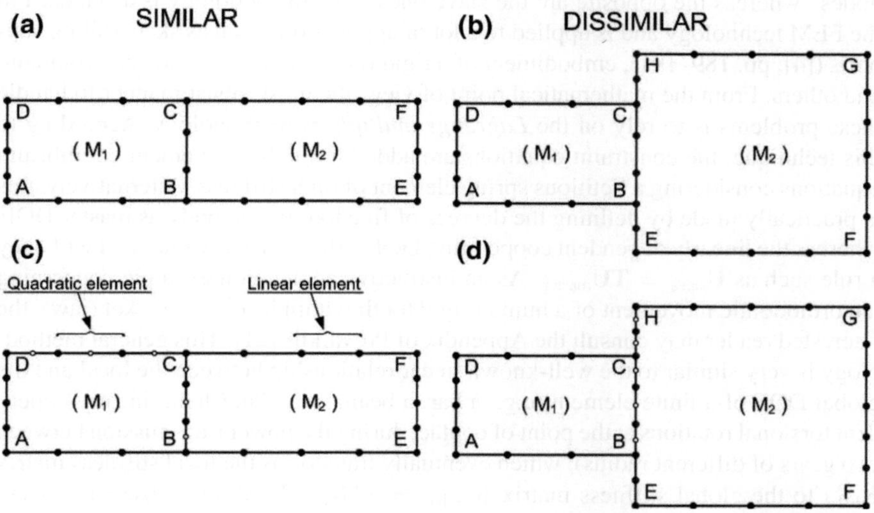

Fig. 13.3 Merging two similar or dissimilar macroelements: **a** and **b** similar and dissimilar with piecewise linear interface; **c** and **d** similar and dissimilar with piecewise quadratic approximation along the interface

Shortcomings actually appear when using Lagrange polynomials, Bernstein ones, or B-splines. For example, with respect to Fig. 13.3b, the side BC of the left macroelement needs a Lagrange polynomial of third degree, whereas the side EH of the right one needs a polynomial of seventh degree.

Nowadays, similar problems are covered under the umbrella of the so-called *mortar* finite element methods, which is the lovely topic of many applied mathematicians and some engineers mostly using *Lagrange multipliers* (e.g., [1, 2, 5, 6, 10–12, 16–18, 20, 24–28]). Closely related are the cell method [19] and the cut-cell or cut-FEM method [7] and papers therein.

13.1.2 *Usual Treatment of Incompatibility in Conventional Finite Elements Through Constraints*

The need to merge incompatible bodies is common in the conventional finite elements. For example, when dealing with surface contact analysis it is usual that the opposite elements will not have absolutely of the same size, so there is a need to match this discrepancy. The usual methodology is to apply a master–slave scheme, in which one set of nodal points (of the same body) are considered to be the "mater nodes" whereas the opposite are the slave ones. This methodology is a standard in the FEM technology and is applied to a lot of applications such as skew rolling supports ([4], pp. 189–191), embodiment of a rigid body within an elastic environment, and others. From the mathematical point of view, the most robust manner to handle these problems is to rely on the *Lagrange multipliers* methodology. According to this technique, the constraint equations are added to the finite element equilibrium equations considering a fictitious spring element of high stiffness. Alternatively, this is practically made by defining the degrees of freedom in one body as master DOF, whereas the linearly dependent cooperating DOF will be the slave ones and will obey a rule such as $\mathbf{U}_{\text{slave}} = \mathbf{T}\mathbf{U}_{\text{master}}$. As an instructive application example concerning the orthodontic movement of a human rigid tooth within its elastic socket (jaw), the interested reader may consult the Appendix of Provatidis [21]. This general methodology is very similar to the well-known linear relationship between the local and the global DOF of a finite element (e.g., a bar, a beam or a plate/shell, linearly dependent torsional rotations at the point of contact during the power transmission between two gears of different radius), which eventually transforms the local stiffness matrix $\mathbf{K}_{\text{local}}$ to the global stiffness matrix $\mathbf{K}_{\text{global}} = \mathbf{T}^{\text{T}}\mathbf{K}_{\text{local}}\mathbf{T}$. Similar observations are valid for the mass matrix as well.

Therefore, for the master part, we can write the following equations of motion (note that upper left index $^{(1)}$ refers to the body No. 1):

$$\begin{bmatrix} {}^{(1)}\mathbf{M}_{11} & {}^{(1)}\mathbf{M}_{12} \\ {}^{(1)}\mathbf{M}_{21} & {}^{(1)}\mathbf{M}_{22} \end{bmatrix} \begin{Bmatrix} {}^{(1)}\ddot{\mathbf{u}}_1 \\ {}^{(1)}\ddot{\mathbf{u}}_2 \end{Bmatrix} + \begin{bmatrix} {}^{(1)}\mathbf{K}_{11} & {}^{(1)}\mathbf{K}_{12} \\ {}^{(1)}\mathbf{K}_{21} & {}^{(1)}\mathbf{K}_{22} \end{bmatrix} \begin{Bmatrix} {}^{(1)}\mathbf{u}_1 \\ {}^{(1)}\mathbf{u}_2 \end{Bmatrix} = \begin{Bmatrix} {}^{(1)}\mathbf{f}_1(t) \\ {}^{(1)}\mathbf{f}_2(t) \end{Bmatrix}, \tag{13.1}$$

whereas for the slave part (note that upper left index $^{(2)}$ refers to the body No. 2):

$$\begin{bmatrix} {}^{(2)}\mathbf{M}_{11} & {}^{(2)}\mathbf{M}_{12} \\ {}^{(2)}\mathbf{M}_{21} & {}^{(2)}\mathbf{M}_{22} \end{bmatrix} \begin{Bmatrix} {}^{(2)}\ddot{\mathbf{u}}_1 \\ {}^{(2)}\ddot{\mathbf{u}}_2 \end{Bmatrix} + \begin{bmatrix} {}^{(2)}\mathbf{K}_{11} & {}^{(2)}\mathbf{K}_{12} \\ {}^{(2)}\mathbf{K}_{21} & {}^{(2)}\mathbf{K}_{22} \end{bmatrix} \begin{Bmatrix} {}^{(2)}\mathbf{u}_1 \\ {}^{(2)}\mathbf{u}_2 \end{Bmatrix} = \begin{Bmatrix} {}^{(2)}\mathbf{f}_1 \\ {}^{(2)}\mathbf{f}_2 \end{Bmatrix}. \tag{13.2}$$

For both bodies, the subscript "1" refers to DOF that is far from the contact area, while the subscript "2" refers to those DOFs involved in the contact region.

Since the DOF $^{(2)}\mathbf{u}_2$ are generally different than those of the $^{(2)}\mathbf{u}_2$ vector, using the proper interpolation (e.g., linear interpolation for linear elements, quadratic interpolation for quadratic elements) along the interface, we can establish a linear relationship of the form:

$$\{^{(2)}\mathbf{u}_2\} = [\mathbf{T}] \cdot \{^{(1)}\mathbf{u}_2\} \tag{13.3}$$

If the vector $[^{(2)}\mathbf{u}_1, {}^{(2)}\mathbf{u}_2]^{\mathrm{T}}$ in Eq. (13.2) is considered as *local*, whereas the corresponding *global* vector is $[^{(1)}\mathbf{u}_2, {}^{(2)}\mathbf{u}_1]^{\mathrm{T}}$, by virtue of Eq. (13.3) the total transformation matrix for the entire body No. 2 (including all its DOF) is written as

$$\begin{Bmatrix} {}^{(2)}\mathbf{u}_1 \\ {}^{(2)}\mathbf{u}_2 \end{Bmatrix} = \begin{bmatrix} \mathbf{0} & \mathbf{I} \\ \mathbf{T} & \mathbf{0} \end{bmatrix} \cdot \begin{Bmatrix} {}^{(1)}\mathbf{u}_2 \\ {}^{(2)}\mathbf{u}_1 \end{Bmatrix} \tag{13.4}$$

Left multiplying both parts Eq. (13.2) [slave part] by the transpose of the matrix

$$\mathbf{T}^{\mathrm{t}} = \begin{bmatrix} \mathbf{0} & \mathbf{I} \\ \mathbf{T} & \mathbf{0} \end{bmatrix}^{\mathrm{t}} \equiv \begin{bmatrix} \mathbf{0} & \mathbf{T}^{\mathrm{t}} \\ \mathbf{I} & \mathbf{0} \end{bmatrix} \tag{13.5}$$

and then adding the produced result in parts with Eq. (13.1), the internal "forces" are canceled (i.e., ${}^{(1)}\mathbf{f}_2 + \mathbf{T}^{\mathrm{t}} \cdot {}^{(2)}\mathbf{f}_2 = \mathbf{0}$), and therefore one obtains:

$$\begin{bmatrix} {}^{(1)}\mathbf{M}_{11} & {}^{(1)}\mathbf{M}_{12} & \mathbf{0} \\ {}^{(1)}\mathbf{M}_{21} & {}^{(1)}\mathbf{M}_{22} + \mathbf{T}^{\mathrm{t}} \cdot {}^{(2)}\mathbf{M}_{22} \cdot \mathbf{T} & \mathbf{T}^{\mathrm{t}} \cdot {}^{(2)}\mathbf{M}_{21} \\ \mathbf{0} & {}^{(2)}\mathbf{M}_{12} \cdot \mathbf{T} & {}^{(2)}\mathbf{M}_{11} \end{bmatrix} \cdot \begin{Bmatrix} {}^{(1)}\ddot{\mathbf{u}}_1 \\ {}^{(1)}\ddot{\mathbf{u}}_2 \\ {}^{(2)}\ddot{\mathbf{u}}_1 \end{Bmatrix}$$
$$+ \begin{bmatrix} {}^{(1)}\mathbf{K}_{11} & {}^{(1)}\mathbf{K}_{12} & \mathbf{0} \\ {}^{(1)}\mathbf{K}_{21} & {}^{(1)}\mathbf{K}_{22} + \mathbf{T}^{\mathrm{t}} \cdot {}^{(2)}\mathbf{K}_{22} \cdot \mathbf{T} & \mathbf{T}^{\mathrm{t}} \cdot {}^{(2)}\mathbf{K}_{21} \\ \mathbf{0} & {}^{(2)}\mathbf{K}_{12} \cdot \mathbf{T} & {}^{(2)}\mathbf{K}_{11} \end{bmatrix} \cdot \begin{Bmatrix} {}^{(1)}\mathbf{u}_1 \\ {}^{(1)}\mathbf{u}_2 \\ {}^{(2)}\mathbf{u}_1 \end{Bmatrix} = \begin{Bmatrix} {}^{(1)}\mathbf{f}_1(t) \\ \mathbf{0} \\ {}^{(2)}\mathbf{f}_1(t) \end{Bmatrix}. \tag{13.6}$$

Clearly, the overall global vector of independent DOFs in Eq. (13.6) is $[^{(1)}\mathbf{u}_1, {}^{(1)}\mathbf{u}_2, {}^{(2)}\mathbf{u}_1]^{\mathrm{T}}$.

Example 13.1 (Mismatching along the Same Interface) Consider a rectangular cavity of dimensions $a \times b = 2.5 \times 1.5$ m, which is decomposed through an interface line normal to the x-axis, into two assemblages of bilinear finite elements. The first half (subregion (I): on the left) consists of $n_{x1} \times n_{y1} = 5 \times 4$ bilinear elements, and the second half (subregion (II): on the right) consists of $n_{x2} \times n_{y2} = 5 \times 5$ ones. Therefore, the interface at $x = a/2 = 1.25$ includes five nodes for the left and six nodes for the right subregion.

The exercise asks to report about the quality of the numerical solution for the following problems (in the same domain):

(i) Steady-state heat conduction along the x-axis, with given temperature 100 °C at $x = 0$ and 0 °C at $x = a$, while ideal insulation is applied to the rest of the two sides parallel to the x-axis (uniform flow).
(ii) Eigenvalue analysis of an acoustic cavity with ideally hard walls (Neumann boundary conditions : $\partial u/\partial n = 0$).

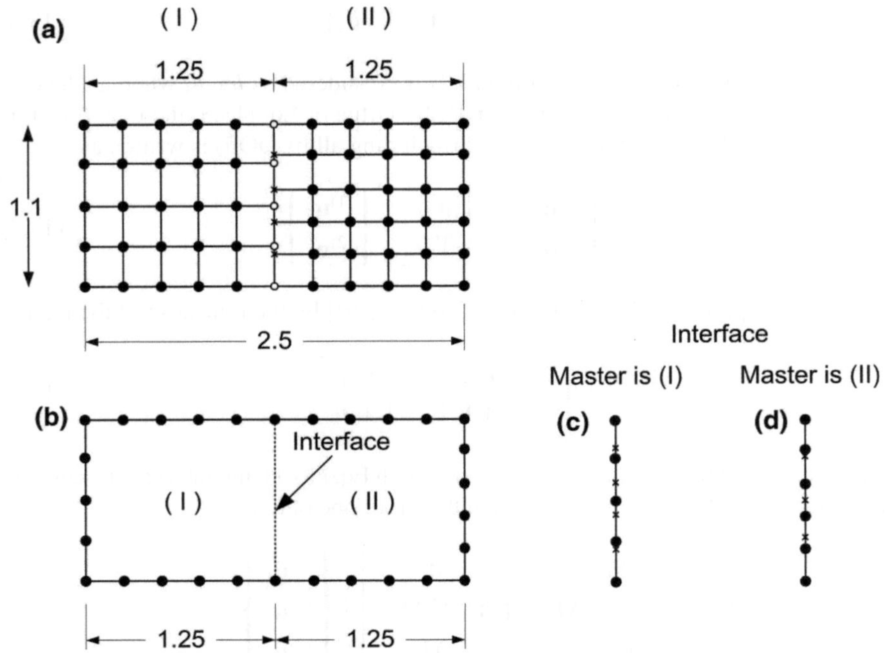

Fig. 13.4 Domain divided into two equal parts using **a** conventional and **b** Coons macroelements, with different mesh densities in the vertical direction, where master element is **c** the left or **d** the right one

Solve this problem in two ways:

- First using conventional finite elements (Fig. 13.4a).
- Second using two Coons macroelements [Fig. 13.4b (cd)].

Solution Consider the two meshes of conventional four-node bilinear quadrilateral elements, as shown in Fig. 13.4a. The left one (I) is the idealization of the left part and consists of 30 nodes (5 rows by 6 columns), whereas the right one (II) is the idealization of the right half and consists of 36 nodes (6 rows by 6 columns).

A. **Conventional Finite Elements**

I. **Heat flow analysis.** Taking as "*master*" degrees of freedom along the interface those that belong to the *left* part (master DOF denoted by white circles), the heat analysis leads to the temperature distribution shown in Fig. 13.5. Clearly, the rectangles represent the assemblage of bilinear finite elements (20 and 25 elements for the left and the right half, respectively), whereas the total number of unconstrained DOF equals to sixty. Although the overall distribution is almost linear as anticipated, however it is characterized by an average absolute error about 1.3 %. It is worthy to point out that the

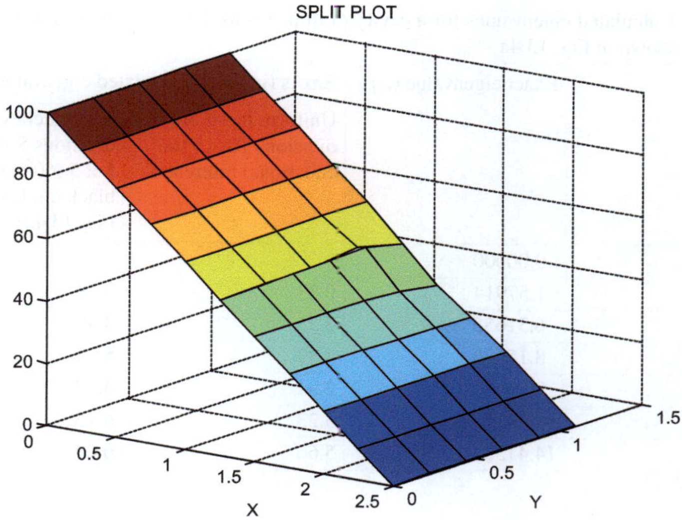

Fig. 13.5 Heat conduction in rectangle of dimensions 2.5 × 1.1

numerical error does not reduce in the neighborhood of the interface but it propagates close to the external boundaries as well. It is reminded that the numerical solution of this particular problem, with full compatibility of the finite elements at the interface, would appear no error at all.

II. **Acoustics**. The problem refers to the same domain and the same discretization as in Fig. 13.4a, which now represents an acoustic cavity with hard walls (Neumann-type boundary conditions). In addition, a uniform mesh of 50 bilinear finite elements (in an arrangement of 10 × 5 subdivisions, i.e., 66 nodes, not shown) is taken as a reference. The calculated eigenvalues are shown in Table 13.1. One may observe that, apart from the first nonzero eigenvalue as ought to be, all the next ones appear an error of acceptable level.

One may observe that even in the case of conventional finite elements, the imposition of constraints between the "master" and "slave" DOF along the interface is a source of some numerical errors, but it eventually offers a reasonable result.

B. Coons and Gordon Macroelements

The same problems are now solved using the two boundary-only (Coons) macroelements shown in Fig. 13.4b, with 18 and 20 nodes, respectively. In addition, a single Coons macroelement with 30 nodes (on the boundary only) is used, as a reference.

Results are presented below for both cases of domain decomposition, where in the first case the master element is on the left whereas in the second case the master

Table 13.1 Calculated eigenvalues for a cavity of dimensions 2.5 × 1.1 using conventional finite elements as shown in Fig. 13.4a

Mode (i)	Exact eigenvalue (ω_i^2)	Errors (in %) of calculated eigenvalues	
		Uniform mesh of divisions 10 × 5 (66 equations) reference	Two patches of divisions 5 × 4 and 5 × 5 (60 equations: black circles in Fig. 13.4a)
1	0.00000	–	–
2	1.57914	0.83	0.79
3	6.31655	3.33	3.49
4	8.15670	3.33	5.25
5	9.73583	2.92	3.70
6	14.2122	5.23	6.45
7	14.4732	5.66	6.09

element is on the right. The total number of active nodes after coupling is 32 and 33, respectively.

Interestingly, in heat flow analysis no error appeared at all.

In the eigenvalue acoustic analysis, the errors of the calculated values are shown in Table 13.2. One may observe that the master–slave procedure does not induce serious numerical errors and in both cases the solution is of high quality (slightly better in the case of 33 DOFs when compared to 32 DOFs). It is worthy to mention that the partitioning of the cavity into two macroelements highly improves the seventh eigenvalue (from 12.16 to only 0.3 %).

13.2 Artificial Nodes

The transfinite interpolation described in the previous chapters may include also "artificial" nodes (internal or external), in order to extend the performance of the interpolation (see [15]). The use of artificial nodes will be explained for the two-dimensional transfinite interpolation case:

$$U(\xi,\eta) = P_\xi\{U\} + P_\eta\{U\} - P_{\xi\eta}\{U\} \tag{13.7}$$

The full expression of transfinite interpolation, given by Eq. (13.7), is applicable to typical cases such as that presented in Fig. 13.6a where auxiliary (artificial) nodes have to be temporarily introduced. The reader may also repeat Example 4.10, which is labeled as "true transfinite element". Following the same concept, artificial nodes have to be introduced for the example of Fig. 13.6b. In contrast, the example of single Fig. 13.6c does not require any auxiliary node at all (it is a tensor product).

Table 13.2 Calculated eigenvalues for a cavity of dimensions 2.5 × 1.1 using one or two Coons macroelements, as shown in Fig. 13.4b (c, d)

Mode (i)	Exact eigenvalue (ω_i^2)	Errors (in %) of calculated eigenvalues using Coons macroelement(s)		
		A single macroelement (30 DOF) reference	Master is (I) 32 DOF (Fig. 13.4c)	Master is (II) 33 DOF (Fig. 13.4d)
1	0.00000	–	–	–
2	1.57914	0.00	0.00	0.00
3	6.31655	0.00	0.00	0.00
4	8.15670	0.00	0.04	0.03
5	9.73583	0.28	0.30	0.12
6	14.2122	0.00	0.03	0.03
7	14.4732	12.16	0.32	0.31
8	22.3689	7.78	7.80	5.61
9	25.2662	0.00	0.73	0.73
10	32.6268	0.73	0.73	0.73

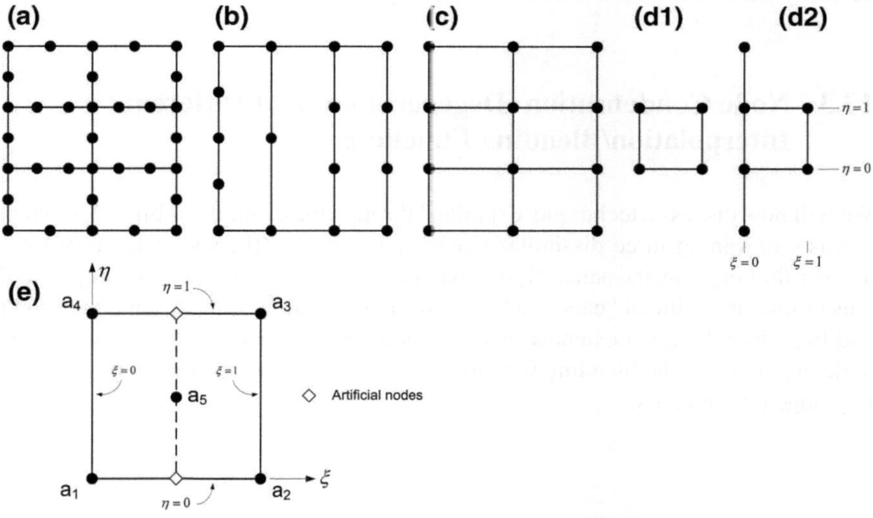

Fig. 13.6 Characteristic cases of Coons interpolation: **a** full formula, **b** blending in ξ-direction, **c** Cartesian product, (d1) real patch to match with c-patch on its left, (d2) modification of d1-patch, **e** artificial nodes

Also, if the patches in Fig. 13.6c, d1 have to be joined together, the element of Fig. 13.6d1 is modified to the element of Fig. 13.6d2 so as to include at $\xi = 0$ the nodes of the element of Fig. 13.6c found at its relevant position $\xi = 1$ ("outer artificial nodes" that look like "ears").

In an analogous manner, the patch in Fig. 13.6e including a central internal node ($\xi = \eta = \frac{1}{2}$) can be easily interpolated after a couple of "inner artificial nodes" is created. Using, for the sake of briefness, the lofting operator $P_\xi\{U\}$ only, the artificial nodes corresponding to the former central node are defined as its projections over both ξ–axes (at $\eta_0 = 0$, $\eta_1 = 1$). Assuming linear interpolation, the projections are $(a_3 + a_4)/2$ and $(a_1 + a_2)/2$, respectively, and the interpolation becomes:

$$\begin{aligned} U(\xi, \eta) = {} & E_0(\xi) \cdot [L_0^1(\eta) \cdot a_1 + L_1^1(\eta) \cdot a_4] \\ & + E_{1/2}(\xi) \cdot [L_0^2(\eta) \cdot (a_1 + a_2)/2 + L_{1/2}^2(\eta) \cdot a_5 \\ & + L_1^2(\eta) \cdot (a_4 + a_3)/2] + E_1(\xi) \cdot [L_0^1(\eta) \cdot a_2 \\ & + L_1^1(\eta) \cdot a_3], \end{aligned} \tag{13.8}$$

where L_j^n denotes the Lagrangian polynomial in η passing through $(n + 1)$ nodes at $0 \leq j = \eta \leq 1$. The full implementation to a 5-node isoparametric element may be found in the solution of Example 4.1 (see Fig. 4.2, Chap. 4).

It is evident that the above procedure applies also to nodes with derivative constraints. It is also remarkable that the former idea of introducing artificial nodes can be easily extended to 3D solid blocks, etc., replacing them with artificial lines, etc.

13.3 Node Condensation (Degeneration) and Different Interpolation/Blending Functions

We will now discuss a technique, explained through the example of Fig. 13.7, which consists in joining three dissimilar Coons' patches (I), (II), and (III). In order to achieve the coupling, the patch (II) is expanded (in an analogous way with Fig. 13.6d2 concerning the artificial "ears"), which is the degeneration of the patch on the right, and the interpolation is extended over the nodes 0, 1/2, and 1. Due to the internal node in patch (II), the blending functions $E(\xi)$ and $E(\eta)$ are chosen to be quadratic Lagrange polynomials.

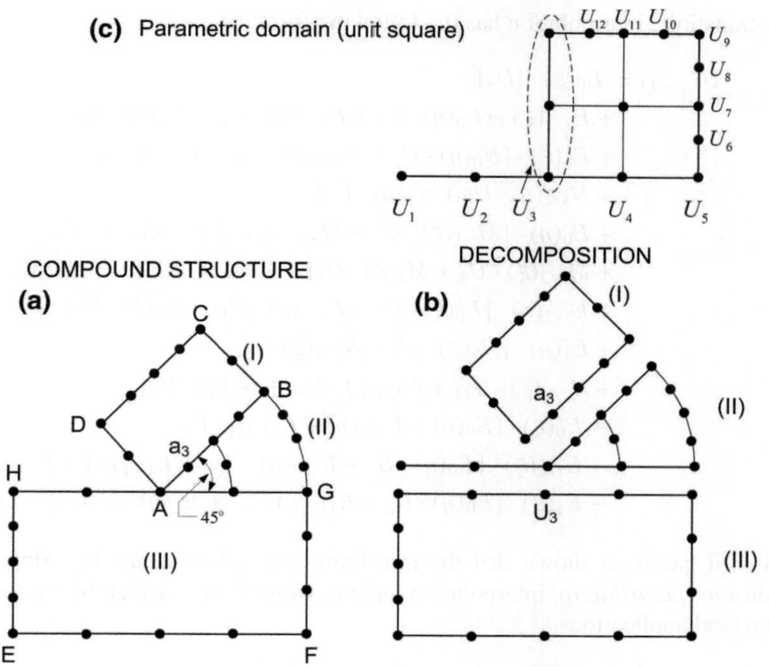

Fig. 13.7 **a** Three dissimilar patches, **b** after decomposition, **c** unit square

For the Lagrange polynomials at the bottom of the unit square (i.e., $[L_i\ (\xi)]_{\eta=0}$), the interpolation is extended over the nodes with $\xi = -1, -\frac{1}{2}, 0, \frac{1}{2}$ and 1. Finally, for the Lagrange polynomials at the top of the unit square (i.e., $[L_i(\xi)]_{\eta=1}$), the interpolation is extended over the nodes 0, ¼, ½, ¾, and 1. Introducing the following Lagrange polynomials:

$L_0, L_{1/2}, L_1$:	Interpolating {0, 0.5, 1}
$M_{-1}, M_{-1/2}, M_0, M_{1/2}, M_1$:	Interpolating {−1, −0.5, 0, 0.5, 1}
$N_0, N_{1/4}, N_{1/2}, N_{3/4}, N_1$:	Interpolating {0, 0.25, 0.5, 0.75, 1}

the transfinite interpolation has the following form:

$$\begin{aligned}
U(\xi,\eta) = {} & E_0(\xi)\cdot[U_3] \\
& + E_{1/2}(\xi)\cdot[L_0(\eta)\cdot U_4 + L_{1/2}(\eta)\cdot U_{13} + L_1(\eta)\cdot U_{11}] \\
& + E_1(\xi)\cdot[N_0(\eta)\cdot U_5 + N_{1/4}(\eta)\cdot U_6 + N_{1/2}(\eta)\cdot U_7 \\
& + N_{3/4}(\eta)\cdot U_8 + N_1(\eta)\cdot U_9] \\
& + E_0(\eta)\cdot[M_{-1}(\xi)\cdot U_1 + M_{-1/2}(\xi)\cdot U_2 + M_0(\xi)\cdot U_3 \\
& + M_{1/2}(\xi)\cdot U_4 + M_1(\xi)\cdot U_5] \\
& + E_{1/2}(\eta)\cdot[L_0(\xi)\cdot U_3 + L_{1/2}(\xi)\cdot U_{13} + L_1(\xi)\cdot U_7] \\
& + E_1(\eta)\cdot[N_0(\xi)\cdot U_3 + N_{1/4}(\xi)\cdot U_{12} \\
& + N_{1/2}(\xi)\cdot U_{11} + N_{3/4}(\xi)\cdot U_{10} + N_1(\xi)\cdot U_9] \\
& - E_0(\xi)\cdot[E_0(\eta) + E_{1/2}(\eta) + E_1(\eta)]\cdot U_3 \\
& - E_{1/2}(\xi)\cdot[E_0(\eta)\cdot U_4 + E_{1/2}(\eta)\cdot U_{13} + E_1(\eta)\cdot U_{11}] \\
& - E_1(\xi)\cdot[E_0(\eta)\cdot U_5 + E_{1/2}(\eta)\cdot U_7 + E_1(\eta)\cdot U_9].
\end{aligned} \tag{13.9}$$

Thus, it has been shown that the transfinite interpolation may be extended to include *non-quadrilateral* interpolation regions, which is of considerable importance for practical applications.

Remark For the interpolation and the blending functions in the η-direction, it is also possible to use a non-polynomial function, e.g., a trigonometric interpolation function of the form:

$$[L(\eta)]_{\xi=1/2} = E(\eta) = b_0 + b_{1/2}\cdot\sin(k\eta) + b_1\cdot\cos(k\eta) \tag{13.10}$$

with

$$k\eta = 0 \quad \text{for } \eta = 0, \quad \text{and} \quad k\eta = 40^\circ \times \frac{\pi}{180}(\text{rad}) \quad \text{for } \eta = 1. \tag{13.11}$$

In Eq. (13.11), the value $k = 40 \times \frac{\pi}{180}$ is chosen according to angle of 40° that is formed in patch (II). The coefficients b_0, $b_{1/2}$ and b_1 can be easily computed by solving a 3×3 linear system obtained by collocating Eq. (13.10) at the nodal point associated to the variables U_5, U_7 and U_9, defining the circular (or elliptic) part.

13.4 Closed Surface Patches

If (instead of Lagrange, Hermite polynomials, or splines) the (periodical) trigonometrical base functions, obtained by Fourier expansion, are used as blending and/or interpolating functions, it is easy to approximate closed-form surfaces such as a helical surface (ξ: trigonometric; η: arbitrary), a sphere (ξ, η: trigonometric), etc. (as shown in Fig. 13.8). Also, axisymmetric surfaces such as part from a torus or the

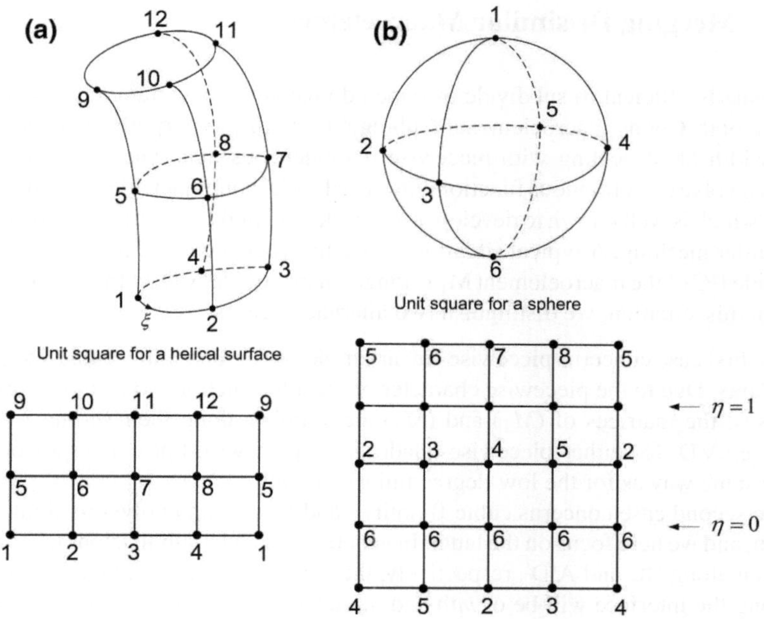

Fig. 13.8 Coons interpolation to closed surfaces: **a** helical and **b** sphere

circular fillet between two coaxial cylinders with different diameters can be approximated by uni-linearly blended interpolation with lofting directed circumferentially.

13.5 Local Control

It is widely accepted that Coons interpolation leads to surfaces of poor quality due to its global nature. Nevertheless, although in most CAD system it is rather obsolete, however it offers certain flexibilities as follows [15]. Let us suppose that after the application of the bilinear Coons interpolation the designer is not satisfied with the produced surface around its central area (e.g., in the strip $3/4 \leq \xi \leq 1$). Keeping unchanged, the left ($0 \leq \xi \leq 1/4$) and the right ($3/4 \leq \xi \leq 1$) subpatches, both perpendicular to the $\xi-$ axis, as well as the linear blending functions, one can now redefine the central subpatch ($1/4 \leq \xi \leq 3/4$) so that it may obtain an additional condition like the second-order derivative $F_{\xi\xi}$ at a specific point (e.g., at $\xi = \eta = 1/2$). This is easily achieved considering for the central subpatch the characteristics (values, derivatives) of the initial patch along the lines $\xi = 1/4$ and $\xi = 3/4$ and applying a process similar to the artificial node technique we met earlier in Sect. 13.2.

13.6 Merging Dissimilar Macroelements

It is usually efficient to subdivide extended domains into a certain number of large and smooth Coons macroelements (subregions) with the purpose of reducing the bandwidth or of dealing with piecewise homogeneous materials. However, if the edges involve discontinuous functions, the usual nodal coupling will give a quadrature error, which is well known to developers of mortar methods for merging domains with dissimilar meshing. A typical situation of two macroelements is shown in Fig. 13.9. The side BC of the macroelement M_1 is longer than the side A′D′ of the macroelement M_2. In this situation, we distinguish two alternative cases:

- The first case concerns piecewise-linear (or piecewise-quadratic, cubic, etc.) interpolants. Due to the piecewise character of the interpolation, no technical problem arises; the matrices of (M_1) and (M_2) are coupled along their shorter interface (here, A′D′ for either piecewise-quadratic or piecewise-linear interpolation), in the same way as for the low-degree finite elements.
- The second case concerns cubic B-splines and Lagrangian polynomial interpolation, and we here focus on the latter. In general, if m and n subdivisions $(n < m)$ are taken along BC and A′D′, respectively, the corresponding Lagrange polynomials along the interface will be of mth and nth degree, respectively. Therefore, continuity condition cannot be fulfilled along the interface. In order to overcome this shortcoming, the degeneration technique mentioned in Sect. 13.3 may be applied. The concept was developed in a Diploma Work by Dionysios Grekas (1988) and was published by Kanarachos et al. [15]. Also, an example was published by Provatidis ([22], pp. 330–331), and here a correction is made in *Example* 13.2.

Example 13.2 (Merging two dissimilar macroelements) The boundary value problem in Fig. 13.9 represents the uniaxial heat flow along a composite structure that consists of a large square (macroelement M_1) coupled with a shorter one (M_2). The temperatures are given along the vertical left (nodes 1, 13, 14, 15, and 16) and vertical right edges (nodes 18, 19, and 20), with values 100 and 40 °C, respectively, while the flux at the junction ($x = 1.0$) is given equally to $q = \partial U/\partial x = -40$. The top side (nodes 9–13 and 21) and the bottom side (nodes 1–5 and 17) are fully insulated ($\partial U/\partial x = 0$). The question is to determine the temperatures at the rest nodes.

Solution The macroelement ABCD (M_1) is chosen as the *master* element because it contains five nodes along the junction at $x = 1$, whereas M_2 includes only three. Therefore, M_1 is handled as a usual boundary-only Coons element. In this example, we choose Lagrange polynomials, so M_1 becomes a member of the well-known Serendipity family. Obviously, the corresponding stiffness matrix, $\mathbf{K}_1$, will be of order 16×16.

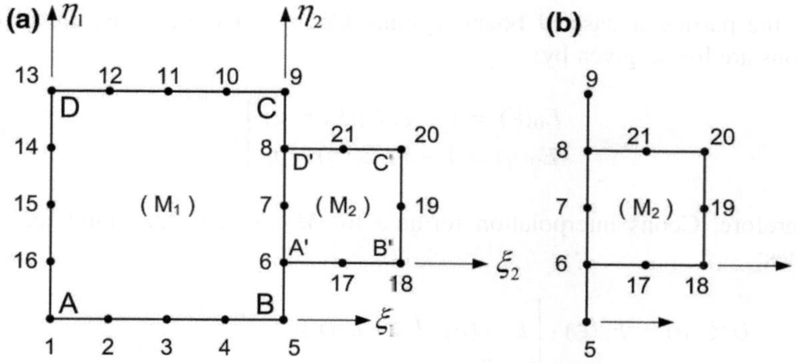

Fig. 13.9 **a** Merging two dissimilar macroelements M_1 and M_2 based on **b** enhanced M_2

The macroelement A′B′C′D′ (M_2) is an eight-node boundary-only Coons element, but it is not always offered for a reliable coupling as is. To build the enhanced model for this macroelement, we consider that information along the line $\xi_2 = 0$ in the Coons patch M_2 is not known strictly along the actual side A′D′ but also along the longer edge BC (of M_1). This means that data should be considered along $\eta_2 = -\frac{1}{2}, 0, \frac{1}{2}, 1$ and $\frac{3}{2}$. In other words, in a sense we have "introduced" two external outer nodes, i.e., 5 and 9, which primarily belong to the master M_1 (see Fig. 13.8b). In this way, both macroelements will have the *same* degrees of freedom along their interface. Then, the enhanced Coons formula for M_2 should include the interpolation along the *entire* side (with local $\xi_2 = 0$) as a polynomial of *fourth* degree, interpolating the five nodal values through the points 5–9.

To this purpose, three sets of Lagrange polynomials are introduced as follows:

- Quartic polynomials along the vertical line 5-6-7-8-9:

$$L_{-\frac{1}{2}}(\eta), L_0(\eta), L_{\frac{1}{2}}(\eta), L_1(\eta), L_{\frac{3}{2}}(\eta)$$

- Cubic polynomials along the vertical line 18-19-20:

$$M_0(\eta), M_{\frac{1}{2}}(\eta), M_1(\eta)$$

- Cubic polynomials along the horizontal line 6-17-18:

$$M_0(\xi), M_{\frac{1}{2}}(\xi), M_1(\xi).$$

The fractional and integer indices in the abovementioned Lagrange polynomials correspond to the ξ_2- or η_2- value of the nodal point to which they are associated. For example, the polynomial $L_{-\frac{1}{2}}(\eta)$ is associated to the node "5", for which $\eta_2 = -\frac{1}{2}$. A similar convention holds for the functions E_0 and E_1 as well (see below).

For the particular case of boundary-only Coons macroelements, the blending functions are linear, given by:

$$\left.\begin{array}{l} E_0(\xi) = 1 - \xi,\ E_1(\xi) = \xi, \\ E_0(\eta) = 1 - \eta,\ E_1(\eta) = \eta \end{array}\right\}. \tag{13.12}$$

Therefore, Coons interpolation formula for M_2 (where ξ, η stand for ξ_2, η_2) becomes:

$$\begin{aligned} U(\xi, \eta) = E_0(\xi) \cdot & \Big[L_{-\frac{1}{2}}(\eta) \cdot U_5 + L_0(\eta) \cdot U_6 \\ & + L_{\frac{1}{2}}(\eta) \cdot U_7 + L_1(\eta) \cdot U_8 + L_{\frac{3}{2}}(\eta) \cdot U_9 \Big] \\ & + E_1(\xi) \cdot \Big[M_0(\eta) \cdot U_{18} + M_{\frac{1}{2}}(\eta) \cdot U_{19} + M_1(\eta) \cdot U_{20} \Big] \\ & + E_0(\eta) \cdot \Big[M_0(\xi) \cdot U_6 + M_{\frac{1}{2}}(\xi) \cdot U_{17} + M_1(\xi) \cdot U_{18} \Big] \\ & + E_1(\eta) \cdot \Big[M_0(\xi) \cdot U_8 + M_{\frac{1}{2}}(\xi) \cdot U_{21} + M_1(\xi) \cdot U_{20} \Big] \\ & - E_0(\xi) \cdot E_0(\eta) \cdot U_6 - E_1(\xi) \cdot E_0(\eta) \cdot U_{18} \\ & - E_0(\xi) \cdot E_1(\eta) \cdot U_8 - E_1(\xi) \cdot E_1(\eta) \cdot U_{20} \end{aligned} \tag{13.13}$$

Equation (12.13) is a correction of a previous formula presented by Provatidis ([23], Eq. (23), p. 331). Each term in a bracket [...] of Eq. (13.13) represents the proper function $U(\xi, \eta)$ along the boundary. For example, the term that is found in the first bracket and represents the variable $U(0, \eta)$ along the side BC of the Coons patch M_1 is of most interest. Here, $U(0, \eta)$ is taken on purpose along the *entire* side of M_1 and not along the shorter side A′D′ of M_2 where it actually refers to. This particular issue is the *trick* to properly couple the two elements (previously presented by Kanarachos et al. [15] and Provatidis [22]).

Moreover, Eq. (13.13) is rearranged by generating common factors by the nodal values, as follows:

$$
\begin{aligned}
U(\xi,\eta) = {} & U_6 \cdot [E_0(\xi)L_0(\eta) + E_0(\eta)M_0(\xi) - E_0(\xi)\cdot E_0(\eta)] \\
& + U_{17} \cdot \left[E_0(\eta)M_{\frac{1}{2}}(\xi)\right] \\
& + U_{18} \cdot [E_0(\eta)M_1(\xi) + E_1(\xi)M_0(\eta) - E_1(\xi)\cdot E_0(\eta)] \\
& + U_{19} \cdot \left[E_1(\xi)M_{\frac{1}{2}}(\eta)\right] \\
& + U_{20} \cdot [E_1(\xi)M_1(\eta) + E_1(\eta)M_1(\xi) - E_1(\xi)\cdot E_1(\eta)] \\
& + U_{21} \cdot \left[E_1(\eta)M_{\frac{1}{2}}(\xi)\right] \\
& + U_8 \cdot [E_0(\xi)L_1(\eta) + E_1(\eta)M_0(\xi) - E_0(\xi)\cdot E_1(\eta)] \\
& + U_7 \cdot \left[E_0(\xi)L_{\frac{1}{2}}(\eta)\right] \\
& + U_5 \cdot \left[E_0(\xi)L_{-\frac{1}{2}}(\eta)\right] \\
& + U_9 \cdot \left[E_0(\xi)L_{\frac{3}{2}}(\eta)\right]
\end{aligned}
\tag{13.14}
$$

Obviously, each term in a bracket [...] of Eq. (13.14) represents the shape function $N_i(\xi,\eta)$ associated to the value of the variable U_i, accordingly.

Note that the quartic polynomials along the vertical line 5-6-7-8-9: $L_{-\frac{1}{2}}(\eta)$, $L_0(\eta)$, $L_{\frac{1}{2}}(\eta)$, $L_1(\eta)$, $L_{\frac{3}{2}}(\eta)$, influence *only* these nodal points.

The sequence of the nodal values in Eq. (13.14) starts from U_6 at the corner A′ of the patch M_2 (Fig. 13.9) and continues (in the counterwise direction) until the eighth nodal value U_7 of this sequence. In addition, the two outer nodes (5 and 9) with nodal values U_5 and U_9 are included. Therefore, the stiffness matrix $\boldsymbol{K}_2$ of the subregion M_2 will be of dimension 10×10.

In summary, in all cases the solution $U(\xi,\eta)$ inside the macroelement is approximated by

$$
U(\xi,\eta) = \sum_{j=1}^{q_e} N_j(\xi,\eta)\cdot U_j \tag{13.15}
$$

with U_j denoting the nodal degrees of freedom appearing at the boundaries of the macroelement as well as in its interior area, and q_e is the total number of its nodes. Obviously, the same coupling concept is applicable to transfinite elements as well as to tensor-product ones.

As usual, the stiffness matrix $\boldsymbol{K}$ of the assemblage will be of dimension 21×21, where the overlapping elements between $\boldsymbol{K}_1$ and $\boldsymbol{K}_2$ refer to all five nodal values of the set $\{U_5, U_6, U_7, U_8, U_9\}$.

The major "difficulty" is the correct calculation of the nonzero part of the external force vector, which is as follows. Since the nonzero flux is given along the segments

56 and 89, boundary integration has to be performed along these two parts. It is worthy to mention that *all* nodes along the interface, i.e., 5–9, influence the aforementioned segments 56 and 89. Therefore, the external forces which are required to be calculated and then be given as input data will be:

$$F_i = q\int_{56} N_i(\xi,\eta)\mathrm{d}y + q\int_{89} N_i(\xi,\eta)\mathrm{d}y, \qquad (13.16)$$

or

$$F_i = q\left[\int_{-\frac{1}{2}}^{0} L_{-\frac{1}{2}}(\eta)\cdot\tfrac{1}{2}\mathrm{d}\eta + \int_{1}^{\frac{3}{2}} L_{\frac{3}{2}}(\eta)\cdot\tfrac{1}{2}\mathrm{d}\eta\right], \quad i = 5,6,7,8,9.$$

The factor $\frac{1}{2}$ is due to the fact that $y = \frac{1}{2}\eta$, with the axis origin at the node 6. The calculation gives the results shown in Table 13.3.

Based on these data, the solution of the linear system (with 13 unknowns, after the deletion of rows and columns associated to given temperatures in $\mathbf{KU} = \boldsymbol{f}$) gives the nodal temperatures shown in Table 13.4. One may observe that the calculated values coincide with those of the exact solution given by:

$$U(x) = 100 - 40x, \quad 0 \le x \le \tfrac{3}{2}.$$

In order to appreciate the high accuracy shown in Table 13.4, let us now see what happens if the two macroelements are coupled as is, without considering the above trick of artificial external nodes. In the latter case, the obtained results are shown in Table 13.5, where one may observe minor changes. Nevertheless, in other problems the difference may be larger.

Table 13.3 Nodal values to be entered as data concerning the external forces

F_5	F_6	F_1	F_1	F_1
−3.2222	−10.4444	7.3333	−10.4444	−3.2222

Table 13.4 Calculated temperature at the points under boundary conditions of Neumann type (10-node enhanced macroelement M_2)

Temperature	Nodes and coordinates				
	2, 12 ($x = 0.25$)	3, 11 ($x = 0.50$)	4, 10 ($x = 0.75$)	5, 6, 7, 8, 9 ($x = 1.00$)	17, 21 ($x = 1.25$)
Calculated	90.0000	80.0000	70.0000	60.0000	50.0000
Exact	90.0000	80.0000	70.0000	60.0000	50.0000

Table 13.5 Calculated temperature at the points under boundary conditions of Neumann type (8-node macroelement M_2)

Nodes						
2, 12	3, 11	4, 10	5, 9	6, 8	7	17, 21
89.9999	79.9999	69.9998	59.9997	60.0261	59.9660	50.0131

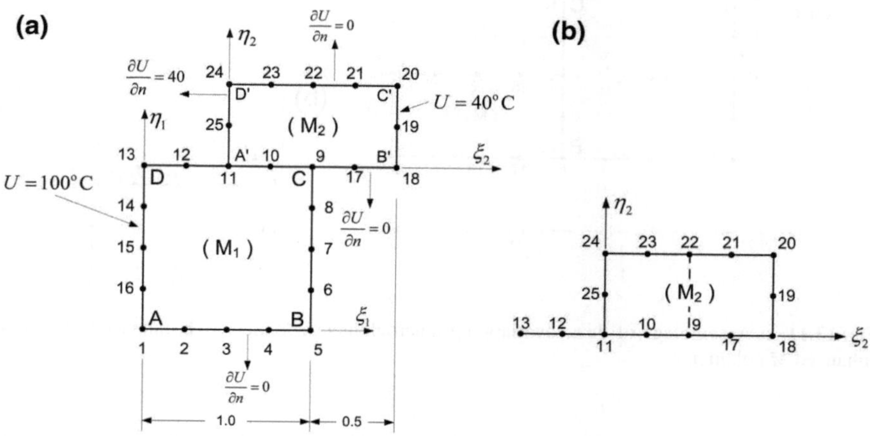

Fig. 13.10 **a** Overlapping elements and **b** enhanced M_2 element

Example 13.3 (Heat Flow in Partially Overlapping Areas) The boundary value problem in Fig. 13.10 represents the uniaxial heat flow over two *partially overlapping* elements. The difference with the previous example is that here the quartic polynomial through the nodes (13, 12, 11, 10, and 9) does not match with the (again) quartic polynomial through the nodes (11, 10, 9, 17, and 18). Nevertheless, in this particular case, when the 16-node element M_1 is coupled with the 12-node element M_2 in the usual way, an *accurate* numerical solution is obtained. This is due to the fact that (i) the internal forces equal to zero because the flux is uniaxial in the direction of the horizontal axis, and (ii) the exact temperature distribution is linear.

Under different boundary values, the accurate coupling between M_1 and M_2 requires an enhanced consideration of M_2 according to Fig. 13.10b, provided the M_1 has been considered as the master element. Nevertheless, the aforementioned enhancement is not sufficient because from the side A′B′ of M_2 only the nodes (11, 10, and 9) can be considered as a subset of the nodes required to define the quartic polynomials along the side DC of M_1. Therefore, there is need to consider a *composite* univariate function along the side A′B′ of M_2 (i.e., of $U(\xi, 0)$ involved in Coons interpolation). An apparent solution to this shortcoming is to revise the subroutine that calculates the global shape functions of M_2, considering an internal separation by the straightline segment defined by the nodes 9 and 22. In this way, regarding M_2 the part between the nodes 11 and 9 will be interpolated by a polynomial of

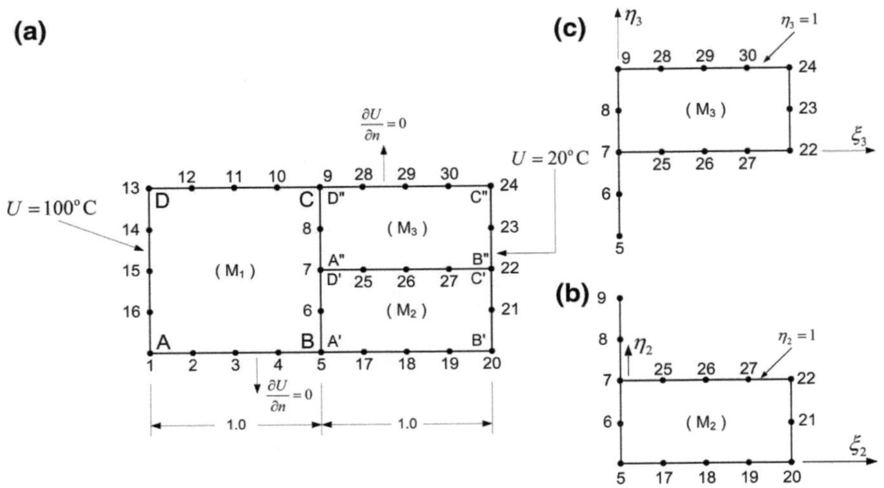

Fig. 13.11 **a** Assemblage of three overlapping macroelements, treated by **b** enhanced M_2 and **c** enhanced M_3 elements

fourth degree, whereas the rest part (9–18) could be interpolated, for example, by a quadratic polynomial.

It is noted that, in a different case that another structure (say M_3) similar to M_1 existed at the right and below M_2 (e.g., consider the symmetric of M_1 with respect to the line BC), we would not have the freedom to choose the quadratic function for the part (9–18). Instead, we had to follow the variation along the corresponding side of the adjacent element M_3, which dictates the use of a quartic polynomial for a second time. Clearly, the variation along the nodes (9, 17, and 18) would be quartic (instead of quadratic).

Example 13.4 (Heat Flow) The boundary value problem in Fig. 13.11 represents the uniaxial heat flow over three macroelements. Following the abovementioned concept of enhancing the two small elements (M_2 and M_3) according to Fig. 13.11b, c, the obtained results are shown in Table 13.6.

One may observe that when using the standard coupling based on 12-node elements, the nodes in the couples (17, 25), (18, 26), and (19, 27) do not have the same calculated temperatures as they ought but they differ from each other.

In contrast, when using two enhanced elements, $M1$ and $M2$, each of 14 nodes, the numerical results coincide with the exact solution.

Example 13.5 (Acoustic Cavity) The boundary value problem in Fig. 13.12 represents a rectangular acoustic cavity of dimensions 2.5×1.1, with hard walls.

Table 13.6 Calculated temperature at the nodes

	Nodes					
	17, 28	25	18, 29	26	19, 30	27
	Temperature					
Calculated as is (each of M_2 and M_3: 12 nodes)	49.9755	49.9563	39.9794	39.9751	29.9893	29.9879
Calculated using enhanced elements (each of M_2 and M_3: 14 nodes)	50.0000	50.0000	40.0000	40.0000	30.0000	30.0000
Exact	50.0000		40.0000		30.0000	

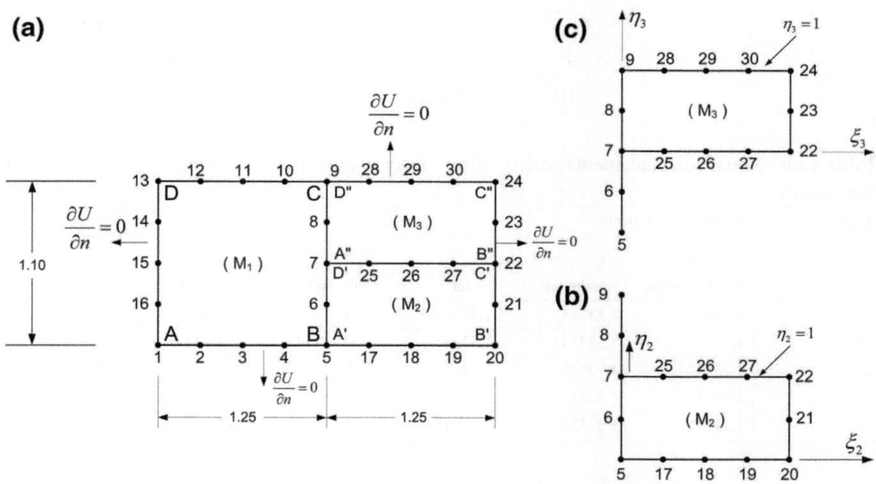

Fig. 13.12 Rectangular acoustic cavity split into three Coons macroelements

The exact solution is given by:

$$\omega_{ij}^2 = \pi^2 c^2 \left[\left(\frac{i}{a} \right)^2 + \left(\frac{j}{b} \right)^2 \right], \quad i, j = 0, 1, 2, \ldots \tag{13.17}$$

In order to test the errors involved by the domain decomposition, this problem is solved using the same type of elements as in the previous example. The numerical results are shown in Table 13.7 for two procedures. In the first procedure, no care

Table 13.7 Calculated eigenvalues of the rectangular acoustic cavity

Mode	Harmonics (i, j)	Exact eigenvalue (ω^2)	Calculated eigenvalue error (in %)	
			Usual macroelements (each of M_2 and M_3: 12 nodes)	Enhanced elements (each of M_2 and M_3: 14 nodes)
1	(0, 0)	0.00000	–	–
2	(1, 0)	1.57914	−0.01	0.00
3	(2, 0)	6.31655	0.06	0.06
4	(0, 1)	8.15670	0.40	0.14
5	(1, 1)	9.73583	0.62	0.61
6	(3, 0)	14.2122	0.51	0.52
7	(2, 1)	14.4732	0.54	0.58
8	(3, 1)	22.3689	8.08	8.11
9	(4, 0)	25.2662	0.73	0.73
10	(0, 2)	32.6268	3.21	1.53

Table 13.8 Nodal displacements using three macroelements, discontinuous or continuous (enhanced)

Node	Nodal displacements					
	$U_x(\times 10^4)$			$U_y(\times 10^6)$		
	Discontinuous	Continuous	Exact	Discontinuous	Continuous	Exact
20	1.33301	1.33333	1.33333	4.97339	5.00000	5.00000
21	1.33351	1.33333	1.33333	2.48832	2.50000	2.50000
22	1.33322	1.33333	1.33333	0	0	0
23	1.33351	1.33333	1.33333	−2.48832	−2.50000	−2.50000
24	1.33301	1.33333	1.33333	−4.97339	−5.00000	−5.00000

was taken for kinematic compatibility, whereas in the second procedure enhanced macroelements (shown in Fig. 13.12b, c) were implemented. One may observe a very small improvement in the numerical solution when the enhanced (fully compatible) elements are used.

Example 13.6 (Plane Beam in Tension) A two-dimensional beam of dimensions 48×12 in plane stress conditions is subjected to a tensile uniform stress (Fig. 13.13a, b). The calculated nodal displacements are shown in Table 13.8. One may observe that when each of the elements M_2 and M_3 consists of 14 nodes (i.e., they are continuous, previously called as "enhanced" elements), the numerical result (for elongation and contraction) coincides with the exact solution.

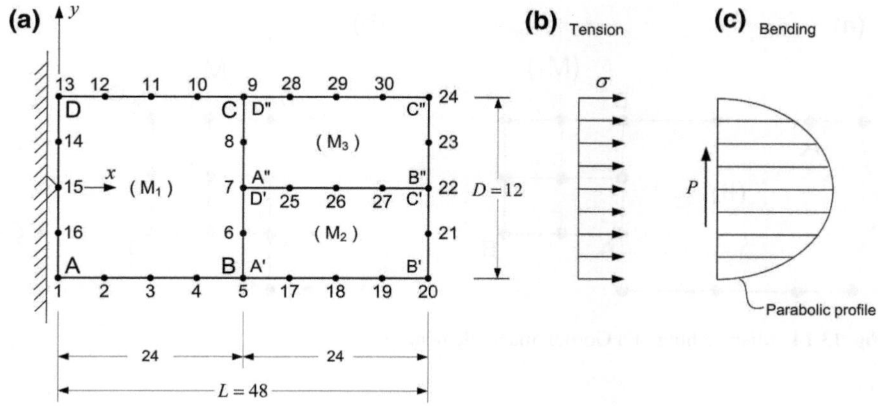

Fig. 13.13 **a** Rectangular plane beam (not in scale) under **b** tension or **c** bending loads

Table 13.9 Nodal displacements using three macroelements, discontinuous or continuous (enhanced)

Node	Nodal Displacements					
	$U_x(\times 10^4)$			$U_y(\times 10^6)$		
	Discontinuous	Continuous	Exact	Discontinuous	Continuous	Exact
20	15.9967	15.9946	16.0000	8.90318e-03	8.89961	8.90000
21	7.93110	7.92627	7.92813	8.90346e-03	8.89969	8.90000
22	0.00000	0.00000	0.00000	8.90359e-03	8.89971	8.90000
23	−7.93110	−7.92627	−7.92813	8.90346e-03	8.89969	8.90000
24	−15.9967	−15.9946	−16.0000	8.90318e-03	8.89961	8.90000
7	0.0000000	0.0000000	0.000000	2.65493e-03	2.85034e-03	0.00285

Example 13.7 (Plane Beam in Bending) The last example of this section refers to the bending of a clamped beam shown in Fig. 13.13a, c, which is supported at its left vertical side, whereas the right vertical side is loaded by a shear load P (with parabolic distribution of the shear stress τ). The following parameters were used: Young's modulus $E = 3.0\times10^7$ kPa; Poisson's ratio $\nu = 0.30$, diameter $D = 12$ m, length $L = 48$ m, shear force $P = 1000$ kN, and plane stress conditions.

The exact solution is given by Timoshenko and Goodier (1970) as well as by Eqs. (3.89–3.91) of Chap. 3.

For the same assemblage of 30 nodes shown in Fig. 13.13a, the results are shown in Table 13.9. One may observe a slight improvement of the numerical solution when using the enhanced elements. It is noted that the accurate solution requires polynomials of third degree in y, which means that in order to obtain the exact solution the right vertical sides of the elements M_2 and M_3 had to be divided into three segments (instead of the current two segments).

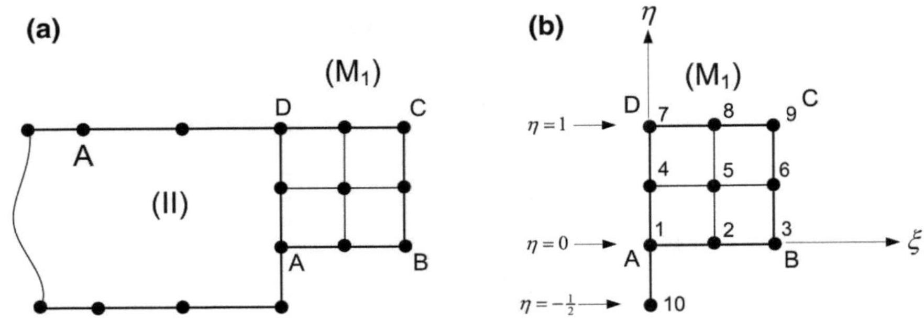

Fig. 13.14 Mismatching of a Gordon macroelement

13.7 Gordon Transfinite Macroelements

Most of the abovementioned results were obtained using (boundary-only) Coons macroelements. For the sake of completeness, a trivial case of the most easy tensor product element will be discussed and a general rule will be extracted.

Let us consider the case shown in Fig. 13.14, where the side DA of the 9-node (quadratic) element (M_1) of Lagrange type is coupled with a cubic element of another type (M_2, either Coons or Gordon).

To this purpose, three sets of Lagrange polynomials are introduced as follows:

- Cubic polynomials along the vertical line 10-1-4-7:

$$M_{-\frac{1}{2}}(\eta),\ M_0(\eta),\ M_{\frac{1}{2}}(\eta),\ M_1(\eta)$$

- Quadratic polynomials along the vertical lines 2-5-8 and 3-6-9:

$$L_0(\eta),\ L_{\frac{1}{2}}(\eta),\ L_1(\eta)$$

- Quadratic polynomials along the horizontal lines 1-2-3, 4-5-6, and 7-8-9:

$$L_0(\xi),\ L_{\frac{1}{2}}(\xi),\ L_1(\xi).$$

The transfinite interpolation includes the following projections:

$$\begin{aligned} P_\xi = E_0(\xi)\Big[M_{-\frac{1}{2}}(\eta)\cdot U_{10} + M_0(\eta)\cdot U_1 + M_{\frac{1}{2}}(\eta)\cdot U_4 + M_1(\eta)\cdot U_7\Big] \\ + E_{\frac{1}{2}}(\xi)\Big[L_0(\eta)\cdot U_2 + L_{\frac{1}{2}}(\eta)\cdot U_5 + L_1(\eta)\cdot U_8\Big] \\ + E_1(\xi)\Big[L_0(\eta)\cdot U_3 + L_{\frac{1}{2}}(\eta)\cdot U_6 + L_1(\eta)\cdot U_9\Big] \end{aligned} \tag{13.18}$$

Also,

$$P_\eta = E_0(\eta)\left[L_0(\eta)\cdot U_1 + L_{\frac{1}{2}}(\eta)\cdot U_2 + L_1(\eta)\cdot U_3\right] + E_{\frac{1}{2}}(\eta)\left[L_0(\eta)\cdot U_4 + L_{\frac{1}{2}}(\eta)\cdot U_5 + L_1(\eta)\cdot U_6\right] + E_1(\eta)\left[L_0(\eta)\cdot U_7 + L_{\frac{1}{2}}(\eta)\cdot U_8 + L_1(\eta)\cdot U_9\right] \tag{13.19}$$

and finally, the corrective term is:

$$P_{\xi\eta} = E_0(\xi)E_0(\eta)U_1 + E_{\frac{1}{2}}(\xi)E_0(\eta)U_2 + E_1(\xi)E_0(\eta)U_3 + E_0(\xi)E_{\frac{1}{2}}(\eta)U_4 + E_{\frac{1}{2}}(\xi)E_{\frac{1}{2}}(\eta)U_5 + E_1(\xi)E_{\frac{1}{2}}(\eta)U_6 + E_0(\xi)E_1(\eta)U_7 + E_{\frac{1}{2}}(\xi)E_1(\eta)U_8 + E_1(\xi)E_1(\eta)U_9 \tag{13.20}$$

Since the blending function is identical with the quadratic Lagrange polynomials:

$$E_0 = L_0,\ E_{\frac{1}{2}} = L_{\frac{1}{2}},\ E_1 = L_1 \tag{13.21}$$

one easily can validate that

$$P_\eta = P_{\xi\eta} \tag{13.22}$$

Therefore, Eq. (13.7) implies:

$$U(\xi,\eta) = P_\xi \tag{13.23}$$

Substituting Eq. (13.18) into Eq. (13.23), by virtue of Eq. (13.21) the latter becomes:

$$U(\xi,\eta) = [L_0(\xi)M_0(\eta)]\cdot U_1 + \left[L_{\frac{1}{2}}(\xi)L_0(\eta)\right]\cdot U_2 + [L_1(\xi)L_0(\eta)]\cdot U_3 + \left[L_0(\xi)M_{\frac{1}{2}}(\eta)\right]\cdot U_4 + \left[L_{\frac{1}{2}}(\xi)L_{\frac{1}{2}}(\eta)\right]\cdot U_5 + \left[L_1(\xi)L_{\frac{1}{2}}(\eta)\right]\cdot U_6 + [L_0(\xi)M_1(\eta)]\cdot U_7 + \left[L_{\frac{1}{2}}(\xi)L_1(\eta)\right]\cdot U_8 + [L_1(\xi)L_1(\eta)]\cdot U_9 + \left[L_0(\xi)M_{-\frac{1}{2}}(\eta)\right]\cdot U_{10} \tag{13.24}$$

One may observe that:

(1) Those nodal points that do not belong to the interface (i.e., nodes 2, 3, 5, 6, 8, and 9) are associated to shape functions that are tensor products of the standard Lagrange polynomials.
(2) The nodes on the interface (i.e., 1, 4, and 5), including that outside the primary macroelement (i.e., 10), are also tensor products with the following difference. The $\xi-$ dependent function is $L_0(\xi)$ (because the interface lies along the isoline $\xi = 0$), whereas the $\eta-$ dependent function is the corresponding enhanced polynomial $M_j(\eta),\ j = -\frac{1}{2}, 0, \frac{1}{2}, 1$.

13.8 Non-rational Bernstein–Bézier, B-Splines, and NURBS

One may start directly from the expression (13.24) and then replace the quadratic Lagrange polynomials with the quadratic Bernstein ones using the identities (see Eq. (6.9a), in Chap. 6]:

$$\underbrace{\begin{bmatrix} L_0(x) \\ L_{\frac{1}{2}}(x) \\ L_1(x) \end{bmatrix}}_{L} = \underbrace{\begin{bmatrix} 1 & -1/2 & 0 \\ 0 & 2 & 0 \\ 0 & -1/2 & 1 \end{bmatrix}}_{T} \cdot \underbrace{\begin{bmatrix} B_{0,2}(x) \\ B_{1,2}(x) \\ B_{2,2}(x) \end{bmatrix}}_{B}, \tag{13.25}$$

where x stands for either of ξ or of η.

Moreover, by virtue of Eq. (6.23) of Chap. 6, the cubic polynomials along the vertical line 10-1-4-7: $M_{-\frac{1}{2}}(\eta)$, $M_0(\eta)$, $M_{\frac{1}{2}}(\eta)$, $M_1(\eta)$ are interrelated with the Bernstein polynomials as follows:

$$\begin{bmatrix} M_{-\frac{1}{2}}(\eta) \\ M_0(\eta) \\ M_{\frac{1}{2}}(\eta) \\ M_1(\eta) \end{bmatrix} = \begin{bmatrix} 1 & -\frac{5}{6} & \frac{1}{3} & 0 \\ 0 & 3 & -\frac{3}{2} & 0 \\ 0 & -\frac{3}{2} & 3 & 0 \\ 0 & \frac{1}{3} & -\frac{5}{6} & 1 \end{bmatrix} \cdot \begin{bmatrix} B_{0,3}(\eta') \\ B_{1,3}(\eta') \\ B_{2,3}(\eta') \\ B_{3,3}(\eta') \end{bmatrix}, \quad -\frac{1}{2} \le \eta \le 1 \tag{13.26}$$

In Eq. (13.26), the Bernstein polynomials should be taken as:

$$\begin{bmatrix} B_{0,3}(\eta') \\ B_{1,3}(\eta') \\ B_{2,3}(\eta') \\ B_{3,3}(\eta') \end{bmatrix} = \begin{bmatrix} (1-\eta')^3 \\ 3(1-\eta')^2\eta' \\ 3(1-\eta')\eta'^2 \\ \eta'^3 \end{bmatrix}, \tag{13.27}$$

with

$$\eta' = \tfrac{2}{3}\eta + \tfrac{1}{3}, \tag{13.28}$$

which ensures that $-\frac{1}{2} \le \eta \le 1$ and $0 \le \eta' \le 1$. ■

An alternative way is to start the procedure directly from Eq. (13.23), i.e., from the relationship $U(\xi,\eta) = P_\xi$, and then continue with Eq. (13.18). Then, we have:

$$\begin{aligned} U(\xi,\eta) = {} & E_0(\xi)\big[B_{0,3}(\eta')\cdot a_{10} + B_{1,3}(\eta')\cdot a_1 + B_{2,3}(\eta')\cdot a_4 + B_{3,3}(\eta')\cdot a_7\big] \\ & + E_{\frac{1}{2}}(\xi)\big[B_{0,2}(\eta)\cdot a_2 + B_{1,2}(\eta)\cdot a_5 + B_{2,2}(\eta)\cdot a_8\big] \\ & + E_1(\xi)\big[B_{0,2}(\eta)\cdot a_3 + B_{1,2}(\eta)\cdot a_6 + B_{2,2}(\eta)\cdot a_9\big] \end{aligned} \tag{13.29}$$

Furthermore, we consider that the three blending functions are quadratic Lagrange polynomials; thus, they fulfill similar to Eq. (13.25) identity:

$$\begin{bmatrix} E_0(\xi) \\ E_{\frac{1}{2}}(\xi) \\ E_1(\xi) \end{bmatrix} = \begin{bmatrix} 1 & -1/2 & 0 \\ 0 & 2 & 0 \\ 0 & -1/2 & 1 \end{bmatrix} \cdot \begin{bmatrix} B_{0,2}(\xi) \\ B_{1,2}(\xi) \\ B_{2,2}(\xi) \end{bmatrix}. \tag{13.30}$$

Finally, substituting Eq. (13.30) into Eq. (13.29), the tensor-product approximation of the variable $U(\xi,\eta)$ into the expanded Bézier patch (with ten coefficients: $a_1,\ldots,a_{10}$ instead of only nine) is straightforward.

Remark In a similar way, the univariate function $U(0,\eta)$ through the points 10, 1, 4, and 7 can be approximated using B-splines or NURBS, and the same holds for $U(\frac{1}{2},\eta)$ and $U(1,\eta)$ as well. Substituting the aforementioned expressions again in Eq. (13.18), the enhanced global shape functions are derived.

13.9 Recapitulation

It was shown that Coons macroelements work well even when the domain is decomposed into a number of them. The most essential matter is the discontinuity between the kinematic degrees of freedom along the interfaces of these elements. A standard methodology was revealed which may be used to improve the accuracy. The so obtained elements were called "enhanced" Coons elements. Obviously, several types of transfinite elements, including also those of Lagrange type, may be included. For the sake of brevity, the tensor-product macroelements based on Bézier and B-splines were discussed only theoretically, but the provided details are sufficient so the reader may develop his/her in-house computer program.

References

1. Abdoulaev GS, Achdou Y, Kuznetsov YA, Prud'Homme C (1999) On a parallel implementation of the mortar element method. Mathematical Modelling and Numerical Analysis 33(2):245–259
2. Akula BR, Yastrebov VA, Vignollet J, Cailletaud G (2017) Domain tying across virtual interfaces: coupling X-FEM with the Mortar method. In: Proceedings CSMA 2017, 13ème colloque National en Calcul des structures, 15–19 Mai 2017, Presqu'île de Giens (Var), 7 pages
3. Apostolatos A, Schmidt R, Wüchner R, Bletzinger K-U (2014) A Nitsche-type formulation and comparison of the most common domain decomposition methods in isogeometric analysis. Int J Numer Meth Eng 97:473–504
4. Bathe K-J (1996) Finite element procedures. Prentice Hall, Upper Saddle River, New Jersey
5. Belgacem FB (1999) The mortar finite element method with Lagrange multipliers. Numer Math 84:173–197
6. Belgacem FB, Maday Y (1997) The mortar finite element method for three dimensional finite elements. Math Model Numer Anal 31:289–302
7. Burman E, Claus S, Hansbo P, Larson MG, Massing A (2014) CutFEM: discretizing geometry and partial differential equations. Int J Numer Meth Eng 104(7):472–501
8. Cavendish JC, Gordon WJ, Hall CA (1977) Substructured macro elements based on locally blended interpolation. Int J Numer Meth Eng 11:1405–1421
9. Cavendish JC, Hall CA (1984) A new class of transitional blended finite elements for the analysis of solid structures. Int J Numer Meth Eng 20:241–253
10. Dornisch W, Stöckler J, Müller R (2018) Coupling of NURBS patches with the isogeometric dual mortar method. In: 6th European conference on computational mechanics (ECCM 6), Glasgow, UK, 11–15 June 2018
11. Dryja M (2005) A Neumann-Neumann algorithm for a mortar discretization of elliptic problems with discontinuous coefficients. Numer Math 99:645–656
12. Egger H (2009) A class of hybrid mortar finite element methods for interface problems with non-matching meshes. Preprint: AICES-2009-2, Aachen Institute for Advanced Study in Computational Engineering Science. Available from: https://www.aices.rwth-aachen.de/files/Inhalte%20Webseite/pdf/Preprints/AICES-2009-2.pdf
13. Gordon WJ, Thiel LC (1982) Transfinite mappings and their application to grid generation. In: Thompson JF (ed) Numerical grid generation, applied mathematics and computation, vol 10–11. pp 171–233
14. Hesch C, Betsch P (2012) Isogeometric analysis and domain decomposition methods. Comput Methods Appl Mech Eng 213–216:104–112
15. Kanarachos A, Grekas D, Provatidis C (1995) Generalized formulation of Coons' interpolation. In: Kaklis PD, Sapidis NS (eds) Computer aided geometric design: from theory to practice, NTUA, Ship-Design Laboratory/CAGD Group, Chapter 7, pp 65–76. ISBN 960-254-068-0
16. Lamichhane BP (2006) Higher order mortar finite elements with dual Lagrange multiplier spaces and applications. Dissertation, Faculty of Mathematics and Physics, University of Stuttgart
17. Maday Y, Mavriplis C, Patera AT (1989) Nonconforming mortar element methods: application to spectral discretizations. In: Domain decomposition methods (Los Angeles, CA, 1988), SIAM, Philadelphia, AP, pp 392–418
18. Marcinkowski L (2001) Domain decomposition methods for mortar finite element discretizations of plate problems. SIAM J Numer Anal 39(4):1097–1114
19. Parvizian J, Düster A, Rank E (2007) Finite cell method: h- and p-extension for embedded domain problems in solid mechanics. Comput Mech 41:121–133
20. Popp A (2012) Mortar methods for computational contact mechanics and general interface problems. Dissertation, Technical University of Munich, Faculty of Mechanical Engineering, June 2012
21. Provatidis C (2000) A comparative FEM-study of tooth mobility using isotropic and anisotropic models of the periodontal ligament. Med Eng Phys 22(5):359–370

22. Provatidis CG (2006) Coons-patch macroelements in two-dimensional parabolic problems. Appl Math Model 30:319–351
23. Provatidis CG (2006) Three-dimensional Coons macroelements: application to eigenvalue and scalar wave propagation problems. Int J Numer Meth Eng 65:111–134
24. Schwedler M (2016) Integrated structural analysis using isogeometric finite element methods. Dissertation, Fakultät Bauingenieurwesen der Bauhaus-Universität Weimar
25. Seshaiyer P, Suri M (1998) Convergence results for non-conforming hp methods: the mortar finite element method. Contemp Math 218:453–459
26. Stefanica D (1999) Domain decomposition methods for mortar finite elements. Dissertation, Courant Institute of Mathematical Sciences, New York University
27. Wassouf Z (2009) Die mortar methode für finite elemente hoher ordnung. Dissertation, Technische Universität München, Lehrstuhl für Computation in Engineering
28. Wohlmuth BI (2000) A mortar finite element method using dual spaces for the Lagrange multiplie. SIAM J Numer Anal 38:989–1012

Chapter 14
Review and Epilogue

Abstract In this chapter, we summarize the most important issues which we expect that the reader should have learned from the previous chapters. We repeat the six most important CAGD interpolations, and then we sketch a draft picture on the evolution of the computational methods which are based on the first five of them (older than isoGeometric Analysis). As the author happened to have been involved in these older CAD-based methodologies since 1982 and has somehow contributed in a large part of the whole CAE spectrum (FEM, BEM, and global collocation), the short history of this research is given from his point of view; a full list of the sixty papers, properly classified (in FEM, BEM, and collocation categories), is given in Appendix. In some places, the text is inspired by self-criticism. We believe that the compact information provided in this chapter, as well as the details in the whole book, will strengthen and broaden the horizon of researchers and postgraduate students in the field of Computational Mechanics. In more detail, we anticipate that the gap between CAD and CAE approaches and communities will be further reduced, bridging the older with the contemporary ideas.

Keywords Review · Coons–Gordon · Personal view · Tensor-product CAD/FEA integration · Overall findings · Epilogue

14.1 Introduction

The Theory of Computer-Aided Geometric Design (CAGD) has been expressed over the years through several different formulations (as mentioned in the previous chapters) until it has been viewed almost the same as, the nonuniform rational B-spline (NURBS). Within this context, this book has tried to present the topic of integration between CAD and FEA for most of the "Precursors" of the isoGeometric Analysis (the latter introduced by Hughes et al. [32]), in the sense that *all* CAGD interpolation formulas may lead to large *isoparametric* finite elements (macroelements) that share the same functional set of basis functions with that inherent in the CAGD representation of surface patches or volume blocks. Not only that, but even a *new* CAGD

C. G. Provatidis, *Precursors of Isogeometric Analysis*, Solid Mechanics and Its Applications 256, https://doi.org/10.1007/978-3-030-03889-2_14

interpolation emerges, it is probably certain that again a *pertinent* FEM method (as well as BEM and Collocation) will be developed (a job for the next generations!).

In chronological sequence, we could say that the theory of CAGD has passed through the following stages (the dates are indicative):

(1) COONS interpolation (1964–1967).
(2) GORDON interpolation (~1969).
(3) BEZIER (~1970).
(4) B-splines (~1972).
(5) BARNHILL interpolation (1973).
(6) NURBS (theoretically after 1975, but practically before 1990).

Surely the above catalog is *incomplete*. For example, the excellent works of Ferguson are missing. According to the distinguished Professor R. E. Barnhill, the "Gregory patch" has to be included (e.g., Gregory [28], among others). Upon finishing the text of this book I find out that works cited in the References of my first papers, such as those regarding Wachspress and pertinent large elements [16, 17, 57] are missing. The author apologizes for these and certainly other numerous omissions he has made, but the reduced space of this volume and his reduced capacity did not allow including numerical examples with all CAGD interpolations.

To make the point of CAD/CAE integration clear to everybody, let us focus in 2D analysis. In this case, the reason for the abovementioned utilization of the CAD representation in the FEM solution (and relevant methods: BEM, Collocation, Spectral, etc.) is quite obvious as follows. In a two-dimensional boundary value problem which is defined in the unit (parametric) square $(\xi, \eta) \in [0, 1] \times [0, 1]$, the *graph* of the approximate solution $U(\xi, \eta)$ represents a *surface* patch provided it is smooth and regular. Therefore, since surface patches can be approximated through the established CAGD interpolations, the relationship between CAGD and CAE (FEM, BEM, Collocation, etc.) becomes obvious. Furthermore, let me give another example from the point of view of a CAD/CAGD researcher. In October, 2001, during a two-day's Workshop I attended a lecture given by the late distinguished Professor J. Hoschek (Technical University of Darmstadt, Germany), who was talking about the fitting of a CAD surface by demanding the minimum strain energy of the corresponding thin elastic plate passing between given points [22]. Since one of the formulations on which the finite element analysis may be based is the minimization of total potential energy (functional), the relationship between CAGD and FEM becomes again evident.

Below we present some critical remarks on CAD formulas, in a simple and instructive way.

14.2 Coons and Gordon Interpolation Formulas

The famous Report written by the late MIT professor Coons [13] is concerned with arbitrary surfaces of space form as well as with hypersurfaces in higher dimensions (e.g., volumes). The Coons primary formula refers to C^0-continuity and is practically an *extension* of the one-dimensional linear interpolation $[U(\xi) = (1-\xi)U_1 + \xi U_2]$ with $\xi \in [0, 1]$ to the two-dimensional unit square with $(\xi, \eta) \in [0, 1] \times [0, 1]$. In more details, for each point $P(\xi, \eta)$ two linear interpolations are performed, the first in the ξ- and the second in the η-direction, and then a corrective term is applied. From the practical point of view, the four corners of the unit square are chosen on the real quadrilateral boundary *ABCD* of the domain, and then a mapping is automatically performed. Moreover, in the same Report [13], Steven Coons studied the circle, the quasi- and the exact sphere, ruled surfaces, and rational polynomial functions. Between others, his formula was extended so as to include derivatives (slopes) in order to achieve C^1-continuity between the adjacent patches. Note that Coons C^1-continuity formula is an extension of the 1D Hermite polynomial interpolation (well known as elementary beam element theory in FEM) to the 2D terrain. Concerning additional data (e.g., given lines, probably normal to the ξ- and η-axes) within a domain, he proposed the splitting in smaller patches to include this extra information. As far as the book in hand is concerned, among others, Steven Coons studied the quadrilateral (*ABCD*) (with C^0- and C^1-continuity) and the hexahedral (*ABCDEFGH*) domains, while he mentioned that the triangular patch can be treated as a degenerated quadrilateral [13]. In conclusion, according to Coons Report, the obtained formulas were exclusively concerned with the geometry $\boldsymbol{x}(\xi, \eta, \zeta)$ only.

Later, William (Bill) Gordon (at General Motors Research Laboratories, Warren, Michigan, USA) extended Coons theory and went a step further, considering a special treatment of *internal* data apart from the boundary of a surface patch or a volume block [23, 24]. And this was achieved without necessarily splitting the domain in subpatches (as Coons had previously proposed). Practically, Gordon took Coons formula $U = (P_\xi + P_\eta - P_{\xi\eta})\{U\}$ and enriched the meaning of the three involved projections by adding data of internal nodes as well. Anyway, the consideration of the internal data allowed for describing more controlled quadrilateral surfaces (patches *ABCD*). In his single author paper [24], the title includes the terms "interpolation and approximation", which means that Bill Gordon had deeply understood that his enhanced formula could be used for the solution of boundary value problems as well. Actually, Bill Gordon (jointly with Charles Hall) two years later, first used his transfinite formula to construct meshes (grids) useful for a further (conventional) finite element analysis [25], and second he proposed novel *transfinite* elements in a *cornerstone* paper [26]. Since the central idea in the FEM is the isoparametric mapping, the aforementioned transfinite elements obviously used the same blending approximation for both the geometry and the physical variable. If the reader is not yet convinced (after having read the aforementioned Gordon-Hall's paper, which has a very mathematical style), even the title "Ritz-Galerkin" of the work by Cavendish et al. [8] is the evidence. Clearly, they used the *same* approximation for the geometry and the physical quantity. The *practical* value of this work appears in one more paper [9] that refers to "substructured macro elements based on locally blending interpolation". Then Bill Gordon became a professor of mathematics and computer science at Drexel University in Philadelphia since 1979. A further paper on the grid generation was published by Gordon and Thiel [27] and, in one sense, could be considered as a precursor work in the so-called isogeometric analysis. Rapidly, Gordon's work influenced other FEM researchers as well [44].

From his published papers, it seems that James Cavendish (at General Motors) has played a great role in CAD/CAE integration. He conducted a Ph.D. thesis in the collocation method to elliptic and parabolic boundary value problems [5] and also continued with the Galerkin method [6]. Concerning the topic of CAD/FEA integration, except of the abovementioned study (jointly with Bill Gordon) on domain decomposition using transfinite elements [9], he also later published [7], on a new class of transitional blended finite elements for the analysis of solid structures, a paper of *intensively CAD character* using the in-house computer program GMSOLID at General Motors in which transfinite elements had been incorporated. Interestingly, he also published jointly with O. C. Zienkiewicz on the blended *infinite* elements for parabolic boundary value problems [10]. The author hypothesizes that researchers working for the industry have many restrictions to reveal all their thoughts (just think of the two de Casteljau reports that still remain classified since 1959), and also they are under pressure to solve the daily problems that appear. A collection of Cavendish's multidisciplinary published work at General Motors may be found in the Internet: (https://www.researchgate.net/scientific-contributions/2040282878_James_C_Cavendish). It is worthy to mention that concerning very complex geometries, Cavendish came back to the classical finite elements of small size for which feature-based algorithms were developed [11].

Also, the third partner, Hall [29] published on the application of the same idea (of the large isoparametric element) in fracture mechanics as well.

Although it was never claimed so, based on the abovementioned facts, it is the author's independent opinion that the idea of "CAD/CAE integration" [in the sense of using the *same* approximation for the geometry $\boldsymbol{x}(\xi, \eta, \zeta)$ and the physical quantity $U(\xi, \eta, \zeta)$] was *de facto* formed in early 1970s by the team of Bill Gordon at General Motors and was continued by late James Cavendish (1942–2010). Again, based on the papers available in the public domain, Gordon and Hall [26] developed the so-called transfinite elements, which, however, initially remained at the theoretical level until James ('Jim') Cavendish implemented them in GMSOLID [7], as mentioned above.

Before going on with the tensor-product CAD-based methods (isoparametric macroelements) that chronologically follow, I feel the need to give some personal facts so as the reader better understands author's position.

14.3 My Personal View and Experience on CAD/CAE Integration

14.3.1 General Conditions

I was graduated in October 1979 from the Mechanical Engineering Department at the National Technical University of Athens (NTUA), Greece. After my military service, in February 1982, I was appointed as a Research Assistant at the CAD/CAE Group in Mechanical Engineering Department (NTUA). As a student (1974–1979) I had heard about spline interpolation, which was taught in the very neighboring Naval Engineering Department. However, it was 1982 when I came in contact with the CAGD methods in a systematic way. The Group leader, Prof. A. E. Kanarachos, had served in Germany for many years, the last seven years as an Associate Professor at Ruhr University Bochum. Among others, he had considerable experience on CAD theory and finite element methodologies, particularly on the influence of the parameterization of the domain in the grid formation (see, Kanarachos and Röper [39] and a pertinent Dissertation by Röper [51]). To give a picture of the dominating mentality in the Group, one of the exercises I was teaching from the blackboard (to undergraduate student of the seventh semester) was concerned with the "derivation of the bilinear isoparametric (4-node) element in terms of Coons interpolation" (see Chap. 3 of this book). The Lecture Notes is still published in the form of a printed book, in which some pages are referring to Coons interpolation formula and their applications (Kanarachos [35, pp. 150–174]).

In 1983, I was involved in the systematic evaluation of the Boundary Element Method (BEM) and I investigated the possibility of using mass matrices as add-on to the static matrices, as a whole or through decomposing the domain into large BEM macroelements [37, 38, 49].

In summer 1984, a new Ph.D. student started his thesis entitled "On the implementation of Coons interpolation in Engineering Analysis", aiming at developing a computer code (called SFICA: structural finite element coons analysis) and also to compare Coons macroelements ("C-elements") with BEM macroelements. Since I was permanent staff, I was responsible for his technical support, and gave him a FORTRAN 77 version of FEAP [59], which was substantially expanded with beam and plate elements including buckling as well as eigenvalue–eigenvector (subspace iteration) software and time-integration subroutines [1]. The computer code was already operating on a UNIX platform (micro VAX II, 4MB RAM).

Being informed from the literature that Lagrange polynomials suffer from numerical oscillations, the Ph.D. student decided to replace them by univariate *natural* cubic B-splines along each of the four sides of a Coons patch *ABCD* [36]. Most of the test problems included in the aforementioned paper had been solved one year earlier using boundary elements [38], except of the L-shaped domain in which the Coons macroelement failed to give better results than the conventional FEM solution. In the lack of a rigorous explanation by the authors till today, I was motivated to include this issue in Chap. 2 of the book in hands. Some additional shortcomings have been commented in Chap. 7 as well.

In 1990, jointly with three coworkers, I had a six-weeks training (in two-course periods) in the French CAD system EUCLID_IS at the premises of Matra Datavision (Palaiseau, Les Ulis, near Paris, France). During this period, I came in direct contact with the Bézier, B-splines and NURBS curves and surfaces. As was reasonable, coming back to Greece I started thinking of replacing Coons interpolation with rational Bézier interpolation and NURBS.

Between the abovementioned two course periods on EUCLID_IS, jointly with the abovementioned coworkers, I participated in a VDI-daily Conference ("Tagung") in Essen-Germany, where my desk (demonstrating the Coons-based computer code SFICA and an in-house BEM code) happened to be very close to that of Professors Szabó and Babuška (they were promoting their PROBE® code, and I still remember a lot of visitors around them to admire the higher-order Legendre polynomials of sixth or seventh degree; in a sense, not quite different than the C-elements). I had the chance to discuss with many other exhibitors either from academia (e.g., Hinton and Owen) or from computer companies (ASKA, LASSO, SDRC/I-DEAS, PERMAS, to mention a few). Therefore, in 1990, I had formed a rather good idea about the state-of-the-art on computational codes and CAD.

Returning back to NTUA, all the team had formed the opinion that the future in computer methods is the triplet (Bézier, B-splines, and NURBS). A Diploma Work by Dimitriou [21] focused in the use of 3D Bézier surfaces in design, and the next year the title of his Ph.D. thesis was determined as "Finite elements based on analytical CAD expressions" (the official document is attached at the end of his Diploma Work: Dimitriou [21]).

So what?

14.3.2 Author's "Odyssey" and "Tutankhamun's Curse"

Most of you are surely aware about the meaning of "Odyssey" (one of two major ancient Greek epic poems attributed to Homer concerning the adventures of Odysseus to return at home after the Trojan War). On the other point of view, the reader may be also aware about the meaning of "Tutankhamun's Curse". Pharaoh's tomb was opened on November 29, 1922. Everyone who was present at the tomb's opening died after a short period. The same is hypothesized to have occurred to the grave robbers (https://en.wikipedia.org/wiki/Curse_of_the_pharaohs).

Sometimes I have the feeling that my attempt at NTUA to develop and promote the idea of CAD-based macroelements was a personal Odyssey in the course toward Ithaca, as will be exposed below. In more details, it should be reported that any student who undertook to deal with the subject of CAD-based macroelements "failed". Not even the first Ph.D. student did *not* complete and did not defend his thesis (officially started in January 1985) because in 1990 he succeeded as an owner of a private company in mechanical engineering, but also the second one (D. G.) stopped very soon in 1989, the third (D. P.) stopped in 1992 after six months, the fourth (J. D.) had a serious car accident in 1993, a few months after he had officially started conducting his Ph.D. thesis and remained in bed for a long period. The fifth Ph.D. student (Dejan D., from Serbia) was very capable in programming but stopped very early, the sixth one (N. Ph.) stopped two years after he begun (however funded from a grant in a quite different topic) as he was employed elsewhere, the seventh one (G. S.) stopped six months after he started. Also, in 2009, an eighth Ph.D. student (S. I.) stopped after two years due to the lack of funding (this fact is commented in the collocation methods below). I do not remember if there are some others, but in any case in my eyes it was a "Tutankhamun's Curse". Now, being distant, I can understand that this particular research topic requires full spiritual devotion, without financial concerns, and (first of all) culture of synergy.

After the abovementioned disaster, I was urged to publish two papers in the First European Conference on Computational Mechanics (ECCM'99), with the suggestion to make no reference to the older paper by Kanarachos and Deriziotis [36]. The first paper was concerned with 2D elastostatics and eigenvalue analysis (C12: Kanarachos et al. 1999), and the main idea was exposed (by the author of the book in hands) in a video-recorded Session chaired by Prof. R. L. Taylor (Berkeley University), in front of many distinguished attendees (Zienkiewicz, Samuelson, etc.). Despite the fact that we did not write it in the paper "expressis verbis", I still remember that I pointed out the possibility to implement any other CAD-based formulas as well. The second paper (abstract) was concerned with 3D potential and acoustic problems (i.e., was the continuation of the first paper in three dimensions), but it was rejected only due to the tight schedule (one paper per presenter).

Upon my return from Munich, I decided to take the subject on my shoulders, in my spare time. The overall result within these 30 years was almost: 30 journal papers and 30 papers in conference proceedings with review, which reflects about 15% of my total publication volume. The full list is given in Appendix 1. In order to avoid duplication in References, my own papers are denoted by the prefix "J" (for Journals) or "C" (for Conferences) followed by the ascending number.

The publication of the first journal papers was a real "Odyssey". Only recently I found the time to safely store most of my email or airmail communications. The first journal paper was much focused and was published rather regularly (J17: Provatidis and Kanarachos 2001). However, when in February 2002, I submitted a second paper entitled "Three-dimensional Coons macroelements in Laplace and acoustic problems" to another journal (of higher impact factor), despite the fact I had provided a long list of *eleven* reviewers, after a few months the assistant Editor (colleague N. L.) called me unofficially by phone and told me that they could not find a willing reviewer! So, in this way I was "forced" to withdraw my work and submit it elsewhere [eventually was published elsewhere in July 2005 (J36)].

In the meantime, several papers of mine already published in conference proceedings, created much material for a paper entitled "CAD-FEA integration using Coons interpolation". To check the impression this idea would have on a skilled CAD researcher, in January 2003, I asked my classmate Prof. P. Kaklis (Naval Engineering Dept. NTUA, now at the University of Strathclyde) to review a two-page abstract I had prepared and he found it very interesting (provided it was written well). Moreover, he suggested changing the title from "CAD-FEA marriage" (an expression I had already used at least twice in Conference papers: C26 twice, and C36, p. 409, probably influenced by the famous paper "marriage a la mode" by Zienkiewicz et al. [60]) to "CAD-FEA *integration*", and adding a somehow related paper of Elain Cohen in the literature [45].

The central idea is summarized as follows:

- The **2D COONS interpolation** could be considered as the connective tissue in many topics such as:
  - CAD interpolation in the geometric patch.
  - Mesh (grid) generation in 2D domains.
  - "C-element" for engineering analysis in 2D problems.
  - 3D BEM analysis (in conjunction with Coons patches).
- Moreover, the **3D COONS** interpolation was the connective tissue for:
  - CAD volume block interpolation.
  - Mesh generation in 3D domains.
  - "C-element" for engineering analysis in 3D problems.
- In summary, 2D and 3D "C-elements" use the nodal points along the minimum number of CAD entities which are required to determine the plane or the solid (geometric) model.
- Regarding the large isoparametric elements for analysis, COONS could be replaced by NURBS (the word "isogeometric" was still unknown).

However, when the idea of the so-defined CAD/CAE integration was subjected in 2003 to a CAD/CAGD journal, it did not even pass the initial editorial screening (was found out of scope). In September 2003, the overall idea (see, Provatidis [50]), where the CAD/FEA integration was implemented using Coons interpolation as the connective tissue (also an extension to NURBS was included), was sent to another journal dealing with similar topics, and after thirteen months it was rejected. Now, being distant, I recognize that trying to pass many ideas in a single paper, this "CAD-FEA integration" report had some weaknesses but the central idea was:

(i) *"Use the same basis functions for both the solid model and the variable"* and also
(ii) *"Put the degrees of freedom along those lines that are absolutely necessary to define the geometric model".*

In reality, the letter of the editor *never* arrived to my office and the decision became known in July 2005 as a response to my email (attached at the end of the Technical Report by Provatidis [50]). In a previous submission of another paper (June 2002), it took also two whole years until major review was asked for the first time, plus half a one year to complete it, and it finally was published in 2006 (paper J41).

It is worthy to mention that, in 2003, a call of proposals from the Committee of Basic Research at NTUA opened, so in May 2003 I submitted a proposal entitled

"DIDYMO (in English: TWIN): *Investigation on the possibility to integrate CAD and CAE methodologies*". After a year of peer review, the proposal was approved (with a high score 9.33/10.00, first in the Mechanical Engineering Department) and the program took the official Contract No. 65/138800; officially it started at June 1, 2004 (with two years duration). Nevertheless, the small amount of only 10,000 Euros was not sufficient to attract new willing Ph.D. students, so only part-time contribution of older ones was put on it.

Concerning the collocation method which was mentioned above, I have the need to say that the idea came in summer 2004 (year of Olympic Games in Athens), while swimming in Skyros island where I had short holidays with the family. The reason is that a lot of reviewers were commenting on the high computer effort required in the Coons macroelement analysis. Although this issue was successfully faced with detail in (J28: Provatidis 2004a), I was highly motivated to find a better solution. As my origin in computer methods was the boundary element analysis, I made the simple thought as follows. In the same "lines" with the indirect BEM where the solution is expanded in a series of *global* fundamental solutions in conjunction with unknown source densities, ϕ, I can preserve the *global* functional set (i.e., the shape functions coming from the CAD-based interpolation: Coons, Gordon, etc.) and follow the same procedure as in IBEM. That means, that Dirichlet and Neumann conditions should be applied to the series expansion of the numerical solution, and then, the obtained equations system is solved. But even this submission, in May 2005, had a successful outcome. In May 2006 (one year later), the editor requested to transfer the submission to another journal in which he was also the editor-in-Chief, but although I accepted his offer I never had a reply. Due to this disaster, I found the opportunity to propose the possibility of implementing CAD-based collocation within "A note in recent progress" at the end of an already accepted paper (J47: Provatidis 2006d, p. 6704) to somehow preserve the "priority" (in patent application terms). My first collocation paper was published in conference proceedings (C122: Provatidis 2007) and then some other papers followed from 2008 onwards, which the reader may find in Chap. 11 as well as in Appendix 1.

But the Odyssey does not stop here. A second research proposal submitted in early 2009 (this time of sufficient budget so as to ensure survival of the Ph.D. student for three years) entitled "Collocation Techniques using CAD-based Approximations" was rejected under the skeptic that "these methods are not offered for nonlinear problems". It seems that the reviewers had not read the book of De Boor [20]. Unfortunately, the financial crisis in the Greek economy since the fall of the year 2009, a few months later forced the Ph.D. student to interrupt his career on CAD-based collocation methods for ever.

Nevertheless, the final outcome of this (CAD/CAE integration) attempt was approximately 30 journal and 30 Conference papers, which are classified in Appendix 1, following the general numbering of my all papers in my CV (http://users.ntua.gr/cprovat).

14.4 Tensor-Product Methods

Tensor-product methods were developed as an extension of tensor-product Lagrange polynomials. They may include nonrational Bernstein–Bézier polynomials and B-splines. They also may include rational Bézier and rational B-splines (NURBS). The theory may be found in standard CAD textbooks and handbooks such as Farin et al. [22].

Regarding the *nonrational* Bernstein–Bézier polynomials, in Chap. 6 it was shown that the two aforementioned sets (Lagrange and Bézier polynomials) are mathematically equivalent. Not only rectangles but also other primitive shapes such as circles and ellipses can be handled as a single tensor-product Lagrangian or Bézierian macroelement. Similarly, a sphere or an ellipsoidal can be handled as a single tensor product (trivariate) 3D macroelement. In these cases, there is a deviation between the mapped object and the ideal solid, but as the number of degrees of freedom increases this deviation becomes extremely small (see, J135: Provatidis 2018).

In contrast, *rational* Bernstein–Bézier polynomials and rational B-splines (NURBS) are not true tensor products, in the sense that the tensor product refers only to the numerator, whereas the denominator is the sum of all combinations of terms in the numerator. It is worthy to mention that concerning a circle (or an ellipse), a tensor product of nine (quadratic) rational Bernstein–Bézier polynomials can establish an ideal mapping between the parametric unit square and the circle (or the ellipse). Regarding the sphere, each of the six spherical caps requires $5 \times 5 = 25$ control points, as has been explained in Chap. 10. In general, the accurate representation of a conical section or surface does not ensure higher accuracy (compared to Lagrange polynomials or simple B-splines) unless additional degrees of freedom are introduced by (i) degree elevation, (ii) knot insertion, or (iii) combination of them (k-refinement). As far as I have understood, I take the risk to say that the contribution of the modern IGA was a catalyst on this particular issue [32]. Of course, in the rational B-splines formulation there are several tricks one can follow; for example, one may define the one-quarter contour of an axial cross section of a sphere and then create an octant by 90° rotation. Details may be found in standard IGA books [4, 14].

Probably since late 1980s, many researchers have worked with 1D, 2D, and 3D approximation using Bézier, B-splines and NURBS. Early CAD-based papers did not rely on control points but they used the prior to Schoenberg [52] truncated power series expressions (e.g., Leung and Au [43]; Kong and Cheung [40], among others). For a fair review, the difficulty is to identify those works that use the CAD patch as an analytical background on which to build a grid of conventional finite elements (e.g., Lee and Hobbs [42]; Heege and Alart [30], among others), or to develop large isoparametric/isogeometric elements with the DOF associated to the control points of the CAD patch [3]. Reasonably, since Bézier is a subclass of B-splines, the state of the art should have been covered by Höllig [31] but, in author's opinion, a detailed review is still missing, as papers from the *electromagnetics* field have to be considered as well (e.g., Cendes [12]; Valle et al. [56]; Martini et al. [46], among others).

Additional information is given in the next subsection.

14.5 The CAD/FEA Integration

A literature survey reveals that many researchers have commented and worked in the matter of so-called CAD/CAE integration, i.e. on the simultaneous use of the same interpolation/approximation for the Solid Model and the Engineering Analysis. Some of them may be found in the book by Cottrell et al. [14].

Apart from the pioneers at General Motors (Gordon, Cavendish) and their co-worker Charles Hall mentioned in the Sect. 14.2, other significant contributions are due to Schramm and Pilkey "in the coupling of geometric description and finite elements" [53] as well as in boundary element applications [54], as well as Schramm et al. [55].

At the same time, many other researchers were used to apply conventional FEM analysis in conjunction with given analytical curves (Boender et al. [2]; Lau and Lo [41], among others), but this topic is out the spirit of this book.

Similar NURBS-based works considering the analytical geometry but now using meshless techniques (somehow similar to IGA), are due to Natekar et al. [47], Zhang and Subbarayan [58], Inoue et al. [33] (focused on shape optimization), and papers therein.

On similar lines, Pourazady and Xu [48], applied the finite element method in conjunction with B-spline and NURBS curves, to minimize the deformation energy functional and solve for the deformed shape of curves subjected to constraints.

Again, a lot of works using B-splines have been documented in the book by Höllig [31], as well as in Chap. 11 (pp. 283–307) of the well-known CAD handbook [22].

14.6 Coding the Findings Documented in Previous Chapters

The core message of the book is that every mathematical formula used in CAGD interpolation hides a functional set (basis functions). Thus, merely using or extending the concept of the *isoparametric* element, the latter functions can be used for computational analysis as well. Not only IGA (based on NURBS) but all previous CAD interpolation formulas, as well as all future ones, can or could be used to form CAD-based macroelements, useful for every type of engineering analysis. The aforementioned engineering analysis does not restrict to the FEM only, but it also includes the "CAD-based Collocation" and the "CAD-based Boundary Element Method" as well as the Spectral Methods, and so on.

Between the several alternative CAD interpolation formulas, the book has focused on (a) Coons interpolation (1964–1967), (b) Gordon transfinite interpolation [23], (c) Bezier interpolation (1970), (d) B-splines interpolation using both the older than Schoenberg [52] and the later Curry and Schoenberg [15] (De Boor [19]) formulation, and (e) Barnhill–Gordon interpolation (1973) in triangular domains.

The overall findings are classified as follows:

- First, following a very standard procedure (easily computerized) all classical quadrilateral finite elements of the "Serendipity family" are derived directly from Coons interpolation formula, which is the oldest one in CAD history. The additional value of Coons interpolation is that it can easily handle all types of trial functions (not only Lagrange polynomials).
- Second, the same happens for all classical triangular elements, which can be derived directly from Barnhill's interpolation formula.
- Third, Gordon transfinite interpolation (historically the second in CAD theory) may lead to (i.e., includes) the conventional family of Lagrange type elements. The possibility to use less or more internal nodes, in several setups within a Gordon macroelement was shown and pertinent numerical results were reported in static and dynamic analysis.
- Fourth, Bézier interpolation (historically the third in CAD theory) may be applied as a tensor product, similar to that used using Lagrange polynomials. It was proven that nonrational Bezier interpolation is equivalent with Lagrange polynomial interpolation, since the same functional set of monomials is used in both cases.
- Fifth, B-spline interpolation (historically the fourth in CAD theory) is applicable in many ways. The simplest way is to apply a tensor product, similar to that used in tensor-product Lagrange polynomials. The meaning of C^m-continuity was explained for partial polynomials of degree p, using either the truncated power series (older than Schoenberg [52]) or the alternative (and more efficient) Curry and Schoenberg [15] (De Boor's formulation [19]). The way of lowering the continuity, either in Schoenberg's or in de Boor's formulation, was explained by examples.
- Sixth, the applicability of all the abovementioned formulations (Coons, Gordon, Bezier, B-splines) has been shown in three-dimensional (3D) problems. Most primitives (cuboid, sphere, ellipsoidal) can be handled with a single macroelement. Application examples in this book include potential problems, whereas elastodynamic analyses may be found in the literature included the author's papers (see Appendix 1).
- Seventh, the same concept (Coons–Gordon, Bézier and B-splines) has been presented for plate bending analysis, theoretically and by presenting numerical results in standard static and dynamic test problems. Hermite macroelements were found to be equivalent with Bézier macroelements.
- Eighth, the applicability of all the abovementioned CAD-based functional sets in "CAD-based Collocation" analysis was briefly outlined. Potential and elasticity problems (2D, 3D) as well as plate bending analysis have been discussed. Lumped masses were developed by shifting the nodal points toward Gaussian points and safely performing nodal collocation.

- Ninth, the inclusion of similar ideas to "Boundary Element Method" (Integral-Equation) analysis has been briefly outlined, mostly in conjunction with Coons—Gordon interpolations. Future research topics related to Bézier and B-splines have been identified.
- Tenth, many 2D problems defined in simple primitive shapes such as a rectangle, a circle, an ellipse can be treated as a single macroelement of any kind (Coons, Gordon, Bézier, B-splines) provided the solution is smooth. A particular class of problems with a great success is the eigenanalysis.
- Eleventh, many 3D problems that refer to solid primitives such as a cube, a sphere, or an ellipsoidal can be successfully solved using a single macroelement.
- Twelfth, the decomposition of the domain into overlapping patches is possible through the Coons and Gordon interpolations. A particular master–slave procedure was reported in Chap. 13 for the first time in such an extent. Artificial nodes, in the form of single or double "ears" are added at the interface to the macroelement having the smallest edge. While for the tensor-product Gordon element the procedure was quite straightforward, it was shown that for Bernstein–Bézier requires some more elaboration.

I would like to add a particular note for the *spline collocation* methods. Spline collocation is a very old technique and it is Carl de Boor, through his Ph.D. thesis [18], who promoted the use of spline collocation to the solution of the two-point boundary value problem. Furthermore, he worked on the Curry and Schoenberg [15] formulation of B-splines and proposed a robust and efficient algorithm [19] to compute their values and derivatives at any specified position. This algorithm is the famous function "`spcol`" in MATLAB®. Interestingly, four years before the publication of De Boor [19] algorithm, in October 7–9, 1968, a team by six spline-oriented mathematicians (Greville, Jerome, Loscalzo, Schoenberg, Schumaker, and Varga) made a proposal to the US Army for funding their activities. I was impressed finding a paper by Jerome and Varga [34] dealing with spline collocation to nonlinear boundary value and eigenvalue problems as well. In other words, applied mathematicians make the beginning and then (sometimes after many years) applied scientists and engineers extend these ideas to the 2D and 3D terrain.

14.7 The Epilogue

This book was an excursion to some of the most dominating CAGD interpolations, from the oldest (Coons) to the most recent one (NURBS), for each of which the corresponding isoparametric macroelement was developed. It was definitely shown that since early 1970s many researchers shared the idea to integrate CAD with CAE, with the objective to handle and optimize those parameters that primarily control the geometric model. Probably the reader will see that there are many issues which have been revisited by contemporary researchers, of course in a modern environment. Nevertheless, many of the central ideas remain rather unaltered.

We hope that this self-contained book will contribute to further reduce the gap between CAD and CAE approaches and communities, bridging the older with the contemporary ideas.

Appendix 1: Author's Papers Related with CAD/CAE Integration

The overall output, including the first paper of the NTUA Group, is shown below. The prefix "J" refers to Journal whereas "C" stands for Conference. Numbering is according to the chronological date, and the existing gaps are due to author's work in different fields, which may be found at:

http://users.ntua.gr/cprovat/index_en.htm

I. **Galerkin Method**

Journal Papers

J0. Kanarachos A, Deriziotis D (1989) On the solution of Laplace and wave propagation problems using 'C-elements'. Finite Elem Anal Des 5:97–109

J17. Provatidis C, Kanarachos A (2001) Performance of a macro-FEM approach using global interpolation (Coons') functions in axisymmetric potential problems. Comput Struct 79:1769–1779

J19. Provatidis CG (2003) Analysis of axisymmetric structures using Coons' interpolation. Finite Elem Anal Des 39:535–558

J24. Provatidis CG (2002) Coons-patch macroelements in potential Robin problems. Forsch Ingenieurwes 67:19–26

J28. Provatidis CG (2004a) Coons-patch macroelements in two-dimensional eigenvalue and scalar wave propagation problems. Comput Struct 82(4–5):383–395

J29. Provatidis CG (2004b) Solution of two-dimensional Poisson problems in quadrilateral domains using transfinite Coons interpolation. Commun Numer Methods Eng 20(7):521–533

J36. Provatidis CG (2005) Three-dimensional Coons macroelements in Laplace and acoustic problems. Comput Struct 83(4–5):1572–1583

J37. Provatidis CG (2005) Analysis of box-like structures using 3-D Coons' interpolation. Commun Numer Methods Eng 21(8):443–456

J41. Provatidis CG (2006a) Coons-patch macroelements in two-dimensional parabolic problems. Appl Math Model 30(4):319–351

J42. Provatidis CG (2006b) Three-dimensional Coons macroelements: application to eigenvalue and scalar wave propagation problems. Int J Numer Methods Eng 65:111–134

J44. Provatidis CG (2006c) Free vibration analysis of two-dimensional structures using Coons-patch macroelements. Finite Elem Anal Des 42(6):518–531

J47. Provatidis CG (2006d) Transient elastodynamic analysis of two-dimensional structures using Coons-patch macroelements. Int J Solids Struct 43(22–23):6688–6706

J69. Provatidis CG (2009) Eigenanalysis of two-dimensional acoustic cavities using transfinite interpolation. J Algorithms Comput Technol 3(4):477–502

J92. Provatidis CG (2011) Equivalent finite element formulations for the calculation of eigenvalues using higher-order polynomials. Appl Math 1(1):13–23. (https://doi.org/10.5923/j.am.20110101.02). Open access. Available online at: http://article.sapub.org/10.5923.j.am.20110101.02.html

J93. Provatidis CG (2012) Two-dimensional elastostatic analysis using Coons-Gordon interpolation. Meccanica 47(4):951–967

J108. Provatidis CG (2013) A review on attempts towards CAD/CAE integration using macroelements. Comput Res 1(3):61–84. Open access. Online at: http://www.hrpub.org/download/20131107/CR2-13801513.pdf

J111. Provatidis CG, Angelidis D (2014) Performance of Coons' macroelements in plate bending analysis. Int J Comput Methods Eng Sci Mech 15:110–125

J114. Provatidis CG (2014) Bézier versus Lagrange polynomials-based finite element analysis of 2-D potential problems. Adv Eng Softw 73:22–34

J119. Provatidis CG (2014) Finite element analysis of structures using C^1-continuous cubic B-splines or equivalent Hermite elements. J Struct 2014(754561):9 pp. Online at: http://www.hindawi.com/journals/jstruc/2014/754561/

J135. Provatidis CG (2018) Engineering analysis with CAD-based macroelements. Arch Appl Mech 88(1–2):121–140. Free online access at: http://rdcu.be/uoQh

Conference Papers

C4. Kanarachos A, Grekas D, Provatidis C (1995) Generalized formulation of Coons' interpolation (chap 7). In: Kaklis P, Sapidis N (eds) Computer aided geometric design: from theory to practice. NTUA, pp 65–76. ISBN: 960-254-068-0

C12. Kanarachos A, Provatidis C, Deriziotis D, Photeas N (1999) A new approach of the FEM analysis of two-dimensional elastic structures using global (Coons') interpolation functions. In: Wunderlich W (ed) ECCM'99—CD proceedings European conference on computational mechanics (ECCM'99): CD proceedings, Munich, Germany, 28 Aug–3 Sept 1999, 763.pdf: pp 1–19. Available online at: http://users.ntua.gr/cprovat/yliko/763.pdf

C19. Provatidis C, Kanarachos A (2000) On the use of Coons' interpolation in CAD/CAE systems. MCME 2000: 2nd international conference on mathematics and computers in mechanical engineering, Vouliagmeni, Greece, 10–15 July 2000. In: Mastorakis N (ed) Systems and control: theory and applications. World Scientific and Engineering Society Press, pp 343–348. ISBN: 960-8052-11-4. Available online at: http://www.wseas.us/e-library/conferences/athens2000/Papers2000/547.pdf

C22. Provatidis CG (2001) A global approximation technique for noise-control analysis. In: Tsahalis DT (ed) Proceedings of EuroNoise-2001, 4th European conference on noise control, 14–17 Jan 2001, vol 1, pp 23–31. University of Patras, Greece. Available online at: http://users.ntua.gr/cprovat/yliko/C22.pdf

C25. Provatidis C (2001) Acoustic analysis of two-dimensional mufflers using large finite elements derived from Coons' interpolation. Proceedings of ASME—Greek section, Patras, Greece, 17–20 Sept 2001. In: Drakatos PA (ed) CD proceedings, paper ANG1/P029, pp 1–6. Available online at: http://users.ntua.gr/cprovat/yliko/C25.pdf

C40. Provatidis CG (2002) 'ACOUSYS': a computer code for the prediction of acoustical behavior in cavities and sound radiation from elastic structures. In: Trohidis A (ed) Proceedings acoustics 2002, Patras, Greece, 30 Sept–1 Oct 2002, paper no. AcP049, pp 333–338. ISBN: 960-7620-25-5

C48. Provatidis C (2003) Frequency analysis and transient response of two-dimensional structures using Coons-patch macroelements. In: Brennan MJ et al (eds) Proceedings VIII international conference on recent advances in structural dynamics, organized by the Institute of Sound & Vibration Research University of Southampton, 14–16 July 2003, paper no. 24. ISBN: 0-854-32-7894

C53. Provatidis CG (2003) Free vibrations of two-dimensional structures using Coons-patch macroelements, FEM and BEM. In: Atluri SN (ed) Proceedings ICCES' 03, Corfu, Greece, 24–29 July 2003

C57. Provatidis CG (2003) Coons-patch macroelements for boundary-value problems governed by fourth-order partial differential equation. In: Lipitakis EA (ed) Proceedings Hellenic-European conference on computational mathematics (HERCMA) 2003, Athens, Greece, 25–27 Sept 2003, pp 634–643. ISBN: 960-87275-1-0

C78. Provatidis CG (2005) Performance of large Gordon-Coons finite elements in 2-D potential problems. In: Georgiou G, Papanastasiou P, Papadrakakis M (eds) Proceedings GRACM 05, Limassol, Cyprus, 29 June–1 July 2005, vol 2, pp 545–552. ISBN: 9963-644-52-Xv.2

C89. Provatidis CG (2006) A global method in biomechanical modeling: application to dental mechanics. In: Middleton J (ed) Proceedings 7th international symposium on computer methods in biomechanics and biomedical engineering, Antibes, Juan Les Pins, Côte d'Azur, France, 22–25 Mar 2006.

C96. Provatidis CG, Vossou CG, Theodorou EG (2006) On the CAD/CAE integration using Coons interpolation. In: Tsahalis D (ed) Proceedings 2nd international conference "From scientific computing to computational engineering", Athens, Greece, 5–8 July 2006, vol III, pp 991–998. ISBN: 960-530-083-4

C108. Provatidis CG (2006) Transient elastodynamic analysis of plane structures using Coons-patch macroelements and modal superposition. In: The eighth international conference on computational structures technology, Las Palmas de Gran Canaria—Spain, 12–15 Sept 2006. (https://doi.org/10.4203/ccp.83.197. You can find paper webpage in the repository by following the link http://dx.doi.org/10.4203/ccp.83.197)

C136. Provatidis C (2008) Plate bending analysis using transfinite interpolation. In: Talaslidis D, Manolis G (eds) Proceedings 6th GRACM international congress on computational mechanics, Thessaloniki, Greece, 19–21 June 2008

C173. Provatidis CG (2009) Higher order Galerkin/Ritz approximations in 1-D eigenvalue problems. In: Proceedings 3rd international conference on experiments/process/system modeling/simulation & optimization (3rd IC-EpsMsO), Athens, 8–11 July 2009

C191. Provatidis CG (2011) Some issues on CAD/CAE integration: global interpolation using isoparametric and isogeometric techniques. In: Proceedings 7th GRACM international congress on computational mechanics, Athens, Greece, 30 June–2 July 2011

C192. Provatidis CG (2011) Global versus local interpolation in the FEM free vibration analysis of prismatic bars. In: Proceedings 7th GRACM international congress on computational mechanics, Athens, Greece, 30 June–2 July 2011

C236. Provatidis C (2016) Engineering analysis using CAD-based macroelements. In: Katsikadelis T, Stavroulakis GE (eds) Recent advances in mechanics. Book of abstracts, 9th German-Greek-Polish symposium, Orthodox Academy of Crete, Kolympari, Greece, 4–9 Sept 2016. Publisher: Technical University of Crete, Chania, Greece, 2016 (e-book), pp 51–52. ISBN: 978-960-8475-25-0

II. **Boundary Element Method**

Conference Papers

C26. Provatidis C (2001) Stress analysis of 3D solid structures using large boundary elements derived from 2D-Coons' interpolation. Proceedings of ASME—Greek section, Patras, Greece, 17–20 Sept 2001. In: Drakatos PA (ed) CD proceedings, paper ANG1/P129, pp 1–6. Available online at: http://users.ntua.gr/cprovat/yliko/C26_asme_3d_elasticity.pdf

C35. Provatidis CG (2002) A comparative study between Coons-patch macroelements and Boundary elements in two-dimensional potential problems. In: Tsahalis DT (ed) Proceedings 4th GRACM congress on computational mechanics, Patras, 27–29 June 2002 (CD proceedings); also: vol I, pp 43–50. Available online at: http://users.ntua.gr/cprovat/yliko/C35.pdf

C36. Provatidis CG (2002) Analysis of three-dimensional sound radiation problems using trimmed patch boundary elements. In: 4th GRACM congress on computational mechanics, Patras, 27–29 June 2002 (CD proceedings); also: vol I, pp 402–409. Available online at: http://users.ntua.gr/cprovat/yliko/C36.pdf

C55. Provatidis CG, Zafiropoulos NK (2003) Free-vibration analysis of three-dimensional solids using Coons-patch boundary superelements. In: Fifth European solids mechanics conference, Thessaloniki, Greece, 17–22 Aug 2003. Extended version, online at: http://users.ntua.gr/cprovat/yliko/C55.pdf

C56. Provatidis CG, Zafiropoulos NK (2003) Determination of eigenfrequencies in three-dimensional acoustic cavities using Coons-patch boundary superelements. In: Fifth European solids mechanics conference, Thessaloniki, Greece, 17–22 Aug 2003. Extended version, online at: http://users.ntua.gr/cprovat/yliko/C56.pdf

Global Collocation Method

Journal Papers

J59. Provatidis CG (2008) Free vibration analysis of elastic rods using global collocation. Arch Appl Mech 78(4):241–250

J60. Provatidis CG (2008) Global collocation method for 2-D rectangular domains. J Mech Mater Struct 3(1):185–194

J63. Provatidis CG (2008) Time- and frequency-domain analysis using lumped mass global collocation. Arch Appl Mech 78(11):909–920

J67. Provatidis CG (2009) Integration-free Coons macroelements for the solution of 2-D Poisson problems. Int J Numer Meth Eng 77:536–557

J70. Provatidis CG, Ioannou KS (2010) Static analysis of two-dimensional elastic structures using global collocation. Arch Appl Mech 80(4):389–400

J99. Provatidis CG, Isidorou SK (2012) Solution of one-dimensional hyperbolic problems using cubic B-splines collocation. Int J Comput Sci Appl (IJCSA) 1(1):12–18. Open access. Available online: http://dpi-journals.com/index.php/IJCSA/article/view/795

J117. Provatidis CG (2014) B-splines collocation eigenanalysis of 2D acoustic problems. J Mech Mater Struct 9(3):259–285

J121. Provatidis CG (2014) Lumped mass acoustic and membrane eigenanalysis using the global collocation method. Cogent Eng 1(1):981366 (http://www.tandfonline.com/doi/abs/10.1080/23311916.2014.981366#.VIKnImc-eSo)

J125. Provatidis C, Fillipatos A (2015) Numerical determination of eigenfrequencies in two-dimensional acoustic cavities using a global collocation method. Rom J Acoust Vibr XII(2):111–115. Open access. Available online: http://sra.ro/Arhiva/2015/nr2/Paper_4_page132-138.pdf

J126. Provatidis C (2015) Lumped mass collocation method for 2D elastodynamic analysis. Int J Adv Eng Technol Comput Sci 2(2):1–15. http://www.irosss.org/ojs/index.php/IJAETCS/article/view/676/207

J128. Provatidis CG (2017) CAD-based collocation eigenanalysis of 2-D elastic structures. Comput Struct 182:55–73

J129. Provatidis CG (2017) B-splines collocation for plate bending eigenanalysis. J Mech Mater Struct 12(4):353–371

Conference Papers

C122. Provatidis CG (2007) Performance of a Lagrange based global finite element collocation method for eigenvalue structural analysis. In: Proceedings 8th Hellenic society of theoretical and applied mechanics (HSTAM) congress, Patras, Greece, 12–14 July 2007, pp 153–160

C135. Provatidis C (2008) Analysis of one-dimensional wave propagation problems using global collocation. In: Proceedings 6th GRACM international congress on computational mechanics, Thessaloniki, Greece, 19–21 June 2008

C172. Isidorou SK, Provatidis CG (2009) Comparison of advanced collocation methods for the solution of ordinary differential equations. In: 3rd international conference on experiments/process/system modeling/simulation & optimization (3rd IC-EpsMsO), Athens, 8–11 July 2009

C193. Provatidis CG, Isidorou SK (2011) B-splines collocation eigenvalue analysis of 1-D problems. In: Proceedings 7th GRACM international congress on computational mechanics, Athens, Greece, 30 June–2 July 2011

References

1. Bathe K-J (1982) Finite element procedures in engineering analysis. Prentice-Hall, New Jersey
2. Boender E, Bronsvoort WF, Post FH (1994) Finite-element mesh generation from constructive-solid-geometry models. Comput Aided Des 26(5):379–392
3. Borden MJ, Scott MA, Evans JA, Hughes TJR (2011) Isogeometric finite element data structures based on Bézier extraction of NURBS. Int J Numer Meth Eng 87:15–47
4. Buffa A, Sangalli G (2012) Isogeometric analysis: a new paradigm in the numerical approximation of PDEs: Cetraro, Italy 2012. Lecture notes in mathematics. Springer, Berlin
5. Cavendish JC (1972) Collocation methods for elliptic and parabolic boundary value problems. Ph.D. dissertation, University of Pittsburgh, USA
6. Cavendish JC, Hall CA (1974) L∞-convergence of collocation and Galerkin approximations to linear two-point parabolic problems. Aequationes Math 11(2–3):230–249
7. Cavendish JC, Hall CA (1984) A new class of transitional blended finite elements for the analysis of solid structures. Int J Numer Methods Eng 28:241–253
8. Cavendish JC, Gordon WJ, Hall CA (1976) Ritz-Galerkin approximations in blending function spaces. Numer Math 26:155–178
9. Cavendish JC, Gordon WJ, Hall CA (1977) Substructured macro elements based on locally blending interpolation. Int J Numer Methods Eng 11:1405–1421
10. Cavendish JC, Hall CA, Zienkiewicz OC (1978) Blended infinite elements for parabolic boundary value problems. Int J Numer Methods Eng 12(12):1841–1851
11. Cavendish JC, Frey WH, Marin SP (1991) Feature-based design and finite element mesh generation for functional surfaces. Adv Eng Softw 13(5–6):226–237
12. Cendes ZJ (1991) Vector finite elements for electromagnetic field computation. IEEE Trans Magn 27(5):3958–3966
13. Coons SA (1964) Surfaces for computer aided design of space form. Project MAC, MIT (1964), revised for MAC-TR-41 (1967). National Technical Information Service (CFSTI), Springfield, VA, USA. Available online: http://publications.csail.mit.edu/lcs/pubs/pdf/MIT-LCS-TR-041.pdf

14. Cottrell J, Hughes TJR, Bazilevs Y (2009) Isogeometric analysis: toward integration of CAD and FEA. Wiley, New York
15. Curry HB, Schoenberg IJ (1966) On Pólya frequency functions IV: the fundamental spline functions and their limits. J Anal Math 17(1):71–107
16. Dahmen W, Dikshit HP, Ojha A (2000) On Wachspress quadrilateral patches. Comput Aided Geom Des 17:879–890
17. Dasgupta G (2003) Integration within polygonal finite elements. ASCE J Aerosp Eng 16(1):9–18
18. De Boor C (1966) The method of projections as applied to the numerical solutions of two-point boundary value problems using cubic splines. Ph.D. thesis, University of Michigan. ftp://cs.wisc.edu/Approx/
19. De Boor C (1972) On calculating with B-splines. J Approx Theory 6:50–62
20. De Boor C (2001) A practical guide to splines, revised edn. Springer, New York (Former edn: Applied mathematical sciences, vol 27. Springer, New York, 1978)
21. Dimitriou J (1992) Modeling and discretization of three-dimensional surfaces using Bézier algorithm. Diploma work (MSc thesis), Mechanical Engineering Department, National Technical University of Athens, Greece, Oct 1992. Online available at: http://users.ntua.gr/cprovat/yliko/Dimitriou_Diploma_Work.pdf
22. Farin G, Hoschek J, Kim M-S (2002) Handbook of computer aided geometric design. Elsevier, North Holland
23. Gordon WJ (1969) Spline-blended surface interpolation through curve networks. Indiana Univ Math J 18(10):931–952. Online available at: http://www.iumj.indiana.edu/IUMJ/FULLTEXT/1969/18/18068
24. Gordon WJ (1971) Blending functions methods of bivariate and multivariate interpolation and approximation. SIAM J Numer Anal 8:158–177
25. Gordon WJ, Hall CA (1973) Construction of curvilinear co-ordinate systems and application to mesh generation. Int J Numer Methods Eng 7:461–477
26. Gordon WJ, Hall CA (1973) Transfinite element methods: blending-function interpolation over arbitrary curved element domains. Numer Math 21:109–112
27. Gordon WJ, Thiel LC (1982) Transfinite mappings and their application to grid generation. Appl Math Comput 11–12:171–233. In: Thompson JP (ed) Numerical grid generation. Elsevier, Amsterdam
28. Gregory JA (1983) N-sided surface patches. In: Gregory JA (ed) Mathematics of surfaces. Clarendon Press, pp 217–232
29. Hall CA, Raymund M, Palusamy S (1979) A macro element approach to computing stress intensity factors for three dimensional structures. Int J Fract 15(3):231–245
30. Heege A, Alart P (1996) A frictional contact element for strongly curved contact problems. Int J Numer Methods Eng 39:165–184
31. Höllig K (2003) Finite element methods with B-splines. SIAM, Philadelphia
32. Hughes TJR, Cottrell JA, Bazilevs Y (2005) Isogeometric analysis: CAD, finite elements, NURBS, exact geometry and mesh refinement. Comput Methods Appl Mech Eng 194:4135–4195
33. Inoue K, Kikuchi Y, Masuyama T (2005) A NURBS finite element method for product shape design. J Eng Des 16(2):157–171
34. Jerome JW, Varga RS (1969) Generalizations of spline functions and applications to nonlinear boundary value and eigenvalue problems. In: Greville TNE (ed) Theory and applications of spline functions. Proceedings of an advanced seminar conducted by the mathematics research center, United States Army, at the University of Wisconsin, Madison, 7–9 Oct 1968. Academic press, New York
35. Kanarachos AE (1985) Dynamics and mechanical vibrations-finite elements. Plaisio-Fountas Publishing Company, Athens (in Greek). http://www.fountasbooks.gr/home/
36. Kanarachos A, Deriziotis D (1989) On the solution of Laplace and wave propagation problems using 'C-elements'. Finite Elem Anal Des 5:97–109

37. Kanarachos A, Provatidis C (1987) Performance of mass matrices for the BEM dynamic analysis of wave propagation problems. Comput Methods Appl Mech Eng 63:155–166
38. Kanarachos A, Provatidis C (1988) On the symmetrization of the BEM formulation. Comput Methods Appl Mech Eng 71:151–165
39. Kanarachos A, Röper O (1979) Rechnerunterstützte Netzgenerierung mit Hilfe der Coonsschen Abbildung. VDI-Z 121:297–303
40. Kong J, Cheung YK (1993) Application of the spline finite strip to the analysis of shear-deformable plates. Comput Struct 46(6):985–988
41. Lau TS, Lo SH (1996) Finite element mesh generation over analytical curved surfaces. Comput Struct 59(2):301–309
42. Lee CK, Hobbs RE (1998) Automatic adaptive finite element mesh generation over rational B-spline surfaces. Comput Struct 69:577–608
43. Leung AYT, Au FTK (1990) Spline finite elements for beam and plate. Comput Struct 31(5):717–729
44. Marshall JA, Mitchell AR (1973) An exact boundary technique for improved accuracy in the finite element method. J Inst Math Appl 12:355–362
45. Martin W, Cohen E (2001) Representation and extraction of volumetric attributes using trivariate splines: a mathematical framework. In: Proceedings of the solid modelling symposium 2001. (http://www.cs.utah.edu)
46. Martini E, Pelosi G, Selleri S (2003) A hybrid finite-element–modal-expansion method with a new type of curvilinear mapping for the analysis of microwave passive devices. IEEE Trans Microw Theory Tech 51(6):1712–1717
47. Natekar D, Zhang X, Subbarayan G (2004) Constructive solid analysis: a hierarchical, geometry-based meshless analysis procedure for integrated design and analysis. Comput Aided Des 36(5):473–486
48. Pourazady M, Xu X (2000) Direct manipulations of B-spline and NURBS curves. Adv Eng Softw 31(2000):107–118
49. Provatidis CG (1987) On the application of the boundary element method in the analysis of field and dynamic problems. Doctoral dissertation, National Technical University of Athens, Department of Mechanical Engineering, Athens, Nov 1987 (in Greek). Available online at: https://www.didaktorika.gr/eadd/handle/10442/0616
50. Provatidis CG (2003) CAD-FEA integration using Coons interpolation. Technical report, submitted to two journals. Available online: http://users.ntua.gr/cprovat/yliko/Provatidis_CAD_CAE_Integration.pdf
51. Röper O (1978) Ein geometrieprozessor für die rehnerinterstützte auslegung von maschinenbauteilen mit hilfe der methode der finite elemente. Dissertation, Ruhr-Uiversität Bochum
52. Schoenberg IJ (1946) Contributions to the problem of approximation of equidistant data by analytic functions. Part A. On the problem of smoothing or graduation. A first class of analytic approximation formulae. Q Appl Math 4(1):45–99
53. Schramm U, Pilkey WD (1993) The coupling of geometric descriptions and finite elements using NURBs—a study in shape optimization. Finite Elem Anal Des 15(1):11–34
54. Schramm U, Pilkey WD (1994) Higher order boundary elements for shape optimization using rational B-splines. Eng Anal Bound Elem 14(3):255–266
55. Schramm U, Pilkey WD, DeVries RI, Zebrowski MP (1995) Shape design for thin-walled beam cross sections using rational B splines. AIAA J 33(11):2205–2211
56. Valle L, Rivas F, Citedra MF (1994) Combining the moment method with geometrical modelling by NURBS surfaces and Bézier patches. IEEE Trans Antennas Propag 42(3):373–381
57. Wachspress EL (1975) A rational finite element basis. Academic Press, Boston
58. Zhang X, Subbarayan G (2006) jNURBS: an object-oriented, symbolic framework for integrated, meshless analysis and optimal design. Adv Eng Softw 37(5):287–311
59. Zienkiewicz OC (1977) The finite element method, 3rd edn. McGraw-Hill, London
60. Zienkiewicz OC, Kelly DW, Bettess P (1977) "Marriage a la mode": the best of both worlds (finite elements and boundary integrals) (chap 5). In: Glowinski R, Rodin EY, Zienkiewicz OC (eds) Energy methods in finite element analysis. Wiley, New York

Index

C. G. Provatidis, *Precursors of Isogeometric Analysis*, Solid Mechanics and Its Applications 256, https://doi.org/10.1007/978-3-030-03889-2

MIX
Papier aus verantwortungsvollen Quellen
Paper from responsible sources
FSC® C105338

If you have any concerns about our products,
you can contact us on
ProductSafety@springernature.com

In case Publisher is established outside the EU,
the EU authorized representative is:
Springer Nature Customer Service Center GmbH
Europaplatz 3, 69115 Heidelberg, Germany

Printed by Libri Plureos GmbH
in Hamburg, Germany